Clymer Collection Series

VINTAGE

SNOWMOBILES

VOLUME I

➢ ARCTIC CAT, 1974-1979

KAWASAKI, 1976-1980

JOHN DEERE, 1972-1977

The world's finest publisher of mechanical how-to manuals

Business Magazines & Media

P.O. Box 12901, Overland Park, KS 66282-2901

FIRST EDITION
First Printing June, 1996
Second Printing May, 1999
Third Printing August, 2002

Printed in U.S.A.

ISBN: 0-89287-677-8

Library of Congress: 96-75970

The following books and guides are published by PRIMEDIA Business Directories & Books.

The Electronics Source Book

More information available at *primediabooks.com*

CONTENTS

QUICK REFERENCE DATA

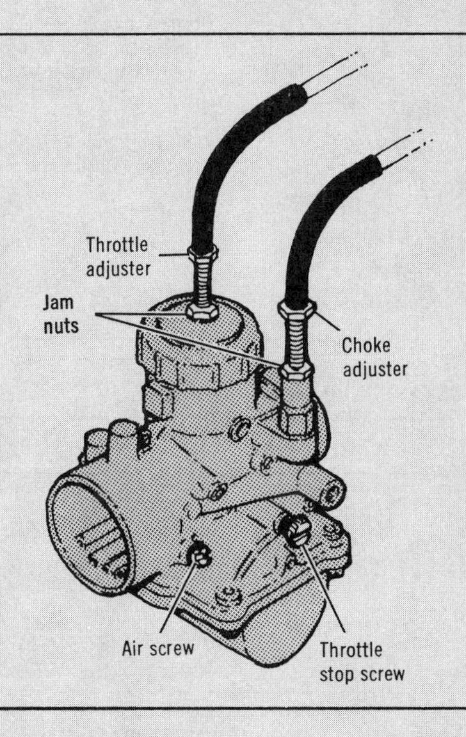

BASIC AIR SCREW SETTING— MIKUNI CARBURETORS

Model/Engine	Turns Out
El Tigre	
1974-1975	
All models	1
1976-1977	
4000	1½
5000	1
1978-1979	1½
Jag All models	1½
Lynx All models	1½
Pantera	
All models thru 1977	1
1978 Fan	1½
1978 Free-air	1
1979 all	1¾
Panther/Cheetah	
4000	¾
5000 thru 1977	1
5000 1978-1979	1¼

BASIC AIR SCREW SETTING — WALBRO CARBURETORS

Model/Engine	Setting
Low speed mixture needle	1 turn out
High speed mixture needle	1⅛ turns out
Throttle cable slack	$\frac{1}{16}$ in.

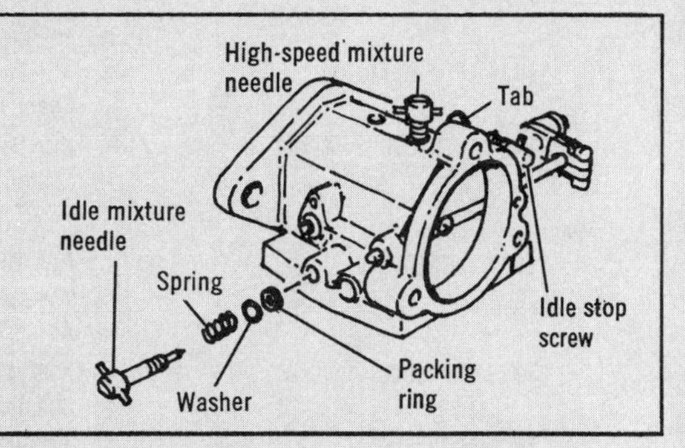

DRIVE BELT SPECIFICATIONS

Model	Year	Outside Circumference	Width
Lynx	All years	43¼ in.	1⅛ in.
Pantera	Through 1975	43¼ in.	1⅛ in.
	1976 on	46¾ in.	1⅛ in.
Jag	Through 1977	44¾ in.	1⅛ in.
	1978-1979	45½ in.	
Cheetah	All years	43¼ in.	1⅛ in.
El Tigre	Through 1977	46¾ in.	1⅛ in.
	1978-1979	43¼ in.	1⅛ in.
Panther	All years	43¼ in.	1⅛ in.

LYNX IGNITION SPECIFICATIONS

Contact breaker gap	0.012-0.016 in. (0.3-0.4mm)
Flywheel bolt torque	65-80 ft.-lb. (9-11 mkg)
Timing	18° or 0.073 in. (1.86mm) BTDC

ADJUSTMENTS

Track tension	¾-1 in. (19-25mm) slack between slide and inside surface of track
Brake adjustment	½-¾ in. (12-18mm) between brake control lever and stop

FUEL, LUBRICATION, AND CAPACITIES

Fuel	Octane rating 88 or higher. Use premium in all high performance machines.
Engine oil	
Type	Arctic Cat Purple Powerlube II-CCI or Chemilube
Mixture ratio	
Except Wankel	20:1
Wankel	30:1
Chaincase oil	Arctic Cat Lube

SPARK PLUG TYPE AND GAP

Model/Engine	Type*	Gap
Lynx		
1974-1975 2000S	NGK BR9HS Champion L77J	0.026-0.030 in. (0.66-0.76mm)
1976-1979 2000S	NGK BR9ESA Champion RN2	0.018-0.022 in. (0.46-0.56mm)
1976-1979 2000T	NGK BR8ESA Champion RN3	0.018-0.022 in. (0.46-0.56mm)
El Tigre		
1974-1975 all	NGK BR9EVA Champion RN2	0.016-0.020 in. (0.41-0.51mm)
1976-1977 4000	NGK BR8ESA Champion RN3	0.018-0.022 in. (0.46-0.56mm)
1976-1979 5000/6000	NGK BR9ESA Champion RN2	0.018-0.022 in. (0.46-0.56mm)
Panther/Cheetah		
1974-1975 all	NGK B8ESA Champion N3 Bosch W215T 28	0.020-0.032 in. (0.51-0.81mm)
1976-1977 4000	NGK BR8ESA Champion RN3	0.018-0.022 in. (0.46-0.56mm)
1976-1979 5000	NGK BR9ESA Champion RN2	0.018-0.022 in. (0.46-0.56mm)
Pantera		
All fan-cooled (except 1978-1979)	NGK BR8ESA Champion RN3	0.018-0.022 in. (0.46-0.56mm)
All free-air and 1978-1979 fan-cooled	NGK BR9ESA Champion RN2	0.018-0.022 in. (0.46-0.56mm)
Jag		
1975	NGK BR8ESA Champion RN3	0.030 in. (0.76mm)
1976-1979	NGK BR8ESA Champion RN3	0.018-0.022 in. (0.46-0.56mm)

*Heat ranges shown are for normal operation. If a colder plug is required because machine is operated at sustained high speeds, change one increment at a time (e.g., if BR8ESA is specified, change to BR9ESA). If a hotter plug is required because machine is operated at low speeds, change one increment at a time (e.g., if BR8ESA is specified, change to BR7ESA).

NGK spark plugs are recommended by the manufacturer; however, an equivalent plug from another manufacturer can be substituted as long as its heat range corresponds to the NGK spark plug specified in this table.

ARCTIC CAT

1974-1979

SERVICE • REPAIR • MAINTENANCE

CHAPTER ONE

GENERAL INFORMATION

Snowmobiling has in recent years become one of the most popular outdoor winter recreational past times. It provides an opportunity for an entire family to experience the splendor of winter and enjoy a season previously regarded by many as miserable.

Snowmobiles also provide an invaluable service in the form of rescue and utility vehicles in areas that would otherwise be inaccessible.

As with all sophisticated machines, snowmobiles require specific periodic maintenance and repair to ensure their reliability and usefulness.

MANUAL ORGANIZATION

This manual provides periodic maintenance, tune-up and general repair procedures for Arctic Cat snowmobiles (except those with Wankel engines) manufactured in 1974 and later.

The supplement at the end of the book contains specific service information for 1978-1979 models. If a procedure is not mentioned in the Supplement, it remains the same as on earlier models.

This chapter provides general information and hints to make all snowmobile work easier and more rewarding. Additional sections cover snowmobile operation, safety and survival techniques.

Chapter Two provides all tune-up and periodic maintenance required to keep your snowmobile in top running condition.

Chapter Three provides numerous methods and suggestions for finding and fixing troubles fast. The chapter also describes how a 2-cycle engine works, to help you analyze troubles logically. Troubleshooting procedures discuss symptoms and logical methods to pinpoint the trouble.

Subsequent chapters describe specific systems such as engine, fuel system, and electrical system. Each provides disassembly, repair, and assembly procedures in easy-to-follow, step-by-step form. If a repair is impractical for home mechanics, it is so indicated. Usually, such repairs are quicker and more economically done by an Arctic Cat dealer or other competent snowmobile repair shop.

Some of the procedures in this manual specify special tools. In all cases, the tool is illustrated in actual use or alone.

The terms NOTE, CAUTION, and WARNING have specific meaning in this book. A NOTE provides additional information to make a step or procedure easier or clearer. Disregarding a NOTE could cause inconvenience, but would not cause damage or personal injury.

A CAUTION emphasizes areas where equipment damage could result. Disregarding a CAUTION could cause permanent mechanical damage; however, personal injury is unlikely.

A WARNING emphasizes areas where personal injury or death could result from negligence.

Mechanical damage may also occur. WARNINGS are to be taken seriously. In some cases serious injury or death has been caused by mechanics disregarding similar warnings.

MACHINE IDENTIFICATION AND PARTS REPLACEMENT

Each snowmobile has a serial number applicable to the machine and a model and serial number for the engine.

See **Figure 1** for the location of the machine serial number on the right side of the tunnel. See **Figure 2** for the location of the VIN (vehicle identification number). **Figure 3** illustrates the location of engine numbers (intake side of block) for most engines. On some engines the number is stamped on the magneto cover (**Figure 4**). Unless specifically noted, all serial numbers called out in procedures in this manual are machine serial numbers and not engine numbers.

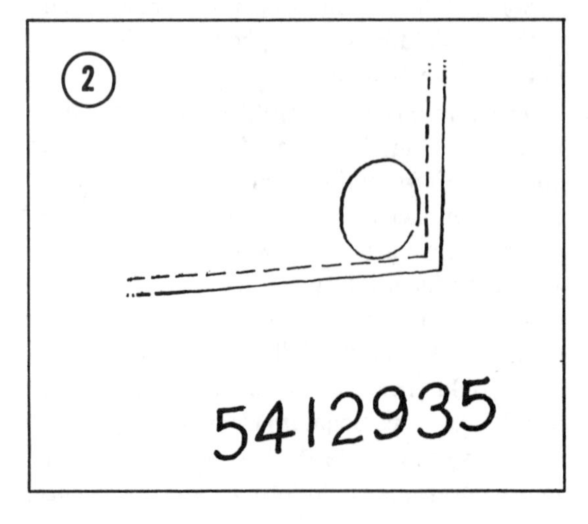

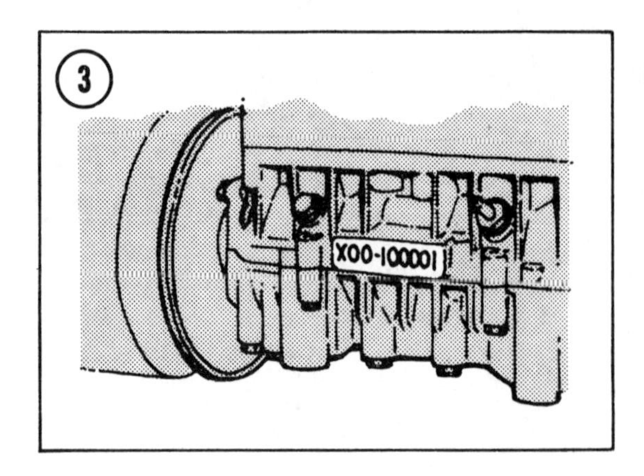

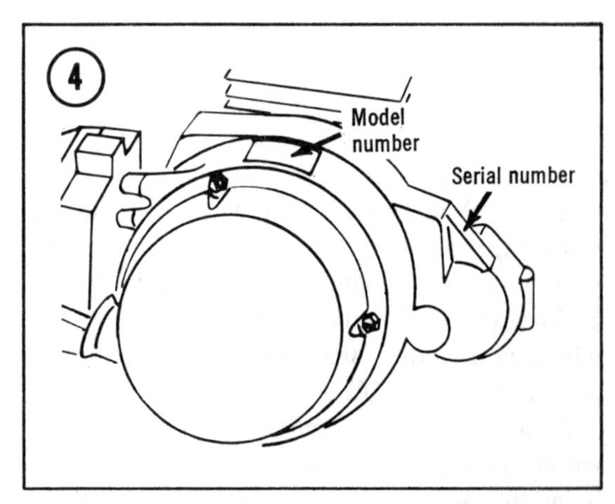

Write down all serial and model numbers applicable to your machine and carry the numbers with you. When you order parts from a dealer, always order by year and engine and machine numbers. If possible, compare old parts to the new ones before purchasing them.

If parts are not alike, have the parts manager explain the difference; the new part may be improved, but it could also be the wrong one.

OPERATION

Fuel Mixing

WARNING

Serious fire hazards always exist around gasoline. Do not allow any smoking in areas where fuel is mixed or when refueling your snowmobile.

Always use fresh fuel. Gasoline loses its potency after sitting for a period of time. Old fuel can cause engine failure and leave you stranded in severe weather.

Proper fuel mixing is very important for the life and efficiency of the engine. All engine lubrication is provided by the oil mixed with the gasoline. Always mix fuel in exact proportions. A "too lean" mixture can cause serious and expensive damage. A "too rich" mixture can cause poor performance and fouled spark plugs which can make an engine difficult or impossible to start.

Use a gasoline with an octane rating of 88 or higher. Use premium grade gasoline in all high-performance racing machines. Mix gasoline in a separate tank, not the snowmobile fuel tank. Use a tank with a larger volume than necessary to allow room for the fuel to agitate and mix completely.

Use Arctic Cat Purple Powerlube II-CCI or Chemilube. Unless absolutely necessary, *do not* use regular brand snowmobile oils. If they have to be used, follow the oil manufacturer's instructions. Never use outboard motor oil, regular mineral oils, or automobile oil; they are almost certain to promote engine damage. All engines except the Wankel use a 20:1 ratio. The Wankel uses a 30:1 ratio.

1. Pour ½ of the required gasoline into a clean container.
2. Add the required amount of oil and mix thoroughly.
3. Add remainder of gasoline and mix entire contents thoroughly.
4. Always use a funnel equipped with a fine screen when adding fuel to snowmobile.

Pre-Start Inspection

The few minutes necessary to prepare the snowmobile, as well as yourself, before starting it, may prevent a breakdown or an accident.

1. Check the cooling system. Insure the cooling fins are clean and free from obstructions.
2. Make sure the exhaust system and carburetors are securely fastened.
3. Check the operation of the throttle control. It should depress without excessive effort and return freely to idle. Ensure the throttle safety switch is operating correctly. To adjust the throttle lever, see *Throttle Adjustment*, Chapter Two. To adjust the throttle safety switch, see *Throttle Safety Switch Adjustment*, Chapter Two.
4. Check the brake control. The brake should fully engage when the lever is depressed about ¾ in. and disengage freely when released. If more than ¾ in. of lever travel is required to engage the brake, or if the brake lever bottoms out on the handle control, adjust it as described in Chapter Two—*Brake Adjustment*.
5. Check the steering to ensure the skis turn freely. If they turn with difficulty, remove ice or snow from around steering mechanism.
6. Check the fuel supply. Never take extended trips without a full fuel tank and a ready reserve for possbile emergencies. For proper fuel mixture ratio, see *Fuel Mixing*, Chapter Two.
7. Check the toolbox. The tool kit and necessary spare parts should be carried at all times. In addition to the tools and equipment supplied with the machine, carry a spare drive belt and extra spark plugs.
8. Check to make sure the headlight, taillights and brakelights are working properly.
9. Make sure all nuts and bolts are tight. A loose nut or bolt could cause serious damage.
10. Clean the windshield with a clean damp cloth. *Do not* use gasoline, solvents, or abrasive cleaners.
11. Check track tension.
12. When the engine is started in extremely cold weather, prop up the back of the machine on a Quik-Jak or tilt the machine to one side, and open the throttle slightly. Allow the track to turn several revolutions to allow the bearings, track, and drive belt proper "warm up" before subjecting them to full load.

WARNING

Before starting engine, be sure no bystanders are in front of or behind the snowmobile as a sudden lurch may cause serious injuries.

13. Start engine and test operation of emergency kill switch and "tether" switch. Check that all lights are working.

Emergency Starting

Always carry a small tool kit with you. Carry an extra starting rope for emergency starting or use the recoil starter rope.

1. Remove hood.
2. Unscrew the bolts and remove recoil starter (**Figure 5**).

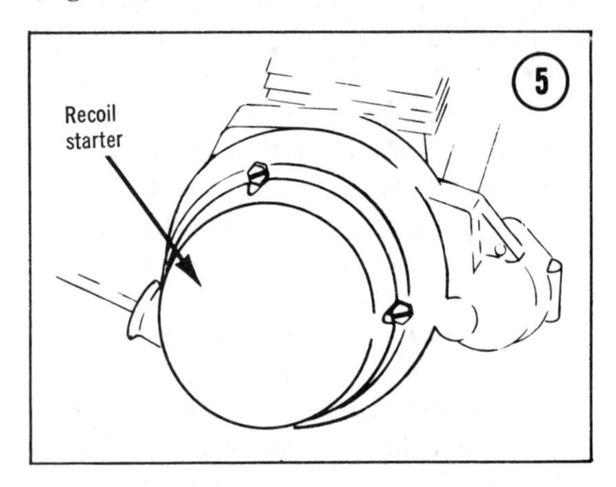

3. Wind rope around starter pulley and pull to crank engine (**Figure 6**).

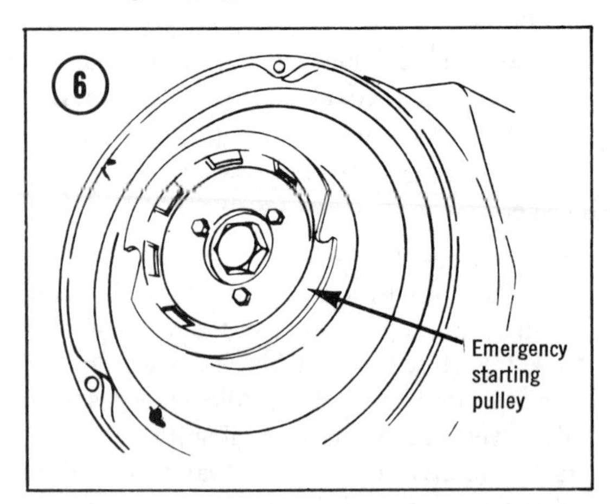

Emergency Stopping

To stop the engine in case of an emergency, pull the tether string or switch the emergency kill switch to the STOP or OFF position.

Towing

When preparing for a long trip, pack extra equipment in a sled, don't try to haul it on the snowmobile. A sled is also ideal for transporting small children.

WARNING

Never tow a sled with ropes or pull straps; always use a solid tow bar. Use of ropes or flexible straps could result in a tailgate accident, when the snowmobile is stopped, with subsequent serious injury.

If it is necessary to tow a disabled snowmobile, securely fasten the disabled machine's skis to the hitch of the tow machine. Remove the drive belt from the disabled machine before towing.

Clearing the Track

If the snowmobile has been operated in deep or slushy snow, it is necessary to clear the track after stopping or the track may freeze, making starting the next time difficult.

WARNING

Always be sure no one is behind the machine when clearing the track. Ice and rocks thrown from the track can cause serious injury.

Tip the snowmobile on its side until the track clears the ground *completely*. Run the track at a moderate speed until all the ice and snow is thrown clear.

CAUTION

If track does freeze, it must be broken loose manually. Attempting to force a frozen track with the engine running will burn and damage the drive belt.

Proper Clothing

Warm and comfortable clothing are a must to provide protection from frostbite. Even mild temperatures can be very uncomfortable and dangerous when combined with a strong wind or when traveling at high speed. See **Table 1** for wind chill factors. Always dress according to what the wind chill factor is, not the temperature. Check with an authorized dealer for suggested types of snowmobile clothing.

WARNING

To provide additional warmth as well as protection against head injury, always wear an approved helmet when snowmobiling.

Table 1 WIND CHILL FACTORS

Estimated Wind Speed in MPH	Actual Thermometer Reading (° F)											
	50	40	30	20	10	0	—10	—20	—30	—40	—50	—60
	Equivalent Temperature (° F)											
Calm	50	40	30	20	10	0	—10	—20	—30	—40	—50	—60
5	48	37	27	16	6	—5	—15	—26	—36	—47	—57	—68
10	40	28	16	4	—9	—21	—33	—46	—58	—70	—83	—95
15	36	22	9	—5	—18	—36	—45	—58	—72	—85	—99	—112
20	32	18	4	—10	—25	—39	—53	—67	—82	—96	—110	—124
25	30	16	0	—15	—29	—44	—59	—74	—88	—104	—118	—133
30	28	13	—2	—18	—33	—48	—63	—79	—94	—109	—125	—140
35	27	11	—4	—20	—35	—49	—67	—82	—98	—113	—129	—145
40	26	10	—6	—21	—37	—53	—69	—85	—100	—116	—132	—148
*	**Little Danger** (for properly clothed person)				**Increasing Danger**			**Great Danger**				
					• Danger from freezing of exposed flesh •							

*Wind speeds greater than 40 mph have little additional effect.

SERVICE HINTS

All procedures described in this book can be performed by anyone reasonably handy with tools. Special tools are required for some procedures; their operation is described and illustrated. These may be purchased at Arctic Cat dealers. If you are on good terms with the dealer's service department, you may be able to borrow from them; however, it should be borne in mind that many of these tools will pay for themselves after the first or second use. If special tools are required, make arrangements to get them before starting. It is frustrating and sometimes expensive to get under way and then find that you are unable to finish up.

Service will be far easier if the machine is clean before beginning work. There are special cleaners for washing the engine-related parts. Just brush or spray on the cleaning solution, let it stand, then rinse it away with a garden hose. Clean all oily or greasy parts with cleaning solvent as they are removed.

Never use gasoline as a cleaning agent, as it presents an extreme fire hazard. Be sure to work in a well-ventilated area when using cleaning solvent. *Keep a fire extinguisher handy.*

Observing the following practices will save time, effort, and frustration as well as prevent possible expensive damage.

1. Tag all similar internal parts for location and mark all mating parts for position. Small parts such as bolts can be identified by placing them in plastic sandwich bags and sealing and labeling the bags with masking tape.

2. Frozen or very tight bolts and screws can often be loosened by soaking with penetrating oil such as WD-40 R, then sharply striking the bolt head a few times with a hammer and punch (or screwdriver for screws). A hammer-driven impact tool can also be very effective. However, make sure the tool is seated squarely on the bolt or nut before striking. Avoid heat unless absolutely necessary, since it may melt, warp, or remove the temper from many parts.

3. Avoid flames or sparks when working near flammable liquids such as gasoline.

4. No parts, except those assembled with a press fit, require unusual force during assembly. If a part is hard to remove or install, find out why before proceeding.

5. Cover all openings after removing parts to keep dirt, small tools, etc., from falling in.

6. Clean all parts as you go along and keep them separated into sub-assemblies. The use of trays, jars, or cans will make reassembly much easier.

7. Make diagrams whenever similar-appearing parts are found. You may *think* you can remember where everything came from—but mistakes are costly. There is also the possibility you may be sidetracked and not return to work for days or even weeks—in which interval carefully laid out parts may have become disturbed.

8. Wiring should be tagged with masking tape and marked as each wire is removed. Again, do not rely on memory alone.

9. When reassembling parts, be sure all shims and washers are replaced exactly as they came out. Whenever a rotating part butts against a stationary part, look for a shim or washer. Use new gaskets if there is any doubt about the condition of old ones. Generally, you should apply gasket cement to only one mating surface so the parts may be easily disassembled in the future. A thin coat of oil on gaskets helps them seal effectively.

10. Heavy grease can be used to hold small parts in place if they tend to fall out during assembly. However, keep grease and oil away from electrical and brake components.

11. High spots may be sanded off a piston with sandpaper, but emery cloth and oil do a much more professional job.

12. Carburetors are best cleaned by disassembling them and soaking the parts in a commercial carburetor cleaner. Never soak gaskets and rubber parts in these cleaners. Never use wire to clean out jets and air passages; they are easily damaged. Use compressed air to blow out the carburetor only if the float has been removed first.

13. Take your time and do the job right. Do not forget that a newly rebuilt snowmobile engine must be broken in the same as a new one. Keep rpm within the limits given in your owner's manual when you get back on the snow.

14. Work safely in a good work area with adequate lighting and allow sufficient time for a repair task.

15. When assembling 2 parts, start all fasteners, then tighten evenly.

Before undertaking a job, read the entire section in this manual which pertains to it. Study the illustrations and text until you have a good idea of what is involved. Many procedures are complicated and errors can be disastrous. When you thoroughly understand what is to be done, follow the prescribed procedure step-by-step.

TOOLS

Every snowmobiler should carry a small tool kit to help make minor adjustments as well as perform emergency repairs.

A normal assortment of ordinary hand tools is required to perform the repair tasks outlined in this manual. The following list represents the minimum requirement:

a. American and metric combination wrenches
b. American and metric socket wrenches
c. Assorted screwdrivers (slot and Phillips type)
d. Pliers
e. Feeler gauges
f. Spark plug wrench
g. Small hammer
h. Plastic or rubber mallet
i. Parts cleaning brush

If purchasing tools, always buy quality tools. They cost more initially, but in most cases will last a lifetime. Remember, the initial expense of new tools is easily offset by the money saved on 1 or 2 repair jobs.

Tune-ups and troubleshooting require a few special tools. All of the following special tools are used in this manual; however, all tools are not necessary for all machine. Read the procedures applicable to your machine to determine what your special tool requirements are.

Ignition Gauge

This tool combines round wire spark plug gap gauges with narrow breaker point feeler gauges (**Figure 7**). The device costs about $3 at auto accessory stores.

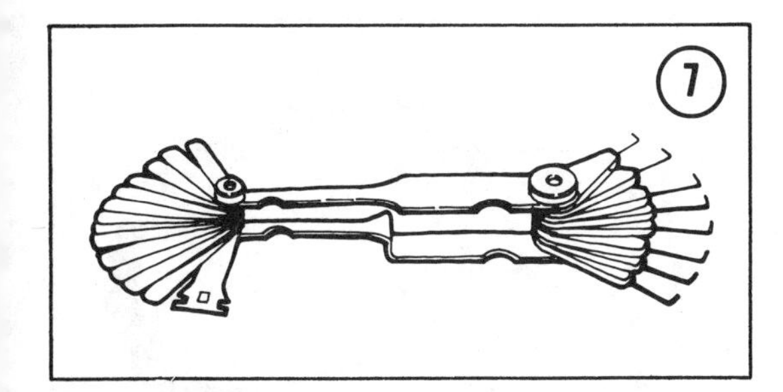

Impact Driver

The impact driver (**Figure 8**) might have been designed with the snowmobiler in mind. It makes removal of screws easy, and eliminates damaged screw slots. Good ones run about $12 at larger hardware stores.

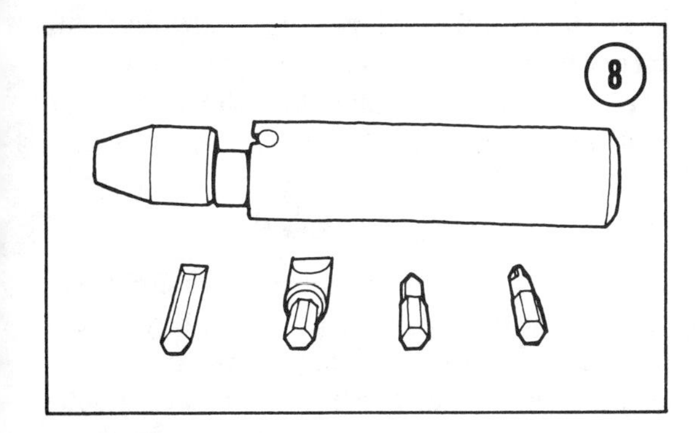

Hydrometer

The hydrometer measures state of charge of the battery and tells much about battery condition. See **Figure 9**. Such an instrument is available at any auto parts store and through most larger mail order outlets. Satisfactory ones cost as little as $3.

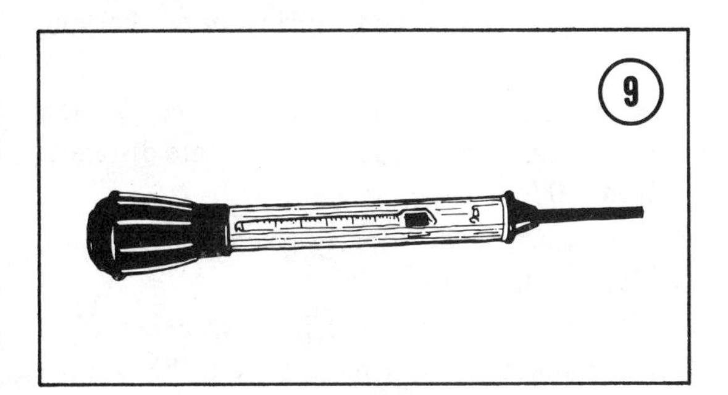

Multimeter (VOM)

A VOM is invaluable for electrical system troubleshooting and service. See **Figure 10**. A

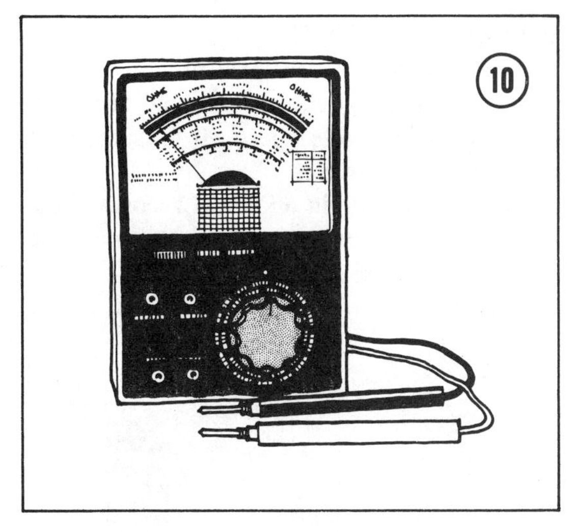

few of its functions may be duplicated by locally fabricated substitutes, but for the serious hobbyist, it is a must. Its uses are described in the applicable sections of this book. Prices start at around $10 at electronics hobbyist stores and mail order outlets.

Timing Gauge

A timing gauge is used to precisely locate the position of the piston before top dead center to achieve the most accurate ignition timing. The instrument is screwed into the spark plug hole and indicates inches and/or millimeters.

The tool shown in **Figure 11** costs about $20 and is available from most dealers and mail order houses. Less expensive tools, which use a vernier scale instead of a dial indicator, are also available.

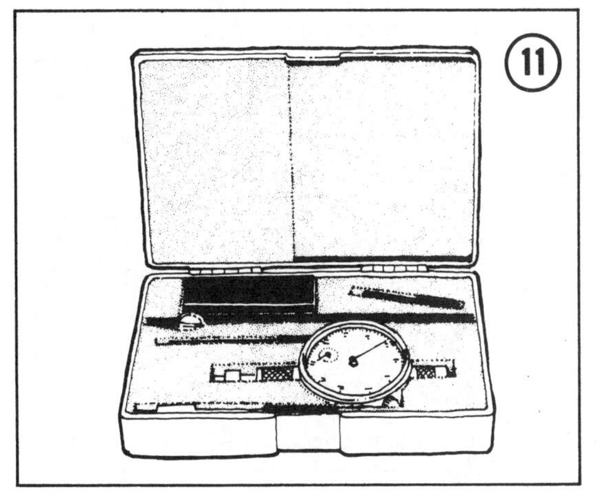

Carburetor Synchronizer

A carburetor synchronizer is used on engines with multiple carburetors to fine tune the synchronization and idle speed. It is sometimes called an air flow meter.

The tool shown in **Figure 12** costs about $10-15 at most dealers, auto parts stores, and mail order houses.

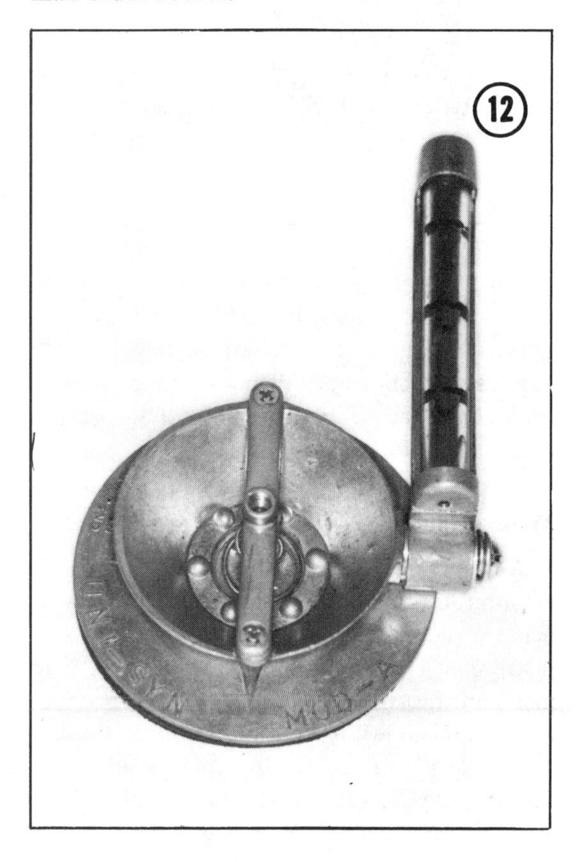
12

Compression Gauge

The compression gauge measures the compression pressure built up in each cylinder. The results, when properly interpreted, indicate general piston, cylinder, ring, and head gasket condition.

Gauges are available with or without the flexible hose. Prices start around $5 at most auto parts stores and mail order outlets. See **Figure 13**.

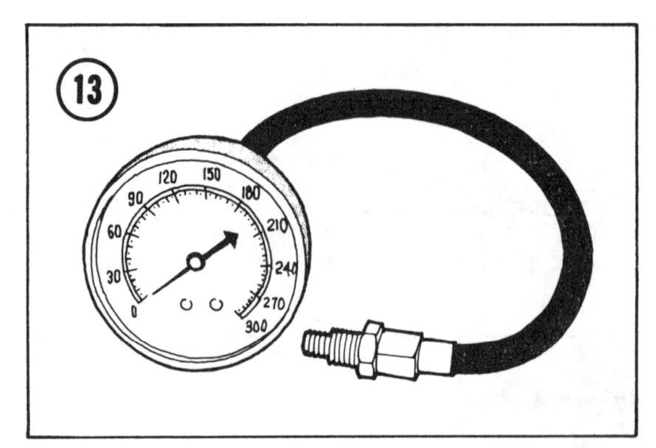

13

EXPENDABLE SUPPLIES

Certain expendable supplies are also required. These include grease, oil, gasket cement, wiping rags, cleaning solvent, and distilled water. Solvent is available at many service stations. Distilled water, required for the battery, is available at every supermarket. It is sold for use in steam irons, and is quite inexpensive. An increasing number of mechanics clean oily parts with a solution of common household detergent or laundry powder.

WORKING SAFELY

Professional mechanics can work for years without sustaining serious injury. If you observe a few rules of common sense and safety, you can enjoy many safe hours servicing your own machine. You can also hurt yourself or damage the machine if you ignore these rules.

1. Never use gasoline as a cleaning solvent.
2. Never smoke or use a torch in the area of flammable liquids, such as cleaning solvent in open containers.
3. Never smoke or use a torch in an area where batteries are charging. Highly explosive hydrogen gas is formed during the charging process.
4. If welding or brazing is required on the machine, remove the fuel tank to a safe distance, at least 50 feet away.
5. Be sure to use properly sized wrenches for nut and bolt turning.
6. If a nut is tight, think for a moment what would happen to your hand should the wrench slip. Be guided accordingly.
7. Keep your work area clean and uncluttered.
8. Wear safety goggles in all operations involving drilling, grinding, or use of a chisel.

9. Never use worn tools.

10. Keep a fire extinguisher handy. Be sure it is rated for gasoline and electrical fires.

SNOWMOBILING CODE OF ETHICS

When snowmobiling, always observe the following code of ethics as provided by the International Snowmobile Industry Association.

1. I will be a good sportsman. I recognize that people judge all snowmobile owners by my actions. I will use my influence with other snowmobile owners to promote sportsmanlike conduct.
2. I will not litter trails or camping areas. I will not pollute streams or lakes.
3. I will not damage living trees, shrubs, or other natural features.
4. I will respect other people's property and rights.
5. I will lend a helping hand when I see someone in distress.
6. I will make myself and my vehicle available to assist search and rescue parties.
7. I will not interfere with or harass hikers, skiers, snowshoers, ice fishermen, or other winter sportsmen. I will respect their rights to enjoy our recreation facilities.
8. I will know and obey all federal, state, provincial, and local rules regulating the operation of snowmobiles in areas where I use my vehicle. I will inform public officials when using public lands.
9. I will not harass wildlife. I will avoid areas posted for the protection and feeding of wildlife.
10. I will stay on marked trails or marked roads open to snowmobiles. I will avoid cross-country travel unless specifically authorized.

SNOWMOBILE SAFETY

General Tips

1. Read your owner's manual and know your machine.
2. Check throttle and brake controls before starting the engine. Frozen controls can cause serious injury.
3. Know how to make an emergency stop.
4. Know all state, provincial, federal, and local laws concerning snowmobiling. Respect private property.
5. Never add fuel while smoking or when the engine is running. Always use fresh, correctly mixed fuel. Incorrect fuel mixtures can cause engine failure, and can leave you stranded in severe weather.
6. Wear adequate clothing to avoid frostbite. Never wear loose scarves, belts, or laces that could catch in moving parts or on tree limbs.
7. Wear eye and head protection. Wear tinted goggles or face shields to guard against snowblindness. Wear yellow eye protection only during white-outs or flat-light conditions.
8. Never allow anyone to operate the snowmobile without proper instruction.
9. Use the "buddy system" for long trips. A snowmobile travels farther in 30 minutes than you can walk in a day.
10. Take along sufficient tools and spare parts for emergency field repairs.
11. Use a sled with a stiff tow bar for carrying extra supplies. Do not overload your snowmobile.
12. Carry emergency survival supplies when going on long trips. Notify friends and relatives of your destination and expected arrival time.
13. Never attempt to repair your machine while the engine is running.
14. Check all machine components and hardware frequently, especially skis and steering.
15. Never lift the rear of machine to clear the track. Tip the machine on its side and be sure no one is behind it.
16. Winch the snowmobile onto a tilt-bed trailer—never drive it on. Secure machine firmly to trailer and ensure the trailer lights operate.

Operating Tips

1. Never operate the machine in crowded areas, or steer toward persons.

2. Avoid avalanche areas and other unsafe terrain.

3. Cross highways (where permitted) at a 90-degree angle after stopping and looking in both directions. Post traffic guards if crossing in groups.

4. Do not ride the snowmobile on or near railroad tracks. The snowmobile engine can drown out the sound of an approaching train. It is difficult to maneuver the snowmobile from between the tracks.

5. Do not ride the snowmobile on ski slopes or other areas with skiers.

6. Always check the thickness of the ice before riding on frozen lakes or rivers. Do not panic if you go through the ice—conserve energy.

7. Keep the headlight and taillight areas free of snow and never ride at night without lights.

8. Do not ride the snowmobile with shields, guards and protective hoods in place.

9. Do not attempt to open new trails at night. Follow established trails or unseen barbed wire or guy wires may cause serious injury or death.

10. Always steer with both hands.

11. Be aware of terrain and avoid operating the snowmobile at excessive speed.

12. Do not panic if the throttle sticks. Pull the "tether" string or push the emergency stop switch.

13. Drive more slowly when carrying a passenger, especially a child.

14. Always allow adequate stopping distance based on ground cover conditions. Ice requires a greater stopping distance to avoid skidding. Apply brakes gradually on ice.

15. Do not speed through wooded areas. Hidden obstructions, hanging limbs, unseen ditches, and even wild animals can cause accidents.

16. Do not tailgate. Rear end collisions can cause injury and machine damage.

17. Do not mix alcoholic beverages with snowmobiling.

18. Keep feet on footrests at all times. Do not permit feet to hang over sides or attempt to stabilize the machine with feet when making turns or in near-spill situations; broken limbs could result.

19. Do not stand on the seat, stunt, or show-off.

20. Do not jump the snowmobile. Injury or machine damage could result.

21. Always keep hands and feet out of the track area when the engine is running. Use extra care when freeing the snowmobile from deep snow.

22. Check the fuel supply regularly. Do not travel further than your fuel will permit you to return.

23. Whenever you leave your machine unattended, remove the "tether" switch and ignition key.

Preparing for a Trip

1. Check all bolts and fasteners for tightness. Do not operate your snowmobile unless it is in top operating condition.

2. Check weather forecasts before starting out on a trip. Cancel your plans if a storm is possible.

3. Study maps of the area before the trip and know where help is located. Note locations of phones, resorts, shelters, towns, farms, and ranches. Know where fuel is available. If possible use the buddy system.

4. Do not overload your snowmobile. Use a sled with a stiff tow bar to haul extra supplies.

5. Do not risk a heart attack if your snowmobile gets stuck in deep snow. Carry a small block and tackle for such situations. Never allow anyone to manually pull on the skis while you attempt to drive machine out.

6. Do not ride beyond one-half the round trip cruising range of your fuel supply. Keep in mind how far it is home.

7. Always carry emergency survival supplies when going on long trips or traveling in unknown territory. Notify friends and relatives of your destination and expected arrival time.

8. Carry adequate eating and cooking utensils (small pans, kettle, plates, cups, etc.) on longer trips. Carry matches in a waterproof container, candles for building a fire, and easy-to-pack food that will not be damaged by freezing. Carry dry food or "space" energy sticks for emergency rations.

9. Pack extra clothing, a tent, sleeping bag,

hand axe and compass. A first aid kit and snow shoes may also come in handy. Space age blankets (one side silverfoil) furnish warmth and can be used as heat reflectors or signalling devices for aerial search parties.

Emergency Survival Techniques

1. Do not panic in the event of an emergency. Relax, think the situation over, then decide on a course of action. You may be within a short distance of help. If possible repair your snowmobile so you can drive to safety. Conserve your energy and stay warm.

2. Keep hands and feet active to promote circulation and avoid frostbite while servicing your machine.

3. Mentally retrace your route. Where was the last point where help could be located? Do not attempt to walk long distances in deep snow. Make yourself comfortable until help arrives.

4. If you are properly equipped for your trip you can turn any undesirable area into a suitable campsite.

5. If necessary, build a small shelter with tree branches or evergreen boughs. Look for a cave or sheltered area against a hill or cliff. Even burrowing in the snow offers protection from the cold and wind.

6. Prepare a signal fire using evergreen boughs and snowmobile oil. If you can not build a fire, make an S-O-S in the snow.

7. Use a policeman's whistle or beat cooking utensils to attract attention or frighten off wild animals.

8. When your camp is established, climb the nearest hill and determine your location. Observe landmarks on the way, so you can find your way back to your campsite. Do not rely on your footprints. They may be covered by blowing snow.

NOTE: If you own a 1978 or later model, first check the Supplement at the back of the book for any new service information.

CHAPTER TWO

PERIODIC MAINTENANCE AND TUNE-UP

Regular maintenance is the best guarantee of a trouble-free, long-lasting snowmobile. An afternoon spent now, cleaning, inspecting, and adjusting, can prevent costly mechanical problems in the future and unexpected breakdowns on the trail.

The procedures presented in this chapter can be easily carried out by anyone with average mechanical skills. The operations are presented step-by-step and if they are followed it is difficult to go wrong.

SERVICE INTERVALS

Service frequency depends on use but as a rule of thumb, the items that follow should be checked and corrected if necessary prior to a long ride or anticipated use of several hours. Though many of the items will not require actual service, even after many hours, a systematic check is a good habit to develop. It can serve to warn of impending trouble and it will allow you to become familiar with the machine and gain a measure of confidence in the bargain. A day of riding is much more pleasant if you are not worrying about the condition of your machine.

FUEL FILTER

The manufacturer recommends that the in-line fuel filter be checked once a month but there is no harm in checking it visually more often to be sure that it has not become clogged by dirty fuel. A typical filter installation is shown in **Figure 1**. If the filter is dirty, disconnect the lines from the filter body and plug the fuel line from the tank to prevent fuel from leaking into the machine.

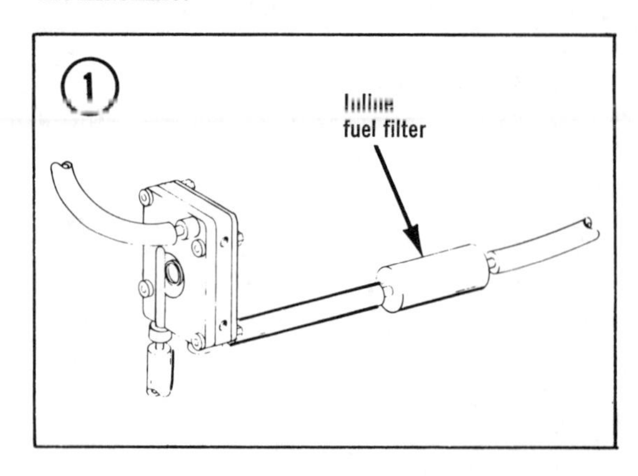

If the filter is not damaged, it can be cleaned by flushing it with fresh gasoline, poured in through the arrow end (fuel pump end) of the filter. Cover both spigots and shake the filter to loosen particles. Then, shake out the gasoline and particles through the fuel tank end. Install the filter with the arrow pointing toward the fuel pump. Make sure connections are tight and leak-free. If the filter care is damaged, replace it with a new filter, obtained from an Arctic Cat dealer.

If this is not convenient, a small automotive filter, the size of the original, may be substituted.

CHAINCASE OIL LEVEL

Check level of chaincase oil at intervals as specified in **Table 1**.

Table 1 SERVICE INTERVALS

Daily
- Cooling fins
- Carburetor and exhaust systems
- Throttle system
- Brake system
- Steering system
- Fuel supply
- Taillight, brakelight, and headlight
- Emergency shut-off switch

Weekly
- Ski alignment
- Wear bar

40 Operating Hours
- Lubrication

Monthly
- In-line filter
- Chain tension
- Track tension and alignment

100 Operating Hours or End of Season
- Drive clutch/driven pulley service check
- Engine service check

Remove the check plug (**Figure 2**) and check that oil level is at the bottom of the access hole.

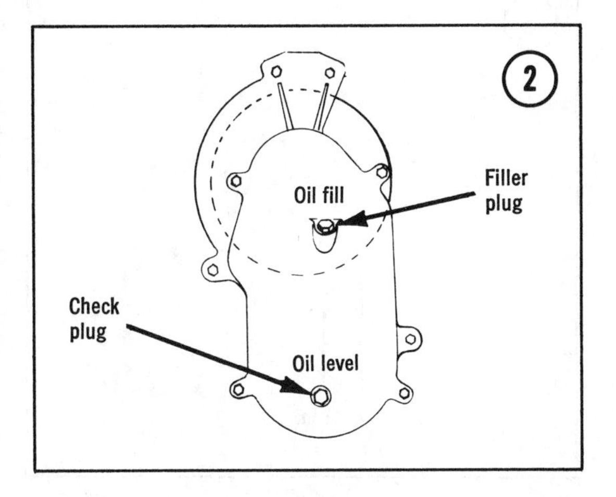

Top up oil level if necessary with Arctic Chain Lube.

Use a syringe or oil suction device to remove old oil when changing oil or for machine storage preparation.

DRIVE BELT

Inspect the drive belt (**Figure 3**) for wear, cracking, or other damage or deterioration. If the belt is not in good condition, full power will not be transmitted from the drive clutch to the driven pulley. Measure the belt (monthly) and compare it with the service specifications in **Table 2**. If the belt is not within specifications, replace it.

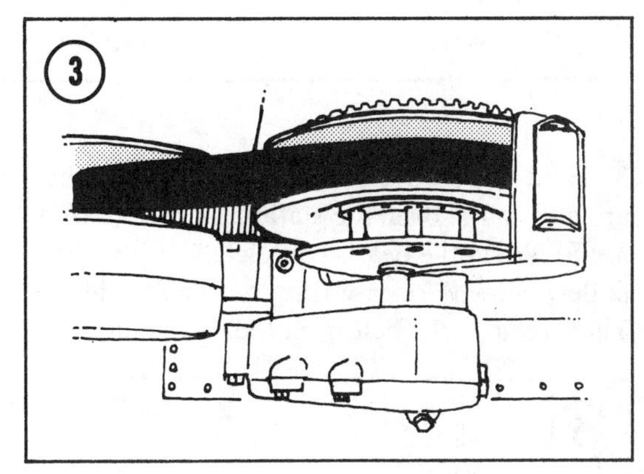

1. Remove the pins from the clutch guard (**Figure 4**) and raise the guard. Set the parking brake.
2. Push against the moveable sheave and rotate it clockwise until the sheaves are apart.

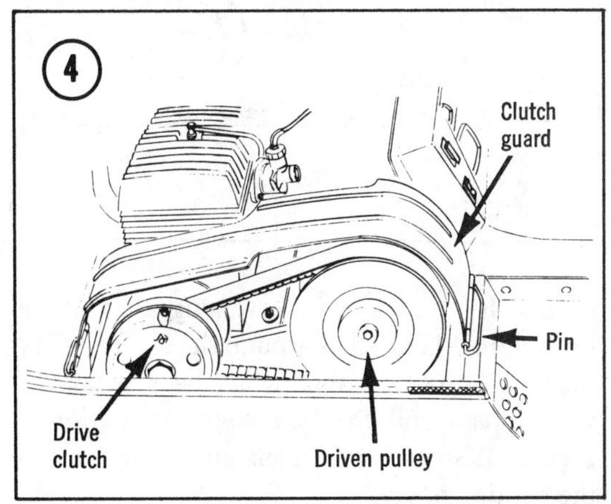

Table 2 DRIVE BELT SPECIFICATIONS

Model/Year	Outside Circumference	Width
Lynx		
All years	43 1/16 to 43 7/8 in.	1 1/8 to 1 9/32 in.
Pantera		
Through 1975	43 1/16 to 43 7/16 in.	1 1/8 to 1 9/32 in.
1976 on	46 1/2 to 46 7/8 in.	1 1/8 to 1 9/32 in.
Jag		
All years	44 5/16 to 44 11/16 in.	1 1/8 to 1 9/32 in.
Cheetah		
All years	43 1/16 to 43 7/16 in.	1 1/8 to 1 9/32 in.
El Tigre		
All years	46 1/2 to 46 7/8 in.	1 1/8 to 1 9/32 in.
Panther		
All years	43 1/16 to 43 7/16 in.	1 1/8 to 1 9/32 in.

3. Hold the sheaves apart, pull the drive belt up, and roll it over the stationary sheave (**Figure 5**). When the belt is completely off the drive pulley, slowly release the moveable sheave. Then, remove the belt from the drive clutch.

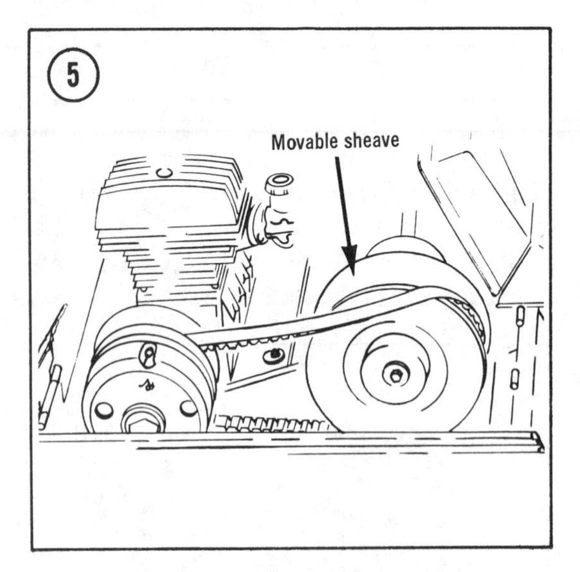

4. Install a new belt by reversing the removal procedure. Place it first around the clutch pulley, then, push the moveable sheave back as far as it will go and roll the belt over the stationary sheave. Reinstall the clutch guard and release the parking brake.

BRAKE

The brake system should be checked daily or each time the machine is ridden. Check for wear and damage to the lever, cable, caliper, pads, and disc. Check also to see that the brake operates smoothly and releases completely when the lever is released.

Pull the lever all the way back and check the distance between the lever and the stop (**Figure 6**). It should be ½-¾ in. (12-18mm). If adjustment is required, turn the adjusting nut (**Figure 7**) clockwise to decrease brake lever travel, and counterclockwise to increase it.

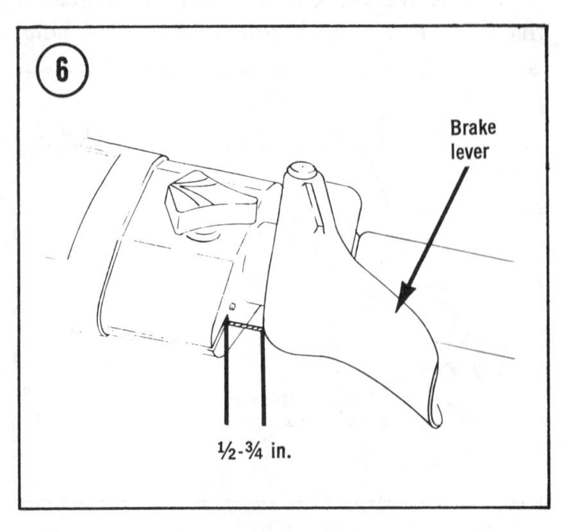

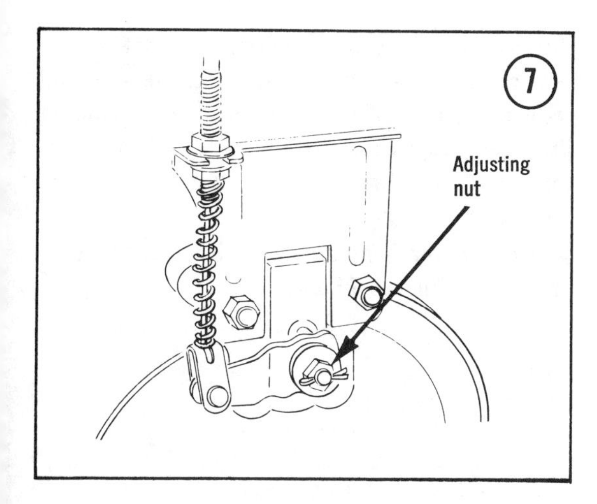

WARNING

Never operate the snowmobile when there is any doubt about the condition of the brake system. Brake failure could cause serious injury.

DRIVE CLUTCH AND DRIVEN PULLEY

The drive clutch and driven pulley need no maintenance other than disassembly and cleaning after 100 hours or at the end of the season. This work should be entrusted to a dealer. In addition, if premature belt failure or wear indicate misalignment of the drive clutch and driven pulley, the problem should be referred to a dealer because special tools are required.

TRACK TENSION

Correct track tension is important. If the track is loose, it will slap the tunnel and can ratchet on the drive sprockets. If it is too tight, the Hi-Fax slides and rear idler will wear rapidly. Tension should be check and corrected if necessary once a month.

1. Clean ice and snow from the track and from inside the skid frame. Raise the rear of the snowmobile so the track is off the ground and free to rotate.

2. Pull down on the track at midpoint (**Figure 8**) with moderate force and measure the distance. It should be ¾-1 in. (19-25mm).

3. If adjustment is required, loosen the jam nuts on the track adjusting bolts (**Figure 9**). If the distance is greater than 1 in. tighten the adjusting bolts and then lock them with the jam nuts to prevent them from turning further. If the distance is less than ¾ in. the adjusting bolts must be loosened and the jam nuts tightened when the tension is correct.

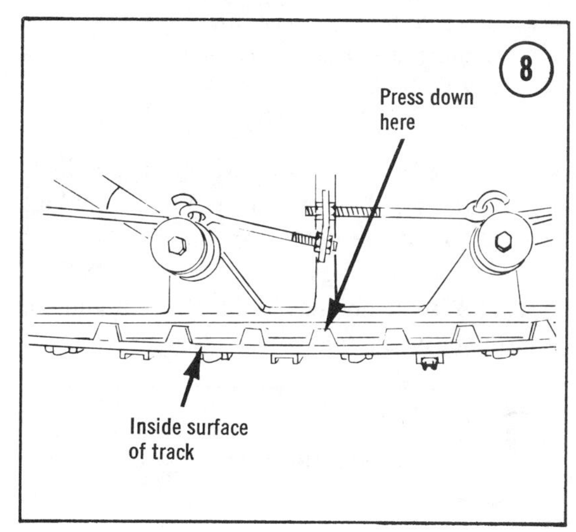

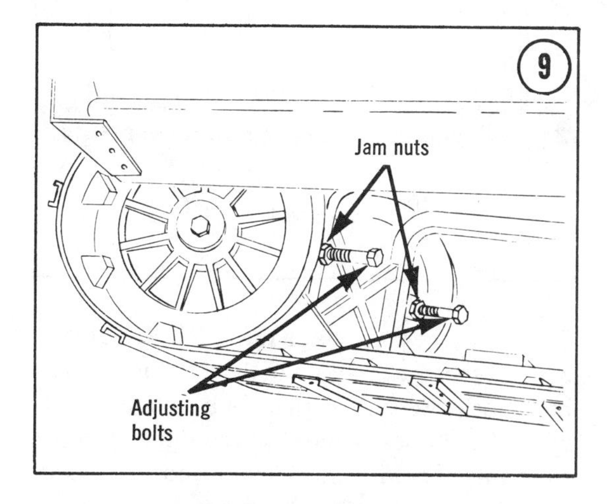

TRACK ALIGNMENT

Track alignment is related to track tension and should be checked and adjusted at the time track tension is checked and adjusted. If the track is misaligned the rear idler wheels, drive lugs, and track will wear rapidly.

1. Clean ice and snow from the track and from inside the skid frame. Raise the rear of the snowmobile so the track is free to rotate. The tips of the skis must be against a wall or other immoveable barrier.

2. Start engine and apply just enough throttle to turn the track several times. Then shut off the engine and allow the track to coast to a stop; do not stop it with the brake.

WARNING

Do not stand in front or to the rear of the snowmobile when the engine is running, and make certain hands, feet, and clothing are kept away from the track when it is turning.

3. After the track has stopped check the alignment of the rear idler wheels and the track lugs (**Figure 10**). If the idlers are centered between the lugs, the alignment is correct. However, if they are offset to one side or the other, alignment adjustment is required.

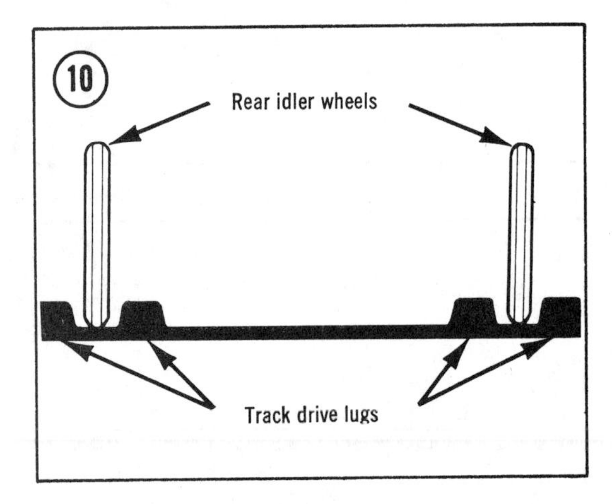

4. Loosen the jam nuts on the rear idler adjuster bolts. If the track is offset to the left, turn the left bolt clockwise or the right bolt counterclockwise. If the track is offset to the right, turn the right bolt clockwise or the left bolt counterclockwise. Also make certain the correct track tension is maintained during the alignment correction. When the adjustment is correct, tighten the jam nuts to prevent the adjuster bolts from turning further.

Test drive the snowmobile then recheck the alignment as described above and make any corrections that are indicated.

SUSPENSION ADJUSTMENT

The suspension can be adjusted to accomodate rider weight (rear spring) and driving conditions (front spring). Increased front spring tension decreases the pressure on the skis and reduces turning effort. Decreased spring tension increases the pressure on the skis and increases turning effort, although steering will be quicker and more responsive.

1. Loosen the jam nuts on either suspension adjusters (**Figures 11 and 12**).

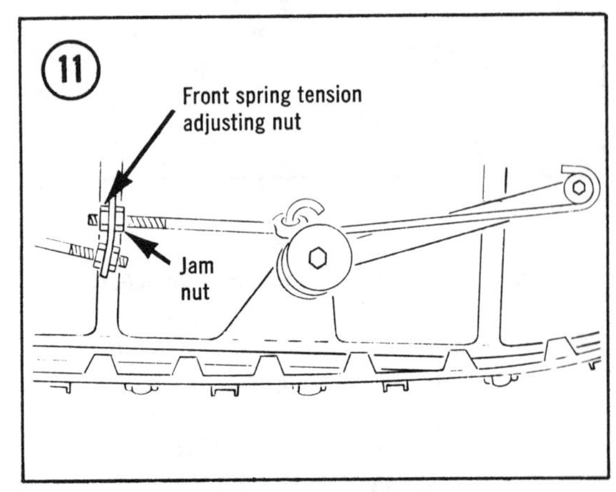

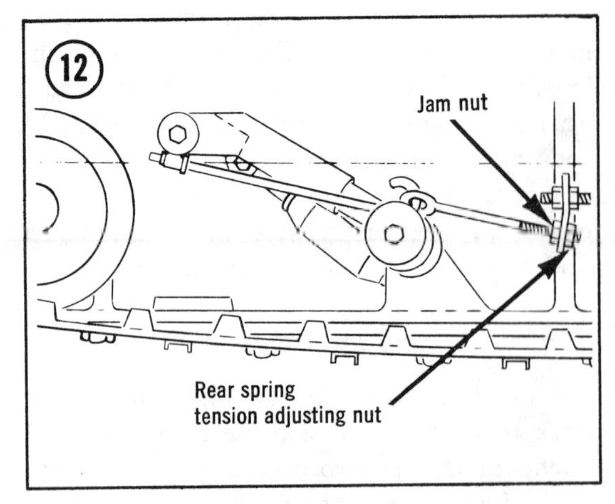

2. Increase spring tension by tightening the adjusting nuts, and decrease it by loosening the adjusting nuts. The rear spring should be adjusted so that it will prevent the machine from bottoming on all but the most severe bumps.

3. When the suspension has been satisfactorily adjusted, tighten the jam nuts.

TUNE-UP

A complete tune-up should be performed after every 100 hours of operation or at the end of the season. Expendable ignition parts (spark plugs for all models, plus points and condenser for Lynx) should be routinely replaced during the tune-up. Have the plugs on hand before you begin.

Always check the condition of spark plug wires, ignition wires, and fuel lines for splitting, loose connections, hardness, and other signs of deterioration. Check that all manifold nuts and carburetor nuts are tight and no crankcase leaks are present. A small air leak can make a good tune-up impossible as well as affect performance. A small air leak can also cause serious damage by allowing the engine to run on a "too-lean" fuel mixture.

The following list of general hints will help make a tune-up easier and more successful.

a. Always use good tools and tune-up equipment. The money saved from 1 or 2 home tune-ups will more than pay for good tools; from that point you are money ahead. Refer to Chapter One for suitable types of tune-up/test equipment.
b. The purchase of a small set of ignition wrenches and 1 or 2 "screwholding" or magnetic screwdrivers will ease the work in replacing breaker points and help eliminate losing small screws.
c. Always purchase quality ignition components.
d. When using a feeler gauge to set breaker points, ensure the blade is wiped clean before inserting between the points.
e. Ensure points are fully open when setting gap with a feeler gauge.
f. Be sure feeler gauge is not tilted or twisted when it is inserted between the contacts. Closely observe the points and withdraw the feeler gauge slowly and carefully. A slight resistance should be felt; however, the movable contact point must *not* "spring back" even slightly when the feeler gauge blade is removed.
g. If breaker points are only slightly pitted they can be dressed lightly with a small ignition point file. *Do not* use sandpaper as it leaves a residue on the points.
h. After points have been installed, always ensure they are properly aligned, or premature pitting and burning will result (see **Figure 13**). Only bend the *fixed* half of the points, not the movable arm.

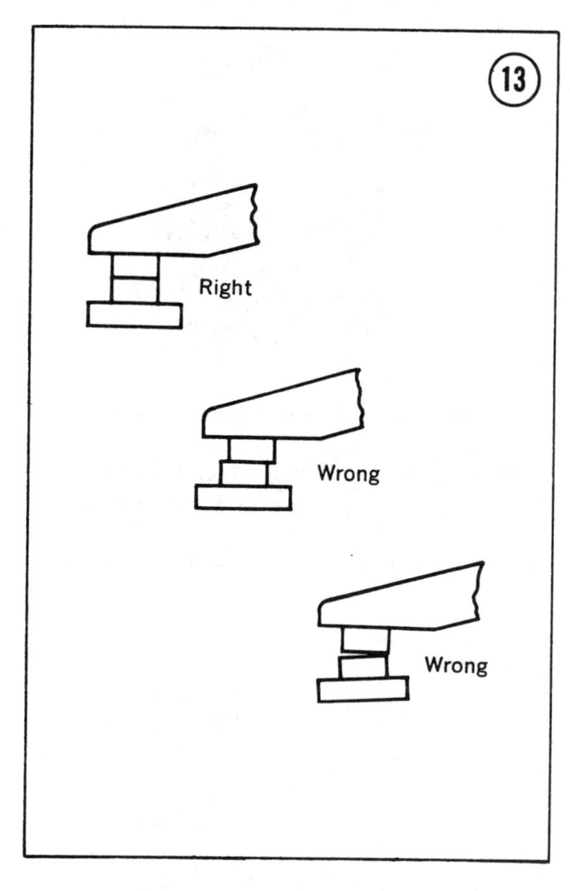

i. When point gap has been set, spring points open and insert a piece of clean paper or board between the contacts. Wipe the contact a few times to remove any trace of oil or grease. A small amount of oil or grease on the contact surfaces will cause the points to prematurely burn or arc.
j. When connecting a timing light or timing tester always follow the manufacturer's instructions.

Because different systems in the engine interact, the procedures should be done in the following order:

1. Check and tighten cylinder head bolts
2. Work on ignition system
3. Adjust carburetor

CYLINDER HEAD BOLTS

1. On Panther and Cheetah models remove the upper engine cooling cowl (**Figure 14**).

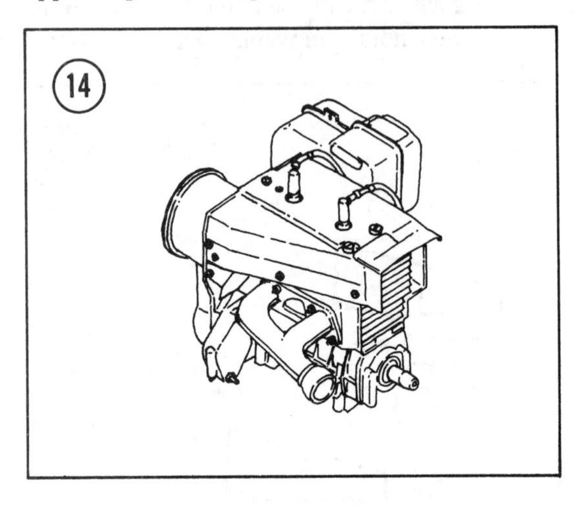

2. Refer to **Figure 15** and tighten the head bolts to the correct torque value, in the pattern shown.

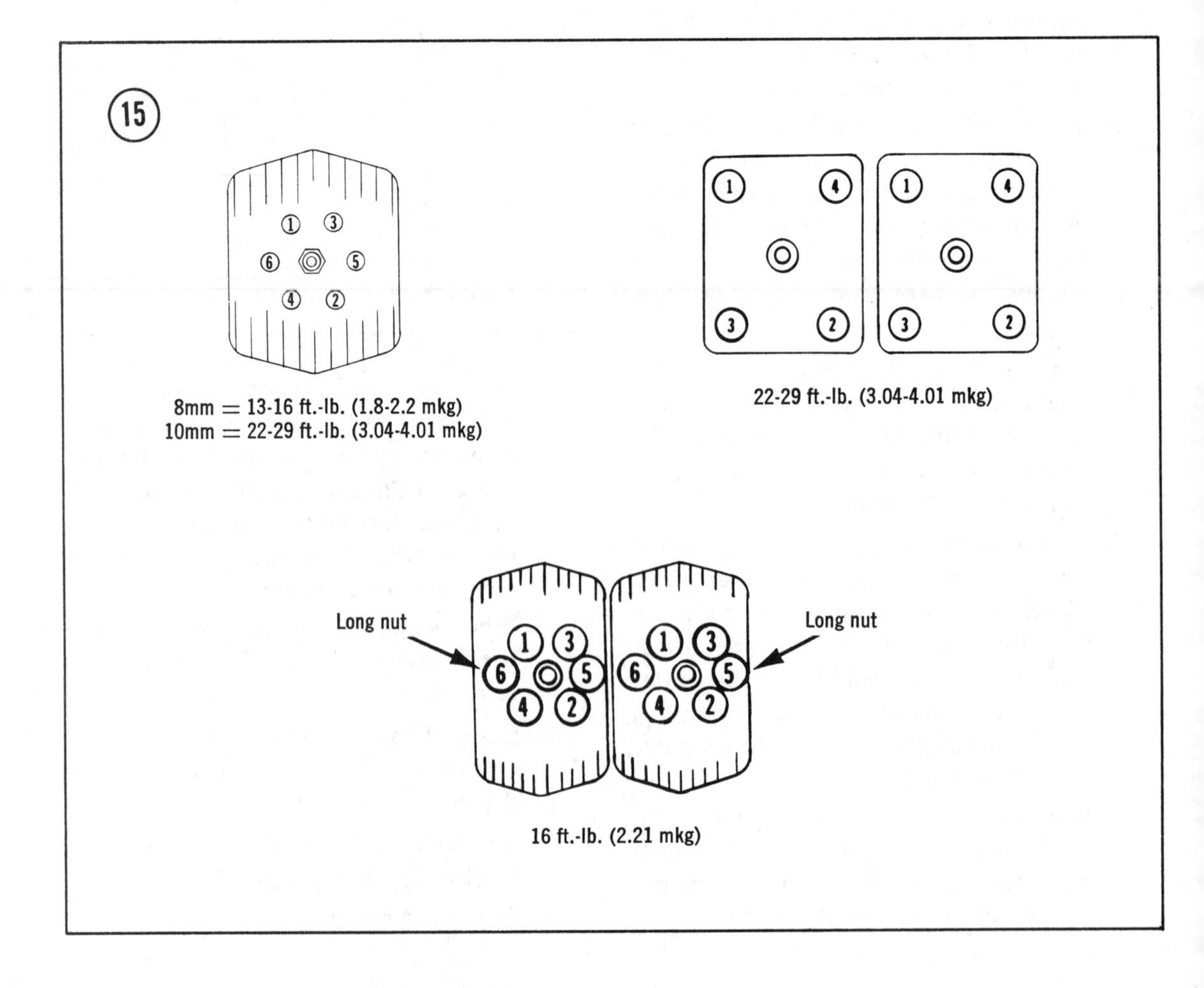

IGNITION SYSTEM

With the exception of the Lynx, Arctic Cat snowmobiles use electronic capacitor discharge ignition (CDI). The CDI is virtually trouble-free and should last the life of the machine without service or adjustment. Air gap and timing are preset and will not change. If trouble is suspected, the unit should be entrusted to a dealer; special skills and test equipment are required to correctly evaluate the condition of the CDI, and incorrect test connections could damage an otherwise good unit and render it unserviceable.

Spark Plugs

Spark plugs should be routinely replaced at each tune-up. In addition, they should be removed and inspected if poor performance indicates that the plugs may be dirty or faulty (see Chapter Three, *Troubleshooting*). In any case, the plugs should be compared to **Figure 16** to

16

Normal plug appearance noted by the brown to grayish-tan deposits and light electrode wear. This plug indicates the correct plug heat range and proper air/fuel ratio.

Red, brown, yellow, and white coatings caused by fuel and oil additives. Such additives should not be used or damage will result.

Carbon fouling distinguished by dry, fluffy black carbon deposits which may be caused by an over-rich air/fuel mixture, excessive hand choking, clogged air filter, or excessive idling.

Shiny yellow glaze on insulator cone is caused when the powdery deposits from fuel and oil additives melt. Melting occurs during hard acceleration after prolonged idling. This glaze conducts electricity and shorts out the plug. Avoid the use of additives at all times.

Oil fouling indicated by wet, oily deposits caused by too much oil in the mix. A hotter plug temporarily reduces oil deposits, but a plug that is too hot leads to preignition and possible engine damage.

Overheated plug indicated by burned or blistered insulator tip and badly worn electrodes. This condition may be caused by preignition, cooling system defects, lean air/fuel ratios, low octane fuel, or over advanced ignition timing.

Spark plug condition photos courtesy of AC Spark Plug Division, General Motors Corporation.

determine their suitability (heat range) and the condition of the engine. Recommended spark plugs are shown in **Table 3**.

Carefully pull the spark plug caps from the plugs by grasping the caps; do not pull on the spark plug wires. Clean the area around the base. Unscrew the plugs from the cylinder heads and compare them to **Figure 16**. If a change in heat range is indicate, it should be only one increment (see **Table 3**).

1. *Normal Condition*—If plugs have a light tan or gray colored deposit and no abnormal gap wear or erosion, good engine, carburetion, and ignition conditions are indicated. The plug in use is of the proper heat range, and may be serviced and returned to use.

2. *Carbon Fouled*—Soft, dry sooty deposits are evidence of incomplete combustion and can usually be attributed to rich carburetion. This condition is also sometimes caused by weak ignition, retarded timing, or low compression. Such a plug may usually be cleaned and returned to service, but the condition which causes fouling should be corrected.

3. *Oil Fouled*—This plug exhibits a black insulator tip, damp oil film over the firing end, and a carbon layer over the entire nose. Electrodes will not be worn. Common causes for this condition are listed in **Table 4**.

Table 4 CAUSES OF FOULED PLUGS

• Improper fuel/oil mixture	• Weak ignition
• Wrong type of oil	• Excessive idling
• Idle speed too low	• Wrong spark plugs (too cold)
• Clogged air silencer	

Oil fouled spark plugs may be cleaned in a pinch, but it is better to replace them. It is important to correct the cause of fouling before the engine is returned to service.

4. *Gap Bridging*—Plugs with this condition exhibit gaps shorted out by combustion chamber deposits used between electrodes. On 2-stroke engines any of the following may be the cause:

a. Improper fuel/oil mixture

b. Clogged exhaust

Be sure to locate and correct the cause of this spark plug condition. Such plugs must be replaced with new ones.

5. *Overheated*—Overheated spark plugs exhibit burned electrodes. The insulator tip will be light gray or even chalk white. The most common cause for this condition is using a spark plug of the wrong heat range (too hot). If it is known that the correct plug is used, other causes are lean fuel mixture, engine overloading or lugging, loose carburetor mounting, or timing advanced too far. Always correct the fault before putting the snowmobile back into service. Such plugs cannot be salvaged; replace with new ones.

6. *Worn Out*—Corrosive gases formed by combustion and high voltage sparks have eroded the electrodes. Spark plugs in this condition require more voltage to fire under hard acceleration—often more than the ignition system can supply. Replace them with new spark plugs of the same heat range.

7. *Preignition*—If electrodes are melted, preignition is almost certainly the cause. Check for carburetor mounting or intake manifold leaks, also over-advanced ignition timing. It is also possible that a plug of the wrong heat range (too hot) is being used. Find the cause of preignition before placing the engine back into service.

The spark plugs recommended by the factory are usually the most suitable for your machine. If riding conditions are mild, it may be advisable to go to spark plugs one step hotter than normal. Unusually severe riding conditions may require slightly colder plugs. See **Table 3**.

CAUTION

Ensure spark plugs used have the correct thread reach. A thread reach too short will cause the exposed threads in the cylinder head to accumulate carbon, resulting in stripped cylinder head threads when the proper plug is installed. A thread reach too long will cause the exposed spark plug threads to accumulate carbon resulting in stripped cylinder head threads when the plug is removed.

It may take some experimentation to arrive at the proper plug heat range for your type of

Table 3 SPARK PLUG TYPE AND GAP

Model/Engine	Type*	Gap
Lynx		
1974-1975 2000S	NGK BR9HS Champion L77J	0.026-0.030 in. (0.66-0.76mm)
1976-1977 2000S	NGK BR9ESA Champion RN2	0.018-0.022 in. (0.46-0.56mm)
1976-1977 2000T	NGK BR8ESA Champion RN3	0.018-0.022 in. (0.46-0.56mm)
El Tigre		
1974-1975 all	NGK BR9EVA Champion RN2	0.016-0.020 in. (0.41-0.51mm)
1976-1977 4000	NGK BR8ESA Champion RN3	0.018-0.022 in. (0.46-0.56mm)
1976-1977 5000	NGK BR9ESA Champion RN2	0.018-0.022 in. (0.46-0.56mm)
Panther/Cheetah		
1974-1975 all	NGK B8ESA Champion N3 Bosch W215T 28	0.020-0.032 in. (0.51-0.81mm)
1976-1977 4000	NGK BR8ESA Champion RN3	0.018-0.022 in. (0.46-0.56mm)
1976-1977 5000	NGK BR9ESA Champion RN2	0.018-0.022 in. (0.46-0.56mm)
Pantera		
All fan-cooled	NGK BR8ESA Champion RN3	0.018-0.022 in. (0.46-0.56mm)
All free-air	NGK BR9ESA Champion RN2	0.018-0.022 in. (0.46-0.56mm)
Jag		
1975	NGK BR8ESA Champion RN3	0.030 in. (0.76mm)
1976-1977	NGK BR8ESA Champion RN3	0.018-0.022 in. (0.46-0.56mm)

*Heat ranges shown are for normal operation. If a colder plug is required because machine is operated at sustained high speeds, change one increment at a time (e.g., if BR8ESA is specified, change to BR9ESA). If a hotter plug is required because machine is operated at low speeds, change one increment at a time (e.g., if BR8ESA is specified, change to BR7ESA).

NGK spark plugs are recommended by the manufacturer; however, an equivalent plug from another manufacturer can be substituted as long as its heat range corresponds to the NGK spark plug specified in this table.

riding. As a general rule, use as cold a spark plug as possible without fouling. This will give the best performance.

Remove and clean spark plugs at least once a season. After cleaning, inspect them for worn or eroded electrodes. Replace them if in doubt about their condition. If the plugs are serviceable, file the center electrodes square, then adjust the gaps by bending the outer electrodes only. Measure the gap with a round wire spark plug gauge only; a flat gauge will yield an incorrect reading.

Set the gap on the new plugs by bending only the side electrode. Correct gap is also shown in **Table 3**. Screw the plugs into the cylinder heads using new washers. Tighten the plugs snugly but do not tighten so much that there is risk of damaging the threads in the heads. Reconnect the high-tension leads.

> NOTE: *On Lynx models, do not install the spark plugs until the timing has been checked (see below).*

Lynx Ignition

The Lynx is equipped with a conventional contact breaker ignition system that requires point replacement and timing adjustment. The procedure that follows applies only to the Lynx.

1. Hold the flywheel with a flywheel spanner (**Figure 17**—part No. 0144-007) and unscrew the large center bolt from the flywheel and the crankshaft.

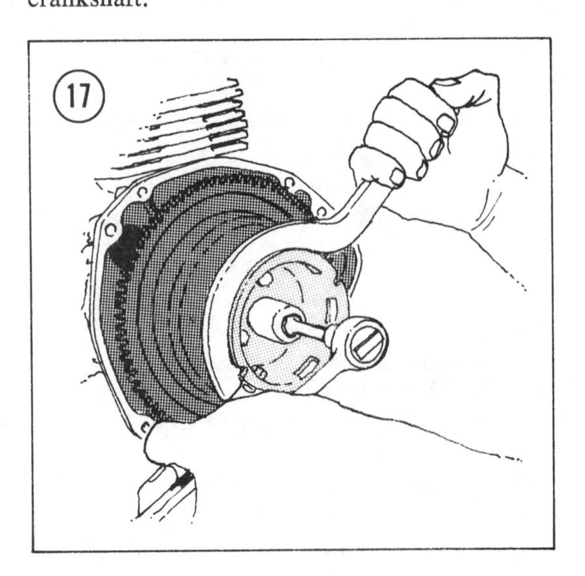

2. Install a flywheel puller (**Figure 18**—part No. 0144-064 or a standard puller like the one shown in **Figure 19**). Tighten the large puller bolt to break the flywheel loose. It may be necessary to rap sharply on the head of the bolt to break the wheel loose. Remove the flywheel and set it aside with the open side up.

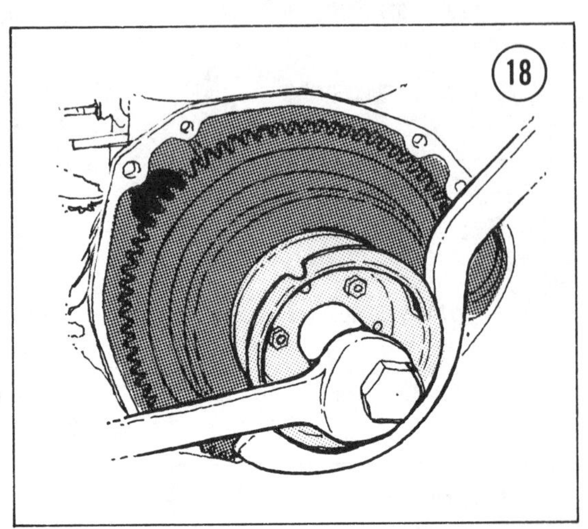

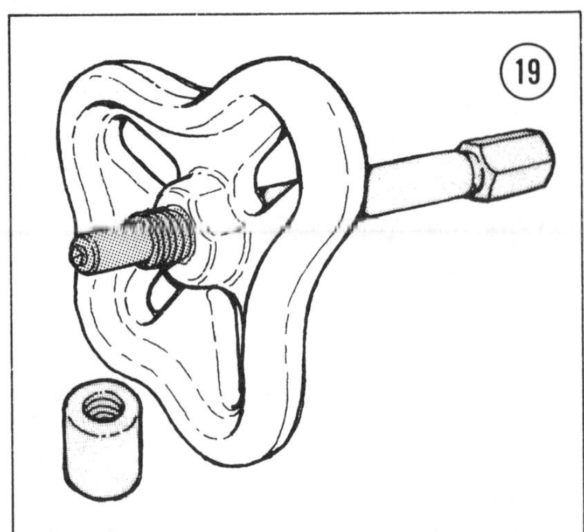

CAUTION
Do not rap on the flywheel.

3. Unscrew the screws that attach the contact breaker assembly to the baseplate (**Figure 20**). Unscrew the nut that attaches the electrical lead to the post on the contact breaker and remove the breaker.

4. Unscrew the screw that attaches the condenser and remove the condenser.

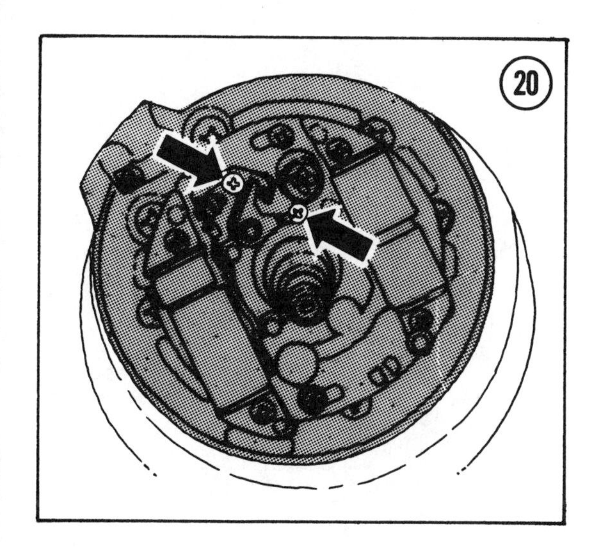

5. Reverse the above to install the contact breaker assembly and the condenser. Make certain the electrical lead is connected and the nut is tight.

6. Before tightening the contact breaker assembly screws, check the gap with a flat feeler gauge. The gap should be 0.012-0.016 in. (0.3-0.4mm). When the gap is correct, tighten the screws securely and recheck to make sure the gap did not change. Reinstall the flywheel and tighten the bolt to 65-80 ft.-lb. (9-11 mkg).

7. Screw a timing gauge (**Figure 21**) into the spark plug hole and rotate the crankshaft until the piston(s) are at TDC (top dead center) (PTO [power take-off] side cylinder on twins).

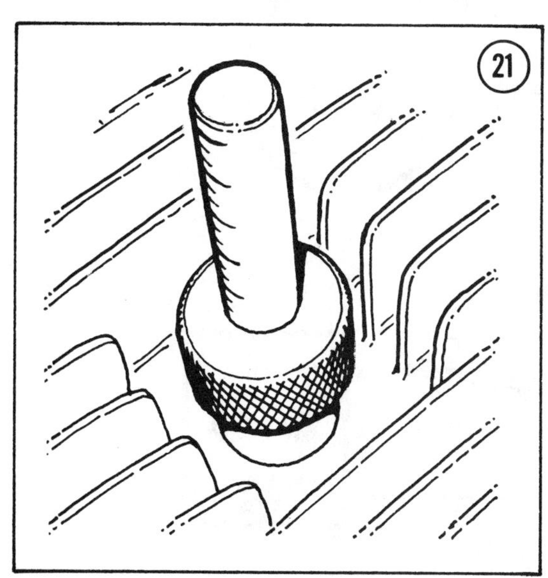

8. Bolt a degree wheel to the flywheel and make corresponding marks on the flywheel and the crankcase to indicate TDC (**Figure 22**). Then rotate the PTO end of the crankshaft 18° clockwise (0.073 in. or 1.86mm movement of the dial indicator). Make another mark on the flywheel adjacent to the first mark on the crankcase. This is the basic timing mark. Remove the degree wheel and the dial indicator. Install spark plugs.

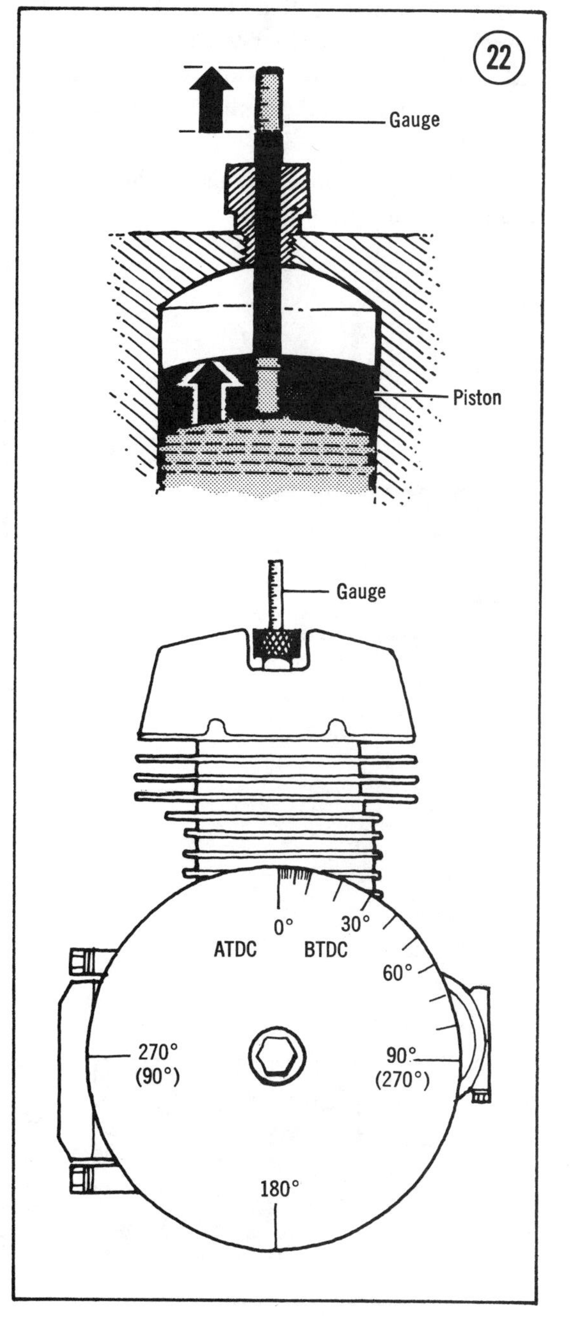

9. Remove the drive belt (see *Drive Belt,* this chapter). Connect a tachometer and timing light as shown in **Figure 23**.

10. Start the engine and warm it up to operating temperature. Direct the timing light at the timing marks and slowly increase the engine speed to a constant 6,000 rpm. The timing marks should align. If they do not, shut off the engine, remove the flywheel, recoil starter, and pulleys. Slightly loosen the 2 screws that attach the magneto baseplate to the case (**Figure 24**) and rotate the plate as required to obtain correct timing. Clockwise rotation retards the timing and counterclockwise advances it.

11. Tighten the screws, reinstall the flywheel and starter, and recheck the timing.

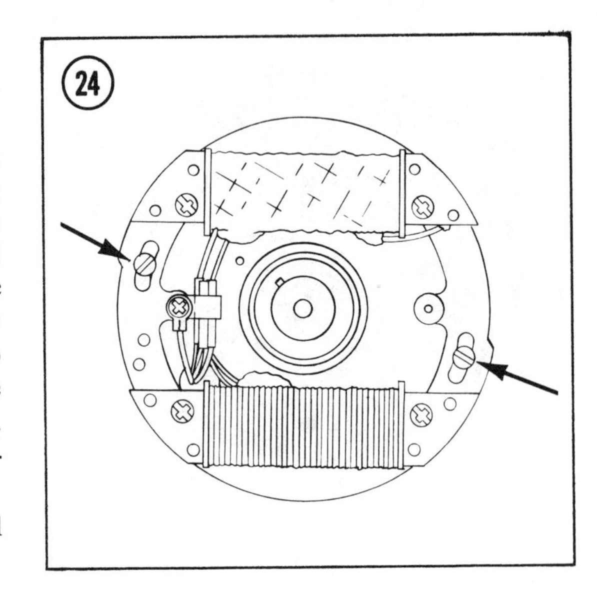

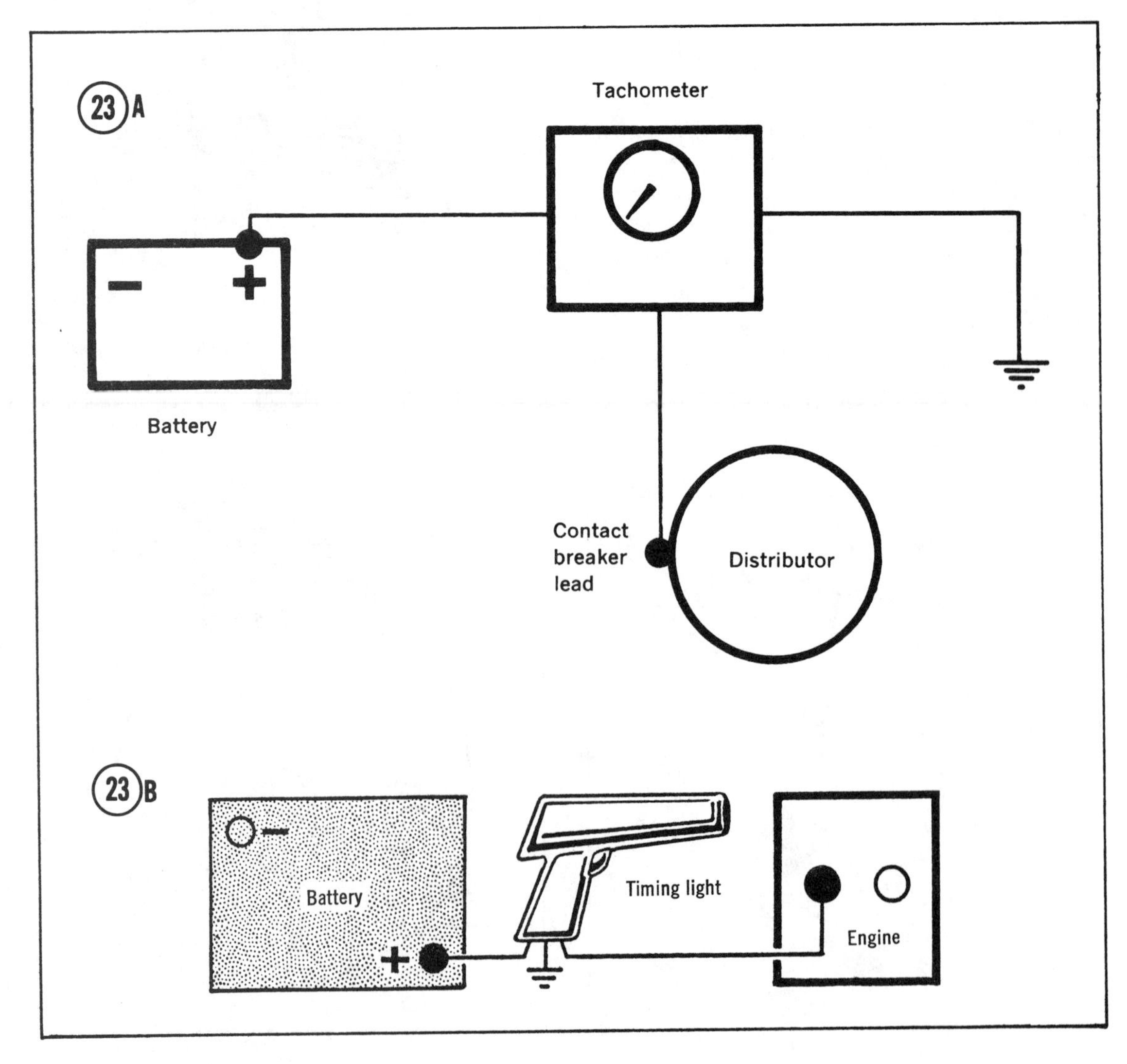

CARBURETION

Two carburetor types are used on Arctic Cat snowmobiles—Walbro and Mikuni. On both types, the main jet is preset at the factory. Extremes of altitude may necessitate a change in main jet; this should be entrusted to a dealer. The adjustments covered here are low-speed and high-speed needle adjustments for the Walbro carburetor and idle speed mixture for the Mikuni carburetor.

WARNING
Carburetor adjustments must be made with the engine off to prevent inadvertent engagement of the drive clutch.

Walbro Carburetor Adjustment

The adjustment needles and seats in the Walbro carburetor can be damaged if they are forced; turn them slowly and feel for the point at which the needles bottom in the seats. For both needles, clockwise turning results in a leaner mixture and counterclockwise turning produces a richer mixture. A too lean mixture results in overheating that could cause serious engine damage; a too-lean mixture will seem to actually improve performance—at the cost of engine life. A too-rich mixture will result in generally poor performance, hard starting, and plug fouling.

1. Turn the low-speed mixture needle in until it just contacts the seat. Then, turn it open—counterclockwise—one turn. This is the basic setting and its correctness can be verified by running the snowmobile at low throttle settings for a short distance and then removing and checking the spark plugs as described earlier. If the plugs indicate a too-lean condition, open the needle slightly—about ½ turn—and run the machine and check the plugs again. Continue to adjust and test until the plug color and condition are satisfactory.

2. Turn the high-speed mixture needle (**Figure 25**) in (clockwise) until it just touches the seat. Then, turn it open (counterclockwise) 1⅛ turns. This is the basic setting and its correctness can be verified by running the machine at full throttle with a full load. As with the low-speed mixture needle, check the color and condition

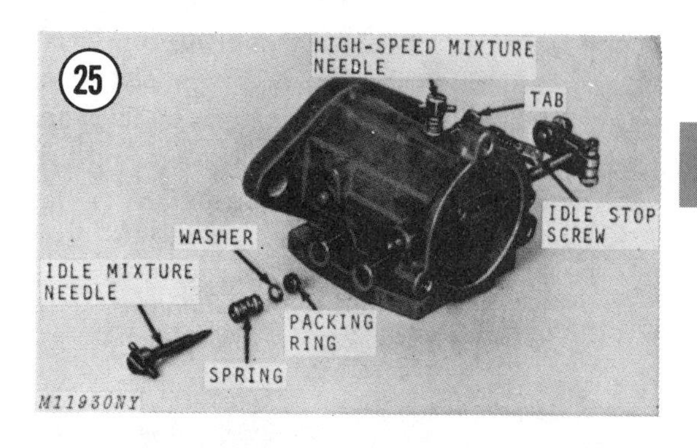

of the spark plugs and adjust accordingly. If the mixture is too lean, serious engine damage is very likely at high speeds, and if the mixture is too rich, the engine will tend to "four-cycle" at top speed with plug fouling a likely result.

3. After the low-speed mixture needle has been adjusted it will be necessary to adjust the idle mixture screw. The 2 adjustments are interdependent and if one is changed the other must also be checked. The correct idle speed occurs just below the point where the clutch engages. To increase the speed, turn the screw clockwise, counterclockwise to decrease it.

4. Adjust the throttle cable by loosening the cable retaining screw (**Figure 26**). Pull the cable until there is no slack, then pull it an additional 1/16 in. and tighten the cable retaining screw.

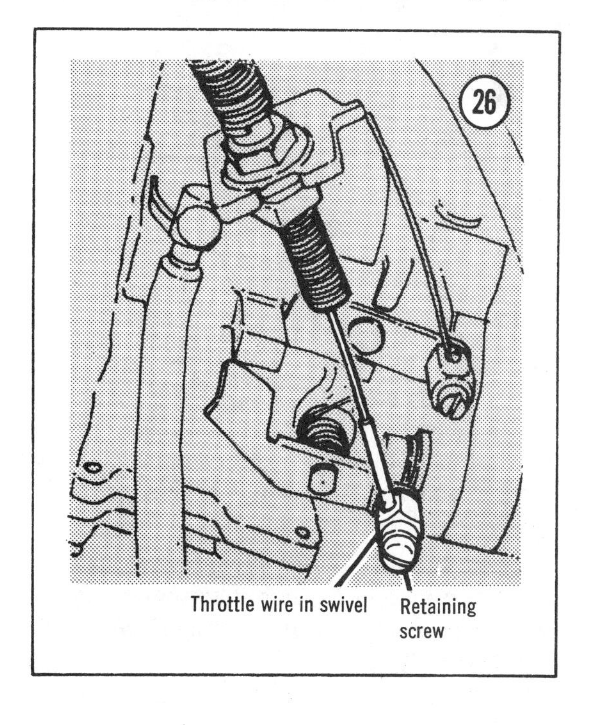

If the adjustment is correct, the throttle lever on the carburetor should be completely open when the hand throttle just contacts the hand grip.

CAUTION

Never attempt to adjust needle with track off the ground. The engine must be adjusted in a "loaded" condition or a too-lean fuel mixture with subsequent serious engine damage will result.

NOTE: *Above 5,000 feet altitude turn high-speed needle one turn open.*

5. Recheck machine performance. If engine stumbles or hesitates on acceleration, it may be necessary to lean or enrichen low-speed needle slightly.

Mikuni Carburetor Starter (Choke) Adjustment

The starting (choke) system on Mikuni carburetors (**Figure 27**) is controlled by a starter plunger with separate metering of fuel/air mixture through independent jets. When the engine is started, the throttle valve must be closed or the starting mixture will be made too lean for engine starting.

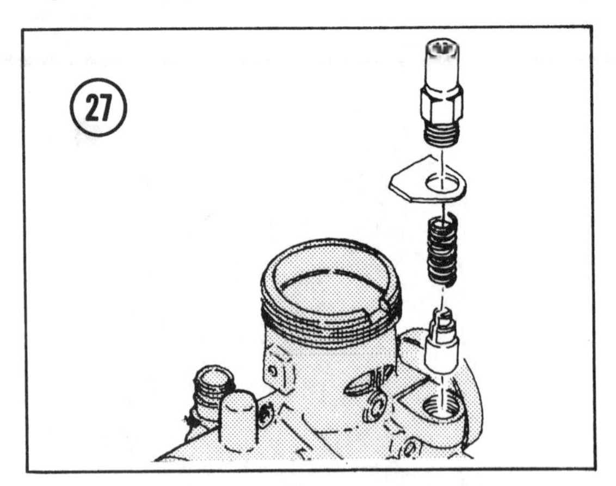

1. Place starter lever on instrument panel in the down position. Lever should have slight free-play.
2. Look through starter plunger air hole (3 o'clock position in the carburetor bore). Check that starter plunger is all the way down in its bore (**Figure 28**).

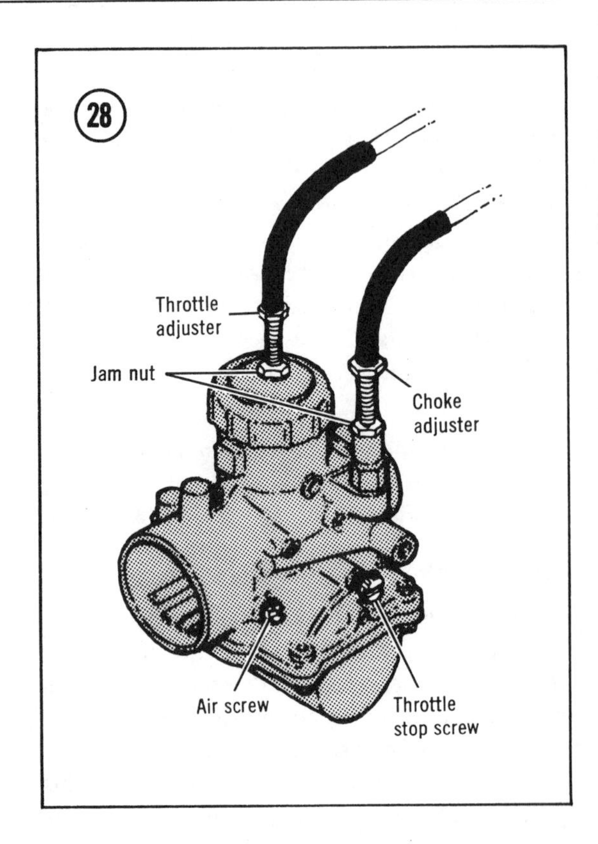

3. To adjust starter plunger perform the following:
   a. Loosen jam nut securing adjusting nut (**Figure 28**).
   b. Rotate adjusting nut clockwise to lower the starter plunger down in its bore. Tighten the jam nut.

   NOTE: *If the starter plunger is not fully down in its bore, the carburetor will run rich and affect entire engine performance level.*

Mikuni Carburetor Adjustment and Synchronization

This procedure includes throttle cable adjustments and idle speed adjustments for all models equipped with Mikuni carburetors.

On models equipped with 2 carburetors a more precise synchronization can be achieved with an air flow meter as described in Chapter One. If such a device is available, perform the following procedure as a preliminary adjustment and proceed to *Mikuni Carburetor Air Flow Meter Synchronization* for the final fine tuning.

Refer to **Figure 29** for this procedure.

1. Remove the air intake silencer.

2. Use strong rubber band and clamp throttle lever to handlebar grip in the wide-open throttle position.

3. Loosen the jam nut securing the adjusting sleeve. Feel inside the carburetor bore and turn the adjusting sleeve until the cutout portion of the throttle valve is flush with the inside of the carburetor bore.

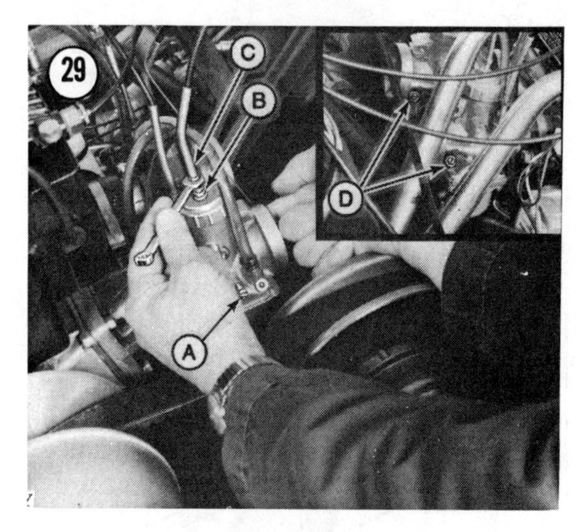

A. Throttle stop screw
B. Jam nut
C. Adjusting sleeve
D. Pilot air screw

4. Turn the adjusting sleeve counterclockwise several additional turns to position the backside of the throttle valve flush with the carburetor bore (**Figure 30**).

> NOTE: *The additional turns on the adjusting sleeve should position the throttle valve flush with or slightly above the carburetor bore. If any part of the throttle slide protrudes into the carburetor bore, turn adjusting sleeve until throttle slide is flush.*

5. Rotate the throttle stop screw (**Figure 29**) counterclockwise until the tip is flush with inside of carburetor bore.

6. Remove rubber band clamp from handlebar and allow throttle to return to idle position.

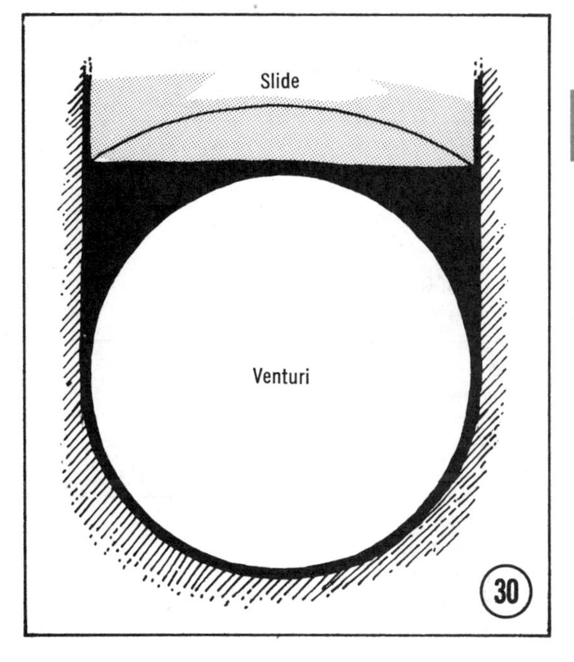

7. Turn in throttle stop screw until tip just contacts throttle slide valve. Turn in stop screw 2 additional turns for a preliminary idle setting.

8. Slowly operate throttle lever on handlebar and observe that throttle valve begins to rise. On models with 2 carburetors, ensure that throttle valves move an equal amount together. Readjust throttle cables if necessary.

9. Slowly turn in pilot air screw until light seating is felt. *Do not* force it or the air screw may be damaged. Back out the pilot air screw number of turns specified in **Table 5**.

10. Install air intake silencer and start engine. Warm up engine to operating temperature and check that idle speed is 1,800 to 2,400 rpm. Adjust throttle stop screw as necessary for specified idle speed. On 2 carburetor models ensure that both throttle stop screws are adjusted an equal amount.

> NOTE: Do not *use pilot air screws to attempt to set engine idle speed. Pilot air screws must be set as specified in* **Table 5**.

Mikuni Carburetor Air Flow Meter Synchronization

To obtain a precise synchronization of twin carburetor models, use an air flow meter device

Table 5 BASIC AIR SCREW SETTING—MIKUNI CARBURETORS

Model/Engine	Turns Out
El Tigre	
1974-1975	
All models	1
1976-1977	
4000	1½
5000	1
Jag	
All models	1½
Lynx	
All models	1½
Pantera	
All models	1
Panther/Cheetah	
4000	¾
5000	1

as described in Chapter One. Perform *Mikuni Carburetor Adjustment and Synchronization* to obtain proper preliminary adjustments.

Refer to **Figure 31** for this procedure.

WARNING

The following procedure is performed with the engine running. Ensure that arms and clothing are clear of drive belt or serious injury may result.

1. Raise and support the rear of the snowmobile so track is clear of the ground.
2. Start the engine. Bind and wedge the throttle lever to maintain engine speed at 4,000 rpm (**Figure 32**).
3. Open the air flow control of air flow meter and place the meter over right carburetor throat. The tube on meter must be vertical.
4. Slowly close the air flow control until float in tube aligns with a graduated mark on tube.
5. Without changing th eadjustment of the air flow control, place the air flow meter on the left carburetor. If the carburetors are equal, no adjustment is necessary.

A. Idle adjusting screw
B. Unisyn synchronizing gauge
C. Tube in vertical position
D. Float
E. Air flow control

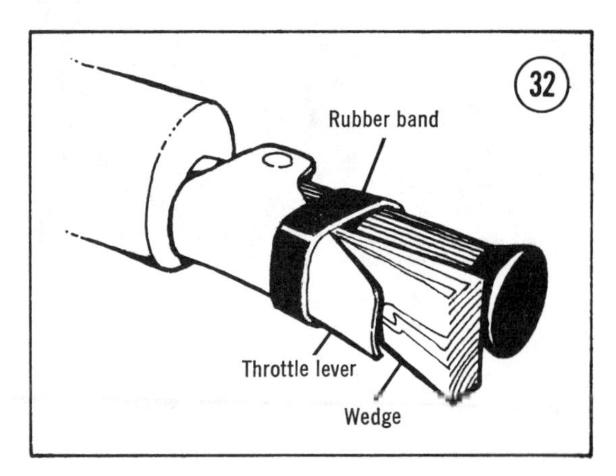

6. If adjustment is necessary, loosen the jam nut on the carburetor with the lowest float level and turn the adjusting sleeve until the float level matches that of the other carburetor.
7. Return the engine to idle and repeat Steps 3, 4, and 5. Adjust the throttle stop screws as necessary for a balanced idle.

Mikuni Carburetor Main Jet Selection

The main jet controls the fuel metering when the carburetor is operating in the ½ to full throttle range. Because temperature and altitude affect the air density, each snowmobile owner will have to perform the following trial and error method of jet selection to obtain peak engine efficiency and performance for his own particular area of operation.

CAUTION
Air intake silencer must be installed during the following procedure or a "too-lean" mixture may result. A "too-lean" fuel mixture can cause engine overheating and subsequent serious damage.

NOTE: *The snowmobile must be operated on a flat, well-packed area for best results.*

1. Operate the machine at wide open throttle for several minutes. If peak rpm cannot be achieved or the engine appears to be laboring, the main jet needs to be changed.
2. Make another trial run and shut off the ignition while the throttle is still wide open. Examine the exhaust and spark plugs to determine if the mixture is too rich or too lean. The mixture is too rich if exhaust manifold or spark plug insulator is dark brown or black. Refer to *Spark Plugs* in this chapter. Decrease the jet size if the mixture is too rich.

NOTE: *Change jet sizes* one *increment at a time and test after each change to obtain best results.*

If the manifold or spark plug insulator is a very light color, the mixture is too lean. Correct by increasing jet size.
3. If the state of fuel/air mixture cannot be determined by color of exhaust manifold or spark plug insulator, assume the mixture is too lean and increase jet size. If operation improves, continue increasing jet size until maximum performance is achieved. If operation gets worse, decrease jet size until best results are obtained.

OFF-SEASON STORAGE

Proper storage techniques are essential to help maintain your snowmobile's life and usefulness. The off season is also an excellent time to perform any maintenance and repair tasks that are necessary.

Placing in Storage

1. Use soap and water to thoroughly clean the exterior of your snowmobile. Use a hose to remove rocks, dirt and debris from the track area. Clean all dirt and debris from the hood and console areas.

CAUTION
Do not spray water around the carburetor or engine. Be sure you allow sufficient time for all components to dry.

2. Use a good automotive type cleaner wax and polish the hood, pan and tunnel. Use a suitable type of upholstery cleaner on the seat. Touch up any scratched or bare metal parts with paint. Paint or oil the skis to prevent rust.
3. Drain the fuel tank. Start the engine and run it at idle to burn off all fuel left in the carburetor. Check the fuel filter and replace it if it is contaminated.
4. Wrap up carburetor(s) and intake manifold in plastic and tape securely (**Figure 33**).

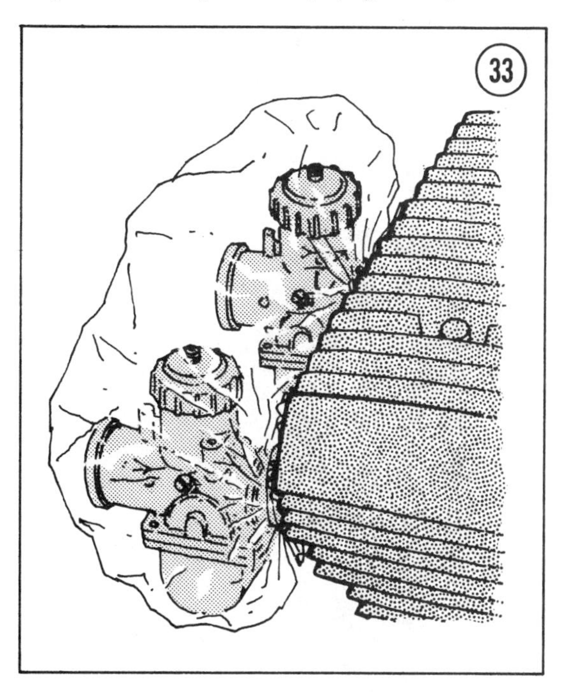

5. Remove the spark plugs and add a teaspoon of Arctic Cat Engine Preserver to each cylinder. Pull the engine over several times with the starter rope to spread the oil over the cylinder walls. Replace the spark plugs.

6. Remove the drive belt. Apply a film of light grease to drive and driven sheaves to prevent rust and corrosion.

7. Change the chaincase oil.

8. Raise the rear of snowmobile off the ground. Loosen the track adjusting screws to remove any tension on the track.

9. Carefully examine all components and assemblies. Make a note of immediate and future maintenance and repair items and order the necessary parts. Check and tighten all nuts and bolts.

10. Cover the snowmobile and store it inside if possible.

Removing From Storage

1. Check and tighten all nuts and bolts.

2. Completely remove grease from the drive and driven sheaves and then clean them with an acetone type cleaner. Install the drive belt.

WARNING

Acetone type cleaners are extremely volatile; keep flame, heat, and cigarettes away.

3. Fill the fuel tank with fresh gasoline/oil mixture. Refer to Chapter One.

4. Check the throttle and brake controls for proper operation and adjust them if necessary.

5. Adjust the track tension.

6. Familiarize yourself with all safety and operating instructions.

7. Start the engine and check the operation of the emergency stop switch and "tether" switch. Check that all lights and switches operate properly. Replace any burned out bulbs.

8. Start out slowly on short rides until you are sure your machine is operating correctly and is dependable.

CHAPTER THREE

TROUBLESHOOTING

Diagnosing snowmobile ills is relatively simple if you use orderly procedures and keep a few basic principles in mind.

Never assume anything. Do not overlook the obvious. If you are riding along and the snowmobile suddenly quits, check the easiest most accessible problem spots first. Is there gasoline in the tank? Has a spark plug wire fallen off? Check the ignition switch. Maybe that last mogul caused you to accidentally switch the emergency switch to OFF or pull the emergency stop "tether" string.

If nothing obvious turns up in a cursory check, look a little further. Learning to recognize and describe symptoms will make repairs easier for you or a mechanic at the shop. Describe problems accurately and fully. Saying that "it won't run" isn't the same as saying "it quit at high speed and wouldn't start", or that "it sat in my garage for three months and then wouldn't start".

Gather as many symptoms together as possible to aid in diagnosis. Note whether the engine lost power gradually or all at once, what color smoke (if any) came from the exhaust, and so on. Remember that the more complicated a machine is, the easier it is to troubleshoot because symptoms point to specific problems.

You do not need fancy equipment or complicated test gear to determine whether repairs can be attempted at home. A few simple checks could save a large repair bill and time lost while the snowmobile sits in a dealer's service department. On the other hand, be realistic and do not attempt repairs beyond your abilities. Service departments tend to charge heavily for putting together disassembled components that may have been abused. Some won't even take on such a job—so use common sense; don't get in over your head.

OPERATING REQUIREMENTS

An engine needs 3 basics to run properly; correct gas/air mixture, compression, and a spark at the right time. If one or more are missing, the engine will not run. The electrical system is the weakest link of the three. More problems result from electrical breakdowns than from any other source. Keep that in mind before you begin tampering with carburetor adjustments.

If the snowmobile has been sitting for any length of time and refuses to start, check the battery (if the machine is so equipped) for a charged condition first, and then look to the gasoline delivery system. This includes the tank, fuel petcocks, lines, and the carburetor. Rust may have formed in the tank, obstructing fuel flow. Gasoline deposits may have gummed up carburetor jets and air passages. Gasoline tends to lose its potency after standing for long periods. Condensation may contaminate it with water. Drain old gas and try starting with a fresh tankful.

Compression, or the lack of it, usually enters the picture only in the case of older machines. Worn or broken pistons, rings, and cylinder bores could prevent starting. Generally a gradual power loss and harder and harder starting will be readily apparent in this case.

PRINCIPLES OF 2-CYCLE ENGINES

The following is a general discussion of a typical 2-cycle piston-port engine.

Figures 1 through 4 illustrate operating principles of piston port engines. During this discussion, assume that the crankshaft is rotating counterclockwise. In **Figure 1**, as the piston travels downward, a transfer port (A) between the crankcase and the cylinder is uncovered. Exhaust gases leave the cylinder through the exhaust port (B), which is also opened by downward movement of the piston. A fresh fuel/air charge, which has previously been compressed slightly by the descending piston, travels from the crankcase (C) to the cylinder through transfer ports (A) as the ports open. Since the incoming charge is under pressure, it rushes into the cylinder quickly and helps to expel exhaust gases from the previous cycle.

Figure 2 illustrates the next phase of the cycle. As the crankshaft continues to rotate, the piston moves upward, closing the exhaust and transfer ports. As the piston continues upward, the air/fuel mixture in the cylinder is compressed. Notice also that a low pressure area is created in the crankcase by the ascending piston at the same time. Further upward movement of the piston uncovers the intake port (D). A fresh fuel/air charge is then drawn into the crankcase through the intake port because of the low pressure created by the upward piston movement.

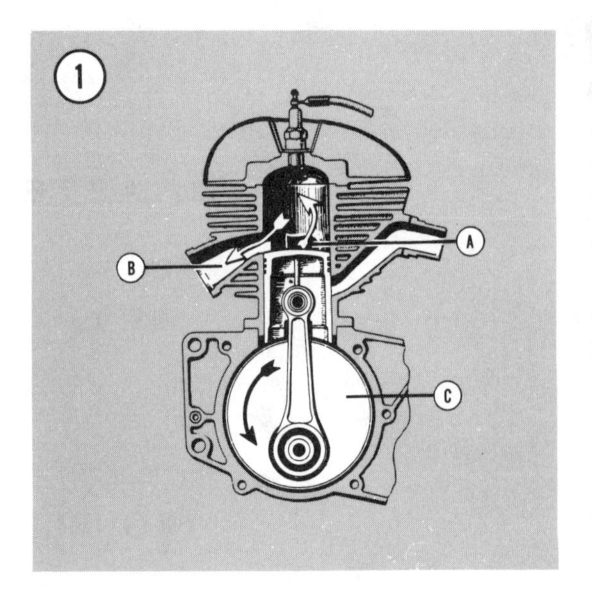

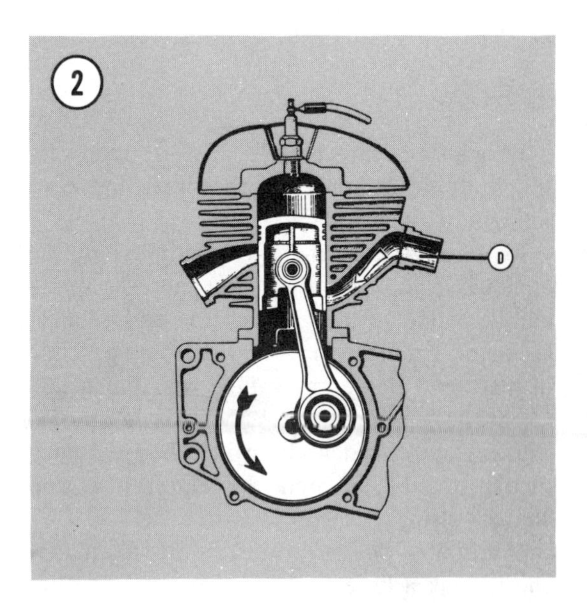

The third phase is shown in **Figure 3**. As the piston approaches top dead center, the spark plug fires, igniting the compressed mixture. The piston is then driven downward by the expanding gases.

When the top of the piston uncovers the exhaust port, the fourth phase begins, as shown in **Figure 4**. The exhaust gases leave the cylinder through the exhaust port. As the piston continues downward, the intake port is closed and the mixture in the crankcase is compressed in preparation for the next cycle. Every downward stroke of the piston is a power stroke.

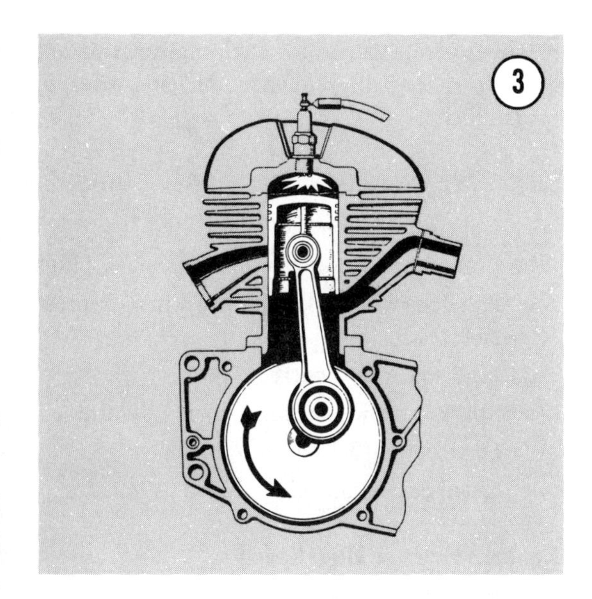

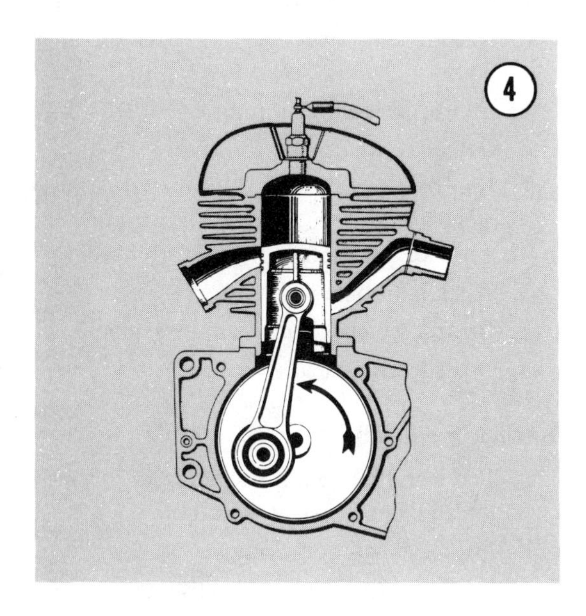

ENGINE STARTING

An engine that refuses to start or is difficult to start can try the patience of anyone. More often than not, the problem is very minor and can be found with a simple and logical troubleshooting approach.

The following items provide a beginning point from which to isolate engine starting problems.

Engine Fails to Start

Perform the following spark test to determine if the ignition system is operating properly.

1. Remove a spark plug.
2. Connect the spark plug connector to the spark plug and clamp the base of spark plug to a good grounding point on the engine. A large alligator clip makes an ideal clamp. Position the spark plug so you can observe the electrode.
3. Turn on the ignition and crank the engine over. A fat blue spark should be evident across the spark plug electrode.

WARNING

On machines equipped with CDI (capacitor discharge ignition), do not hold spark plug, wire, or connector or a serious electrical shock may result.

4. If the spark is good, check for one or more of the following possible malfunctions:
   a. Fouled or defective spark plugs
   b. Obstructed fuel filter or fuel line
   c. Defective fuel pump
   d. Leaking head gasket—perform compression test.
5. If spark is not good, check for one or more of the following:
   a. Burned, pitted, or improperly gapped breaker points
   b. Weak ignition coil or condenser
   c. Loose electrical connections
   d. Defective CDI components—have CDI system checked by an authorized dealer.

Engine Difficult to Start

Check for one or more of the following possible malfunctions:

a. Fouled spark plugs
b. Improperly adjusted choke
c. Defective or improperly adjusted breaker points
d. Contaminated fuel system
e. Improperly adjusted carburetor
f. Weak ignition coil
g. Incorrect fuel mixture
h. Crankcase drain plugs loose or missing
i. Poor compression—perform compression test.

Engine Will Not Crank

Check for one or more of the following possible malfunctions:

a. Defective recoil starter
b. Seized piston
c. Seized crankshaft bearings
d. Broken connecting rod

Compression Test

Perform a compression test to determine condition of piston ring sealing qualities, piston wear, and condition of head gasket seal.

1. Remove the spark plugs. Insert a compression gauge in one spark plug hole (**Figure 5**). Refer to Chapter One for a suitable type of compression tester.

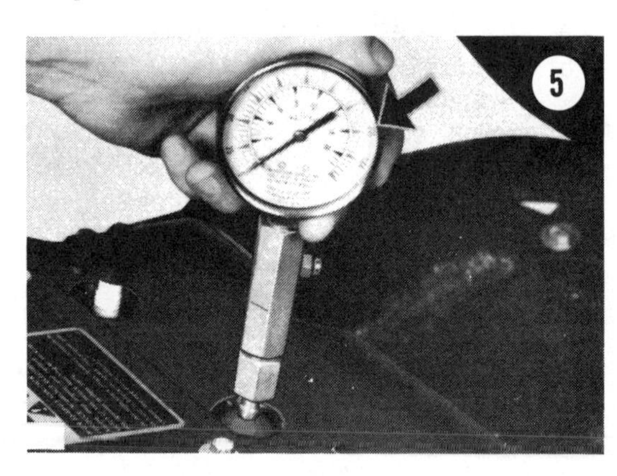

2. Crank the engine vigorously and record compression reading. Repeat for other cylinder. Compression readings should be from 120 to 175 psi (8.44-12.30 kg/cm²). Maximum allowable variation between cylinders is 10 psi (0.70 kg/cm²).

3. If compression is low or variance between cylinder is excessive, check for defective head gaskets, damaged cylinders and pistons, or stuck piston rings.

ENGINE PERFORMANCE

The following items are a starting point from which to isolate a performance malfunction. It is assumed the engine runs but is not operating at peak efficiency.

The possible causes for each malfunction are listed in a logical sequence and in order of probability.

Engine Will Not Idle

a. Carburetor incorrectly adjusted
b. Fouled or improperly gapped spark plugs
c. Head gasket leaking—perform compression test.
d. Fuel mixture incorrect
e. Spark advance mechanism not retarding
f. Obstructed fuel pump impulse tube
g. Crankcase drain plugs loose or missing

Engine Misses at High Speed

a. Fouled or improperly gapped spark plugs
b. Defective or improperly gapped breaker points
c. Improper ignition timing
d. Defective fuel pump
e. Improper carburetor high-speed adjustment (Walbro Carburetor) or improper main jet selection (Mikuni Carburetor)
f. Weak ignition coil
g. Obstructed fuel pump impulse tube
h. Obstructed fuel filter

Engine Overheating

a. Too lean fuel mixture—incorrect carburetor adjustment or jet selection
b. Improper ignition timing
c. Incorrect spark plug heat range
d. Intake system or crankcase air leak
e. Cooling fan belt broken or slipping
f. Cooling fan defective
g. Damaged or blocked cooling fins

Engine Smokes and Runs Rough

a. Carburetor adjusted incorrectly—mixture too rich
b. Incorrect fuel/oil mixture
c. Choke not operating properly
d. Obstructed muffler
e. Water or other contaminates in fuel

Engine Loses Power

a. Carburetor incorrectly adjusted
b. Engine overheating
c. Defective or improperly gapped breaker points
d. Improper ignition timing
e. Incorrectly gapped spark plugs
f. Weak ignition coil
g. Obstructed muffler
h. Dragging brake

Engine Lacks Acceleration

a. Carburetor mixture too lean
b. Defective fuel pump
c. Incorrect fuel/oil mixture
d. Defective or improperly gapped breaker points
e. Improper ignition timing
f. Dragging brake

ENGINE FAILURE ANALYSIS

Overheating is the major cause of serious and expensive engine failures. It is important that each snowmobile owner understand all the causes of engine overheating and take the necessary precautions to avoid expensive overheating damage. Proper preventive maintenance and careful attention to all potential problem areas can often prevent a serious malfunction before it happens.

Fuel

All Arctic Cat snowmobile engines rely on a proper fuel/oil mixture for engine lubrication. Always use an approved oil and mix the fuel carefully as described in Chapter One.

Gasoline must be of sufficiently high octane (88 or higher) to avoid "knocking" and "detonation".

Fuel/Air Mixture

Fuel/air mixture is determined by carburetor adjustment (Walbro) or main jet selection (Mikuni). Always adjust carburetors carefully and pay particular attention to avoid a "too-lean" mixture.

Heat

Excessive external heat on the engine can be caused by the following:

a. Hood louvers plugged with snow
b. Damaged or plugged cylinder and head cooling fins
c. Slipping or broken fan belt
d. Damaged cooling fan
e. Operating snowmobile in hot weather
f. Plugged or restricted exhaust system

See **Figures 6 and 7** for examples of cylinder and piston scuffing caused by excessive heat.

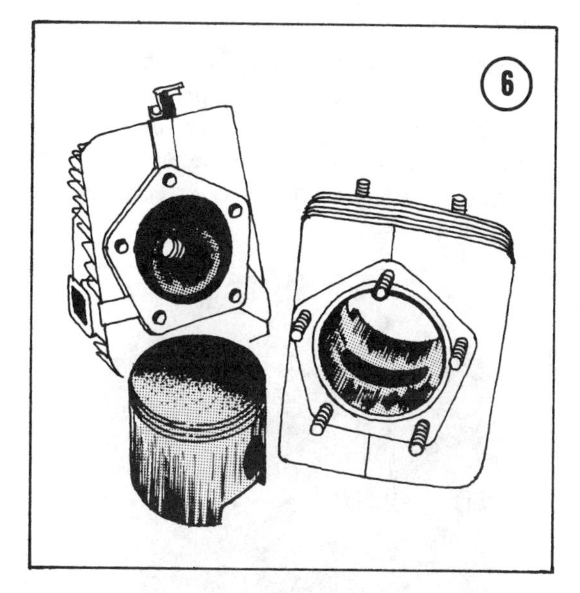

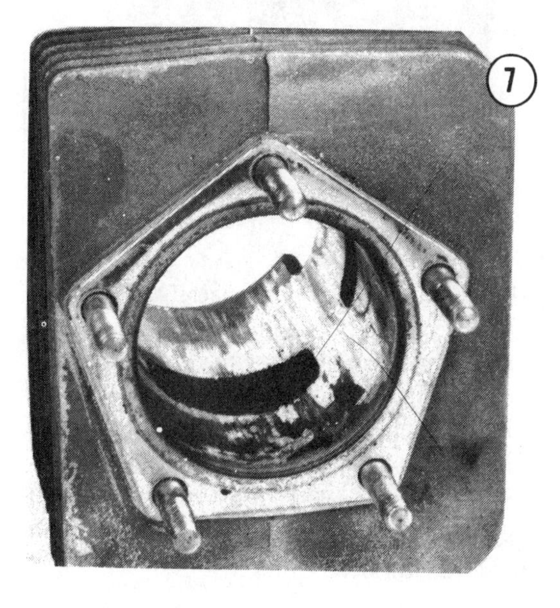

Dirt

Dirt is a potential problem for all snowmobiles. The air intake silencers on all models are not designed to filter incoming air. Avoid running snowmobiles in areas that are not completely snow covered.

Ignition Timing

Ignition timing that is too far advanced can cause "knocking" or "detonation". Timing that is too retarded causes excessive heat buildup in the cylinder exhaust port area.

Spark Plugs

Spark plugs must be of a correct heat range. Too hot a heat range can cause pre-ignition and detonation which can ultimately result in piston burn-through as shown in **Figure 8**.

Refer to Chapter Two for recommended spark plugs.

Pre-Ignition

Pre-ignition is caused by excessive heat in the combustion chamber due to a spark plug of improper heat range and/or too lean a fuel mixture. See **Figure 9** for an example of a melted and scuffed piston caused by pre-ignition.

Detonation (Knocking)

Knocking is caused by a too lean fuel mixture and/or too low octane fuel.

ELECTRICAL SYSTEM

The following items provide a starting point from which to troubleshoot electrical system malfunctions. The possible causes for each malfunction are listed in a logical sequence and in order of probability.

Ignition system malfunctions are outlined under *Engine Starting* and *Engine Performance*.

Lights Will Not Light

a. Bulbs are burned out
b. Loose electrical connections
c. Defective switch
d. Defective lighting coil or alternator
e. Defective voltage regulator
f. Defective battery (electric-start models)

Bulbs Burn Out Rapidly

a. Incorrect bulb type
b. Defective voltage regulator

Lights Too Bright or Too Dim

a. Defective voltage regulator
b. Defective alternator

Discharged Battery (Electric-Start Models)

a. Defective battery
b. Low electrolyte level
c. Dirty or loose electrical connections
d. Defective voltage regulator
e. Defective lighting coil
f. Defective rectifier
g. Defective circuit breaker

Cracked Battery Case

a. Discharged battery allowed to freeze
b. Improperly installed hold-down clamp
c. Improperly attached battery cables

Starter Motor Does Not Operate

a. Loose electrical connections
b. Discharged battery
c. Defective starter solenoid
d. Defective starter motor
e. Defective circuit breaker
f. Defective ignition switch

Poor Starter Performance

a. Commutator or brushes worn, dirty, or oil soaked
b. Binding armature
c. Weak brush springs
d. Armature open, shorted, or grounded

POWER TRAIN

The following items provide a starting point from which to troubleshoot power train malfunctions. The possible causes for each malfunction are listed in a logical sequence and in order of probability. Also refer to *Drive Belt Wear Analysis*.

Drive Belt Not Operating Smoothly in Drive Sheave

a. Face of drive sheave is rough, grooved, pitted, or scored
b. Defective drive belt

Uneven Drive Belt Wear

a. Misaligned drive and driven sheaves
b. Loose engine mounts

Glazed Drive Belt

a. Excessive slippage
b. Oil or grease on sheave surfaces

Drive Belt Worn Narrow in One Place

a. Excessive slippage caused by stuck track
b. Too high engine idle speed

Drive Belt Too Tight at Idle

a. Engine idle speed too fast
b. Distance between sheaves incorrect
c. Belt length incorrect

Drive Belt Edge Cord Failure

a. Misaligned sheaves
b. Loosen engine mounting bolts

Brake Not Holding Properly

a. Incorrect brake cable adjustment
b. Brake lining or pucks worn
c. Oil saturated brake lining or pucks
d. Sheared key on brake pulley or disc
e. Incorrect brake adjustment

Brake Not Releasing Properly

a. Weak or broken return spring
b. Bent or damaged brake lever
c. Incorrect brake adjustment

Leaking Chaincase

a. Gaskets on driveshaft bearing flanges or secondary shaft bearing flanges damaged
b. Damaged O-ring on driveshaft or secondary shaft
c. Cracked or broken chaincase

Rapid Chain and Sprocket Wear

a. Insufficient chaincase oil
b. Misaligned sprockets
c. Broken chain tension blocks

DRIVE BELT WEAR ANALYSIS

Frayed Edge

A rapidly wearing drive belt with a frayed edge cord indicates the drive belt is misaligned (see **Figure 10**). Also check for loose engine mounting bolts.

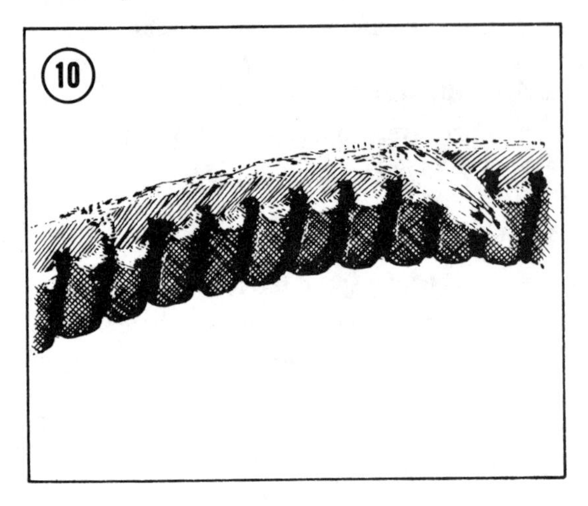

Worn Narrow in One Section

Excessive slippage due to a stuck track or too high an engine idle speed will cause the drive belt to be worn narrow in one section (see **Figure 11**).

Belt Disintegration

Drive belt disintegration is usually caused by misalignment. Disintegration can also be caused by using an incorrect belt or oil or grease on sheave surfaces (see **Figure 12**).

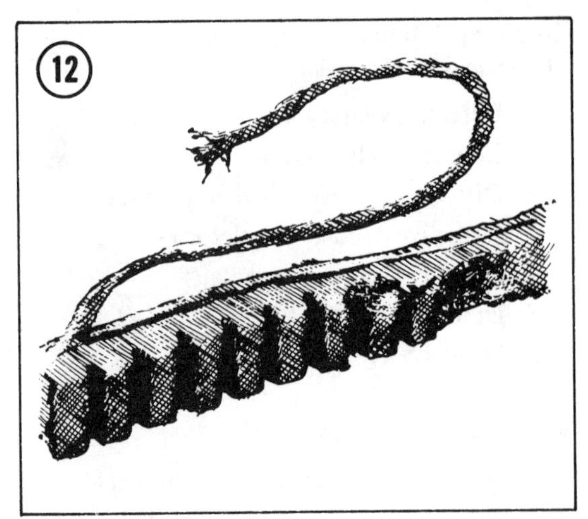

Sheared Cogs

Sheared cogs as shown in **Figure 13** are usually caused by violent drive sheave engagement. This is an indication of a defective or improperly installed drive sheave.

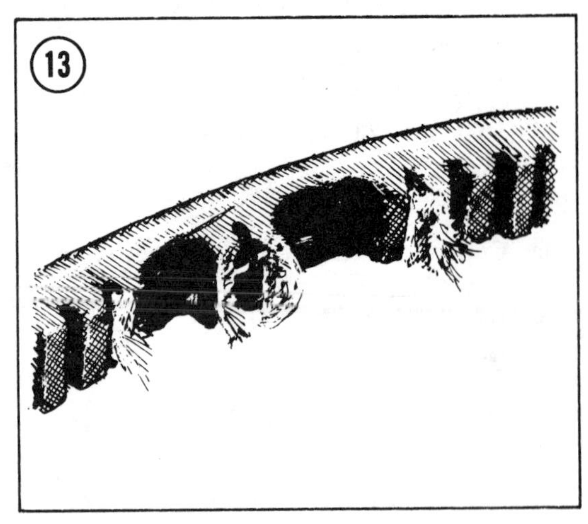

SKIS AND STEERING

The following items provide a starting point from which to troubleshoot ski and steering malfunctions. The possible causes for each malfunction are listed in a logical sequence and in order of probability.

Loose Steering

a. Loose steering post bushing
b. Loose steering post or steering column cap screw

c. Loose tie rod ends
d. Worn spindle bushings
e. Stripped spindle splines

Unequal Steering

a. Improperly adjusted tie rods
b. Improperly installed steering arms

Rapid Ski Wear

a. Skis misaligned
b. Worn out ski wear rods (Skags)
c. Worn out spring wear plate

TRACK ASSEMBLY

The following items provide a starting point from which to troubleshoot track assembly malfunctions. The possible causes for each are listed in a logical sequence and in order of probability. Also refer to *Track Wear Analysis*.

Frayed Track Edge

Track misaligned

Track Grooved on Inner Surface

a. Track too tight
b. Frozen rear idler shaft bearing

Track Drive Ratcheting

Track too loose

Rear Idlers Turning on Shaft

Frozen rear idler shaft bearings

TRACK WEAR ANALYSIS

The majority of track failures and abnormal wear patterns are caused by negligence, abuse, and poor maintenance. The following items illustrate typical examples. In all cases the damage could have been avoided by proper maintenance and good operator technique.

Obstruction Damage

Cuts, slashes, and gouges in the track surface are caused by hitting obstructions such as broken glass, sharp rocks or buried steel (see **Figure 14**).

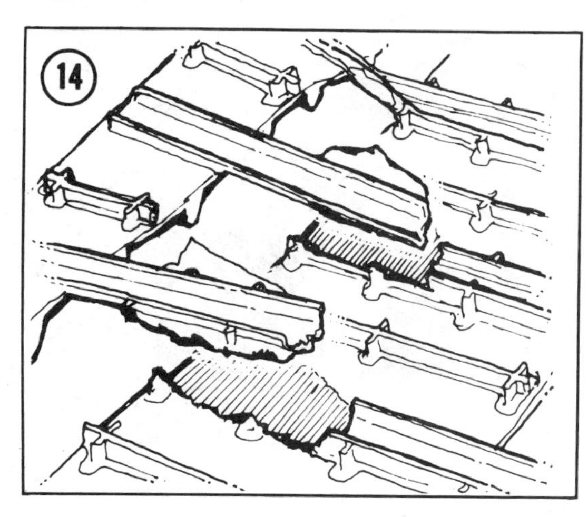

Worn Grouser Bars

Excessively worn grouser bars are caused by snowmobile operation over rough and non-snow covered terrain such as gravel roads and highway roadsides (**Figure 15**).

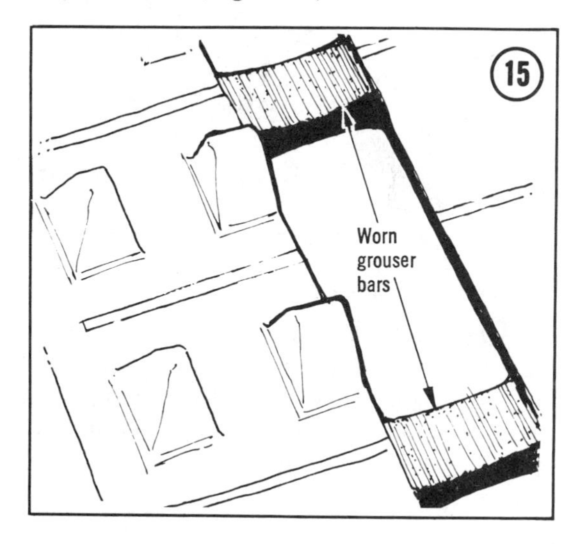

Lug Damage

Lug damage as shown in **Figure 16** is caused by lack of snow lubrication.

Ratcheting Damage

Insufficient track tension is a major cause of ratcheting damage to the top of the lugs (**Figure 17**). Ratcheting can also be caused by too great a load and constant "jack-rabbit" starts.

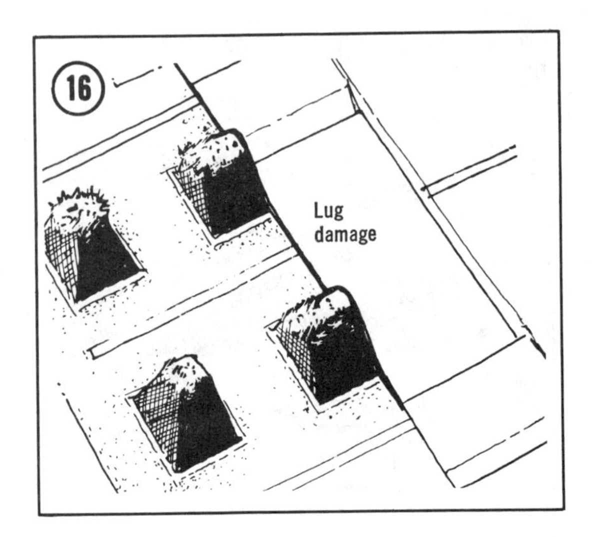

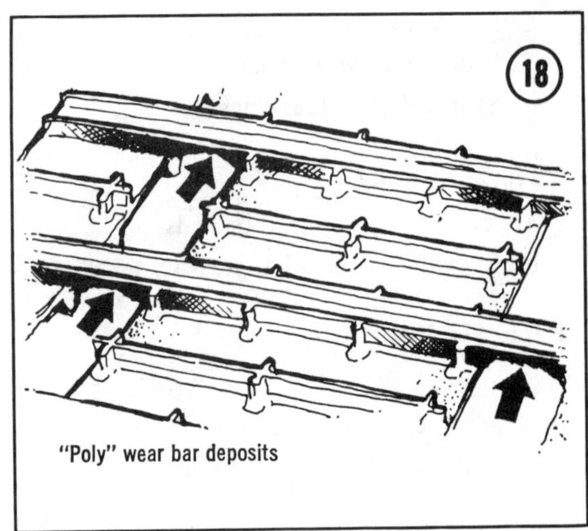

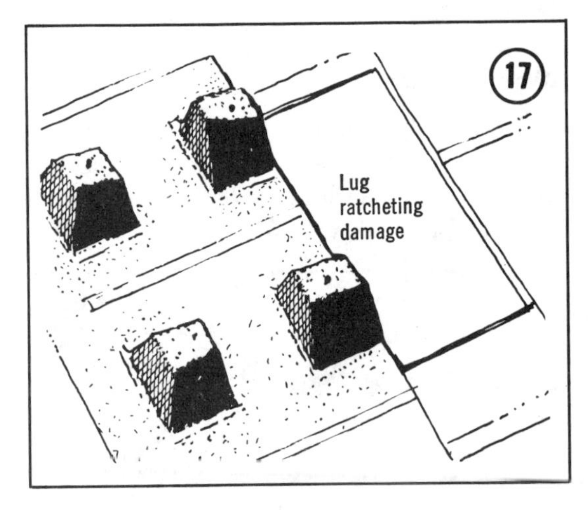

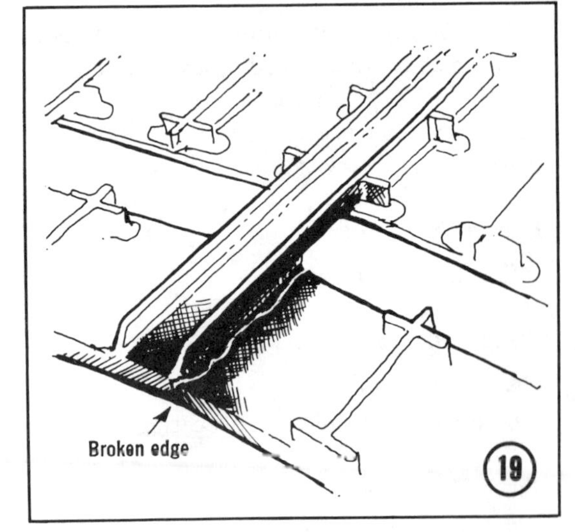

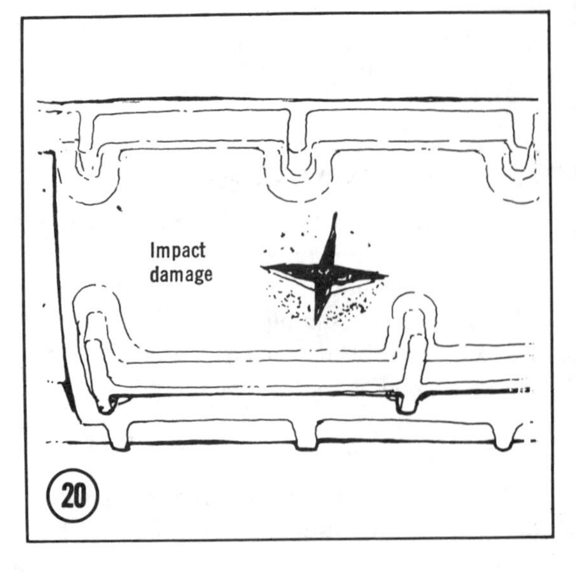

Over-tension Damage

Excessive track tension can cause too much friction on the wear bars. This friction causes the wear bars to melt and adhere to the track grouser bars. See **Figure 18**. An indication of this condition is a "sticky" track that has a tendency to "lock up."

Loose Track Damage

A track adjusted too loosely can cause the outer edge to flex excessively. This results in the type of damage shown in **Figure 19**. Excessive weight can also contribute to the damage.

Impact Damage

Impact damage as shown in **Figure 20** causes the track rubber to open and expose the cord.

This frequently happens in more than one place. Impact damage is usually caused by riding on rough or frozen ground or ice. Also, insufficient track tension can allow the track to pound against the track stabilizers inside the tunnel.

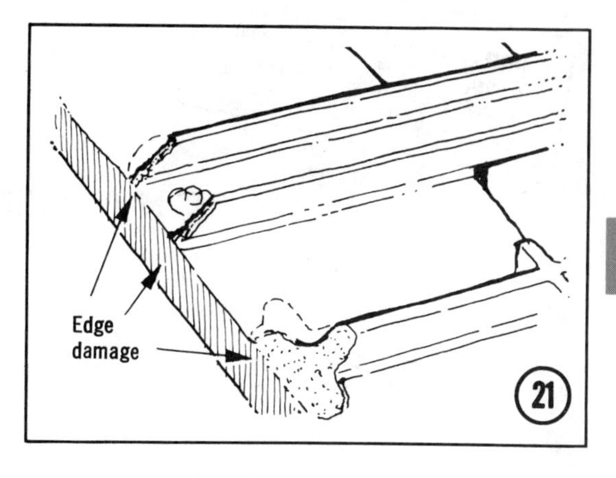

Edge Damage

Edge damage as shown in **Figure 21** is usually caused by tipping the snowmobile on its side to clear the track and allowing the track edge to contact an abrasive surface.

NOTE: If you own a 1978 or later model, first check the Supplement at the back of the book for any new service information.

CHAPTER FOUR

ENGINE

All Arctic Cat snowmobiles are powered with either 2- or single-cylinder 2-stroke engines (except for those equipped with Wankel engines—not covered in this manual). Refer to **Table 1** to determine which engine is equipped with your machine. See Chapter Three for the *Principles of Operation* of 2-cycle piston-port engines.

All Arctic Cat engines are equipped with ball-type main crankshaft bearings and needle bearings on both ends of the connecting rods.

On 1974-1975 models, the crankshaft, lower end bearings, crank pins, connecting rods, interior bearings, and seals are available *only* as a complete assembly. For 1976-1977 models, the crankshaft components are available individually. For all models and years, the outer bearings and seals are available separately. In all cases, however, it is recommended that lower end work be entrusted to a dealer; experience and special measuring equipment, along with a press, *are required* to service the lower end. Engine removal and upper end disassembly are covered to reduce the cost of lower end service.

UPPER END OVERHAUL

An upper end overhaul is within the abilities of the average hobbyist mechanic equipped with a reasonable range of hand tools, and inside and outside micrometers. Before beginning work, read Chapter One, and particularly the headings *Service Hints, Tools, Expendable Supplies,* and *Working Safely*. The information they contain will contribute to the efficiency, effectiveness, and safety of your work.

Also, read the procedures in this chapter carefully and completely before picking up a wrench. It is also a good idea to physically compare the instructions with the actual machine beforehand to familiarize yourself with the procedure and the equipment.

A complete upper end overhaul can be performed without removing the engine from the machine. However, you may find it more convenient to have the engine on a workbench, and should a small part or tool or dirt fall into the crankcase, it is more easily removed if the engine can be inverted. For these reasons, engine removal is described first.

The removal/installation procedures, as well as most service procedures, are virtually identical for all Arctic Cat engines. Exceptions are noted where they apply.

When measuring wear surfaces for critical dimensions, be sure to consult the appropriate tables for your engine. This holds true also for critical torque values.

Table 1 ACCEPTABLE CYLINDER DIAMETERS

Model/Engine	Dimensions*
Jag	
2000	2.126-2.127 in. (54.000-54.025mm)
3000	2.362-2.363 in .(60.000-60.020mm)
Panther, 1974-1975	
340	2.362-2.368 in. (60.000-60.150mm)
440	2.677-2.683 in. (68.000-68.150mm)
Panther/Cheetah, 1976-1977	
4000	2.559-2.560 in. (65.000-65.015mm)
5000	2.756-2.757 in. (70.000-70.015mm)
Pantera	
5000 (all)	2.755-2.756 in. (70.000-70.015mm)
Lynx	
2000S (single)	2.755-2.756 in. (70.000-70.015mm)
2000T (twin)	2.126-2.127 in. (54.000-54.020mm)
El Tigre, 1974-1975	
250	2.008-2.015 in. (51.000-51.190mm)
295	2.205-2.211 in. (56.000-56.150mm)
340	2.362-2.370 in. (60.000-60.190mm)
400	2.560-2.567 in. (65.000-65.190mm)
440	2.677-2.684 in. (68.000-68.184mm)
El Tigre, 1976-1977	
4000	2.559-2.560 in. (65.000-65.015mm)
5000	2.755-2.756 in. (70.000-70.015mm)

*Maximum deviation between cylinders = 0.002 in. (0.051mm).

Engine Removal/Installation

1. Open the hood and unplug the headlight harness. Remove the E-rings from the hood hinge pins, disconnect the hood cables, and remove the hood. Set it out of the way.

2. Remove the handle from the recoil starter. Grasp the starter rope firmly and untie the safety knot that prevents it from passing through the dashboard. Pull the rope through the dashboard and retie the knot to prevent the rope from being drawn all the way into the starter.

3. Unscrew the bolts that attach the recoil starter assembly to the engine (**Figure 1**) and collect the washers. Remove the starter assembly and set it out of the way.

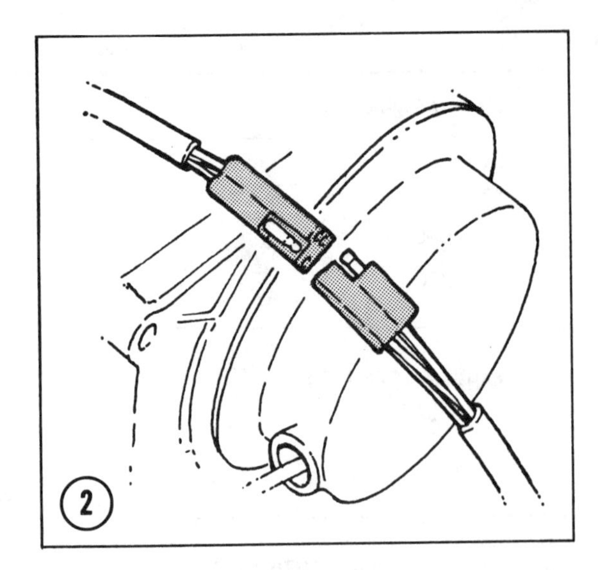

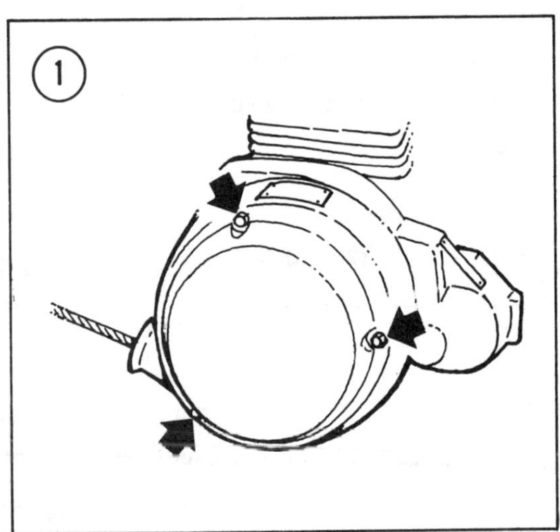

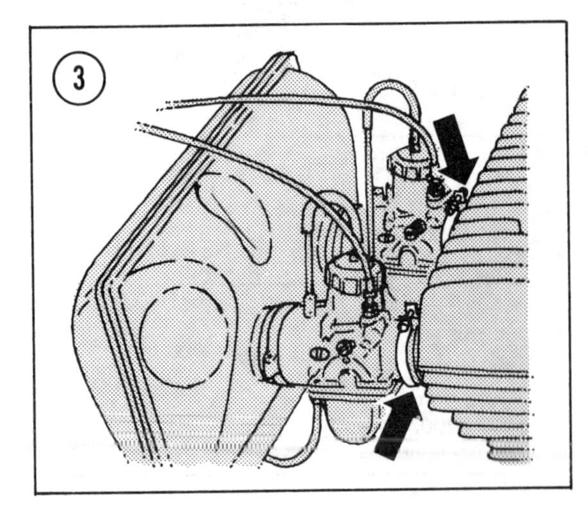

4. Unplug the main engine wiring harness at the engine connector (**Figure 2**).

5. Remove the belt guard, then remove the drive belt as described in Chapter Two, *Drive Belt*.

6. For Mikuni carburetors, loosen the carburetor flange clamp (**Figure 3**) and pull the carburetor out of the flange. For Walbro carburetors, unscrew the nuts that attach the carburetor to the manifold and collect the washers (**Figure 4**). Loosen the clamp on the silencer connector and disconnect the carburetor. Set it out of the way, with the cables attached.

7. Disconnect the fuel pump impulse line from the crankcase.

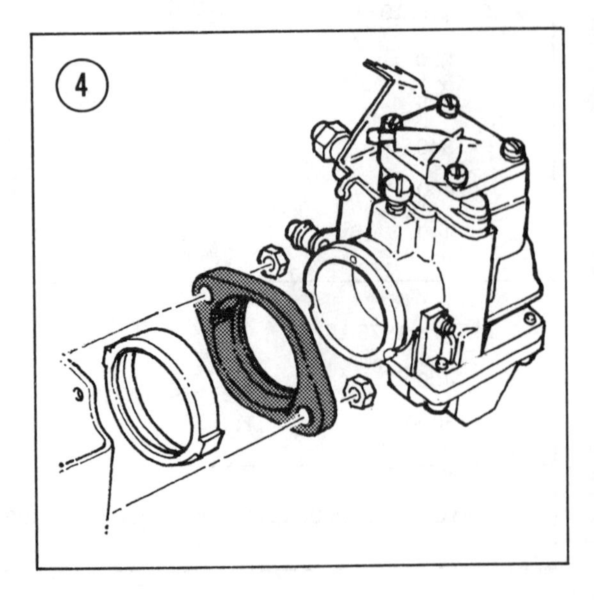

8. Place a piece of cardboard on the floor on the MAG side of the machine, and tile the snowmobile on its side. Unscrew the locknut from the rear motor mounts (**Figure 5**). Set the machine upright and unscrew the locknuts from the front motor mounts. Collect the washers.

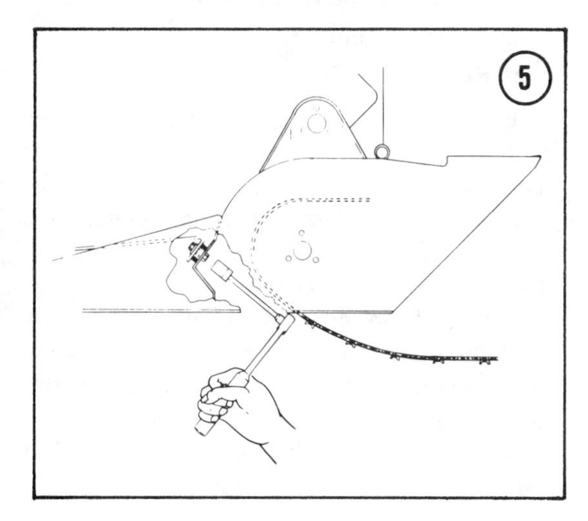

9. Unscrew the drive clutch bolt and collect the washer (**Figure 6**). On some models it is first necessary to remove the rubber plug from the belly pan to gain access to the bolt. Remove the drive clutch with the special Arctic puller

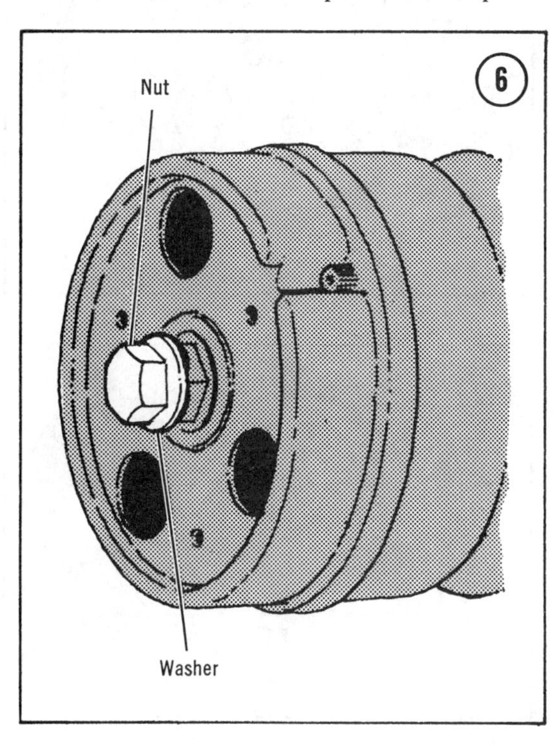

(**Figure 7**, part No. 0144-104 for Panther/Cheetah models, No. 0144-064 for others).

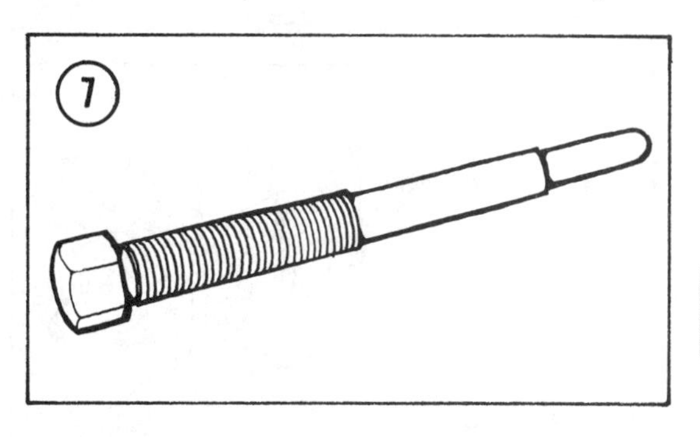

10. Disconnect the muffler from the exhaust manifold. Some models are fitted with a clip (**Figure 8**), and others use springs (**Figure 9**).

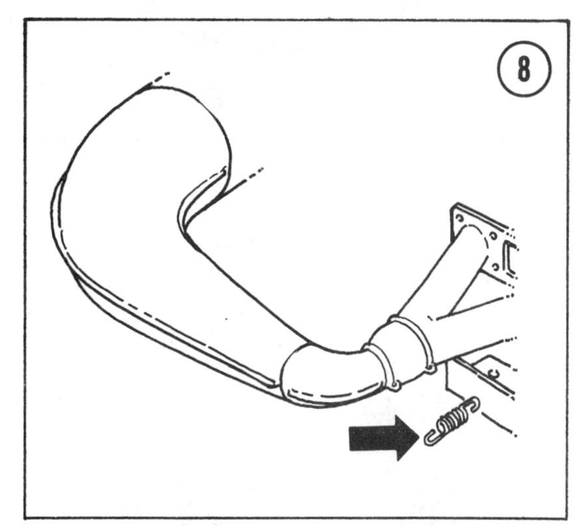

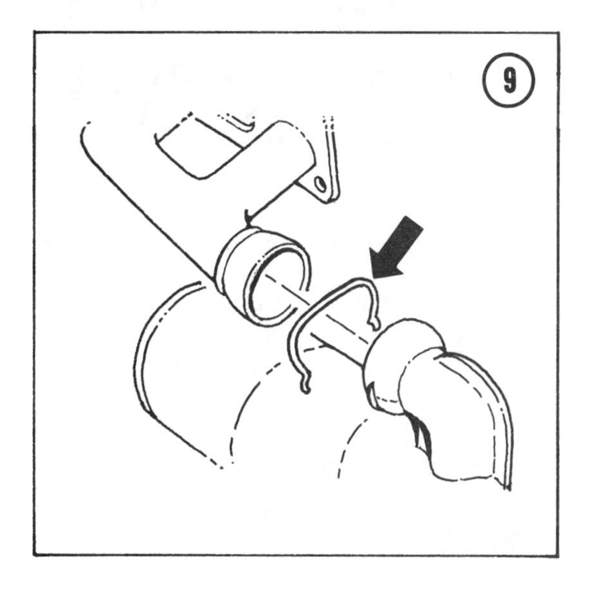

11. Unscrew the bolts that attach the engine plate to the engine (**Figure 10**) and collect the washers. Set the engine plate assembly aside.

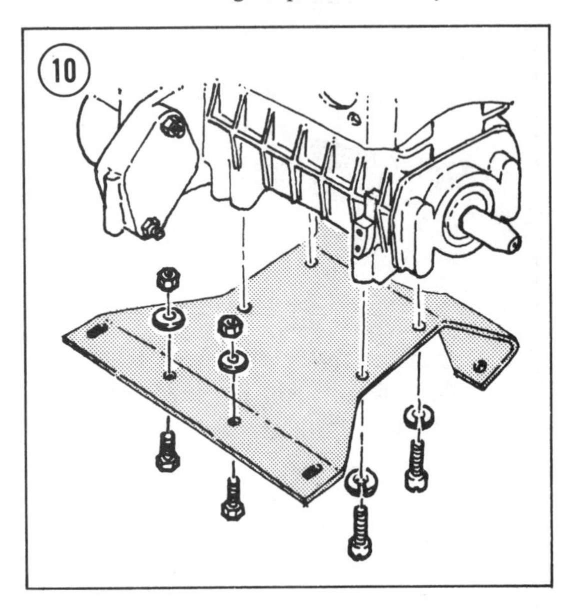

12. Unscrew the nuts that attach the exhaust manifold to the engine (**Figure 11**) and collect the washers. Remove the manifold.

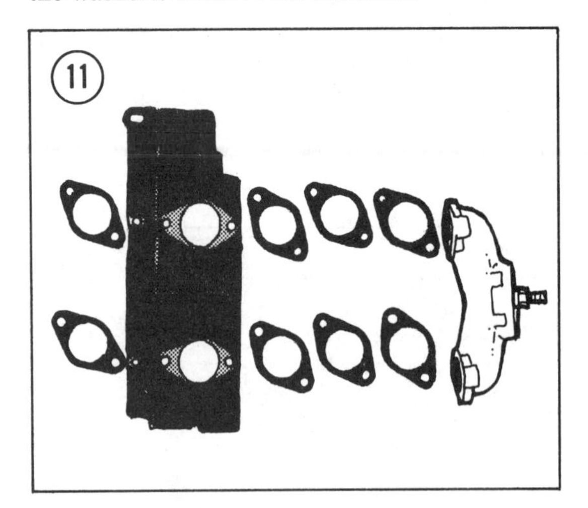

13. With assistance, lift the engine out of the machine after making certain that all attachments and connections have been removed or disconnected.

Upper End Disassembly

An orderly sequence should be followed to efficiently and correctly disassemble the engine upper end once the engine has been removed from the machine.

a. Remove exterior components (cooling shroud, coil, CDI "black box," etc.).
b. Remove spark plugs from cylinder head.
c. Remove cylinder heads and gaskets.
d. Remove cylinders.
e. Remove pistons.

1. On engines equipped with fan cooling, the cooling fan assembly and the shroud must be removed. First unscrew the bolts that attach the starting pulley (**Figure 12**) and collect the washers. Remove the starter and fan drive pulleys from the flywheel (**Figure 13**). Disengage the fan belt from the fan driven pulley. Unscrew the bolts that attach the fan housing assembly to the engine, collect the washers, and remove the fan assembly. Unscrew the bolts that attach the fan shroud (or cowling) to the engine, collect the washers, and remove the shroud.

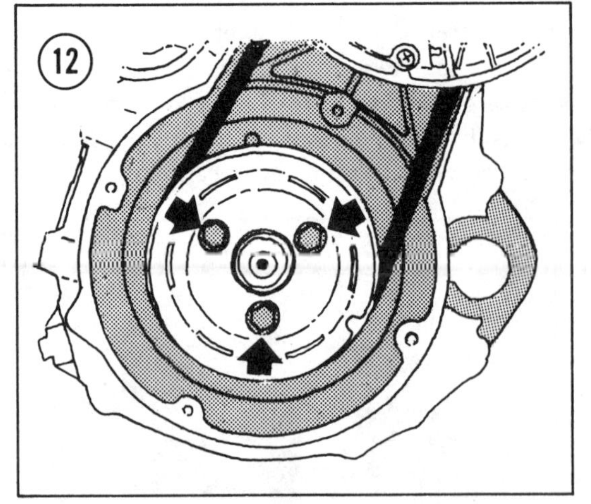

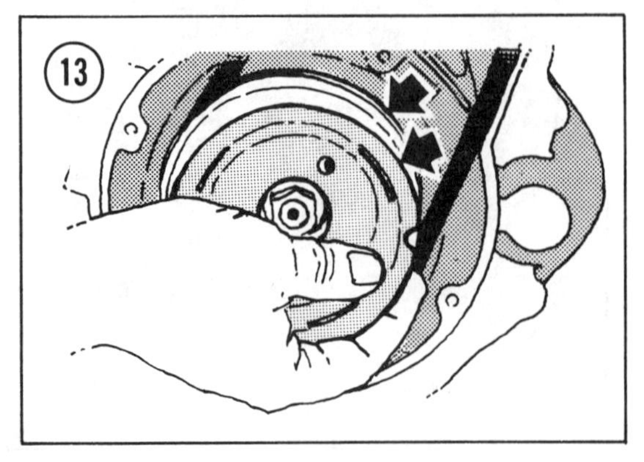

2. On all engines, unscrew the cylinder head nuts progressively in the appropriate pattern shown in **Figure 14**. Begin with ¼ turn for each nut until all nuts have been loosened 2 full turns. While this may seem tedious, it will help to prevent warping of the cylinder head. When all the nuts have been loosened, unscrew them and remove the cylinder head. If the head does not release from the cylinder with upward pressure, tap lightly around the bottom of the head with a soft mallet. Remove the cylinder head gasket.

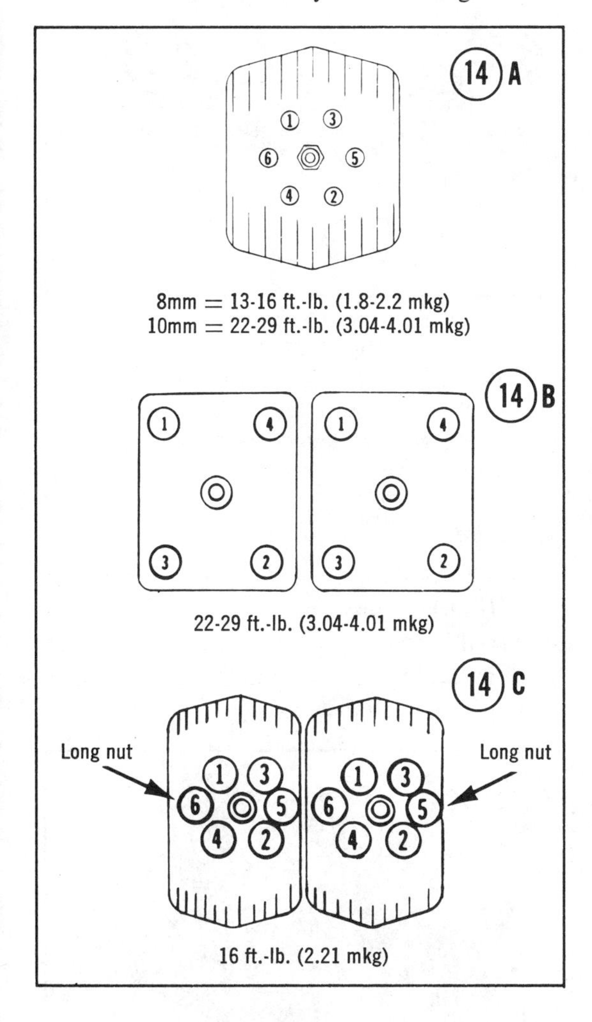

3. Unscrew the exhaust manifold nuts, collect the washers, and remove the exhaust manifold (**Figure 15**).

4. Unscrew the cylinder base nuts (**Figure 16**), collect the washers, and remove the cylinders by lifting straight up; don't rotate the cylinders as they are raised or you may damage otherwise good piston rings. When the cylinders have been removed, place a clean shop cloth in the top of each half of the crankcase to prevent small parts and foreign matter from falling into the crankcases.

5. Remove the circlips from both sides of each piston (**Figure 17**) and push the pin out. An

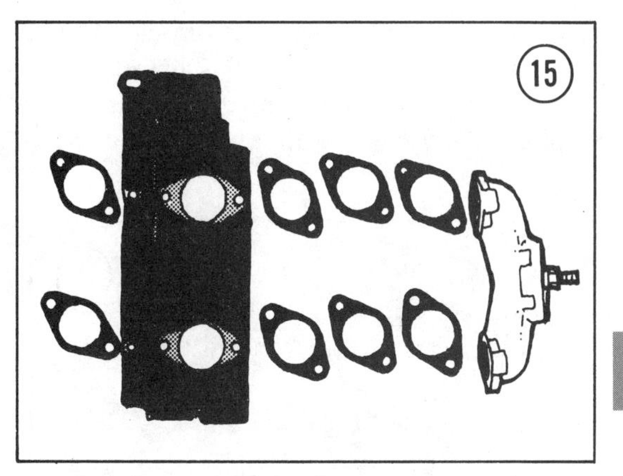

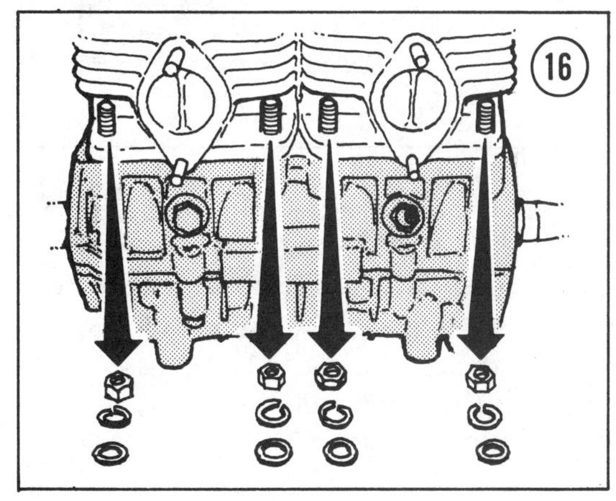

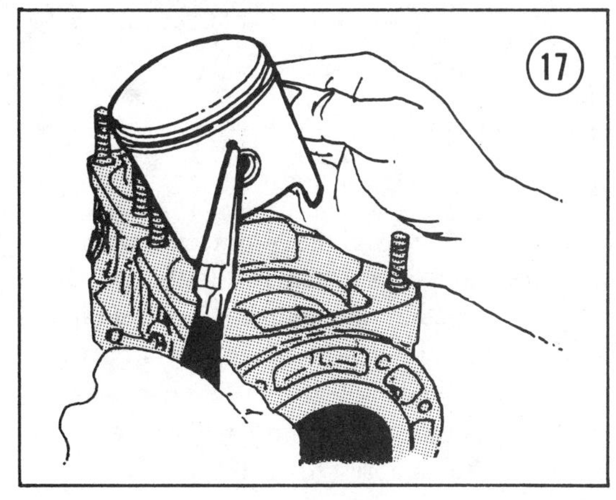

extractor (**Figure 18**) is available through Arctic Cat dealers (part No. 0144-003). In most cases, it will be necessary to remove the pins after wrapping the pistons with a cloth heated in boiling water.

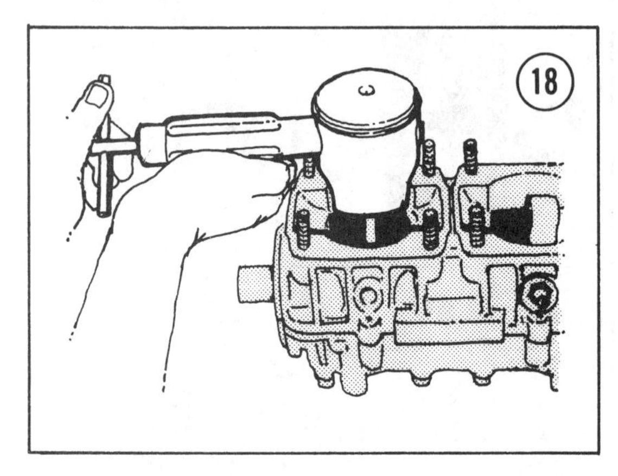

6. Remove the small end bearing from the connecting rod. For twin cylinder engines, keep the piston sets separated and do not mix the pieces.
7. Secure the connecting rods with rubber bands stretched over the cylinder base studs (**Figure 19**) to prevent the rods from striking the crankcase and damaging it.

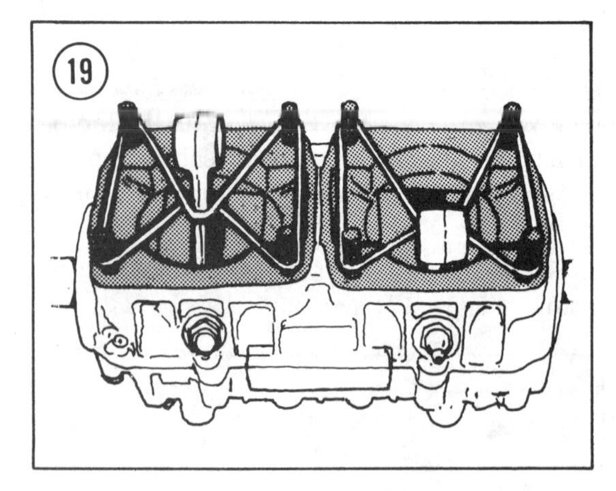

Cleaning

1. Scrape carbon from the combustion chamber in the head and the exhaust port in the cylinder, using a soft metal (aluminum) or wood scraper. Do not use a hard metal scraper; it will burr the surfaces and create hot spots.
2. Clean the head and cylinder with solvent and dry them with compressed air if possible.
3. Remove the rings from the piston and clean the dome with a soft scraper. Clean the ring grooves with a ring groove scraper or a piece of old piston ring (**Figure 20**). Clean the piston with solvent and dry it with compressed air.

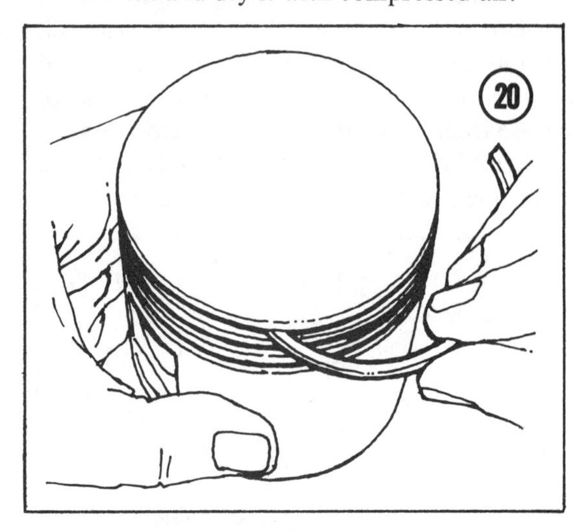

CAUTION

Pistons fitted with keystone cross-section rings must not be cleaned with a ring groove scraper. Instead, use a piece of old ring.

Inspection

1. Check the flatness of the cylinder head on either a surface plate or a piece of glass (**Figure 21**). If the head does not make contact over the entire sealing surface it must be trued. This is a job for a specialist.

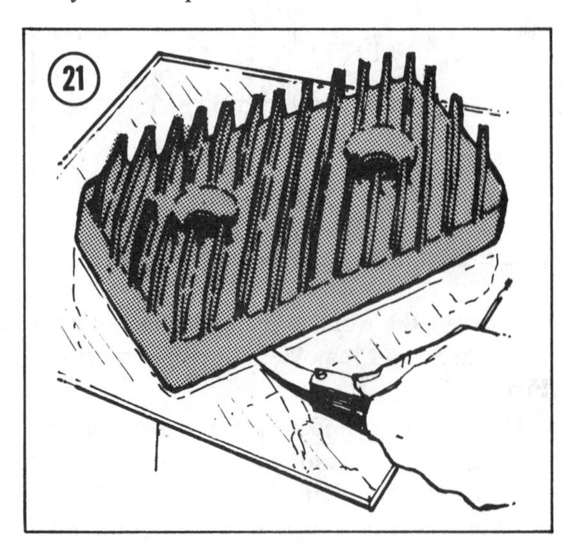

2. Check the cylinder and the piston for wear, galling, scuffing, or burning. Minor irregularities can be removed from the piston with crocus cloth and light oil. The cylinder may be cleaned up with a light honing, provided it is within specifications as described below.

3. Measure the cylinder bore with an inside micrometer or cylinder gauge. Measure ⅜ in. below the top of the cylinder, in 2 locations, 90° apart (**Figure 22**). If the 2 measurements differ by more than 0.002 in. (0.05mm), the cylinder is out-of-round beyond specification and must be bored or replaced. Measure again in 2 locations 90° apart just above the intake port. If these 2 measurements differ by more than 0.002 in. (0.05mm), the cylinder is out of specification and must be bored or replaced. Acceptable cylinder diameters are shown in **Table 1**.

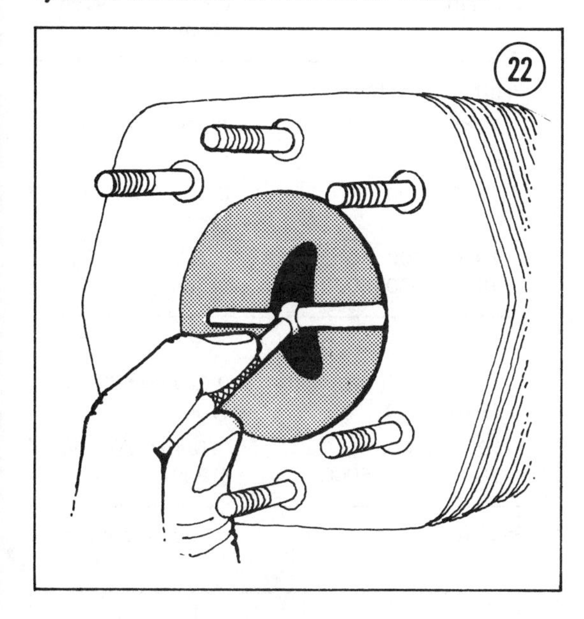

4. Check piston skirt-to-cylinder clearance by first measuring the base of the cylinder bore, front to back (**Figure 23**). With an outside micrometer, measure the piston skirt, front and rear, about ¼ in. (6mm) from the bottom (**Figure 24**). Subtract the piston measurement from the cylinder measurement to determine actual piston skirt-to-cylinder clearance. Refer to **Table 2** for acceptable clearances. If the bore is okay, but the clearance is excessive, the piston must be replaced with a new unit that includes a matched pin and bearing, as well as rings.

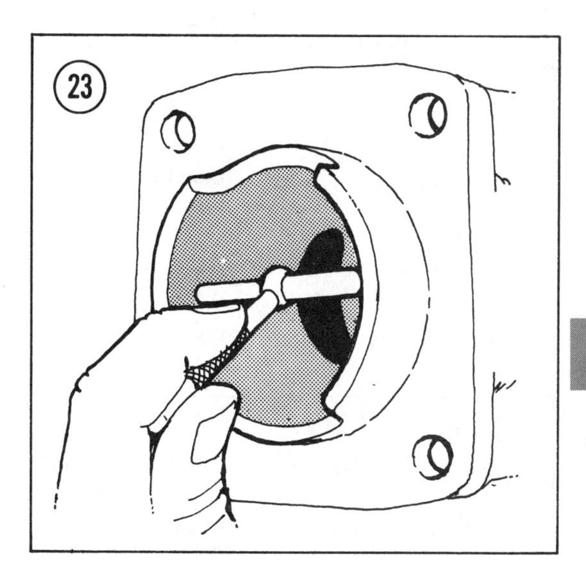

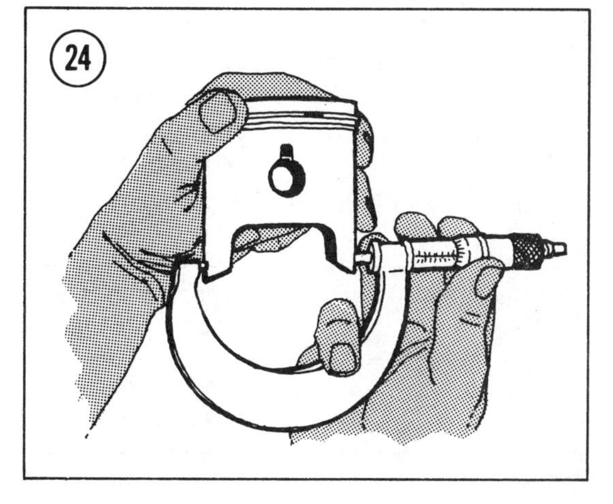

5. Measure the piston ring end gap. Place a ring into the cylinder, ⅜ in. from the top. Use the piston to square the ring with the cylinder by pressing down on the ring with the skirt. Measure the gap as shown in **Figure 25**. Refer to **Table 3** for acceptable end gap. If the gap is too large for either ring, replace them as a set. Check new rings in the manner just described, and if the gap is too small, file material off the ends of the rings.

6. Measure the piston pin ¼ in. from each end (**Figure 26**) and compare the actual measurement to **Table 4**. If the pin measurement is not acceptable, replace the pin and bearing.

7. Measure the pin bore in the piston (**Figure 27**) and compare the results with **Table 5**. If the bore measurement is not acceptable, re-

Table 2 PISTON-TO-CYLINDER CLEARANCE

Model/Engine	Clearance
Jag (all)	0.0015-0.0020 in. (0.04-0.05mm)
Panther/Cheetah	
1974-1975 (all)	0.0008-0.0020 in. (0.020-0.050mm)
1976-1977 (all)	0.0020-0.0025 in. (0.050-0.064mm)
El Tigre	
1974-1975 295/340	0.0028-0.0042 in. (0.071-0.106mm)
1974-1975 400/440	0.0033-0.0048 in. (0.084-0.122mm)
1974-1975 340 FR1	0.0030-0.0045 in. (0.076-0.114mm)
1974-1975 440 FR1	0.0022-0.0037 in. (0.056-0.094mm)
1974-1975 250	0.0026-0.0041 in. (0.066-0.104mm)
1976-1977 (all)	0.0020-0.0025 in. (0.050-0.064mm)
Pantera (all)	0.0020-0.0025 in. (0.050-0.064mm)
Lynx (all)	0.0015-0.0025 in. (0.038-0.064mm)

25

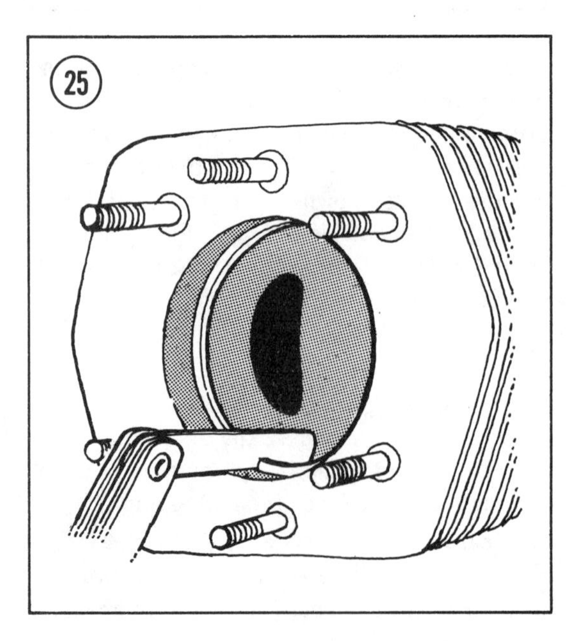

26

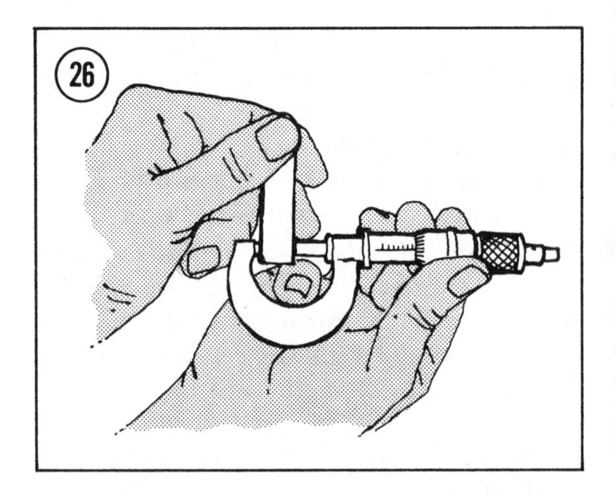

Table 4 PISTON PIN DIAMETER

Model/Engine	Diameter
Jag (all)	0.6297-0.6299 in. (16mm)
Panther/Cheetah	
1974-1975 340/440	0.6297-0.6299 in. (16mm)
1976 4000	0.6297-0.6299 in. (16mm)
1976 5000, 1977 4000/5000	0.7085-0.7087 in. (18mm)
El Tigre	
1974-1975 (all), 1976 4000	0.6297-0.6299 in. (16mm)
1976 5000, 1977 4000/5000	0.7085-0.7087 in. (18mm)
Pantera (all)	0.7085-0.7087 in. (18mm)
Lynx	
Single	0.7085-0.7087 in. (18mm)
Twin	0.6297-0.6299 in. (16mm)

place the piston, pin, and bearing as a set, along with new rings.

8. Measure the small end bore of the connecting rod with a snap gauge (**Figure 28**), then measure the snap gauge with an outside micrometer (**Figure 29**). Compare the measure-

Table 3 PISTON RING END GAP

Model/Engine	Gap	Wear Limit
Jag (all)	0.006-0.014 in. (0.15-0.36mm)	0.031 in. (0.79mm)
Panther/Cheetah		
340	0.006-0.014 in. (0.15-0.36mm)	0.031 in. (0.79mm)
440	0.008-0.016 in. (0.20-0.40mm)	0.033 in. (0.85mm)
4000	0.006-0.014 in. (0.15-0.36mm)	0.031 in. (0.79mm)
5000	0.008-0.016 in. (0.20-0.40mm)	0.033 in. (0.85mm)
El Tigre		
1974-1975 295/340	0.006-0.014 in. (0.15-0.36mm)	0.031 in. (0.79mm)
1974-1975 400/440	0.008-0.016 in. (0.20-0.40mm)	0.033 in. (0.85mm)
1974-1975 340 FR1/ 440 FR1/250	0.012-0.019 in. (0.30-0.48mm)	0.035 in. (0.89mm)
1976-1977 4000	0.006-0.014 in. (0.15-0.36mm)	0.031 in. (0.79mm)
1976-1977 5000	0.008-0.016 in. (0.20-0.40mm)	0.033 in. (0.85mm)
Pantera (all)	0.006-0.014 in. (0.15-0.36mm)	0.031 in. (0.79mm)
Lynx		
Single	0.008-0.016 in. (0.20-0.40mm)	0.033 in. (0.84mm)
Twin	0.006-0.014 in. (0.15-0.36mm)	0.031 in. (0.79mm)

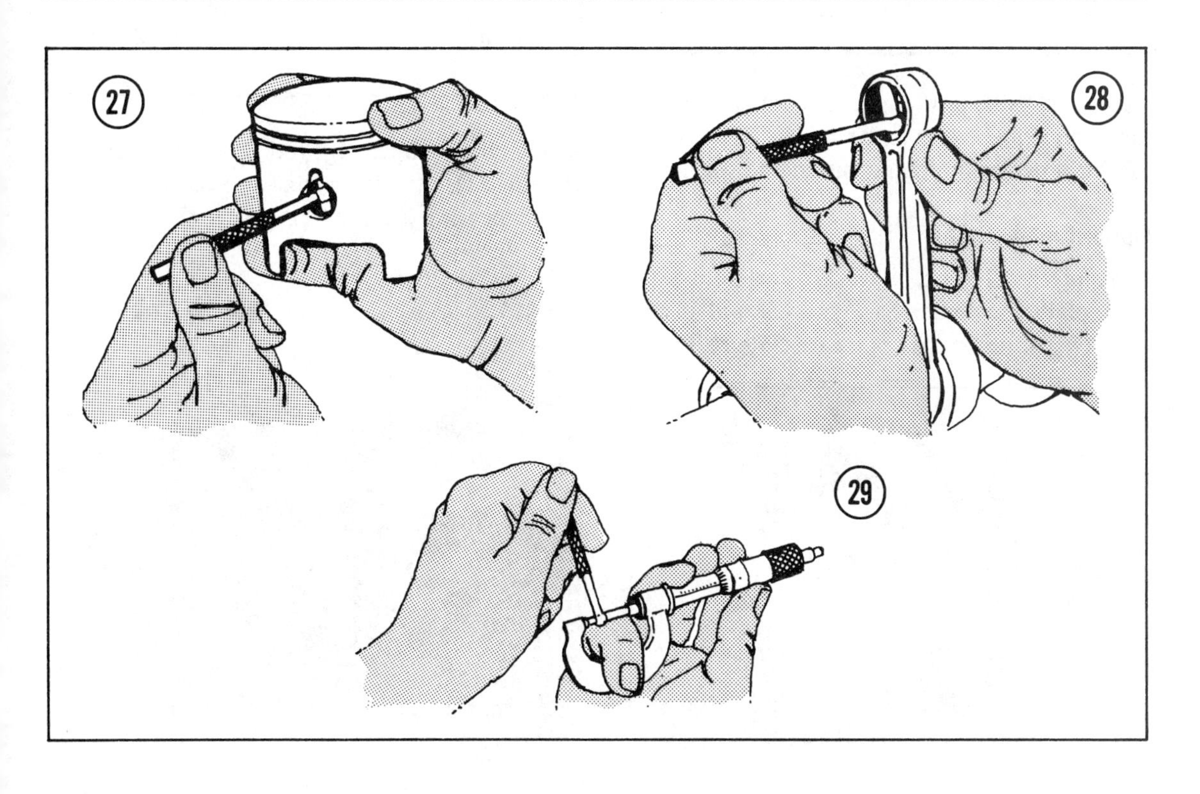

Table 5 PISTON PIN BORE DIAMETER

Model/Engine	Diameter
Jag (all)	0.6298-0.6301 in. (16mm)
Panther/Cheetah	
1974-1975 340/440	0.6298-0.6301 in. (16mm)
1976 4000	0.6298-0.6301 in. (16mm)
1976 5000, 1977 4000/5000	0.7086-0.7089 in. (18mm)
El Tigre	
1974-1975 (all), 1976 4000	0.6298-0.6301 in. (16mm)
1976 5000, 1977 4000/5000	0.7086-0.7089 in. (18mm)
Pantera (all)	0.7086-0.7089 in. (18mm)
Lynx	
Single	0.7086-0.7089 in. (18mm)
Twin	0.6298-0.6301 in. (16mm)

ment with **Table 6**. If the bore is not acceptable, the connecting rod must be replaced. This is a job for a dealer or competent engine specialist.

Upper End Assembly

1. Apply a thin coat of gasket sealer to both sides of new cylinder base gaskets and install them over studs on the crankcase (**Figure 30**).

Table 6
CONNECTING ROD SMALL END DIAMETER

Model/Engine	Diameter
Jag (all)	0.8268-0.8273 in. (21.00-21.01mm)
Panther/Cheetah	
1974-1975 340/440	0.7875-0.7880 in. (20.00-20.01mm)
1976 4000	0.8268-0.8273 in. (21.00-21.01mm)
1976 5000, 1977 4000/5000	0.9055-0.9060 in. (23.00-23.01mm)
El Tigre	
1974-1975 (all)	0.7875-0.7880 in. (20.00-20.01mm)
1976 4000	0.8268-0.8273 in. (21.00-21.01mm)
1976 5000, 1977 4000/5000	0.9055-0.9060 in. (23.00-23.01mm)
Pantera (all)	0.9055-0.9060 in. (23.00-23.01mm)
Lynx	
Single	0.9055-0.9060 in. (23.00-23.01mm)
Twin	0.8268-0.8273 in. (21.00-21.01mm)

2. Oil the connecting rod big end bearing (**Figure 31**), the piston pin, and the piston pin needle bearing (**Figure 32**).

3. Install the pistons with the arrow on the piston crown facing the exhaust port. This is essential so the ring end gaps will be correctly positioned and will not snag in the ports. See **Figure 33**. Install the piston pin clips, making sure they are completely seated in their grooves with the open ends of the clips facing down (**Figure 34**).

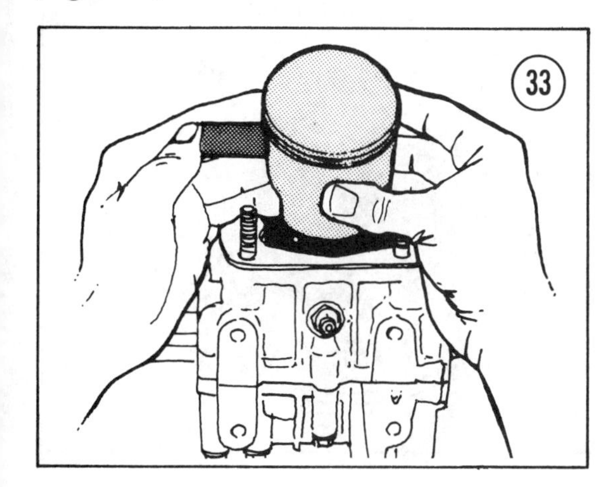

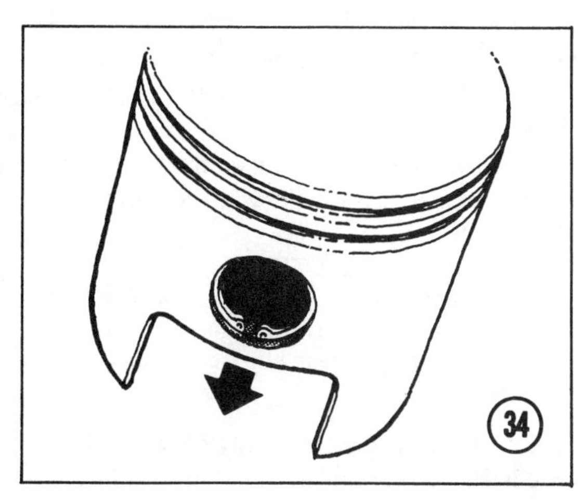

4. Coat the piston and the cylinder bore lightly with oil. Line up the piston ring ends with the locating pins in the ring grooves (**Figure 35**). Install a ring compressor or a large hose clamp to compress the rings.

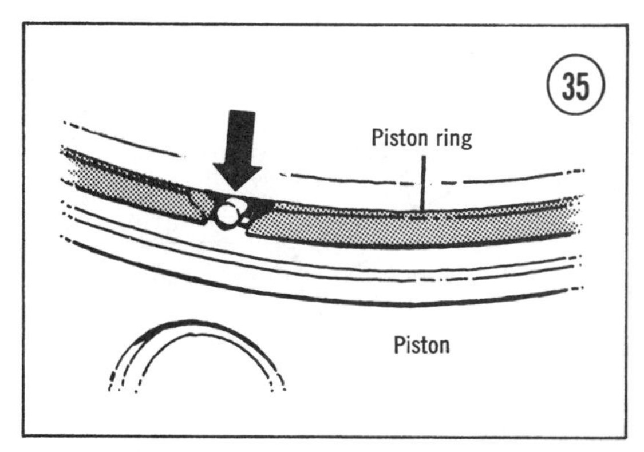

5. Line up the cylinder with the piston, hold the piston to prevent it from rocking, and push the cylinder down over the piston. Some resistance will be felt, but if binding is experienced, remove the cylinder and locate and correct the problem before proceeding; don't force the cylinder down over the piston.

6. When the cylinder is in place, screw on and tighten the cylinder base nuts to 22-29 ft.-lb. (3.04-4.01 mkg) in a crisscross pattern. See **Figure 36**.

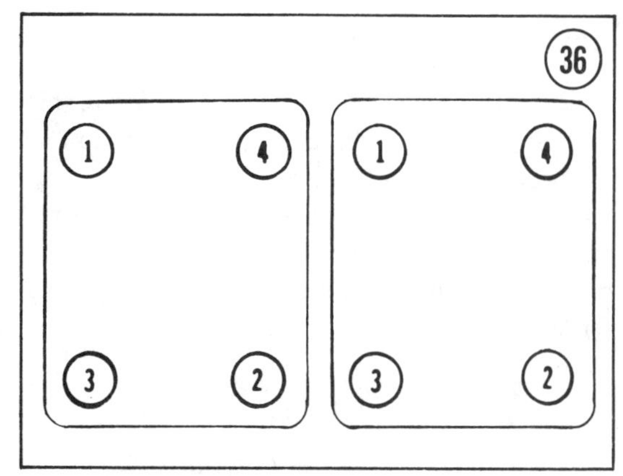

7. Set new head gaskets in place, then install the cylinder heads. Tighten the cylinder head nuts to 22-29 ft.-lb. (3.04-4.01 mkg) for JAG engines, or 16 ft.-lb. (2.21 mkg) for all other engines in the appropriate pattern shown in **Figure 37**.

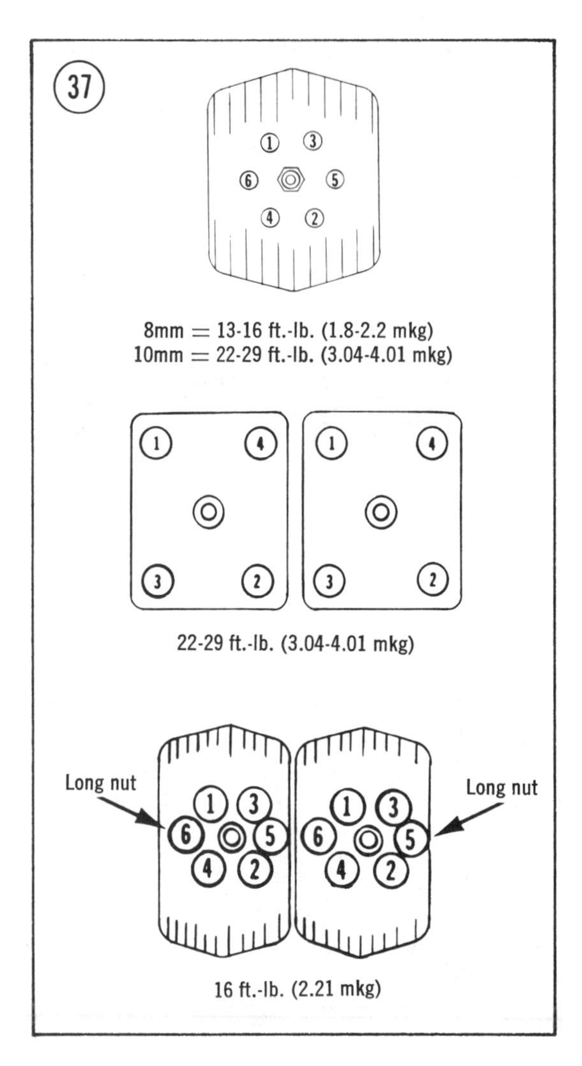

8. Refer to *Engine Removal/Installation* in this chapter and install the engine and the support hardware. Refer to Chapter Two, *Tune-up,* and tune the engine as described. Remember, a rebuilt upper end must be broken-in in the same manner as a new engine if it is to enjoy long service life.

LOWER END OVERHAUL

Lower end overhaul as described here consists of removal and installation of the crankshaft assembly. A surface plate, V-blocks, and dial indicator stands and bases are required to accurately check the condition of the crankshaft assembly. In addition, early crankshaft assemblies cannot be serviced and must be replaced as an assembly, while the latest crankshafts require a hydraulic press and considerable experience to disassemble, assemble, and align them.

It is recommended that the crankcase halves be parted and the crankshaft assembly entrusted to a dealer for inspection and repair.

Disassembly

1. Refer to *Engine Removal/Installation* and *Upper End Disassembly* and remove the engine from the machine and disassemble upper end.

2. Remove the flywheel, rotor, and stator as described in Chapter Six. Unscrew the screws that attach the magneto base to the crankcase (**Figure 38**). An impact driver should be used to loosen the screws and prevent damage to the screw slots. Tap the magneto base off the crankcase with a soft mallet.

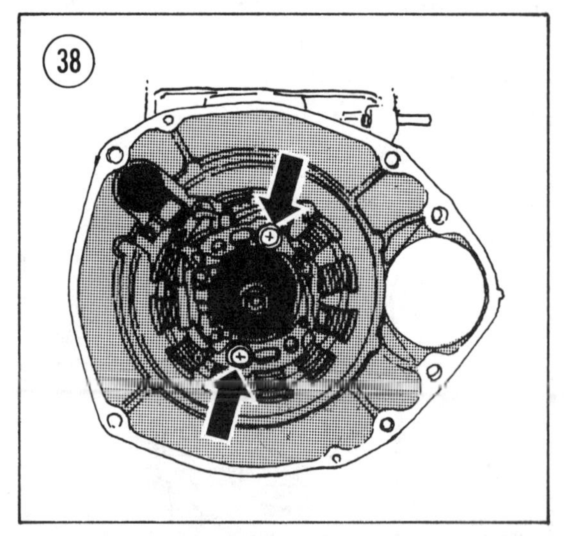

3. Unscrew the bolts from the bottom of the crankcase in the appropriate pattern shown in **Figure 39** to reduce the possibility of warping the crankcase halves.

4. Part the cases by tapping on the large bosses on the upper case half with a soft mallet (**Figure 40**).

CAUTION

Do not pry the cases apart with a screwdriver or any other similar tool; damage to the sealing surfaces is likely to result.

5. Lift the crankshaft assembly out of the lower case half. Pay particular attention to shims and

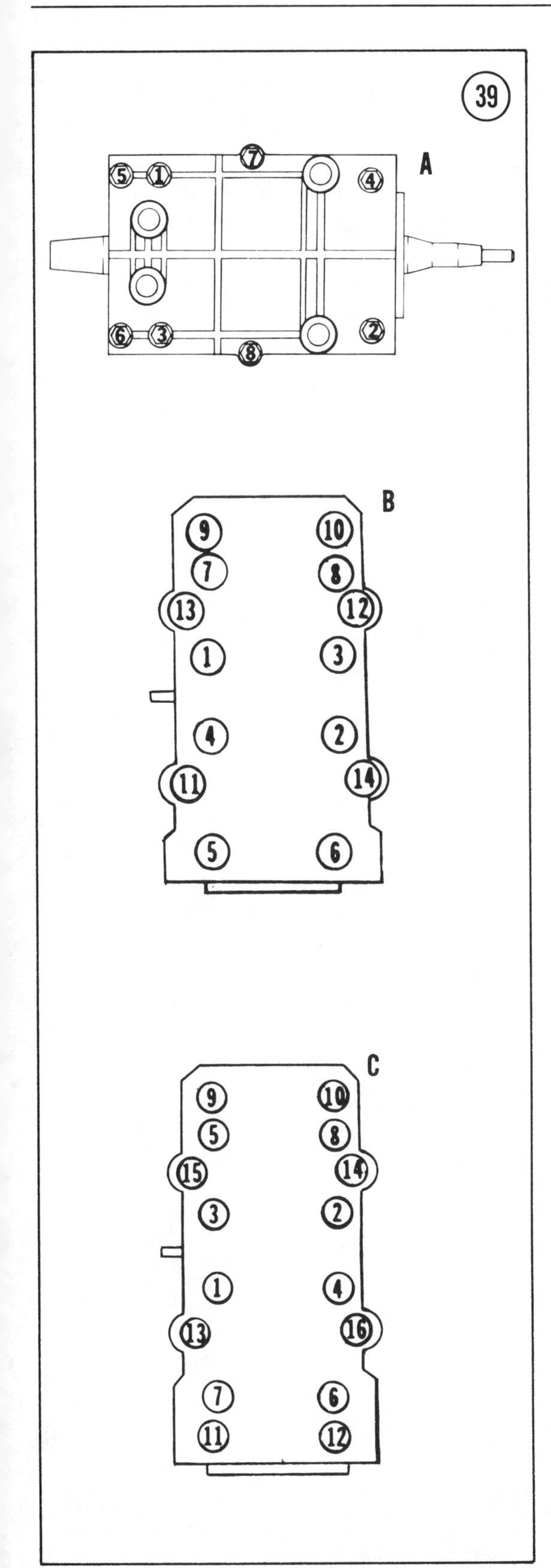

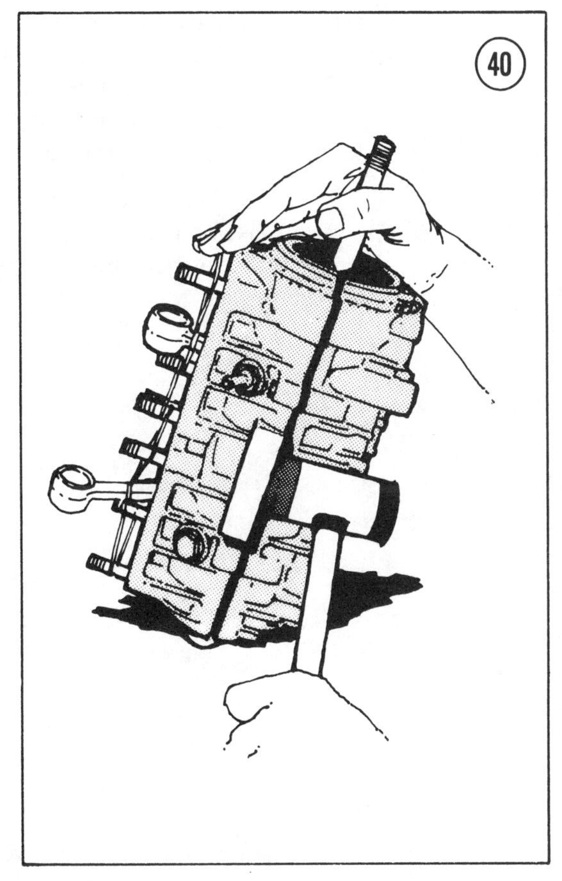

half rings in the bottom case, and make a detailed drawing to indicate their locations.

6. Clean the crankcase halves and the crankshaft assembly with solvent and dry them with compressed air. Have the crankshaft assembly inspected and serviced by an Arctic Cat dealer or engine specialist.

Assembly

1. Fill the open face of each crankshaft seal with grease and install the seals on the ends of the shaft with the open faces inward. Oil the connecting rod and main bearings.

2. Lightly coat the sealing surface of both crankcase halves with a pliable crankcase sealer such as Liquid Gasket, RTV Silicone Seal, or Permatex Form-A-Gasket.

3. Set the crankshaft assembly in the bottom case half, making sure the half ring is installed in the center on twins, and the dowels in the case engage the alignment holes in the bearings.

4. Set the top case half in place and press it down by hand until it contacts the sealing surface of the lower case half. If it does not, retrace the above steps to find the reason; no force is required to assemble the case halves if everything is lined up.

5. Install all of the crankcase bolts finger-tight. Then, tighten them in the sequence shown in **Figure 39** to the correct torque:

6mm	6-7 ft.-lb. (0.82-0.96 mkg)
8mm	13-16 ft.-lb. (1.79-2.21 mkg)
10mm	15-22 ft.-lb. (2.07-3.04 mkg)

6. Reverse the remaining disassembly steps to assemble the engine. Refer to the instructions for assembling the upper end and installing the engine in this chapter. Refer to Chapter Two for engine tune-up.

NOTE: If you own a 1978 or later model, first check the Supplement at the back of the book for any new service information.

CHAPTER FIVE

FUEL SYSTEM

The fuel system consists of a fuel tank, fuel line or lines, in-line fuel filter and carburetor.

The Walbro carburetor (**Figure 1**) has an integral fuel pump. An auxiliary impulse pump (**Figure 2**) is provided on models equipped with Mikuni carburetors (**Figure 3**). The fuel pumps operate off differential pressure in the engine crankcase.

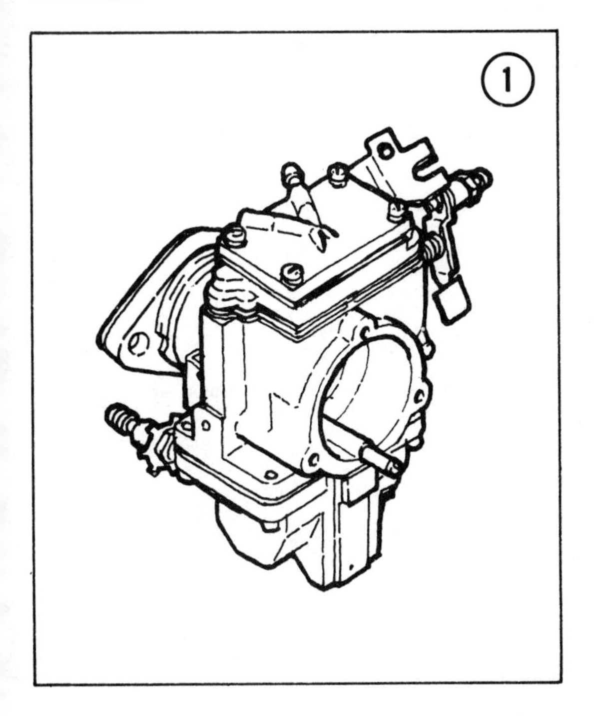

1

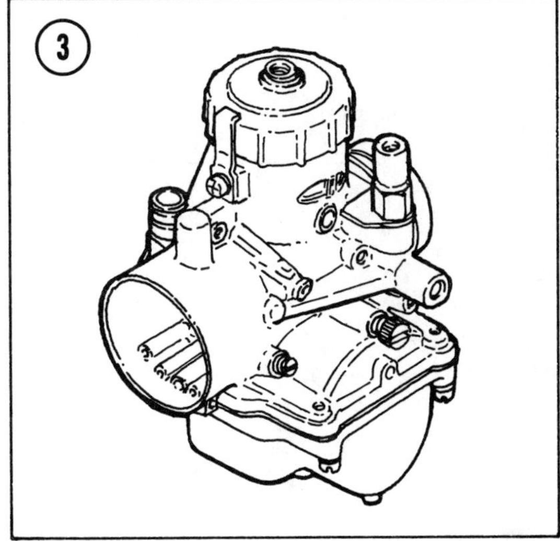

3

An air intake silencer is used on all models to quiet incoming air and to catch fuel that may spit back out of the carburetor.

This chapter covers removal, installation, and replacement and/or repair of carburetors, fuel pumps, in-line filters and fuel tanks. Carburetor tuning is covered in Chapter Two.

WALBRO CARBURETOR

Removal

1. Remove the windshield and console.

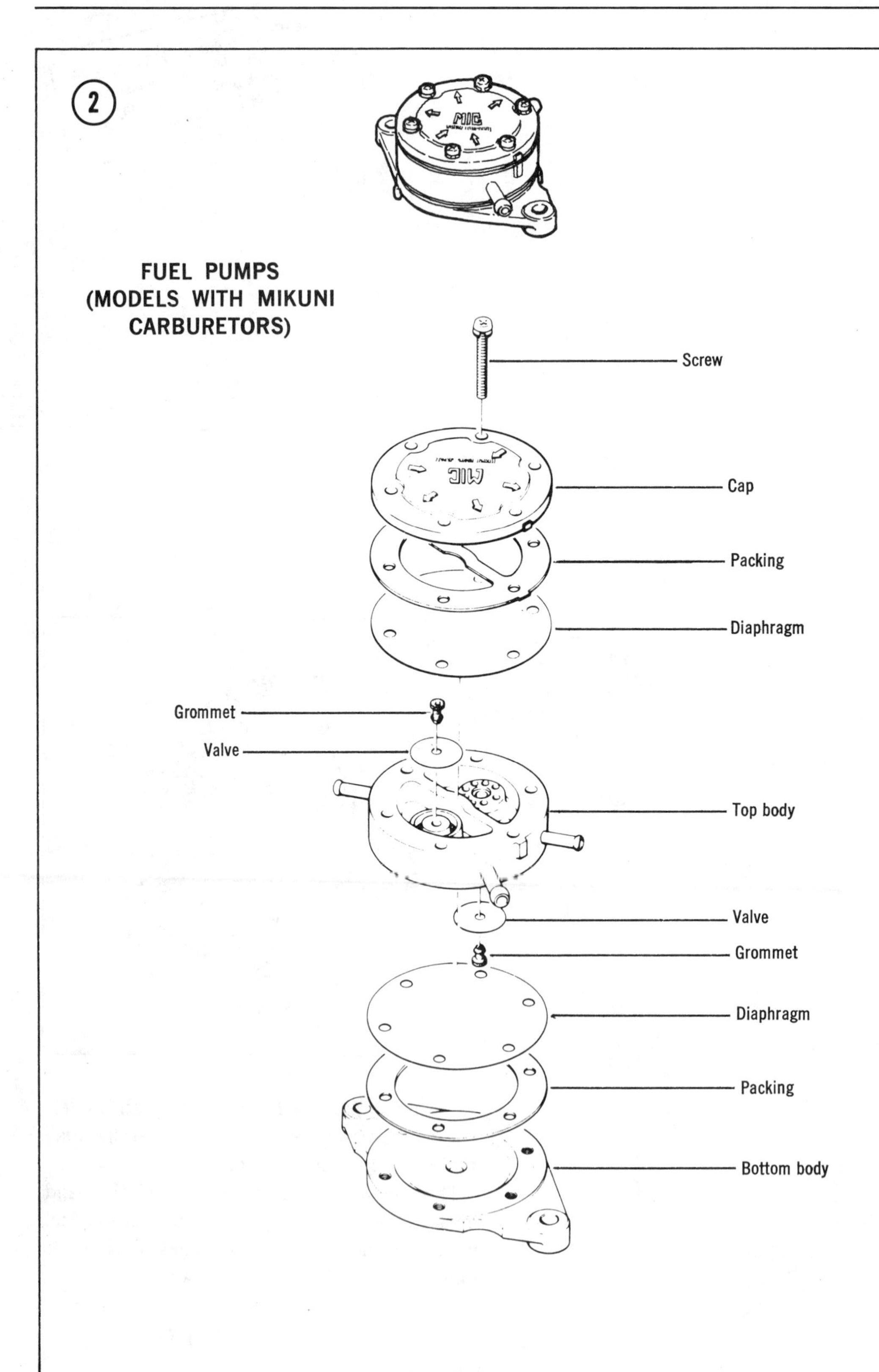
2
FUEL PUMPS
(MODELS WITH MIKUNI
CARBURETORS)
Screw
Cap
Packing
Diaphragm
Grommet
Valve
Top body
Valve
Grommet
Diaphragm
Packing
Bottom body

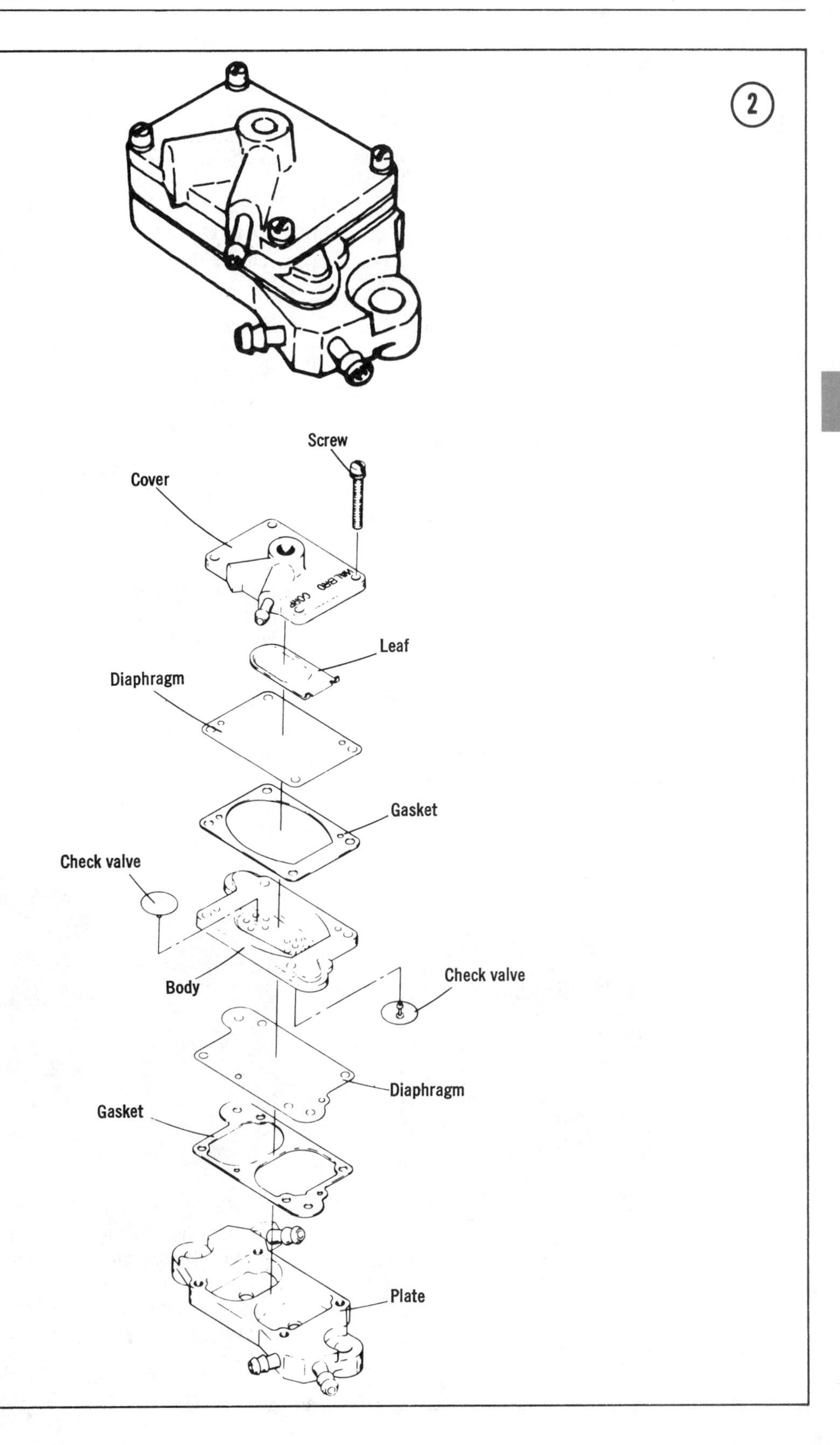
2
Screw
Cover
WALBRO CORP
Leaf
Diaphragm
Gasket
Check valve
Body
Check valve
Diaphragm
Gasket
Plate

2. Disconnect the choke and throttle cables. Disconnect the fuel line and plug it. Remove the vapor return lines, and fuel pump impulse tube from carburetor.

3. Remove the silencer and throttle and choke brackets. Remove the carburetor from intake manifold.

Disassembly

1. Clean exterior of carburetor with a nonflammable solvent.

CAUTION

Never use compressed air to clean an assembled carburetor or the diaphragm may be damaged.

2. Carefully disassemble carburetor according to **Figure 4**. Pay particular attention to the location of different sized screws and springs.

3. Perform *Cleaning and Inspection.*

Cleaning and Inspection

A special carburetor cleaning solution, tank and basket (**Figure 5**) can be purchased at auto parts stores for a few dollars. When not in use the tank can be sealed to prevent the reuseable solution from evaporating.

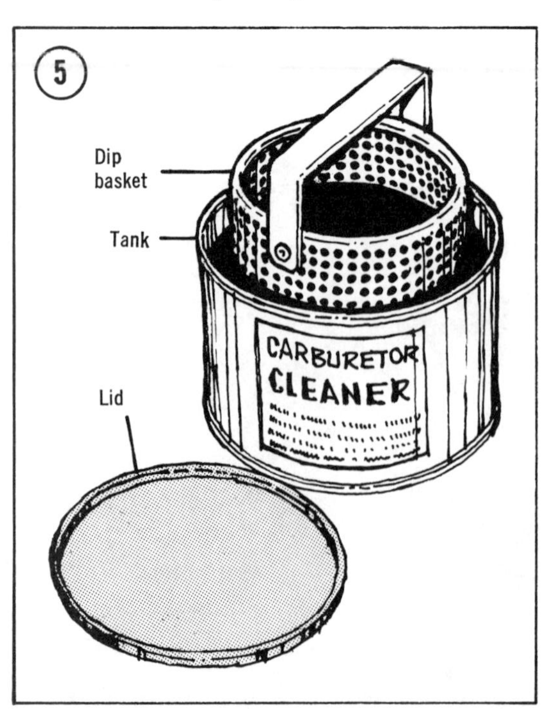

WARNING

Most carburetor cleaners are highly caustic. They must be handled with extreme care or skin burns and possible eye injury may result.

1. Clean all metallic parts in carburetor cleaning solvent. Do not place gaskets or diaphragms in solvent or they will be destroyed.

CAUTION

Never clean holes or passages with small drill bits or wire or a slight enlargement or burring of holes will result and drastically affect carburetor performance.

2. After cleaning carburetor parts, dry them with compressed air. Make sure all holes are open and free of carbon and dirt.

WARNING

Wear safety glasses when drying the parts with compressed air to keep solvent out of your eyes.

NOTE: *Do not use rags or wastepaper to dry parts. Lint may plug jets or channels and affect carburetor operation.*

3. Inspect shaft bearing surfaces in the carburetor body (**Figure 6**) for excessive wear.

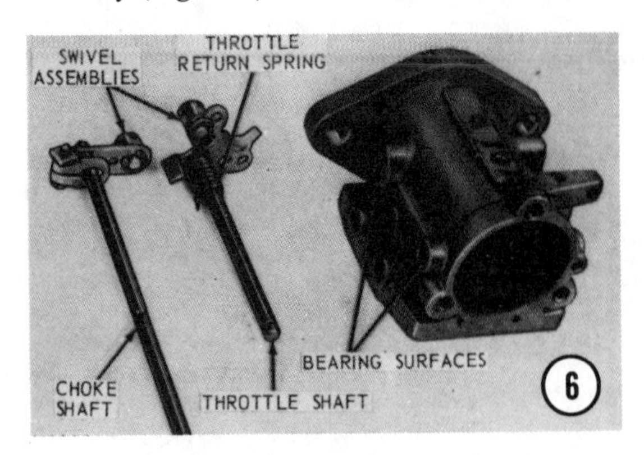

CAUTION

If excessive clearance is found between shafts and carburetor body, worn parts must be replaced. Excessive clearance will allow air to enter causing a damaging too-lean mixture.

4. Inspect the choke and throttle plates for damage. Inspect the swivel assemblies on choke

and throttle levers for wear. Check the condition of throttle return spring. Replace all worn parts.

5. Inspect the mixture needles and needle valve seating surfaces for pitting or wear (**Figure 7**), and replace them if they are worn or damaged.

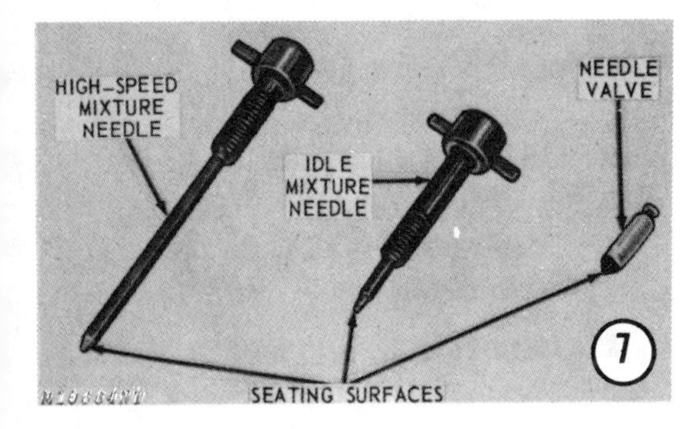

6. Inspect the diaphragms for distortion, cracks or punctures (**Figure 8**) and replace them if they are less than perfect.

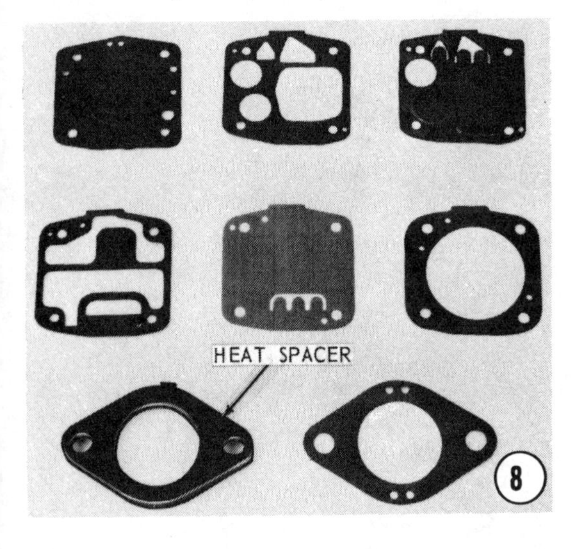

7. Inspect the carburetor mounting gasket and heat spacer gasket.

Assembly

Refer to **Figure 4** for this procedure.

1. Install the throttle shaft, with spring, in the carburetor body and secure it with the retaining ring.

2. Insert the throttle plate as shown in **Figure 9** with numbers facing out and below the shaft. Close the throttle plate to center it in the carburetor body before tightening the screw, and

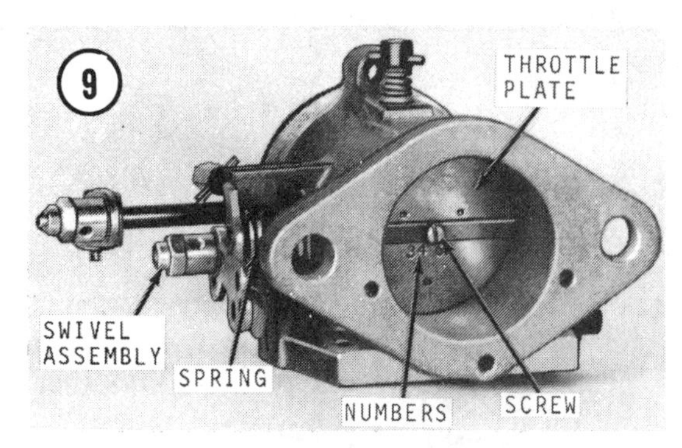

secure screw with Loctite Lok-N-Seal medium strength fastener locking compound.

3. Install the detent spring and detent ball in hole in carburetor body (**Figure 10**).

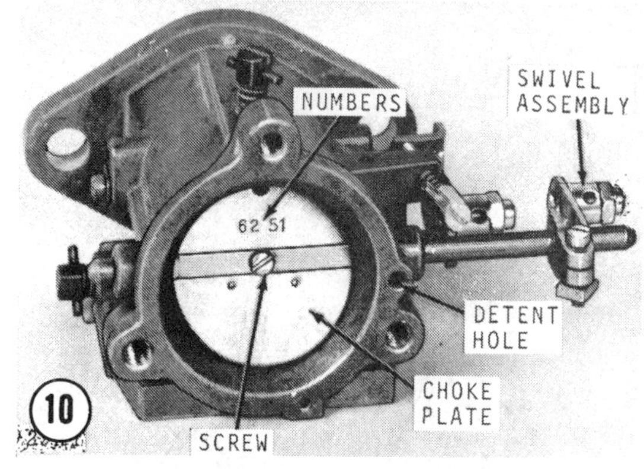

4. Install the choke shaft in the carburetor body. Insert the choke plate as shown in **Figure 10** with numbers facing out and the notch on top. Close the choke plate to center it in the carburetor body before tightening the screw. Secure screw with Loctite (see Step 2).

5. Install the spring, washer, and packing rings on the mixture needles (**Figure 11**). Install the needles in carburetor body. Install the idle stop screw and spring.

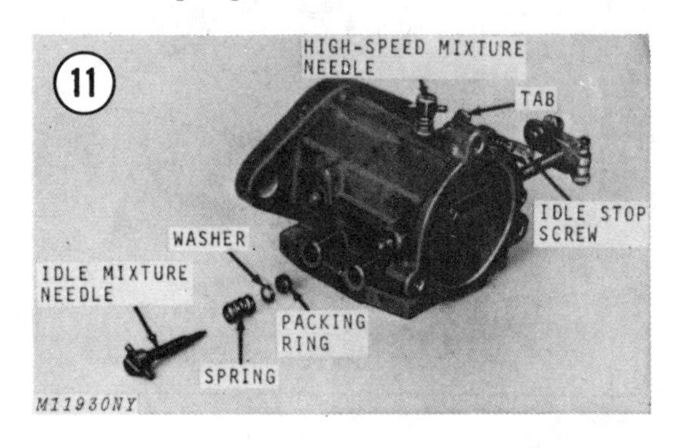

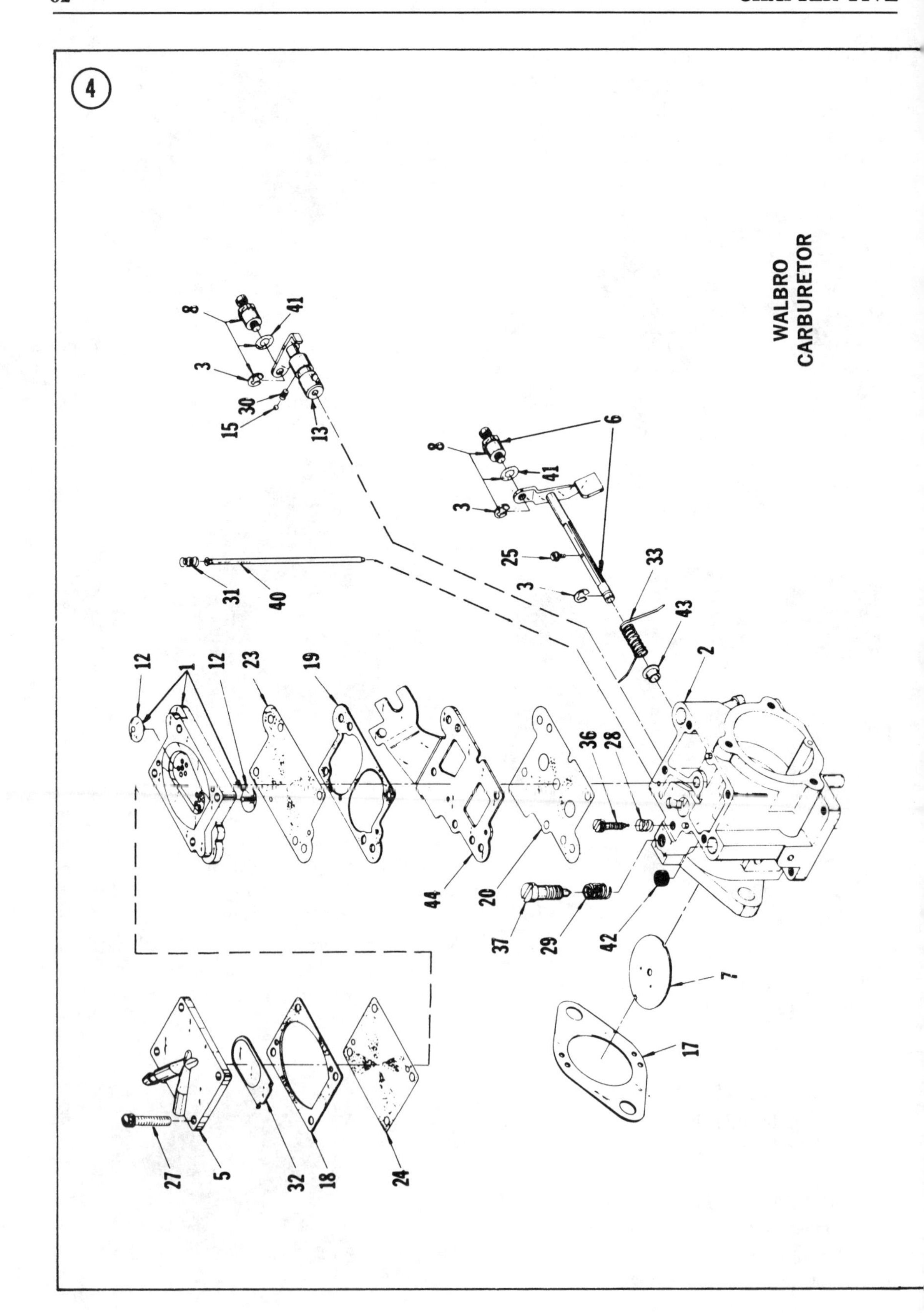
4
WALBRO
CARBURETOR
1
2
3
5
6
7
8
12
13
15
17
18
19
20
23
24
25
27
28
29
30
31
32
33
36
37
40
41
42
43
44

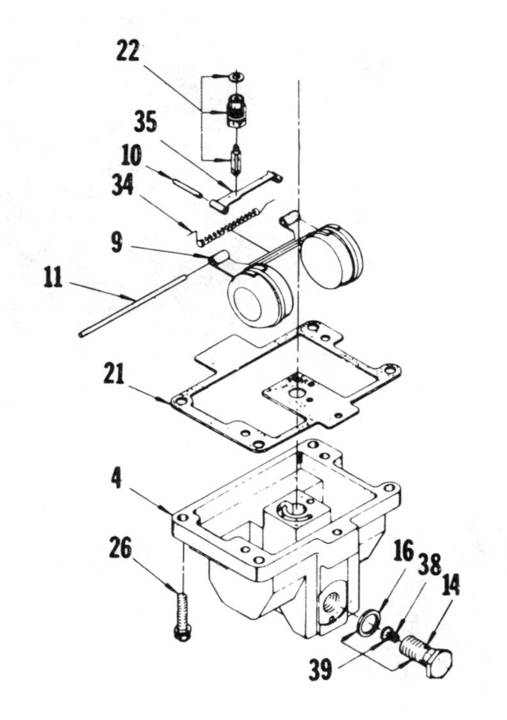

*Parts in repair kit (also includes gasket/diaphragm kit)

**Parts in gasket/diaphragm kit

(4)

1. Fuel pump body
2. Carburetor body
3. Retaining ring*
4. Fuel bowl
5. Fuel pump cover
6. Throttle shaft and swivel
7. Throttle valve
8. Swivel assembly
9. Float
10. Pin-float valve lever*
11. Float pin
12. Check valve
13. Enrichment valve
14. Jet limiting plug
15. Friction ball*
16. Jet plug gasket**
17. Intake flange gasket**
18. Fuel pump gasket**
19. Surge chamber gasket**
20. Gasket**
21. Fuel bowl gasket**
22. Float valve*
23. Surge chamber diaphragm**
24. Fuel pump diaphragm**
25. Throttle valve screw*
26. Fuel bowl screw*
27. Fuel pump screw*
28. Idle fuel screw springs
29. Idle air screw spring
30. Enrichener friction spring*
31. Idle tube pressure spring*
32. Fuel pump leaf spring
33. Throttle return spring
34. Float support spring
35. Float valve lever
36. Idle fuel screw
37. Idle air screw
38. High speed jet
39. Jet assembly
40. Idle fuel feed tube
41. Swivel
42. Fuel inlet screen
43. Throttle bushing
44. Throttle bracket plate

6. Carefully seat both needles. Use finger pressure only. Do not force the needle or damage to needle and/or seat will result.

7. Open the high-speed mixture needle ⅞ turn. Open the idle mixture 1 turn.

8. Turn the idle stop screw in until it contacts the tab on the throttle lever, then turn it in one additional turn.

> NOTE: *The preceding steps are preliminary carburetor adjustments. Complete carburetor adjustment must be performed as described in Chapter Two after the carburetors are installed.*

9. Install the thin and thick gaskets, circuit diaphragm, and circuit plate as shown in **Figure 12**. Secure them with 3 screws.

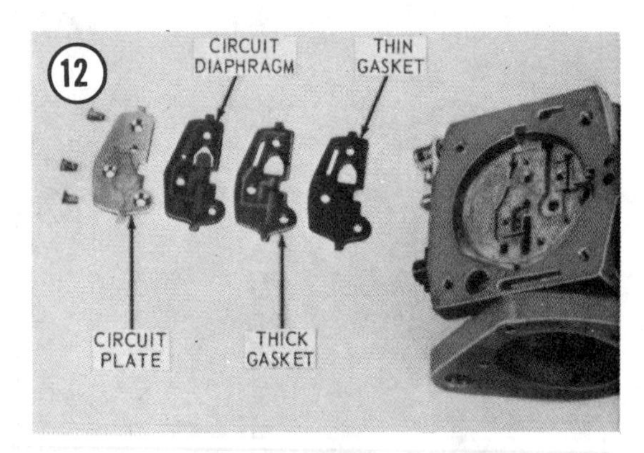

10. Place the spring in the cavity in the carburetor body, and the needle valve in the tab on the metering lever. Install the lever pin through the metering lever (**Figure 13**).

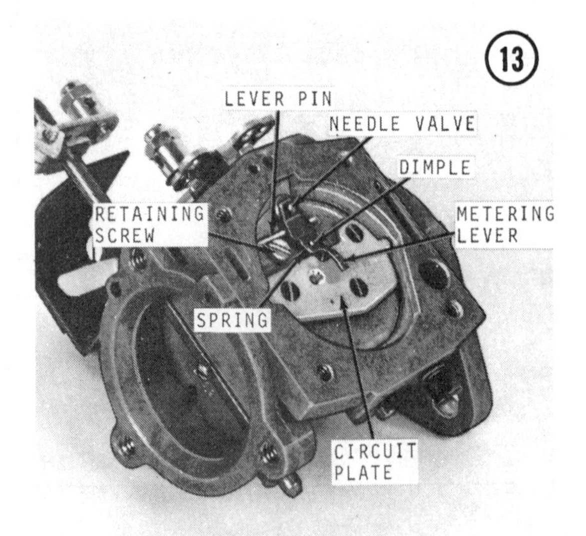

11. Install the assembly in the carburetor body and secure it with the screw. Ensure that the spring is positioned around the dimple in the metering lever.

12. Invert the carburetor and lay a straightedge across the carburetor body. Adjust the needle valve tab so the lever is 0.005-0.020 in. (0.127-0.508mm) above the carburetor body as shown in **Figure 14**.

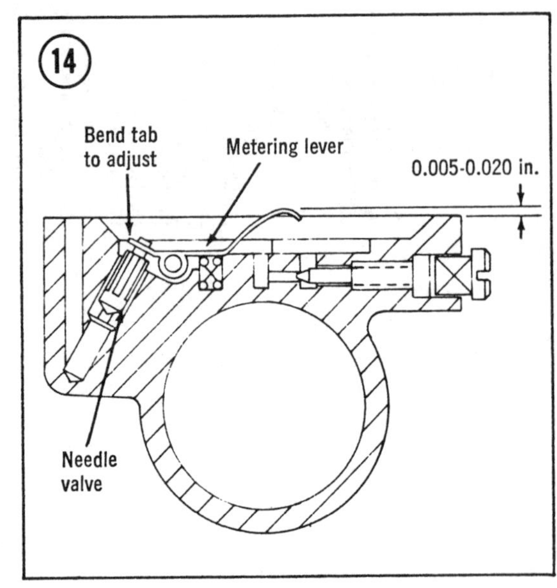

13. Install the valve springs, pressure spring, check valve, and fuel pump spring in the metering diaphragm plate as shown in **Figure 15**.

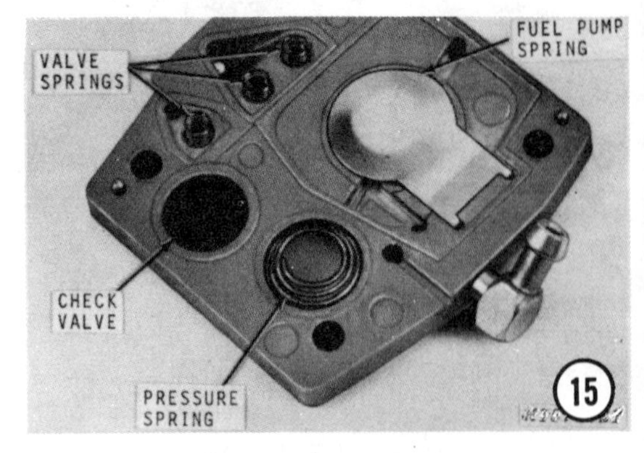

14. Install the small screen (32, **Figure 4**) in other side of metering diaphragm plate.

15. Install diaphragms, gaskets, filter screen, and plates in carburetor body as shown in **Figure 16**. Install and tighten the 4 cover screws.

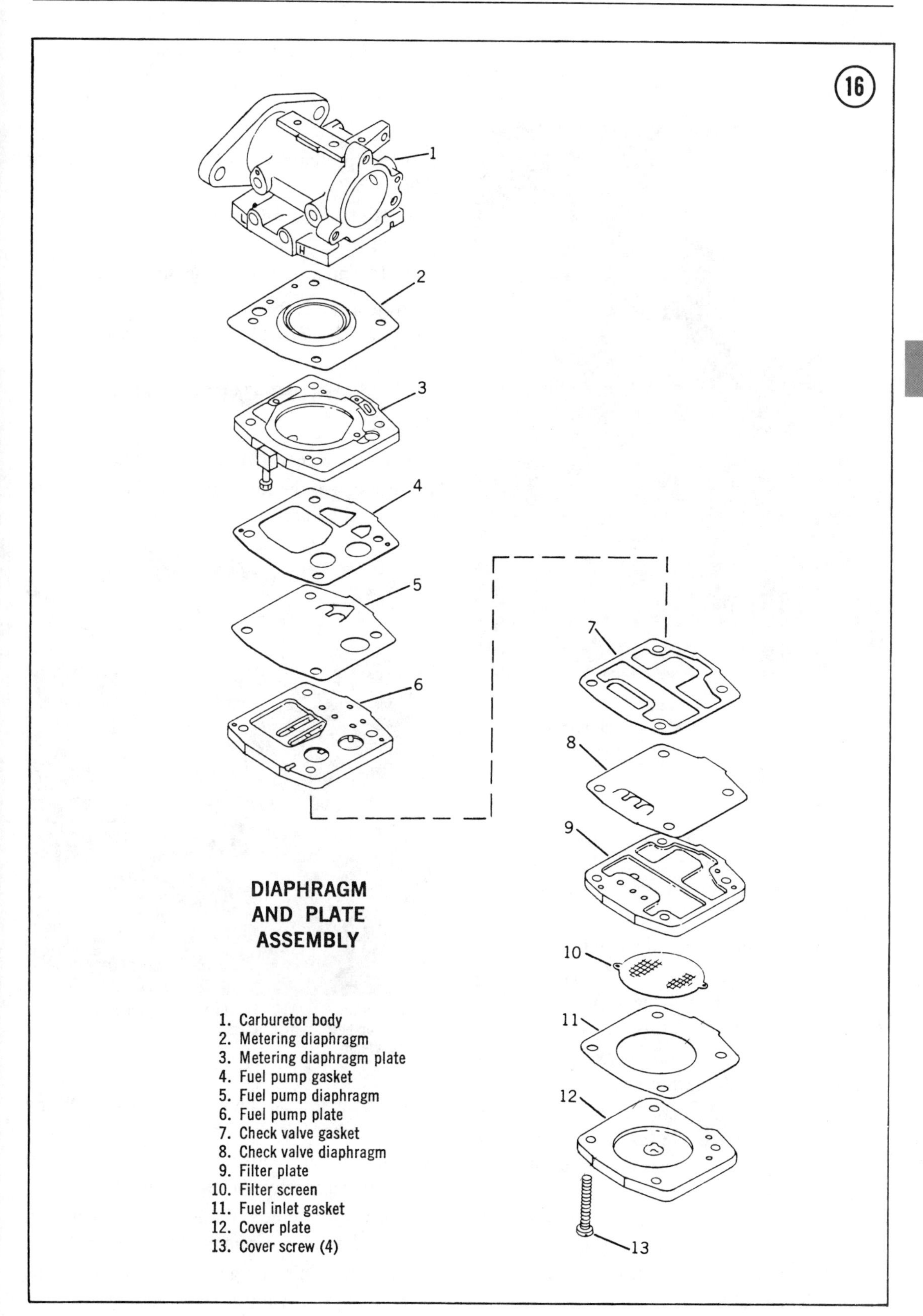
16
1
2
3
4
5
6
7
8
9
10
11
12
13
DIAPHRAGM
AND PLATE
ASSEMBLY
1. Carburetor body
2. Metering diaphragm
3. Metering diaphragm plate
4. Fuel pump gasket
5. Fuel pump diaphragm
6. Fuel pump plate
7. Check valve gasket
8. Check valve diaphragm
9. Filter plate
10. Filter screen
11. Fuel inlet gasket
12. Cover plate
13. Cover screw (4)

5

Installation

Refer to **Figure 17** for this procedure.

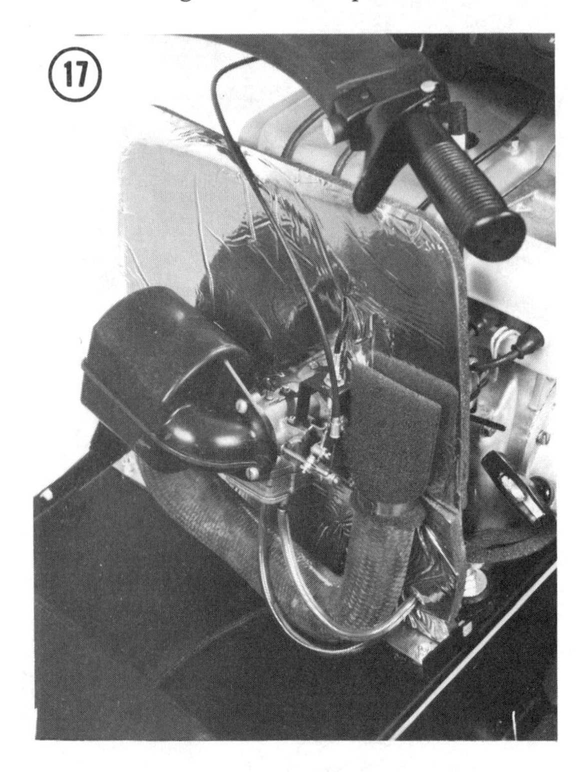

1. Choke cable
2. Silencer
3. Heat spacer and gaskets
4. Throttle bracket
5. Fuel line (green)
6. Fuel line (clear)

1. Install a new gasket, heat spacer, another new gasket, small square heat shield, a third new gasket and the carburetor, and tighten the nuts securely.

2. Install the silencer and throttle bracket.

3. Connect the fuel line to the center fitting under carburetor. Install the vapor return line (clear) on other lower fitting.

4. Connect the impulse tube to the angle fitting on the left side of the carburetor.

5. Connect the choke cable to the choke lever. Ensure that the choke is in the full-open position when the choke lever on the instrument panel is pushed down.

6. Connect the throttle cable to the throttle lever and adjust it so the throttle lever fully opens when the throttle control on the handlebar is fully actuated.

NOTE: *Make sure the dowel on the end of the throttle cable is correctly positioned in the recess of the handlebar throttle control.*

7. Make preliminary carburetor adjustments as follows:

   a. High-speed adjustment needle ⅞ turn open.

   b. Idle mixture needle one turn open.

8. Adjust the carburetor as described in Chapter Two—*Walbro Carburetor Adjustment.*

MIKUNI CARBURETOR

Removal

Refer to **Figure 18** for this procedure.

1. Disconnect the fuel line from the carburetor and plug the line.

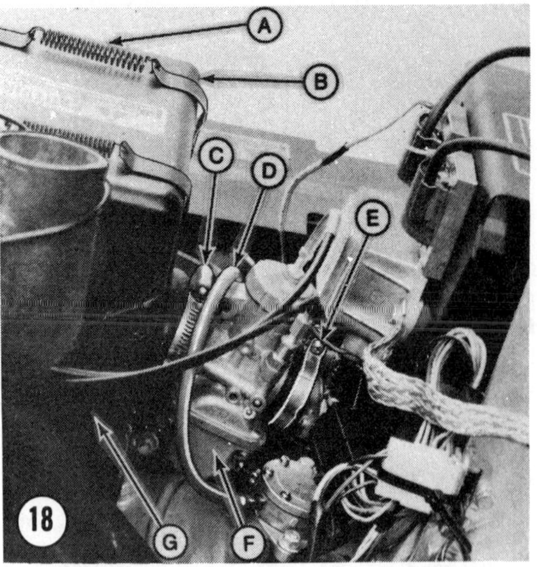

A. Spring
B. Top half
C. Clamp
D. Fuel line
E. Clamp
F. Carburetor
G. Lower half

2. Unscrew the carburetor top (**Figure 19**) and withdraw the slide assembly.

3. Unscrew the choke assembly (**Figure 20**) from the carburetor and withdraw it.

4. Loosen the screws in the clamping bands on the air silencer and remove the silencer.

5. Loosen the clamping bands on the manifolds (**Figure 21**) and remove the carburetors.

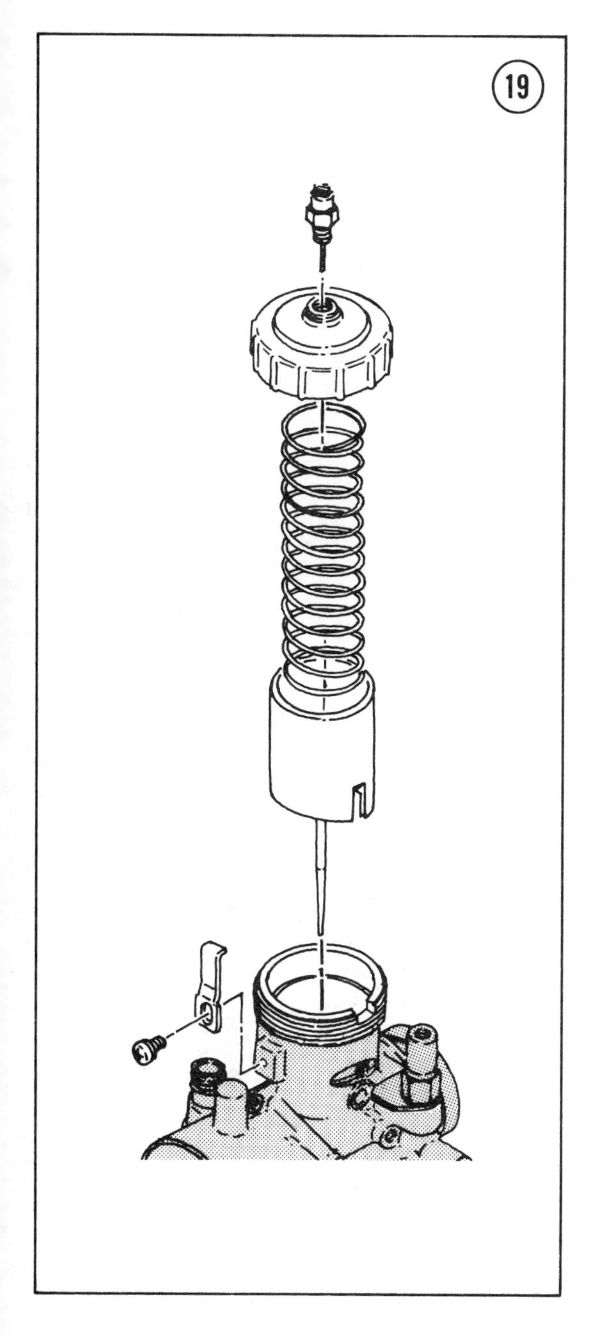

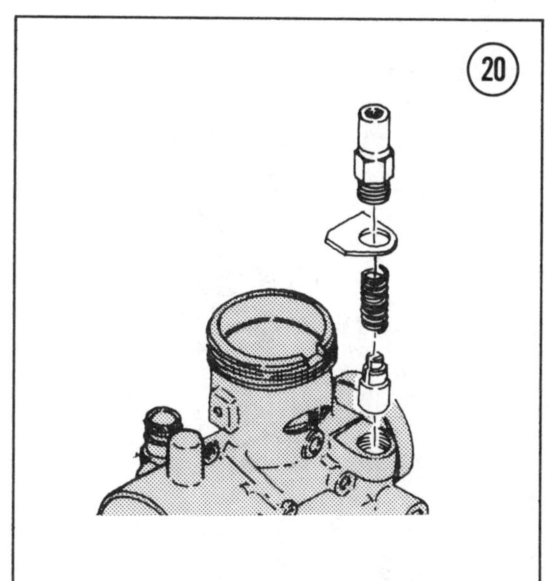

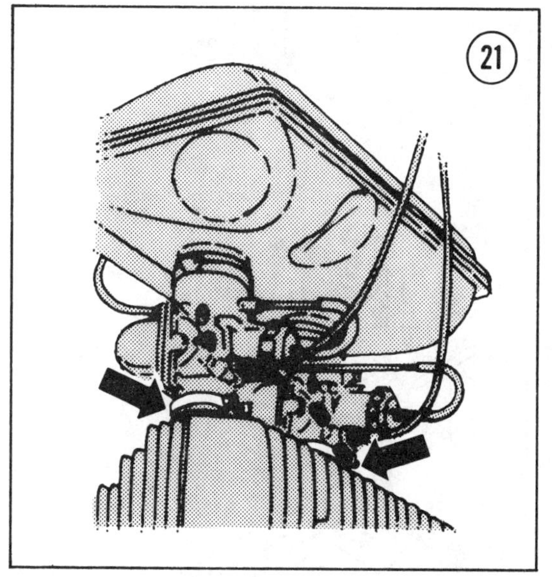

Disassembly

Refer to **Figure 22** for this procedure.

1. Remove the drain plug from bottom of the float chamber. Drain the fuel into a suitable container. Install the drain plug.

WARNING

Handle and dispose of drained fuel carefully or a serious or fatal fire may occur.

2. Remove the throttle stop screw and spring.
3. Remove the air screw and spring.
4. Remove the float chamber as shown in **Figure 23**. Gently lift out floats from the carburetor body.
5. Using a 6mm socket or box end wrench, gently remove the main jet and ring.
6. Remove the float arm pin and float arm. Lift off the baffle plate and gaskets (**Figure 23**).
7. Gently remove the inlet needle valve assembly with washer.

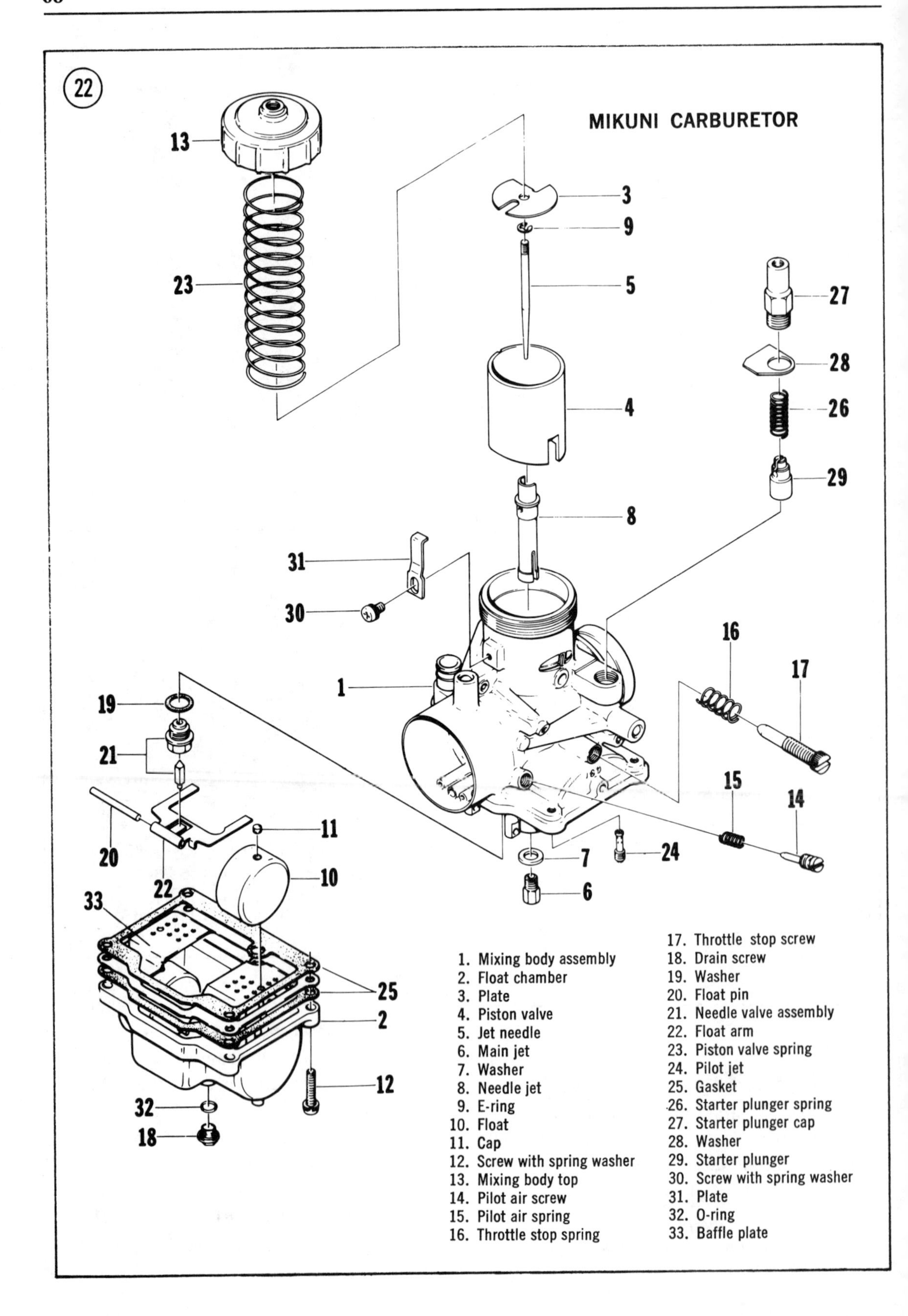
22
MIKUNI CARBURETOR
13
23
3
9
5
4
8
27
28
26
29
31
30
1
16
17
19
21
15
14
20
22
11
10
7
24
6
33
25
2
12
32
18
1. Mixing body assembly
2. Float chamber
3. Plate
4. Piston valve
5. Jet needle
6. Main jet
7. Washer
8. Needle jet
9. E-ring
10. Float
11. Cap
12. Screw with spring washer
13. Mixing body top
14. Pilot air screw
15. Pilot air spring
16. Throttle stop spring
17. Throttle stop screw
18. Drain screw
19. Washer
20. Float pin
21. Needle valve assembly
22. Float arm
23. Piston valve spring
24. Pilot jet
25. Gasket
26. Starter plunger spring
27. Starter plunger cap
28. Washer
29. Starter plunger
30. Screw with spring washer
31. Plate
32. O-ring
33. Baffle plate

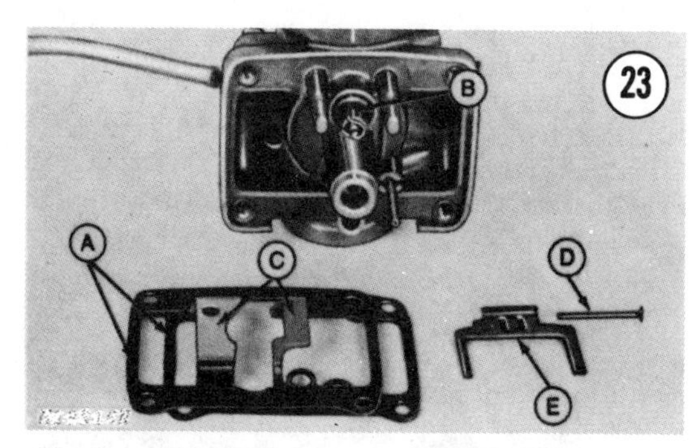

A. Gaskets
B. Inlet needle valve assembly
C. Baffle plate
D. Float air pin
E. Float arm

8. Gently push the needle jet from mixing chamber using an awl or similar sharp pointed device. See **Figure 24**.

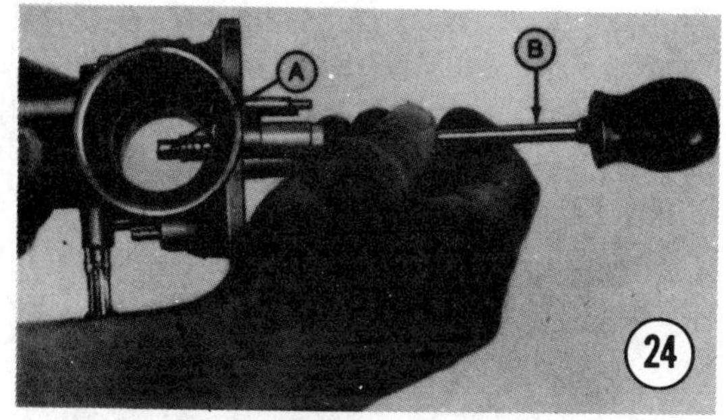

A. Needle jet B. Awl

9. Compress the throttle slide spring and remove the cable lock plate from the top of the slide. Disconnect the end of the cable from the throttle slide. Remove the needle valve from the slide. Note the position of the E-ring on the needle valve; the E-ring locates the needle with respect to the needle valve and a change in position can affect fuel mixture and performance. If the carburetor has been performing correctly, you will want to reinstall the E-ring on the notch from which it was removed.

Cleaning and Inspection

WARNING

Most carburetor cleaners are highly caustic. They must be handled with extreme care or skin burns and possible eye injury may result.

1. Clean all metallic parts in carburetor cleaning solvent. Do not place gaskets in solvent or they will be destroyed.

CAUTION

Never clean holes or passages with small drill bits or wire or a slight enlargement or burring of hole will result and drastically affect carburetor performance.

2. Inspect the float chamber and carburetor body for fine cracks or evidence of fuel leaks.
3. Check the throttle spring for distortion or damage.
4. Inspect the air screw and throttle stop screw for surface damage or stripped threads.
5. Inspect the pilot jet and main jet for damage or stripped threads.

CAUTION

The pilot jet and main jet must be scrupulously clean and shiny. Any burring, roughness, or abrasion will cause a lean fuel and air mixture and possible engine damage.

6. Remove the retainer and inlet valve from the valve seat. Carefully examine the seating surface on the inlet valve and seat for damage. Make sure that the retainer does not bind and hinder movement of the inlet valve.
7. Inspect the jet needle and needle jet for damage. The jet needle must slide freely within the needle jet.
8. Install the float guides in the float chamber. Move the floats up and down several times to ensure that they are not binding on the float guides.
9. Inspect the float arm and float pin to ensure the float arm does not bind on the pin.
10. Inspect the choke plunger. The plunger must move freely in the passage of carburetor body.
11. Install the throttle valve in the carburetor body and move it up and down several times to check for sticking motion or looseness. Ensure that the guide pin in the carburetor body is not broken.

Assembly

Refer to **Figure 22** for this procedure.

1. Using a small screwdriver, install the pilot jet in the carburetor body as shown in **Figure 25**.

2. Install the gaskets and baffle plate on the carburetor body (**Figure 26**). Install the second gasket on top of baffle plate.
3. Place the washer on the inlet nedle valve seat and install the seat in the carburetor body (**Figure 26**). Install the inlet valve (point down) and retainer.

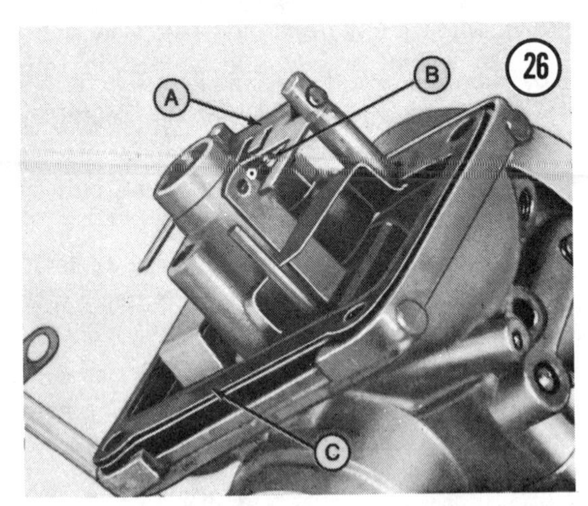

A. Float arm B. Inlet valve C. Baffle plate and gaskets

4. Install the float arm and secure it with the float arm pin.
5. Invert the carburetor body and check the float level. The float bowl sealing surface of the carburetor body (**Figure 27**) must be parallel with the float arm. If necessary, adjust it by bending the float arm actuating tab.
6. Install the needle jet. Make sure the notch on

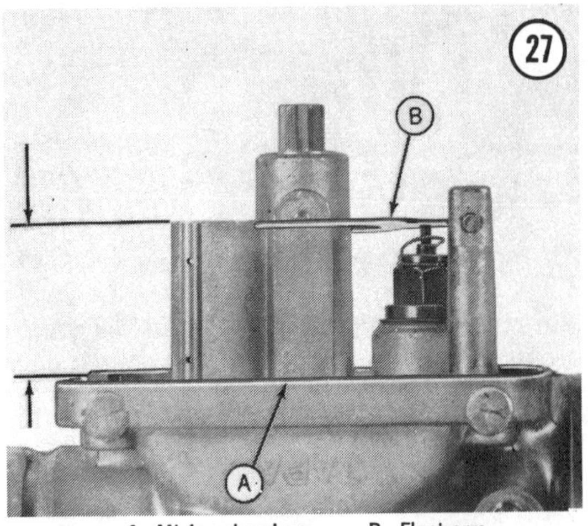

A. Mixing chamber B. Float arm

the needle jet is correctly aligned with the pin in bore of the carburetor body (**Figure 28**). Install the ring over the needle jet bore (recess in ring next to bore) and screw the main jet into the needle jet.

A. Pin B. Notch C. Needle jet

7. Slide the floats over the float pins. The pins on the float must be down and point to the inside of float chamber as shown in **Figure 29**.
8. Install the float chamber on the carburetor body and secure it with 4 screws.
9. Slide the air screw spring over air screw and screw in air screw gently.

CAUTION

Do not force the air screw or seat damage may occur.

10. Install the throttle stop screw and spring. Install the screw until it is just flush with the inside of the bore.

Installation

1. Position the carburetor in rubber mount and tighten the clamp screw.

2. Connect the fuel line from the pump to the carburetor.

3. Install the E-ring in the groove in the needle valve from which it was removed. Install the needle valve in the throttle slide (**Figure 30**).

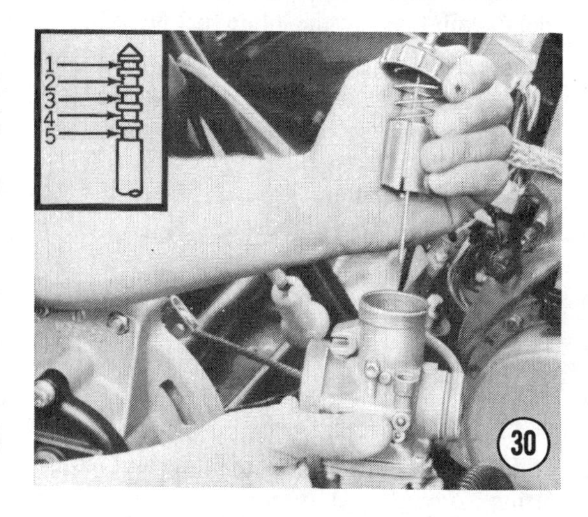

4. Route the throttle cable end button through cap, spring, lock plate, and slot in throttle slide as shown in **Figure 30**.

5. Slide the cable into the narrow part of the slot in the throttle valve. Install the lock plate between the spring and throttle slide with the tab on the plate in the slot of the throttle slide.

6. Install the throttle assembly in the carburetor body, making sure the needle engages the needle valve and the slot in the throttle engages the knob in the throttle bore (**Figure 31**). Compress the throttle valve spring and tighten the cap on the carburetor body.

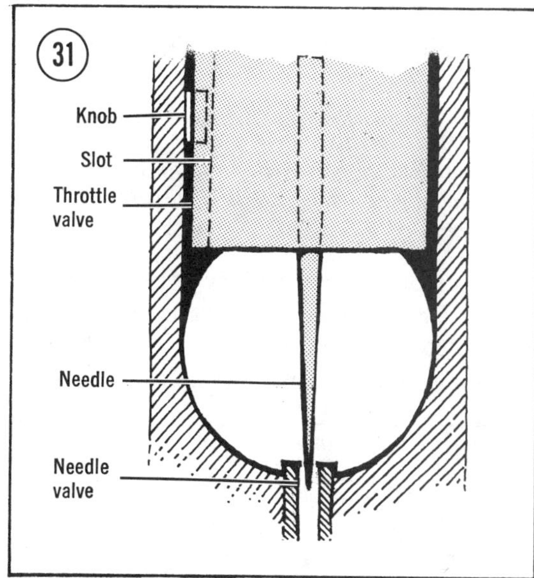

7. Move the choke lever on the instrument panel to the OFF position. Route the choke cable end button through the cap and spring. Hook the end button in the choke plunger as shown in **Figure 32** and place the washer on the carburetor body. Install the assembly and tighten the cap.

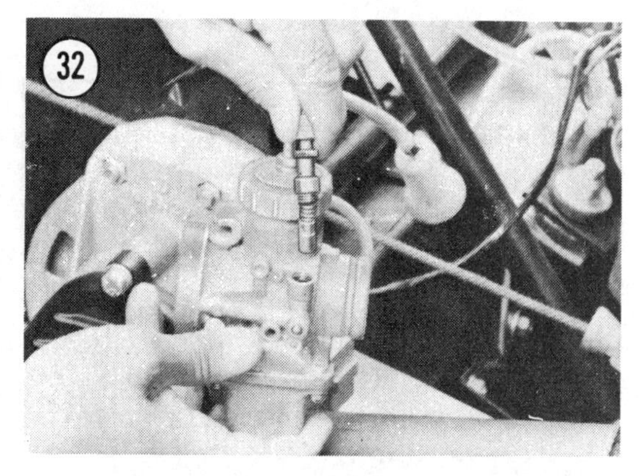

8. Perform carburetor adjustments as described in Chapter Two, *Mikuni Carburetor Adjustment and Synchronization*.

9. Install the air intake silencer.

10. Install windshield and console if removed.

5

INTAKE SILENCERS

Intake silencers are installed on snowmobiles to quiet the sound of rushing air and to catch fuel that spits back out of the carburetor throat.

The silencer is not intended to filter incoming air. Operate snowmobiles only in clean, snow-covered areas.

CAUTION

Never operate the snowmobile with the silencer removed. Loss of power and engine damage will result from a lean fuel mixture.

Service of air intake silencers is limited to removal and cleaning of components. Refer to **Figures 33 through 35** for air intake silencers used on various models.

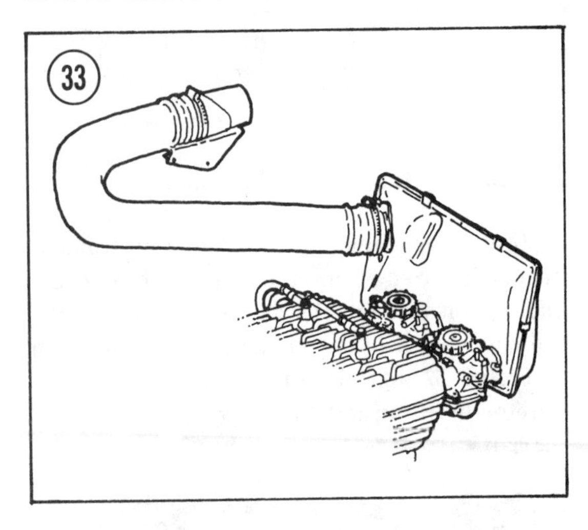

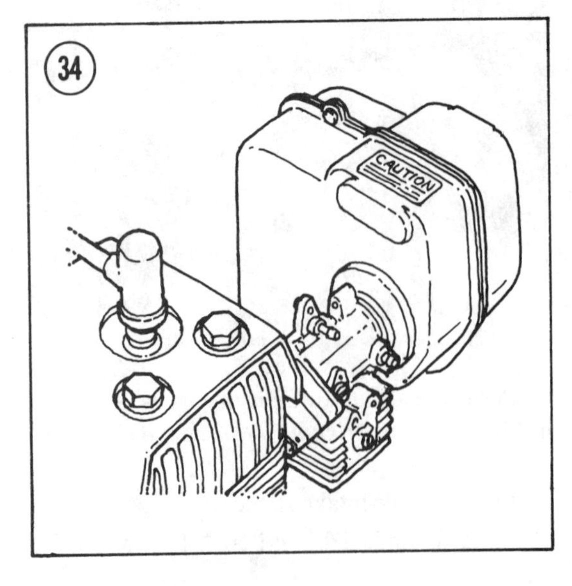

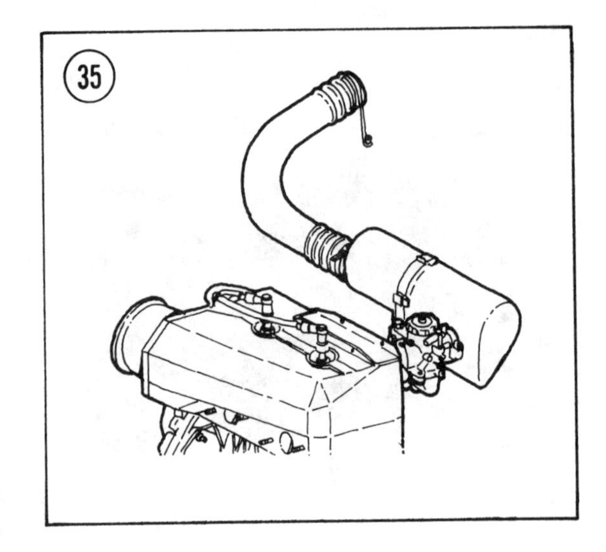

FUEL TANK

The fuel tank (**Figure 36**) incorporates a fuel gauge in the filler cap and a spill ledge to prevent spilled fuel from spilling on to the seat. The fuel tank cap is sealed and the tank is vented by a line to the top of the tank.

A fuel shut-off valve is located between the fuel pickup line and the inline fuel filter.

NOTE: *If snowmobile has been transported on a trailer without the fuel valve shut off, engine may be flooded.*

Removal/Installation

1. Disconnect fuel lines and vent lines.
2. Remove seat and tank hold-on clip.
3. Remove seat by sliding tank rearward.
4. Installation is the reverse of these steps.

Pickup Screen Cleaning

1. Disconnect fuel lines from fitting and remove fitting from fuel tank (**Figure 37**).
2. Remove pickup screen from end of line.
3. Rinse screen carefully in solvent and blow dry with compressed air. Replace screen if damaged. Replace gasket on fuel line fitting if necessary.

INLINE FUEL FILTER

Service of inline fuel filter is limited to annual replacement or replacement when contamina-

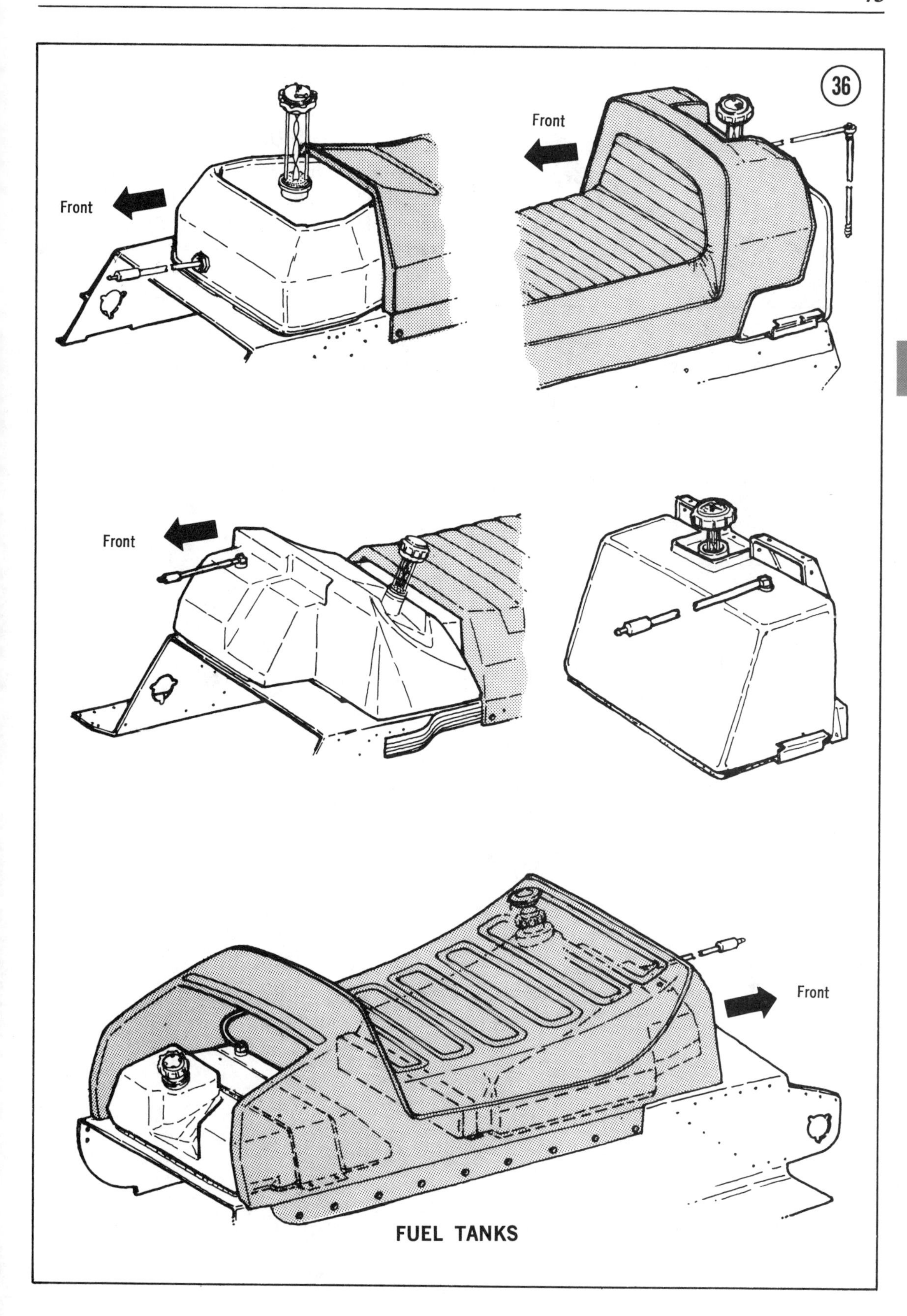

FUEL TANKS

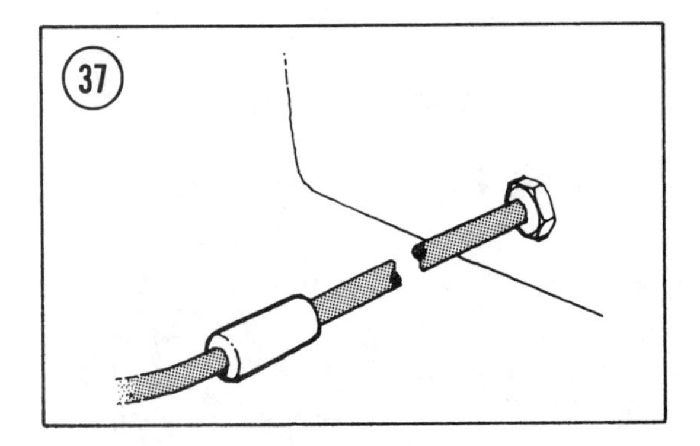

tion builds up at the base of the cone in the filter unit.

FUEL PUMP

To check fuel pump operation, disconnect fuel line from pump to carburetor at the carburetor. Make sure ignition switch is off and pull recoil starter handle and check for fuel flow at fuel line. If fuel flow from pump is unsatisfactory, replace pump.

CHAPTER SIX

ELECTRICAL SYSTEM

The electrical system on Arctic Cat snowmobiles consists of an ignition system, lighting system, and an electric starting system on some models.

Two types of ignition systems are used; a magneto (Lynx I and 2000S) and (capacitor discharge ignition) CDI (others).

The lighting system consists of a headlight, brake/taillight, and console lights.

The electric starting system on some models consists of a battery, a starter and solenoid, and charging equipment.

This chapter includes testing and repair of some components of the ignition, lighting, and charging systems. Some testing and repair tasks referenced in this chapter require special testing equipment and tools. These tasks are best accomplished by an authorized dealer or competent auto electric shop.

Refer to Chapter Two for ignition timing and breaker point adjustment.

CAPACITOR DISCHARGE IGNITION (CDI)

The capacitor discharge ignition system (Figure 1) consists of a permanent magnet flywheel, alternator, and solid-state capacitor. The system supplies high voltage for ignition and generates current required for the lighting system.

The flywheel incorporates a magnet and is mounted on the engine crankshaft. The flywheel and magnet revolve around the stator assembly, which is fixed to the engine. Current is generated in the 12 pole windings of the stator.

Nine poles supply power for the lighting system and 3 poles supply power for the ignition.

The ignition timing ring, alternator stator, and electronic pack require special test equipment to troubleshoot malfunctions and to monitor performance. If trouble exists in any of these units, refer the testing and repair to an authorized dealer.

Testing Engine Emergency Shutoff Switch

Refer to the appropriate wiring diagram for this procedure.

1. Disconnect the black wires leading to the terminal block.

2. Connect test leads of a continuity test light to the black wires.

3. Turn the ignition switch to ON position.

4. With the shutoff switch in the OFF position,

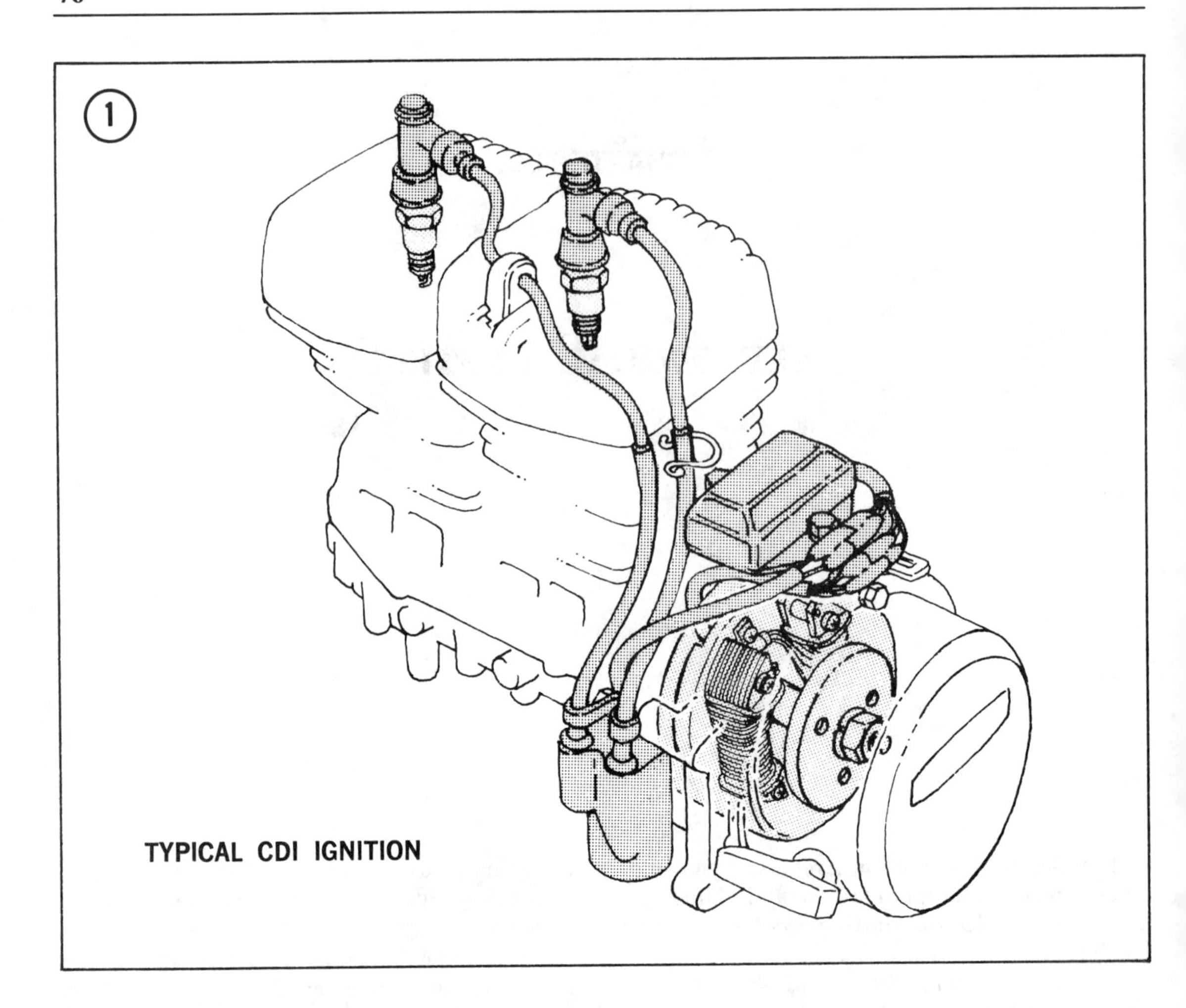

the test light should light. The test light should not light with the shutoff switch in the ON position. Replace the switch if it is defective.

Flywheel, Alternator Stator, and Trigger Removal/Installation

It is not necessary to remove engine from snowmobile to perform the following procedure.

1. Remove the recoil starter, flywheel housing, and lower fan sheave.
2. Install special flywheel holding tool (part No. 0144-007), or locally fabricated equivalent, to flywheel (**Figure 2**).
3. While holding flywheel, remove retaining nut.
4. Install puller (part No. 0144-112 for panther and Cheetah, 0144-064 for others) to flywheel as shown in **Figure 3**. Tighten puller and remove flywheel. If the flywheel does not release, rap sharply on the puller bolt with a hammer.

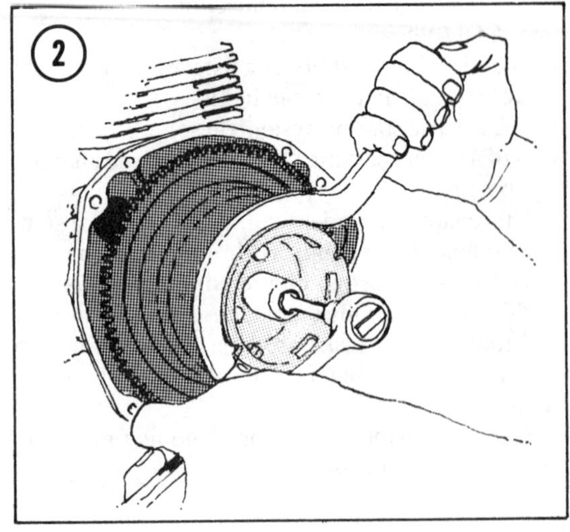

CAUTION

Do not strike flywheel with a steel hammer or serious damage to flywheel may occur.

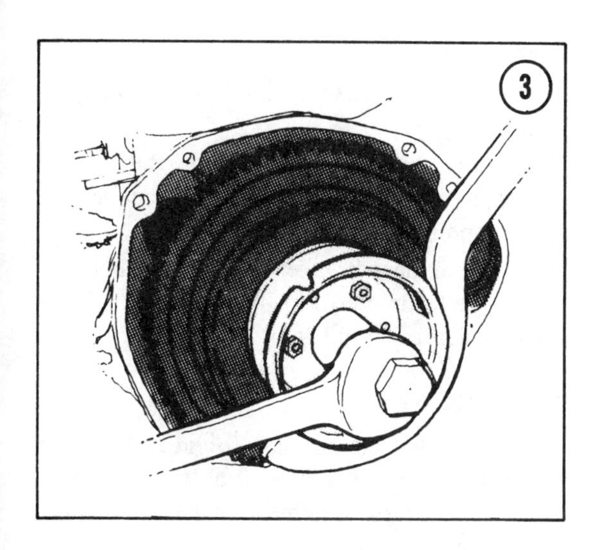

5. Remove the stator and trigger assembly (**Figure 4**).

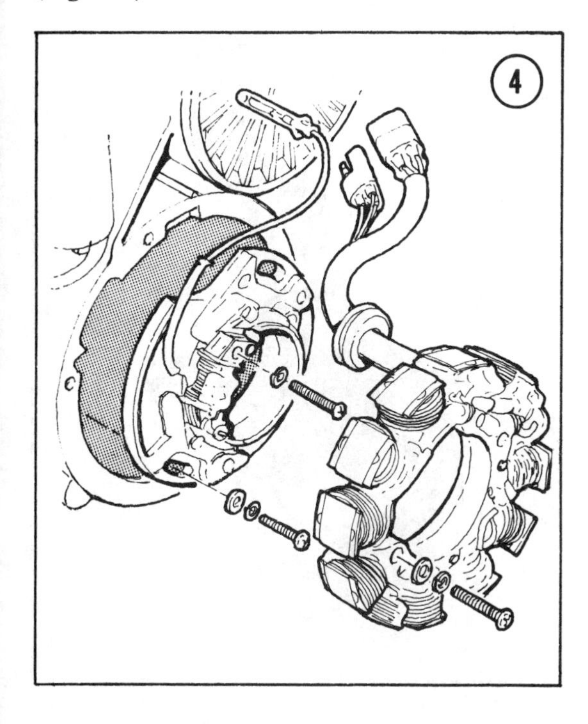

NOTE: *Screws securing stator and trigger assembly are secured with Loctite. Tap screwdriver with hammer to help break Loctite loose.*

6. Installation is the reverse of these steps. Keep the following points in mind:

   a. Use Loctite to secure stator and trigger screws.

   b. Position 4 screws on trigger assembly in the center of their slots for preliminary ignition timing.

   c. Torque flywheel retaining nut to 60 ft.-lb. (8.3 mkg).

   d. Refer to Chapter Two and perform ignition timing.

MAGNETO IGNITION

The magneto equipped Lynx features an energy transfer ignition system (**Figure 5**).

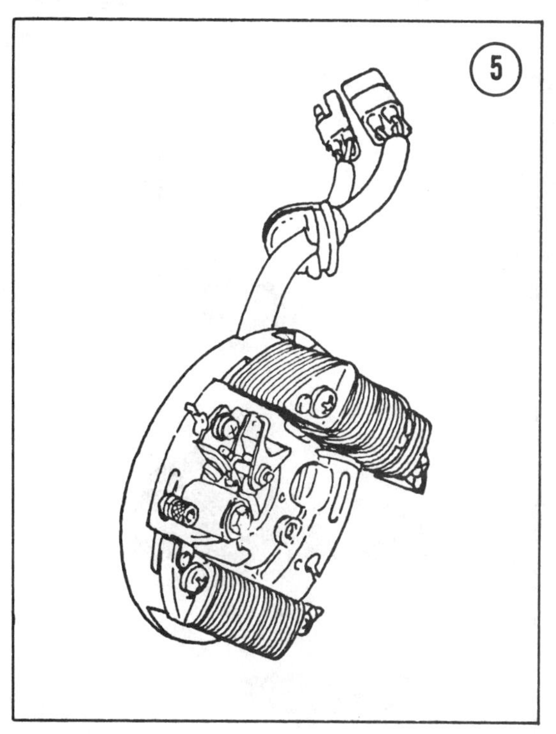

The ignition system consists of an ignition generating coil, a set of breaker points, a condenser, and an ignition coil.

Refer to Chapter Two for breaker point and timing adjustments.

Emergency Stop Switch Test

1. Disconnect the coupler to emergency stop switch.

2. Connect a light-type continuity tester between the terminals in coupler.

3. The test light must light when the switch is pressed and go out when the stop switch is in the operating position.

Ignition Switch Test

Check the ignition switch with a light-type continuity tester. Refer to the schematic and check for continuity between switch terminals.

Stator Assembly Removal

Remove recoil starter, fan cover, and flywheel to gain access to stator assembly. Remove the backing plate if the stator assembly is to be removed.

Refer to **Figure 6** for the following procedures.

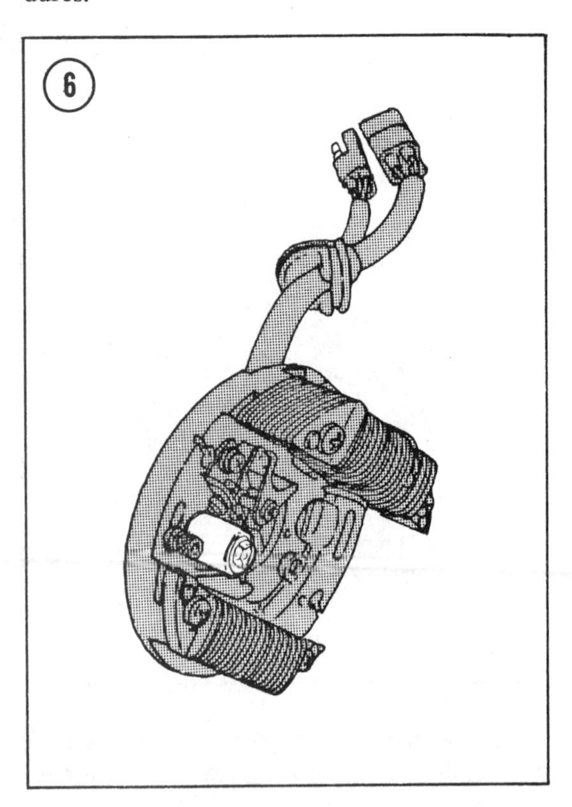

Condenser

1. Loosen the soldered leads on condenser terminal with a soldering iron.

2. Remove screw that secures the condenser to the stator plate and remove condenser.

3. Install a new condenser and solder the leads to terminal.

CAUTION

Exercise care when resoldering wire to the condenser; too much heat can destroy the condenser.

Breaker Points

1. Loosen the breaker point terminal and disconect the leads.

2. Remove the screw that attaches the breaker points to the stator plate and remove the breaker point assembly.

3. Install a new breaker point assembly and attach the leads. Perform breaker point adjustment as described in Chapter Two.

Felt Oil Pads

Replace the felt oil pads (**Figure 7**) if their lubricating capacity is questionable. Oil the pads with 1 or 2 drops of light oil whenever breaker points are replaced.

LIGHTING SYSTEM

The lighting system consists of a headlight and brake/taillight unit, instrument lights, and an AC (alternating current) generating device. Switches control all lighting circuits.

On Panther/Cheetah models equipped with an electric starter, AC is converted to DC (direct current) by a rectifier and then used to keep the battery charged.

TESTING

If the lights fail to work, do not immediately assume the worst—a major failure in the alternator, regulator, or lighting coils. Very often the problem can be found in the lamps; look for burned out filaments or bulbs that are loose in their sockets. In addition, check the harness connectors to ensure that they are clean and tight and have not come disconnected.

Four things are required for the lights to function. You must have current and an uninterrupted, non-shorted path for it to follow. The switch must work correctly. A good ground is required. The light bulbs must be in good condition and correctly installed in clean, dry sockets.

A lighting troubleshooting chart is provided at the end of this chapter to speed the diagnosis of lighting system trouble. The tests that follow can confirm both good and poor conditions.

To effectively trace electrical troubles, you will need an ohmmeter and voltmeter or a multimeter like that described in Chapter Two.

Lighting Coil Resistance Test

1. Unplug the main wiring harness connector at the engine (**Figure 8**).

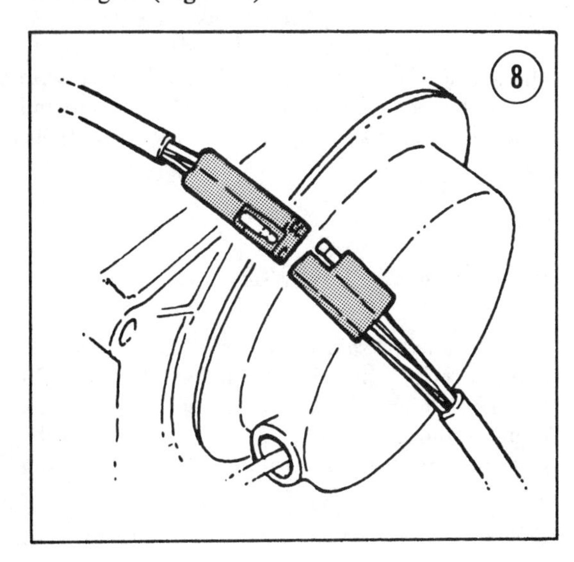

2. Connect an ohmmeter to the 2 yellow wires on the engine side of the connector (**Figure 9**) and measure the resistance. It should be 0.145 ohm for all engines except the Lynx 2000S

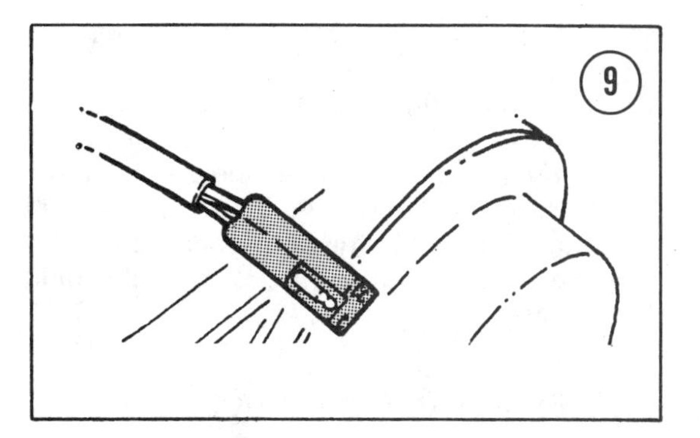

which is 0.45 ohm. If the resistance is correct, proceed to check the voltage regulator as described below.

6

Voltage Regulator Test

1. Raise the rear of the snowmobile so the track is clear of the floor and support it securely, making sure the track is clear and will rotate without interference.

2. Reconnect main wiring harness connector.

3. Remove the voltage regulator but do not disconnect it. Make sure that it is not in contact with any part of the machine.

4. Select the 100 VAC scale on a voltmeter or multimeter. Connect one of the meter leads to the yellow wire on the voltage regulator and the other lead to ground.

CAUTION

The voltage tester must have a test capacity in excess of 30 VAC; if it does not, high voltage during testing may damage it.

5. Start the engine and run it at a fast idle. The voltmeter should indicate 12-15 VAC.

6. Increase the engine speed and watch the meter. It should indicate 15-25 VAC as the engine speed increases.

7. If the voltmeter indicates the ranges shown in Step 5 (12-15 VAC at idle), and Step 6 (15-25 VAC at speed), the alternator is operating satisfactorily and the voltage regulator can now be tested as described in Step 8. If the readings are not as shown, trouble may exist in the alternator or in the wiring. Inspect and clean all connections and make sure they are tight and dry. If this fails to produce good readings, the alter-

nator should be tested by a dealer or competent automotive electrical specialist.
8. Attach the voltage regulator to the chassis. Connect the voltmeter as described in Step 4 and start the engine. Read the meter first at idle and then at increased engine speed. It should indicate 12-15 VAC for both conditions. If it does not, the regulator is defective and should be replaced.

Electric Start System (Panther/Cheetah)

The electric starting system consists of a 12-volt battery, starter motor, starter solenoid, rectifier, and fuse.

The starter motor engages a ring gear on the flywheel to turn the engine over. The battery is kept charged by the alternator. A fuse is used to protect the system from overloads or short circuits.

If difficulty is experienced with the optional electric start system installed on some Panther/Cheetah models, check the obvious-and most likely cause of trouble; the battery. Make certain the battery and starter connections are clean and tight. Check and service the battery as described below. If the battery is satisfactory, proceed to test the starter and solenoid as described.

Battery Removal/Installation

1. Disconnect the negative (-) battery cable. Then disconnect the positive (+) cable.
2. Loosen hold-down bolts and unhook bolts form battery box. Remove hold-down clamp.
3. Disconnect vent tube from battery. Carefully lift battery out of battery box.
4. Installation is the reverse of these steps. Keep the following points in mind:

CAUTION
Be sure battery connections are correct or serious damage to electrical components will occur.

a. Be sure exterior of battery and terminals are clean and free from corrosion.
b. Connect positive (+) cable to battery first.

Battery Cleaning and Service

Electrolyte level in the battery should be checked periodically, especially during periods of regular operation. Use only distilled water and top off the battery to bottom of ring (filler neck) so the tops of the plates are covered. *Do not* overfill.

Battery corrosion is a normal reaction ; however, it should be cleaned off periodically to keep battery deterioration to a minimum.

Remove battery and wire brush terminals and cable ends. Wash terminals and exterior of battery with about a 4:1 solution of warm water and baking soda.

CAUTION
Do not allow any baking soda solution to enter battery cells or serious battery damage may result.

Wash battery box and hold-down bolts with baking soda solution. Rinse all parts in clear water and wipe dry.

In freezing weather never add water to a battery unless the machine will be operated for a period of time to mix electrolyte and water.

CAUTION
Keep battery fully charged . A discharged battery will freeze, causing battery case to break.

Remove the battery from the machine during extended non-use periods and keep battery fully charged. Perform periodic specific gravity tests with a hydrometer to determine the level of charge and how long charge stays up before it starts to deteriorate.

Battery Specific Gravity Test

Determine the state of charge of the battery with a hydrometer. To use this instrument, place the suction tube (Figure 10) into the filler opening and draw in just enough electrolyte to lift the float. Hold the instrument in a vertical position and take the reading at eye level.

Specific gravity of electrolyte varies with temperature, so it is necessary to apply a temperature correction to the reading you obtain. For each 10° that the battery temperature exceeds 80° F, add 0.004 to the indicated specific

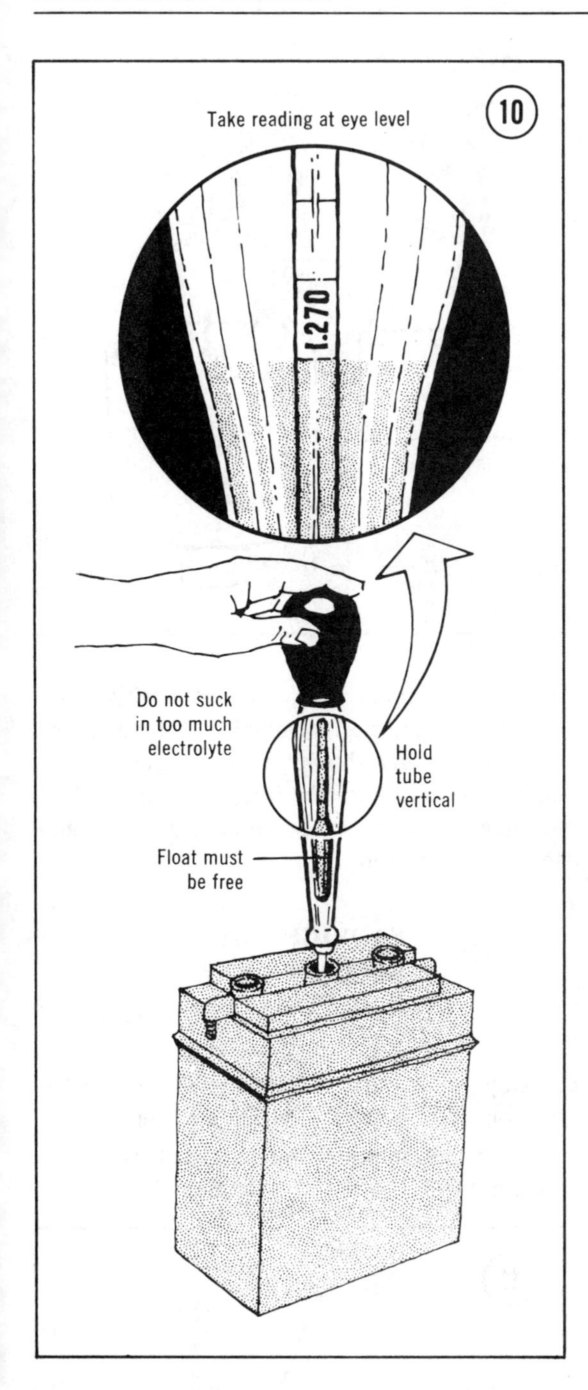

gravity. Subtract 0.004 from the indicated value for each 10° that the battery temperature is below 80°F.

WARNING

Do not smoke or permit any open flame in any area where batteries are being charged. Highly explosive hydrogen gas is formed during the charging process.

The specific gravity of a fully charged battery is 1.260. If the specific gravity is below 1.220, recharge the battery (**Figure 11**).

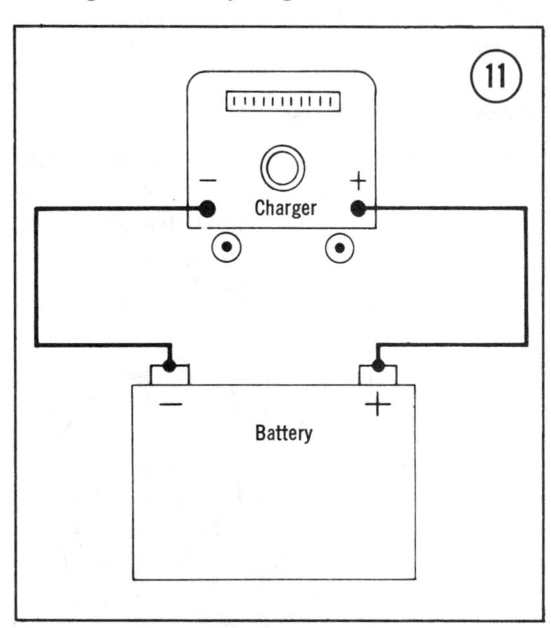

Starter Test

If starter fails to crank engine or cranks engine very slowly, perform the following:

1. Inspect cranking circuit wiring for loose or bady corroded connections or damaged wiring.
2. Perform *Battery Specific Gravity Test* to be certain battery is charged and not defective.
3. Crank engine with recoil starter to make sure engine turns freely and is not seized.

NOTE: *Remove spark plug wires. The following bypasses the ignition switch.*

4. If starter still will not crank engine, place a heavy jumper lead from positive (+) battery terminal directly to starter terminal (**Figure 12**).

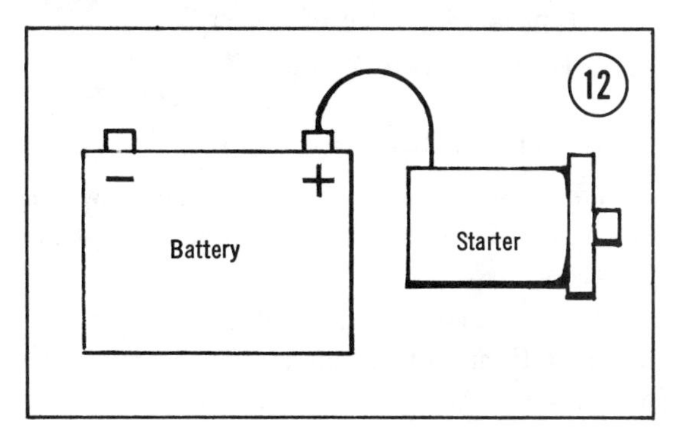

This bypasses the ignition switch, circuit breaker and starter solenoid. If starter now cranks the engine, then one of these items is defective. If starter will still not crank engine, starter is defective.

Starter Removal/Installation

Refer to **Figure 13** for this procedure. Starter repair consists of armature and/or brush replacement. It is recommended that all starter service and repair be referred to an authorized dealer or competent auto electric shop.

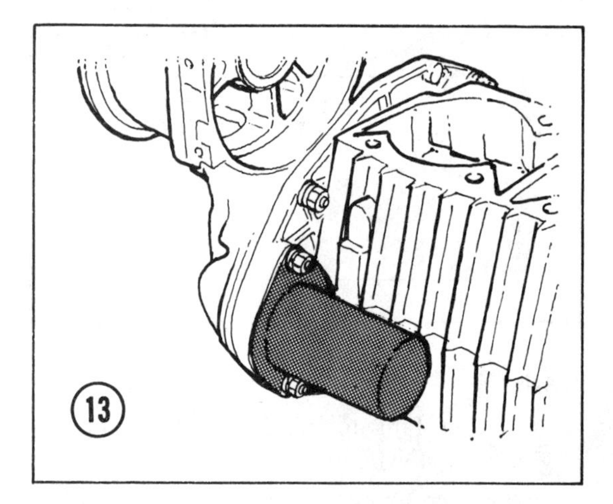

1. Disconnect ground cable from battery.
2. Disconnect solenoid-to-starter cable from starter terminal.
3. Remove mounting bolts securing starter to engine and mounting bracket to engine. Remove starter and mounting bracket.
4. Remove mounting bracket from starter.
5. Installation is the reverse of these steps. Keep the following points in mind:
   a. Tighten bolts securing bracket last to prevent misalignment of starter.
   b. Torque bolts to 6 ft.-lb. (0.8 mkg).

Starter Solenoid Test

1. Starter solenoid is a sealed magnet switch and cannot be repaired. If defective, it must be replaced.
2. Remove and insulate cable from starter terminal. Connect test light across 2 large terminals, **Figure 14**, of starter solenoid.

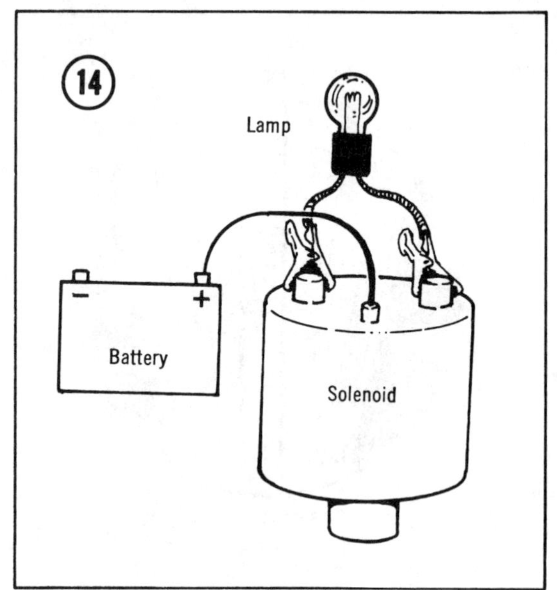

3. With a jumper lead, connect positive (+) battery post to small terminal on solenoid. The solenoid plunger should snap in, light the test lamp, and hold until the jumper is removed. If this does not occur, the solenoid is defective and should be replaced.

HEADLIGHT BULB REPLACEMENT

1. Raise the hood to gain access to the rear of the headlight.
2. Unplug the harness connector by pulling on the connector—not the wiring harness.
3. Turn the retaining ring counterclockwise to release it from the housing (**Figure 15**).

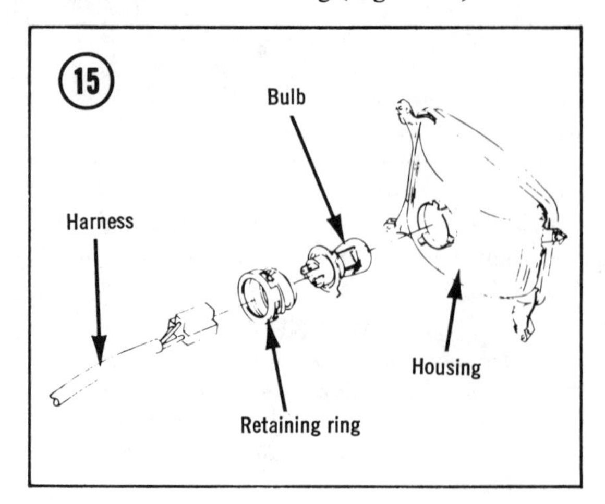

4. Remove the old bulb and install a new one of the same rating. The rating appears on the bulb. Set the retaining ring in place and turn it clockwise to lock it. Reconnect wiring harness.

HEADLIGHT AIMING

1. Refer to Chapter Two, *Suspension Adjustment*, and check (and adjust if necessary) the suspension.
2. Set the snowmobile on a level surface, 25 feet (7.6m) from a vertical surface such as a wall (**Figure 16**).
3. Measure the distance from the floor to the center of the headlight lens. Make a mark on the vertical surface the same distance from the floor. For instance, if the center of the headlight lens is 24 in. above the floor, mark "A" in **Figure 16** should also be 24 in. above the floor.
4. Start the engine, turn on the headlight and set the beam selector at HIGH; do not adjust the light with the selector set at LOW.
5. The most intense area of the beam on the wall must be 2 in. below the "A" mark (see **Figure 17**).
6. If the beam aim is not correct, move the headlight up or down as required by turning the spring loaded adjuster screws in or out (**Figure 18**).

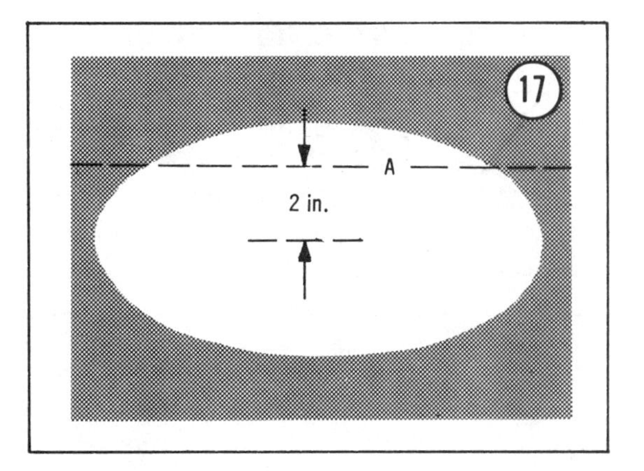

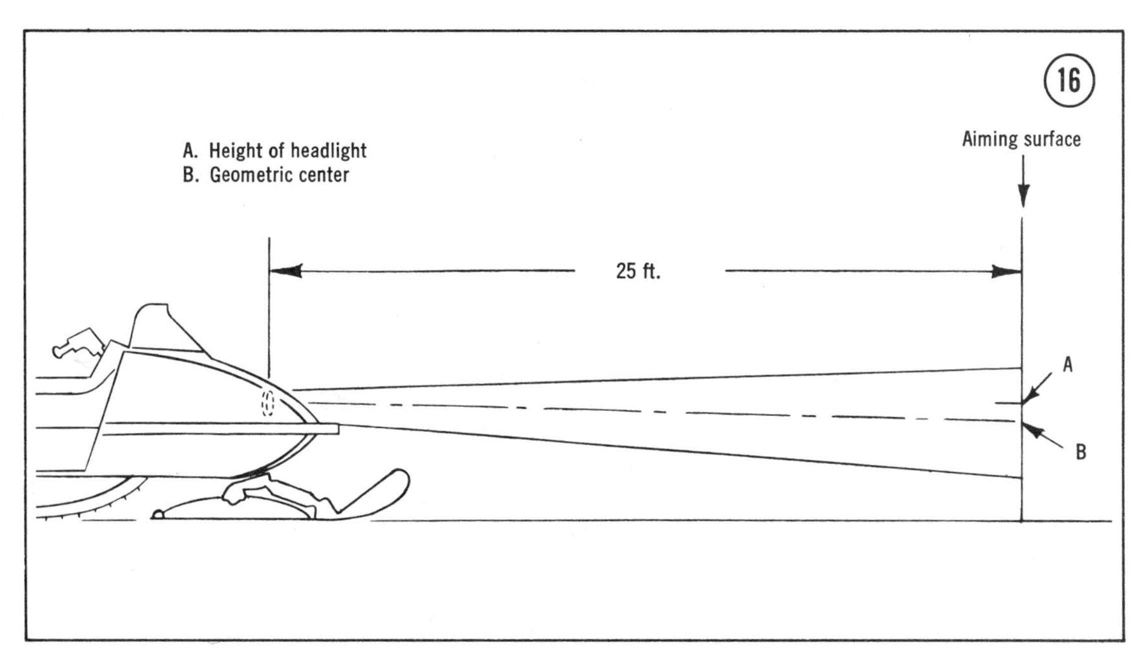

6

BRAKE/TAILLIGHT BULB REPLACEMENT

1. Refer to **Figure 19** (typical lens) and unscrew the screws that attach the lens to the rear housing or toolbox. Remove the lens and remove the defective bulb by pressing in on it and turning it counterclockwise.

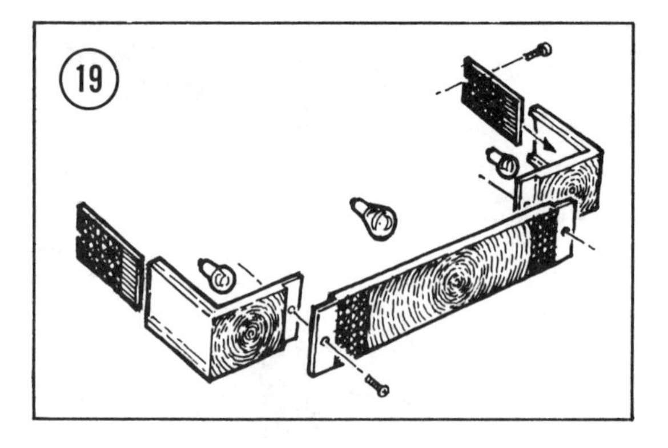

2. Clean the socket to remove any corrosion, dirt, or moisture.

3. Line up the guide pins on the new bulb with the slots in the socket and press the bulb in and turn in clockwise to lock it in place.

> NOTE: *On dual filament bulbs (combination brake/taillight) the pins are at different distances from the base of the bulb; make sure they correspond with the different length slots in the socket.*

4. Clean the inside of the lens with a mild detergent and warm water before installing. Install all of the screws finger-tight before tightening, then tighten them securely but not so tight that they damage the lens.

NOTE: If you own a 1978 or later model, first check the Supplement at the back of the book for any new service information.

CHAPTER SEVEN

POWER TRAIN

The power train consists of a drive belt, drive and driven sheaves, drive chain and sprockets, secondary shaft, drive shaft and a brake assembly. See **Figure 1** for a typical example of drive train components.

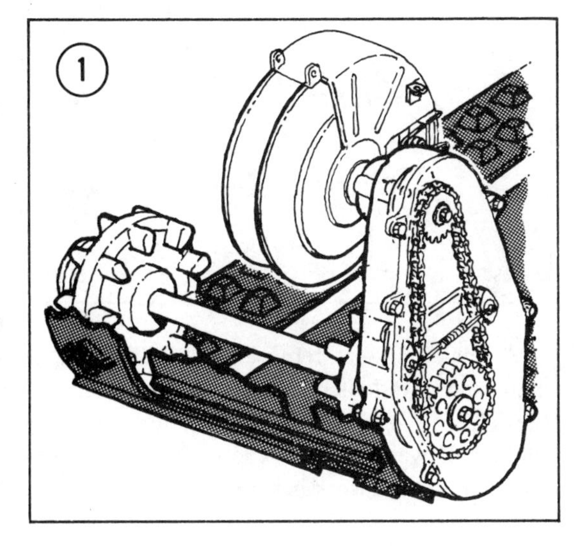

All machines are equipped with a mechanical disc brake.

Some procedures in this chapter require the use of special tools for removal and repair work. If such tools are not available, and substitutes cannot be locally fabricated, refer the removal and repair work to an authorized dealer.

DRIVE BELT

Inspect the drive belt (**Figure 2**) for wear, cracking, or other damage or deterioration. If the belt is not in good condition, full power will not be transmitted from the drive clutch to the driven pulley. Measure the belt (monthly) and compare it with the service specifications in **Table 1**. If the belt is not within specifications, replace it.

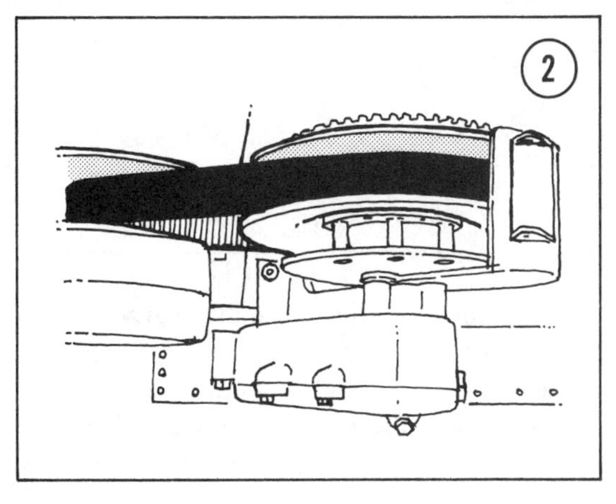

1. Remove the pins from the clutch guard (**Figure 3**) and raise the guard. Set the parking brake.
2. Push against the moveable sheave and rotate it clockwise until the sheaves are apart.

Table 1 DRIVE BELT SPECIFICATIONS

Year/Model	Outside Circumference	Width
Lynx		
All years	43 1/16 to 43 7/8 in.	1 1/8 to 1 9/32 in.
Pantera		
Through 1975	43 1/16 to 43 7/16 in.	1 1/8 to 1 9/32 in.
1976 on	46 1/2 to 46 7/8 in.	1 1/8 to 1 9/32 in.
Jag		
All years	44 5/16 to 44 11/16 in.	1 1/8 to 1 9/32 in.
Cheetah		
All years	43 1/16 to 43 7/16 in.	1 1/8 to 1 9/32 in.
El Tigre		
All years	46 1/2 to 46 7/8 in.	1 1/8 to 1 9/32 in.
Panther		
All years	43 1/16 to 43 7/16 in.	1 1/8 to 1 9/32 in.

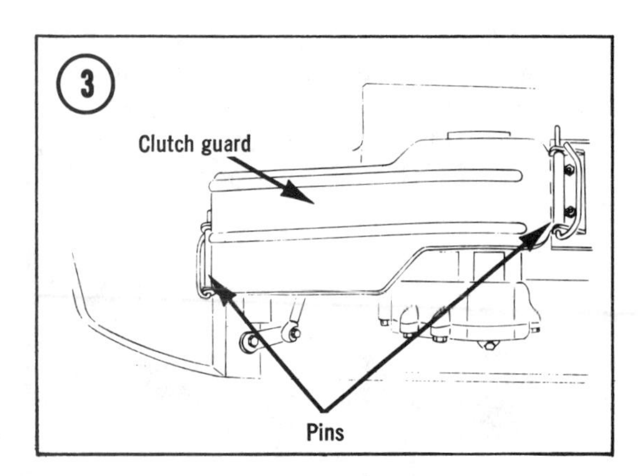

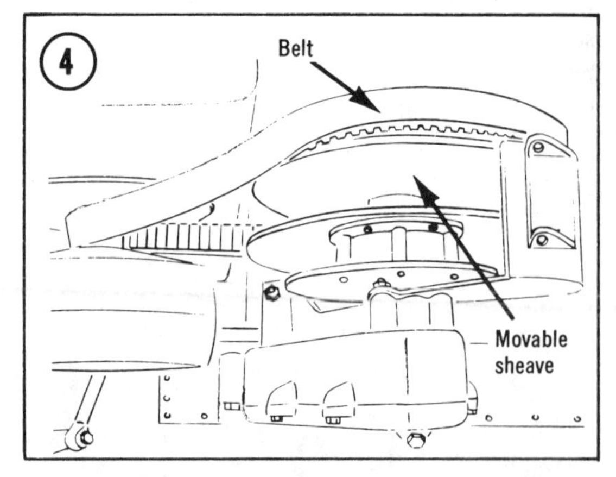

3. Hold the sheaves apart, pull the drive belt up, and roll it over the stationary sheave (**Figure 4**). When the belt is completely off the drive pulley, slowly release the moveable sheave. Then, remove the belt from the drive clutch. Reverse these steps to install the belt.

Belt Alignment

The alignment and center-to-center distances of the drive clutch and the driven pulley were set at the time the snowmobile was built, and adjustment is rarely required other than when the clutch and driven pulley are removed and reinstalled.

Special tools are required to measure alignment and parallelism and to remove and install the clutch. If misalignment is indicated by rapid belt wear or loss of power, it is recommended that the work be entrusted to a dealer.

Drive Clutch Removal/Installation

Service of the drive clutch should be entrusted to a dealer. The cost of service can be reduced by removing and installing it yourself.

1. Raise the clutch shield (**Figure 3**) and remove the drive belt as described above.

Remove the rubber plug from the left side of the bellypan if so equipped (**Figure 5**).

2. Unscrew the bolt that attaches the drive pulley to the crankshaft and collect the washer.

3. Screw the Arctic Cat clutch puller bolt into the clutch (**Figure 6**). Tighten the bolt to withdraw the clutch from the crankshaft. If an impact wrench is being used for this task, the clutch can be held by hand; however, if a breaker bar or conventional wrench is used, it will also be necessary to hold the clutch with a chain wrench. If the clutch is reluctant to release from the shaft, occasionally rap sharply on the puller bolt head with a hammer and continue to tighten it.

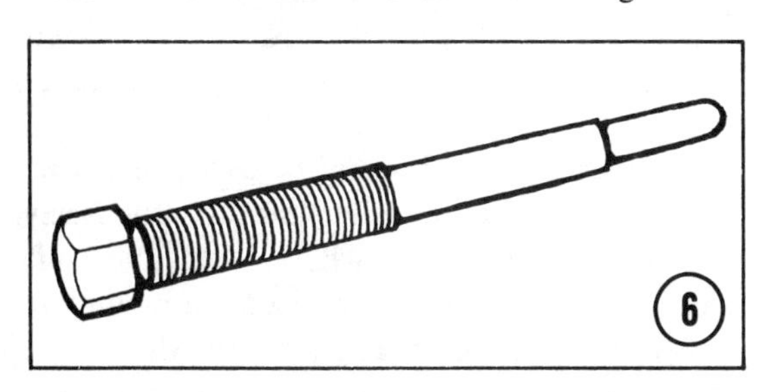

4. To install the drive clutch, install the drive belt on the driven pulley then loop it around the drive clutch.

5. Set the drive clutch on the shaft and screw in the bolt, with a washer installed, and tighten it to 55-60 ft.-lb. (7.6-8.3 mkg).

6. Have the alignment of the drive clutch and driven pulley checked by a dealer.

Driven Pulley Removal/Installation

As with the drive clutch, service of the driven pulley should be entrusted to a dealer.

1. Raise the clutch shield and remove the drive belt as described above.

2. Unscrew the pulley bolt and remove the washer (**Figure 7**).

3. Remove the pulley from the shaft and note the number of shims and washers. Remove the key from the shaft.

4. To install the driven pulley, return all the shims that were removed from the shaft.

5. Install the pulley on the shaft and rotate it to line up its keyway with the keyway on the shaft. Tap the key into place. Screw in the pulley bolt, with a washer and any shims that were removed from the outer face of the pulley, and tighten it to 17 ft.-lb. (2.4 mkg).

6. Have the alignment of the driven pulley and the drive clutch checked by a dealer. Install the drive belt as described in Chapter Two, *Drive Belt*. Install the clutch shield.

Drive Chain and Sprockets Removal/Installation

1. Drain the fuel tank. Turn the machine onto its left side (power take-off—PTO) to prevent oil from the chaincase spilling into the belly pan.

2. Unscrew the bolts that hold the chaincase cover in place and remove the washers. Tap the cover to break it loose and remove it and the gasket.

3. Remove the cotter key and washer from the tensioner post (**Figure 8**) and disconnect the spring. Discard the cotter key.

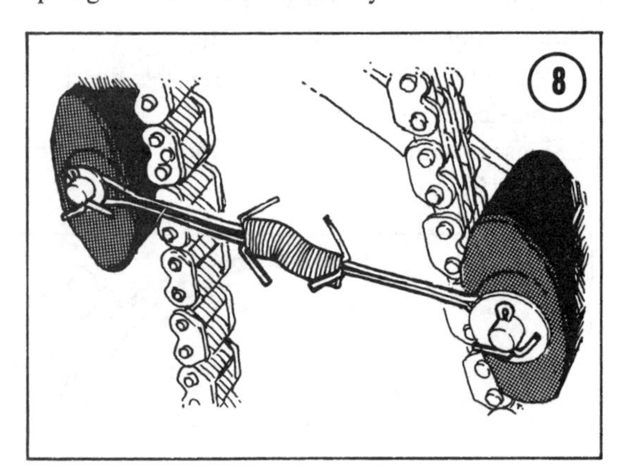

4. Unscrew the sprocket bolts and remove the washers. Slide both sprockets off their shafts with the chain in place.

5. To install the sprockets and chain, assemble them with chain looped over both sprockets, line them up with the shafts, and slide them on.

6. Screw in the bolts, with the washers installed, and tighten them to 24 ft.-lb. (3.3 mkg).

7. Place a straightedge across the sprockets (**Figure 9**) and check their alignment. Shim the sprocket that is closest to the inside of the chaincase. A gap of 0.030 in. (0.76mm) is allowable if a shim of the precise thickness to make the sprockets parallel is not available.

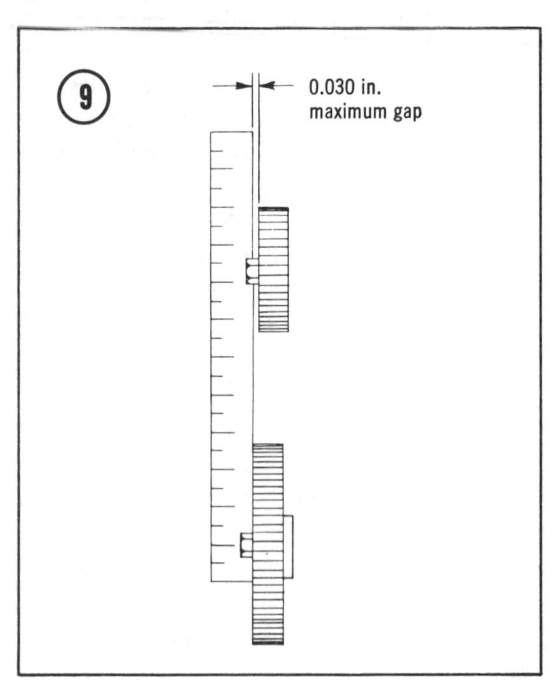

8. Install the tensioner spring and secure it with a washer and a *new* cotter key.

9. If the gasket is in good condition, install it. If there is any question about its condition, replace it with a new one. Install the chaincase cover and thread in all 4 bolts before tightening them.

Driven Shaft
Removal/Installation

1. Refer to *Driven Pulley Removal/Installation* and remove the driven pulley. Refer to *Drive Chain and Sprockets Removal/Installation* and remove the chain and sprockets.

2. Unscrew the nuts from the MAG side bearing flange (**Figure 10**). Remove the flange, O-ring and bearing.

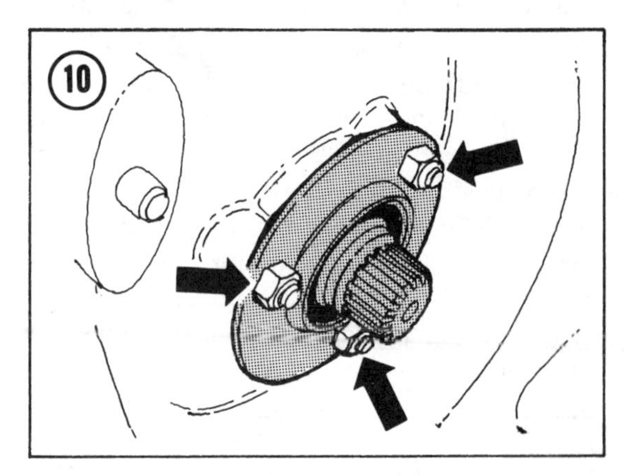

3. Loosen the set screw in the PTO side bearing collar and rotate the collar until it turns freely.

4. Move the shaft toward the PTO side while rotating it to prevent the brake from binding on the shaft. Remove the key from the shaft.

5. Loosen the nuts that attach the MAG side bearing holder (**Figure 11**). They do not have to be removed to allow the shaft to be withdrawn.

6. Tilt the MAG end of the shaft and remove it.

7. Before installing the shaft, coat it lightly with oil. Insert it through the PTO side bearing, splines first, and install the bearing collar with the large ID toward the bearing.

NOTE: *If the shaft was burred by the set screw in the lock collar, dress it with a fine file.*

8. Install the key in the shaft and set the brake disc and hub in place, with the large flange facing the PTO side. Make sure the key does not fall out.

9. Install the MAG side bearing, O-ring, and flange. Screw the locknuts on finger-tight. Rotate the shaft 2 or 3 revolutions and tighten the flange nuts to 12 ft.-lb. (1.7 mkg).

10. Tap on the PTO end of the shaft to seat it against the MAG side bearing. Turn the bearing collar in the direction of shaft rotation to lock the bearing. Tighten the setscrew in the collar.

11. Install the chains and sprocket and the chaincase cover, and the driven pulley as described earlier in this chapter. Have the alignment of the pulley and the drive clutch checked by a dealer.

Track Drive Shaft and Track Removal/Installation

1. Refer to *Drive Chain and Sprockets Removal/Installation* (this chapter) and remove the drive chain and sprockets. Refer to *Suspension Removal/Installation*, Chapter Nine and remove the skid frame.

2. Refer to **Figure 12** and remove the speedometer drive head. Unscrew the nuts from the MAG side bearing flange for the drive shaft and remove the flange, bearing, and O-ring.

3. Move the drive shaft to the MAG side and pull the PTO end out of the front mounting hole. Tilt the shaft away from the tunnel and remove the track. Withdraw the drive shaft from the chaincase.

4. Clean all of the parts thoroughly. Use solvent for metal parts and soap and water for rubber and plastic pieces such as O-rings. Dry them thoroughly, preferably with compressed air.

5. Inspect the track and replace any missing rivets and damaged cleats. Rotate the bearings by hand to check for roughness and excessive radial play. Replace them if there is any doubt about their condition. Inspect the drive shaft for damaged threads and fretted splines and replace it if it is not in good condition.

6. To install the track and drive shaft, set the shaft assembly in the track with the splined end of the shaft on the MAG side.

7. On the unsplined end of the shaft, install the lock collar with the large ID end out. Install the retainer plate (with the lock flange toward the lock collar), bearing, and outer retainer plate (with the flange toward the end of the shaft).

8. Set the track and drive shaft assembly in place and guide the splined end of the shaft into the chaincase.

9. Align the PTO end of the shaft with the holes in the tunnel. Install the 3 carriage bolts through the tunnel from the inside, through the retainer plates, the tunnel, front end, and the speedometer drive head.

10. Install the MAG side bearing on the shaft with the sealed side toward the sprockets. Then install the O-ring and the retainer plate. Screw on the nuts finger-tight.

11. Align the track drive sprockets equidistant from the inside edges of the tunnel. When the distance is equal, tighten the bearing lock collar.

12. Tighten the nuts for each of the bearing retainers to 20 ft.-lb. (2.8 mkg).

13. Refer to *Suspension Removal/Installation*, Chapter Nine, and install the skid frame. Refer to *Drive Chain and Sprockets Removal/Installation*, this chapter.

14. Refer to *Track Tension, Track Alignment*, and *Suspension Adjustment* in Chapter Two and adjust and align the suspension and track.

CHAPTER EIGHT

FRONT SUSPENSION AND STEERING

The front suspension and steering consists of spring mounted skis on spindles connected to the steering column by tie rods (**Figure 1**).

The skis have replaceable wear rods (skags). All machines except the early Lynx I are equipped with mono-leaf springs. Some are equipped with shock absorbers. The Lynx I has multi-leaf springs.

The handlebar steering column on some models is equipped with Zerk-type fittings that must be lubricated at 40-hour intervals. All snowmobiles use a one piece steering column.

The tie rod and drag link ends have both right-hand and left-hand threads. The tie rods, drag link, and spindles are designed to bend rather than break if extreme shock loads are encountered.

SKIS AND STEERING

Ski Wear Bars

The ski wear bars, or skags, aid turning the snowmobile, and they protect the bottoms of the skis from wear caused by road crossings and bare terrain. The bars are expendable and should be checked weekly to ensure that they have not worn down to the point where they no longer afford adequate protection to the running surfaces of the skis and cease to aid turning. If the wear bars are excessively worn, turning will be imprecise and control marginal.

Removal/Installation

1. The fuel tank should be no more than ¼ full to prevent spillage of fuel. Turn the machine on its side and protect the finish with cardboard placed on the floor.
2. Remove ice and snow from the skis.
3. Unscrew the locknut and remove the flat washer from each ski (**Figure 2**).

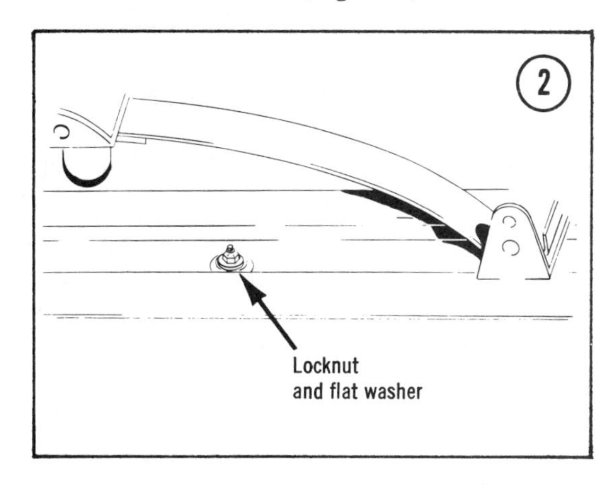

4. Pry the wear bar off the bottom of each ski until the stud is out of the hole. Place a block of wood between the ski and the wear bar, *behind*

8

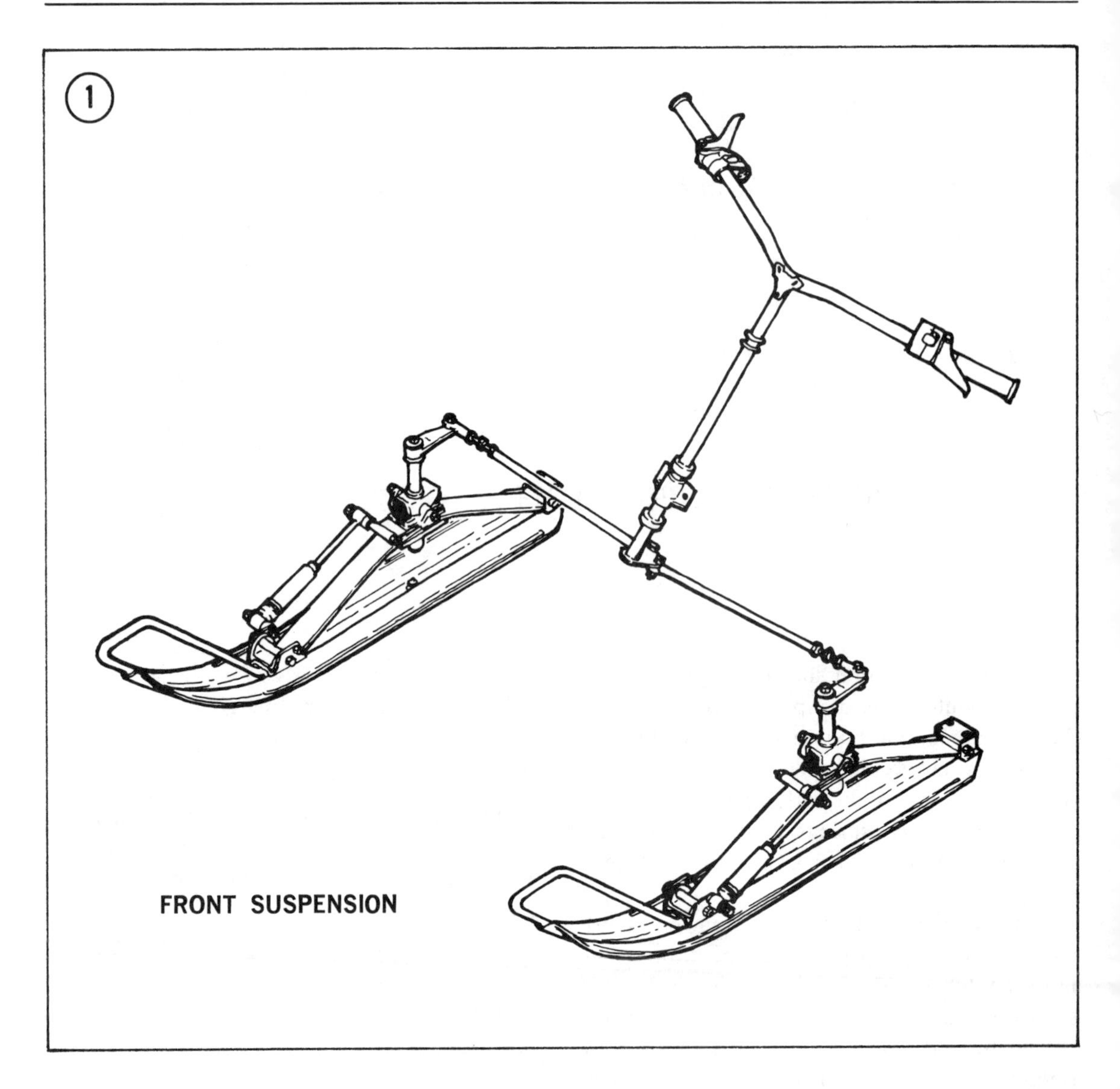

the stud, and rap it with a hammer to drive the rear of the wear bar out of the hole in the rear of the ski (**Figure 3**). Then, pull the wear bar out of the hole in the front of the ski.

5. Install a new wear bar by first inserting the forward end into the hole in the front of the ski. Insert the wooden block between the ski and the wear bar *ahead* of the stud. Tap on the block to drive the wear bar to the rear, while at the same time guide the rear end of the wear bar into the hole in the rear of the ski until the stud lines up with the hole in the top of the ski.

6. Remove the wooden block and push the stud into the hole. Install the washer and screw on and tighten the locknut on the stud.

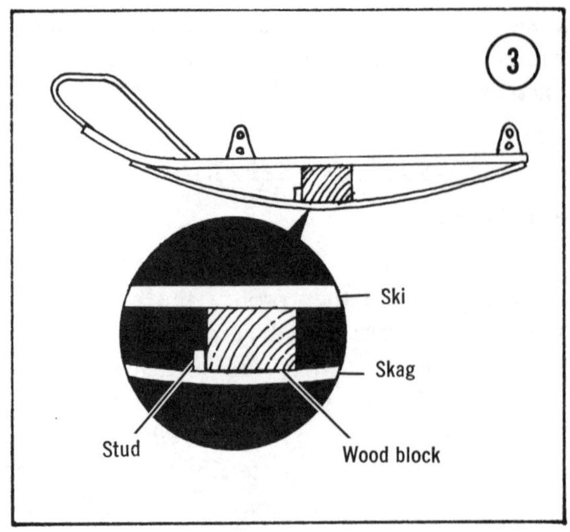

Ski Removal/Disassembly (Spindle Shock Mount)

1. Turn the machine on its side. The fuel tank must be nearly empty to prevent spillage. Refer to **Figure 4** and unscrew the locknut from the bolt that attaches the ski assembly to the spindle.

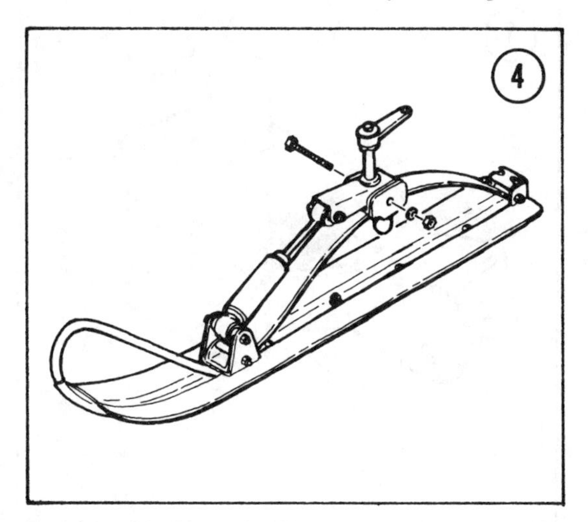

2. Unscrew the locknut from the shock absorber top bolt. Withdraw both bolts and remove the ski.

3. Unscrew the locknut from the shock absorber bottom bolt, withdraw the bolt and remove the shock absorber.

4. With a vise, compress the spring about an inch and remove the nut and bolt from the front spring mount. Carefully release the pressure and then remove the nut and bolt from the rear spring mount. Remove the spring and the rear bumper. (The bumper can be left in place if it is in good condition.)

Ski Assembly/Installation (Spindle Shock Mount)

1. Attach the rear bumper to the ski with pop-rivets (**Figure 5**).

2. Install the saddle, liner, and block on the ski. Tighten the nuts and bolts to 35 ft.-lb. (4.84 mkg).

> NOTE: *The saddles are interchangeable, from one ski to the other. However, they must be arranged so that each has the threaded hole on the inside.*

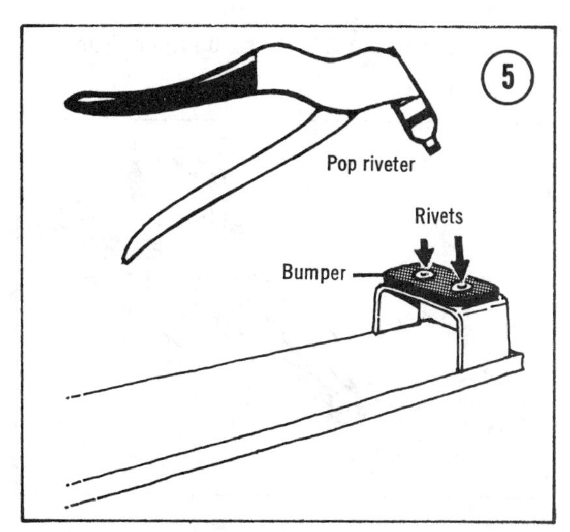

3. Connect the rear of the spring to the rear mount and tighten the nut and bolt to 20 ft.-lb. (2.77 mkg). The locknut must be on the inside.

4. Place the spring slide saddle and the spring in the front mount, then install the sleeve in the slide saddle. Compress the spring with a vise to line up the holes in the front mount with the holes in the slide saddle and the sleeve. Install the nut and bolt, with the nut on the inside, and tighten them to 35 ft.-lb. (4.84 mkg). Release the pressure and remove the ski from the vise.

5. Attach the lower end of the shock absorber to the ski. Use the long sleeve and install a plastic bushing on either end with the rounded side of each bushing facing out. Tighten the shock absorber bolt and nut to 50 ft.-lb. (6.9 mkg). As for the other bolts, the locknut must be on the inside.

6. Attach the skis to the spindles with the threaded hole in each saddle on the inside. Tighten the bolt in the saddle to 30 ft.-lb. (4.15 mkg). Then install the locknut and tighten it to 30 ft.-lb.

7. Connect the upper end of the shock absorber to the spindle and tighten the nut and bolt, with the nut on the inside, to 50 ft.-lb. (6.9 mkg).

Ski Removal/Disassembly (Saddle Shock Mount)

1. Unscrew the locknut and bolt that attach the ski to the spindle and remove the ski (**Fig-**

ure 6). Remove the rubber damper from the spring saddle.

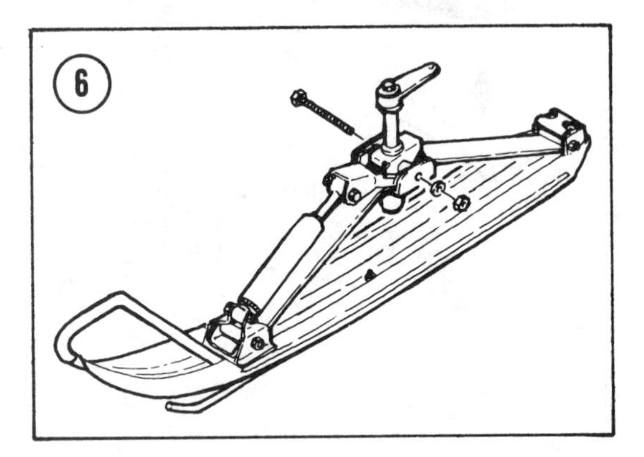

2. Unscrew the nuts and bolts from the shock absorber mounts and remove the shock absorber and the bushings.
3. With a vise, compress the spring about an inch and remove the nut and bolt from the front spring mount. Collect the sleeve and saddle. Carefully release the pressure on the spring.
4. Unscrew the nut and bolt from the rear spring mount and remove the spring. Remove the rear ski bumper. (The bumper can be left in place if it is in good condition.)

Ski Assembly/Installation (Saddle Shock Mount)

1. Rivet a new rear bumper to the ski if the old one was removed. Place the rear of the spring in the rear mount and screw in and tighten the bolt and locknut to 20 ft.-lb. (2.8 mkg). The nut must be on the inside of the ski.
2. Place the spring slide saddle and the spring in the front mount, then install the sleeve in the slide saddle. Compress the spring with a vise to line up the holes in the front mount with the holes in the slide saddle and sleeve. Install the nut and bolt, with the nut on the inside of the ski, and tighten them to 35 ft.-lb. (4.8 mkg). Release the pressure and remove the ski from the vise.
3. Set 2 plastic bushings into each eye of the shock absorber. Install the shock absorber, body down, and tighten the nuts and bolts, with the nuts on the inside, to 30 ft.-lb. (4.2 mkg).
4. Set the rubber damper in the spring saddle (**Figure 7**) and attach the ski assembly to the spindle. Make sure the threaded hole in the saddle is to the inside. Tighten the bolt in the saddle to 30 ft.-lb. (4.2 mkg). Then, screw on the locknut and tighten it to 30 ft.-lb.

Tie Rod Removal/Installation

The tie rods are designed to bend rather than break upon severe ski impact. If they are severely bent, they should be replaced rather than straightened. In either case, the rods must be removed.

1. Refer to Chapter Four, *Engine Removal*, and remove the engine from the machine.
2. Remove the nuts and bolts that attach the tie rods to the spindle arms and steering post (**Figure 8**).
3. Disassemble the tie rods by loosening the jam nuts on either end and unscrewing the adjuster bolts.
4. To assemble and install the tie rods reverse the above. Note that the adjuster bolts have both right-hand and left-hand threads. The end of the bolt with the brass colored jam nut screws into the tie rod, and the silver nut end screws into the tie rod end.
5. Install the tie rods with the adjuster end at the spindles.
6. Reinstall the engine as described in Chapter Four, *Engine Installation*, and align the skis as described below.

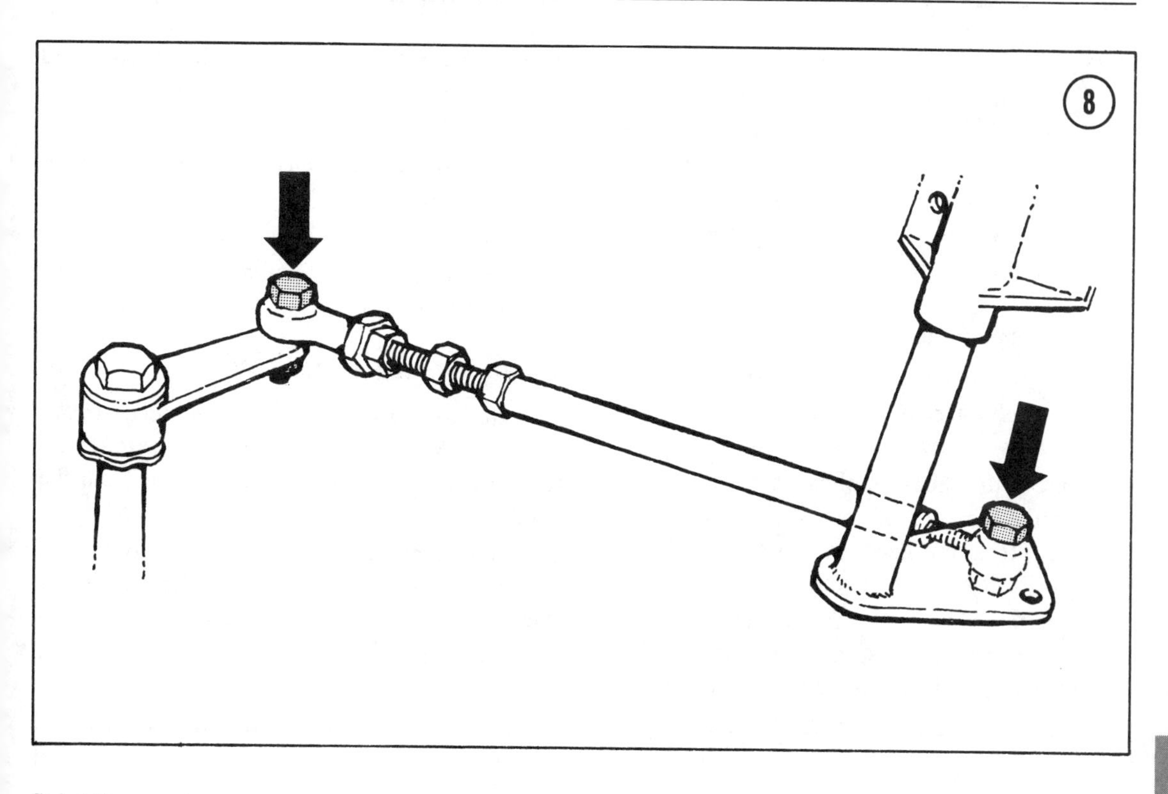

Ski Alignment

Ski alignment must be checked and corrected if necessary when any of the steering components (steering post, tie rods, spindles) are disconnected or replaced. It should also be checked during major service such as at the end of the season.

1. Set the handlebar in the straight ahead position and loosen the jam nuts on both tie rod ends (**Figure 9**).

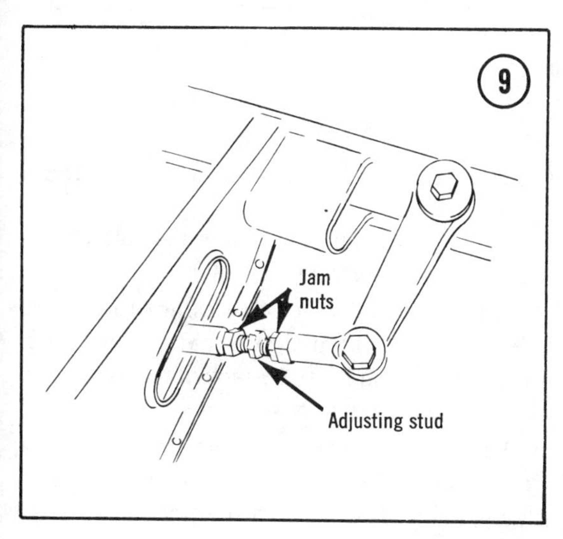

2. Raise and block the machine about 6 in. off the floor. Determine which ski is most closely aligned with the machine and measure the distance between the skis at two locations (**Figure 10**). The measurements should be no more than ¼ in. greater at the front than the rear (skis toed-out).

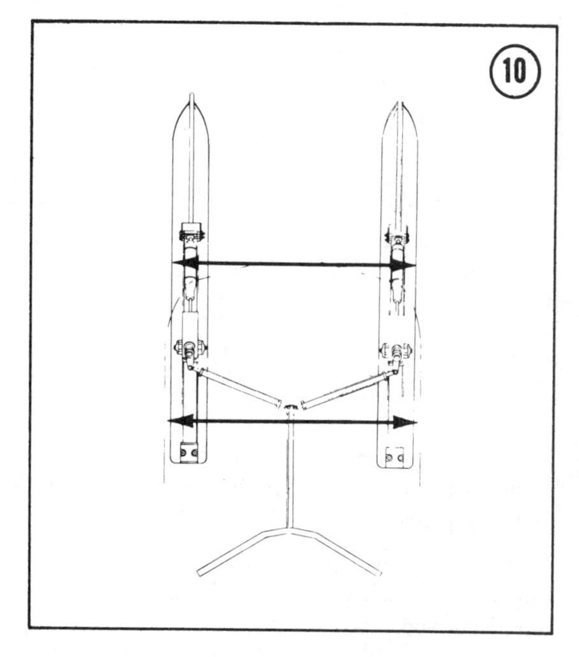

3. If adjustment is required one or both adjusting studs in the tie rods must be turned to make the skis parallel and square with the handlebars.
4. When adjustment is correct, tighten the jam nuts against the tie rods and the tie rod ends, taking care not to turn the adjusting stud further after the adjustment is correct.

Spindle Removal/Installation

On some models it is necessary to remove the pulse charger and drive clutch from the engine. This is described in Chapter Four.

1. Remove the ski assemblies from the spindles as described earlier under *Ski Removal/Disassembly*. It is first necessary to raise and support the front of the machine.
2. Remove the bolt and flat washer that hold the spindle arm on the spindle (**Figure 11**). Remove the arm and washers from the spindle.

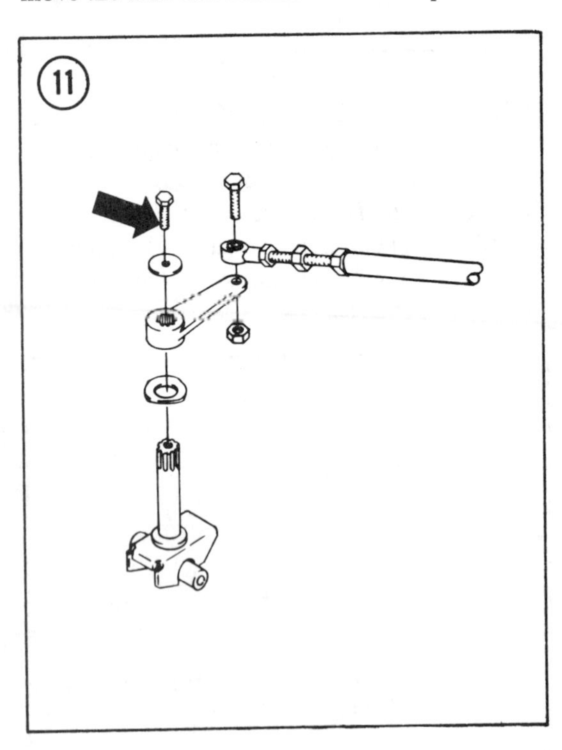

3. Pull down on the spindle to remove it. It may be necessary to tap the spindle out of its mount with a soft drift.
4. To install the spindle, slide it up into the mount and install the washers in the order they were removed.
5. Install the spindle arm and flat washer and screw in and tighten the bolt to 20 ft.-lb. (2.8 mkg).
6. Refer to *Ski Assembly/Installation* and install the skis.
7. Refer to *Ski Alignment* and align the skis.
8. Reinstall the pulse charger and drive clutch if they were removed.

Steering Post Removal/Installation

1. Refer to *Throttle Handle Removal/Installation* and *Brake Handle Removal/Installation* (this chapter) and remove the controls from the handlebar.
2. Remove the front and rear motor mount nuts and lift the engine just enough to clear the lower steering post bracket. Unscrew the nuts and bolts that attach the bracket to the frame (**Figure 12**).

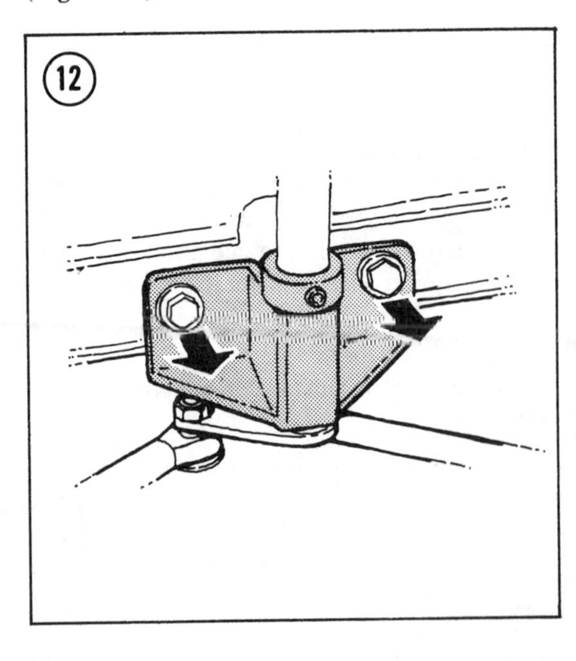

3. Unscrew the nuts and bolts that attach the tie rods to the steering quadrant.
4. On some models it is necessary to remove the seat, move the fuel tank back, and detach the dashboard so it can be moved to gain access to the upper steering post bracket.
5. Unscrew the nuts from the upper steering post bracket (**Figure 13**) and remove the cable clamp, column clamp, and bushing halves.

6. Remove the steering post from the machine.

7. Install the steering post by reversing the removal steps. Lubricate the lower steering post bracket and install the lower bracket bolts and nuts but do not tighten them yet.

8. On models equipped with bushings for the upper steering post mount, install the bushing halves, the upper clamp, and the cable clamp. Screw on the locknuts and tighten them to 10 ft.-lb. (1.4 mkg). Tighten the lower steering post bracket nuts.

9. If the dash was removed, reinstall it. Reinstall the fuel tank and the seat cushion. Reconnect the tie rods as described earlier (*Tie Rod Removal/Installation*).

10. Set the engine back in place and screw on the nuts but do not tighten them until they have all been installed. Then, tighten them to 35 ft.-lb. (4.8 mkg).

11. Install the controls (see *Throttle Handle Removal/Installation* and *Brake Handle Removal/Installation* in this chapter).

12. Refer to *Ski Alignment* and check and correct the alignment of the skis if necessary.

Throttle Handle Removal/Installation

1. Remove the pad from the center of the handlebar after removing the clips.

2. Remove the circlip from the throttle lever pin and then remove the pin (**Figure 14**). Disconnect the cable from the throttle lever.

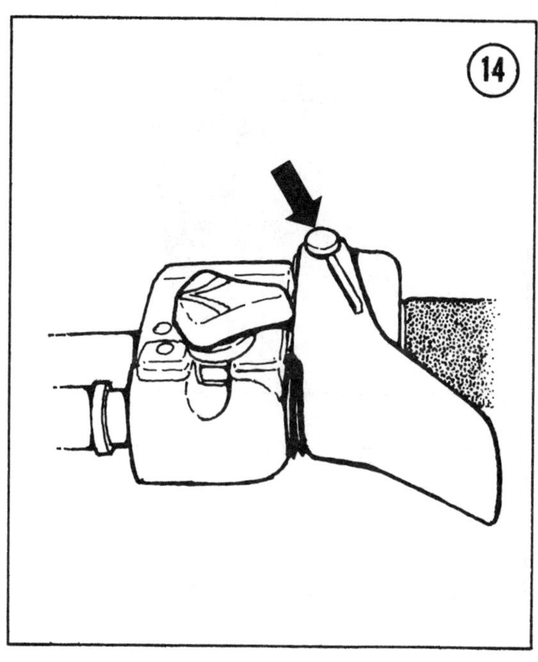

3. Unscrew the screws that hold the cap to the throttle handle (**Figure 15**). Slide the cap toward the center of the handlebar and remove the safety stop switch (**Figure 16**).

4. Remove the spring pin from the throttle handle (**Figure 17**). Pull the throttle handle and the cap off the handlebar.

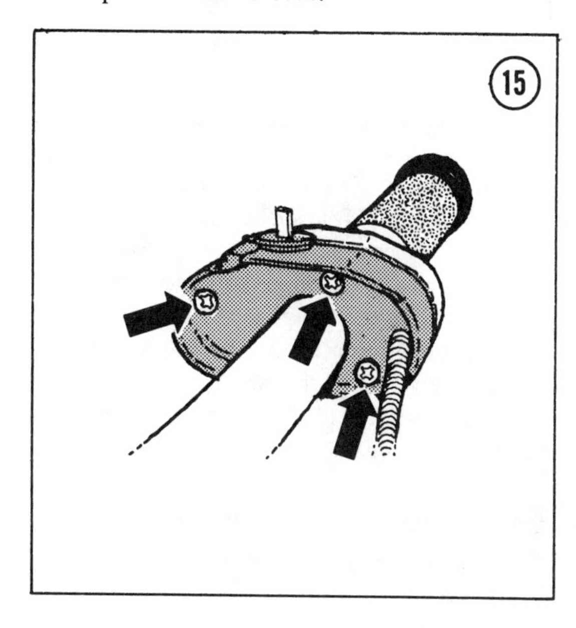

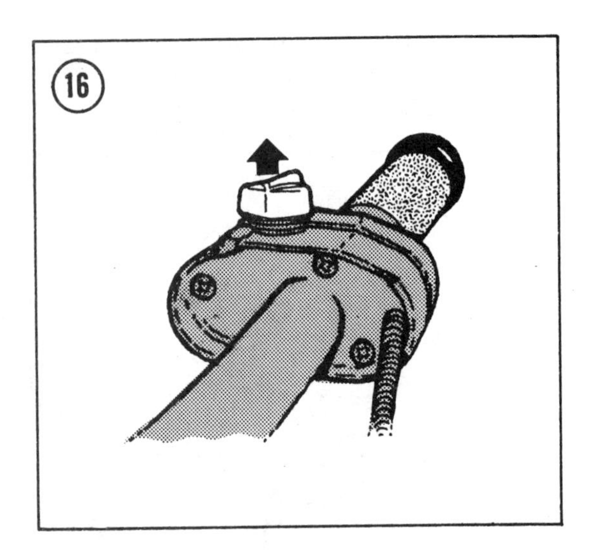

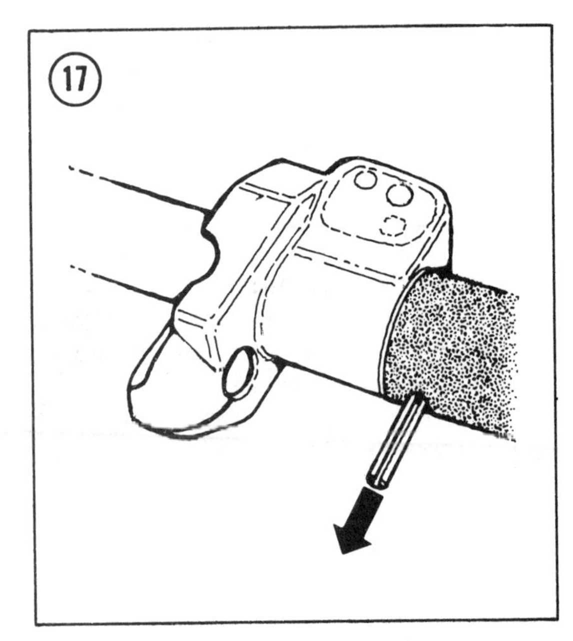

5. Reverse the above steps to install the throttle handle. Make sure the socket for the emergency stop switch faces the rear. Check also that the safety switch and tether are correctly positioned. Before installing the pad on the handlebar, check the operation of the throttle to ensure it moves freely and returns when released. If it is not satisfactory, find the cause and correct it before operating the machine.

Brake Handle Removal/Installation

1. Remove the pad from the center of the handlebar after removing the clips.

2. Remove the circlip from the brake lever pin and then remove the pin (**Figure 18**). Remove the lever, spring, and actuator.

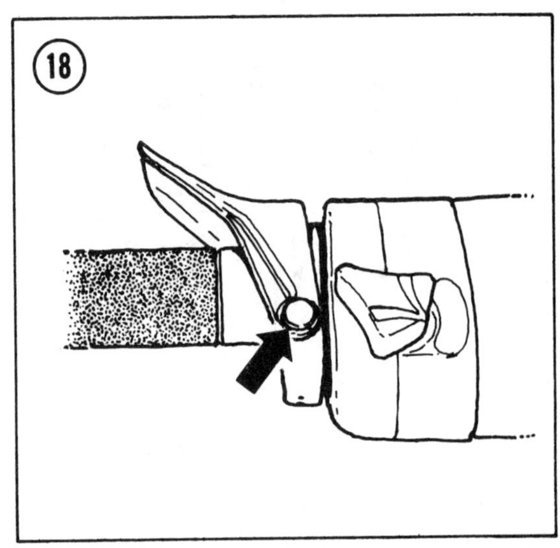

3. Slide the cable and lever toward the center of the handlebar and disconnect the cable (**Figure 19**).

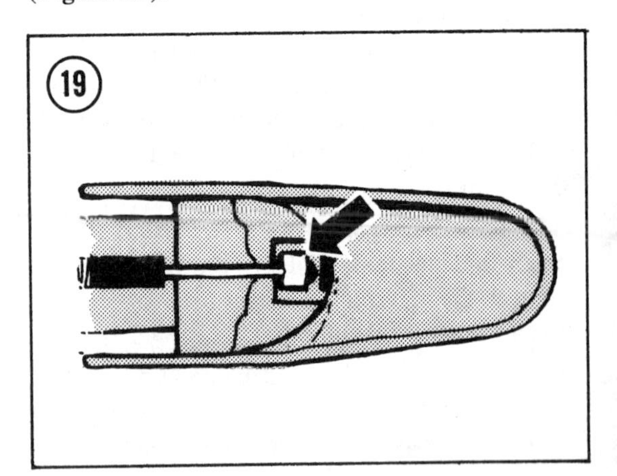

4. Unscrew the screws from the cap (**Figure 20**). Move the cap toward the center of the handlebar and remove the dimmer and brake light switches from the handle.

5. Remove the spring pin from the brake handle (**Figure 21**). Pull the brake handle and cap off the handlebar.

6. Reverse the above to assemble and install the brake lever assembly. Test the operation of the brake and the lights when assembly is complete and correct any unsatisfactory conditions.

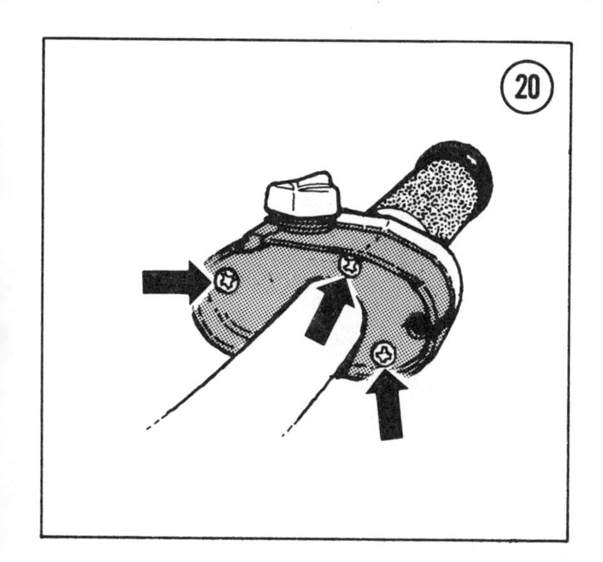
20

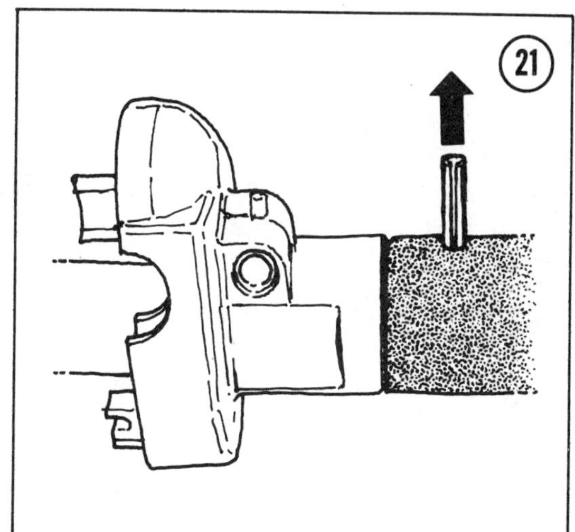
21

CHAPTER NINE

REAR SUSPENSION AND TRACK

Arctic Cat snowmobiles are equipped with slide rail rear suspension which utilizes a rear idler assembly.

The slide rail suspension (**Figure 1**) is fitted with adjustable springs, shock absorbers, and replaceable wear bars. The suspension also includes a weight transfer adjustment to vary the amount of pressure on the skis.

This chapter includes removal and installation procedures for suspension components and tracks. Refer to Chapter Two for suspension and track adjustments and Chapter Three for *Track Wear Analysis.*

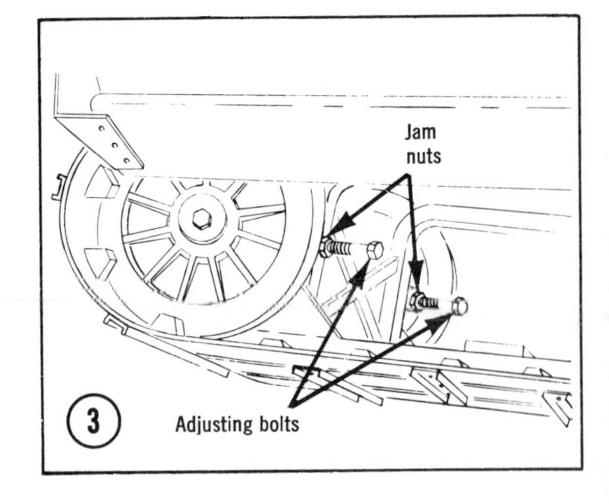

Suspension Removal/Installation

1. Unscrew the bolts that attach the skid frame to the tunnel (**Figure 2**).

2. Loosen the track adjusting screws (**Figure 3**).

3. Raise the rear of the machine a couple of feet and support it. Pull the skid frame out of the track and remove the axles from the front and rear arms.

4. Thoroughly wash the skid frame with detergent and water and dry it completely with clean rags and compressed air. Lightly sand the paint in chipped areas and repaint them. Inspect the Hi-Fax slides and replace them if they are severely worn (see *Hi-Fax Slide Replacement,* this chapter).

5. Lightly coat the inner axles with low-temperature grease.

6. Turn the snowmobile on its side and pull the track out of the tunnel. Set the skid frame in the track and then install the front and rear axles.

7. Hold the track and frame at a 45° angle to the tunnel and line up the forward holes in the skid frame with the forward holes in the tunnel. Screw in the bolts, with a flat washer installed, but do not tighten them until the rear bolts have been installed. It will be necessary to tip the

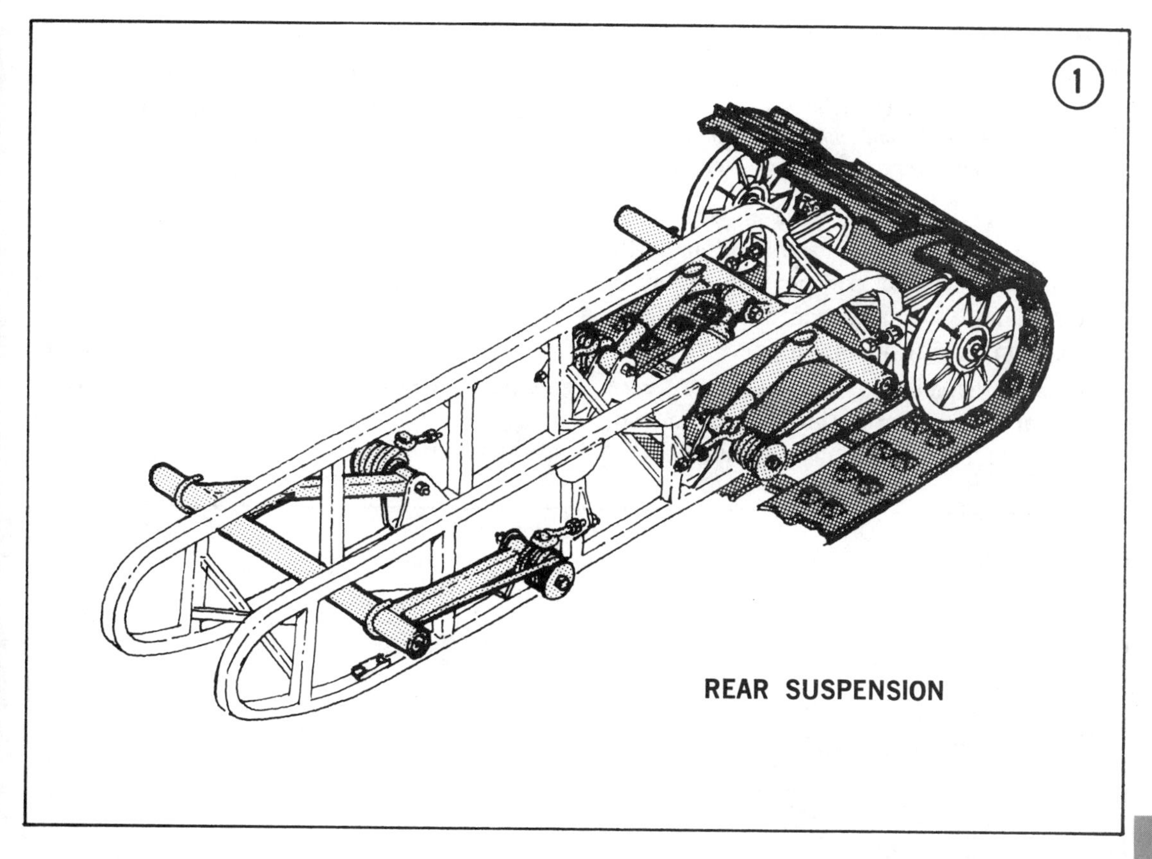
1
REAR SUSPENSION

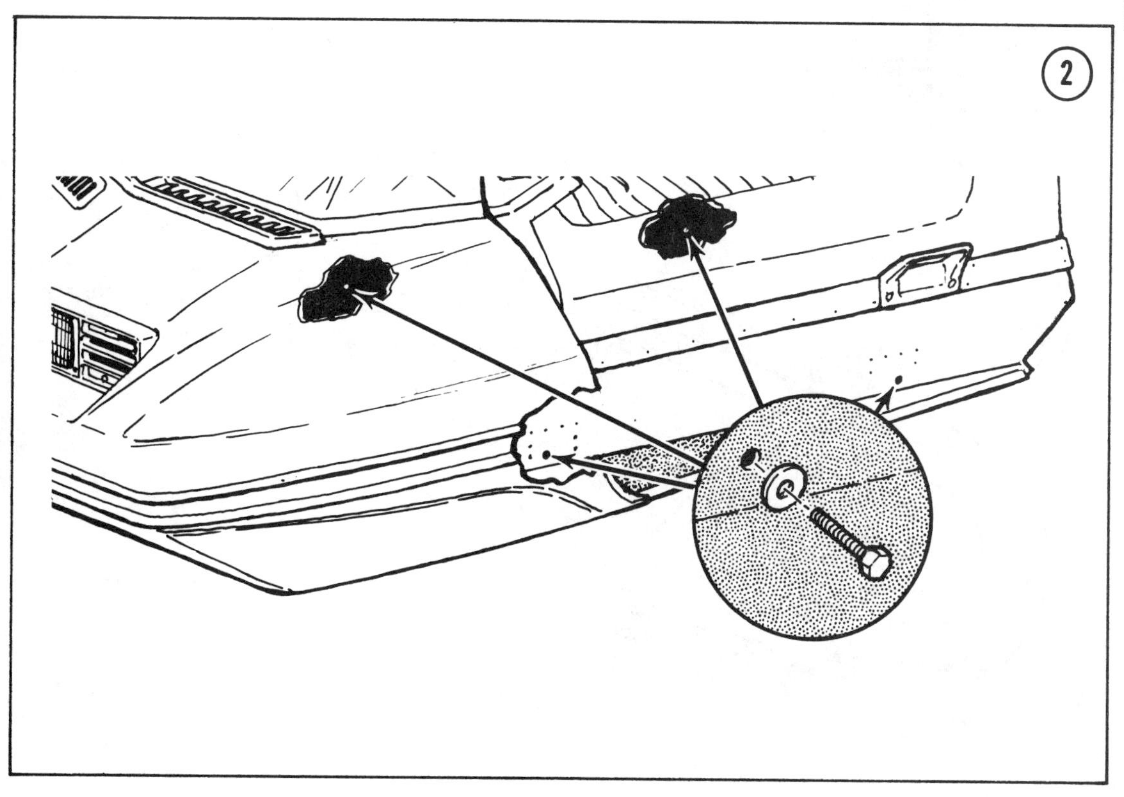
2

snowmobile from one side to the other to install both bolts.

CAUTION

The fuel tank and chaincase should be drained before performing Step 6.

8. Push the skid frame and track into the tunnel and line up the rear holes in the skid frame with the holes in the tunnel. Do not tighten any of the bolts until all 4 have been installed. Then, tighten them to 35 ft.-lb. (4.8 mkg).

9. Refer to Chapter Two, *Track Tension, Track Alignment,* and *Suspension Adjustment* and adjust and align the suspension and track.

Hi-Fax Slide Replacement

1. Invert the skid frame on a clean surface. Remove the rivets from the Hi-Fax slides, beginning at the rear and working forward, with a chisel and hammer (**Figure 4**). Remove all of the rivet ends from the rails.

2. Turn the skid frame over and remove the wear strips (**Figure 5**) in the manner just described.

3. The new Hi-Fax strips must be at room temperature (70°F) before being installed. Rivet the end of a slide to the forward (curved) end of the skid frame (**Figure 6**).

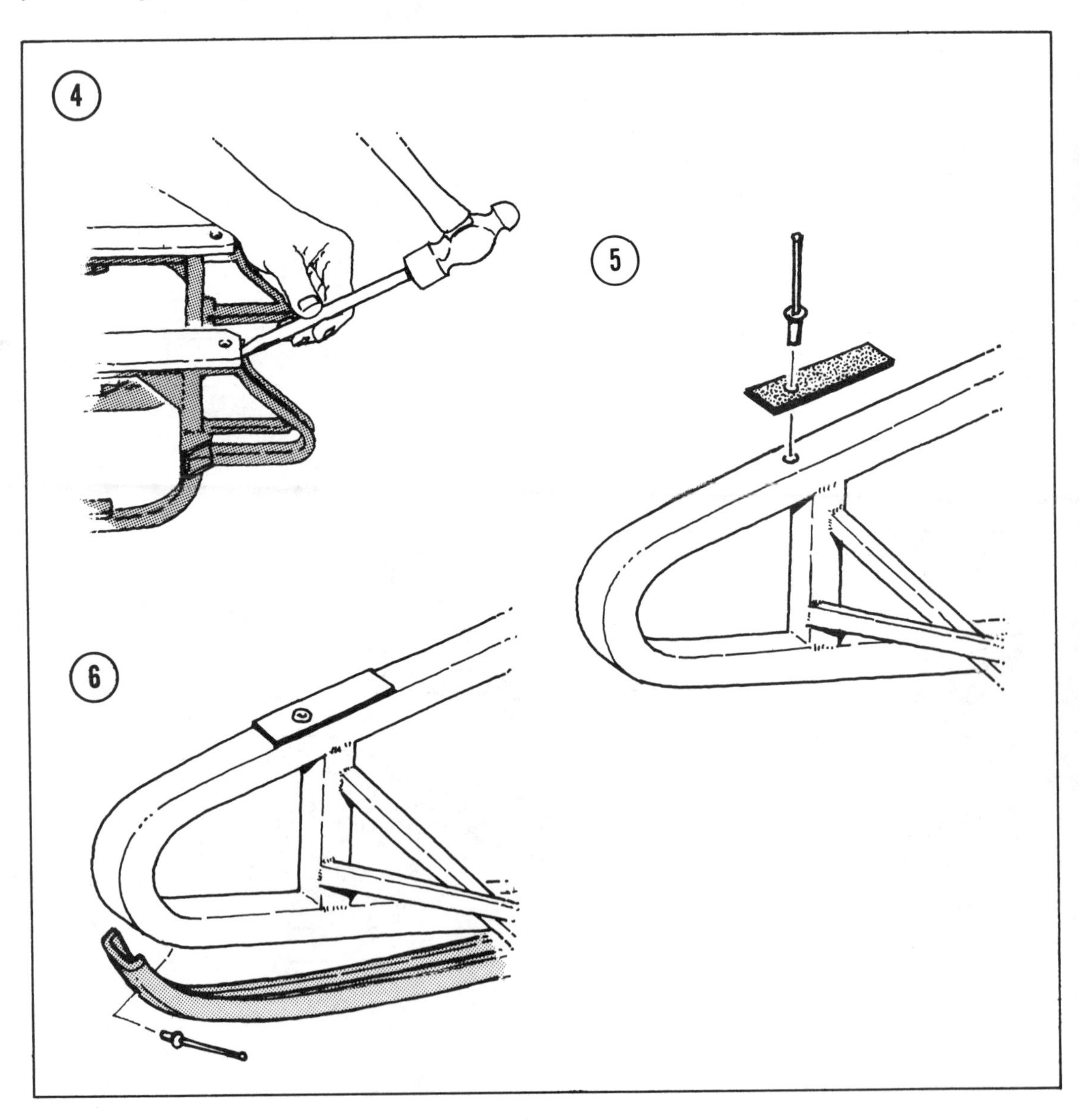

4. Heat the Hi-Fax with a propane torch, bend it around the frame, and rivet it in place in the next hole. Continue to heat, bend and rivet the strip in place until the last rivet has been installed in the end of the frame. Saw off excess length with a hacksaw. Do not attempt to bend the Hi-Fax without heating it; breakage is likely to occur. Install the other strip in the manner just described.

5. Invert the frame and rivet new wear strips to the upper rails.

6. Refer to *Suspension Removal/Installation* (this chapter) and install the skid frame.

SUPPLEMENT

1978 AND LATER SERVICE INFORMATION

The following supplement provides procedures and specifications unique to 1978-1979 Arctic Cat snowmobiles. All other procedures are identical to those for earlier models, contained in the main body of the book.

The chapter headings in this supplement correspond to those in the main body of this book. If a chapter is not included in the supplement, there are no changes affecting 1978-1979 models.

CHAPTER ONE

GENERAL INFORMATION

Arctic Cat models for 1978-1979 include:

Pantera FC (fan-cooled)
Pantera FA (free air)
Panther 4000
Panther 5000
Cheetah 5000
Jag 2000
Jag 3000
Lynx T
Lynx S
El Tigre 5000
El Tigre 6000
El Tigre Cross Country 340

With the exception of the liquid-cooled engine introduced in 1978 for the El Tigre 6000, procedures remain the same as for early models.

CHAPTER TWO

PERIODIC MAINTENANCE AND TUNE-UP

Essential maintenance and tune-up information for 1978-1979 models is contained in **Table 1** *(Drive Belt Specifications),* **Table 2** (*Ignition Specifications*), and **Table 3** (*Basic Air Screw Setting — Mikuni Carburetors*).

SUSPENSION ADJUSTMENT— EL TIGRE MODELS

Front spring adjustment affects steering by altering the weight on the skis. Rear spring adjustment compensates for rider weight.

Front Spring

1. To increase pressure on the skis, loosen the self-locking spring adjuster nuts (**Figure 1**). Turn both nuts the same amount.
2. To decrease the pressure on the skis, tighten the self-locking spring adjuster nuts.

Rear Spring

1. Use the spark plug wrench handle to turn the spring adjustment cam mounted on the tunnel (**Figure 2**).

2. The cam positions are marked 1, 2, 3, and 4 and correspond to changes in tension, with No. 1 representing the lowest tension setting.

> NOTE: *If the tension setting is being increased, move the cam numerically; if it's at "1," don't move it directly to "4." Instead, move it first to "2" then to "3."*

3. Make certain both cams are at the same setting.

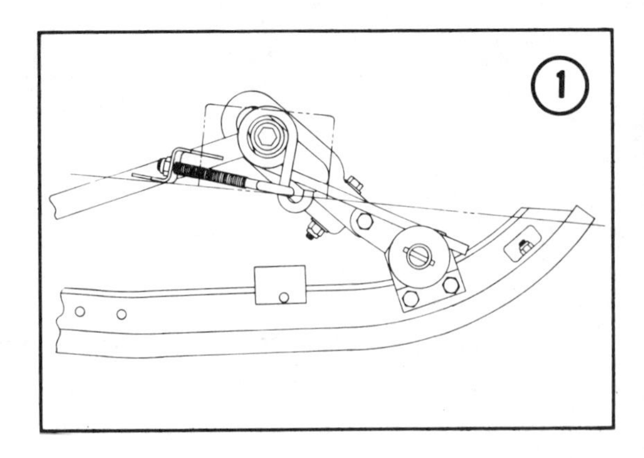

Table 1 DRIVE BELT SPECIFICATIONS

Model	Outside Circumference	Width
Lynx	43.25 in. (109.86cm)	1.25 in. (31.8mm)
Pantera	46.7 in. (118.62cm)	1.25 in. (31.8mm)
Jag	45.5 in. (115.57cm)	1.25 in. (31.8mm)
Cheetah	43.25 in. (109.86cm)	1.25 in. (31.8mm)
El Tigre	43.25 in. (109.86cm)	1.25 in. (31.8mm)
Panther	43.25 in. (109.86cm)	1.25 in. (31.8mm)

Table 2 IGNITION SPECIFICATIONS

Model	Ignition Timing @ 6,000 rpm	Spark Plug[1] NGK	Champion
Lynx			
Single (1978)	22° BTDC	BR9ES	N2
Single (1979)	18° BTDC	BR8ES	N3
Twin (1978) —			
Kokusan CDI	16° BTDC	BR8ES	N3
Nippon-Denso CDI	18° BTDC	BR8ES	N3
Twin (1979)	18° BTDC	BR8ES	N3
El Tigre			
5000	20° BTDC	BR9ES	N2
6000	22° BTDC	BR9ES	N2
Panther/Cheetah	18° BTDC	BR9ES	N2
Pantera			
Fan-cooled	18° BTDC	BR9ES	N2
Free air	20° BTDC	BR9ES	N2
Jag			
1978	16° BTDC	BR8ES	N3
1979	18° BTDC	BR8ES	N3

1. Spark plug gap (all models) = 0.020 in. (0.5mm).

NOTE: Spark plug heat ranges shown are for average operation (a combination of low- and high-speed running). If a colder plug is required because the machine is operated at sustained high speed, change one increment at a time; for instance, if a BR8ES is recommended, change to a BR9ES. If a hotter plug is required because of sustained low-speed operation, the change would be to a BR7ES.

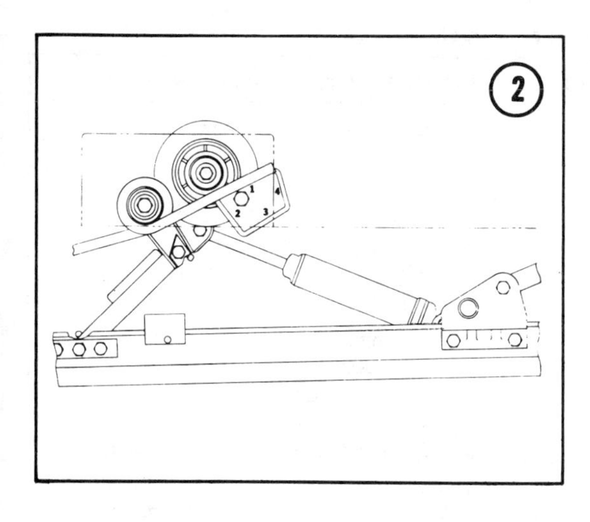

Table 3 BASIC AIR SCREW SETTING MIKUNI CARBURETORS

Model	Turns out
El Tigre (5000, 6000, & Cross Country)	1½
Jag (all models)	1½
Lynx (all models)	1½
Pantera	
1978 fan-cooled	1½
1978 free-air	1
1979	1¾
Panther	
4000	¾
5000	1¼
Cheetah	1¼

CHAPTER FOUR

ENGINE

Critical engine specifications for 1978-1979 models are shown in **Table 4**. Tightening torques are shown in **Table 5**.

LIQUID-COOLED ENGINE (EL TIGRE 6000)

Complete service and repair of the liquid-cooled engine used in the El Tigre 6000 is presented below. Upper-end rebuild (described below) can be carried out without removing the engine from the sled; however, it is much easier to work on a bench.

Removal/Installation

1. Raise the hood. Remove the pins from the clutch guard and remove the guard. Remove the drive belt from the sheaves.
2. Disconnect the springs that connect the expansion chamber to the manifold and remove the chamber.
3. Place a container beneath the joint of the hose and the water inlet manifold. Loosen the clamp and disconnect the hose. Allow the coolant to drain. Disconnect the other hose from the thermostat cover. Disconnect the temperature gauge lead from the cylinder head.
4. Disconnect the high-tension leads from the spark plugs.
5. Remove the air box from the carburetors. Loosen the carburetor clamps and slip the carburetors out of the intake adapters. Disconnect the impulse line from the crankcase.
6. Loosen the clamp that connects the heat exchanger hose to the water pump and disconnect the hose from the pump.
7. Unplug the main electrical harness and the ignition harness.
8. Unscrew the bolts that attach the recoil starter and remove the starter.
9. Unscrew the 4 engine mounting bolts that connect the mounting plate to the tunnel and the angle support.
10. Remove the engine from the sled. If the engine is to be completely disassembled, remove the engine mount plate and the drive clutch.

Table 4 ENGINE SPECIFICATIONS

	PANTERA FC	PANTERA FA	PANTHER 4000	PAN/CHE 5000
Bore	70mm (2.756 in.)	70mm (2.756 in.)	65mm (2.559 in.)	70mm (2.756 in.)
Stroke	65mm (2.559 in.)	65mm (2.559 in.)	65mm (2.559 in.)	65mm (2.559 in.)
Piston ring end gap wear limit	0.20-0.80mm (0.008-0.031 in.)	0.20-0.80mm (0.008-0.031 in.)	0.15-0.80mm (0.006-0.031 in.)	0.20-0.80mm (0.008-0.031 in.)
Piston skirt/ cylinder clearance*	0.05-0.06mm (0.0020-0.0025 in.)	0.05-0.06mm (0.0020-0.0025 in.)	0.05-0.06mm (0.0020-0.0025 in.)	0.05-0.06mm (0.0020-0.0025 in.)
Piston pin diameters	18mm (0.7085-0.7087 in.)	18mm (0.7085-0.7087 in.)	18mm (0.7085-0.7087 in.)	18mm (0.7085-0.7087 in.)
Piston pin bore diameters	18mm (0.7085-0.7089 in.)	18mm (0.7085-0.7089 in.)	18mm (0.7085-0.7089 in.)	18mm (0.7085-0.7089 in.)
Connecting rod small end diameter	23.00-23.03mm (0.9055-0.9067 in.)	23.00-23.03mm (0.9055-0.9067 in.)	23.00-23.03mm (0.9055-0.9067 in.)	23.00-23.03mm (0.9055-0.9067 in.)
Connecting rod radial play	0.02-0.03mm (0.0008-0.0012 in.)	0.02-0.03mm (0.0008-0.0012 in.)	0.02-0.03mm (0.0008-0.0012 in.)	0.02-0.03mm (0.0008-0.0012 in.)
Crankshaft end play	0.05-0.10mm (0.002-0.004 in.)	0.05-0.10mm (0.002-0.004 in.)	0.05-0.10mm (0.002-0.004 in.)	0.05-0.10mm (0.002-0.004 in.)
Crankshaft runout	0.05mm (0.002 in.)	0.05mm (0.002 in.)	0.05mm (0.002 in.)	0.05mm (0.002 in.)
	JAG 2000	**JAG 3000**	**LYNX T**	**LYNX S**
Bore	54mm (2.126 in.)	60mm (2.362 in.)	54mm (2.126 in.)	70mm (2.756 in.)
Stroke	60mm (2.362 in.)	60mm (2.362 in.)	60mm (2.362 in.)	65mm (2.559 in.)
Piston ring end gap wear limit	0.15-0.80mm (0.006-0.031 in.)	0.15-0.80mm (0.006-0.031 in.)	0.15-0.80mm (0.006-0.031 in.)	0.20-0.80mm (0.008-0.031 in.)
Piston skirt/ cylinder clearance*	0.04-0.06mm (0.0015-0.0025 in.)	0.04-0.06mm (0.0015-0.0025 in.)	0.04-0.06mm (0.0015-0.0025 in.)	0.05-0.06mm (0.0020-0.0025 in.)
Piston pin diameters	16mm (0.6297-0.6299 in.)	16mm (0.6297-0.6299 in.)	16mm (0.6297-0.6299 in.)	18mm (0.7085-0.7087 in.)
Piston pin bore diameters	16mm (0.6298-0.6301 in.)	16mm (0.6298-0.6301 in.)	16mm (0.6298-0.6301 in.)	18mm (0.7085-0.7089 in.)
Connecting rod small end diameter	21.00-21.01mm (0.8268-0.8272 in.)	21.00-21.01mm (0.8268-0.8272 in.)	21.00-21.01mm (0.8268-0.8272 in.)	23.00-23.03mm (0.9055-0.9067 in.)
Connecting rod radial play	0.02-0.03mm (0.0008-0.0012 in.)	0.02-0.03mm (0.0008-0.0012 in.)	0.02-0.03mm (0.0008-0.0012 in.)	0.02-0.03mm (0.0008-0.0012 in.)
Crankshaft end play	0.05-0.10mm (0.002-0.004 in.)	0.05-0.10mm (0.002-0.004 in.)	0.05-0.10mm (0.002-0.004 in.)	0.05-0.10mm (0.002-0.004 in.)
Crankshaft runout	0.05mm (0.002 in.)	0.05mm (0.002 in.)	0.05mm (0.002 in.)	0.05mm (0.002 in.)

*The specifications in the above chart are manufacturing tolerances.
The piston skirt/cylinder clearance wear limit for all models is 0.15mm or 0.006 of an inch.

Table 4 ENGINE SPECIFICATIONS

	EL TIGRE 5000	EL TIGRE 6000	EL TIGRE CROSS COUNTRY 340
Bore	70mm (2.756 in.)	68mm (2.677 in.)	60mm (2.362 in.)
Stroke	65mm (2.559 in.)	60mm (2.362 in.)	60mm (2.362 in.)
Piston ring end gap wear limit	0.8mm (0.031 in.)	0.8mm (0.031 in.)	0.8mm (0.031 in.)
Piston skirt/ cylinder clearance*	0.04-0.05mm (0.0015-0.002 in.)	0.04-0.05mm (0.0015-0.002 in.)	0.04-0.05mm (0.0015-0.002 in.)
Piston pin diameters	18mm (0.7085-0.7087 in.)	18mm (0.7085-0.7087 in.)	18mm (0.7085-0.7087 in.)
Piston pin bore diameters	18mm (0.7085-0.7089 in.)	18mm (0.7085-0.7089 in.)	18mm (0.7085-0.7089 in.)
Connecting rod small end diameter	23.00-23.03mm (0.9055-0.9067 in.)	23.00-23.03mm (0.9055-0.9067 in.)	23.00-23.03mm (0.9055-0.9067 in.)
Connecting rod radial play	0.02-0.03mm (0.00067-0.0013 in.)	0.02-0.03mm (0.00067-0.0013 in.)	0.02-0.03mm (0.00067-0.0013 in.)
Crankshaft end play	0.05-0.10mm (0.002-0.004 in.)	0.05-0.10mm (0.002-0.004 in.)	0.05-0.10mm (0.002-0.004 in.)
Crankshaft runout	0.05mm (0.002 in.)	0.05mm (0.002 in.)	0.05mm (0.002 in.)

*The specifications in the above chart are manufacturing tolerances. The piston skirt/cylinder clearance wear limit for all models is 0.15mm or 0.006 of an inch.

Table 5 CRITICAL TORQUE VALUES

Fastener	Torque, Ft.-lb. (N•m)
Cylinder head bolts[1]	13-16 (18-22)
Cylinder base nuts[2]	22-29 (30-40)
Crankcase bolts — 6mm	6-7 (8-10)
Crankcase bolts — 8mm	13-16 (18-22)
Crankcase bolts — 10mm[3]	15-22 (20-30)
Flywheel (stator) nut	49-63 (68-87)
Intake manifold nuts	11-14 (15-19)
Exhaust manifold nuts	11-14 (15-19)
Engine mounting bolts	55 (76)
Spark plug	18-20 (25-28)

1. Jag 2000/3000, Lynx T, El Tigre 6000 and Cross Country = 22-29 (30-40)
2. El Tigre 6000 and Cross Country = 16 (23)
3. El Tigre 6000 and Cross Country = 29 (40)

11. Installation is the reverse of removal. If the engine mounting plate was removed, tighten the mounting bolts to 55 ft.-lb. (75 N•m). Set the engine in the sled and install the cupped washers, mounts, and bolts. Tighten the bolts to 23 ft.-lb. (32 N•m).

12. Install the recoil starter. Pull the rope until the pawls engage, then tighten the bolts to 7 ft.-lb. (10 N•m).

13. Connect the impulse line to the crankcase. Connect the main and ignition electrical plugs. Apply thread sealer to the temperature sender and screw it into the cylinder head. Take care not to overtighten and damage it.

14. Connect the water hoses and tighten the clamps securely. Install the carburetors and the air box. Make sure the clamps are tight to prevent air leaks. Connect the high-tension leads to the spark plugs.

15. Install the exhaust manifold and tighten the nuts to 14. ft.-lb. (19 N•m). Install the expansion chamber and connect the springs. Install the clutch and the clutch guard. Check the clutch alignment and then install the belt.

16. Remove the cap from the radiator and fill the system with a mixture of anti-freeze and water; follow the anti-freeze manufacturer's recommendations for mixing. Check for leaks and correct any that are found.

17. Refer to Chapter Two and service, adjust, and tune up the sled as described.

UPPER-END REBUILD

1. Disconnect the coolant bypass hose from the cylinder head. Remove the thermostat cover and the thermostat (**Figure 3**).

2. Unscrew the cylinder head bolts in a crisscross pattern (**Figure 4**). Use a soft mallet to break the head loose (**Figure 5**) and remove it and the gasket.

3. Remove the coolant intake manifold from the cylinders (**Figure 6**).

4. Unscrew the cylinder base nuts and remove the flat washers and lockwashers. Tap around the cylinder bases with a soft mallet to break the cylinders loose, then remove them by pulling straight up; do not twist or rotate the cylinders as they are being removed or the

3

4

5

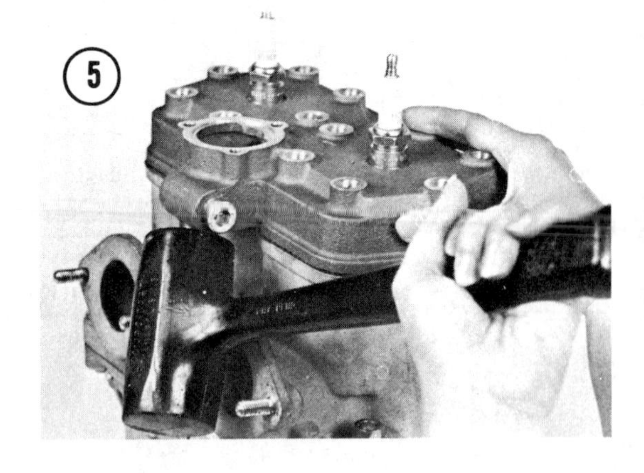

6

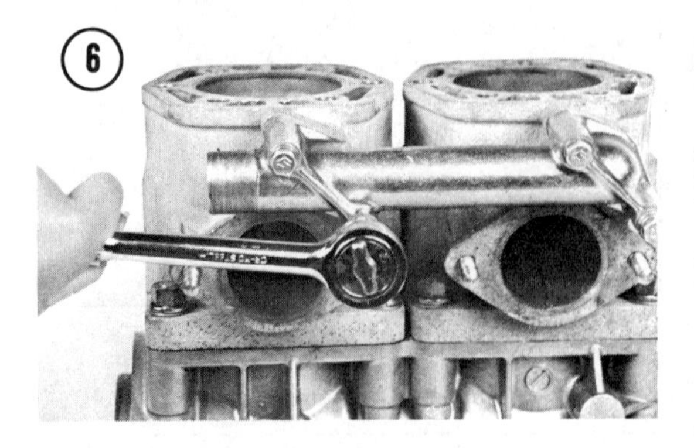

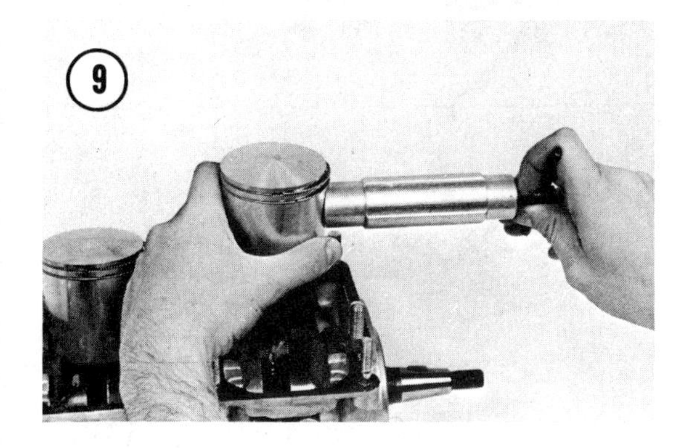

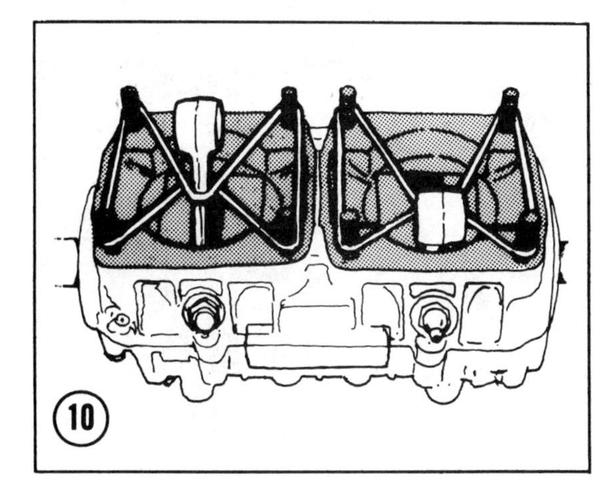

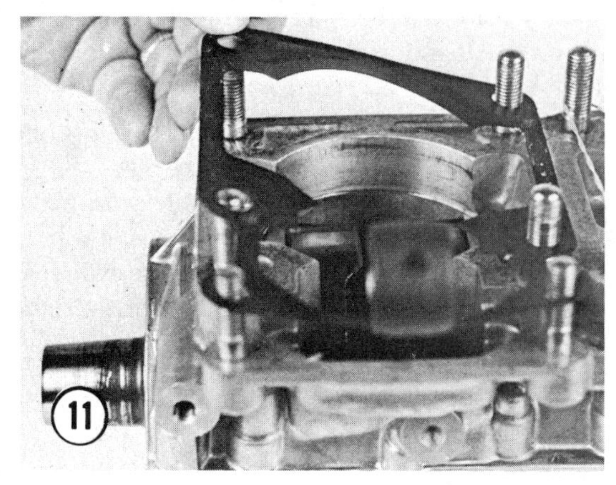

piston rings will likely be broken. Mark the cylinders for location (right or left). See **Figure 7**.

CAUTION

Be careful not to damage the reed valve stop in the cylinder.

5. Place clean shop rags in the crankcase to prevent dirt and small parts from falling in. Remove the circlips from the pistons (**Figure 8**).

6. Mark the pistons for location (right and left) and direction (front). Press out the pins with a pin extractor (**Figure 9**).

7. Secure the connecting rods with rubber bands (**Figure 10**) to prevent them from striking the edges of the crankcase.

8. Refer to *Inspection* which follows, correct any out-of-specification conditions, then proceed with the next step to assemble the upper end.

9. Oil the connecting rod big-end bearing with fresh engine oil. Lightly coat both sides of new cylinder base gaskets with General Electric RTV sealer (or an equivalent). Make certain there are no voids in the sealer around the transfer ports and the reed block. Install the gaskets (**Figure 11**).

10. Oil the connecting rod small end bearings and install them in the rods (**Figure 12**).

11. Install the pistons, making sure location and direction are correct. If the pistons are marked with arrows, they should point to the exhaust port. If they do not have arrows, locate the retaining pin for the top ring toward the intake side. Make sure the pin circlips are com-

pletely seated in their grooves. Install the rings with the letters facing up. Lightly oil the piston and rings.

12. Oil the cylinder bore. Support the piston and slide the cylinder down into place (**Figure 13**). Make sure the rings are lined up with the locating pins in the piston (**Figure 14**), and compress the rings and slowly work the cylinder down over the studs. Install the washers and nuts but do not tighten them yet.

CAUTION

Do not twist or rotate the cylinder while it is being installed; ring damage is likely to result.

13. Coat the water manifold flanges with RTV sealer (**Figure 15**). Install new gaskets and the water manifold. Tighten the nuts to 7 ft.-lb. (10 N•m). Tighten the cylinder base nuts progressively in the pattern shown in **Figure 16**. The small nuts should be tightened to 7 ft.-lb. (10 N•m). Tighten the larger nuts to 16 ft.-lb. (22 N•m).

14. Lightly coat new head gaskets with RTV sealer and install them on the cylinders.

CAUTION

*The large water passage holes (**Figure 17**) must be positioned toward the intake side of the engine. If they are installed incorrectly, serious engine damage is likely to result.*

15. Install the cylinder head, flat washers, lockwashers, bolts, and nuts and tighten them in the pattern shown in **Figure 18** to 29 ft.-lb. (40 N•m).

16. Pressure test the engine as described at the end of this chapter.

17. Install the thermostat with the spring down (**Figure 19**). Coat a new thermostat cover gasket with RTV sealant and install the gasket and cover, with the outlet toward the PTO side of the engine (**Figure 20**). Install the bypass hose on the cylinder head.

18. Continue engine assembly as described in *Removal/Installation* after the engine has been installed in the sled. Refer to Chapter Two and service and tune the engine as described. Fill the cooling system and correct any leaks that are found.

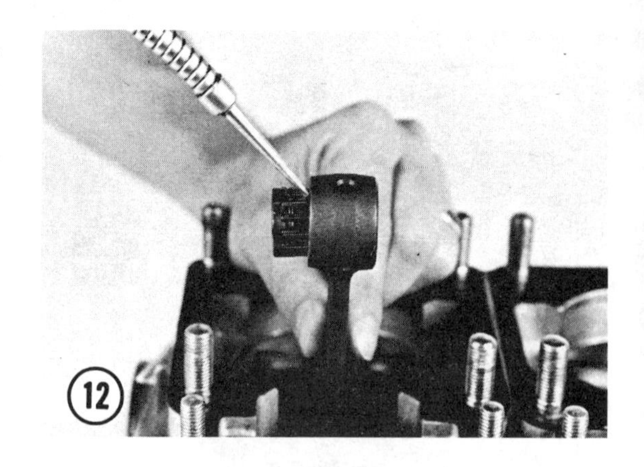

12

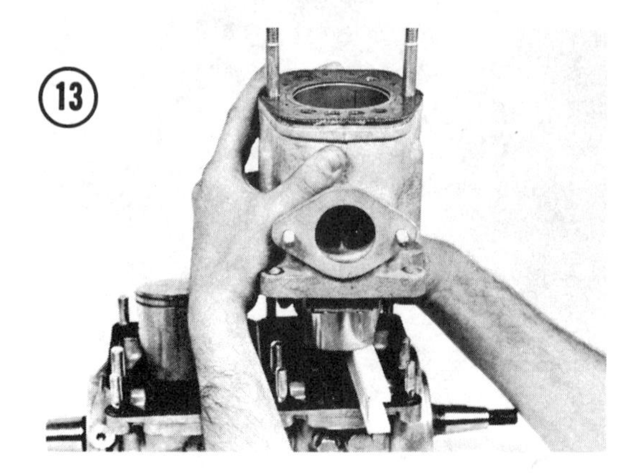

13

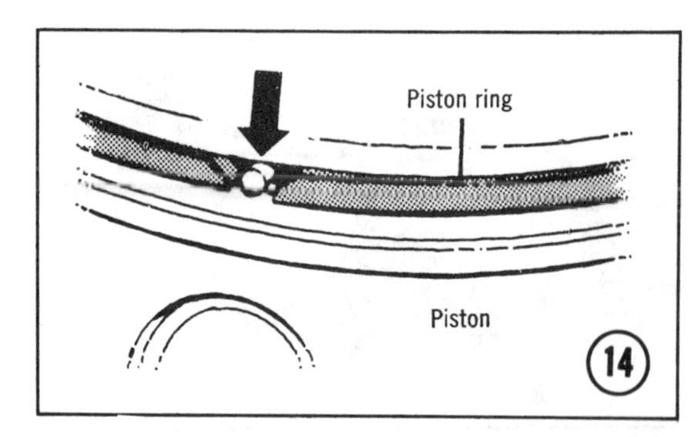

14

15

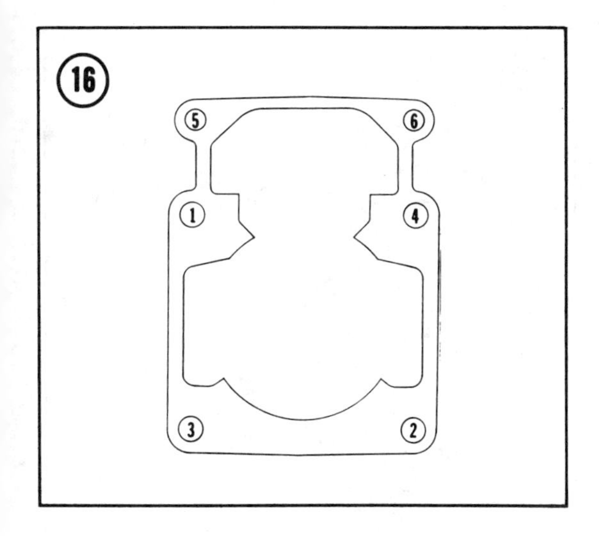

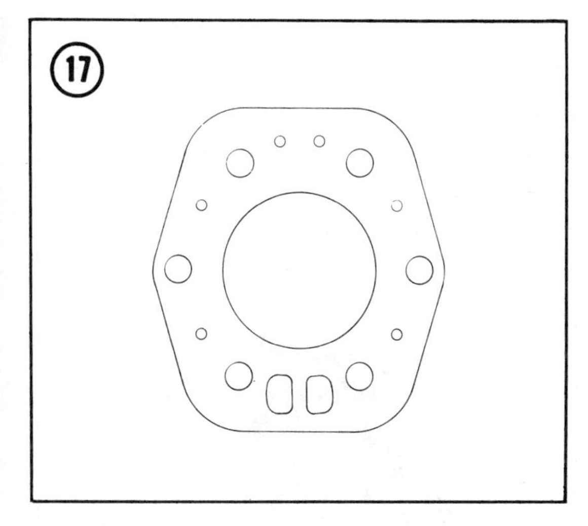

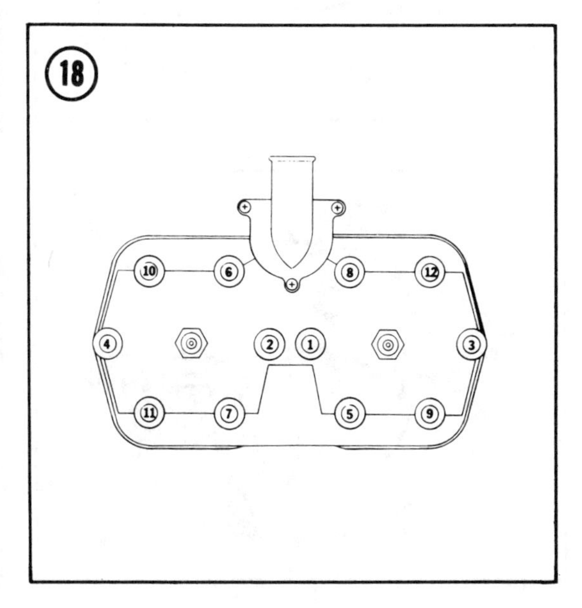

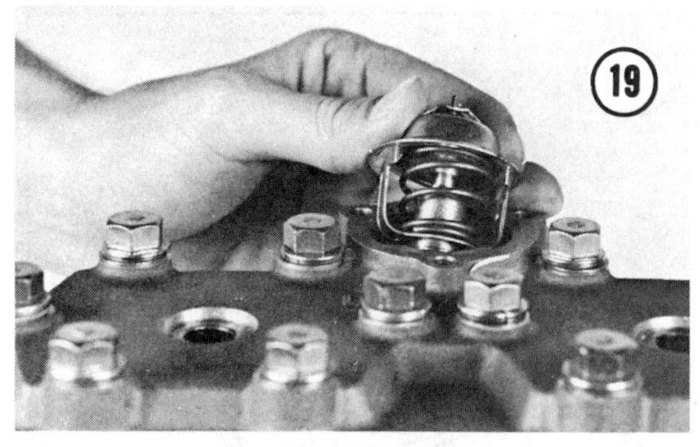

Inspection

1. Scrape carbon from the combustion chambers in the cylinder heads and the exhaust ports in the cylinders using a soft metal (aluminum) or wood scraper. Do not use a hard metal scraper; it will burr the surfaces and create hot spots.

2. Clean the heads and cylinders with solvent and dry them with compressed air if possible.

3. Remove the rings from the pistons and clean the crowns with a soft scraper. Clean the ring grooves with a piece of old piston ring (**Figure 21**). Clean the piston with solvent and dry it with compressed air.

4. Check the flatness of the cylinder head on a surface plate or piece of glass (**Figure 22**). If the head does not make contact over the entire sealing surface, it must be trued; this is a job for an expert.

5. Check the cylinders and pistons for wear, galling, scuffing, and burning. Minor irregularities can be removed from the pistons with crocus cloth and light oil. The cylinders may be cleaned up with a light honing, provided they are within the specifications described below.

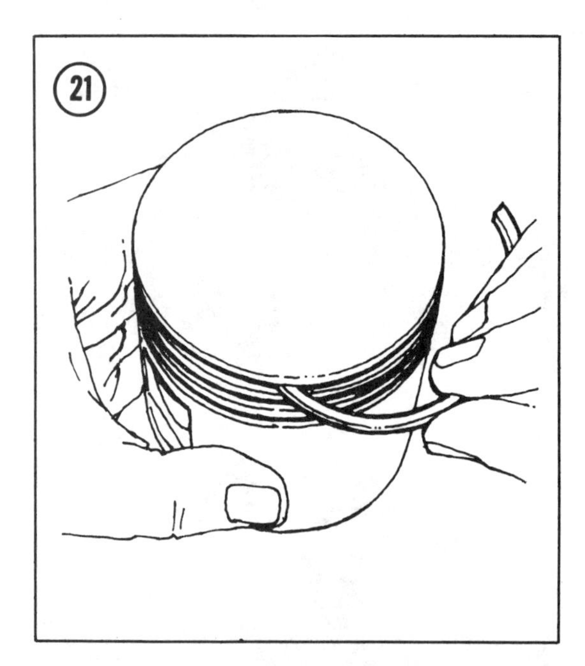

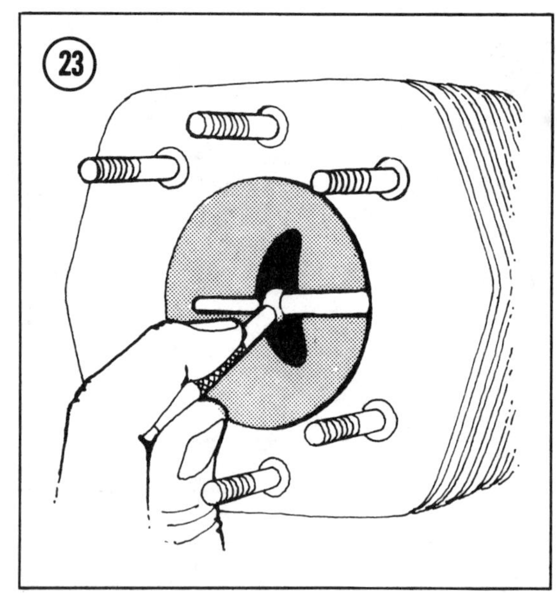

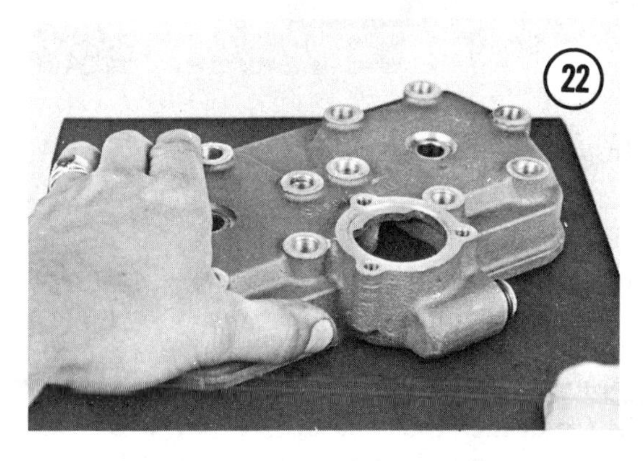

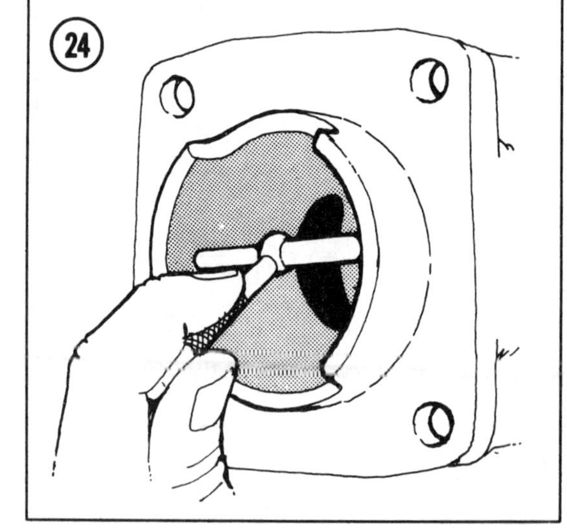

6. Measure the cylinder bore with an inside micrometer or cylinder gauge. Measure ⅜ in. below the top of the cylinder, in 2 locations, 90 degrees apart (**Figure 23**). If the 2 measurements differ by more than 0.002 in. (0.05mm), the cylinder is out of round beyond specification and must be bored or replaced. Measure again in 2 locations 90 degrees apart just above the intake port. If these measurements differ from the first measurements by more than 0.002 in. (0.05mm), the cylinder taper is beyond specification and the cylinder must be bored or replaced. If one cylinder is out of specification, both must be bored or replaced and new pistons fitted.

7. Check the piston skirt-to-cylinder clearance by first measuring the cylinder bore front to

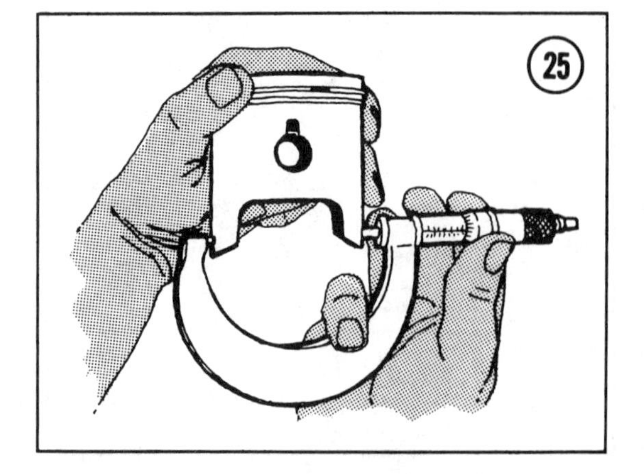

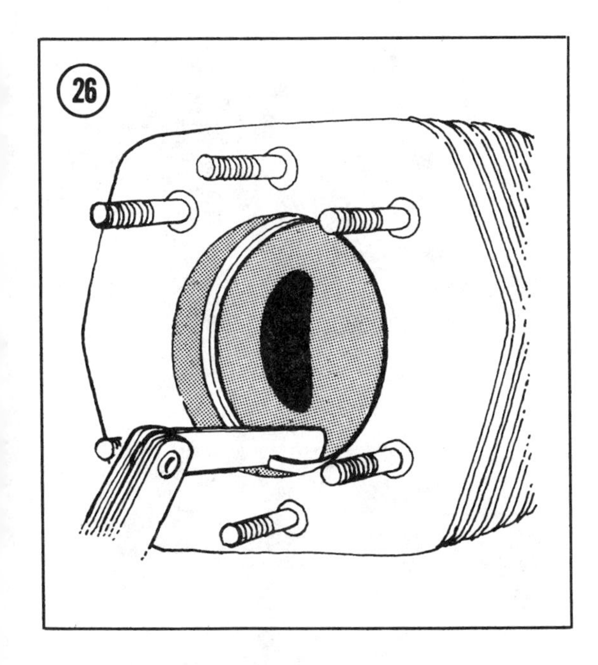

back (**Figure 24**) at the base. Then, measure the piston skirt, front and rear, about ¼ in. from the bottom with an outside micrometer (**Figure 25**). Subtract the piston measurement from the cylinder measurement to determine the actual piston skirt-to-cylinder clearance (see **Table 4**). If the clearance is excessive but the cylinder dimensions are acceptable, the pistons should be replaced with new units that include matched pins and bearings as well as rings.

8. Measure the piston ring end gap. Place each ring into its respective cylinder, ⅜ in. from the top. Use the piston to square the ring with the bore by pressing down on the ring with the piston skirt. Measure the gap as shown in **Figure 26**. See **Table 4**. If the gap is too large for any one ring, replace them all as a set and check the gap of the new rings. If the gap is too small, carefully file the rings (**Figure 27**) until they are correct.

9. Inspect the piston pin for galling at the ends — an indication that the pin is rotating in the piston. If this type of damage is apparent, the piston, pin, and small-end bearing must be replaced as a set. (The pin bores in the piston will no doubt be similarily damaged.)

10. Set the bearing and pin in the small end of each connecting rod and check for rocking movement, back and forth and up and down (**Figure 28**). If movement is apparent, recheck with a new bearing and pin. If movement is still apparent, the rod is worn and must be replaced. This is a job for a dealer.

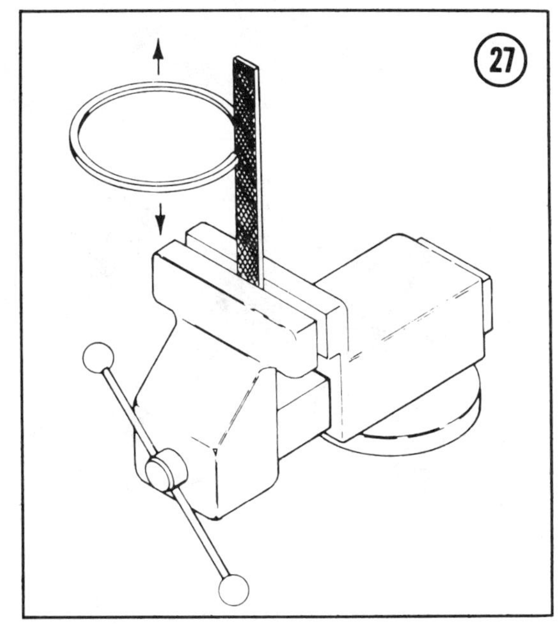

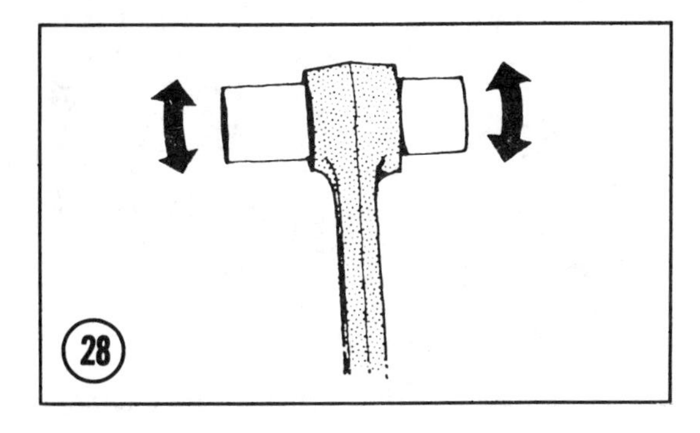

LOWER-END REBUILD

Lower-end rebuild as described here consists of removal and installation of the crankshaft assembly. Sophisticated tools and much experience are required to accurately check the condition of the crankshaft assembly. In addition, a hydraulic press and special knowledge are required to disassemble, assemble, and accurately align the crankshaft assembly. It is recommended that the crankcase halves be parted and the crankshaft assembly be entrusted to a dealer for inspection and repair.

Seal replacement should also be entrusted to a dealer. A leaking crankshaft seal can cause

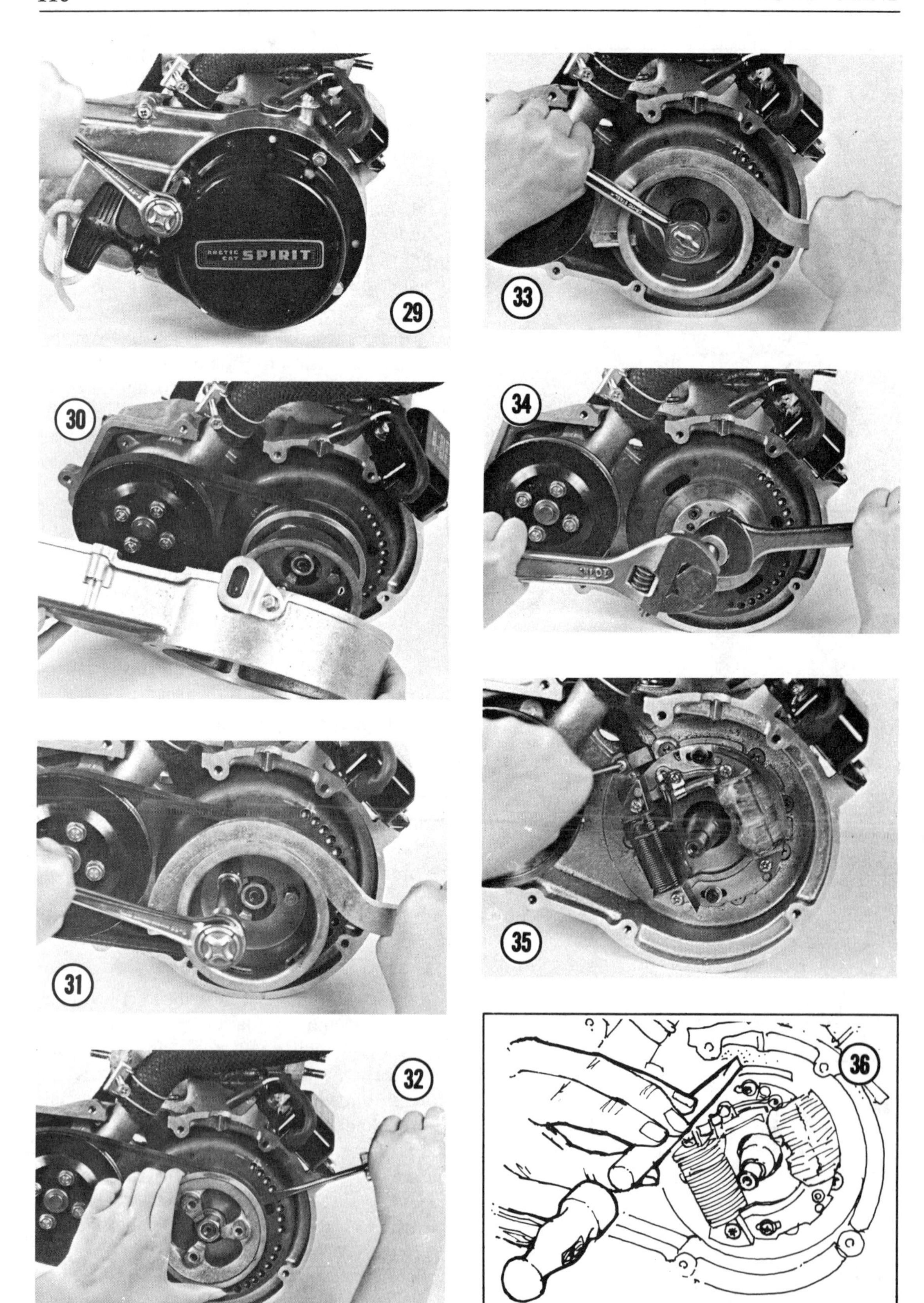
ARCTIC CAT SPIRIT
29
30
31
32
33
34
35
36

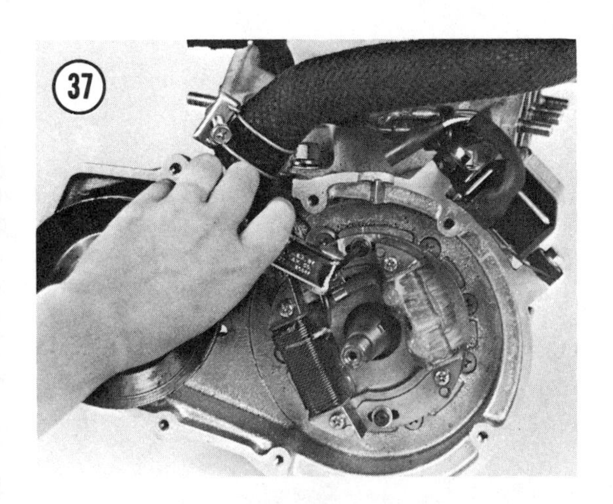

(37)

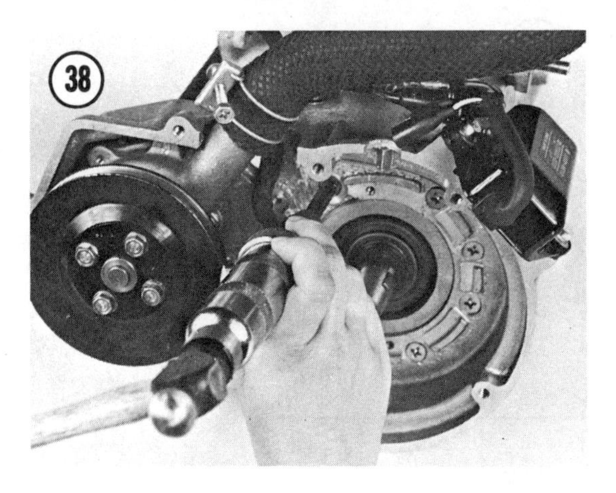

(38)

the engine to run lean and create a situation that could result in severe damage; if it is leaking severely, it will prevent the engine from running at all.

Disassembly

1. Refer to *Removal/Installation* and *Upper-End Rebuild* and remove the engine from the sled and disassemble to upper end.

2. Remove the recoil starter (**Figure 29**). Note the position of the handle relative to the engine so that it can be installed in the same location.

3. Remove the magneto cover (**Figure 30**).

4. Hold the starter drum to prevent it from turning and unscrew the bolts (**Figure 31**). Remove the drum.

5. Carefully pry the pump drive pulley off the magneto rotor (**Figure 32**). Remove the belt.

6. Reinstall the starter drum, install the bolts, hold the drum to prevent it from turning, and

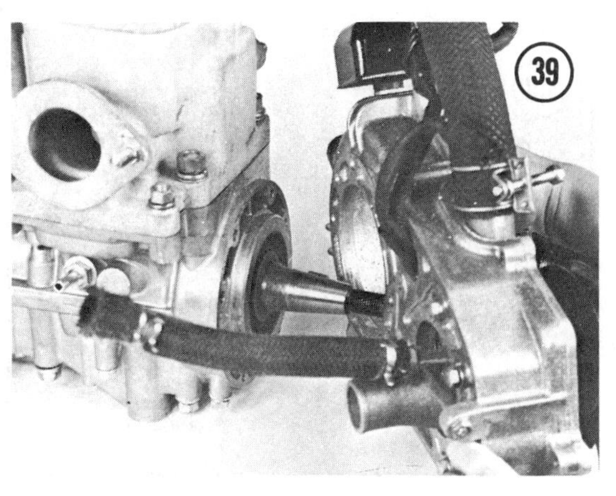

(39)

unscrew the rotor bolt (**Figure 33**). Remove the starter drum.

CAUTION

Do not run the starter drum bolts all the way in; they could damage the stator coils.

7. Install the puller on the flywheel (Arctic Cat part No. 0144-112), hold the body to prevent it from turning, and tighten the puller bolt (**Figure 34**) to remove the flywheel. It may be necessary to rap on the head of the puller bolt with a hammer. Remove the flywheel, wrap it in clean newspaper, and set it out of the way.

8. Remove the wiring harness clamp (**Figure 35**). Scribe an alignment mark on the stator and the magneto housing (**Figure 36**) for reference during assembly. Unscrew the Allen screws that attach the stator to the housing (**Figure 37**) and remove the stator. Wrap it in clean newspaper and set it aside.

9. Loosen the magneto housing screws with an impact driver (**Figure 38**), unscrew them, and remove the housing. It will probably be necessary to tap the housing loose with a rubber mallet (**Figure 39**).

10. Unscrew the 16 bolts that hold the crankcase halves together (**Figure 40**). Carefully tap the crankcase halves apart with a rubber mallet (**Figure 41**).

CAUTION

Do not pry the case halves apart with a chisel, screwdriver, or similar tool; the sealing surfaces are easily damaged.

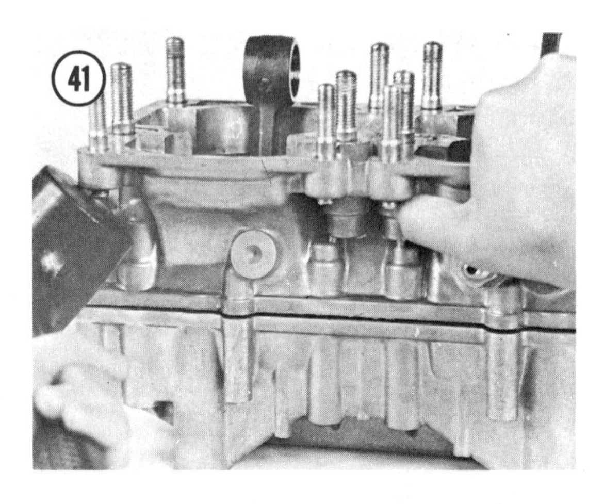

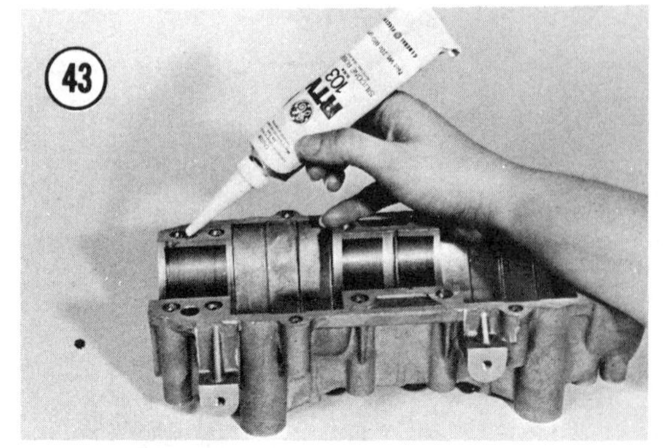

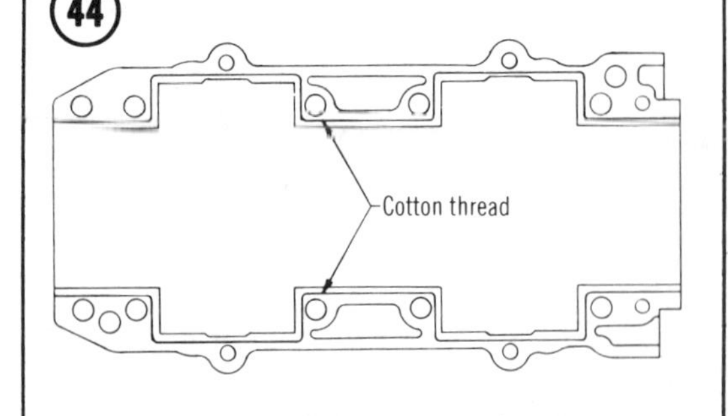

11. Lift the crankshaft assembly out of the lower case half (**Figure 42**). Remove the half-ring from the case; this piece controls crankshaft end-float.

12. Refer the crankcase and crankshaft assembly to a dealer for inspection and repair. Then, assemble the lower end as described below.

Assembly

1. Check the lower crankcase half to make sure all of the bearing locating dowels are in place. Apply a thin, uniform coat of RTV sealer to the mating surface of the crankcase lower half (**Figure 43**). Lay a strand of No. 50 cotton thread along the sealing surface, inside the bolt hole pattern (**Figure 44**). This will ensure a good seal of the 2 halves.

CAUTION

Use only cotton thread; synthetic thread may damage the sealing surface.

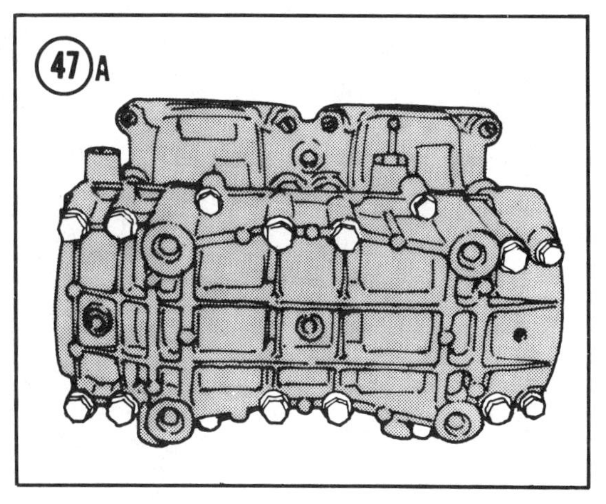

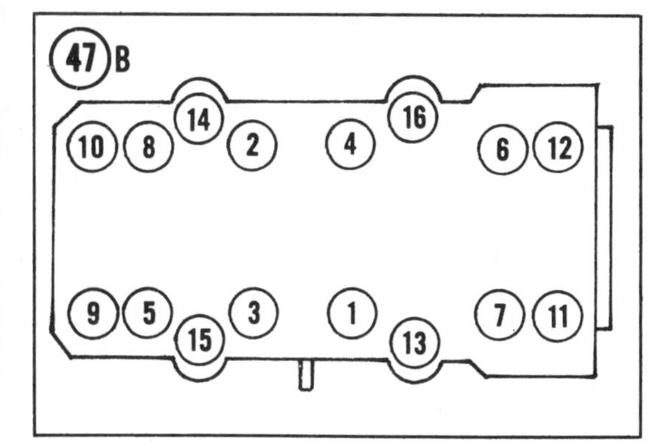

2. Set the half ring in the lower case. Line up the holes in the bearings so they will seal on the locating dowels in the case (**Figure 45**).

3. Grease the recessed side of new end seals and slide them onto the crankshaft with the open side (spring side) facing in toward the bearings.

4. Carefully set the crankshaft assembly in the lower case half (**Figure 46**), making sure the dowels engage the holes in the bearings and the half-ring engages the groove.

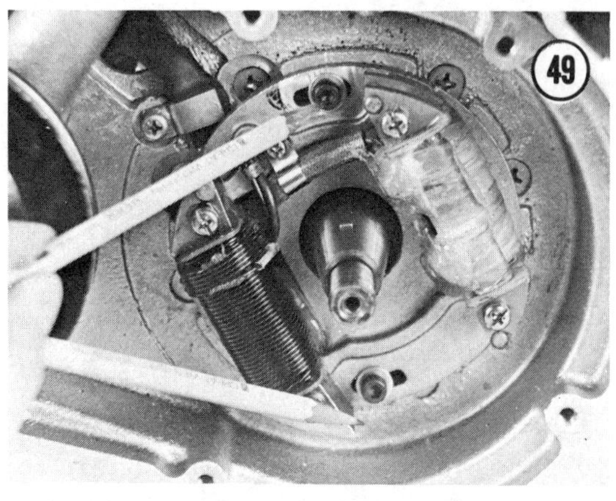

5. Set the top half of the crankcase in place, making sure it is in contact with the lower half, all along the sealing surfaces. Then, turn the assembled cases on their side and install the through-bolts and washers. Screw them evenly, finger-tight (**Figure 47**). Then, tighten them in the pattern shown to the following values:

6mm bolt = 7 ft.-lb. (10 N•m)
8mm bolt = 16 ft.-lb. (22N•m)
10mm bolt = 29 ft.-lb. (40N•m)

6. Install the magneto case. It may be necessary to tap it into place with a rubber mallet. Clean the threads of the 8 case screws and apply Loctite to them. Screw them in and set them with an impact screwdriver (**Figure 48**).

7. Set the stator in position and align the scribe marks made during disassembly. Screw in and tighten the 2 Allen screws (**Figure 49**). Install the wiring harness clamp.

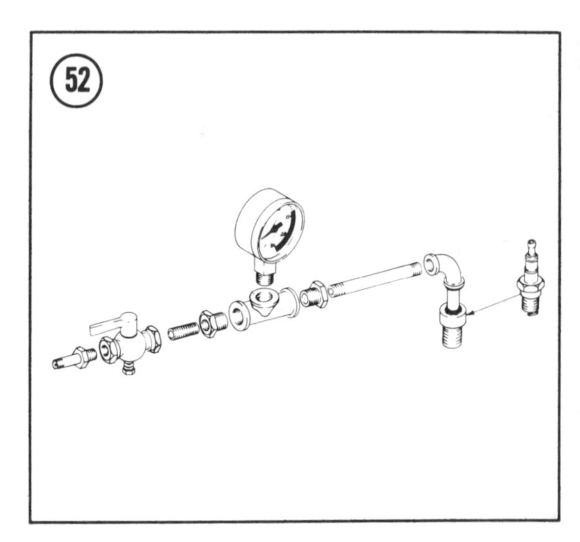

8. Clean the stator coils and the inside of the rotor to ensure there is no dirt to magnetically attract metal particles. Set the key in the crankshaft, install the rotor (flywheel), spacer, lockwasher, and nut. Tighten the flywheel nut to 65 ft.-lb. (90 N•m).

9. Install the pump drive belt and the drive pulley. Install the starter drum and tighten the bolts to 7 ft.-lb. (10 N•m). See **Figure 50**. Check the deflection of the belt; it should be ¼ in. (6mm) midway between the 2 pulleys. If adjustment is required, turn the bolt that is located behind the water pump.

10. Install the magneto cover (**Figure 51**). Install the recoil starter with the handle oriented as it was before.

11. Refer to *Upper-End Rebuild* and *Removal/Installation* described earlier and complete the assembly and installation of the engine. Refer to Chapter Two and service and tune the engine as described.

ENGINE PRESSURE TESTING

It's essential that a two-stroke engine be leak-free. At the least, an air leak will cause poor performance; at the worst, it can cause severe engine damage. Pressure testing is relatively simple but requires some special hardware. Complete kits are available from your Arctic Cat distributor (part No. 0144-127) or Azusa Engineering, Inc., P.O. Box N, Azusa, CA 91702 (part No. 3800). As an alternative, a kit can be assembled from simple plumbing hardware (**Figure 52**). In addition to the pieces shown, you will need plates and gaskets to seal the intake and exhaust flanges.

1. Plug the intake and exhaust ports.

2. Connect the tester to the engine and make sure that all openings, such as spark plug holes, ports, impulse line fitting, etc., are closed.

3. Pressurize the system to 12 psi and close the valve on the inlet line.

CAUTION

Do not pressurize the system beyond 15 psi or the crankcase seals will be damaged.

4. Watch the gauge and time the rate of pressure drop. If it drops faster than 1 pound/minute/cylinder there is an unacceptable leak. Check all sealing surfaces and openings with soapy water and a brush. Bubbles will form in the area of a leak.

CHAPTER FIVE

FUEL SYSTEM

Basic air screw settings for Mikuni carburetors installed on 1978-1979 models are shown in **Table 3**.

REED VALVE (EL TIGRE)

The reed valve assembly should be inspected at least annually or when decreased performance indicates that a reed petal may have broken.

1. Remove the cylinders from the engine as described in Chapter Four of this supplement (*Upper-End Rebuild*).

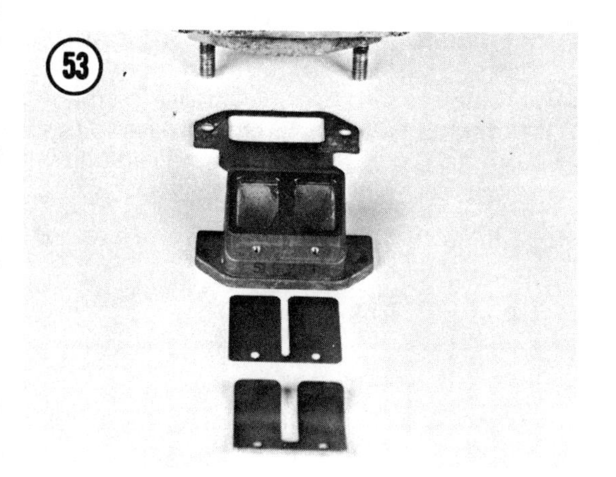

2. Remove the screws that hold the reed block in the cylinder and remove the reed block.

3. Disassemble the reed block, reed, and reed stopper **(Figure 53)**.

4. Inspect the rubber seating area of the block for signs of wear and deterioration. Replace the block if it's not in good condition.

5. Inspect the reed petals for cracks and bending and replace them if they are less than perfect. After they have been inspected, coat them lightly with oil to retard corrosion.

6. Install the reed with the bevelled corner in the lower right corner of the reed block **(Figure 54)**.

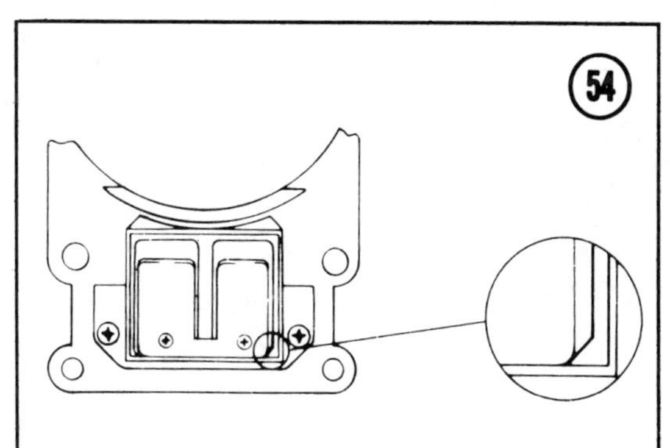

CHAPTER SEVEN

POWER TRAIN

Drive belt specifications for 1978-1979 models are shown in **Table 1.**

INDEX

M

O

P

R

S

T

V

W

NOTES

NOTES

Clymer Collection Series

VINTAGE

SNOWMOBILES

VOLUME I

ARCTIC CAT, 1974-1979

➢ KAWASAKI, 1976-1980

JOHN DEERE, 1972-1977

CONTENTS

QUICK REFERENCE DATA

LUBRICATION INTERVALS

Interval	Item	Lubricant
Every 20 hours	Lubricate control cables	LPS, Break-Free, WD40, Dri-Slide
	Check chaincase oil level	Dexron II automatic transmission fluid
	Check gearcase oil level (SS models)	SAE 10W-30 motor oil
Annually	Change chaincase oil	Dexron II automatic transmission fluid
	Change gearcase oil	SAE 10W-30
	Lubricate steering bushings	Low-temperature grease
	Lubricate suspension axles	Low-temperature grease
	Lubricate recoil starter	Low-temperature grease
	Lubricate ski spindles	Low-temperature grease
	Lubricate driven sheave shaft and sliding sheave	Low-temperature grease
	Lubricate tie-rod ends	Low-temperature grease

ENGINE FASTENER TORQUE

Model	Cylinder Head	Flywheel	Engine Mounts	Drive Sheave
SA and SB	2.2 mkg (16 ft.-lb.)	8.3 mkg (60 ft.-lb.)	2.2 mkg (16 ft.-lb.)	7.6-8.3 mkg (55-60 ft.-lb.)
SS and ST	2.2 mkg (16 ft.-lb.)	8.3 mkg (60 ft.-lb.)	2.2 mkg (16 ft.-lb.)	8.3-1.0 mkg (60-70 ft.-lb.)

IGNITION SPECIFICATIONS

Model	Spark Plug (NGK)	Gap, mm (in.)	Static Timing, mm (in.) BTDC	Dynamic Timing, °BTDC @ rpm
SA340-A1	BR-8ES	0.6 (0.024)	3.53 (0.139)	25° @ 6,500
SA440-A1	BR-8ES	0.6 (0.024)	3.53 (0.139)	25° @ 6,500
SB340-A1	B-9EV	0.6 (0.024)	1.6 (0.063)	17° @ 6,500
SB440-A1	B-9EV	0.6 (0.024)	1.6 (0.063)	17° @ 6,500
SA340-A2	BR-8ESA	0.8 (0.032)	3.53 (0.139)	25° @ 6,500
SB440-A2	BR-9EV	0.6 (0.024)	1.86 (0.073)	18° @ 6,500
SA340-A3 and later	BR-8ESA	0.8 (0.032)	0.145 (0.006)	25° @ 6,500
SB440-A3 and later	BR-9EV	0.6 (0.024)	1.86 (0.073)	18° @ 6,500
SS340 all	BR-9EV	0.6 (0.024)	2.75 (0.108)	22° @ 6,500
ST440 all	BR-9EV	0.6 (0.024)	1.86 (0.073)	18° @ 6,500

CARBURETOR SETTINGS

Model	Idle Speed, rpm	Air Screw, turns out	Throttle Stop Screw, turns out
SA340-A1	2,000	1.5	3
SA440-A1	2,000	1.5	3
SB340-A1	2,000	1.5	3
SB440-A1	2,000	2	6
SA340-A2	2,000	1.5	3
SB440-A2	2,500	1.5	3
SA340-A3 and later	2,000	1.5	6
SB440-A3 and later	2,500	1.5	6
SS340 all	3,000	1.5	3
ST440 all	3,000	1.5	3

DRIVE BELT SPECIFICATIONS

Model	Width (Top Surface)		Outside Circumference
SA and SB	New	31.7 mm (1¼ in.)	1,105 mm (43.5 in.)
	Service limit	26 mm (1.02 in.)	
SS and ST	New	31.7 mm	1,184 mm (46.6 in.)
	Service limit	26 mm (1.02 in.)	

- NOTES -

CHAPTER ONE

GENERAL INFORMATION

Snowmobiling has, in recent years, become one of the most popular outdoor winter recreational pastimes. It provides an opportunity for the entire family to experience the splendor of winter and enjoy a season previously regarded by many as miserable.

Snowmobiles also provide an invaluable service in the form of rescue and utility vehicles in areas that would otherwise be inaccessible.

As with all sophisticated machines, snowmobiles require specific periodic maintenance and repair to ensure their reliability and usefulness.

MANUAL ORGANIZATION

This manual provides information for periodic maintenance, tune-up, and general repair for Kawasaki snowmobiles manufactured in 1976 and later.

This chapter provides general information and hints to make all snowmobile work easier and more rewarding. Additional sections cover snowmobile operation, safety, and survival techniques.

Chapter Two provides all periodic maintenance and tune-up information required to keep your snowmobile in top running condition.

Chapter Three provides methods and suggestions for finding and fixing trouble fast. The chapter also describes how a 2-cycle engine works to help you to analyze trouble logically. Troubleshooting procedures discuss typical symptoms and logical methods to pinpoint trouble.

Subsequent chapters describe specific systems such as the engine, fuel and exhaust systems, electrical system, and power train. Each chapter provides disassembly, inspection, repair, and assembly procedures in easy-to-follow, step-by-step form.

Tables are found at the ends of the chapters.

If a repair is impractical for the owner/mechanic, it is so indicated. Usually, such repairs are quicker and more economically done by a Kawasaki dealer or other competent snowmobile repair shop.

Some of the procedures described in this manual call for special tools. In all cases, the tool is illustrated, often in actual use.

Throughout the manual the words NOTE, CAUTION, and WARNING are used when special emphasis is required; each has a very specific meaning.

A NOTE provides additional information to make a step or procedure easier or clearer. If you disregard a NOTE you may be inconvenienced, but damage to the machine or personal injury would not occur.

A CAUTION emphasizes areas where equipment damage could result. If you disregard a CAUTION mechanical damage may occur but persona injury is unlikely.

A WARNING emphasizes areas where personal injury or death could result from a disregard for the information. Also, mechanical damage is likely to occur. Take all WARNINGS seriously; they could mean your life.

MACHINE IDENTIFICATION AND PARTS REPLACEMENT

Each snowmobile has a serial number applicable to the machine and a model and serial number for the engine.

The machine serial number is located on a plate riveted to right side of the body (**Figure 1**). The engine model and serial number are stamped on the upper crankcase half, on the right side (**Figure 2**).

Write down all serial and model numbers applicable to your machine and carry them with you when you order parts from a dealer. Always order by year and engine and machine numbers. If possible, have the old parts with you so you can compare them with the new parts. If the parts are not alike, have the parts man explain the reason for the difference and insist upon assurance that the new parts are correct and will fit.

OPERATION

Lubrication

Kawasaki SS and ST models are equipped with oil injection lubrication and require no pre-mixing of oil and gasoline. Engine lubrication for SS and ST models is described in Chapter Two (see *Pre-Ride Checks*).

SA and SB models are lubricated with oil mixed with the fuel.

WARNING

Serious fire hazards always exist around gasoline. Do not allow smoking in areas where fuel is mixed or when you are refueling your snowmobile.

Always use fresh gasoline; gasoline loses its potency after sitting for a period of time. Old fuel can cause engine failure and leave you stranded in severe weather.

Correct fuel/oil mixing is essential for the life and efficiency of the engine. Always mix fuel and oil in exact proportions; don't guess. An oil-lean mixture can cause serious and expensive engine damage, and an oil-rich mixture can cause poor performance and fouled spark plugs which make the engine difficult or impossible to start.

Use gasoline with an octane rating of 89 or higher. Mix the gasoline and oil in a separate container — not in the machine's fuel tank. The mixing tank should be larger than the amount of gasoline and oil being mixed to permit room for them to agitate and mix thoroughly.

Use Kawasaki Snowmobile Oil or an equivalent.

NOTE: *Acceptable oil will be marked with* ***B. I. A. Certified C. T. W.*** *on the top of the can.*

Use the appropriate fuel/oil ratio shown in **Table 1 or Table 2.**

1. Pour the required amount of oil into a clean and empty mixing can.
2. Add ½ the gasoline required and mix thoroughly after capping the mixing can. Gently shake the can back and forth, side to side, and up and down for a couple of minutes.
3. Add the remainder of the required amount of gasoline and mix thoroughly once again.

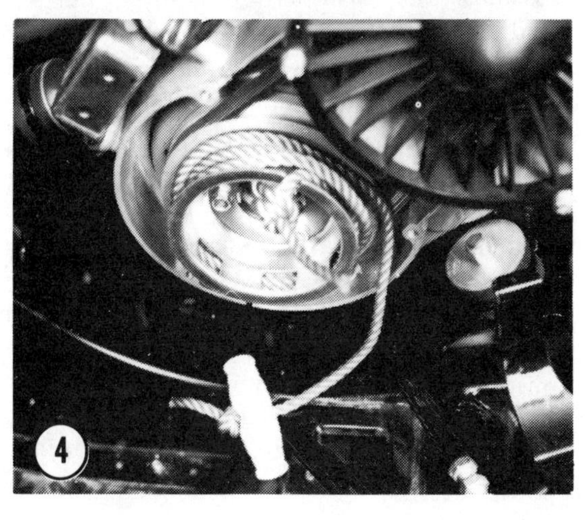

4. Fill the snowmobile tank with "mix" using a funnel equipped with a fine-mesh filter screen.

Emergency Starting

Always carry a small tool kit with you, and carry an extra rope for emergency starting.

Open the hood and remove the recoil starter assembly **(Figure 3)**. Wind the emergency starting rope around the pulley **(Figure 4)** and pull it to crank the engine.

Emergency Stopping

To stop the engine in an emergency, pull the tether switch or, turn the emergency kill switch to STOP or OFF.

Towing

When preparing for a long trip, pack extra equipment in a sled; don't carry it on the machine. A sled is also ideal for transporting small children.

WARNING
Never tow a sled with ropes or pull straps. The use of ropes or flexible straps could result in a tailgate accident if you stop suddenly or travel down long, steep hills.

If it is necessary to tow a disabled snowmobile, securely fasten the disabled machine's skis to the hitch of the tow machine. Remove the drive belt from the disabled machine and tow it at low and moderate speeds. *Don't tow a disabled machine with a rope or strap; with the drive belt removed the brake is inoperative on most machines.*

Clearing the Track

If the machine has been operated in deep or slushy snow, clear the track after stopping to prevent it from freezing and making starting and running difficult later on.

WARNING
Make sure no one is behind or in front of the machine when you are clearing

the track. Ice and rocks thrown from the track can cause injury.

Tip the machine onto its side until the track clears the snow *completely*. Run the engine at a moderate speed until all the ice and snow is thrown clear.

CAUTION
If the track does freeze, break it loose by hand. If you attempt to break it loose with the engine you will burn and damage the drive belt.

Proper Clothing

Warm and comfortable clothing is essential to provide protection from frostbite. Even mild temperatures can be very uncomfortable and dangerous when combined with strong wind or when travelling at high speed (see **Table 3** for wind chill factors). Always dress according to what the wind chill factor is — not for the air temperature alone.

It's best to layer your clothing so that as the temperature and wind chill factor change, clothing can be removed or added to prevent perspiring or to provide additional warmth.

WARNING
Always wear an approved safety helmet. Not only does it protect against head injury, but it will also provide added warmth. And wear a full face shield to prevent frostbite of the face.

SERVICE HINTS

All procedures described in this book can be performed by anyone reasonably handy with tools. Special tools are required for some procedures, and their operation is described or illustrated. These may be purchased through your dealer; however, it's a good idea to anticipate your needs and order the tools beforehand because not many dealers stock special tools in their parts department.

Service will be far easier if the machine is clean before you begin work on it. There are special cleaners for washing the engine and related parts. Just spray on the cleaning solution, let it stand, then rinse it away with clear water. Clean all oily or greasy parts with cleaning solvent as they are removed.

WARNING
Never use gasoline as a cleaning solvent. It represents an extreme fire hazard. Work in a well-ventilated area and keep a fire extinguisher handy, just in case.

Observe the following practices and you will save time, effort, and frustration as well as prevent possible expensive damage.

1. Tag all internal parts for location and mark all mating parts for position. Small parts such as bolts can be identified by placing them in plastic bags and sealing and labeling them with masking tape.

2. Frozen or very tight bolts and screws can often by loosened by soaking the threads with penetrating oil such as Liquid Wrench, Break-Free, or WD-40, then sharply striking the bolt head a few times with a hammer and punch (or an impact driver for screws).

NOTE: *Liquid Wrench and WD-40 are available through most hardware and auto parts stores. Break-Free can be found at many gun shops.*

Avoid using heat on stubborn fasteners unless absolutely necessary; it may melt, warp, or remove the temper from parts.

3. Avoid flames or sparks when working near flammable liquids such as gasoline.

4. No parts except those assembled with a press fit require unusual force during disassembly or assembly. If a part is hard to remove or install, find out why before proceeding.

5. Cover all openings after removing parts to keep dirt, small tools, etc., from falling in.

6. Clean all parts as they are removed and keep them separated into subassemblies. The use of trays, muffin tins, or cans for segregating related parts will make reassembly much easier.

7. Make diagrams whenever similar-appearing parts are found. You may *think* you can remember where everything came from — but mistakes are costly and time consuming. There is also the possibility that you may be side-tracked and not return to work on the machine for days or even weeks, during which time

carefully laid out parts may become disturbed. And as you remove parts that are not quite symmetrical, such as a spacer with a chamfer on one side or a gear that is flat on one side and shouldered on the other, make a note of how the part is oriented to prevent it from being installed incorrectly later on.

8. Pay particular attention to the location of shims and washers as they are removed. Whenever a rotating part butts against a stationary part, look for a shim or washer.

9. Ask your dealer for recommendations about replacing gaskets. Some gaskets can be reused, but if there is any doubt about a gasket's condition, replace it. And have the new gaskets on hand before beginning work. It's a good idea to apply gasket cement to only one side of a gasket (unless otherwise specified) so parts may be easily disassembled in the future. Only a very light coat of gasket cement is required.

10. Heavy grease can be used to hold small parts in place if they tend to fall out during assembly. However, keep grease away from electrical and brake components.

11. High spots and roughness may be sanded off pistons with fine sandpaper, but emery cloth and oil do a more professional job.

12. Wiring should be tagged with masking tape and marked as each wire is removed or disconnected.

13. Carburetors are best cleaned by disassembling them and soaking the metal parts in a commercial carburetor cleaner. Omit gaskets, O-rings, and other non-metal parts from the solution, otherwise they will be damaged. Never use wire or fine drill bits to clean jets and passages; the parts are easily damaged. Use compressed air to blow out the carburetor, but only after the float and needle valve have been removed.

14. Work carefully in a clean, well-lighted area and allow sufficient time to do a good job.

15. Read the entire section in this handbook that pertains to a specific job before beginning work. Study the illustrations and the related text until you have a good idea of what is involved. Many procedures are complicated and errors can be costly. When you fully understand what is to be done, follow the prescribed procedure step-by-step.

16. Remember that a newly rebuilt engine must be broken in just as though it were new. Excessive loads or sustained high engine speeds can damage new parts if they haven't had sufficient time to seat in. After a "fresh" engine has been run for several hours, check critical fasteners for tightness.

TOOLS

Carry a small tool kit in the machine so minor adjustments and emergency repairs can be made on the trail.

A basic assortment of hand tools is required to perform repair and service tasks described in this handbook. The list that follows represents a minimum requirement.

a. Metric combination wrenches
b. Metric socket wrenches
c. Assorted screwdrivers (slot and Phillips types)
d. Pliers
e. Flat feeler gauges
f. Ignition gauge
g. Spark plug wrench
h. Hammer
e. Plastic or rubber mallet
j. Parts cleaning brush
k. Fine wire brush

If you are purchasing tools, always buy good quality tools. They cost more initially but in most cases will last a lifetime. And remember that the initial expense of new tools is easily offset by the money saved on one or two repair jobs.

Tune-up and troubleshooting require a few special tools. All of the special tools described below are used in this manual, however, all tools are not required for all machines. Read the procedures applicable to your machine to determine what your special tool requirements are.

Ignition Gauge

This tool combines round wire spark plug gap gauges with narrow breaker point feeler gauges (**Figure 5**). The gauge costs about $3 at auto parts stores.

Impact Driver

The impact driver (**Figure 6**) may prove to be one of the most valuable tools in your shop. It makes removal of screws easy and eliminates damaged screw slots. When fitted with a socket wrench, it can be used to shock stubborn nuts and bolts loose. A good impact driver costs about $12 at auto parts and hardware stores.

Hydrometer

A hydrometer measures the state of charge of a battery and tells much about the battery condition. See **Figure 7**. A suitable hydrometer costs just a few dollars and is available through auto parts stores.

Multimeter (VOM)

A VOM is invaluable for electrical system troubleshooting and service. See **Figure 8**. A few of its functions may be duplicated by locally fabricated substitutes but for the serious hobbyist/mechanic it is a must. Its uses are described in the applicable sections of this book. Prices start at around $12 at electronics hobbyist stores and mail order outlets.

Carburetor Synchronizer

A carburetor synchronizer is used on engines with dual carburetors to fine-tune the slide synchronization and idle speed. It is sometimes called an airflow meter.

The tool shown in **Figure 10** costs about $15 at auto parts stores and through mail order outlets.

Compression Gauge

The compression gauge (**Figure 11**) measures the compression pressure buildup in each cylinder. When properly interpreted, the results indicate general piston, cylinder, piston ring, and head gasket condition.

Gauges are available with or without a flexible hose and a good one is usually less than $10 through auto parts stores.

Timing Gauge

A timing gauge is used to precisely locate the position of the piston in the cylinder to achieve the most accurate ignition timing. The instrument is screwed into the spark plug hole. Most gauges indicate both inches and millimeters.

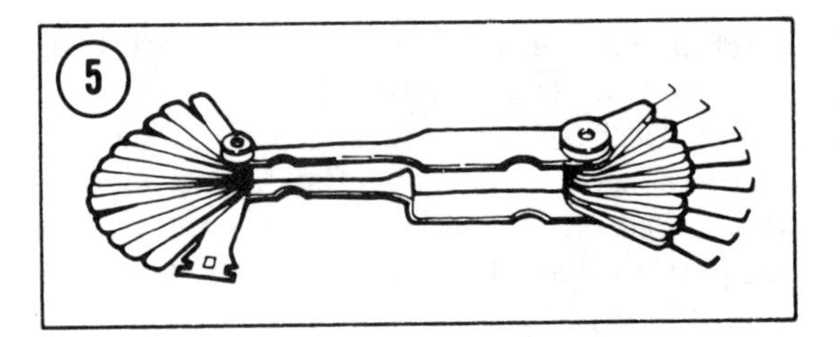
5

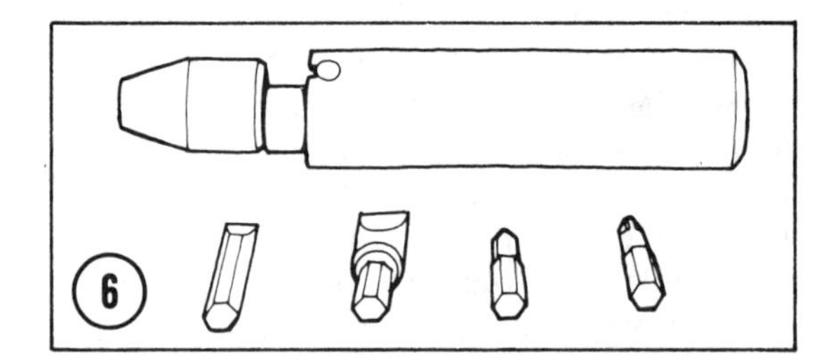
6

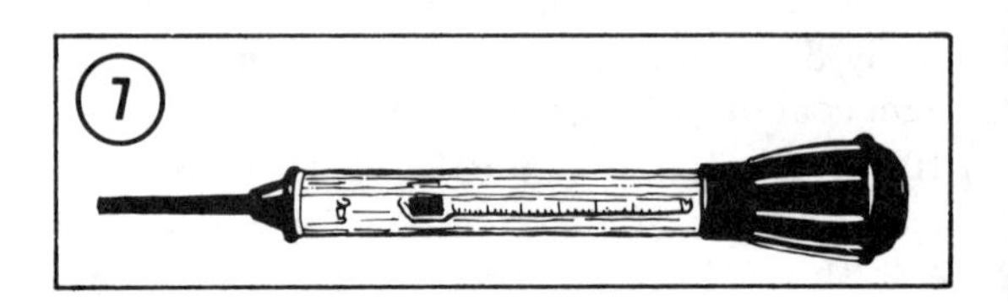
7

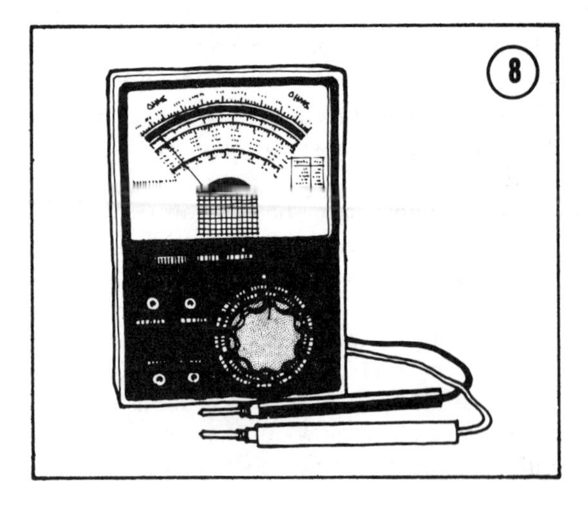
8

The gauge set shown in **Figure 9** costs about $25.

Sheave Gauge

A sheave gauge (**Figure 12**) should be an integral part of your tool inventory. It is used to measure both the center-to-center distance of

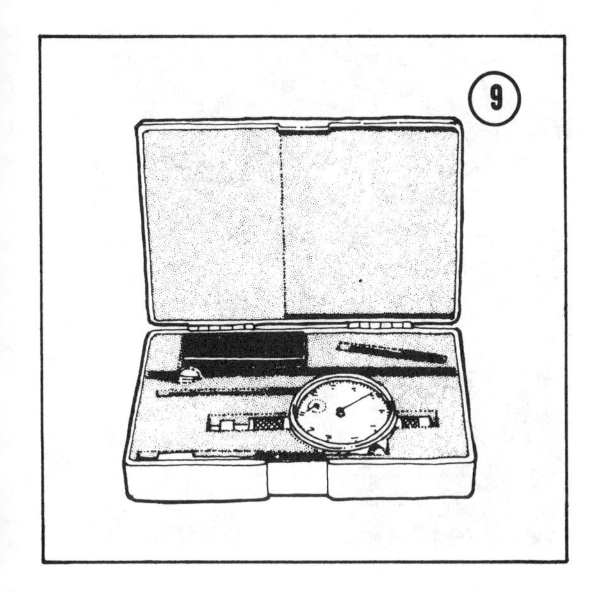

9

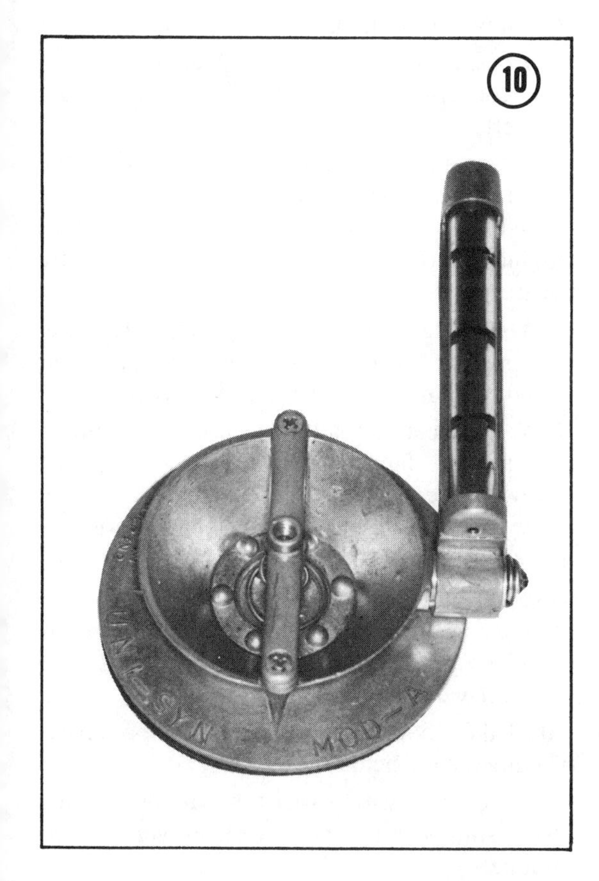

10

the drive and driven sheaves and sheave offset. The gauge is available only through Kawasaki dealers and will most likely have to be specially ordered. Refer to **Table 4** for the correct gauge for your machine.

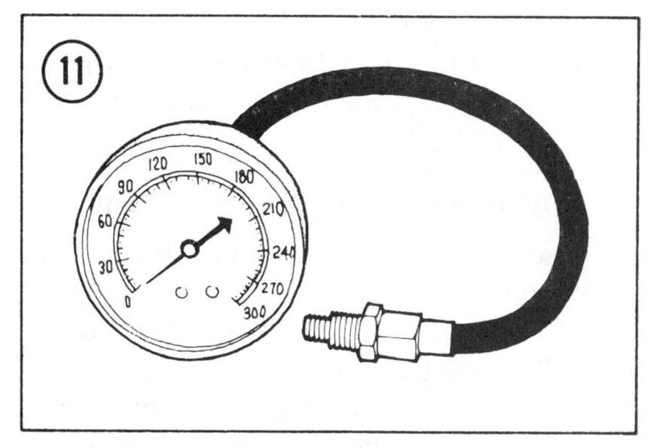

11

EXPENDABLE SUPPLIES

Certain expendable supplies are required for many of the procedures described in this handbook. These include grease, oil, gasket cement, shop rags, cleaning solvent, and distilled water for machines equipped with a battery.

WORKING SAFELY

Professional mechanics can work for years without sustaining serious injury. If you observe a few rules and exercise common sense and safety, you too can enjoy many safe and satisfying hours servicing your own machine. However, if you ignore the rules you can also hurt yourself or damage your machine.

1. Never use gasoline as a cleaning solvent.
2. Never smoke or use a torch around flammable liquids such as cleaning solvent.
3. Never smoke or use a torch in an area where a battery is being charged. Highly explosive hydrogen gas is formed during the charging process.
4. If welding or brazing is required on the machine, remove the fuel tank and set it well out of the way — 25 feet or more.
5. Always use the correct size wrench on nuts and bolts.
6. If a nut or bolt is tight, think for a moment what would happen to your hand should the wrench slip or the fastener suddenly come loose.
7. Keep your work area clean and uncluttered.
8. Wear safety goggles in all operations involving drilling, grinding, or use of a chisel or air hose.

9. Never use worn tools.

10. Keep a fire extinguisher handy and make sure it's rated for gasoline and electrical fires.

SNOWMOBILING CODE OF ETHICS

When snowmobiling, always observe the following code of ethics set down by the International Snowmobile Industry Association.

1. I will be a good sportsman. I recognize that people judge all snowmobile owners by my actions. I will use my influence with other snowmobile owners to promote sportsmanlike conduct.

2. I will not litter trails or camping areas. I will not pollute streams or lakes.

3. I will not damage living trees, shrubs, or other natural features.

4. I will respect other people's property and rights.

5. I will lend a helping hand when I see someone in distress.

6. I will make myself and my vehicle available to assist search and rescue parties.

7. I will not interfere with or harass hikers, skiers, snowshoers, ice fishermen, or other winter sportsmen. I will respect their rights to enjoy our recreation facilities.

8. I will learn and obey all federal, state, provincial, and local rules regulating the operation of snowmobiles in areas where I use my vehicle. I will inform public officials when using public lands.

9. I will not harass wildlife. I will avoid areas posted for the protection and feeding of wildlife.

10. I will stay on marked trails or marked roads open to snowmobiles. I will avoid cross-country travel unless specifically authorized.

SNOWMOBILE SAFETY

General Tips

1. Read your owners manual and know your machine.

2. Check throttle and brake controls before starting the engine. Frozen controls can cause serious injury.

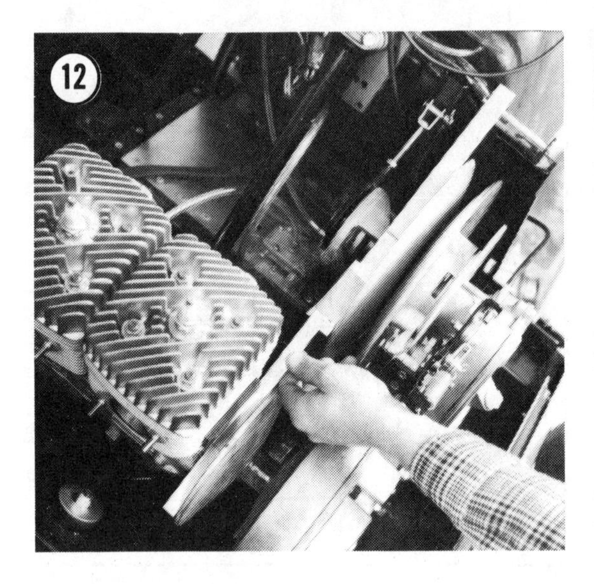

3. Know how to make an emergency stop.

4. Know all state, provincial, federal, and local laws concerning snowmobiling. Respect private property.

5. Never add fuel while smoking or when the engine is running. Always use fresh, correctly mixed fuel. Incorrect fuel mixtures can cause engine failure, and leave you stranded in severe weather.

6. Wear adequate clothing to avoid frostbite. Never wear loose scarves, belts, or laces that could catch in moving parts or on tree limbs.

7. Wear eye and head protection. Wear tinted goggles or face shields to guard against snowblindness. Wear yellow eye protection only during white-outs or flat-light conditions.

8. Never allow anyone to operate the snowmobile without proper instruction.

9. Use the "buddy system" for long trips. A snowmobile travels farther in 30 minutes than you can walk in a day.

10. Take along sufficient tools and spare parts for emergency field repairs.

11. Use a sled with a stiff tow bar for carrying extra supplies. Do not overload your snowmobile.

12. Carry emergency survival supplies when going on long trips. Notify friends and relatives of your destination and expected arrival time.

13. Never attempt to repair your machine while the engine is running.

14. Check all machine components and hardware frequently, especially skis and steering.

15. Never lift the rear of machine to clear the track. Tip the machine on its side and be sure that no one is behind it.

16. Winch the snowmobile onto a tilt-bed trailer — never drive it on. Secure the machine firmly to the trailer and ensure that the trailer lights operate.

Operating Tips

1. Never operate the machine in crowded areas, or steer towards people.
2. Avoid avalanche areas and other unsafe terrain.
3. Cross highways (where permitted) at a 90-degree angle after stopping and looking in both directions. Post traffic guards if crossing in groups.
4. Do not ride the snowmobile on or near railroad tracks. The snowmobile engine can drown out the sound of an approaching train. It is difficult to maneuver the snowmobile from between the tracks.
5. Do not ride the snowmobile on ski slopes or other areas with skiers.
6. Always check the thickness of the ice before riding on frozen lakes or rivers. Do not panic if you go through the ice — conserve energy.
7. Keep the headlight and taillight areas free of snow and never ride at night without lights.
8. Do not ride the snowmobile without shields, guards, and protective hoods in place.
9. Do not attempt to open new trails at night. Follow established trails; unseen barbed wire or guy wires may cause serious injury or death.
10. Always steer with both hands.
11. Be aware of terrain and avoid operating the snowmobile at excessive speed.
12. Do not panic if the throttle sticks. Pull the "tether" string or push emergency stop switch.
13. Drive more slowly when carrying a passenger, especially a child.
14. Always allow adequate stopping distance based on ground cover conditions. Ice requires a greater stopping distance to avoid skidding. Apply brakes gradually on ice.
15. Do not speed through wooded areas. Hidden obstructions, hanging limbs, unseen ditches, and even wild animals can cause accidents.
16. Do not tailgate. Rear end collisions can cause injury and machine damage.
17. Do not mix alcoholic beverages with snowmobiling.
18. Keep feet on footrests at all times. Do not permit feet to hang over sides or attempt to stabilize the machine with feet when making turns or in near-spill situations; broken limbs could result.
19. Do not stand on the seat, stunt, or show-off.
20. Do not jump the snowmobile. Injury or machine damage could result.
21. Always keep hands and feet out of the track area when the engine is running. Use extra care when freeing the snowmobile from deep snow.
22. Check the fuel supply regularly. Do not travel further than your fuel will permit you to return.
23. Whenever you leave your machine unattended, remove the "tether" switch and ignition key.

Preparing for a Trip

1. Check all bolts and fasteners for tightness. Do not operate your snowmobile unless it is in top operating condition.
2. Check weather forecasts before starting out on a trip. Cancel your plans if a storm is possible.
3. Study maps of the area before the trip and know where help is located. Note locations of phones, resorts, shelters, towns, farms, and ranches. Know where fuel is available. If possible use the buddy system.
4. Do not overload your snowmobile. Use a sled with a stiff tow bar to haul extra supplies.
5. Do not risk a heart attack if your snowmobile gets stuck in deep snow. Carry a small block and tackle for such situations. Never allow anyone to manually pull on the skis while you attempt to drive the machine out.

6. Do not ride beyond one-half the round trip cruising range of your fuel supply. Keep in mind how far it is home.

7. Always carry emergency survival supplies when going on long trips or traveling in unknown territory. Notify friends and relatives of your destination and expected arrival time.

8. Carry adequate eating and cooking utensils (small pans, kettle, plates, cups, etc.) on longer trips. Carry matches in a waterproof container, candles for building a fire, and easy-to-pack food that will not be damaged by freezing. Carry dry food or "space" energy sticks for emergency rations.

9. Pack extra clothing, a tent, sleeping bag, hand axe and compass. A first aid kit and snow shoes may also come in handy. Space age blankets (one side silverfoil) furnish warmth and can be used as heat reflectors or signalling devices for aerial search parties.

Emergency Survival Techniques

1. Do not panic in the event of an emergency. Relax, think the situation over, then decide on a course of action. You may be within a short distance of help. If possible repair your snowmobile so you can drive to safety. Conserve your energy and stay warm.

2. Keep hands and feet active to promote circulation and avoid frostbite while servicing your machine.

3. Mentally retrace your route. Where was the last point where help could be located? Do not attempt to walk long distances in deep snow. Make yourself comfortable until help arrives.

4. If you are properly equipped for your trip you can turn any undesirable area into a suitable campsite.

5. If necessary, build a small shelter with tree branches or evergreen boughs. Look for a cave or sheltered area against a hill or cliff. Even burrowing in the snow offers protection from the cold and wind.

6. Prepare a signal fire using evergreen boughs and snowmobile oil. If you can not build a fire, make an S-O-S in the snow.

7. Use a policeman's whistle or beat cooking utensils to attract attention or frighten off wild animals.

8. When your camp is established, climb the nearest hill and determine your location. Observe landmarks on the way so you can find your way back to your campsite. Do not rely on your footprints. They may be covered by blowing snow.

Table 1 25:1 FUEL/OIL MIXING RATIO— SA AND SB A1 SERIES

Fuel	Oil	
U.S. Gallons	Oz.	cc
0.5	2.6	77
1.0	5.2	154
1.5	7.7	258
2.0	10.3	305
2.5	12.8	379
5.0	25.6	757
Liters		cc
1.0		40
2.0		80
3.0		120
4.0		160
5.0		200
6.0		240

Table 2 40:1 FUEL/OIL MIXING RATIO— SA AND SB A2 AND LATER SERIES

Fuel	Oil	
U.S. Gallons	Oz.	cc
0.5	1.6	48
1.0	3.2	95
1.5	4.8	142
2.0	6.4	189
2.5	8.0	237
5.0	16.0	473
Liters		cc
1.0		25
2.0		50
3.0		75
4.0		100
5.0		125
6.0		150

Table 3 WIND CHILL FACTORS

Estimated Wind Speed in MPH	Actual Thermometer Reading (° F)											
	50	40	30	20	10	0	–10	–20	–30	–40	–50	–60
	Equivalent Temperature (° F)											
Calm	50	40	30	20	10	0	–10	–20	–30	–40	–50	–60
5	48	37	27	16	6	–5	–15	–26	–36	–47	–57	–68
10	40	28	16	4	–9	–21	–33	–46	–58	–70	–83	–95
15	36	22	9	–5	–18	–36	–45	–58	–72	–85	–99	–112
20	32	18	4	–10	–25	–39	–53	–67	–82	–96	–110	–124
25	30	16	0	–15	–29	–44	–59	–74	–88	–104	–118	–133
30	28	13	–2	–18	–33	–48	–63	–79	–94	–109	–125	–140
35	27	11	–4	–20	–35	–49	–67	–82	–98	–113	–129	–145
40	26	10	–6	–21	–37	–53	–69	–85	–100	–116	–132	–148
*	Little Danger (for properly clothed person)				Increasing Danger			Great Danger				
					• Danger from freezing of exposed flesh •							

*Wind speeds greater than 40 mph have little additional effect.

Table 4 SHEAVE GAUGE IDENTIFICATION

Model	Gauge (Kawasaki Part No.)
SA and SB—All	205207
SS and ST—All	57001-3503

CHAPTER TWO

PERIODIC MAINTENANCE

A program of regular routine maintenance will ensure that your snowmobile will provide many hours of efficient, trouble-free service. An afternoon or evening spent now, cleaning, inspecting, and adjusting, can prevent costly mechanical problems in the future and unexpected and often dangerous breakdowns on the trail.

The procedures described in this chapter are simple and straightforward. They are arranged by service intervals so that tasks that should be carried out at one time are grouped together to make servicing easy and more enjoyable.

Tables are found at the end of the chapter.

SERVICE INTERVALS

Factory-recommended service intervals are shown in **Table 1**. Service requirements may vary depending on type of use. For example, for extremely hard use, over rough and uncertain terrain, track adjustment and condition should be checked more frequently than recommended in the table.

Although many of the items to be checked won't require any actual service, even after many hours of operation, it is a good idea to follow the recommendations to help you develop the habit of carrying out a systematic inspection. It can serve to warn of impending trouble and it will allow you to become familiar with your machine and gain confidence about its condition and your ability to take care of it. A day of riding is much more pleasant if you're not worrying about the condition of your snowmobile.

PRE-RIDE CHECKS

The checks and inspection that follow should be carried out before each ride.

Fuel Level

Check the fuel level in the tank **(Figure 1)** and top it up if it is not full. Even if you are planning on riding just a short distance it's a good idea to begin with a full tank of gas; an unexpected change in weather or trail conditions could require you to be out longer than you had anticipated.

Oil Level

On machines equipped with oil injection lubrication, check the oil tank and add oil if the level is low. Use Kawasaki Snowmobile Oil or an equivalent.

NOTE

Acceptable oil will be marked with B.I.A. Certified T.C.W. on the top of the can.

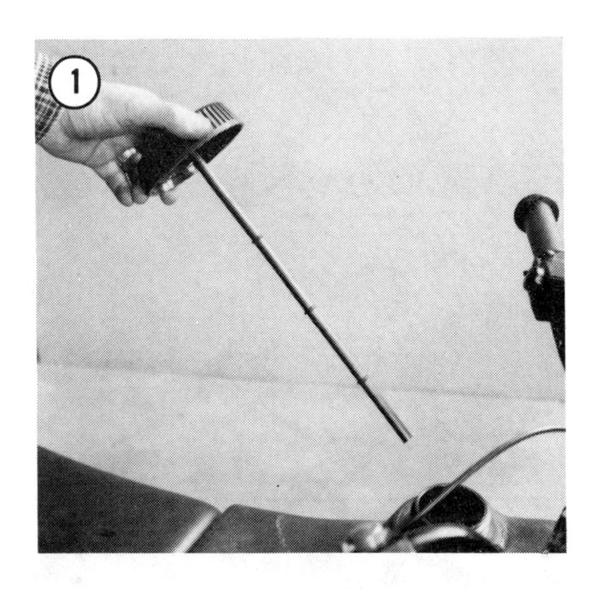

*If fuel and oil mixing are required, refer to **Lubrication** in Chapter One.*

Drive Belt

Check the condition of the drive belt. If it is severely worn, cracked, damaged, or deteriorated, it should be replaced (see Chapter Seven, *Drive Belt*).

Engine Stop Switch

Start the engine and allow it to warm up to the point that it will idle with the throttle unattended. Then, operate the engine STOP switch and if the engine does not shut off, refer to Chapter Six and locate and correct the trouble before riding the machine.

Tether Switch

Start the engine and allow it to idle. Disconnect the tether switch and if the engine does not shut off, refer to Chapter Six and locate and correct the trouble before riding the machine.

Lights

Check the operation of the headlight, taillight, and stoplight and if any are not operating, refer to Chapter Six and locate and correct the trouble before riding the machine.

Track

Check the track to make sure it is not frozen. If it is, lean the machine over and free the track by hand; *do not free it by revving the engine or the drive belt will be damaged.*

Coolant Level

On liquid-cooled machines, the coolant level should be 1/4-1/2 in. (6-12 mm) below the tank top. Use a 55%/45% mixture of ethylene glycol antifreeze and water.

FIRST 5-10 HOURS OR 50-100 MILES (80-160 KM)

The following checks and corrections are recommended for new machines and should also be applied to machines that have undergone major repair work and to all machines at the beginning of the season.

Ignition Timing

Refer to *Tune-up* later in this chapter and check the ignition timing as described.

Starter Cable Free Play

Refer to *Tune-up* later in this chapter and check the starter cable free play as described.

Throttle Cable Free Play

Refer to *Tune-up* later in this chapter and check the throttle cable free play as described.

Carburetor Adjustment

Refer to *Tune-up* later in this chapter and check carburetor adjustment as described.

Track Tension and Alignment

There are 2 track adjustments — tension and alignment — that should be carried out after 5-10 hours of operation for a new machine, or after a new track has been installed, and at 20-hour intervals from then on.

Correct track tension is important because if the track is too loose, it will slap on the bottom of the tunnel and wear both the track and the tunnel. Also, a loose track can ratchet on the drive sprockets and damage both the track and the sprockets.

If the track is too tight, it will rapidly wear the slide runner material and the rubber on the idler wheels, and degrade performance because of increased friction and drag on the drive system.

Track alignment is related to track tension and should be checked and adjusted when the tension is checked and adjusted. If the track is misaligned, the rear idler wheels, drive sprocket lugs, and track lugs will wear rapidly. Also, performance will be poor because of resistance of the track lugs against the sides of the wheels.

1. Turn the machine onto its side and rest it on a piece of cardboard to protect the finish.
2. Clean ice, snow, and dirt from the track and suspension.
3. Pull the track away from the suspension frame at midpoint with moderate force and measure the distance (**Figure 2**). For SS340-A2, SS440-A2, and ST440-A2 models the distance should be 1¼-1½ in. (31-38 mm). For all other models, the distance should ¾ in. (19 mm).
4. If adjustment is required, loosen the locknuts on the track adjuster bolts (**Figure 3**). If the track-to-runner distance is more than specified, tighten the adjuster bolts equally until the distance is correct. If the distance is less than specified, loosen the adjuster bolts equally until the distance is correct. Then, tighten the locknuts.
5. Position the machine on its skis, so the tips of the skis are against a wall or other immoveable barrier. Elevate and support the rear of the machine so the track is completely clear of the ground and free to rotate.
6. Start the engine and apply just enough throttle to turn the track several revolutions. Then, shut off the engine and allow the track to coast to a stop; don't stop it with the brake.

WARNING

Don't stand behind or in front of the machine when the engine is running, and take care to keep hands, feet, and clothing away from the track when it is turning.

7. Check the alignment of the rear idler wheels and the track (**Figure 4**). If the idler wheels are equal distances from the lugs and the edges of the track and the openings in the track are centered with the slide runners (**Figure 5**), the alignment is correct. However, if the track is offset to one side or the other, alignment adjustment is required.

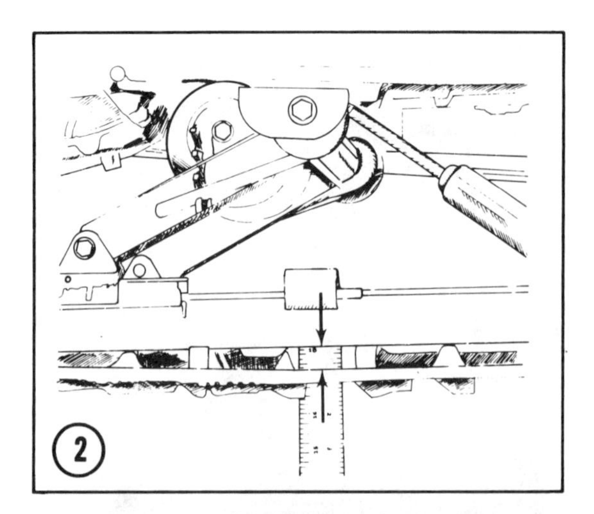

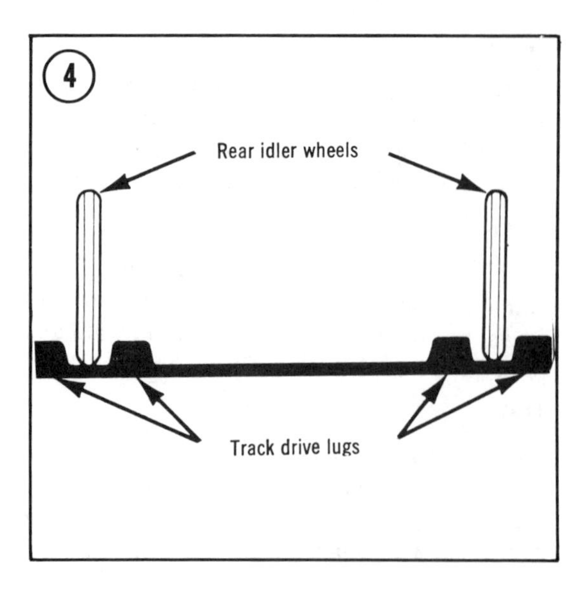

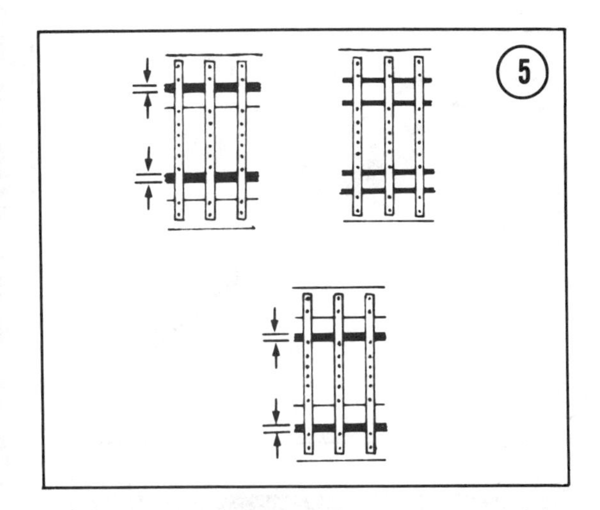

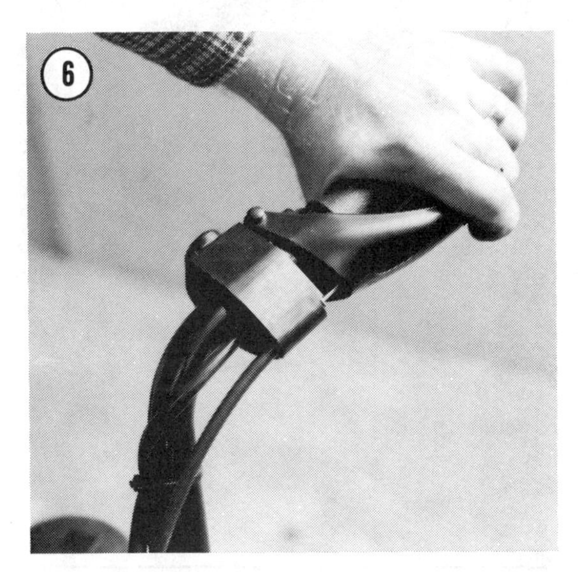

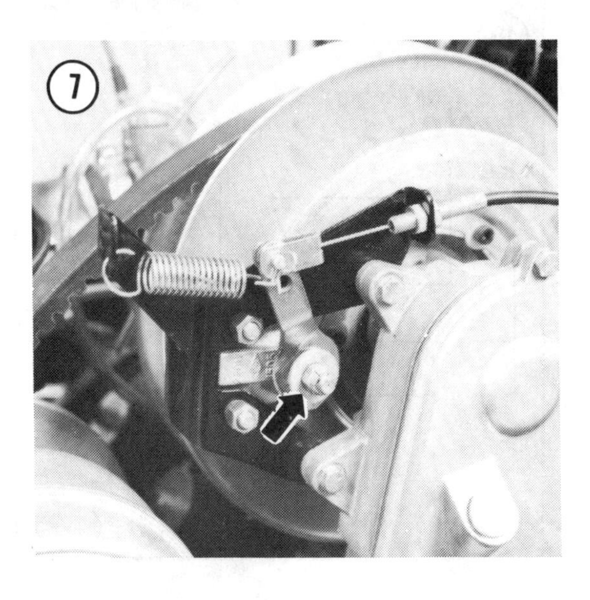

8. Loosen the locknuts on the track adjuster bolts. If the track is offset to the left, tighten the left adjuster bolt and loosen the right one equal amounts until the track is centered. If the track is offset to the right, tighten the right adjuster bolt and loosen the left one equal amounts. Then, tighten the locknuts.

9. Repeat Steps 6 and 7 and if alignment is still not correct, repeat Step 8 and then Steps 6 and 7 again until the track is correctly aligned. Then recheck track tension.

Brake Adjustment (and Pad Wear)

Pull the brake lever (**Figure 6**) so there is about ¾ in. (19 mm) of clearance between the lever and the base. At the same time, rotate the driven sheave back and forth and check for movement of the pads. The sheave should move with some slight resistance and the pads should just begin to move with it.

If adjustment is required, tighten the adjuster nut (**Figure 7**) to move the pads closer to the disc, or loosen it to move them back.

When adjustment is correct, check the serviceability of the pads by measuring the distance the threads on the adjuster bolt protrude from the adjuster nut (**Figure 7**). If the threads protrude $\frac{7}{16}$ in. (11.1 mm) or more, the pads should be replaced as described in Chapter Seven.

EVERY 20 HOURS OR 300 MILES (480 KM)

The following checks and corrections should be made every 20 hours or 300 miles (480 km) of operation.

Track Tension and Alignment

Check track tension and alignment as described earlier (see *Track Tension and Alignment*).

Brake Adjustment (and Pad Wear)

Adjust the brake and check for pad wear as described earlier. See *Brake Adjustment (and Pad Wear)*.

Skag Wear

Inspect the ski runners (skags) for wear (**Figure 8**) and replace them if they are more than half worn down or are cracked at the mounting studs (see *Ski Runner Replacement*, Chapter Eight).

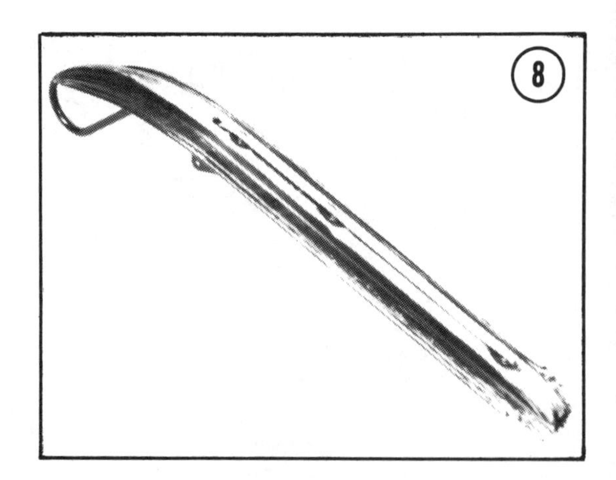

Slide Rail Runners

Measure the slide rail runners (**Figure 9**) and replace them if they are worn past the service limit shown in **Figure 10**. See Chapter Nine, *Slide Runner Replacement*.

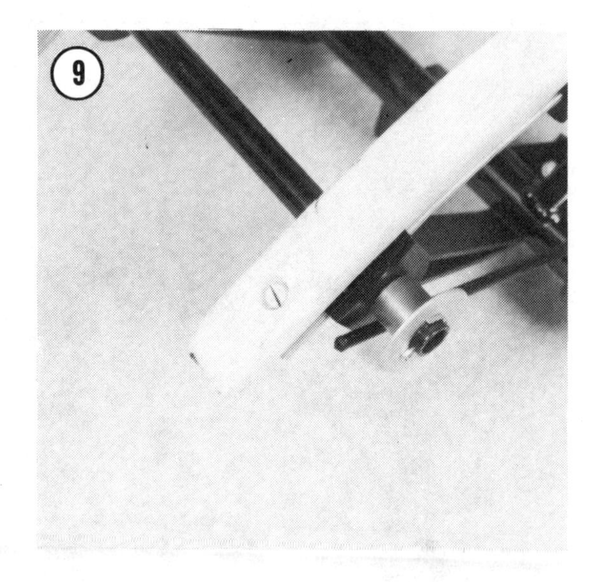

Battery

On models equipped with electric starters, check the electrolyte level in the battery. See *Battery* in Chapter Six.

EVERY 40 HOURS OR 600 MILES (960 KM)

The following checks and corrections should be made every 40 hours or 600 miles (960 km) of operation.

Spark Plugs

Refer to *Tune-up* later in this chapter and clean, regap or replace the spark plugs as described.

Drive Belt

Remove the belt guard (**Figures 11 and 12**) and inspect the drive belt for wear, cracks, damage, or deterioration. Replace the belt if worn beyond the service limit shown in **Table 2**. See *Drive Belt* in Chapter Seven.

Ski Alignment

Refer to *Ski Alignment* in Chapter Eight and check and correct ski alignment as described.

Fasteners

Check all engine, drive train, suspension, and steering fasteners for tightness.

EVERY 60 HOURS OR 900 MILES (1,440 KM)

The following checks and corrections should be made every 60 hours or 900 miles (1,440 km) of operation.

Starter Cable Free Play

Refer to *Tune-up* later in this chapter and check the starter cable free play as described.

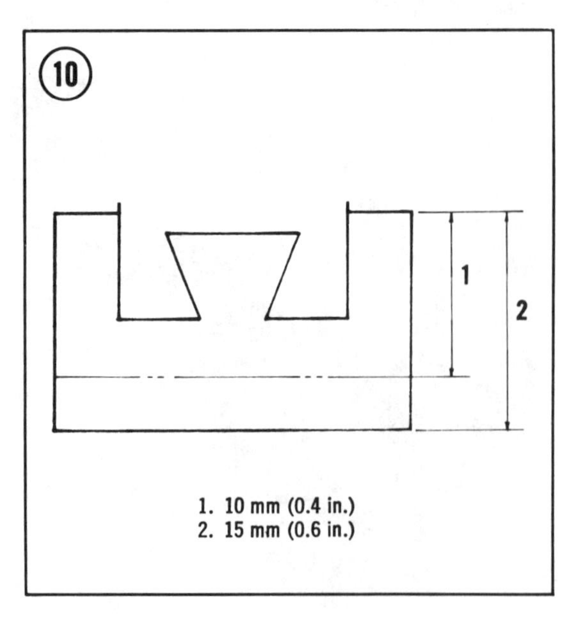

1. 10 mm (0.4 in.)
2. 15 mm (0.6 in.)

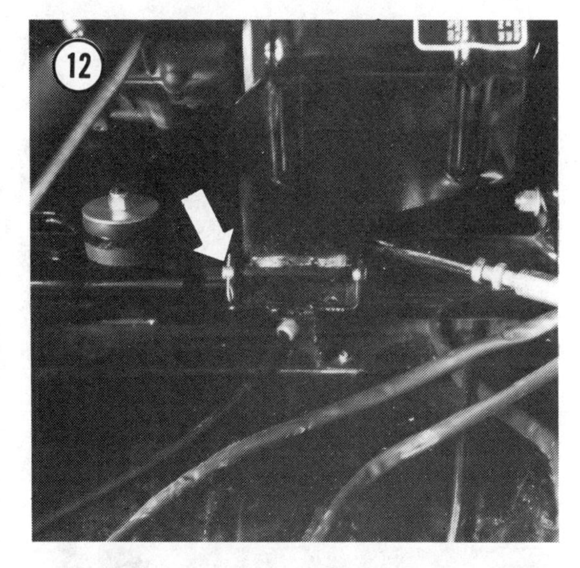

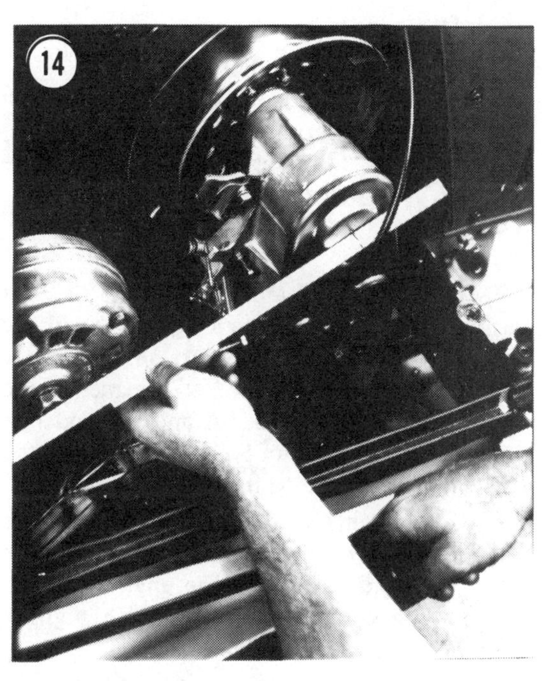

Throttle Cable Free Play

Refer to *Tune-up* later in this chapter and check the throttle cable free play as described.

Fuel Filter

The fuel filter should be routinely replaced at this service interval. Disconnect the lines from the filter (**Figure 13**) and connect them to a new unit. Check the filter case for a flow-direction arrow and install it with the arrow facing in the direction of fuel flow.

Drive Sheave Adjustment

The center-to-center distance from the drive sheave to the driven sheave, and the offset of the sheaves must be correctly maintained for good performance and long belt life.

1. Remove the belt guard (**Figures 11 and 12**).
2. Measure the distance between the sheave centers (**Figure 14**). For SA and SB models, the distance should be 262 mm (10.3 in.). For SS and ST models, the distance should be 304 mm (12 in.).
3. If adjustment is required, loosen the chain-case mounting bolts (**Figure 15** — SA and SB models) and turn the clevis adjuster (**Figure 16**)

in or out until the distance is correct. Then, tighten the chaincase mounting bolts.

For SS and ST models, loosen the bolts on the jackshaft bearing retainer (**Figure 17**), loosen the locknut on the adjuster, and turn the adjuster in or out until the distance is correct. Then, tighten the adjuster locknut and the bearing retainer bolts.

4. Measure the offset of the sheaves. This is easily done with a sheave gauge available through your dealer. See **Figure 18**. The gauge should contact each sheave at 2 places at the same time. The offset for SA and SB models is 11.5 mm (0.454 in.). For SS and ST models, the offset is 14.9 mm (0.588 in.).

5. If adjustment is required, loosen all the engine mounting bolts (**Figure 19**—SA and SB front bolts) and move the engine right or left until the offset is correct. Then, tighten the engine mounting bolts.

For SS and ST models, remove the trim piece from the body (**Figure 20**) and unscrew the bolt that attaches the driven sheave to the jackshaft. Carefully draw the sheave off the shaft and collect the shims from behind it (**Figure 21**). Add or subtract shims as required to correct the offset. Then, install the driven sheave and trim piece.

NOTE: *The sheave gauge for SA and SB models is identified with Kawasaki Part No. 205207. The gauge for SS and ST models is Part No. 57001-3503.*

Sheave Condition

Check the contact surfaces of the sheaves for scoring and wear. They should be smooth and shiny.

ANNUALLY

In addition to the checks and service just described, the following tasks should be carried out annually, preferably at the end of the winter season, before the machine is stored for the summer.

Upper End Decarbonization

Decarbonization of the upper end involves cleaning the piston crowns, combustion chambers, exhaust ports, and exhaust system. Refer

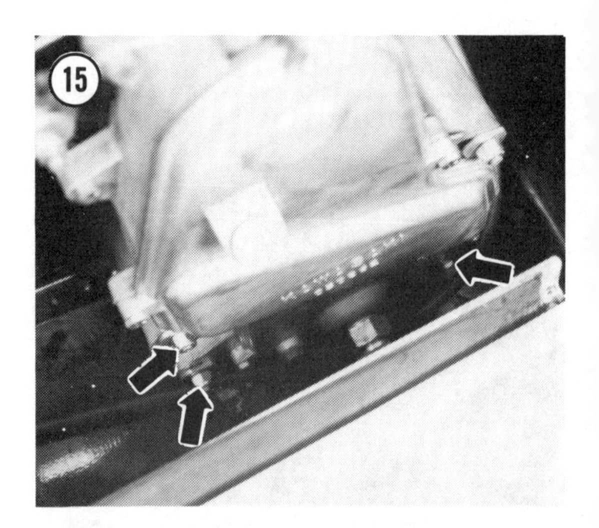

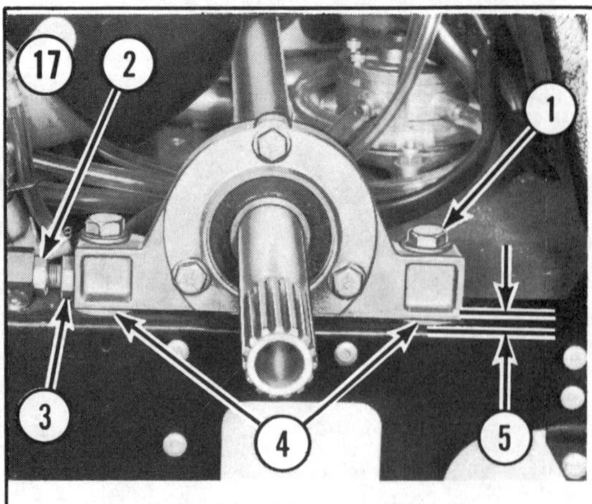

1. Bearing retainer mounting bolts
2. Adjusting bolt locknut
3. Adjusting bolt
4. Bearing retainer shims
5. Parallel

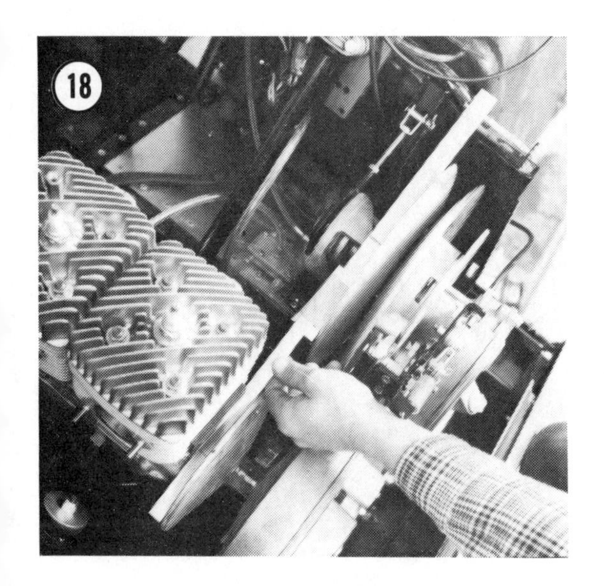

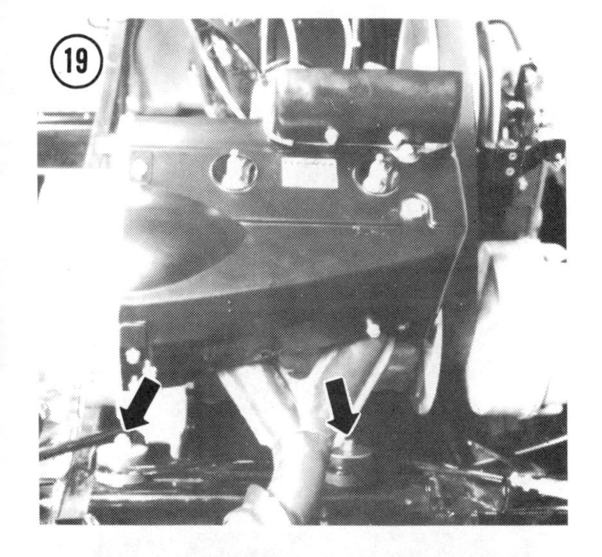

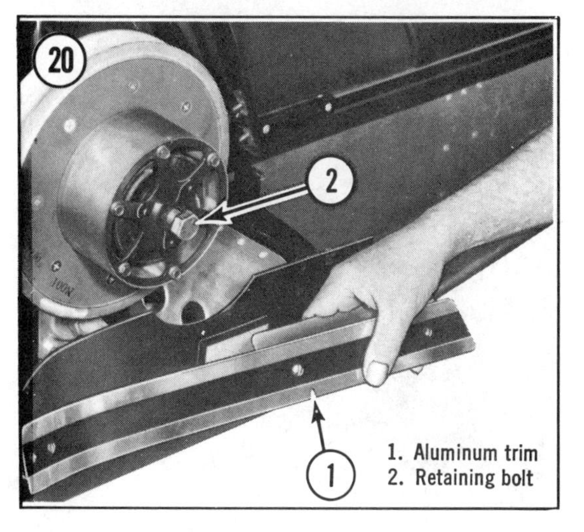

1. Aluminum trim
2. Retaining bolt

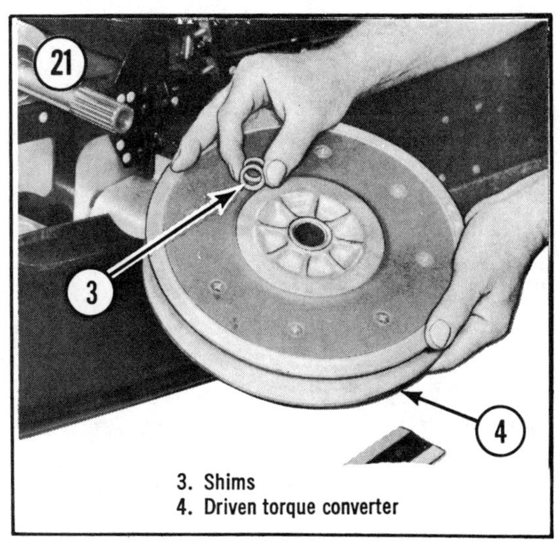

3. Shims
4. Driven torque converter

to Chapter Four, *Upper End Service*, and Chapter Five, *Exhaust System*, for the correct procedures.

Ignition Timing

Refer to *Tune-up* later in this chapter and check the ignition timing as described.

Carburetor Service

Clean, inspect, and adjust the carburetor as described in Chapter Five and under *Tune-up* later in this chapter.

Fuel Tank

Refer to Chapter Five, *Fuel Tank*, and remove, clean, and inspect the fuel tank.

Chain Tensioner Guides

Refer to Chapter Seven, *Chain Tensioner*, and inspect the tensioner guides and replace them if they are worn.

Track

Refer to Chapter Three, *Track Wear Analysis*, and inspect the track for wear and damage.

LUBRICATION

The lubrication services that follow should be carried out at the intervals shown in **Table 3**. The table also lists recommended lubricants.

Control Cables

The control cables should be lubricated with a light, low-temperature lubricant such as WD-40, Break-Free, LPS, or Dri-Slide.

1. Unscrew the top from the carburetor (**Figure 22**), remove the slide assembly, and disconnect the cable. Elevate the end of the cable and slowly pour lubricant down the cable sheath. Work the cable back and forth to aid the flow. When lubricant begins to run out the other end of the cable sheath, reconnect it to the slide and install the slide in the carburetor. Adjust the cable free play as described in *Tune-up* in this chapter.
2. Disconnect the brake cable from the brake (**Figure 23**). Lubricate the cable as described above. After the cable has been reconnected, refer to *Brake Adjustment* in this chapter and adjust the brake as described.

Driven Sheave

Apply low-temperature grease to the ramp sliding surfaces on the driven sheave (**Figure 24**).

Chaincase

The chaincase oil level should be checked at 20-hour intervals. Remove the level plug on the front of the case (**Figure 25**). Oil should just begin to seep out. If necessary, add Kawasaki Chain Lubricant.

The chaincase oil should be changed at the end of each season. The oil can be drained by removing the chaincase cover (**Figure 26**) and

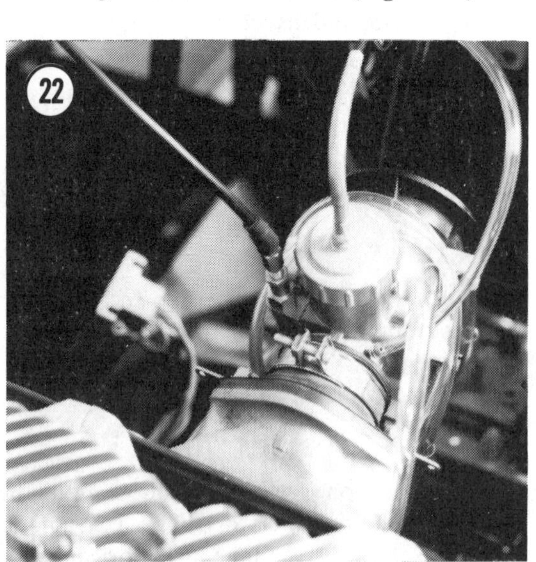

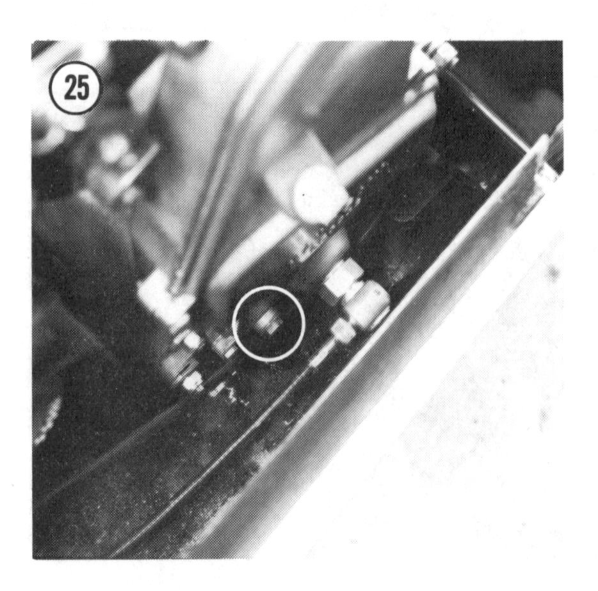

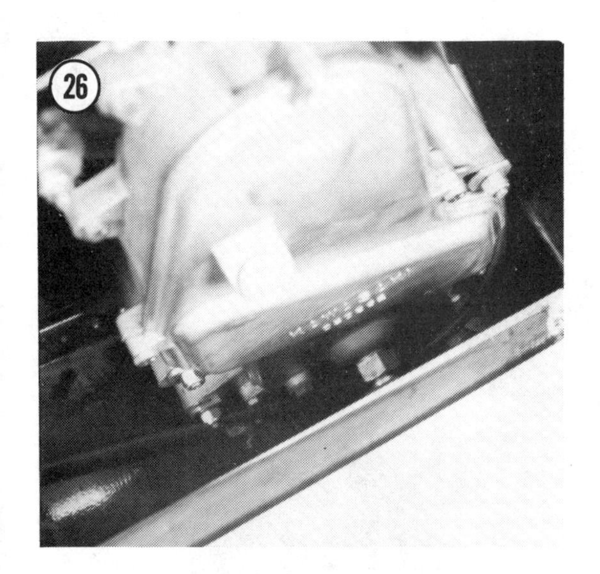

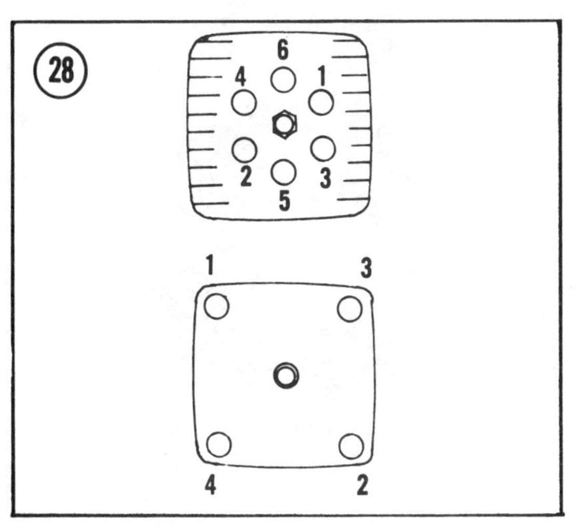

allowing the oil to drain down into the body and out through a hole beneath the chaincase. However, this makes a messy job of it. Instead, turn the machine onto its right side, then remove the chaincase cover and remove the oil with a turkey baster.

Install the cover, set the machine upright, and pour oil into the chaincase through the filler opening until it runs out the level hole. Then, screw the plugs into both holes.

Gear Case

On models with oil injection or liquid cooling, check the gear case oil level every 20 hours of operation, and change it annually. Refer to *Oil Pump Removal* in Chapter Four.

TUNE-UP

The factory-recommended intervals for checking the factors relating to engine tuning (spark plug condition and gap, ignition timing, carburetion) would seem to be at odds with experience. This occurs because experience has shown that required service intervals differ from one system to the next. For example, while spark plug condition and gap should be checked for all models at 20-hour intervals, along with contact breaker gap and ignition timing for flywheel magneto models, ignition timing on CDI equipped models need be checked only if the ignition assembly has been removed and reinstalled.

However, a complete tune-up procedure is presented here for convenience, and only a little extra time is required to ensure that all engine tuning variables are correct.

Because the different systems in the engine interact, the tune-up procedure should be done in the following order:

a. Check and tighten cylinder head fasteners.

b. Work on ignition system.

c. Adjust carburetion.

CYLINDER HEAD FASTENERS

1. On air-cooled models, disconnect the spark plug high-tension leads and remove the cooling shroud (**Figure 27**).

2. On all models, tighten the cylinder head fasteners in the pattern shown in **Figure 28** to 2.2 mkg (16 ft.-lb.).

29

SPARK PLUG CONDITIONS

NORMAL USE

OIL FOULED

CARBON FOULED

OVERHEATED

GAP BRIDGED

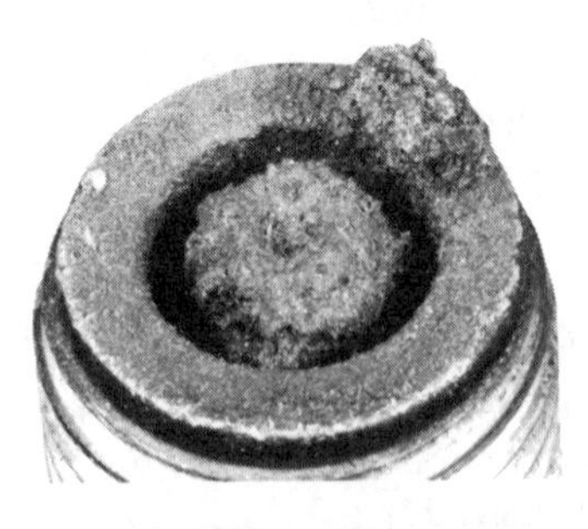

SUSTAINED PREIGNITION

WORN OUT

Photos courtesy of Champion Spark Plug Company.

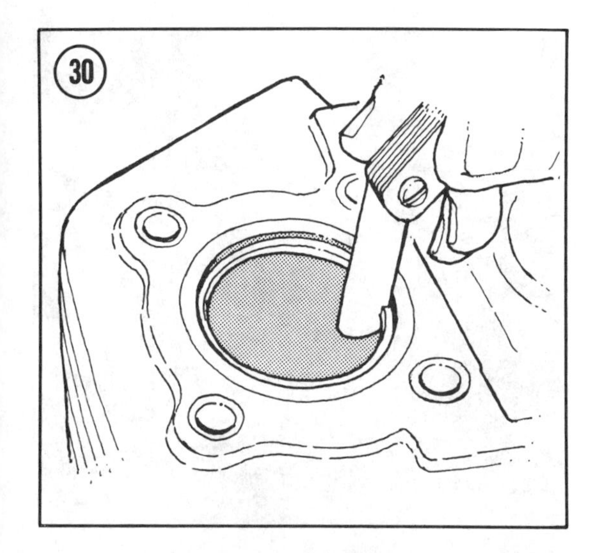

3. Install the cooling shroud on air-cooled models.

IGNITION SYSTEM

Spark Plugs

For all models, spark plugs should be routinely checked and their gap adjusted at 20-hour or 300-mile (480 km) intervals. In addition, they should be checked any time hard starting or poor performance indicates that they may be faulty or that carburetion may be out of adjustment.

1. Carefully disconnect the high-tension leads from the plugs by grasping the caps and turning them slightly as you pull them loose. Don't pull on the wires.
2. Clean the area around the base of the plugs with compressed air.
3. Unscrew the plugs from the cylinder heads and compare their condition to **Figure 29**.
4. Check the heat range of the plugs and compare them with the specifications in **Table 4**. If the heat range is not correct, replace them with correct plugs.
5. If the plugs are in good condition and are the correct heat range, clean them with a fine wire brush and solvent and dry them with compressed air.
6. Check the electrode gap with a round wire gauge and compare it with the gap specified in **Table 4**. If adjustment is required, carefully bend the side electrode (**Figure 30**).
7. Check timing as described under *Contact Breaker Ignition* or *Capacitive Discharge Ignition* which follows.

Screw the plugs into the cylinder heads until they are firmly seated, then tighten them an additional $\frac{1}{4}$ turn.

CONTACT BREAKER IGNITION

Check the timing and contact breaker gap at 100-hour intervals. Replace the condensers and contact breaker assemblies annually, or when the contacts become severely pitted.

Timing/Adjustment

1. Remove the muffler (**Figure 31**), recoil starter assembly (**Figure 32**), and starter pulley (**Figure 33**).

2. Dip a stiff business card in lacquer thinner or a similar non-petroleum base solvent such as ethyl alcohol, place the card between the contact breaker points, and draw it through the points several times, until the points are clean and no discoloration remains on the card.

Inspect the contact surfaces of the points and if they are smooth and parallel, proceed with the next step. If they are not, dress them with a few strokes of a point file and then clean them with lacquer thinner.

3. Rotate the crankshaft until one set of contacts is at its widest point. Measure the gap with a flat feeler gauge. The gap should be 0.35 mm (0.014 in.). If it's not, loosen the breaker mounting screw (**Figure 34**) and move the stationary plate as required. Then, tighten the screw and recheck the gap.

4. Install a timing gauge in the right spark plug hole (**Figure 35**).

5. Disconnect the red and white primary leads from the coil and connect one lead of a buzz-box or continuity light to the red lead from the magneto and the other tester lead to ground (**Figure 36**).

> NOTE: *At the precise instant the light goes out or the buzz-box ceases to make noise, the points are just beginning to open.*

6. Rotate the crankshaft to bring the piston to TDC, as indicated by the dial gauge. Zero the gauge. Then, slowly turn the crankshaft counterclockwise until the gauge indicates the timing shown in **Table 4**. This is the firing point.

7. Slightly loosen the stator mounting screws (**Figure 37**), and slowly turn the stator until the points just begin to open (indicated by the light or buzz-box) with the crankshaft at the firing point. Check the dial gauge to make sure the crankshaft has not moved, then tighten the stator screws.

8. Recheck the contact breaker gap at the widest point. It may not be possible to achieve both correct timing and contact breaker gap if the breaker assembly has been in use for awhile. In such case, it will be necessary to compromise by turning the stator plate slightly to alter the firing point within the limits shown in **Table 4**.

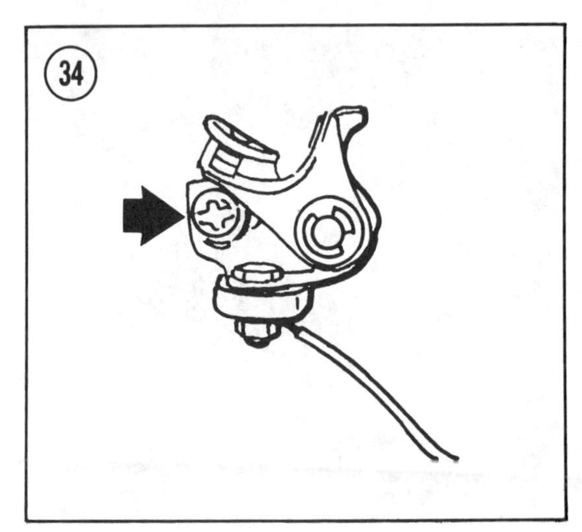

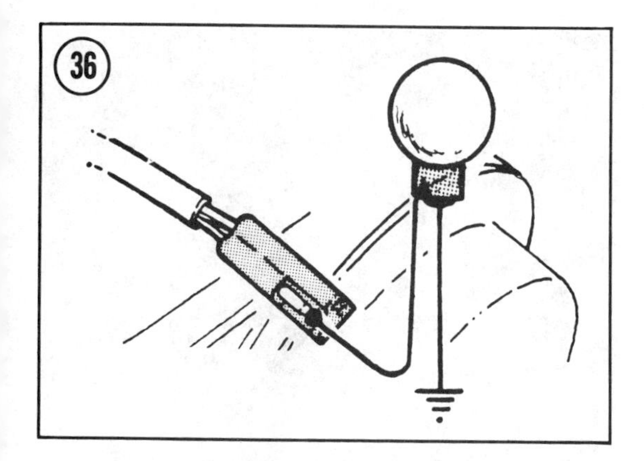

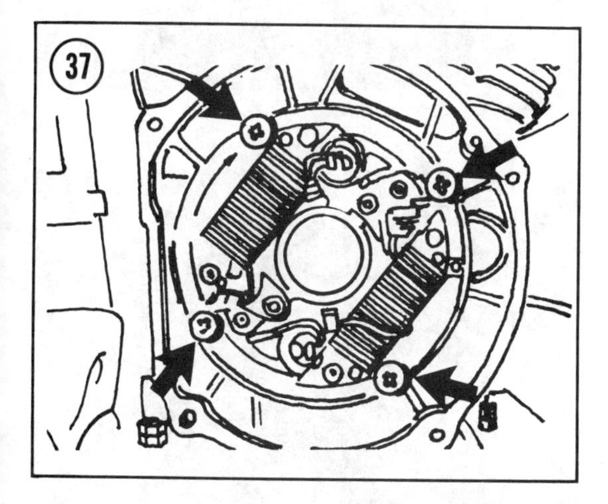

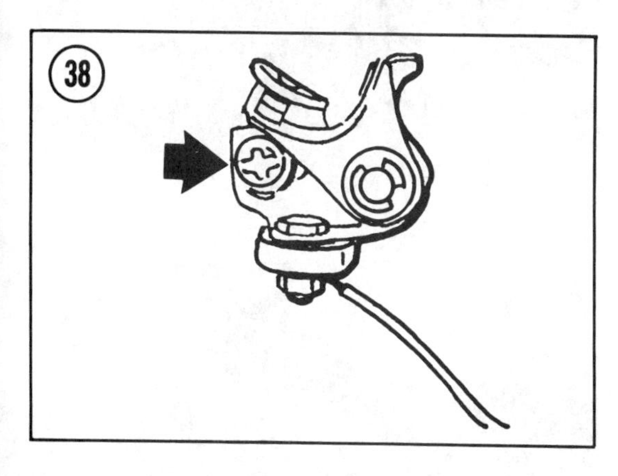

9. Adjust the timing and contact breaker gap for the left cylinder in manner just described. Then, check the dynamic timing as described later under *Strobe Timing*.

Contact Breaker Replacement

The contact breaker assemblies should be replaced annually or when accurate contact breaker gap adjustment is no longer possible because of point wear.

1. Remove the contact breaker mounting screw (**Figure 38**) and remove the breaker assembly.
2. Loosen the terminal nut and disconnect the primary lead.
3. Connect the primary lead to a new breaker assembly and tighten the nut.
4. Install the contact breaker assemblies on the stator.
5. Refer to *Timing/Adjustment* and adjust the contact breaker gap.

Condenser Replacement

The condensers should be replaced routinely when contact breaker assemblies are replaced.

1. Loosen the soldered leads on the condenser terminal with a soldering iron.
2. Remove the screws that attach the condensers to the stator plate and remove the condensers.
3. Install new condensers and solder the leads to the terminals.

CAUTION

Be careful when soldering the leads to the condensers; too much heat can destroy the condensers.

CAPACITIVE DISCHARGE IGNITION (CDI)

Once installed and correctly adjusted, the CDI will remain in time. However, the timing procedure is described and should be carried out if ignition timing troubles are suspected.

1. Remove the muffler (**Figure 39**), recoil starter assembly (**Figure 40**), and starter pulley (**Figure 41**).
2. Install a timing gauge in the right spark plug hole (**Figure 42**).
3. Rotate the crankshaft to bring the piston to TDC as indicated by the gauge. Then zero the gauge.
4. Rotate the crankshaft counterclockwise (viewed from the right) until the gauge needle has made about 3½ revolutions. Then, slowly turn the crankshaft clockwise until the gauge indicates the timing shown in **Table 4**.

5. Check the alignment of the marks on the flywheel and the pulser coil (**Figure 43**). If the marks are not aligned, loosen the 2 screws that attach the pulser coil to the stator and carefully move the coil to line up the marks. Then, without moving the coil any further, tighten the screws.

6. Repeat the above steps to check the timing of the left cylinder. Then, check the dynamic timing of the engine as described below under *Strobe Timing* after restoring the engine to running condition.

STROBE TIMING

Static timing, done with the use of a dial timing gauge, is very accurate and will suit most needs. However, only dynamic timing of an engine warmed up to operating temperature will ensure spot-on ignition performance.

The procedure that follows is not only a valuable tuning aid, but it is fun and rewarding as well because it requires you to do a bit of "re-engineering" on your machine that will increase your understanding of it and your ability to maintain it in top running order.

First, make a rigid pointer from a thin piece of aluminum stock, using the appropriate pattern shown in **Figure 44**. Then, attach the pointer to the engine (**Figure 45**) and use the procedure that follows.

1. Remove the spark plugs and install a timing gauge in the right spark plug hole.

2. Turn the drive sheave counterclockwise to bring the piston to TDC, as indicated by the timing gauge. Zero the gauge.

3. Slowly turn the drive sheave clockwise until the dial gauge indicates the timing shown in **Table 4**. This is the firing point. Make a mark on the drive sheave in line with the fixed pointer. Then, make 4 marks on each side of the firing mark, 1/16 in. (27 mm) from the mark and 1/16 in. apart. Each mark represents 1° of crankshaft rotation, or 1° of timing.

> NOTE: *Make the marks with a fine, felt marker. Don't scribe them on the drive sheave; the sheave is not keyed to the crankshaft, and when it is removed for a subsequent repair job, it is unlikely that it will be installed in the same position.*

39

40

41

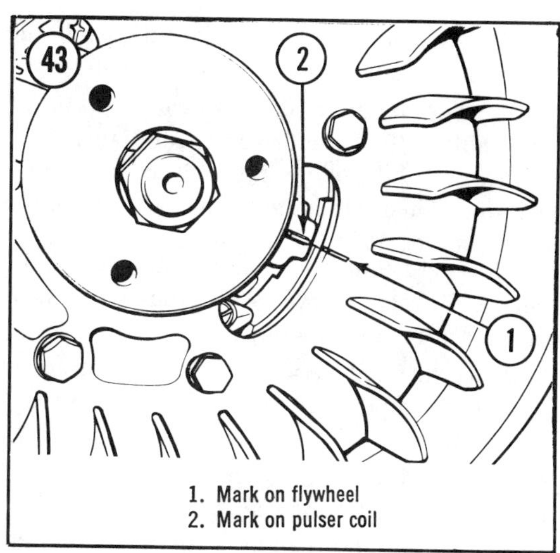

1. Mark on flywheel
2. Mark on pulser coil

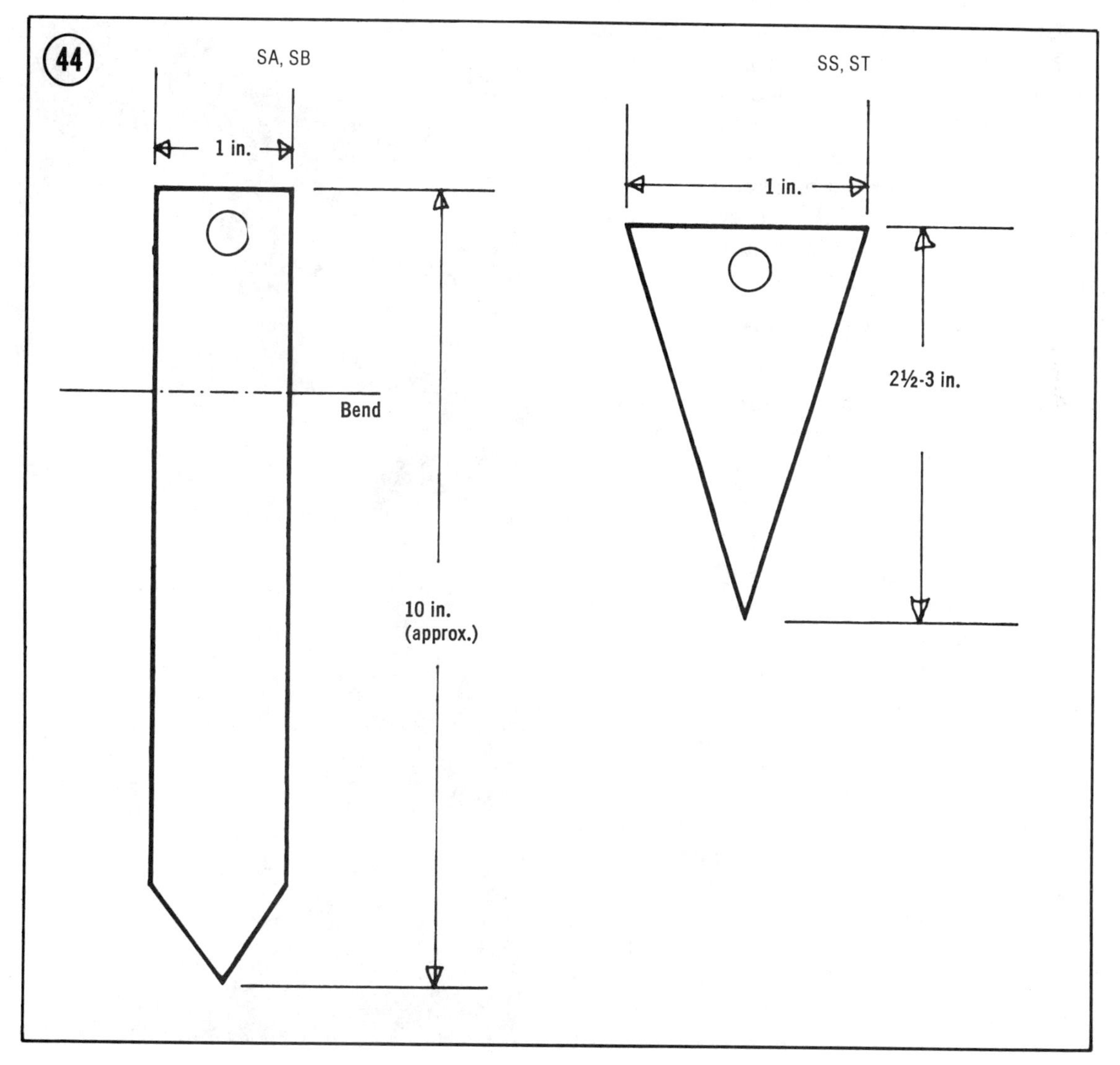

4. Remove the timing gauge and install the spark plugs and connect the high-tension leads.

5. Remove the drive belt by pressing in on the moveable half of the driven sheave, turning it clockwise, and rolling the belt over it (**Figure 46**). Then, disconnect the belt from the drive sheave and install the guard.

6. Connect a timing light and tachometer to the engine in accordance with the instrument manufacturer's instructions.

7. Start the engine and allow it to warm up to operating temperature — 4 or 5 minutes should be sufficient.

8. Slowly increase the engine speed from idle to the dynamic timing speed shown in **Table 4** and direct the timing light at the fixed pointer. If the timing is correct, the firing point mark on the drive sheave will line up with the fixed pointer. If the mark does not line up with the pointer, note how far retarded or advanced it is (in degrees), for reference and readjust the timing as described earlier for static timing. On models with contact breaker points, check the dynamic timing of the left cylinder in the same manner just described for the right, and adjust timing for the left cylinder.

45

46

CARBURETION

There are 5 adjustments possible with either of the 2 carburetors that are used on Kawasaki snowmobiles. The adjustments are interdependent and should be carried out in the order that follows for best results.

a. Starter cable free play
b. Throttle cable free play (and slide synchronization on the dual carburetor models)
c. Idle mixture/throttle stop
d. Slide needle
e. Main jet/metering adjustment

Starter Cable Free Play

With the starter lever down (off), disconnect the boot from the cable adjuster (**Figure 47**) and pull up on the cable sheath to check the free play. It should be 1.5 mm ($\frac{1}{16}$ in.). If it is less, the carburetor will run rich. If adjustment is required, loosen the locknut and turn the cable adjuster in or out until the free play is correct. Then, hold the adjuster to prevent it from turn-

47

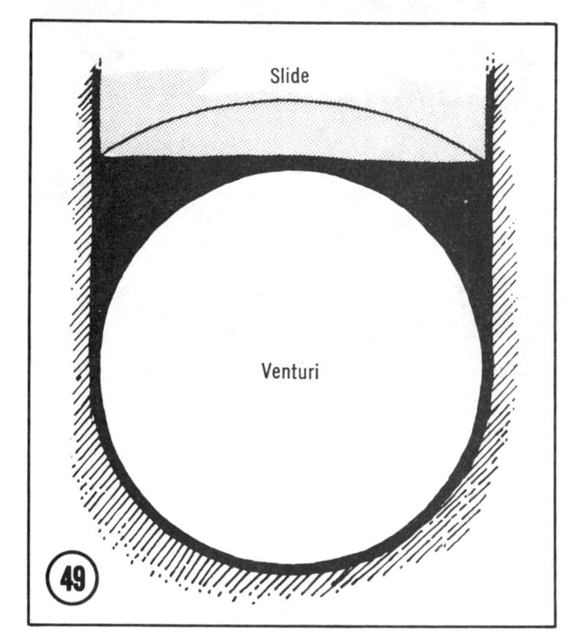

ing further and tighten the locknut and reseat the boot.

Throttle Cable Free Play

1. Remove the air intake silencer (**Figure 48**).

2. Fully open the throttle hand lever and feel inside the carburetor bore to make sure the rear edge of the throttle slide is clear of the venturi (**Figure 49**). If adjustment is required, loosen the locknut on the cable adjuster and turn the adjuster as required. Then, tighten the locknut.

> NOTE: *On dual carburetor models all adjustments must be equal for both carburetors.*

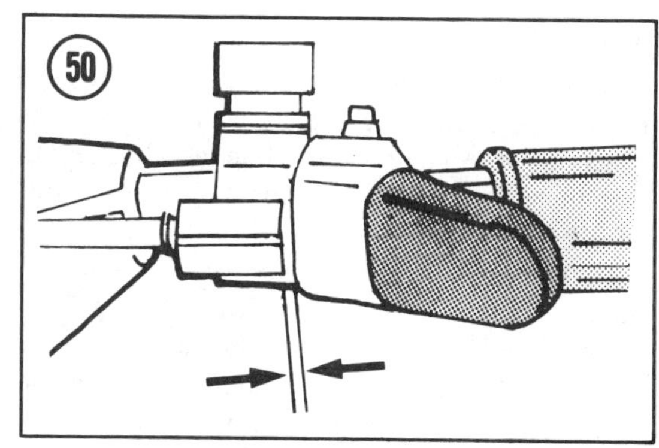

3. Run the throttle stop screw out until its tip is flush with the inside of the slide bore. Then run the screw in the number of turns specified in **Table 5** for the basic idle setting. Close the throttle and allow the slide to rest on the throttle stop screw.

4. Operate the throttle hand control and check for minimal free play (**Figure 50**). If adjustment is required, loosen the locknut and turn the cable adjuster as required. Then tighten the locknut. On dual carburetor models, make sure the throttle slides move equal amounts.

Idle Mixture/Throttle Stop

1. Slowly turn the idle mixture screw (**Figure 51**) in until it just seats. *Don't tighten the screw or it will be damaged.* Then, back the screw out the number of turns specified in **Table 5**. On dual carburetor models, turn the screws out an equal number of turns.

2. Install the air intake silencer. Position the machine with the skis against a wall and raise the rear of the machine and support it so the track is clear of the ground and free to rotate. Connect a tachometer to the engine in accordance with the instrument manufacturer's instructions.

3. Start the engine and allow it to warm up to operating temperature.

4. Set the idle speed with the throttle stop screw (**Figure 52**) as specified in **Table 5**. On dual carburetor models, turn the throttle stop screws equal amounts.

NOTE

On dual carburetor models, synchronize the carburetors as described later (see ***Carburetor Synchronization****) before installing the air intake silencer.*

Slide Needle Position

Mid-range performance is controlled by slide needle position. To check its correctness, first warm up the engine and then accelerate the machine hard over hard-packed snow and "feel" the engine response.

If the engine seems starved or if it knocks, the mid-range mixture is too lean. In this case, remove the slide assembly from the carburetor and raise the needle in the slide by moving the needle clip to the next lower notch (**Figure 53**). Then, recheck the performance and if the mixture is still too lean, raise the needle one more position; move the needle one position at a time.

If the engine "blubbers" or if there is excessive exhaust smoke during hard acceleration, the mid-range mixture is too rich. In this case, lower the needle in the slide by moving the clip to a higher notch. Then recheck the performance and if it is still too rich, lower the needle another notch.

CAUTION

After adjusting the slide needle position, check the operation of the throttle to make sure the slide moves freely in its bore and closes when the throttle is released.

Main Jet/Metering Adjustment

Check the condition of the high-speed mixture ratio by conducting a "plug chop" as described under *Jetting* in Chapter Five. For carburetors equipped with a main jet adjuster screw (**Figure 54**), the mixture can be made leaner by turning the screw clockwise and made richer by turning it counterclockwise.

CAUTION

During testing and adjustment, turn the screw no more than 1/8 turn at a time. Massive changes, such as 1/2 turn, may take the mixture beyond the optimum point and yield inaccurate information, or, at the worst, the mixture could be made so lean that engine damage could result.

If the carburetor is not equipped with a main jet adjusting screw, high-speed mixture corrections must be made by changing the main jet size. In this case, also refer to *Jetting* in Chapter Five and make required adjustments and changes as described.

When all the adjustments discussed have been made, test the machine in all modes of carburetor performance — starting, idle, acceleration, and sustained high-speed operation. If performance is deficient in any mode, readjust

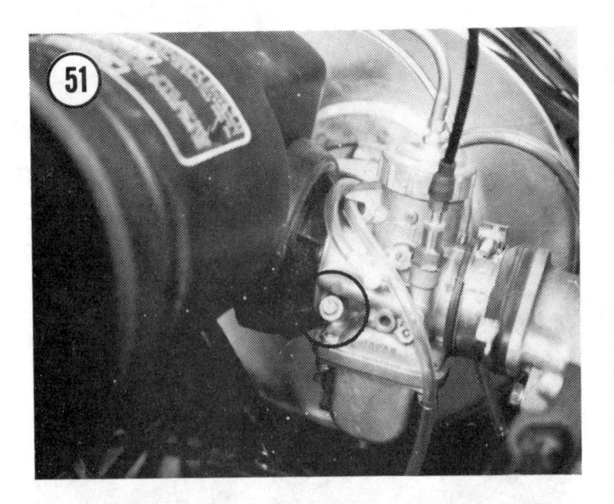

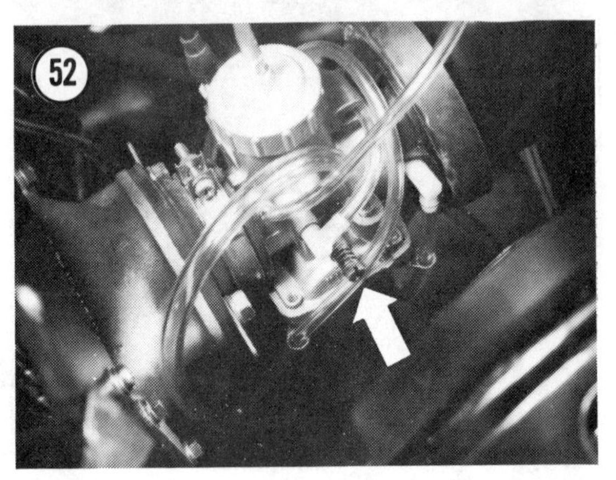

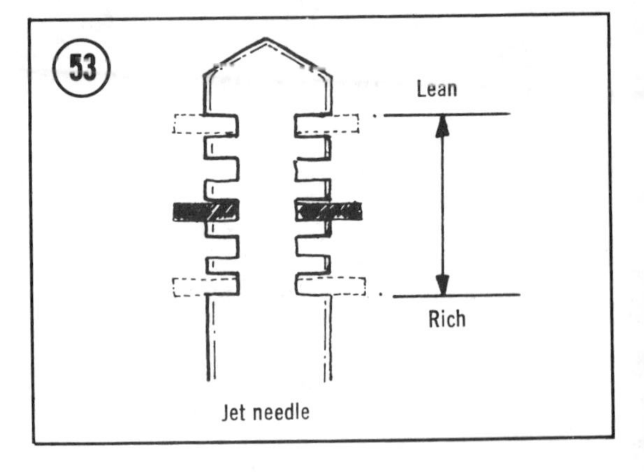

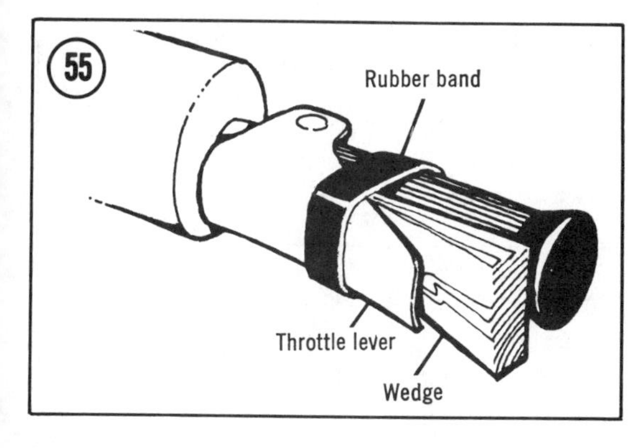

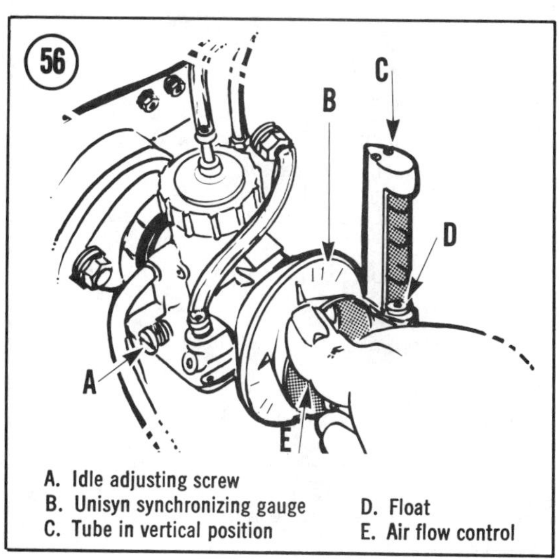

A. Idle adjusting screw
B. Unisyn synchronizing gauge
C. Tube in vertical position
D. Float
E. Air flow control

the carburetor as just described until total performance is satisfactory.

Carburetor Synchronization

On dual carburetor models, the carburetors must be precisely synchronized if maximum performance is to be realized. A carburetor synchronizer like the one described in Chapter One is essential for this work.

1. Make the basic adjustments described earlier (see *Starter Cable Free Play, Throttle Cable Free Play,* and *Idle Mixture/Throttle Stop*).

2. Position the machine with the skis against the wall and raise the rear of the machine and support it so the track is clear of the ground and free to rotate. Connect a tachometer to the engine in accordance with the instrument manufacturer's instructions.

3. Start the engine and allow it to warm up to operating temperature. Bind and wedge the throttle to maintain engine speed at 4,000 rpm (**Figure 55**).

4. Open the air flow control on the air flow meter and place the meter over the throat of the left carburetor, with the tube on the meter positioned vertical (**Figure 56**).

5. Slowly close the air flow control on the meter until the float lines up with a graduated mark on the tube.

6. Without changing the adjustment of the air flow control, move the meter to the other carburetor and observe the float. If its position does not change, the carburetors are synchronized.

7. If adjustment is required, loosen the locknut on the cable adjuster of the carburetor that records the lowest float position on the meter and turn the adjuster out until the float position matches that for the other carburetor.

8. Remove the wedge and rubber band from the throttle lever and allow the engine speed to return to idle. Repeat Steps 4, 5, and 6 and correct the idle synchronization by turning the idle stop screw of the carburetor with the lowest reading in until the float position for that carburetor matches that for the other.

9. When the carburetors have been accurately synchronized, tighten the locknuts on the cable adjusters and install the air intake silencer.

10. Check the engine idle speed and, if necessary, set it to the value shown in **Table 5** by turning the idle stop screws equal amounts.

Table 1 SERVICE INTERVALS

Interval	Item
Before each ride	Check fuel level Check oil level Inspect drive belt Check operation of engine stop switch Check operation of tether switch Check operation of lights Check coolant level
First 5-10 hours or 50-100 miles (80-160 km)	Check ignition timing Check starter cable free play Check throttle cable free play Check carburetor adjustments Check track adjustments Change gear case oil
Every 20 hours or 300 miles (480 km)	Check track adjustments Adjust brake Inspect ski runners (skags) Check battery electrolyte
Every 40 hours or 600 miles (960 km)	Clean, regap spark plugs Inspect drive belt Check ski alignment Check fasteners for tightness
Every 60 hours or 900 miles (1,440 km)	Check starter cable free play Check throttle cable free play Replace fuel filter Adjust drive sheaves Inspect sheaves for wear
Annually	Remove carbon from engine upper end and exhaust system Check and adjust ignition timing Clean, inspect, and adjust carburetor Clean and inspect fuel tank Inspect chain tensioner guides Inspect track

Table 2 DRIVE BELT SPECIFICATIONS

Model	Width (Top Surface)		Outside Circumference
SA and SB	New	31.7 mm (1¼ in.)	1,105 mm (43.5 in.)
	Service limit	26 mm (1.02 in.)	
SS and ST	New	31.7 mm	1,184 mm (46.6 in.)
	Service limit	26 mm (1.02 in.)	

Table 3 LUBRICATION INTERVALS

Interval	Item	Lubricant
Every 20 hours	Lubricate control cables	LPS, Break-Free, WD40, Dri-Slide
	Check chaincase oil level	Dexron II automatic transmission fluid
	Check gearcase oil level (SS models)	SAE 10W-30 motor oil
Annually	Change chaincase oil	Dexron II automatic transmission fluid
	Change gearcase oil	SAE 10W-30
	Lubricate steering bushings	Low-temperature grease
	Lubricate suspension axles	Low-temperature grease
	Lubricate recoil starter	Low-temperature grease
	Lubricate ski spindles	Low-temperature grease
	Lubricate driven sheave shaft and sliding sheave	Low-temperature grease
	Lubricate tie-rod ends	Low-temperature grease

Table 4 IGNITION SPECIFICATIONS

Model	Spark Plug (NGK)	Gap, mm (in.)	Static Timing, mm (in.) BTDC	Dynamic Timing, °BTDC @ rpm
SA340-A1	BR-8ES	0.6 (0.024)	3.53 (0.139)	25° @ 6,500
SA440-A1	BR-8ES	0.6 (0.024)	3.53 (0.139)	25° @ 6,500
SB340-A1	B-9EV	0.6 (0.024)	1.6 (0.063)	17° @ 6,500
SB440-A1	B-9EV	0.6 (0.024)	1.6 (0.063)	17° @ 6,500
SA340-A2	BR-8ESA	0.8 (0.032)	3.53 (0.139)	25° @ 6,500
SB440-A2	BR-9EV	0.6 (0.024)	1.86 (0.073)	18° @ 6,500
SA340-A3 and later	BR-8ESA	0.8 (0.032)	0.145 (0.006)	25° @ 6,500
SB440-A3 and later	BR-9EV	0.6 (0.024)	1.86 (0.073)	18° @ 6,500
SS340 all	BR-9EV	0.6 (0.024)	2.75 (0.108)	22° @ 6,500
ST440 all	BR-9EV	0.6 (0.024)	1.86 (0.073)	18° @ 6,500

Table 5 CARBURETOR SETTINGS

Model	Idle Speed, rpm	Air Screw, turns out	Throttle Stop Screw, turns out
SA340-A1	2,000	1.5	3
SA440-A1	2,000	1.5	3
SB340-A1	2,000	1.5	3
SB440-A1	2,000	2	6
SA340-A2	2,000	1.5	3
SB440-A2	2,500	1.5	3
SA340-A3 and later	2,000	1.5	6
SB440-A3 and later	2,500	1.5	6
SS340 all	3,000	1.5	3
ST440 all	3,000	1.5	3

CHAPTER THREE

TROUBLESHOOTING

Diagnosing snowmobile ills is relatively simple if you use orderly procedures and keep a few basic principles in mind.

Never assume anything. Do not overlook the obvious. If you are riding along and the snowmobile suddenly quits, check the easiest most accessible problem spots first. Is there gasoline in the tank? Has a spark plug wire fallen off? Check the ignition switch. Maybe that last mogul caused you to accidentally switch the emergency switch to OFF or pull the emergency stop "tether" string.

If nothing obvious turns up in a cursory check, look a little further. Learning to recognize and describe symptoms will make repairs easier for you or a mechanic at the shop. Describe problems accurately and fully. Saying that "it won't run" isn't the same as saying "it quit at high speed and wouldn't start," or that "it sat in my garage for three months and then wouldn't start."

Gather as many symptoms together as possible to aid in diagnosis. Note whether the engine lost power gradually or all at once, what color smoke (if any) came from the exhaust, and so on. Remember that the more complicated a machine is, the easier it is to troubleshoot because symptoms point to specific problems.

You do not need fancy equipment or complicated test gear to determine whether repairs can be attempted at home. A few simple checks could save a large repair bill and time lost while the snowmobile sits in a dealer's service department. On the other hand, be realistic and do not attempt repairs beyond your abilities. Service departments tend to charge heavily for putting together disassembled components that may have been abused. Some won't even take on such a job — so use common sense; don't get in over your head.

OPERATING REQUIREMENTS

An engine needs 3 basics to run properly; correct gas/air mixture, compression, and a spark at the right time. If one or more are missing, the engine will not run. The electrical system is the weakest link of the three. More problems result from electrical breakdowns than from any other source. Keep that in mind before you begin tampering with carburetor adjustments.

If the snowmobile has been sitting for any length of time and refuses to start, check the battery (if the machine is so equipped) for a charged condition first, and then look to the gasoline delivery system. This includes the tank, fuel petcock, lines, and the carburetor. Sediment may have formed in the tank, obstructing fuel flow. Gasoline deposits may have gummed up carburetor jets and air passages. Gasoline tends to lose its potency after standing for long periods. Condensation may contaminate it with water. Drain old gas and try starting with a fresh tankful.

Compression, or the lack of it, usually enters the picture only in the case of older machines. Worn or broken pistons, rings, and cylinder bores could prevent starting. Generally a gradual power loss and harder and harder starting will be readily apparent in this case.

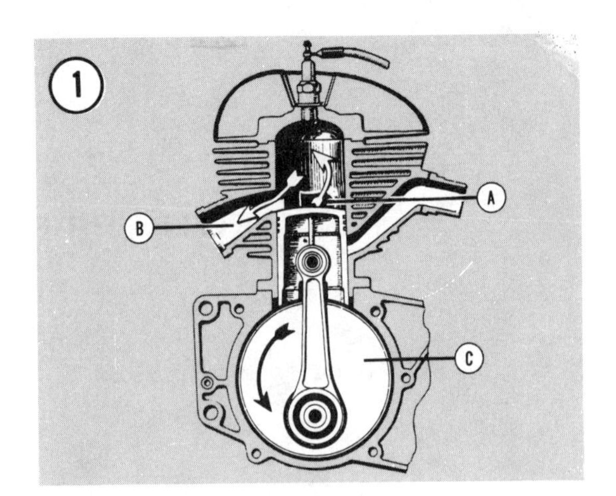

PRINCIPLES OF 2-CYCLE ENGINES

The following is a general discussion of a typical 2-cycle piston-port engine.

Figures 1 through 4 illustrate operating principles of piston port engines. During this discussion, assume that the crankshaft is rotating counterclockwise. In **Figure 1**, as the piston travels downward, a transfer port (A) between the crankcase and the cylinder is uncovered. Exhaust gases leave the cylinder through the exhaust port (B), which is also opened by downward movement of the piston. A fresh fuel/air charge, which has previously been compressed slightly by the descending piston, travels from the crankcase (C) to the cylinder through transfer ports (A) as the ports open. Since the incoming charge is under pressure, it rushes into the cylinder quickly and helps to expel exhaust gases from the previous cycle.

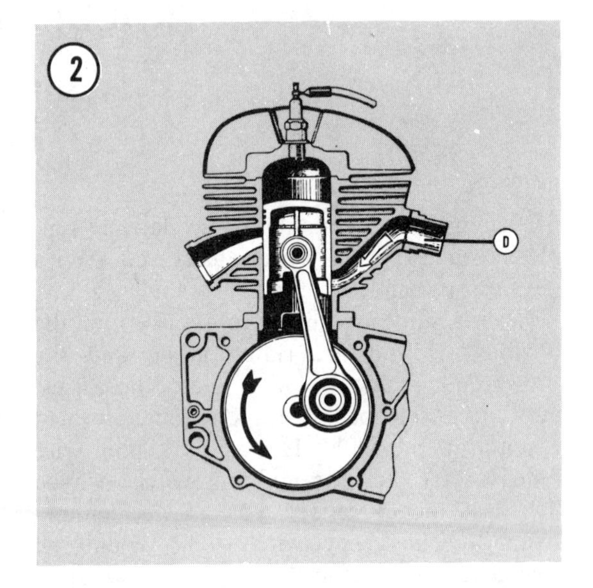

Figure 2 illustrates the next phase of the cycle. As the crankshaft continues to rotate, the piston moves upward, closing the exhaust and transfer ports. As the piston continues upward, the air/fuel mixture in the cylinder is compressed. Notice also that a low pressure area is created in the crankcase by the ascending piston at the same time. Further upward movement of the piston uncovers the intake port (D). A fresh fuel/air charge is then drawn into the crankcase through the intake port because of the low pressure created by the upward piston movement.

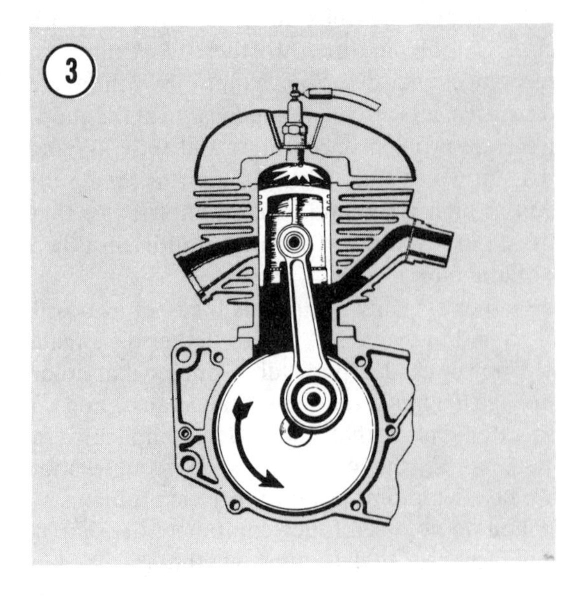

The third phase is shown in **Figure 3**. As the piston approaches top dead center, the spark plug fires, igniting the compressed mixture. The piston is then driven downward by the expanding gases.

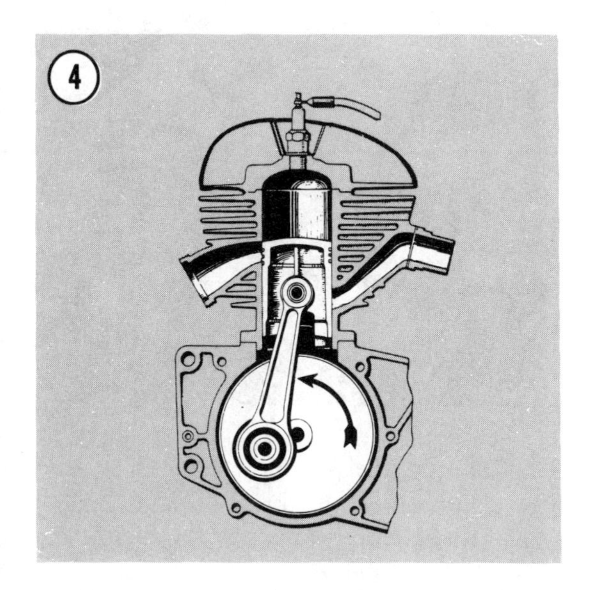

When the top of the piston uncovers the exhaust port, the fourth phase begins, as shown in **Figure 4**. The exhaust gases leave the cylinder through the exhaust port. As the piston continues downward, the intake port is closed and the mixture in the crankcase is compressed in preparation for the next cycle. Every downward stroke of the piston is a power stroke.

ENGINE STARTING

An engine that refuses to start or is difficult to start can try the patience of anyone. More often than not, the problem is very minor and can be found with a simple and logical troubleshooting approach.

The following items provide a beginning point from which to isolate engine starting problems.

Engine Fails to Start

Perform the following spark test to determine if the ignition system is operating properly.

1. Remove a spark plug.
2. Connect the spark plug connector to the spark plug and clamp the base of spark plug to a good grounding point on the engine. A large alligator clip makes an ideal clamp. Position the spark plug so you can observe the electrode.
3. Turn on the ignition and crank the engine over. A fat blue spark should be evident across the spark plug electrode.

WARNING

On machines equipped with CDI (capacitor discharge ignition), do not hold spark plug, wire, or connector or a serious electrical shock may result.

4. If the spark is good, check for one or more of the following possible malfunctions:
   a. Fouled or defective spark plugs
   b. Obstructed fuel filter or fuel line
   c. Defective fuel pump
   d. Leaking head gasket — perform the compression test
5. If spark is not good, check for one or more of the following:
   a. Burned, pitted, or improperly gapped breaker points
   b. Weak ignition coil or condenser
   c. Loose electrical connections
   d. Defective CDI components — have CDI system checked by an authorized dealer.

Engine Difficult to Start

Check for one or more of the following possible malfunctions:

a. Fouled spark plugs
b. Improperly adjusted choke
c. Defective or improperly adjusted breaker points
d. Contaminated fuel system
e. Improperly adjusted carburetor
f. Weak ignition coil
g. Incorrect fuel mixture
h. Crankcase drain plugs loose or missing
i. Poor compression — perform the compression test

Engine Will Not Crank

Check for one or more of the following possible malfunctions:

a. Defective recoil starter
b. Seized piston

c. Seized crankshaft bearings
d. Broken connecting rod

Compression Test

Perform a compression test to determine condition of piston ring sealing qualities, piston wear, and condition of head gasket seal.

1. Remove the spark plugs. Insert a compression gauge in one spark plug hole **(Figure 5)**. Refer to Chapter One for a suitable type of compression tester.

2. Crank the engine vigorously and record compression reading. Repeat for other cylinder. Compression readings should be from 120 to 175 psi (8.44-12.30 kg/cm^2). Maximum allowable variation between cylinders is 10 psi (0.70 kg/cm^2).

3. If compression is low or variance between cylinders is excessive, check for defective head gaskets, damaged cylinders and pistons, or stuck piston rings.

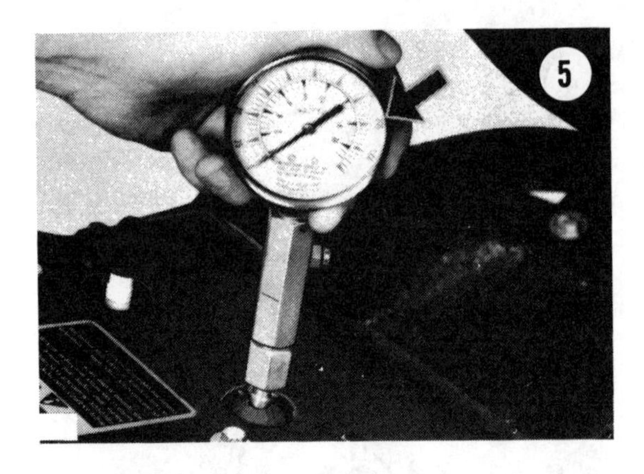

ENGINE PERFORMANCE

The following items are a starting point from which to isolate a performance malfunction. It is assumed the engine runs but is not operating at peak efficiency.

The possible causes for each malfunction are listed in a logical sequence and in order of probability.

Engine Will Not Idle

a. Carburetor incorrectly adjusted
b. Fouled or improperly gapped spark plugs
c. Head gasket leaking — perform the compression test
d. Fuel mixture incorrect
e. Spark advance mechanism not retarding
f. Obstructed fuel pump impulse tube
g. Crankcase drain plugs loose or missing

Engine Misses at High Speed

a. Fouled or improperly gapped spark plugs
b. Defective or improperly gapped breaker points
c. Improper ignition timing
d. Defective fuel pump
e. Improper carburetor high-speed adjustment or improper main jet selection
f. Weak ignition coil
g. Obstructed fuel pump impulse tube
h. Obstructed fuel filter

Engine Overheating

a. Too lean fuel mixture — incorrect carburetor adjustment or jet selection
b. Improper ignition timing
c. Incorrect spark plug heat range
d. Intake system or crankcase air leak
e. Cooling fan belt broken or slipping
f. Cooling fan defective
g. Damaged or blocked cooling fins

Engine Smokes and Runs Rough

a. Carburetor adjusted incorrectly — mixture too rich
b. Incorrect fuel/oil mixture
c. Choke not operating properly
d. Obstructed muffler
e. Water or other contaminates in fuel

Engine Loses Power

a. Carburetor incorrectly adjusted
b. Engine overheating
c. Defective or improperly gapped breaker points
d. Improper ignition timing
e. Incorrectly gapped spark plugs
f. Weak ignition coil

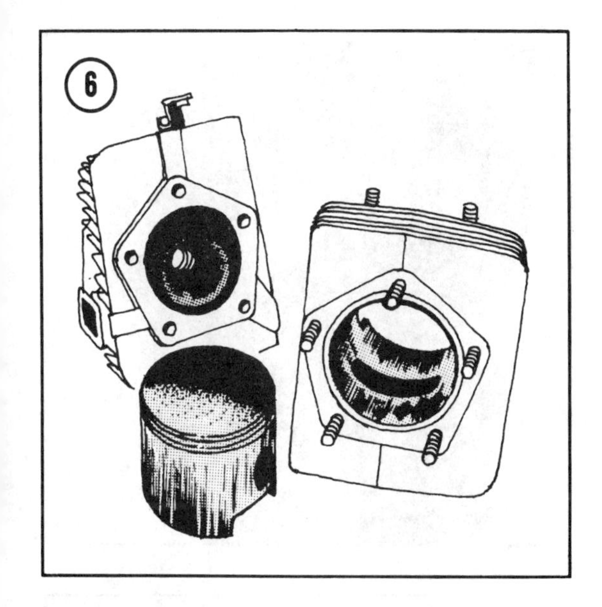

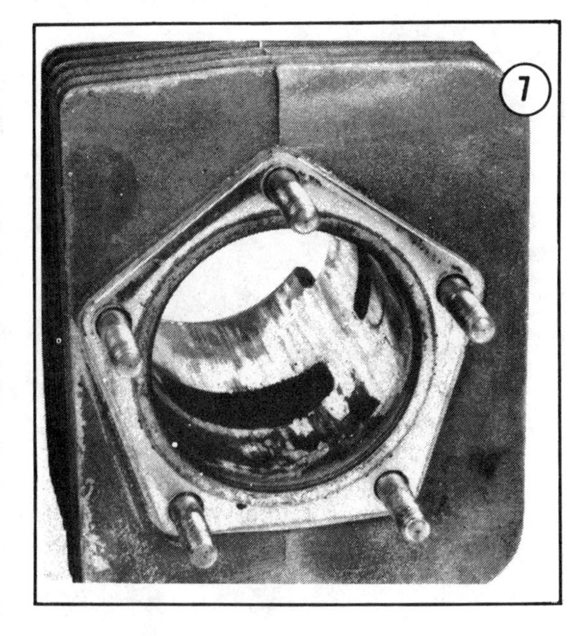

g. Obstructed muffler
h. Dragging brakes

Engine Lacks Acceleration

a. Carburetor mixture too lean
b. Defective fuel pump
c. Incorrect fuel/oil mixture
d. Defective or improperly gapped breaker points
e. Improper ignition timing
f. Dragging brakes

ENGINE FAILURE ANALYSIS

Overheating is the major cause of serious and expensive engine failures. It is important that each snowmobile owner understand all the causes of engine overheating and take the necessary precautions to avoid expensive overheating damage. Proper preventive maintenance and careful attention to all potential problem areas can often prevent serious malfunction before it happens.

Fuel

All snowmobile engines rely on a proper fuel/oil mixture for engine lubrication. Always use an approved oil and mix the fuel carefully as described in Chapter One.

Gasoline must be of sufficiently high octane (88 or higher) to avoid "knocking" and "detonation."

Fuel/Air Mixture

Fuel/air mixture is determined by carburetor adjustment or main jet selection. Always adjust carburetors carefully and pay particular attention to avoid a "too-lean" mixture.

Heat

Excessive external heat on the engine can be caused by the following:

a. Hood louvers plugged with snow
b. Damaged or plugged cylinder and head cooling fins
c. Slipping or broken fan belt
d. Damaged cooling fan
e. Operating snowmobile in hot weather
f. Plugged or restricted exhaust system

See **Figures 6 and 7** for examples of cylinder and piston scuffing caused by excessive heat.

Dirt

Dirt is a potential problem for all snowmobiles. The air intake silencers on all models are not designed to filter incoming air. Avoid running snowmobiles in areas that are not completely snow covered.

Ignition Timing

Ignition timing that is too far advanced can cause "knocking" or "detonation." Timing that is too retarded causes excessive heat buildup in the cylinder exhaust port area.

Spark Plugs

Spark plugs must be of a correct heat range. Too hot a heat range can cause pre-ignition and detonation which can ultimately result in piston burn-through as shown in **Figure 8**.

Refer to Chapter Two for recommended spark plugs.

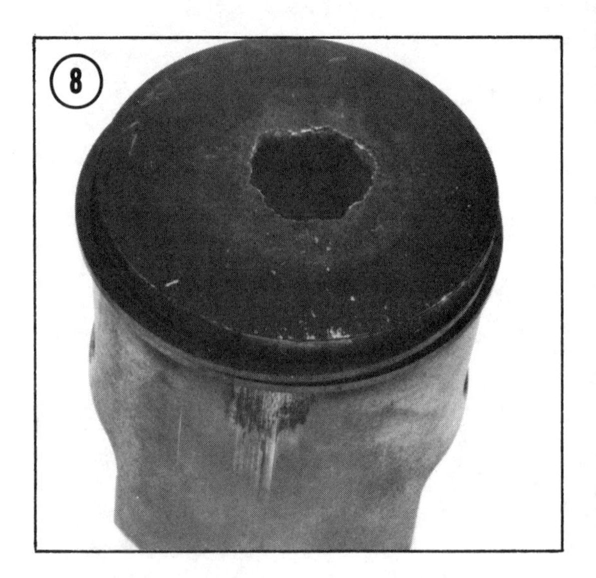

Pre-ignition

Pre-ignition is caused by excessive heat in the combustion chamber due to a spark plug of improper heat range and/or too lean a fuel mixture. See **Figure 9** for an example of a melted and scuffed piston caused by pre-ignition.

Detonation (Knocking)

Knocking is caused by a too lean fuel mixture and/or too low octane fuel.

ELECTRICAL SYSTEM

The following items provide a starting point from which to troubleshoot electrical system malfunctions. The possible causes for each malfunction are listed in a logical sequence and in order of probability.

Ignition system malfunctions are outlined under *Engine Starting* and *Engine Performance*.

Lights Will Not Light

a. Bulbs are burned out
b. Loose electrical connections
c. Defective switch
d. Defective lighting coil or alternator
e. Defective voltage regulator
f. Defective battery (electric-start models)

Bulbs Burn Out Rapidly

a. Incorrect bulb type
b. Defective voltage regulator

Lights Too Bright or Too Dim

a. Defective voltage regulator
b. Defective alternator

Discharged Battery (Electric-start Models)

a. Defective battery
b. Low electrolyte level

c. Dirty or loose electrical connections
d. Defective voltage regulator
e. Defective lighting coil
f. Defective rectifier
g. Defective circuit breaker

Cracked Battery Case

a. Discharged battery allowed to freeze
b. Improperly installed hold-down clamp
c. Improperly attached battery cables

Starter Motor Does Not Operate

a. Loose electrical connections
b. Discharged battery
c. Defective starter solenoid
d. Defective starter motor
e. Defective circuit breaker
f. Defective ignition switch

Poor Starter Performance

a. Commutator or brushes worn, dirty, or oil soaked
b. Binding armature
c. Weak brush springs
d. Armature open, shorted, or grounded

POWER TRAIN

The following items provide a starting point from which to troubleshoot power train malfunctions. The possible causes for each malfunction are listed in a logical sequence and in order of probability. Also refer to *Drive Belt Wear Analysis*.

Drive Belt Not Operating Smoothly in Drive Sheave

a. Face of drive sheave is rough, grooved, pitted, or scored
b. Defective drive belt

Uneven Drive Belt Wear

a. Misaligned drive and driven sheaves
b. Loose engine mounts

Glazed Drive Belt

a. Excessive slippage
b. Oil or grease on sheave surfaces

Drive Belt Worn Narrow in One Place

a. Excessive slippage caused by stuck track
b. Too high engine idle speed

Drive Belt Too Tight at Idle

a. Engine idle speed too fast
b. Distance between sheaves incorrect
c. Belt length incorrect

Drive Belt Edge Cord Failure

a. Misaligned sheaves
b. Loosen engine mounting bolts

Brake Not Holding Properly

a. Incorrect brake cable adjustment
b. Brake lining or pucks worn
c. Oil saturated brake lining or pucks
d. Sheared key on brake pulley or disc
e. Incorrect brake adjustment

Brake Not Releasing Properly

a. Weak or broken return spring
b. Bent or damaged brake lever
c. Incorrect brake adjustment

Leaking Chaincase

a. Gaskets on drive shaft bearing flanges or secondary shaft bearing flanges damaged
b. Damaged O-ring on drive shaft or secondary shaft
c. Cracked or broken chaincase

Rapid Chain and Sprocket Wear

a. Insufficient chaincase oil
b. Misaligned sprockets
c. Broken chain tension blocks

DRIVE BELT WEAR ANALYSIS

Frayed Edge

A rapidly wearing drive belt with a frayed edge cord indicates the drive belt is misaligned (see **Figure 10**). Also check for loose engine mounting bolts.

Worn Narrow in One Section

Excessive slippage due to a stuck track or too high an engine idle speed will cause the drive belt to be worn narrow in one section (see **Figure 11**).

Belt Disintegration

Drive belt disintegration is usually caused by misalignment. Disintegration can also be caused by using an incorrect belt or oil or grease on sheave surfaces (see **Figure 12**).

Sheared Cogs

Sheared cogs as shown in **Figure 13** are usually caused by violent drive sheave engagement. This is an indication of a defective or improperly installed drive sheave.

SKIS AND STEERING

The following items provide a starting point from which to troubleshoot ski and steering malfunctions. The possible causes for each malfunction are listed in a logical sequence and in order or probability.

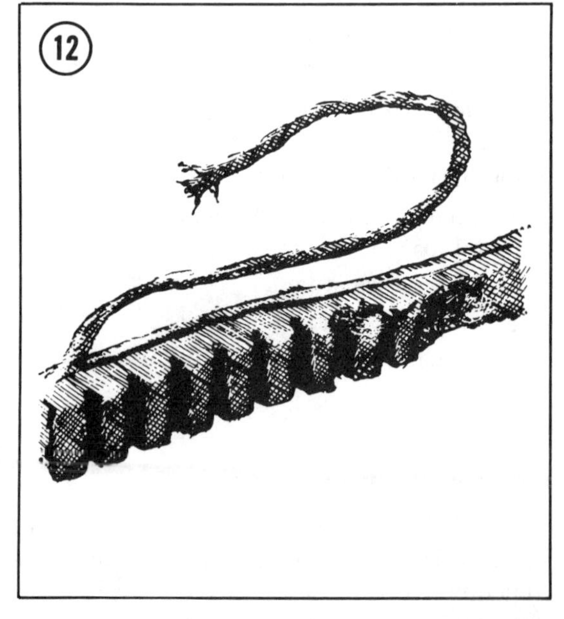

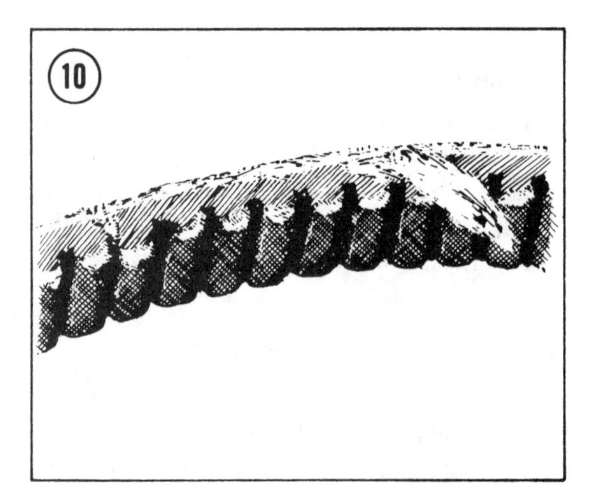

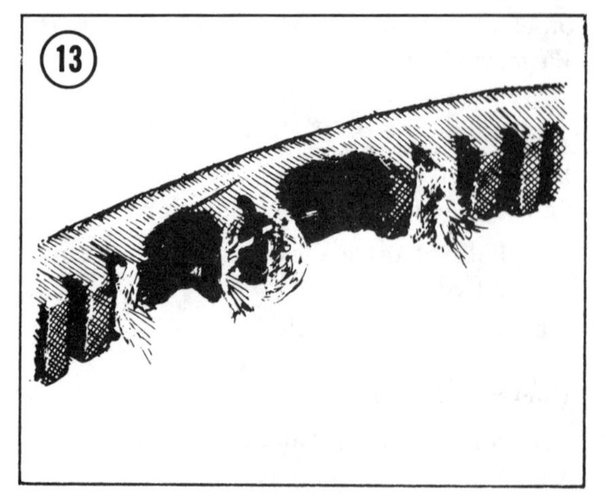

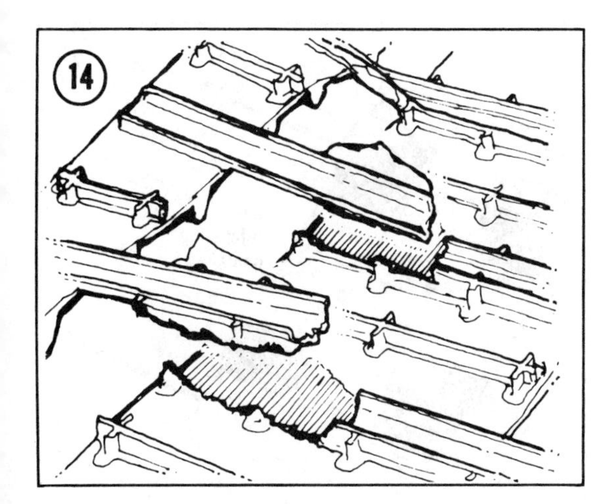

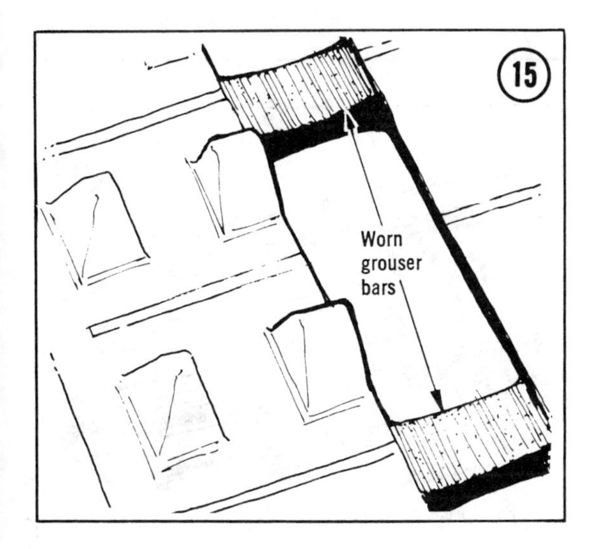

Loose Steering

a. Loose steering post bushing
b. Loose steering post or steering column cap screw
c. Loose tie rod ends
d. Worn spindle bushings
e. Stripped spindle splines

Unequal Steering

a. Improperly adjusted tie rods
b. Improperly installed steering arms

Rapid Ski Wear

a. Skis misaligned
b. Worn out ski wear rods (Skags)
c. Worn out spring wear plate

TRACK ASSEMBLY

The following items provide a starting point from which to troubleshoot track assembly malfunctions. The possible causes for each are listed in a logical sequence and in order of probability. Also refer to *Track Wear Analysis*.

Frayed Track Edge

Track misaligned

Track Grooved on Inner Surface

a. Track too tight
b. Frozen rear idler shaft bearing

Track Drive Ratcheting

Track too loose

Rear Idlers Turning on Shaft

Frozen rear idler shaft bearings

TRACK WEAR ANALYSIS

The majority of track failures and abnormal wear patterns are caused by negligence, abuse, and poor maintenance. The following items illustrate typical examples. In all cases the damage could have been avoided by proper maintenance and good operator technique.

Obstruction Damage

Cuts, slashes, and gouges in the track surface are caused by hitting obstructions such as broken glass, sharp rocks or buried steel (see **Figure 14**).

Worn Grouser Bars

Excessively worn grouser bars are caused by snowmobile operation over rough and non-snow covered terrain such as gravel roads and highway roadsides (**Figure 15**).

Lug Damage

Lug damage as shown in **Figure 16** is caused by lack of snow lubrication.

Ratcheting Damage

Insufficient track tension is a major cause of ratcheting damage to the top of the lugs (**Fig-**

ure 17). Ratcheting can also be caused by too great a load and constant "jack-rabbit" starts.

Over-tension Damage

Excessive track tension can cause too much friction on the wear bars. This friction causes the wear bars to melt and adhere to the track grouser bars. See **Figure 18**. An indication of this condition is a "sticky" track that has a tendency to "lock up."

Loose Track Damage

A track adjusted too loosely can cause the outer edge to flex excessively. This results in the type of damage shown in **Figure 19**. Excessive weight can also contribute to the damage.

Impact Damage

Impact damage as shown in **Figure 20** causes the track rubber to open and expose the cord. This frequently happens in more than one place. Impact damage is usually caused by riding on rough or frozen ground or ice. Also, insufficient track tension can allow the track to pound against the track stabilizers inside the tunnel.

Edge Damage

Edge damage as shown in **Figure 21** is usually caused by tipping the snowmobile on its side to clear the track and allowing the track edge to contact an abrasive surface.

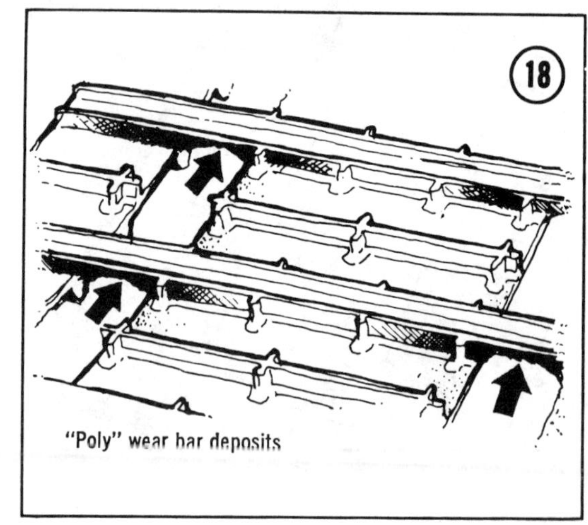

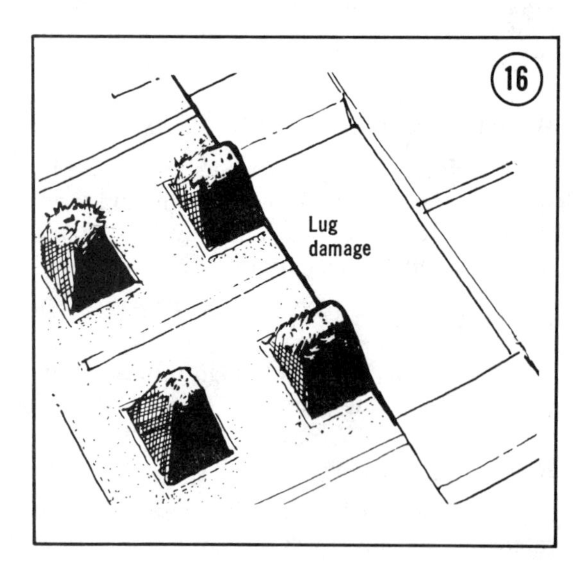

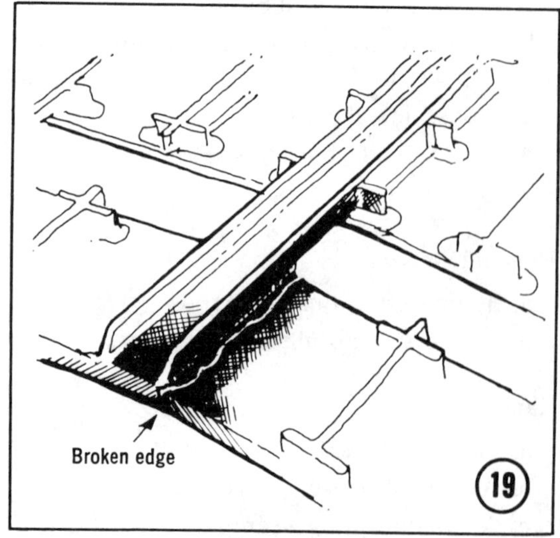

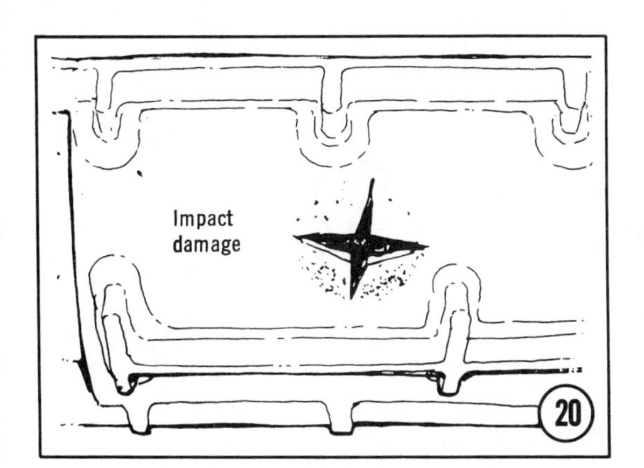
Impact
damage
20

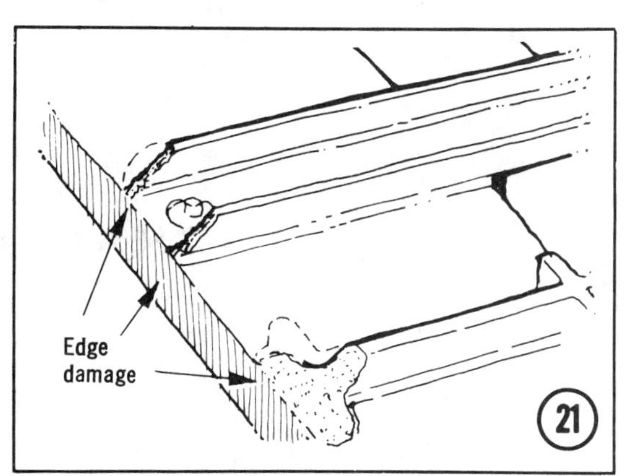
Edge
damage
21

CHAPTER FOUR

ENGINE

All Kawasaki snowmobiles are powered with a twin-cylinder 2-stroke engines. Refer to Chapter Two for *Principles of Operation* of 2-cycle engines.

Work on all engines is essentially the same, although some differences exist between models (air cooling vs liquid cooling, piston-port intake control vs reed valve, etc.), and where differences are important to the work being described they will be pointed out.

All engines are equipped with ball-type main crankshaft bearings and needle bearings on both ends of the connecting rods. Crankshaft components are available as individual parts; however, other than replacement of outer seals, it's recommended that crankshaft work be entrusted to a dealer or other competent engine specialist. Experience and special measuring equipment, along with a hydraulic press, are required to service the crankshaft. Engine removal and disassembly are covered in detail so that cost of lower end service can be minimized.

An upper end overhaul is within the abilities of the average hobbyist mechanic equipped with a reasonable range of hand tools and inside and outside micrometers.

Before beginning any work on the engine, read Chapter One, and particularly the headings *Service Hints*, *Tools, Expendable Supplies,* and *Working Safely*. The information they contain will contribute to the efficiency, effectiveness, and safety of your work.

Also, carefully read and understand the appropriate procedures in this chapter before beginning work. It is a good idea to physically compare the instructions with the actual machine, as far as possible, beforehand to familiarize yourself with the procedures and the equipment.

A complete upper end overhaul can be performed without removing the engine from the machine. However, if you have the time you may find it more convenient to have the engine on a workbench, and should a small part or tool or bit of dirt fal into the crankcase, it is more easily removed if the engine can be inverted. For these reasons, engine removal is described first.

The removal/installation procedures, as well as most service procedures, are virtually identical for all Kawasaki engines. Exceptions are noted where they apply.

When measuring surfaces for critical dimensions, be sure to consult the appropriate table for your particular engine. All tables are at the ends of the chapters. This holds true for critical tightening torques as well.

Finally, if you are disassembling the upper end even for a routine inspection and carbon removal — anticipating no parts replacement

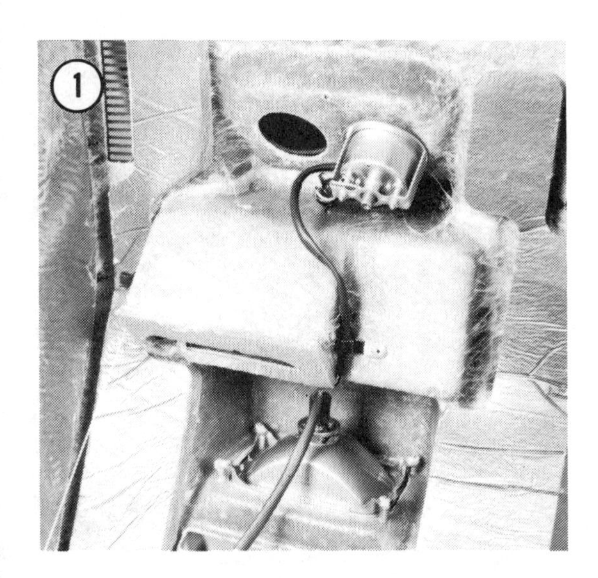

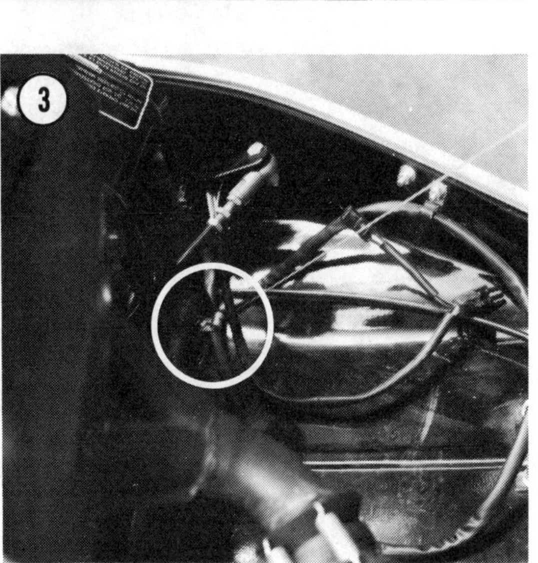

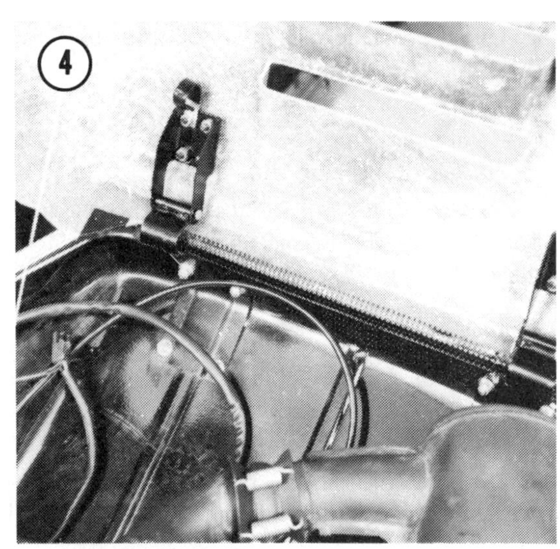

— purchase a gasket set beforehand; you can't reassemble the upper end without it, and if the pistons, rings, and bores are in good condition you will save time by being able to complete the job right away.

ENGINE REMOVAL — AIR-COOLED

Engine removal is virtually identical for all Kawasaki snowmobiles with a few exceptions that are noted as they occur. The machine illustrated is an SA-440-A4 and is the most typical example.

1. Open the hood all the way. Disconnect the speedometer cable and unplug the connector from the headlight (**Figure 1**). Disconnect the wiring harness from the hood clip (**Figure 2**).

2. Disconnect the hood restraining cable (**Figure 3**) and with assistance, support the hood and remove the hood hinge pins (**Figure 4**). Remove the hood and set it well out of the way to prevent it from being damaged.

3. Disconnect the springs that secure the exhaust system headpipe to the manifold (**Figure 5**) and remove the long springs that hold the muffler in its cradle. Remove the exhaust system from the machine. Remove the recoil starter (**Figure 6**).

4. Disconnect the fuel line from the carburetor (**Figure 7**) and plug it to prevent fuel from leaking. Unscrew the starter valve and pull it out of the carburetor. Unscrew the carburetor top and remove the slide assembly.

5. Disconnect the fuel pump pulse line from the crankcase (**Figure 8**). Disconnect the wiring harness connectors.

6. Remove the fan cover (**Figure 9**) and the cooling shroud (**Figure 10**).

7. Unscrew the 3 bolts that connect the starter pulley to the flywheel (**Figure 11**) and remove the pulley and fan belt. Remove the fan case (**Figure 12**).

8. Remove the drive belt guard (**Figure 13**). Press in on the moveable drive sheave, turn it clockwise, and roll the drive belt over it (**Figure 14**). Remove the belt from the drive sheave.

9. Unscrew the engine mounting bolts (**Figure 15** — 2 in front of the engine and 2 behind it).

10. Double check to make sure all hoses, cables, and electrical leads between the engine and the machine have been disconnected. Then, with assistance, lift the engine out of the machine and set it on a workbench.

ENGINE REMOVAL LIQUID-COOLED

Removal of the liquid-cooled engine is virtually the same as for the air-cooled engine. However, the cooling system must first be drained and the coolant hoses disconnected.

WARNING

Do not drain the system or disconnect the coolant hoses when the engine and coolant are hot; serious burns are likely to result.

1. Carefully loosen the radiator cap to relieve system pressure.

2. Place a container (5 quarts or larger) beneath the machine where the coolant overflow hose comes through the body.

3. Disconnect the coolant overflow hose from the reservoir, pass it beneath the engine, and connect it to the coolant drain valve in front of the engine (**Figure 16**).

4. Raise the rear of the machine and support it. This is necessary to aid the coolant to drain from the heat exchanger.

5. Open the drain valve and allow about 10 minutes for the coolant to drain.

5

6

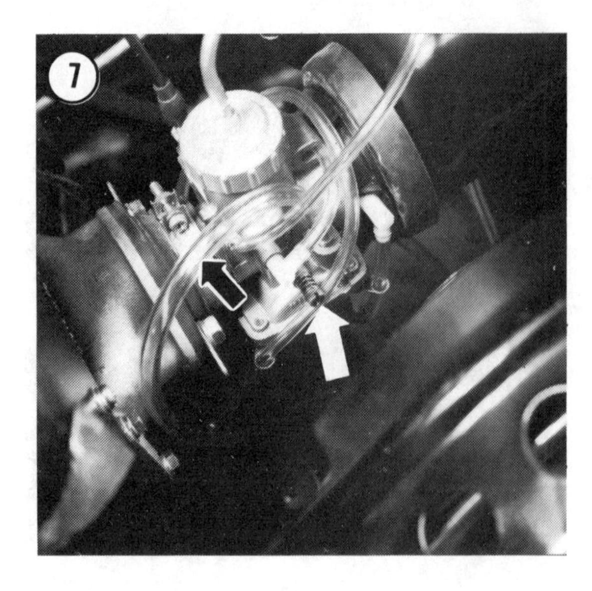
7

8

9

10

11

12

13

6. Disconnect the top coolant hose from the thermostat cover. Disconnect the bottom coolant hose from the water pump (**Figure 17**).

> NOTE: *Some residual coolant will drain from the pump and the lower hose at this point. It's unavoidable and will have to be cleaned up after the engine has been removed from the machine.*

ENGINE INSTALLATION — AIR-COOLED

1. Make sure the rubber engine mount shock absorbers are in place and set the engine into the machine. Tighten the engine mounting bolts finger-tight.
2. With a sheave gauge (**Figure 18**), check and adjust the sheave offset and when it is correct, tighten the engine mounting bolts to the value shown in **Table 1** at the end of this chapter. Then, install the drive belt and the belt guard.
3. Install the fan case (**Figure 19**) and the fan belt and starter pulley (**Figure 20**).
4. Install the cooling shroud (**Figure 21**) and the fan cover (**Figure 22**).
5. Connect the wiring harness connectors and the fuel pump pulse line (**Figure 23**).
6. Install the slide assembly in the carburetor, taking care to line up the nub in the slide bore with the slot in the slide and the needle valve with the needle jet. Screw in the starter valve and connect the fuel line (**Figure 24**).
7. Install the recoil starter assembly (**Figure 25**) and the exhaust system (**Figure 26**). Connect the headpipe-to-manifold springs first, then the muffler hold-down springs.
8. With assistance, align the hood hinges and install the pins (**Figure 27**). Connect the hood restraining cable (**Figure 28**).
9. Route the wiring harness under the hood clip (**Figure 29**) and plug the connector into the headlight (**Figure 30**) and connect the speedometer cable.

ENGINE INSTALLATION — LIQUID-COOLED

Installation of the liquid-cooled engine is essentially the same as that described for air-

14

15

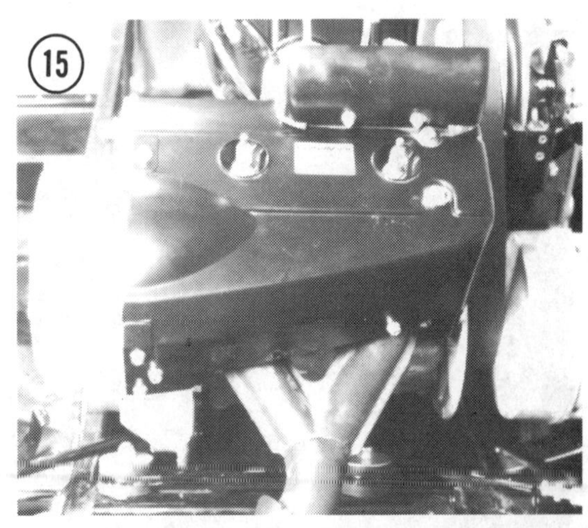

16

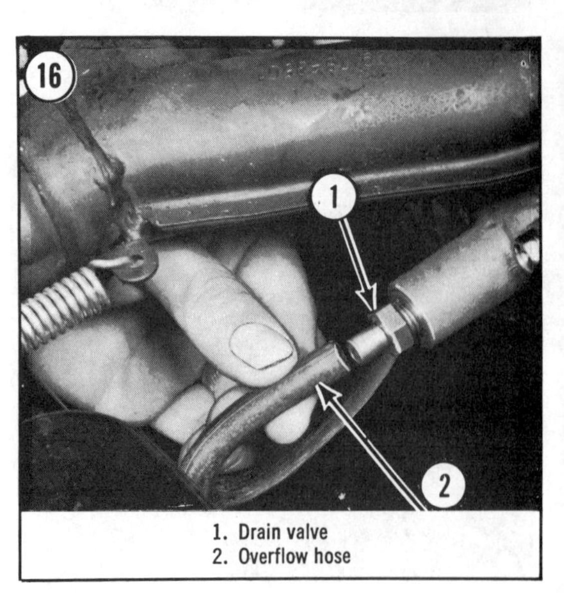

1. Drain valve
2. Overflow hose

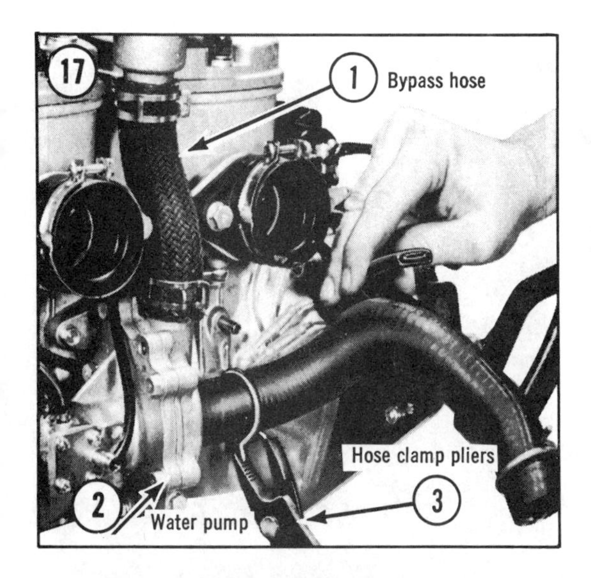
17
1 Bypass hose
Hose clamp pliers
2 Water pump
3

20

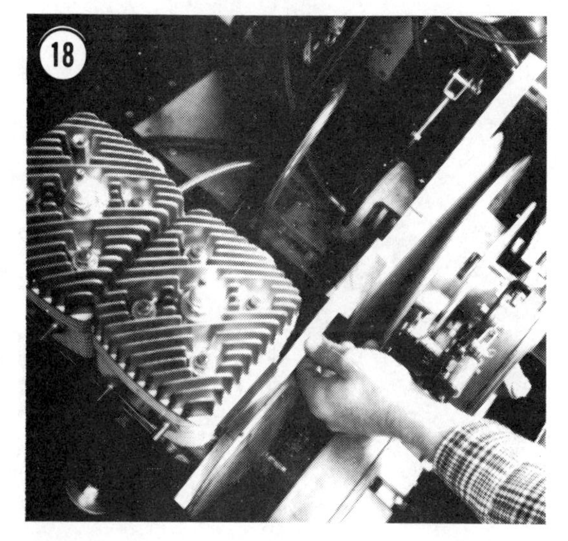
18

21

19

22

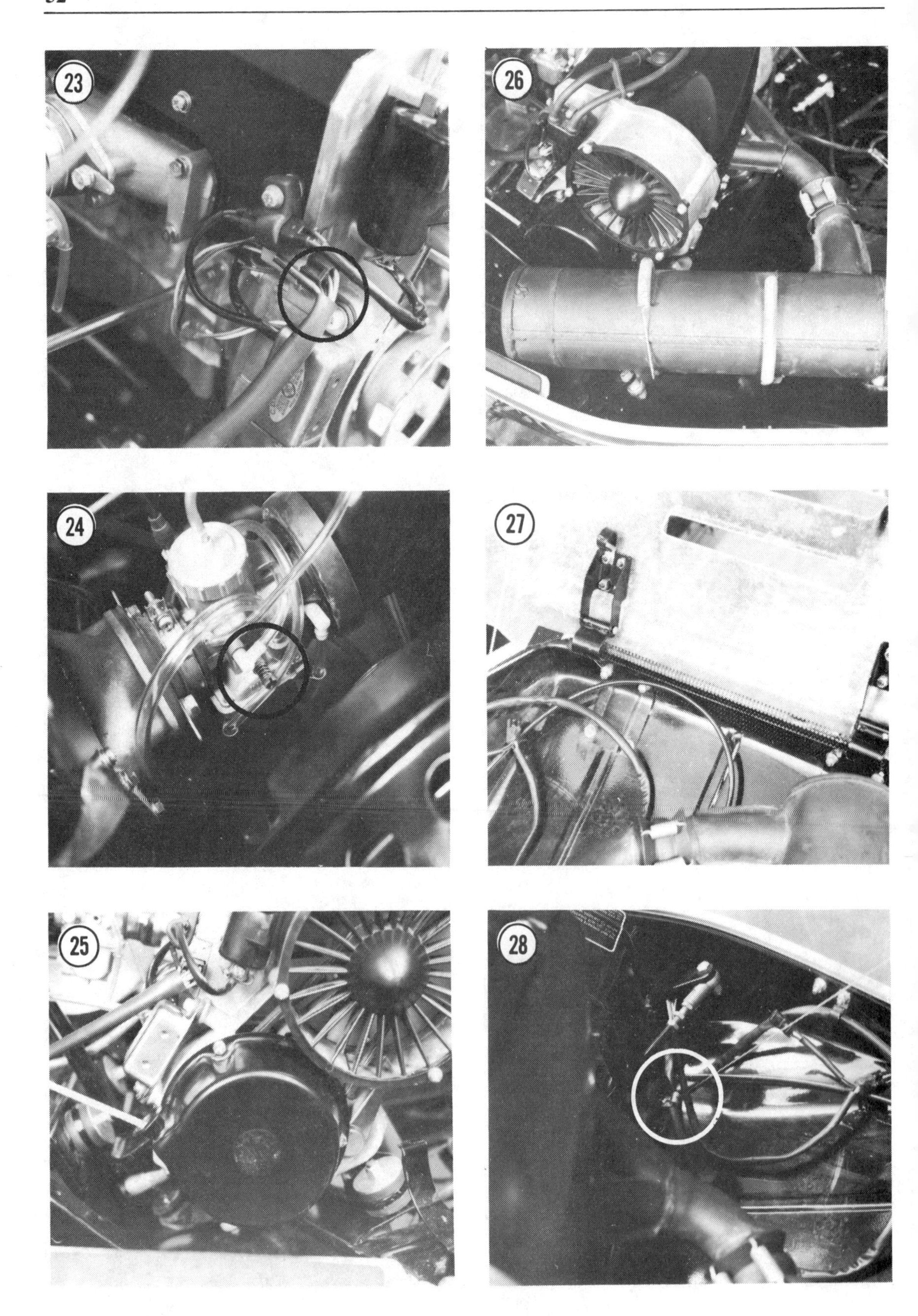
23
26
24
27
25
28

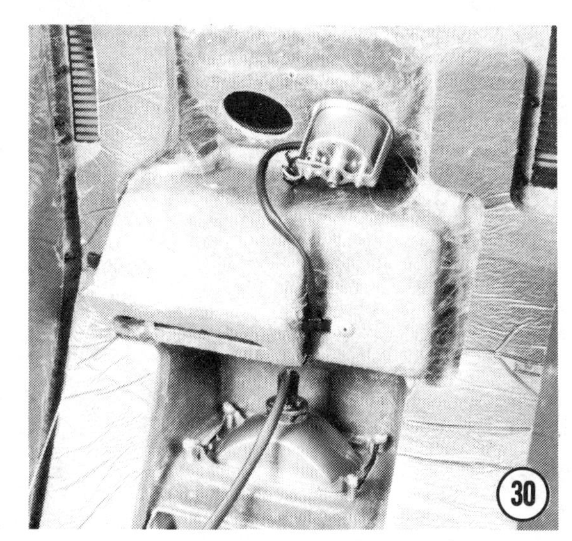

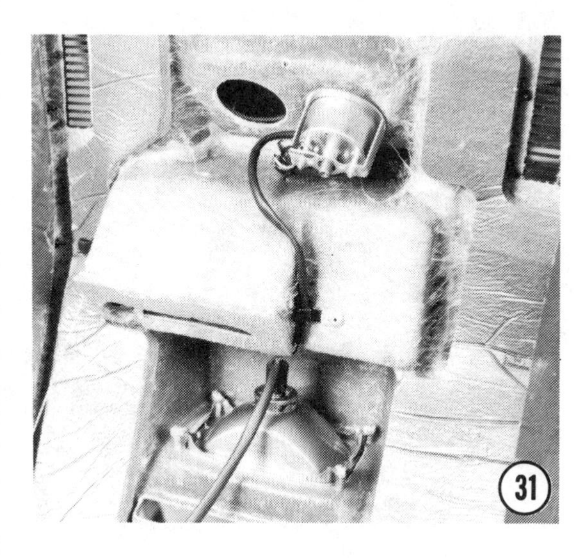

cooled engine above. When the installation steps described have been carried out, proceed as follows.

1. Connect the top and bottom coolant hoses to the engine and tighten the clamps securely.

2. Refer to *Liquid-Cooled Engines* in Chapter Two and fill and bleed the coolant system as described.

UPPER END SERVICE — AIR-COOLED ENGINES

The engine upper end can be serviced (including carbon removal) and repaired with the engine installed in the machine, and while many mechanics find that upper end engine work is easier when the engine has been removed from the machine and worked on on a bench, a great deal of time can be saved with the procedure that follows. This procedure also applies when the engine has been removed from the machine.

Disassembly

1. Open the hood all the way. Disconnect the speedometer cable and unplug the connector from the headlight (**Figure 31**). Disconnect the wiring harness from the hood clip (**Figure 32**).

2. Disconnect the hood restraining cable (**Figure 33**) and with assistance, support the hood and remove the hinge pins (**Figure 34**). Remove the hood and set it well out of the way to prevent it from being damaged.

3. Disconnect the springs that secure the exhaust system headpipe to the manifold (**Figure 35**) and remove the long springs that hold the muffler in its cradle. Remove the exhaust system from the machine. Remove the recoil starter (**Figure 36**).

4. Remove the fan cover (**Figure 37**) and the cooling shroud (**Figure 38**). Disconnect the harness (**Figure 39**).

5. Unscrew the 3 bolts that connect the starter pulley to the flywheel (**Figure 40**) and remove the pulley and the fan belt. Remove the fan case (**Figure 41**).

6. Disconnect the fuel line from the carburetor and plug it to prevent fuel from leaking from the tank. Unscrew the starter valve and the carburetor top and remove the slide assembly (**Figure 42**).

32
35
33
36
34
37

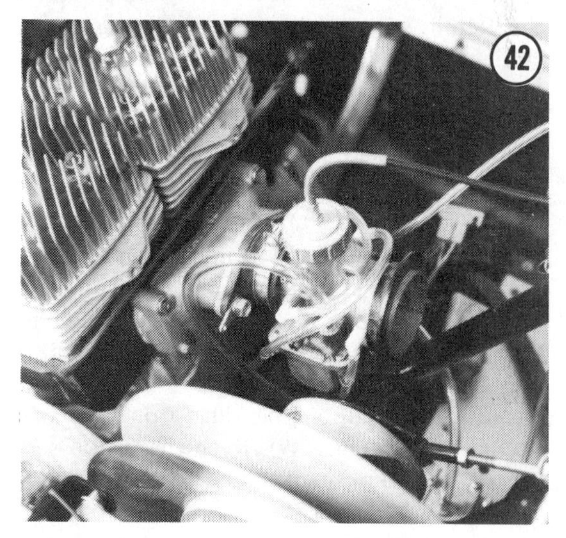

7. Loosen the carburetor clamping band (**Figure 43**) and remove the carburetor.

8. Remove the intake manifold (**Figure 44**) and the exhaust manifold (**Figure 45**).

9. Remove the bottom cooling shroud panels from the cylinders (**Figure 46 and 47**).

10. Mark the cylinder heads for location — left and right — and loosen the nuts 2 full turns, ½ turn at a time, in a crisscross pattern to prevent warping the heads. Then, unscrew the nuts completely and remove the heads and the head gaskets.

11. Remove the cylinders by lifting them straight up — do not twist them, otherwise the piston rings may be damaged. It may be necessary to first tap around the base of each cylinder with a soft mallet to break them loose from the crankcase. Mark the cylinders for location.
12. Place a clean shop rag in the crankcase to prevent dirt and small parts and tools from falling into the engine. Remove the piston pin clip from the outside of each piston (**Figure 48**), press the pins out, and remove the pistons. Remove the small end bearings from the connecting rods (**Figure 49**). Keep the piston sets (piston, pin, clip, small end bearing) separated; do not mix the pieces.
13. Secure the connecting rods with rubber bands stretched over the studs (**Figure 50**) to prevent the rods from striking the crankcase and damaging it.

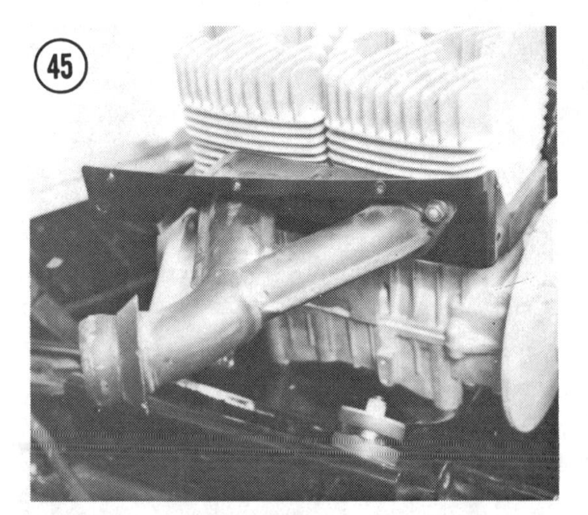

Cleaning

1. Scrape carbon from the combustion chamber in the heads and the exhaust ports in the cylinders with a soft metal (aluminum) or wood scraper. Do not use a hard metal scraper; it will burr the surfaces and create hot spots.
2. Clean the heads and cylinders with solvent and dry them with compressed air.
3. Remove the rings from the pistons and clean the piston crowns with a soft scraper. Clean the

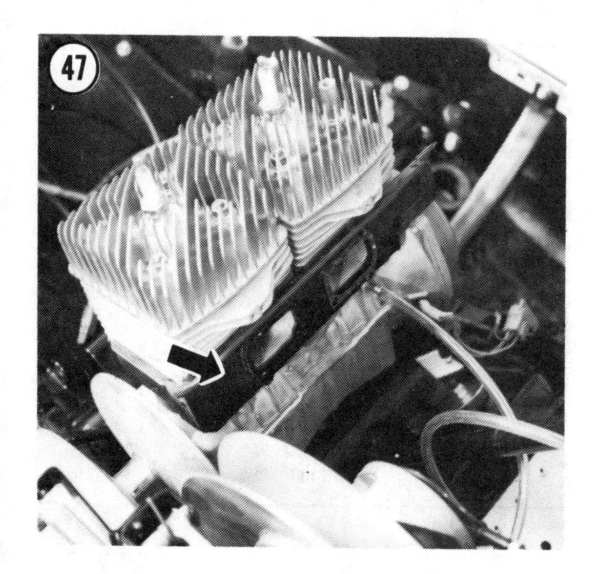

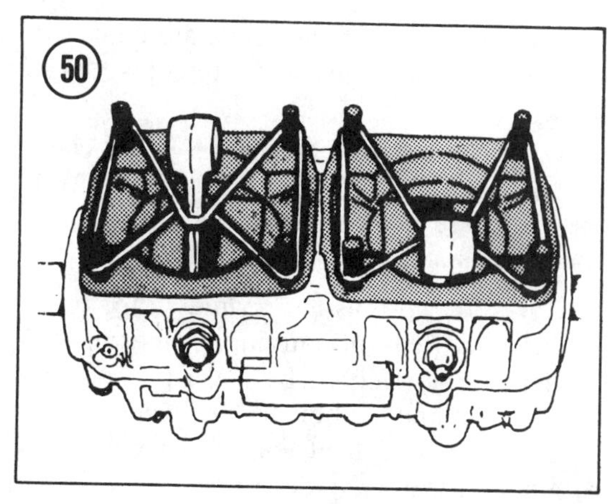

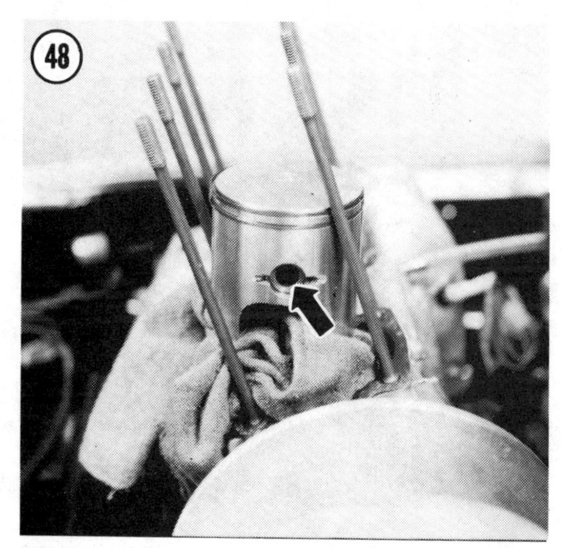

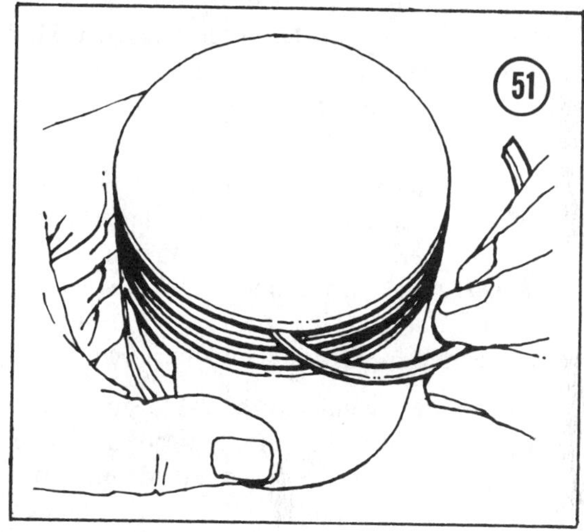

rings grooves with a piece of old piston ring (**Figure 51**). Wash the pistons with solvent and dry them with compressed air.

CAUTION

Pistons fitted with Keystone cross-section rings must be cleaned with a piece of Keystone ring to prevent damage to the ring grooves.

4. Clean the small end bearings with solvent and dry them with compressed air.

5. Remove the old cylinder base gaskets from the bottoms of the cylinders and the top of the crankcase, using a soft scraper. Be careful to prevent any pieces from falling into the crankcase.

Inspection

1. Inspect the reed valve petals (if so equipped) for fatigue cracks. The petals should fit flush

against their seats. If the petals are damaged, replace them as described in Chapter Five, *Reed Valve*.

2. With a straightedge and feeler gauge, check the flatness of the cylinder heads (**Figure 52**). If the heads are not flat, have them trued by a competent machine shop.

3. Check the pistons and cylinders for wear, galling, scuffing, or burning. Minor irregularities can be removed from the pistons with crocus cloth and light oil. The cylinder may be cleaned up with a light honing, provided it is within specifications as described below.

4. Inspect the inside of the pistons for cracks in the corners of the transfer notches (**Figure 53**). Fatigue in this area can cause the piston skirt to collapse. If cracks are visible, replace the pistons.

5. Measure the bore with an inside micrometer or cylinder gauge. Measure 3/8 in. below the top of the cylinder, in 2 locations 90° apart (**Figure 54**). If the 2 measurements differ by more than 0.05 mm (0.002 in.), the cylinder is out-of-round beyond specification and must be bored.

6. Measure the cylinder again in 2 locations 90° apart just above the intake port. If these measurements differ from those made in Step 5 by more than 0.05 mm (0.002 in.), the cylinder is excessively tapered and must be bored. Acceptable cylinder dimensions are shown in **Table 2** at the end of this chapter.

7. Check piston skirt-to-cylinder clearance by first measuring the base of the cylinder bore, front to back (**Figure 55**). With an outside micrometer, measure the piston skirt, front to back, about 6 mm (1/4 in.) from the bottom (**Figure 56**). Subtract the piston measurement from the cylinder measurement to determine actual piston skirt-to-cylinder clearance. Refer to **Table 2** for acceptable clearance. If the bore is OK but the clearance is excessive, install new piston sets, including pins, bearings, and rings.

8. Measure piston ring end gap. Place a ring into the cylinder 6 mm (1/4 in.) from the top. Press down on the ring with the piston skirt to square the ring with the bore. Measure the gap as shown in **Figure 57**. Refer to **Table 2** for ac-

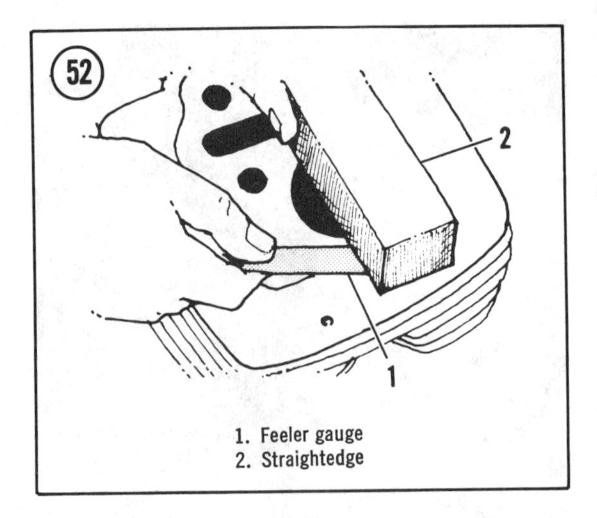

1. Feeler gauge
2. Straightedge

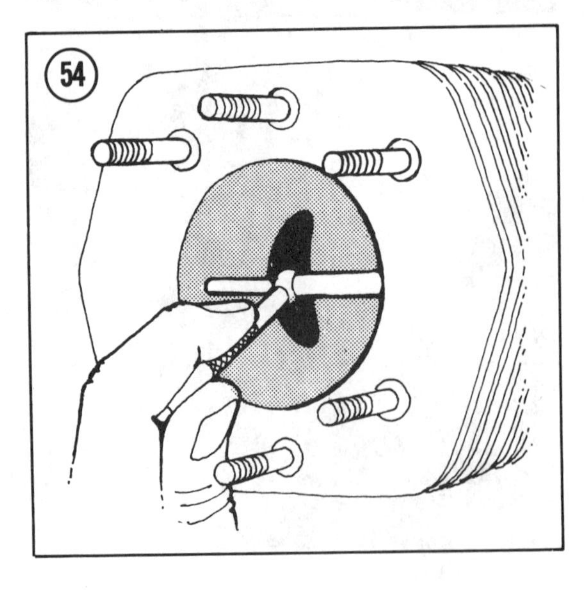

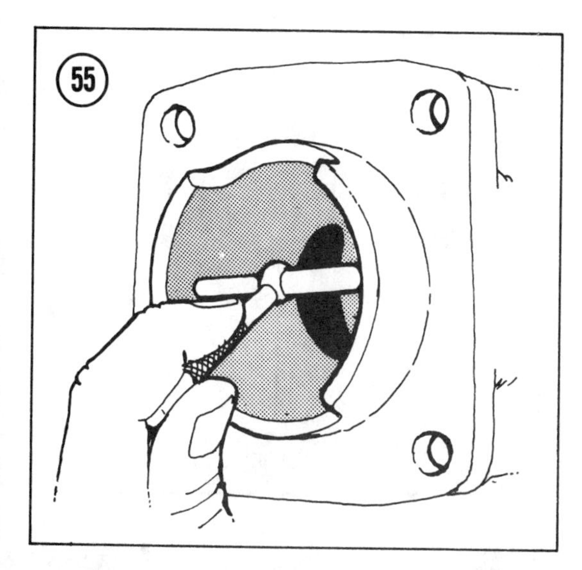
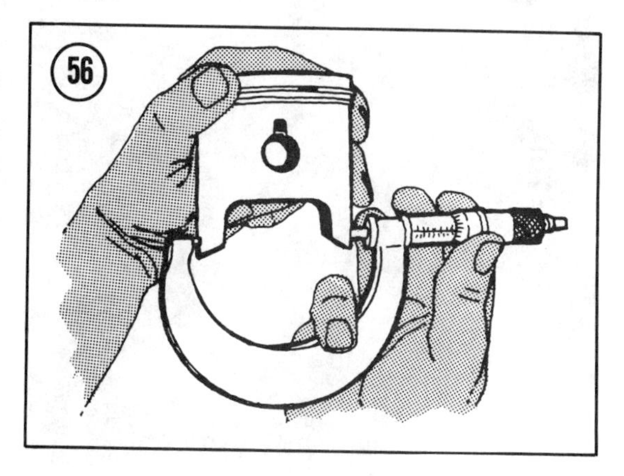
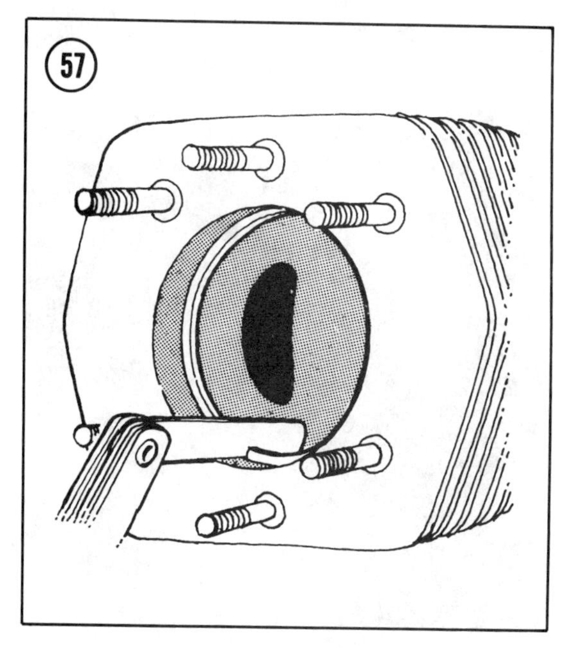
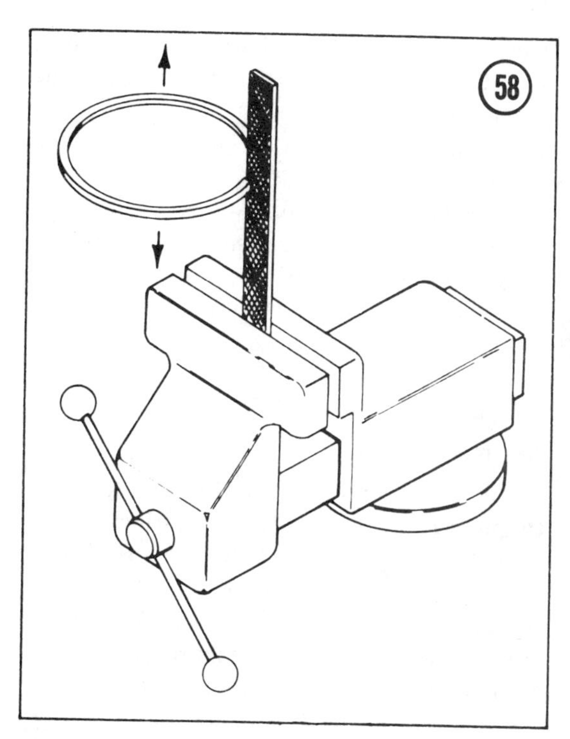
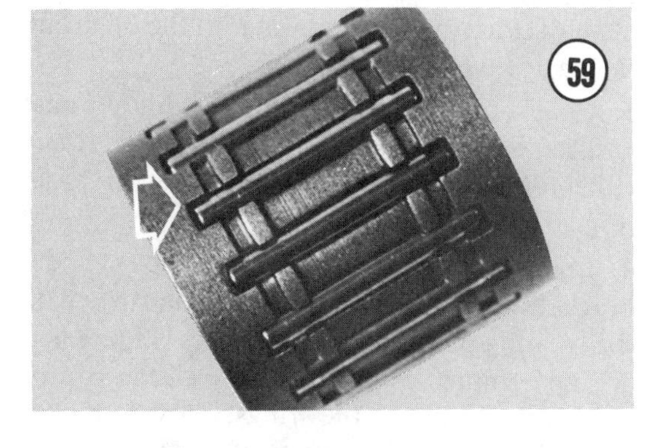

ceptable ring end gap. If the gap is too large for either ring, replace them as a set. Check new rings in the manner just described, and if the gap is too small, carefully file material off the ends of the rings **(Figure 58)**.

9. Inspect the piston pins for wear and discoloration. If the pins are not satisfactory, they must be replaced along with the piston sets.

10. Inspect the bearing cages for cracks at the corners of the roller beds **(Figure 59)**. Inspect the rollers for flat spots and replace the bearing if either condition is apparent.

11. Inspect the bearing bore in each of the connecting rods. It should be smooth. If the bore is

worn or scored, the rod will have to be replaced. This is a job for your dealer or other specialist.

12. Lightly oil the bearing and pin and install them in the rod. Check for play and rotation (**Figure 60**). There should be no vertical play and the pin should rotate smoothly. If play or roughness is apparent, the pin and bearing should be replaced, along with the piston set.

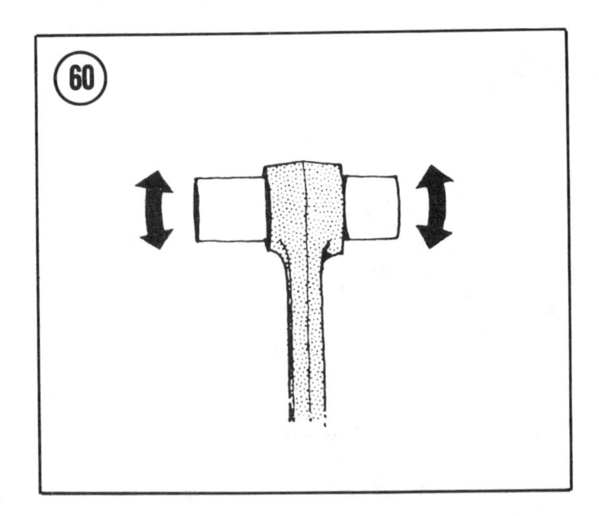

Assembly

1. Oil the connecting rod big end bearings (**Figure 61**). Oil the upper end bearings and install them in the connecting rods (**Figure 62**).

2. Oil the piston pins. Install the pistons with the arrow on the crown facing forward — toward the exhaust port (**Figure 63**). This is essential so the piston ring ends will be correctly positioned and will not snag in the ports. Install the piston pin clips, making sure they are completely seated in their grooves with the open end facing down (**Figure 64**).

3. Coat both surfaces of new cylinder base gaskets with gasket cement and install them either on the crankcase or the cylinder bases (**Figure 65**).

4. Lightly oil the pistons and the cylinder bores. Line up the ends of the piston rings with the locating pins in the ring grooves (**Figure 66**). A ring compressor or large hose clamp can be used to compress the rings; however, the tapered lead on the bottom of the cylinders is generous and the cylinder can be installed without the aid of a compressor.

5. Line up the cylinder with the piston, hold the piston to prevent it from rocking, and push the cylinder down over it. Some resistance is normal but if binding is felt, remove the cylinder and locate and correct the trouble before proceeding. *Don't force the cylinder over the piston.* If the cylinder does bind, it's likely that a ring is not correctly seated in its groove, or its ends may not be lined up with the locating pin in its groove.

6. Install the head gasket, head, washers, and nuts. Run the nuts down finger-tight. *Don't torque them yet.*

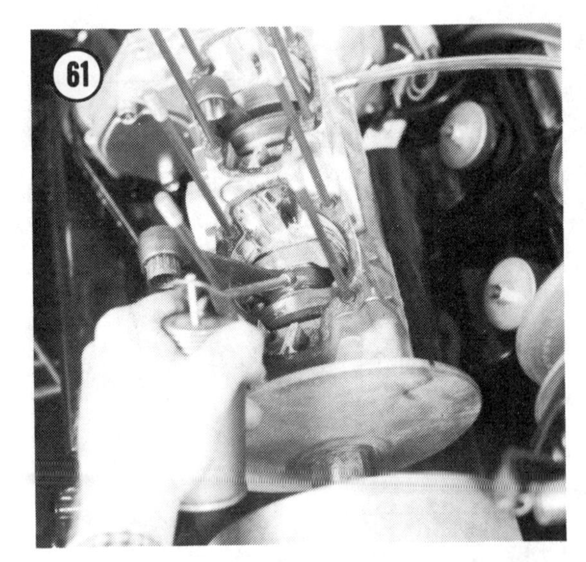

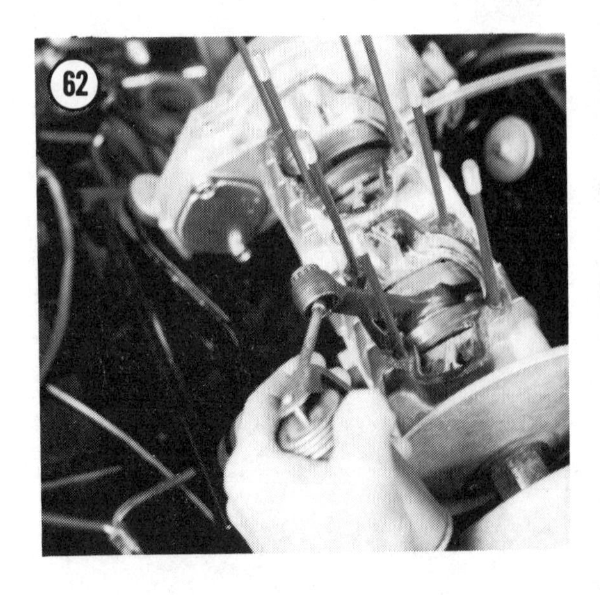

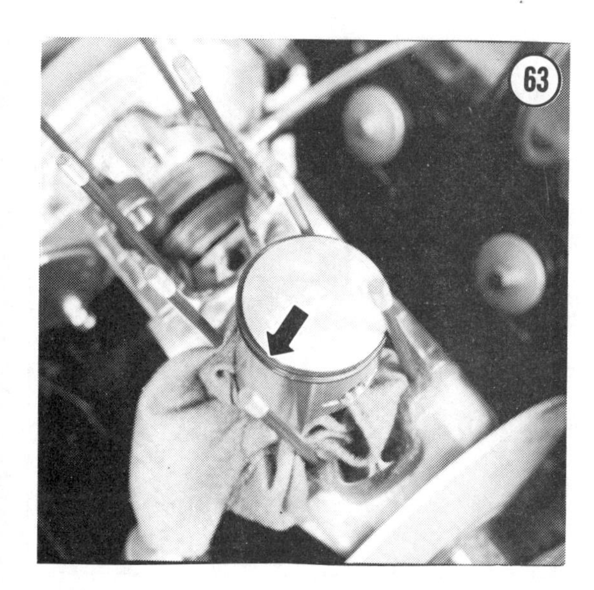

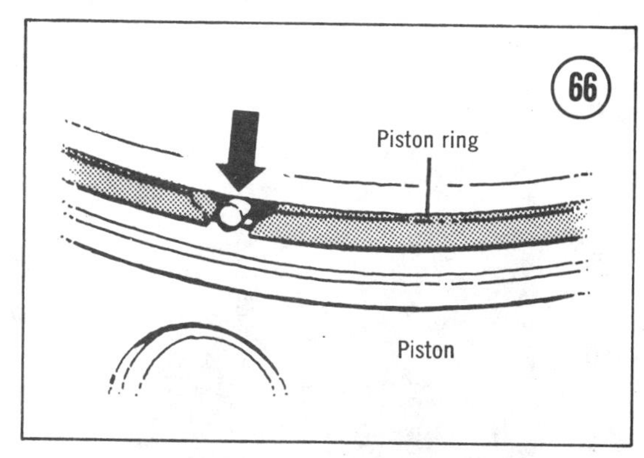

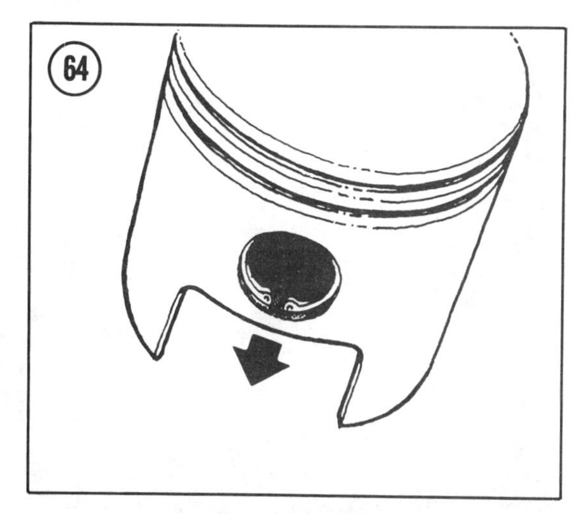

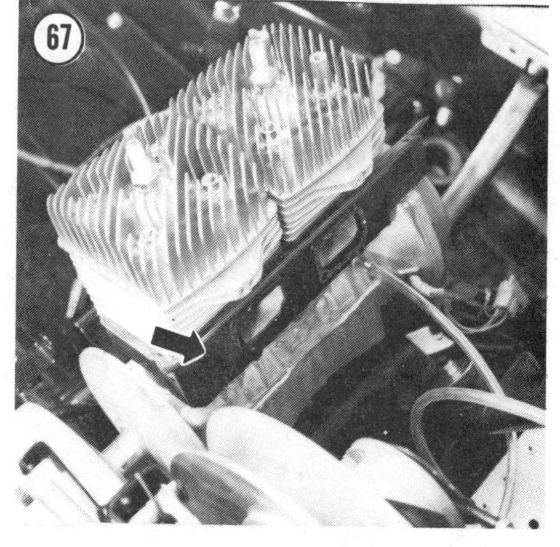

7. Install the lower cooling shroud panels (**Figure 67 and 68**). Install the exhaust manifold (**Figure 69**). Coat both sides of the intake manifold gaskets with gasket cement and install them and the manifold. Run the left bottom bolt in first (**Figure 70**).

> NOTE: *On models with reed valves, set the reed cages into the cylinders before installing the intake manifold.*

8. Tighten the manifold nuts and bolts to align the cylinders, then tighten the cylinder head nuts in a crisscross pattern to the value shown in **Table 1**.

9. Install the carburetor and tighten the clamping band. Connect the fuel line, screw in the starter valve, and install the slide assembly. Make sure the groove in the slide engages the knob inside the slide bore and the needle fits in-

68

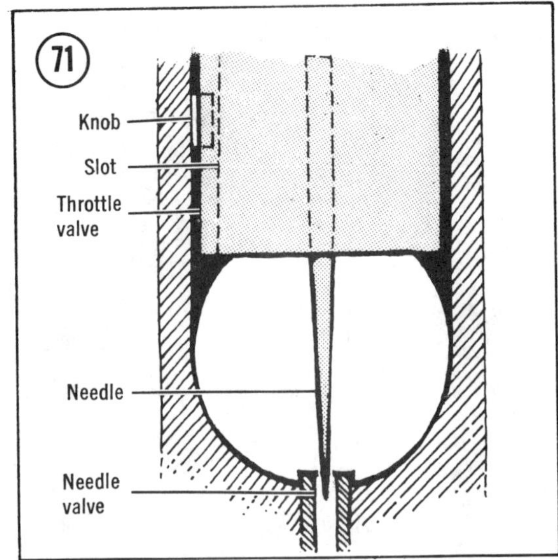
71
Knob
Slot
Throttle valve
Needle
Needle valve

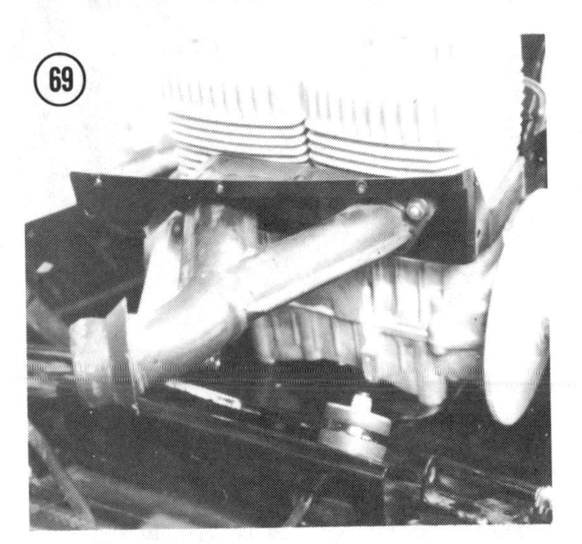
69

72

70

73

to the needle valve. Screw the top on securely (**Figure 71**).

10. Install the fan case (**Figure 72**), and the fan belt and starter pulley (**Figure 73**). Tighten the bolts to the value shown in **Table 1**. Connect the leads (**Figure 74**).

11. Install the cooling shroud (**Figure 75**) and the fan cover (**Figure 76**).

12. Install the starter assembly and the exhaust system (**Figure 77**). Connect the headpipe-to-manifold springs first, then connect the long hold-down springs.

13. With assistance, align the hood with the hinges in the body and install the hinge pins. Attach the restraining cable (**Figure 78**).

14. Route the wiring harness under the hood clip to prevent it from being crimped when the hood is closed or burned on the exhaust system (**Figure 79**). Connect the plug to the headlight (**Figure 80**) and connect the cable to the speedometer.

15. Double check to make sure all lines, cables, hoses, and wires have been reconnected and are correctly routed.

16. Refer to Chapter Two and tune the engine and service the machine as described.

CAUTION

Following upper end service, if any parts were replaced or if the cylinders were honed or bored, the engine should be broken in as though it were new. Also, it's a good idea to retorque the cylinder head nuts after several hours of operation.

UPPER END SERVICE — LIQUID-COOLED

Upper end service for the liquid-cooled engine is essentially the same as that described for the air-cooled engine with the exception of those points noted below.

Disassembly

Before disassembling the upper end, first drain the coolant.

WARNING

Do not drain the system or disconnect coolant hoses when the engine and coolant are hot.

1. Carefully loosen the radiator cap to relieve system pressure.
2. Place a container (5 quarts or larger) beneath the machine where the coolant overflow hose comes through the body.
3. Disconnect the coolant overflow hose from the reservoir, pass it beneath the engine, and connect it to the coolant drain valve in front of the engine (**Figure 81**).
4. Raise the rear of the machine and support it. This is necessary to aid the coolant to drain from the heat exchanger.
5. Open the drain valve and allow about 10 minutes for the coolant to drain.

79

80

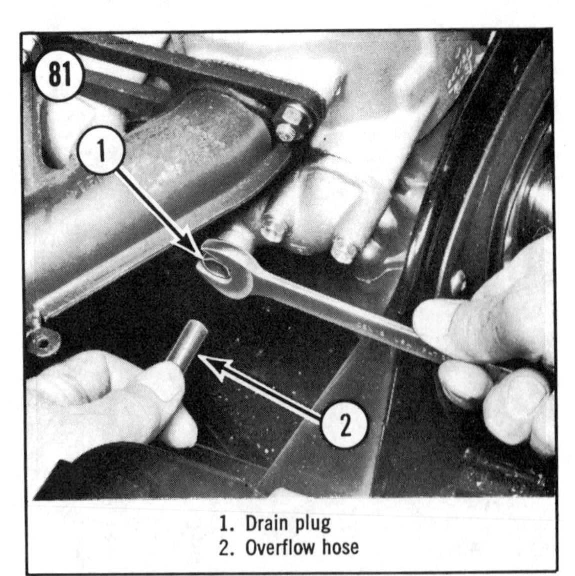

81

1. Drain plug
2. Overflow hose

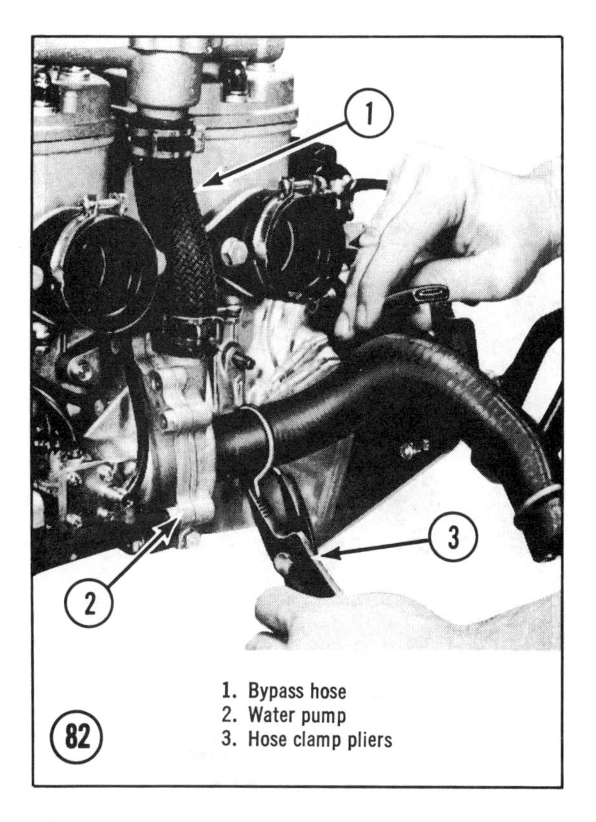

1. Bypass hose
2. Water pump
3. Hose clamp pliers

82

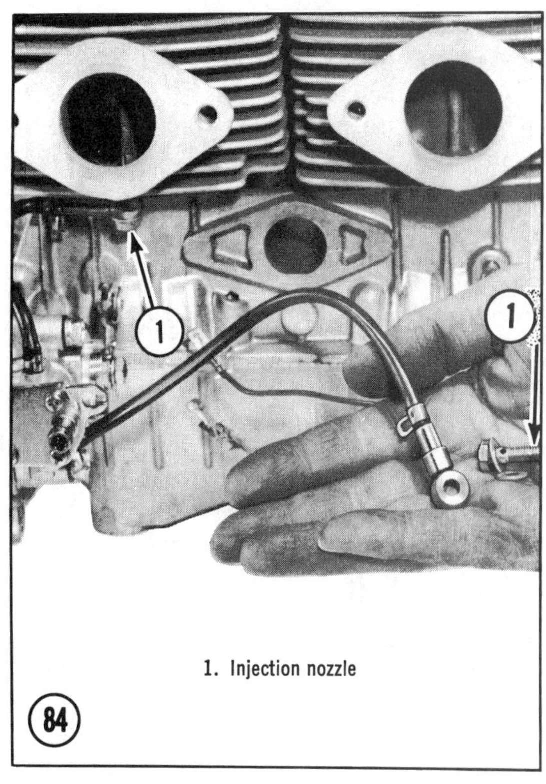

1. Injection nozzle

84

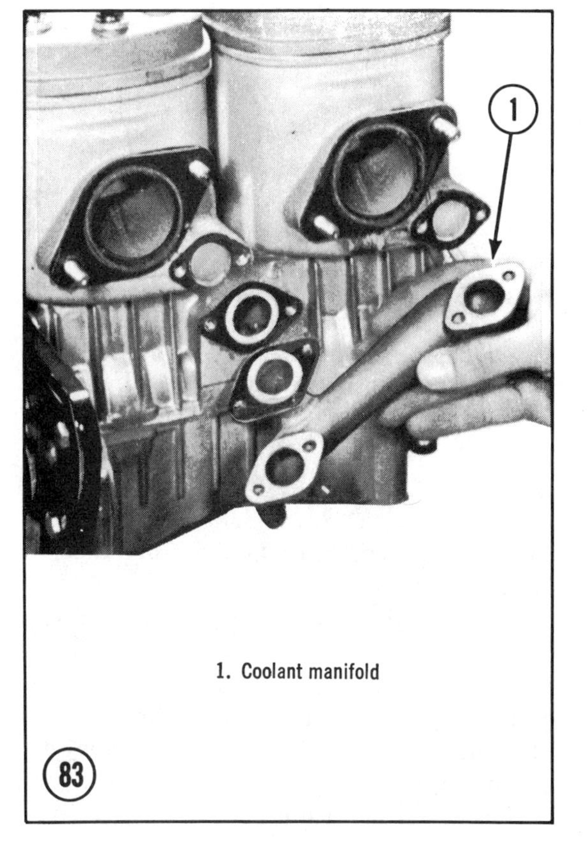

1. Coolant manifold

83

6. Disconnect the top coolant hose from the thermostat cover. Disconnect the bottom hose from the water pump (**Figure 82**).

> NOTE: *Some residual coolant will drain into the body from the pump and the lower hose. Flush it with mild soapy water and then clear water to prevent the coolant from attacking the paint.*

7. Remove the exhaust manifold. Loosen the bypass hose clamp at the water pump, unscrew the bolts that attach the water manifold to the heads, and remove the manifold and bypass hose (**Figure 82**).

8. Remove the water manifold from the front of the engine (**Figure 83**).

9. Disconnect the oil injection lines from the intake flanges (**Figure 84**). Cover the nozzles and the ends of the line with a plastic bag and tape it in place.

10. Mark the cylinders and heads for location (right and left) and position (front and back).

11. Remove the carburetors.

12. Unscrew the cylinder head nuts progressivey ½ turn at a time, in a crisscross pattern to

prevent warping the heads. When the nuts have been loosened 4 full turns, unscrew them all the way. Tap around the base of the heads to break them loose from the cylinders and remove them.

CAUTION

Be careful to keep residual coolant that may remain in the heads from leaking into the cylinders.

13. Tap around the base of the cylinders to break them loose from the gaskets and then slide them up the studs to remove them. Stuff a clean rag under each piston to keep dirt and parts from entering the crankcase.
14. Remove the outer piston pin clip from each piston (**Figure 85**), push out the pins, remove the pistons, and collect the small end bearings.

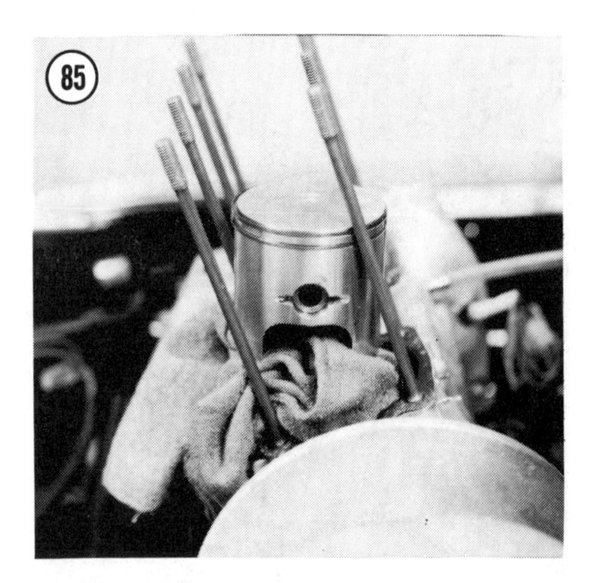

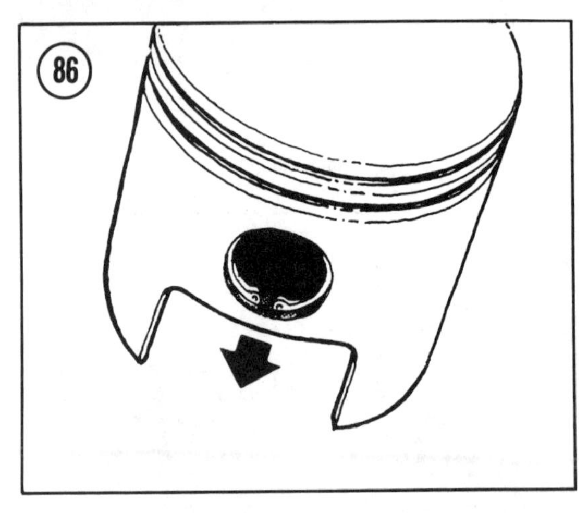

Cleaning

1. Scrape the carbon from the combustion chamber in the heads and the exhaust ports in the cylinder, using a soft metal (aluminum) or wood scraper. Do not use a hard metal scraper; it will burr the surfaces and create hot spots.
2. Scrape old gasket material from the sealing surfaces of the heads and cylinders with a soft metal or wood scraper; taking care not to damage the surfaces. Clean the openings in the water passages and then clean the heads and cylinders with solvent and blow them dry with compressed air.
3. Remove the rings from the pistons and clean the piston crowns with a soft scraper. Clean the ring grooves with a piece of old ring. Clean the pistons with solvent and blow them dry with compressed air.
4. Clean the small end bearings with solvent and dry them with compressed air.
5. Clean the thermostat and the passages in the water manifolds with solvent and dry them with compressed air.

Inspection

The inspection of cylinder heads, cylinders, pistons, and bearings is the same as for air-cooled engines described earlier. However, the cylinder bores in the liquid-cooled engine are chrome plated and must be checked for damage. Check for peeling chrome toward the top of the cylinder and around the edges of the ports. If only slight peeling is apparent, the cylinders are satisfactory. But, if peeling is extensive, replace the cylinders.

Critical dimensions and tolerances are shown in **Table 2** at the end of this chapter.

Assembly

1. Oil the connecting rod bottom end bearings. Oil the upper end bearings and install them in the connecting rods.
2. Oil the piston pins. Install the pistons with the arrow on the crown facing forward — toward the exhaust port. This is essential so the piston ring ends will not snag on the ports. Install the piston pin clips, making sure they are completely seated in their grooves with the open end facing down (**Figure 86**).

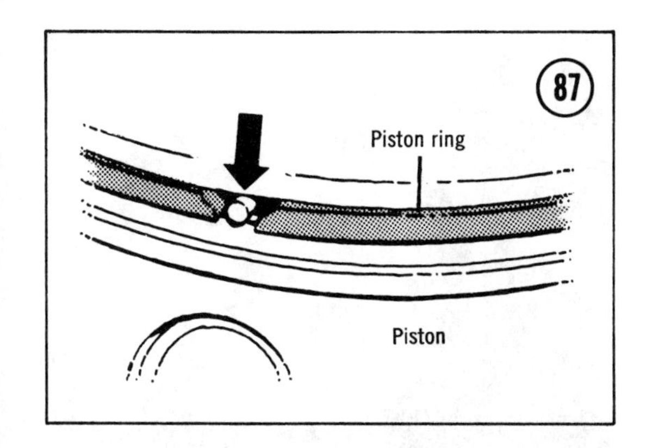

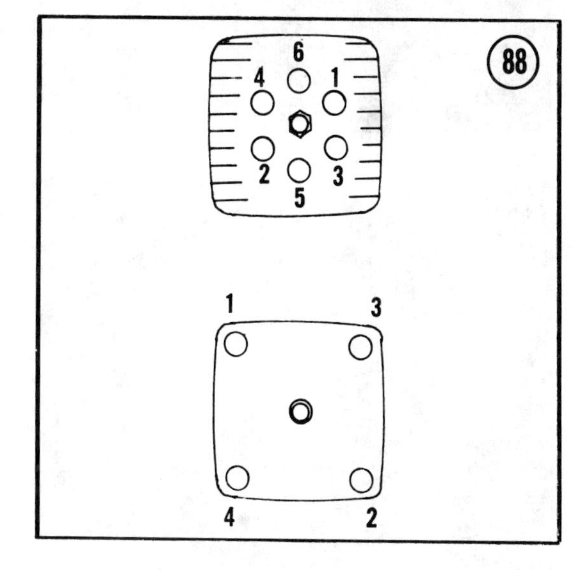

3. Coat both surfaces of new cylinder base gaskets with silicone sealer and install them on the crankcase studs.

4. Lightly oil the pistons and the cylinder bores. Line up the ends of the rings with the locating pins in the ring grooves (**Figure 87**). A ring compressor or large hose clamp can be used to compress the rings.

5. Note the position and location marks made on the cylinders earlier, line up one cylinder with its respective piston, hold the piston to prevent it from rocking, and push the cylinder down over it. Some resistance is normal, but if binding is felt, remove the cylinder and locate and correct the trouble before proceeding. *Don't force the cylinder over the piston.* If the cylinder does bind, it's likely that a ring is not correctly seated in its groove, or its ends may not be lined up with the locating pin in the groove.

NOTE: *After the cylinders have been installed, check for ring breakage by pressing on the rings with your finger, through the exhaust ports. If the rings spring back they are OK.*

6. Coat both surfaces of new head gaskets with silicone sealer. Take care not to get sealer in the water passages. Install the gaskets and the cylinder heads. Screw on the nuts finger-tight. *Don't torque them yet.*

7. Coat the flanges of front coolant manifold and the companion flanges on the cylinders and crankcase with silicone sealer. Install the manifold and screw in and tighten the 4 bolts that connect the manifold to the cylinders. Then, tighten the cylinder head nuts in a crisscross pattern to 2.2 mkg (16 ft.-lb.). See **Figure 88**. Finally, tighten the 2 bolts that attach the coolant manifold to the crankcase.

8. Connect the oil injection lines to the inlet flanges.

9. Connect the bypass hose to the water pump and install the coolant manifold on the cylinder heads. Coat all mating surfaces with silicone sealer.

10. Install the exhaust manifold and the carburetors.

When assembly is complete, refer to Chapter Two and fill and bleed the cooling system, tune the engine, and perform routine service.

NOTE: *Following upper end service, if any of the parts were replaced the engine should be broken in as though it were new. Also, it's a good idea to retorque the cylinder head nuts after several hours of operation.*

LOWER END SERVICE

Lower end service as described here consists of removal, inspection, and installation of the crankshaft assembly. Inspection requires V-blocks (or another suitable centering device such as a lathe), a dial indicator with a base and stand, and a working knowledge of precision measurement. If you lack the skills or equipment, your dealer can inspect the assembly and determine what service is required.

The detailed inspection procedures are included for those equipped for the job. Critical dimensions and tolerances are shown in **Table 3** at the end of this chapter.

Crankshaft service — replacement of unsatisfactory parts or alignment of crankshaft components — should be entrusted to your dealer or an engine specialist. A hydraulic press and considerable experience are necessary to disassemble, assemble, and accurately align the crankshaft assembly, which, in the case of the average twin-cylinder engine, is made up of 7 pressed-together pieces, not counting the bearings, seals, and connecting rods.

Disassembly

1. Refer to *Engine Removal* and *Upper End Service* and remove the engine from the machine and disassemble the upper end.

NOTE

*Before removing the engine, loosen the flywheel/rotor nut or bolt shown removed in **Figure 89** and the drive sheave bolt (**Figure 90**). This is much easier when the engine is held firmly in place.*

2. Unscrew the flywheel/rotor bolt and install a puller (**Figure 91**). See **Table 4** for puller identification.

3. Mark the stator and crankcase for reference during assembly (**Figure 92**). Unscrew the stator screws and remove the stator.

4. Unscrew the bolts that attach the engine mount plate (**Figure 93**) and remove the plate.

5. Unscrew the crankcase bolts in the pattern shown in **Figure 94A or 94B**, depending on your engine.

6. Tap on the large bolt bosses with a soft mallet to break the crankcase halves apart; remove the bottom half.

CAUTION

Do not pry the cases apart with a screwdriver or any other sharp tool, otherwise the sealing surfaces will be damaged.

7. Lift the crankshaft assembly out of the upper case half. Pay particular attention to the locating ring and alignment pins for reference during assembly.

Cleaning

1. Clean the case halves and the crankshaft assembly in *fresh solvent* and blow them dry with compressed air.

89

90

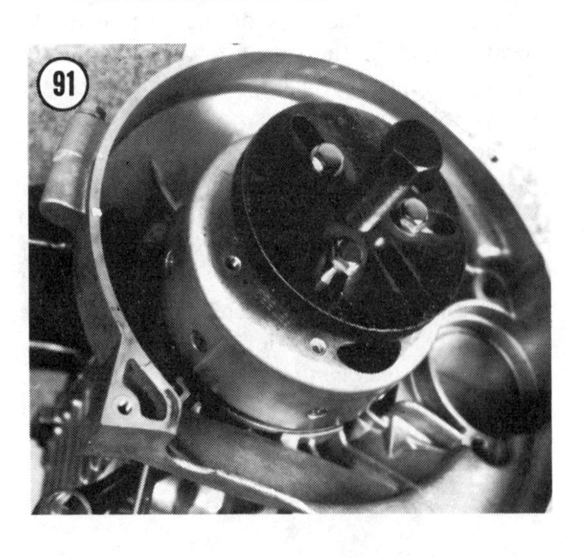

91

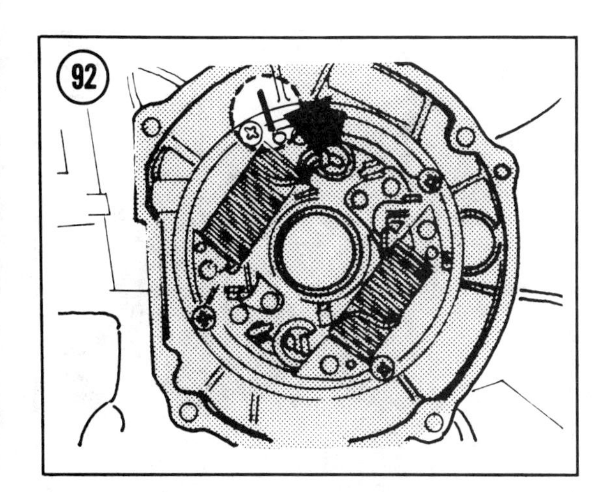
92

93

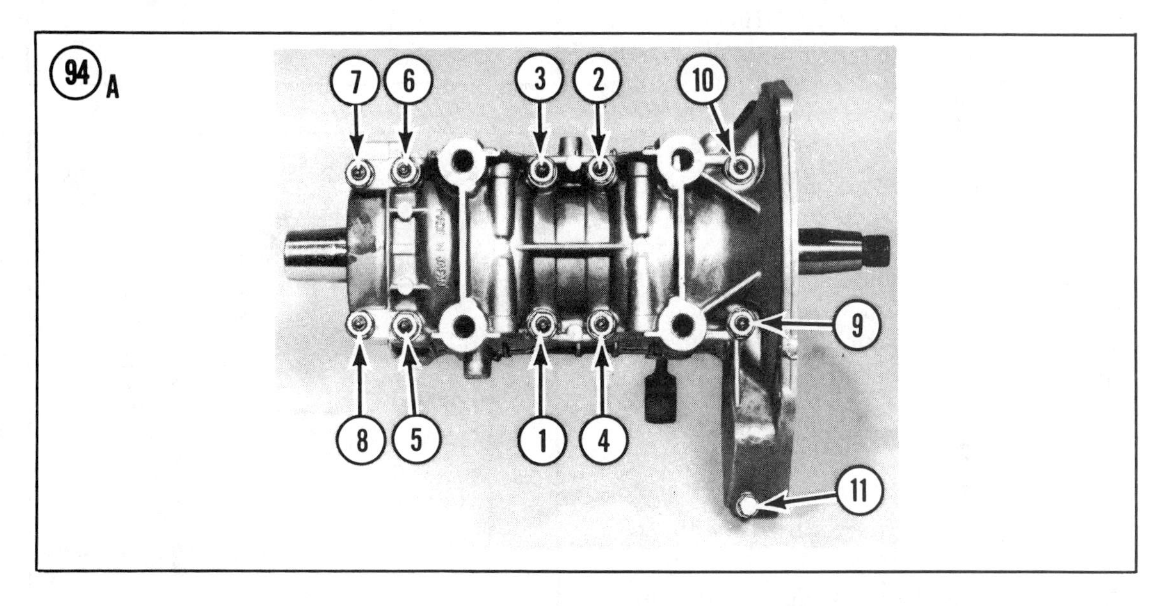
94 A
7
6
3
2
10
9
8
5
1
4
11

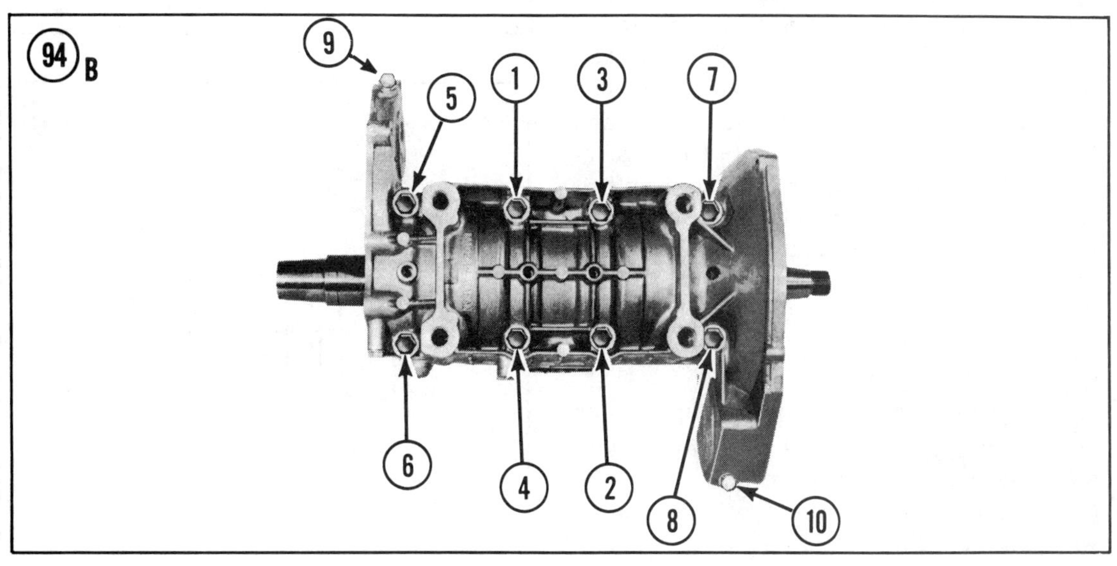
94 B
9
5
1
3
7
6
4
2
8
10

2. Carefully remove old gasket cement from the case mating surfaces, using a soft scraper.

Inspection

1. Rotate the bearings by hand and feel for play and roughness. The bearings should turn freely and smoothly and should have no apparent radial play. If you are uncertain about their condition, have them checked by your dealer or an engine specialist.
2. Check the oil seals **(Figure 95)** for damage and replace them if their condition is doubtful.
3. Check the axial play of the connecting rod by measuring its side-to-side movement at the small end with a dial indicator **(Figure 96)**. If the play exceeds 2 mm (0.08 in.), the big end bearings or the connecting rod or both must be replaced. Refer this work to your dealer or an engine specialist.
4. Support crankshaft assembly in V-blocks or a lathe and measure the runout of the shaft with a dial indicator **(Figure 97)**. Maximum allowable runout for the ends is 0.05 mm (0.002 in.). If the runout is excessive, have the assembly trued by your dealer or an engine specialist.

95

96

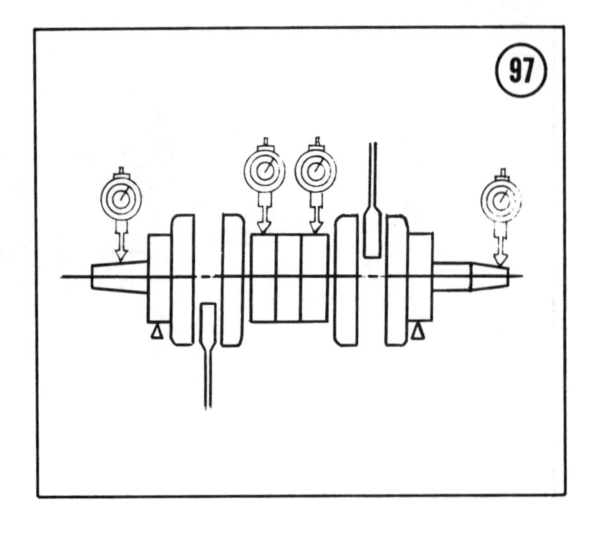

97

Assembly

1. Liberally oil all the bearings with fresh engine oil.
2. Set the crankshaft assembly in the upper case half, taking care to line up the pin in the center seal with the recess in the case **(Figure 98)**. Push the crankshaft as far as it will go toward the sheave end of the engine and measure the end play with a flat feeler gauge inserted between the center bearing and the thrust washer on the magneto end of the shaft. The maximum allowable end play is 0.35 mm (0.014 in.); however, it's recommended that the end play be set at 0.08-0.13 mm (0.003-0.005 in.). If adjustment is required, refer to **Table 5** for the correct shim. Use only one shim.
3. Make certain the oil seal lips line up with the grooves in the case and all the bearings and seals are completely seated.
4. Coat the sealing surfaces of both case halves with gasket cement.

1. Aligning pin

5. Set the lower case in place, making sure it contacts the upper case half along the entire sealing surface.

6. Screw in the crankcase bolts and run them down finger-tight. Rotate the crankshaft several complete revolutions to make sure it is correctly seated and not binding. Then, tighten the bolts progressively in the pattern shown in **Figure 99A or 99B**, depending on your engine. Occasionally rotate the crankshaft and check for binding and resistance. If the shaft binds, stop and locate and correct the trouble before continuing. Finally, tighten the bolts in the pattern shown to the torque specified in **Table 1** at the end of this chapter.

7. Install engine mounting plate (**Figure 100**).

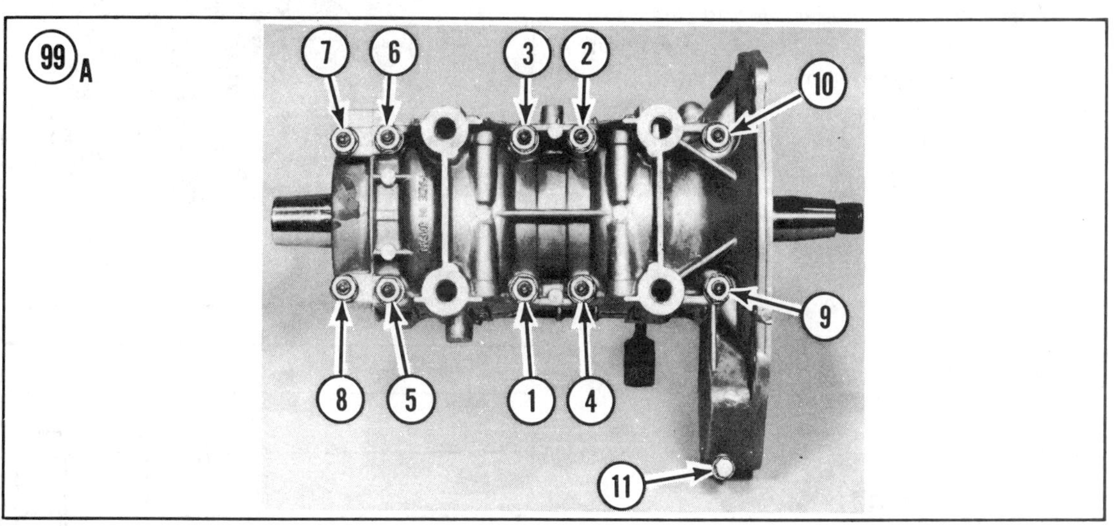

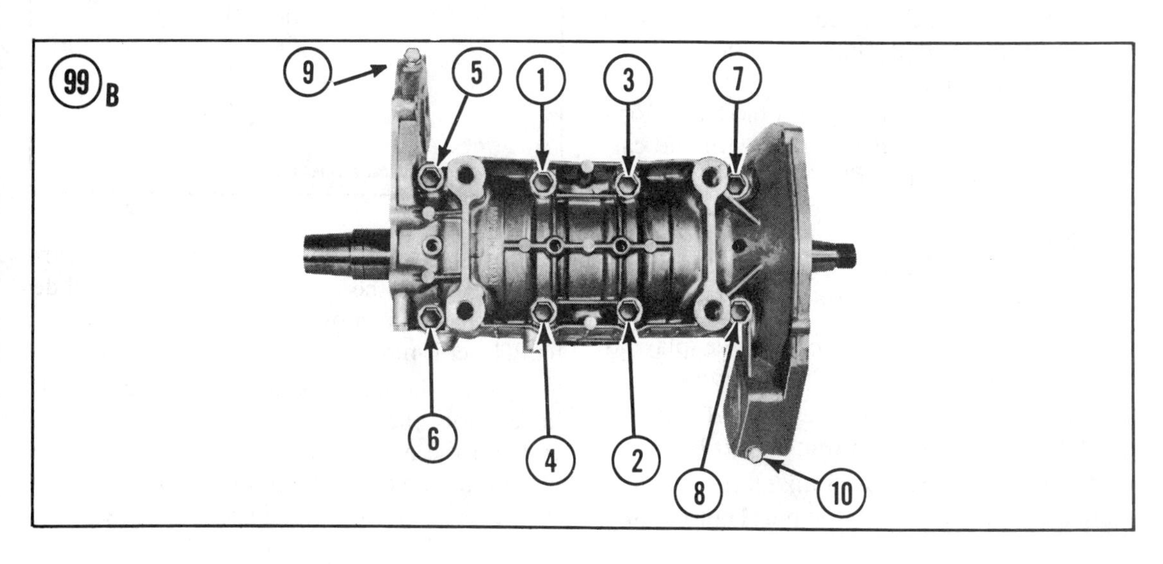

8. Invert the crankcase and oil the big and small end bearings with fresh engine oil.

9. Line up the mark on the stator with the mark on the crankcase (**Figure 101**) and screw in and tighten the stator screws.

10. Set the key in the end of the crankshaft, line it up with the keyway in the rotor, and push the rotor onto the crankshaft. Screw the bolt into the end of the crankshaft and tighten it to the torque specified in **Table 1** at the end of this chapter.

11. Clean the taper on the sheave end of the crankshaft and the taper inside the drive sheave. Push the sheave onto the crankshaft, install the bolt and washers (**Figure 102**), and tighten the bolt to the value specified in **Table 1** at the end of this chapter.

12. Refer to *Upper End Service* and *Engine Installation* in this chapter and complete assembly and installation of the engine as described.

13. When installation and assembly are complete, refer to Chapter Two and service the machine and tune the engine.

CAUTION

If any parts were replaced, break in the engine as though it were new. After several hours of operation, check the engine mounting bolts for tightness and retorque the cylinder head nuts.

OIL PUMP

SS and ST models are equipped with an oil injection system that eliminates the need for premixing fuel and oil.

The oil pump is virtually trouble free. However, it should be synchronized with the carburetor each time the carburetor free play is adjusted to ensure the engine receives an adequate supply of oil.

Adjustment

1. Adjust the carburetor cable free play as described in Chapter Two, *Carburetion*.

2. Back out the idle stop screws equal amounts until the throttle slides bottom in their bores.

3. Push the oil pump bellcrank all the way forward until it contacts the stop pin (**Figure 103**).

100

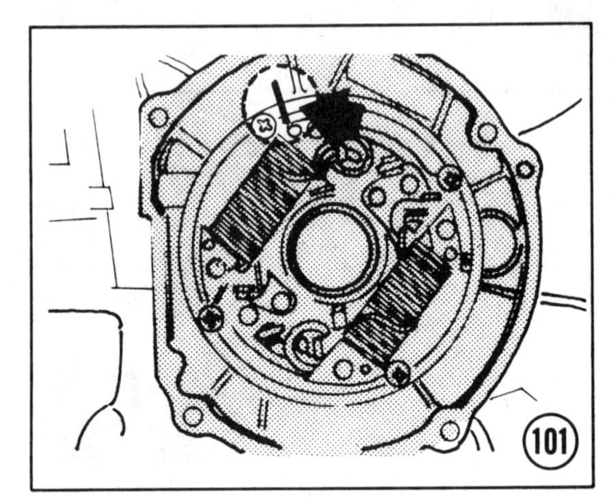

101

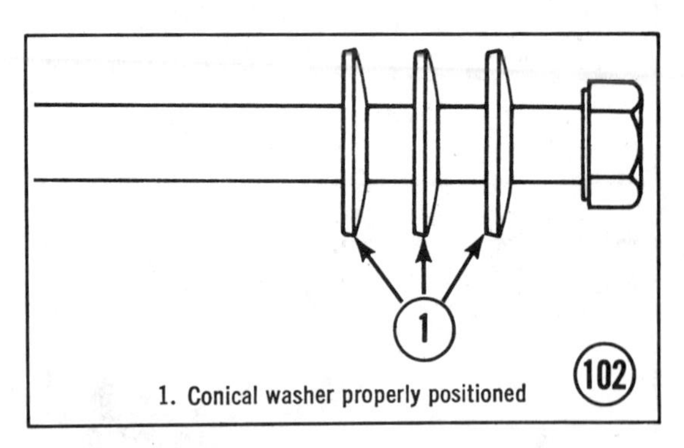

1. Conical washer properly positioned

102

4. Adjust the throttle cable on the left carburetor so that there is no free play — the slide should begin to move as soon as the throttle hand lever is moved.

5. Loosen the locknuts on the pump cable adjuster, hold the bellcrank against the stop pin, and run the adjuster out until there is no free play in the cable. The lever on the pump and the left carburetor slide should move at the same

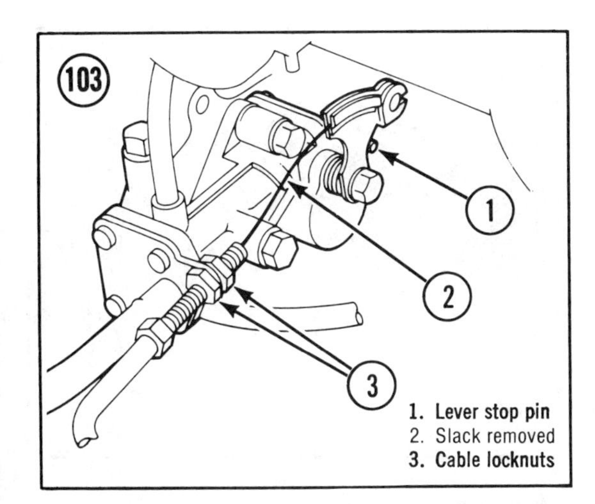

1. **Lever stop pin**
2. Slack removed
3. **Cable locknuts**

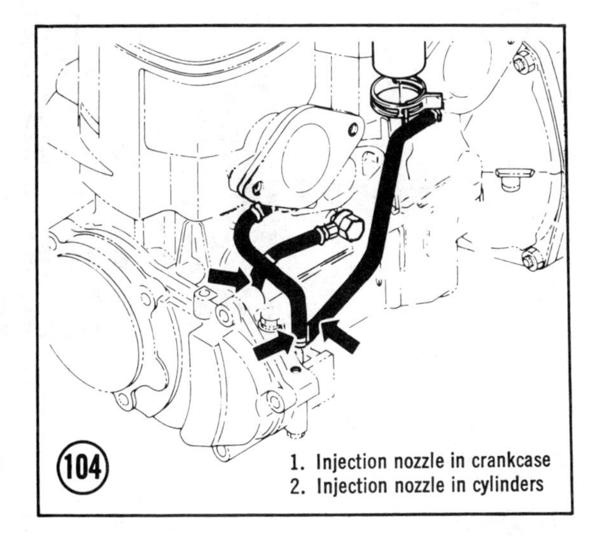

1. **Injection nozzle in crankcase**
2. **Injection nozzle in cylinders**

time, as soon as the throttle hand lever is moved.

6. Adjust the cable free play for the right carburetor in the same manner as for the left. When the hand lever is moved, both carburetor slides and the pump lever should begin to move at the same time.

7. Adjust the carburetors as described in Chapter Two, *Carburetion*, and finally tighten the cable adjuster locknuts on the pump cable.

Output Test

The output of the oil pump should be checked any time the engine is disassembled for major work, or when inadequate oil supply is suspected.

The engine must be run on premixed fuel/oil for this test. A quart of gasoline will be sufficient. Add to it 0.8 oz. (25 cc) of Kawasaki Snowmobile Oil, cap the container, and mix the two thoroughly. Remove and drain the fuel tank as described in Chapter Five, reinstall it, and pour in the fuel/oil mixture.

1. Disconnect the outlet lines from the oil pump. Connect a 2-foot length of tubing to each pump outlet. Place the other end of each tube in separate metric graduates or beakers.

2. Pull the oil pump lever all the way open and hold it in this position with a rubber band.

3. Start the engine and run it at about 3,000 rpm until oil just begins to run out the end of each test tube. Then, allow the engine to run at 3,000 rpm and time it for 1 minute from the point that oil begins to flow into the graduates.

4. Shut off the engine after 1 minute and measure the oil in each of the graduates. Each should contain 3.35-4.00 cc. If the amount is less for either line, replace the pump. If the pump output is satisfactory, remove the test lines and reconnect the oil supply lines and remove the rubber band from the oil pump lever. Fill the fuel tank with fresh gasoline; the small amount of pre-mix that remains should have no appreciable effect on engine performance.

Removal/Installation

The engine must be removed from the machine before the oil pump can be removed. See *Engine Removal* earlier in this chapter.

1. Disconnect the oil supply lines from the pump and plug them (**Figure 104**).

2. Place a container beneath the front of the oil pump gearcase, remove the drain plug (**Figure 105**), and allow the oil to drain.

3. Remove the gearcase cover (**Figure 106**) and collect the spacer from the idler gear and remove the gear (**Figure 107**).

4. Hold the oil pump gear to prevent it from turning, unscrew the nut, and remove the gear.

5. Unscrew the bolts that attach the pump to the gearcase and remove the pump.

6. Reverse the above to install the pump. Install the O-ring in the groove in the flange and coat the flange with silicone sealer. Use new washers on the pump mounting bolts.

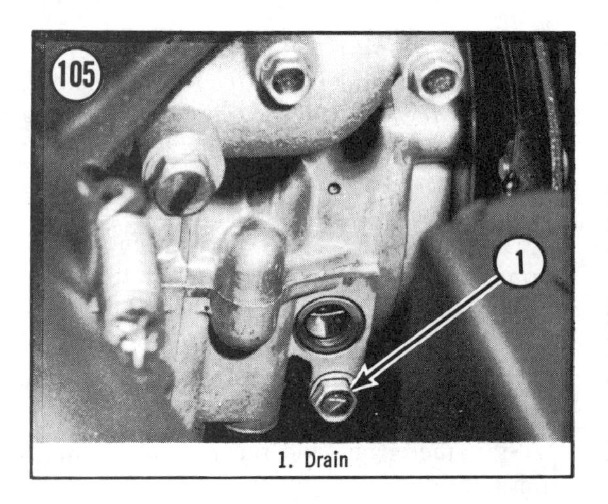

1. Drain

7. Install the pump gear and nut. Install a spacer behind the idler gear and one in front of it.

8. Install the O-ring seal in the groove in the gearcase cover and lightly coat the sealing surface with silicone sealer. Fill the groove in the outer seal in the cover with grease and install the cover.

9. Connect the oil supply lines to the pump.

10. Fill the gearcase with SAE 10W-30 oil, to the center of the sight glass.

NOTE

The gearcase oil should be changed after 10 hours of operation, and annually from then on.

11. Install the engine in the machine as described earlier in this chapter (see *Engine Installation*).

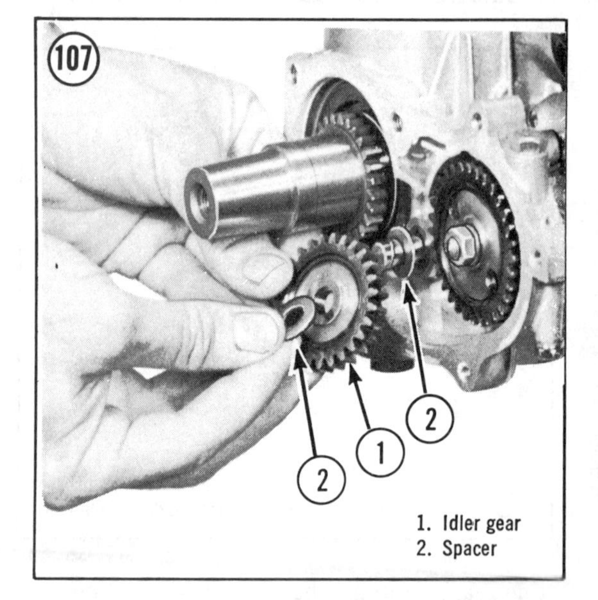

1. Idler gear
2. Spacer

STARTER

Starter service and repair usually involves replacement of the starter rope, and after long use the unit may require complete servicing including cleaning, lubrication, and replacement of broken parts.

Disassembly

1. Remove the exhaust system and the starter assembly (**Figure 108**).

2. Unscrew the locknut from the starter center shaft (**Figure 109**).

3. Remove the cover and the spring.

4. Remove the coil spring, plastic washers, hairpin springs, and pawls (**Figures 110 and 111**).

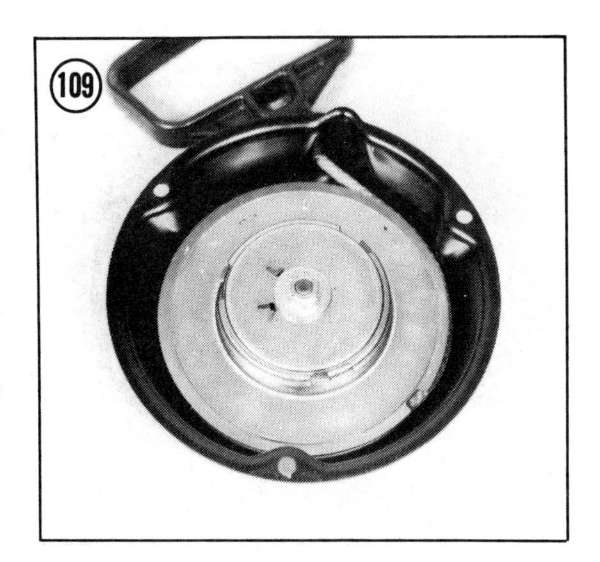
109

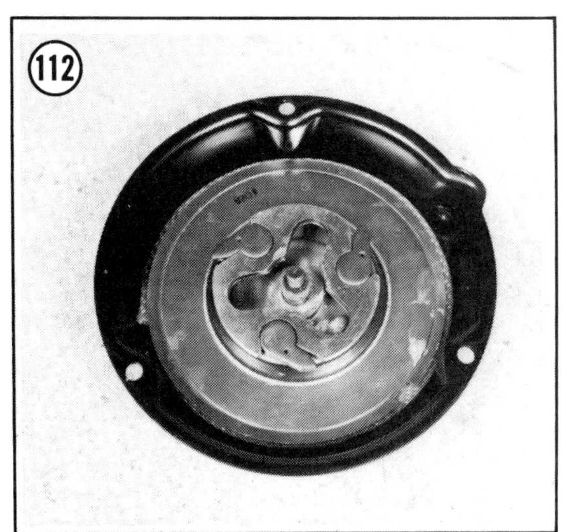
112

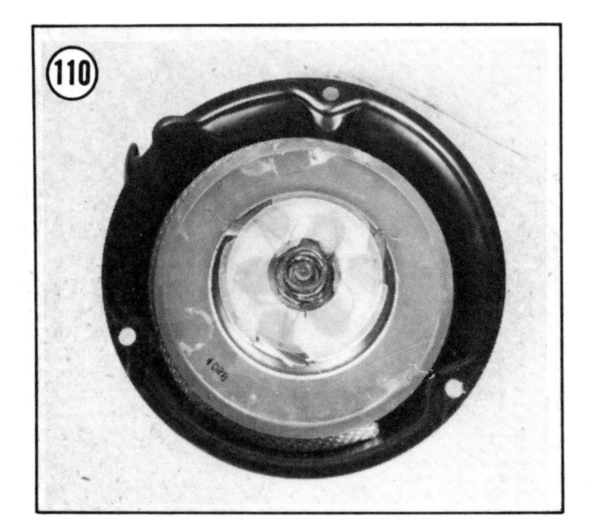
110

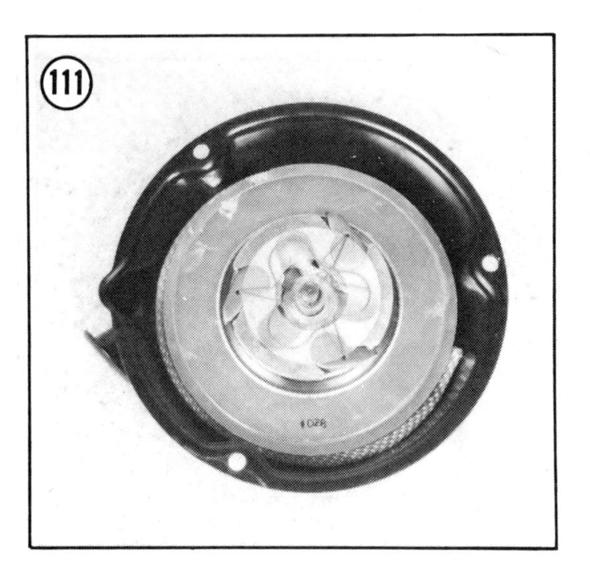
111

5. Remove the reel from the case (**Figure 112**), taking care not to dislodge the spring.

Cleaning/Inspection/Lubrication

1. Clean the parts in solvent and dry them with compressed air.

2. Lubricate the spring and the back of the reel with low-temperature grease (**Figure 113**).

3. Replace any parts that are worn or damaged.

4. Routinely replace the starter rope.

Assembly

1. If the spring was removed, carefully wind it back in with assistance, beginning with the outside of the coil.

2. Wind a new rope around the reel and set the reel into the case, making sure it engages with the inside end of the spring.

3. Set the pawl holder in the reel (**Figure 114**) and install the nylon sleeve (**Figure 115**) on the shaft.

4. Install the pawls (**Figure 116**), the plastic washer and hairpin springs (**Figure 117**), and the outer plastic washer (**Figure 118**).

5. Install the small coil spring and hook the end of it in the slot in the drum.

6. Install the cover, hook the end of the spring in it, and wind it $\frac{1}{3}$ turn clockwise. Install the washer and nut on the shaft.

7. Install the starter handle on the rope and install the starter assembly and exhaust system on the machine.

113

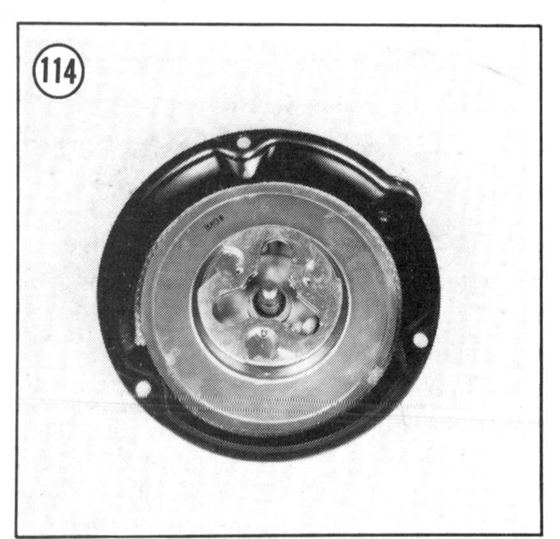
114

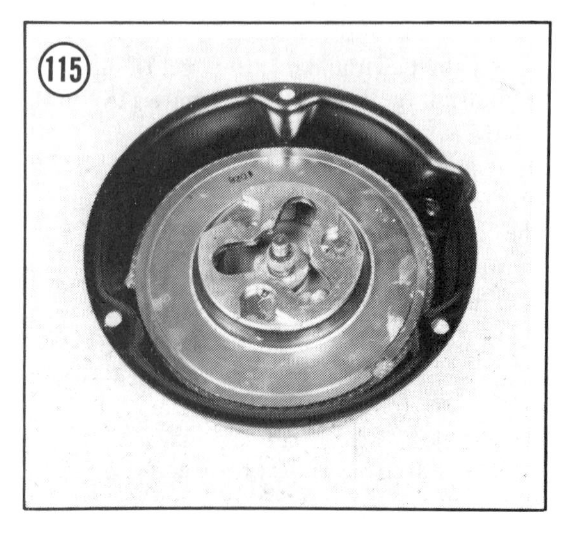
115

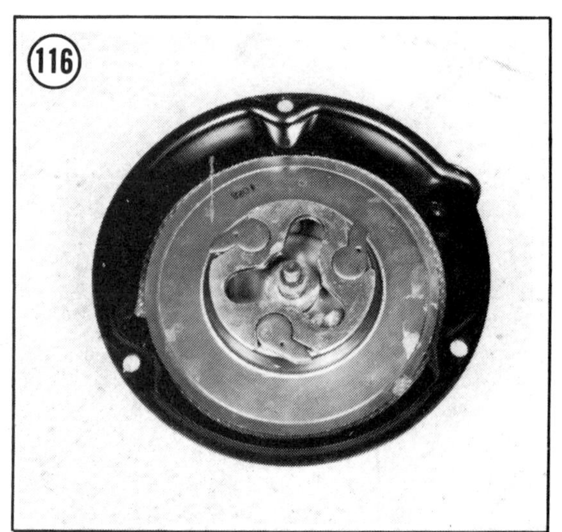
116

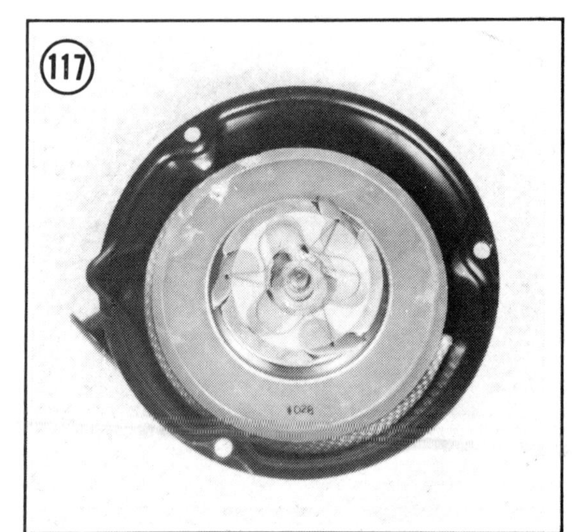
117

118

Table 1 ENGINE FASTENER TORQUE

Model	Cylinder Head	Flywheel	Engine Mounts	Drive Sheave
SA and SB	2.2 mkg (16 ft.-lb.)	8.3 mkg (60 ft.-lb.)	2.2 mkg (16 ft.-lb.)	7.6-8.3 mkg (55-60 ft.-lb.)
SS and ST	2.2 mkg (16 ft.-lb.)	8.3 mkg (60 ft.-lb.)	2.2 mkg (16 ft.-lb.)	8.3-1.0 mkg (60-70 ft.-lb.)

Table 2 UPPER END SPECIFICATIONS

Model	Piston/Cylinder Clearance, mm (in.)	Piston Ring End Gap, mm (in.)	Bore Limits, mm (in.)
SA340 all	0.02-0.06 (0.0008-0.003)	0.15-0.35 (0.006-0.014)	60.10 (2.366)
SA440-A1	0.02-0.06 (0.0008-0.003)	0.20-0.40 (0.008-0.016)	68.10 (2.68)
SB340 all	0.045-0.05 (0.0018-0.0020)	0.35-0.55 (0.014-0.022)	60.10 (2.366)
SB440-A1	0.045-0.05 (0.0018-0.0020)	0.35-0.55 (0.014-0.022)	68.10 (2.68)
SB440-A2 and later	0.05-0.10 (0.002-0.004)	0.2-0.4 (0.008-0.016)	68.10 (2.68)
SS340-A1	0.08-0.12 (0.003-0.005)	0.2-0.4 (0.008-0.016)	60.10 (2.366)
SS340-A2 and later	0.02-0.04 (0.0008-0.0016)	0.2-0.4 (0.008-0.016)	60.10 (2.366)
SS440-A1	0.05-0.10 (0.0020-0.0040)	0.2-0.4 (0.008-0.016)	68.10 (2.68)
SS440-A2 and later	0.02-0.04 (0.0008-0.0016)	0.2-0.4 (0.008-0.016)	68.10 (2.68)
ST440 all	0.05-0.10 (0.002-0.004)	0.2-0.4 (0.008-0.016)	68.10 (2.68)

Table 3 BOTTOM END SPECIFICATIONS

Model	Crankshaft Runout, mm (in.)	Crankshaft End Play, mm (in.)	Connecting Rod Radial Play, mm (in.)
SA and SB	0.05 (0.002)	0.48 (0.019)	0.02-0.03 (0.0008-0.0010)
SS and ST	0.05 (0.002)	0.38 (0.015)	0.02-0.03 (0.0008-0.0010)

Table 4 FLYWHEEL/ROTOR PULLER IDENTIFICATION

Model	Puller (Kawasaki Part No.)
SA340-A1, A2	56019-011
SA340-A3 and later	57001-3509
SA440	56019-011
SB440-A2	56019-011
SB440-A3 and later	57001-3509
ST440	56019-01
SS340-A1	56019-01
SS340-A2 and later	57001-3509
SS440-A1	56019-01
SS440-A2 and later	57001-3509

Table 5 CRANKSHAFT SHIMS

Thickness, mm (in.)
0.7 ± 0.03 (0.0276 ± 0.0012)
0.8 ± 0.03 (0.0315 ± 0.0012)
0.9 ± 0.03 (0.0354 ± 0.0016)
1.0 ± 0.04 (0.03937 ± 0.0016)
1.2 ± 0.05 (0.0472 ± 0.0020)

CHAPTER FIVE

FUEL AND EXHAUST SYSTEMS

The fuel system consists of a carburetor — or dual carburetors on some models — fuel pump, fuel tank, lines, and a filter. On all models, the fuel pump is operated by differential pressure in the crankcase.

An air intake silencer is used to quiet incoming air and catch fuel that may spit back through the carburetor.

The exhaust system consists of an integrated muffler/expansion chamber connected to the engine with an exhaust manifold.

This chapter covers removal, installation, repair or replacement of carburetors, fuel pump, filter, and tank, along with cleaning of the exhaust system. Adjustment of the carburetors is described later in this chapter. Carburetor jetting is also described. Tables are found at the end of the chapter.

CARBURETORS

All Kawasaki snowmobiles use Mikuni slide-type carburetors — either VM Zinc or VM Econo-Jet. Removal and installation procedures are the same for both types. Service is described separately for each type because of significant differences.

Removal/Installation

1. Raise the hood and remove the air intake silencer (**Figure 1**).

2. Unscrew the starter valve from the carburetor (**Figure 2**).
3. Disconnect the primer tube (**Figure 3**).
4. Disconnect the fuel line (**Figure 4**) and plug it to prevent fuel leakage. Unscrew the carburetor top and remove the slide assembly (**Figure 4**).
5. Loosen the clamping band (**Figure 5**) and remove the carburetor from the manifold.
6. Reverse the above to install the carburetor. Carefully clean the inside of the rubber manifold sleeve before inserting the carburetor.

Tighten the clamping band to the point that the carburetor can't be rotated in the sleeve. When installing the slide assembly, line up the groove in the slide with the knob in the carburetor, and the needle with the needle jet (**Figure 6**). Refer to *Adjustment* later in this chapter and adjust the carburetor as described.

MIKUNI VM ZINC CARBURETOR

The VM Zinc carburetor is equipped with an enrichment valve to aid warmup. Intake is controlled by a throttle slide. Idle mixture is controlled by a fixed pilot jet and an adjustable air control needle. Low and mid-range mixture is controlled by the pilot jet and the needle jet. High-speed mixture is controlled by the main jet.

Disassembly

Refer to **Figure 7** for the procedure that follows.

1. Unscrew the 4 screws that attach the float bowl to the carburetor and remove the bowl. Empty the gasoline into a sealable container.
2. Remove the float hinge pin (6) and the float arm (11). Tap out the valve needle and unscrew the valve body (18).
3. Unscrew the main jet (10) and the pilot jet (8).
4. Press the needle jet (28) out of the carburetor.

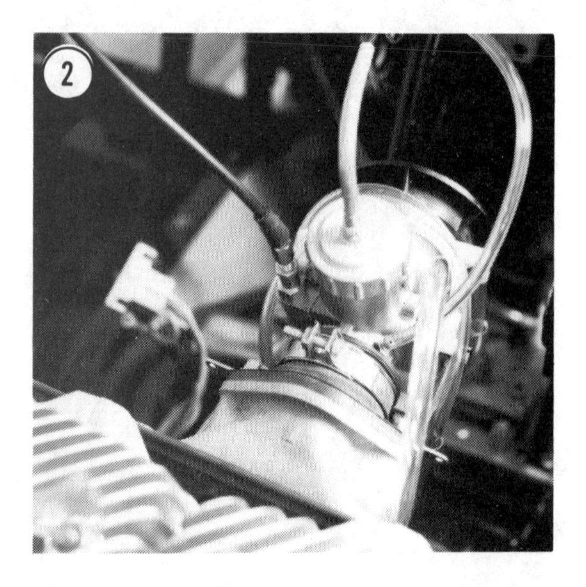

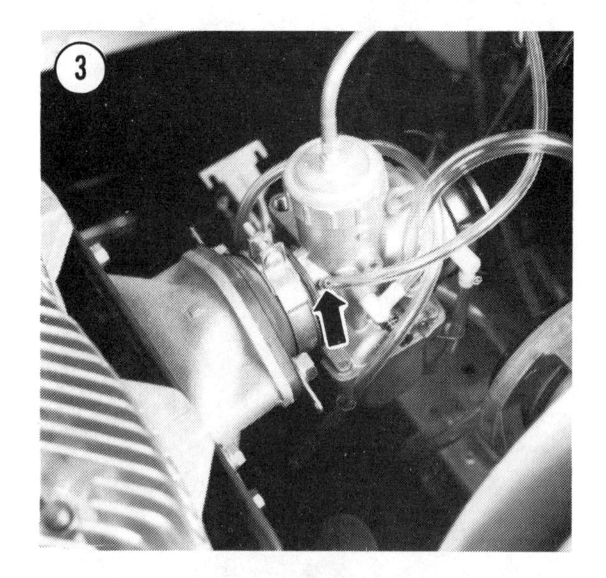

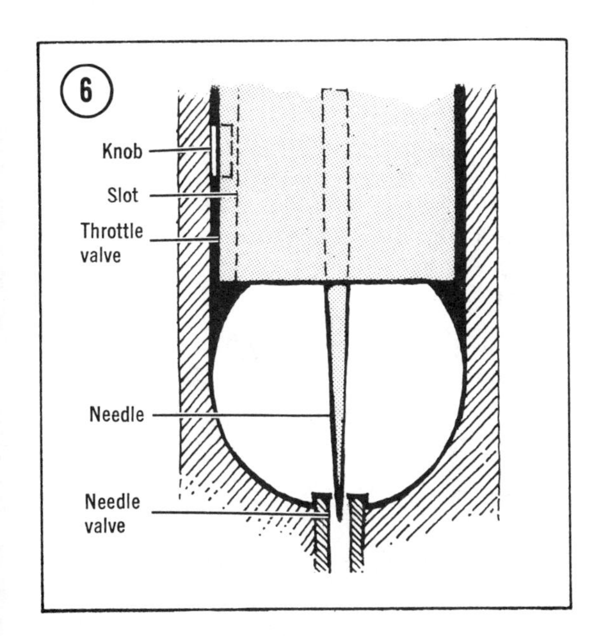

5. Unscrew the throttle stop screw (5) and spring (4).

6. Unscrew the idle mixture screw (2) and spring (3).

7. Disconnect the throttle cable from the slide (27) and disassemble the slide/needle assembly.

Cleaning and Inspection

A special carburetor cleaning solvent, tank, and basket (**Figure 8**) can be purchased at auto parts stores for just a few dollars. When not in use, the tank can be sealed to prevent the reuseable solvent from evaporating.

WARNING

Most carburetor cleaning solvents are highly caustic and can cause skin burns and eye injury.

1. Clean all the metal parts in carburetor cleaning solvent.

CAUTION

Do not place gaskets, O-rings, floats, or other non-metal parts in the solvent or they will be damaged.

2. Dry the parts with compressed air. Blow out all jets and passages to ensure that no solvent or residue remains.

CAUTION

Never clean passages or jets with small drill bits or wire; the soft metal is very easily burred, resulting in changes in the effective size of the passage or jet and altered fuel flow rate.

3. Inspect the carburetor body and float bowl for fine cracks or evidence of fuel leaks. Minor damage can be repaired with epoxy or "liquid aluminum" type fillers.

4. Inspect the taper of the idle mixture screw for scoring and replace it if it is less than perfect. Inspect the threads of the throttle stop screw for damage which could cause it to bind.

5. Inspect the jets for internal damage and damaged threads. Replace any jet that is less than perfect. Make certain replacement jets are the same size as the original.

CAUTION

The jets must be scrupulously clean and shiny. Any burring, roughness, or abrasion could cause a lean mixture that will result in major engine damage.

6. Inspect the float valve seat and needle taper for scoring, wear, or any other signs of damage and replace them as a set if either is less than perfect. A damage float valve will result in flooding of the float bowl and impair performance. In addition, accumulation of raw gasoline in the closed engine compartment presents a severe fire hazard.

7. Inspect the needle jet and needle for scoring, wear, or damage and replace them as a set if either piece is less than perfect. Check the movement of the needle in the jet. It must move freely in and out without binding.

8. Check the movement of the float on the pivot pin. It must move freely without binding.

9. Check the movement of the starter valve in the bore in the carburetor. It should move freely in and out.

10. Check the movement of the throttle slide in the carburetor. It should move freely up and down without binding and it should not be loose. If either the slide or the slide bore appear worn, they should be replaced.

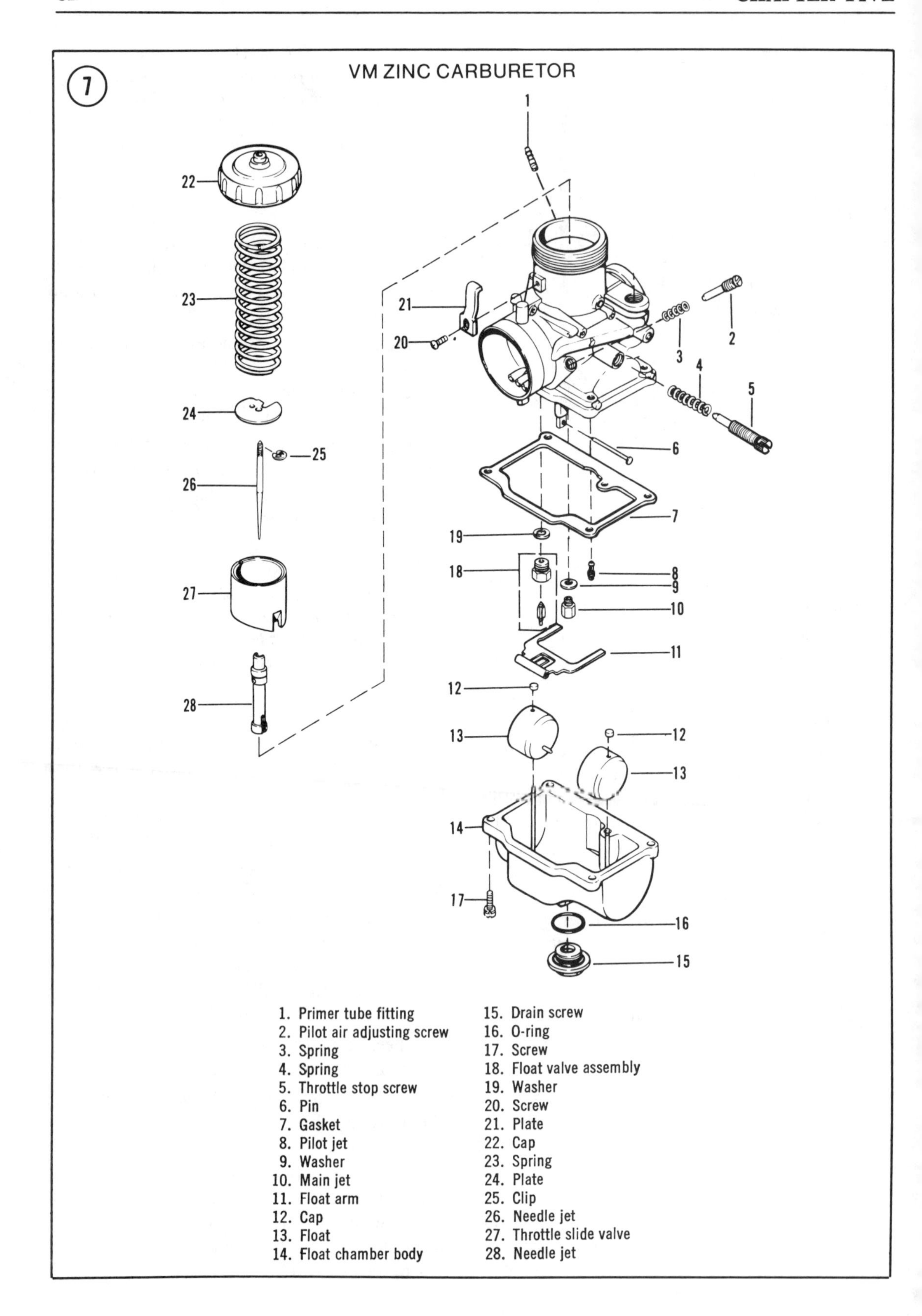
7
VM ZINC CARBURETOR
1
2
3
4
5
6
7
8
9
10
11
12
12
13
13
14
15
16
17
18
19
20
21
22
23
24
25
26
27
28
1. Primer tube fitting
2. Pilot air adjusting screw
3. Spring
4. Spring
5. Throttle stop screw
6. Pin
7. Gasket
8. Pilot jet
9. Washer
10. Main jet
11. Float arm
12. Cap
13. Float
14. Float chamber body
15. Drain screw
16. O-ring
17. Screw
18. Float valve assembly
19. Washer
20. Screw
21. Plate
22. Cap
23. Spring
24. Plate
25. Clip
26. Needle jet
27. Throttle slide valve
28. Needle jet

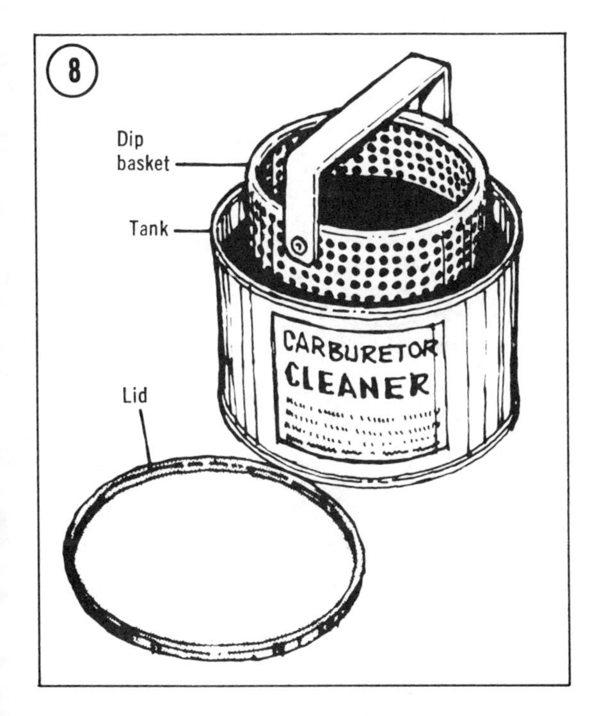

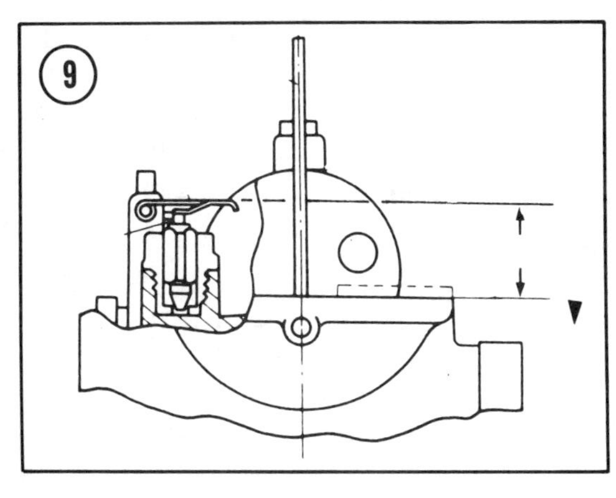

Assembly

Refer to **Figure 7** for the procedure that follows.

1. Assemble the slide (27), needle (26), clip (25), plate (24), spring (23) and cap (22). Make certain the needle clip is in the same slot from which it was removed. When the cable has been engaged in the slide, lock it in place with the plate.
2. Install the idle mixture screw (2) and spring (3). Run the screw in carefully, just until it seats. *Don't tighten it; the screw and seat are easily damaged.* Back the screw out 1½ turns. This basic setting will permit the engine to start and idle well enough to allow correct adjustment to be made.
3. Install the throttle stop screw (5) and spring (4). Run the screw in about 10 turns.
4. Press the needle jet (28) into the carburetor. The shielded portion of the spray nozzle must face the intake side of the venturi.
5. Install the main jet (10) and the pilot jet (8). Run them in until they seat but take care not to overtighten them.
6. Screw in the float valve body and install the needle (18). Set the float arm (11) in place and push in the hinge pin (6). Make sure the pin is completely seated and that the float arm and needle move freely.
7. Measure the float height, and correct it if necessary, as described below.
8. Install the float bowl gasket (7) and float bowl (14). Start all the screws in before tightening them.

Float Level Adjustment

The float level should be routinely checked and adjusted whenever the carburetor is disassembled and cleaned, or when flooding indicates that the level may be incorrect.

Invert the carburetor and allow the float arm to contact the needle valve, but don't compress the spring-loaded plunger in the needle.

The float arm should be parallel to the carburetor base (**Figure 9**). If it is not, carefully bend the tang that contacts the needle valve until the arm is parallel.

MIKUNI ECONO-JET CARBURETOR

The Econo-Jet carburetor is equipped with an enrichment valve to aid warmup. Intake is controlled by a throttle slide. Idle mixture is controlled by a fixed pilot jet and an adjustable air-control needle. Low and mid-range mixture is controlled by the pilot jet and the needle and needle jet. High-speed mixture is controlled by the main jet and Econo-Jet. The Econo-Jet flows at speeds above 5,000 rpm to augment the flow of the main jet, which is relatively small to promote good fuel economy.

5

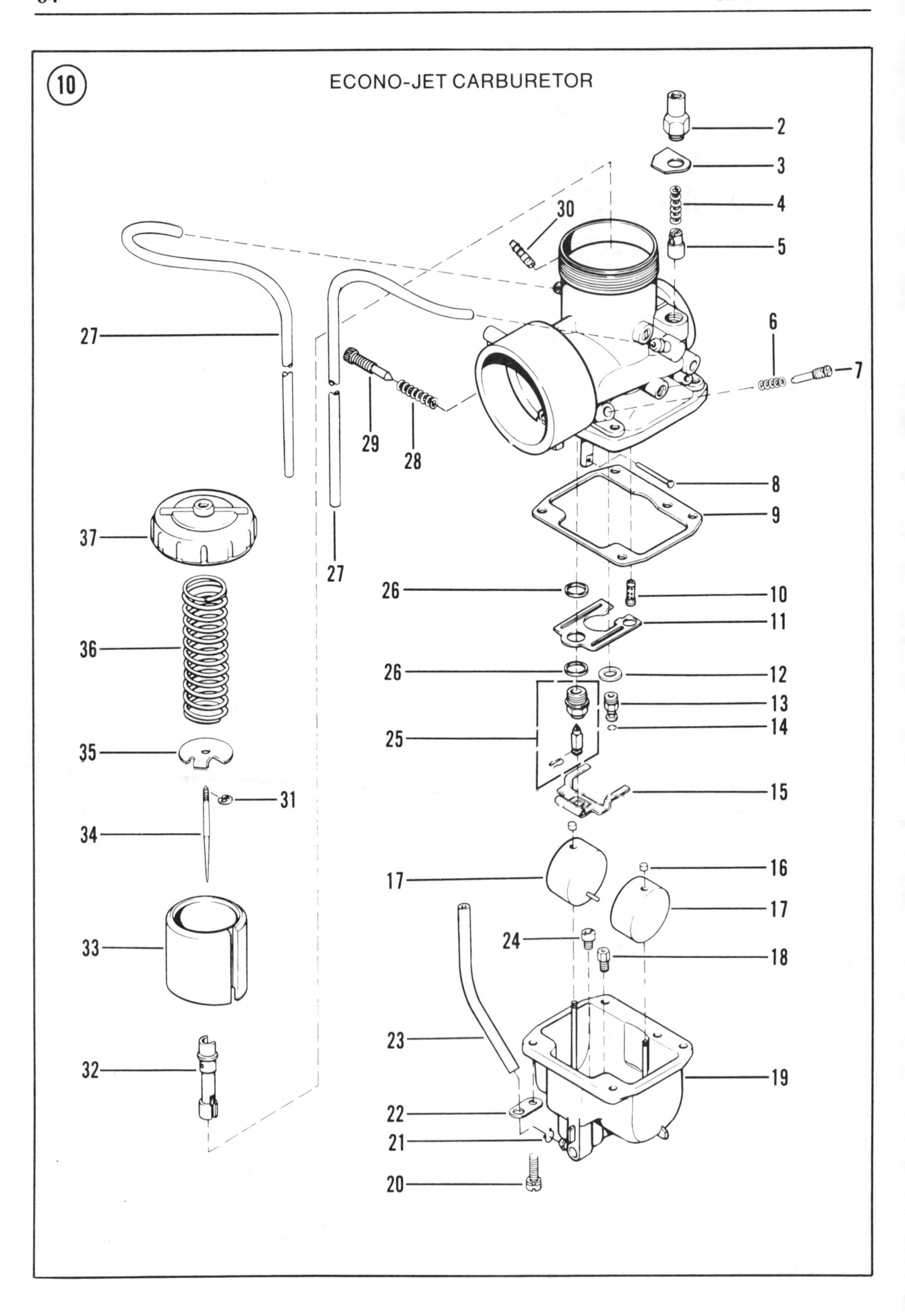
10
ECONO-JET CARBURETOR
2
3
4
5
6
7
8
9
10
11
12
13
14
15
16
17
18
19
20
21
22
23
24
25
26
26
27
27
28
29
30
31
32
33
34
35
36
37

ECONO-JET CARBURETOR

1. Carburetor assembly
2. Enrichener plunger cap
3. Cap washer
4. Enrichener plunger spring
5. Enrichener plunger
6. Air screw spring
7. Pilot air screw
8. Float pin
9. Float chamber gasket
10. Pilot jet
11. Baffle
12. Holder washer
13. Carburetor needle holder
14. Holder O-ring
15. Float arm
16. Float retainer
17. Float
18. Main jet
19. Float chamber
20. Special screw
21. Econo jet tube clamp
22. Tube retainer
23. Tube
24. Econo jet
25. Float valve assembly
26. Float valve gasket
27. Tube
28. Throttle screw spring
29. Throttle stop screw
30. Primer tube fitting
31. Carburetor needle clip
32. Needle jet
33. Throttle valve
34. Needle
35. Spring seat
36. Throttle valve spring
37. Mixing chamber cap

12

Disassembly

Refer to **Figure 10** for the procedure that follows.

1. Unscrew the 4 screws that attach the float bowl to the carburetor (**Figure 11**) and remove the bowl. Empty the gasoline into a sealable container.

2. Remove the float hinge pin (8) and the float arm (15). Remove the clip and needle from the float valve and unscrew the valve body (25) from the carburetor.

3. Unscrew the valve holder (13) and remove the baffle (11).

4. Unscrew the pilot jet (10).

5. From inside the venturi, press the needle jet out of the carburetor (**Figures 12 and 13**).

6. Unscrew the throttle stop screw (29) and remove the spring (28).

7. Unscrew the idle mixture screw (7) and remove the spring (6).

8. Unscrew the main jet and the Econo-Jet (18 and 24, **Figure 14 and Figure 10**).

9. Disconnect the throttle cable from the slide (33) and take apart the slide/needle assembly.

Cleaning and Inspection

1. Clean all metal parts in carburetor cleaning solvent.

CAUTION

Do not place gaskets, O-rings, floats, or other non-metal parts in the solvent or they will be damaged.

WARNING

Most carburetor cleaning solvents are highly caustic and can cause skin burns and eye injury.

2. Dry the parts with compressed air. Blow out all jets and passages to ensure that no solvent or residue remains.

CAUTION

Never clean passages or jets with small drill bits or wire; the soft metal is very easily burred, resulting in the effective size of the passage or jet being changed and altering the fuel flow rate.

3. Inspect the carburetor body and float bowl for fine cracks or evidence of fuel leaks. Minor damage can be repaired with epoxy or "liquid aluminum" type fillers.
4. Inspect the taper of the idle mixture screw for scoring and replace it if it is less than perfect. Inspect the threads of the throttle stop screw for damage which could cause it to bind.
5. Inspect the jets for internal damage and damaged threads. Replace any jet that is less than perfect. Make certain replacement jets are the same size as the original.

CAUTION

The jets must be scrupulously clean and shiny. Any burring, roughness, or abrasion could cause a lean mixture that will result in major engine damage.

6. Inspect the float vale seat and needle taper for scoring, wear, or any other signs of damage and replace them as a set if either is less than perfect. A damaged float valve will result in flooding of the float bowl and impair performance. In addition, accumulation of raw gasoline in the closed engine compartment presents a severe fire hazard.
7. Inspect the needle jet and needle for scoring, wear, or damage and replace them as a set if either piece is less than perfect. Check the movement of the needle in the jet. It must move freely in and out without binding.
8. Check the movement of the float arm on the pivot pin. It must move freely without binding.
9. Check the movement of the starter valve in the bore in the carburetor. It should move freely up and down.

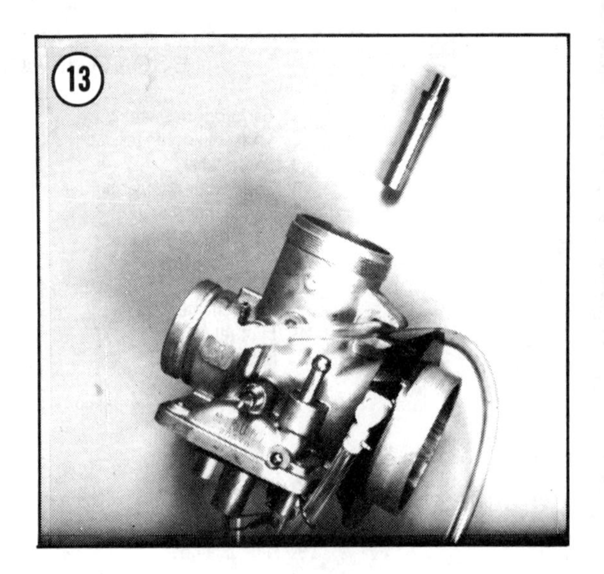

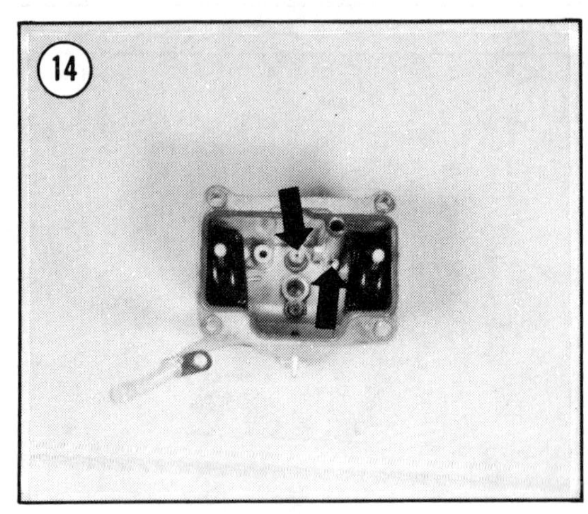

10. Check the movement of the throttle slide in the carburetor. It should move freely up and down without binding and it should not be loose. If either the slide or the slide bore appear worn, they should be replaced.

Assembly

Refer to **Figure 10** for the procedure that follows.

1. Assemble the throttle slide (33), needle (34), clip (31), plate (35), spring (36) and cap (37). Make certain the needle clip is in the same slot from which it was removed. When the cable has been engaged in the slide, lock it in place with the plate.
2. Install the idle mixture screw (7) and spring (6). Run the screw in carefully, just until it

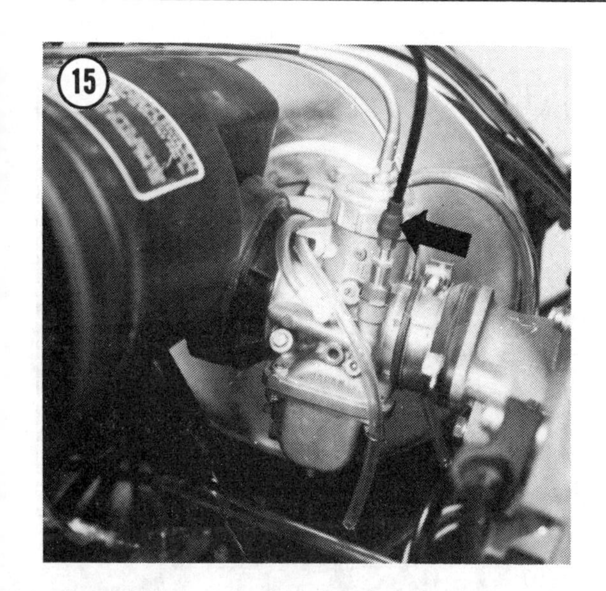

15

16

seats. *Don't tighten it; the screw and seat are easily damaged.* Back the screw out 1½ turns. This basic setting will permit the engine to start and idle well enough to allow correct adjustment to be made.

3. Install the throttle stop screw (29) and spring (28). Run the screw in about 10 turns.

4. Press the needle jet (32) into the carburetor. The shielded portion of the spray nozzle must face the intake side of the venturi.

5. Install the main jet (18) and the Econo-Jet (24) in the float bowl.

6. Install the pilot jet (10).

7. Install the baffle (11) and the needle jet holder (13).

8. Screw in the float valve body, set the valve needle into the valve body, and lock the needle in place with the clip (25).

9. Set the float arm in place (15) and push in the hinge pin (8). Make sure the pin is completely seated and that the arm moves freely.

10. Check the float height and correct it if necessary as described earlier under *Float Level Adjustment*.

11. Install the float bowl gasket (9) and the float bowl (19). Start all the screws in before tightening them.

CARBURETION ADJUSTMENT

There are 5 adjustments possible with either of the 2 carburetors that are used on Kawasaki snowmobiles. The adjustments are interdependent and should be carried out in the order that follows for best results.

a. Starter cable free play
b. Throttle cable free play (and slide synchronization on dual carburetor models)
c. Idle mixture/throttle stop
d. Slide needle
e. Main jet/metering adjustment

Starter Cable Free Play

With the starter lever down (off), disconnect the boot from the cable adjuster (**Figure 15**) and pull up on the cable sheath to check the free play. It should be 1.5 mm ($\frac{1}{16}$ in.). If it is less, the carburetor will run rich. If adjustment is required, loosen the locknut and turn the cable adjuster in or out until the free play is correct. Then, hold the adjuster to prevent it from turning further and tighten the locknut and reseat the boot.

Throttle Cable Free Play

1. Remove the air intake silencer (**Figure 16**).

2. Fully open the throttle hand lever and feel inside the carburetor bore to make sure the rear edge of the throttle slide is clear of the venturi

(**Figure 17**). If adjustment is required, loosen the locknut on the cable adjuster and turn the adjuster as required. Then, tighten the locknut.

> NOTE: *On dual carburetor models all adjustments must be equal for both carburetors.*

3. Run the throttle stop screw out until its tip is flush with the inside of the slide bore. Then run the screw in the number of turns specified in **Table 1** for the basic idle setting. Close the throttle and allow the slide to rest on the throttle stop screw.
4. Operate the throttle hand control and check for minimal free play (**Figure 18**). If adjustment is required, loosen the locknut and turn the valve adjuster as required. Then tighten the locknut. On dual carburetor models, make sure the throttle slides move equal amounts.

Idle Mixture/Throttle Stop

1. Slowly turn the idle mixture screw (**Figure 19**) in until it just seats. *Don't tighten the screw or it will be damaged.* Then, back the screw out the number of turns specified in **Table 1**. On dual carburetor models, turn the screws out an equal number of turns.

2. Install the air intake silencer. Position the machine with the skis against a wall and raise

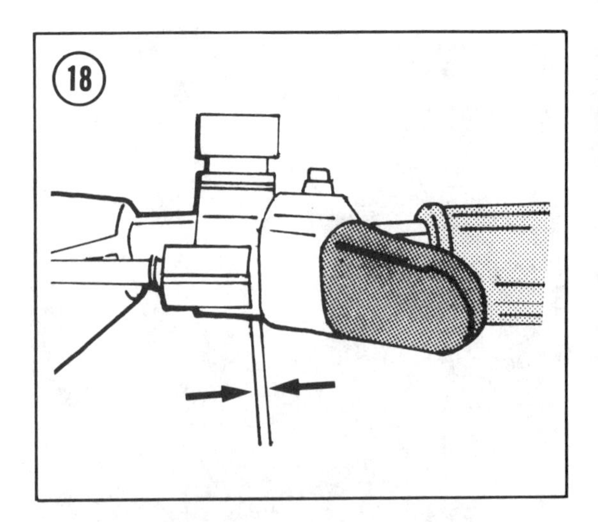

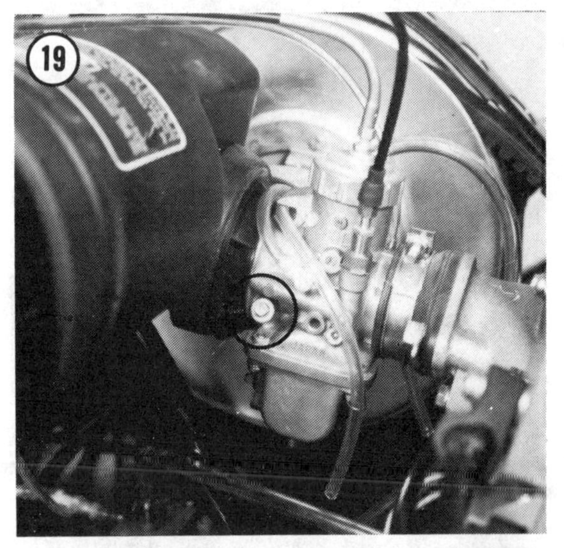

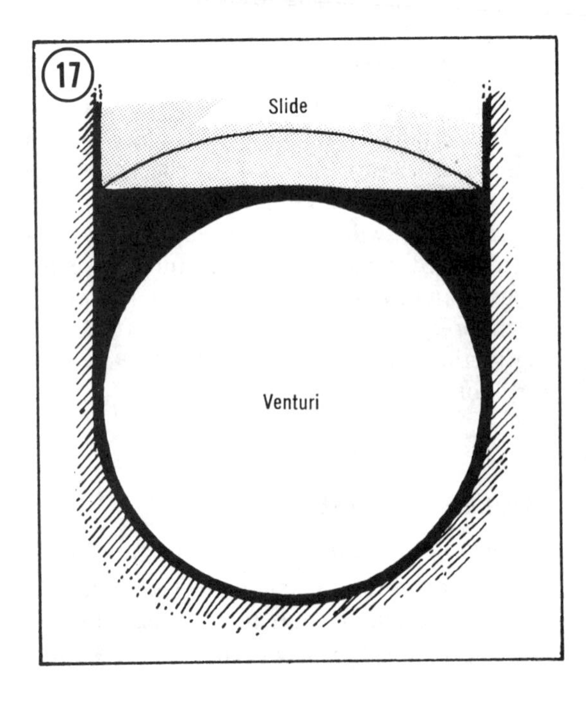

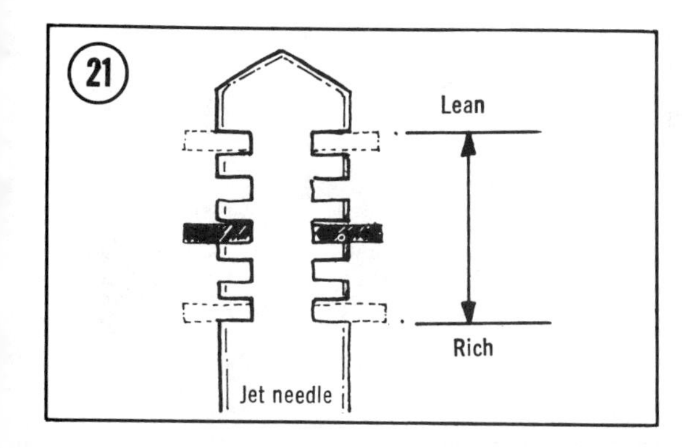

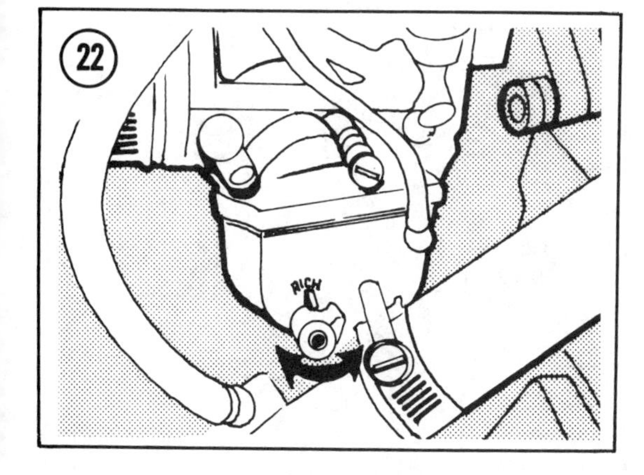

the rear of the machine and support it so the track is clear of the ground and free to rotate. Connect a tachometer to the engine in accordance with the instrument manufacturer's instructions.

3. Start the engine and allow it to warm up to operating temperature.

4. Set the idle speed with the throttle stop screw (**Figure 20**) as specified in **Table 1**. On dual carburetor models, turn the throttle stop screws equal amounts.

> NOTE: *On dual carburetor models, synchronize the carburetors as described later (see **Carburetor Synchronization**) before installing the air intake silencer.*

Slide Needle Position

Mid-range performance is controlled by slide needle position. To check its correctness, first warm up the engine and then accelerate the machine hard over hard-packed snow and "feel" the engine response.

If the engine seems starved or if it knocks, the mid-range mixture is too lean. In this case, remove the slide assembly from the carburetor and raise the needle in the slide by moving the needle clip to the next lower notch (**Figure 21**). Then, recheck the performance and if the mixture is still too lean, raise the needle one more position; move the needle one position at a time.

If the engine "blubbers" or if there is excessive exhaust smoke during hard acceleration, the mid-range mixture is too rich. In this case, lower the needle in the slide by moving the clip to a higher notch. Then recheck the performance and if it is still too rich, lower the needle another notch.

CAUTION

After adjusting the slide needle position, check the operation of the throttle to make sure the slide moves freely in its bore and closes when the throttle is released.

Main Jet/Metering Adjustment

Check the condition of the high-speed mixture ratio by conducting a "plug chop" as described under *Jetting* in this chapter. For carburetors equipped with a main jet adjuster screw (**Figure 22**), the mixture can be made leaner by turning the screw clockwise and made richer by turning it counterclockwise.

CAUTION

During testing and adjustment, turn the screw no more than 1/8 turn at a time. Massive changes, such as 1/2 turn, may take the mixture beyond the optimum point and yield inaccurate information, or, at the worst, the mixture could be made so lean that engine damage could result.

If the carburetor is not equipped with a main jet adjusting screw, high-speed mixture corrections must be made by changing the main jet size. In this case, also refer to *Jetting* in this chapter and make required adjustments and changes as described.

When all the adjustments discussed have been made, test the machine in all modes of carburetor performance — starting, idle, acceleration, and sustained high-speed operation. If performance is deficient in any mode, readjust the carburetor as just described until total performance is satisfactory.

Carburetor Synchronization

On dual carburetor models, the carburetors must be precisely synchronized if maximum performance is to be realized. A carburetor synchronizer like the one described in Chapter One is essential for this work.

1. Make the basic adjustments described earlier (see *Starter Cable Free Play, Throttle Cable Free Play*, and *Idle Mixture/Throttle Stop*).

2. Position the machine with the skis against a wall and raise the rear of the machine and support it so the track is clear of the ground and free to rotate. Connect a tachometer to the engine in accordance with the instrument manufacturer's instructions.

3. Start the engine and allow it to warm up to operating temperature. Bind and wedge the throttle to maintain engine speed at 4,000 rpm (**Figure 23**).

4. Open the air flow control on the air flow meter and place the meter over the throat of the left carburetor, with the tube on the meter positioned vertical (**Figure 24**).

5. Slowly close the air flow control on the meter until the float lines up with a graduated mark on the tube.

6. Without changing the adjustment of the air flow control, move the meter to the other carburetor and observe the float. If its position does not change, the carburetors are synchronized.

7. If adjustment is required, loosen the locknut on the cable adjuster of the carburetor that records the lowest float position on the meter and turn the adjuster out until the float position matches that for the other carburetor.

8. Remove the wedge and rubber band from the throttle lever and allow the engine speed to return to idle. Repeat Steps 4, 5, and 6 and correct the idle synchronization by turning the idle stop screw of the carburetor with the lowest reading in until the float position for that carburetor matches that for the other.

9. When the carburetors have been accurately synchronized, tighten the locknuts on the cable adjusters and install the air intake silencer.

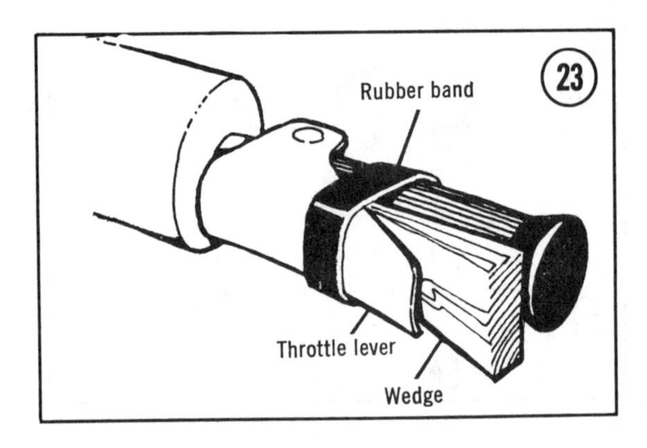

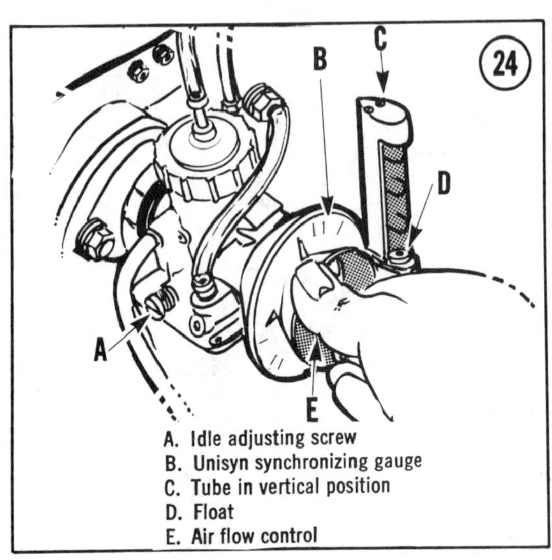

A. Idle adjusting screw
B. Unisyn synchronizing gauge
C. Tube in vertical position
D. Float
E. Air flow control

10. Check the engine idle speed and, if necessary, set it to the value shown in **Table 1** by turning the idle stop screws equal amounts.

JETTING

The air/fuel mixture ratio has a significant effect on engine performance and service life. For purposes of discussion, a ratio of 15 parts air to 1 part fuel is generally accepted. If there are proportionately more parts of fuel — say 12 to 1 — the mixture is fuel rich and there is not enough air to properly oxidize the fuel, allowing some of it to go unburned and provide less work than it's capable of. Not only is fuel efficiency reduced, but, in more immediate terms, engine performance is poor. The engine tends to be sluggish and respond poorly.

On the other hand, if the ratio is perhaps 20 parts air to 1 part fuel, the mixture is fuel-lean. While the engine may not seem sluggish and in fact will often seem to be performing better than normal, a lean condition raises the temperature in the combustion chamber, often to the point of melting the piston crown, combustion chamber, and the insulation and electrodes on the spark plug.

Mixture ratio is affected by changes in air density which varies with changes in temperature and altitude. Air density increases as temperature decreases, and decreases as altitude increases.

The standard jets and settings shown in **Table 1** are the basic factory settings. They are selected for operation in outside air temperatures of – 4° to 32° F (– 20° to 0° C) at sea level.

As a rule of thumb, main jet size should be decreased one size (for example, from a No. 210 to a No. 200 jet) for each 4,000 foot increase in altitude if the outside temperature remains in the basic range (– 4° to 32° F).

However, if the outside temperature is significantly lower than the basic range — say – 36° to – °4 F — an increase in jet size is required. Thus, it can be seen that if the machine is to be operated at higher altitude (4,000 feet) and colder temperature than the basic range, the standard jet may very well be correct.

It's important to understand that changes in altitude and temperature from the basic ranges can offset one another.

NOTE: *The values used above to explain the relationship of temperature and altitude as they affect air density are selected only for illustration; correct jetting can be achieved only by patient experimenting under actual conditions.*

Before making any changes in jetting, make the basic adjustments described under *Adjustment* for your particular carburetor and carry out ignition adjustments described in Chapter Six. Then, test ride the machine and check for the symptoms described in **Table 2**.

Next, conduct a "plug chop." Accelerate the machine at full throttle for about 100 yards on hard-packed snow. Then, kill the engine with the ignition switch and stop the machine with the brake — not with engine compression.

WARNING
Make certain you have clear snow ahead of you before accelerating to make a plug chop. Check the area you will be accelerating over beforehand to make sure there are no hidden obstacles.

Unscrew the spark plugs from the cylinder heads and compare their condition to **Figure 25**. If the spark plugs indicate a lean mixture, the main jet size should be increased, and if a rich condition is indicated, the main jet size should be decreased.

NOTE: *If the carburetor is equipped with a high-speed adjuster screw, turn the screw in to richen the mixture or out to lean it.*

Make jet changes one size at a time and recheck with a plug chop as described above. (If the high-speed circuit has an adjuster screw, turn the screw ⅛ turn at a time.) Be patient and don't be satisfied until the jetting is correct; the little bit of additional time required to do a thorough job can mean the difference between fair and excellent performance — and it may very well prevent expensive engine damage.

REED VALVE

The reed valve should be inspected routinely when the engine upper end is serviced, or when a decline in engine performance that can't be corrected with ignition and carburetion tuning indicates that the reed valve may have broken reed petals.

Removal

1. Loosen the carburetor clamping bands **(Figure 26)** and remove the carburetor. It's not necessary to disconnect the cables and clamps from the carburetor — just hang it out of the way.

2. Unscrew the intake manifold nuts, tap the manifold with a soft mallet to break it loose, and remove it.

3. Remove the reed valve assembly from the cylinder.

Inspection

1. Check the reed petals for fatigue cracks. Check to see if the petals fit flush against the seats. Check the fit by applying suction to the

SPARK PLUG CONDITIONS

NORMAL USE

OIL FOULED

CARBON FOULED

OVERHEATED

GAP BRIDGED

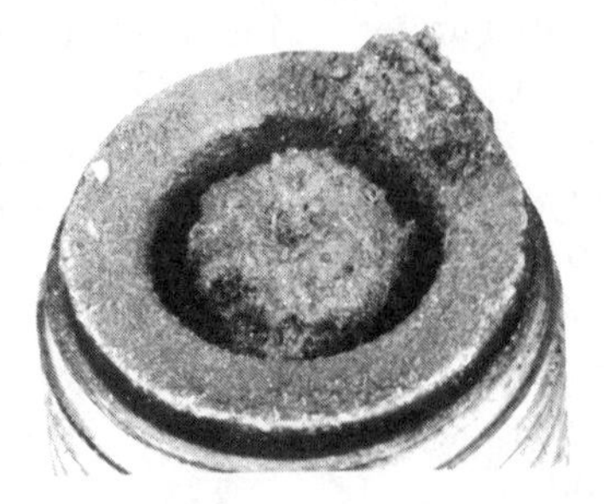

SUSTAINED PREIGNITION

WORN OUT

Photos courtesy of Champion Spark Plug Company.

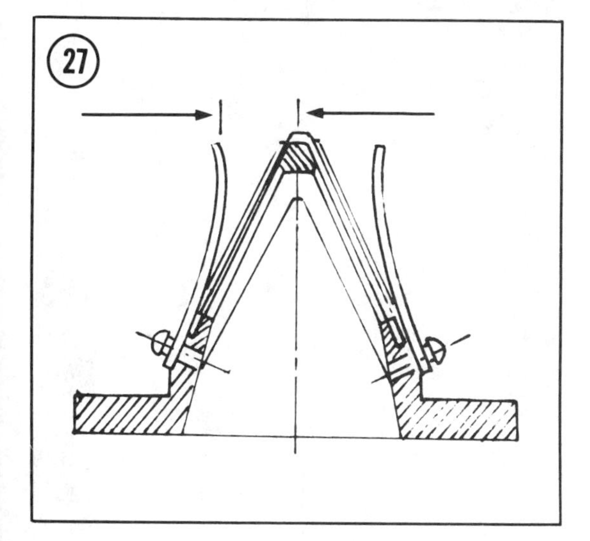

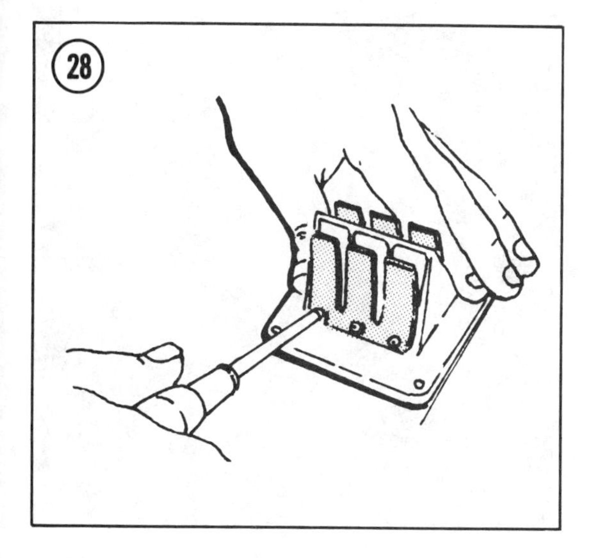

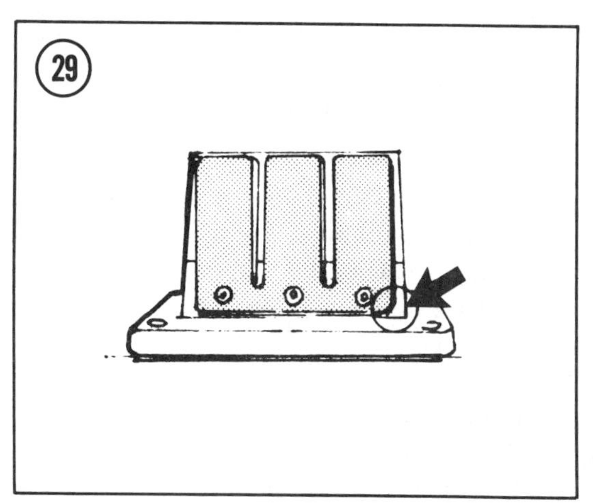

carburetor side of the valve. Replace the petals as described below (*Reed Petal Replacement*) if they are less than perfect.

2. Check the clearance of the valve stoppers (**Figure 27**). It should be 10.2-10.6 mm (0.402-0.418 in.). If the clearance is greater than specified, the reed petals will break, and if it is less the petals won't open completely and performance will be poor.

If the clearance is incorrect, carefully bend the stoppers as required.

Reed Petal Replacement

1. Remove the screws that attach the stopper plates and reeds to the valve block (**Figure 28**). Mark the stoppers and valve so the stoppers can be installed on the sides from which they were removed.
2. Clean the reed block and stopper plates in solvent and dry them.
3. Coat the threads of the screws with Loctite or a similar compound.
4. Position the new reeds and the stoppers with the notched corner positioned as shown (**Figure 29**). Screw in and tighten the screws.
5. Check the stopper clearance as described above.

Installation

1. Install the reed valve assembly and gasket in the cylinder.
2. Install the intake manifold and gaskets. Start all the nuts on and run them down finger-tight before tightening them with a wrench.

3. Install the carburetor and tighten the clamping bands.

FUEL TANK

Removal/Installation

1. Raise the hood. Disconnect the fuel supply line from the bottom front of the tank and plug the line to prevent leakage.
2. Pull back on the tank at the top (**Figure 30**) to disconnect the tank from the steering support loop.
3. Lift the tank out of the vehicle.
4. Reverse the above to install the tank. Make certain it is correctly seated on the tunnel and engaged with the steering support loop. Reconnect the fuel line.

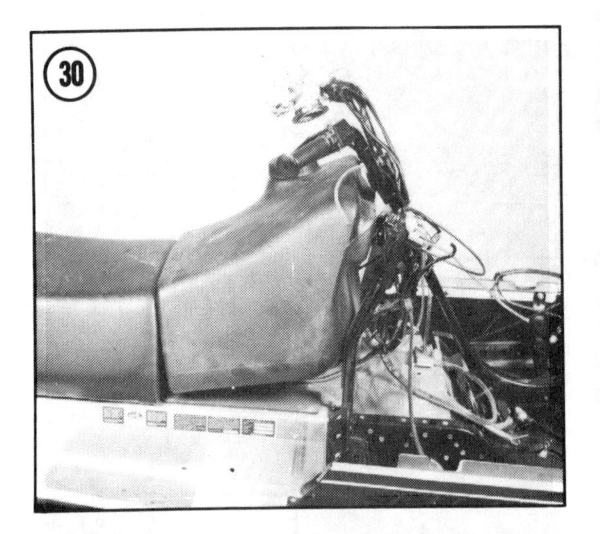

Cleaning/Inspection

1. Pour old gasoline from the tank into a sealable container.
2. Pour about 1 quart of fresh gasoline into the tank and slosh it around for several minutes to loosen sediment. Then pour the contents into a sealable container.
3. Examine the tank for cracks and abrasions, particularly at points where the tank contacts the body. Abraded areas can be protected by cushioning them with a coating of non-hardening silicone sealer and allowing it to dry before installing the tank. However, if abrading is extensive, or if the tank is leaking, replace it.

FUEL FILTER

Replace the fuel filter at the end of the winter season, before storing the machine for the summer, and after the fuel tank has been cleaned as described above.

Disconnect the fuel line from the tank (**Figure 31**) and remove the filter. Install a new filter, making sure the arrow on the filter points in the direction of fuel flow from the tank. Push the hoses all the way onto the filter nipples.

FUEL PUMP

All models are equipped with a remote, pulse-type fuel pump that is located just ahead of the fuel tank (**Figure 32**).

Removal/Installation

1. Raise the hood and remove the air intake silencer (**Figure 33**).

2. Disconnect the lines from the pump and unscrew the mounting bolts (**Figure 32**). Remove the pump.

3. Reverse the above to install the pump. Make sure the lines are correctly connected.

Disassembly/Assembly

Refer to **Figure 34 or Figure 35** for the procedure that follows.

1. Remove the screws that attach the cover to the body.

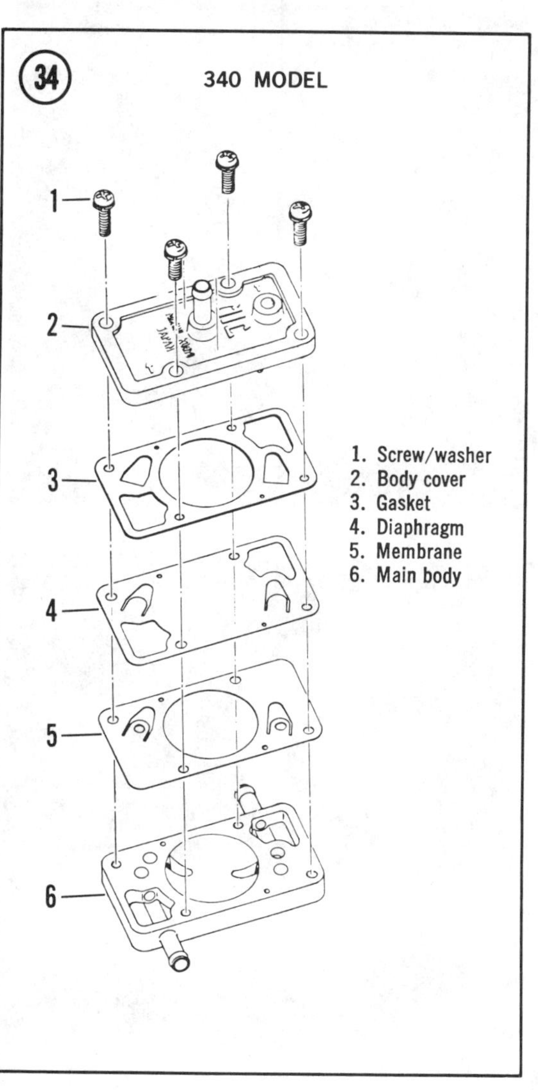

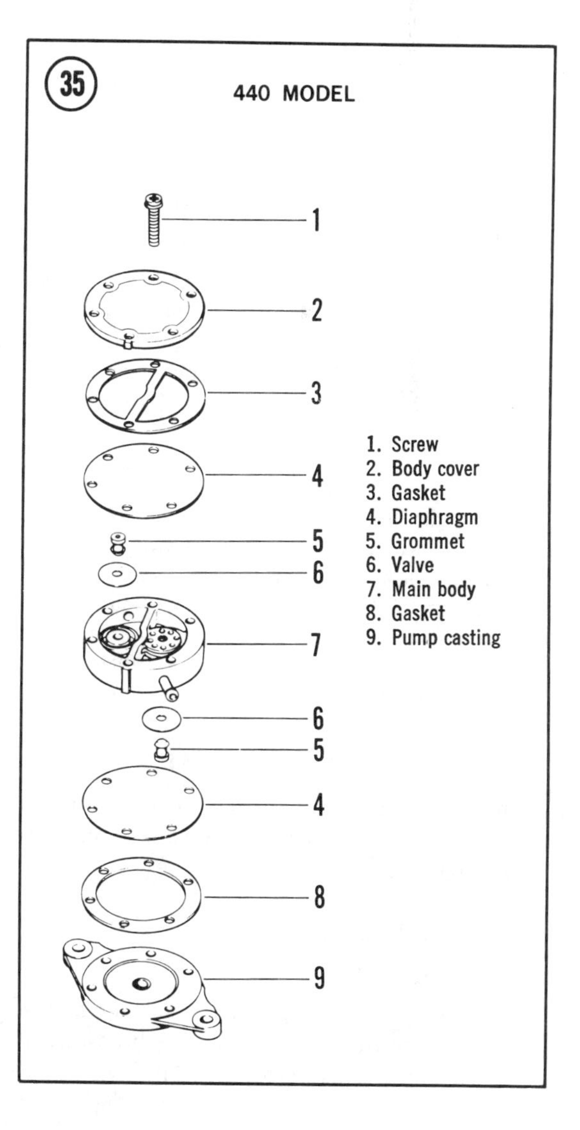

2. Separate the cover from the body, taking care not to damage the gasket or membrane. Clean and inspect the pump as described below.

3. Reverse the above to assemble the pump.

> NOTE: *On 440 models, replace the grommets with new ones, and line up the tabs on the body, diaphragms, and gaskets.*

Cleaning/Inspection

1. Clean all metal parts with solvent and dry them with compressed air. *Do not clean the rubber diaphragms in solvent.*

2. Inspect the diaphragms for damage and replace them if there is any question about their condition.

3. Blow out the ports in the pump body and cover with compressed air.

Pump Output Test

1. Raise the hood and disconnect the output line from the fuel pump (**Figure 36**). Place the end of the line in a graduate or beaker.

2. Start the engine and allow it to idle for 30 seconds. Then, shut it off and measure the fuel in the graduate. For 340 models, the fuel volume should be 192 cc (6.5 oz.), and for 440 models, 266 cc (9 oz.). If the amount is noticeably less, clean and inspect the pump as described above.

WARNING

The fuel pump output test should be conducted outside or in a well-ventilated area; the gasoline fumes that collect during the test are very explosive, particularly if they are exposed to a furnace or water heater pilot light or any other open flame.

EXHAUST SYSTEM

The exhaust system requires no service other than an annual cleaning to remove carbon that builds up in the headpipe.

Removal/Installation

1. Raise the hood.

2. Disconnect the springs that attach the headpipe to the manifold and remove the hold-down springs over the muffler (**Figure 37**).

3. Lift the system out of the machine.

4. Reverse the above to install the system.

Cleaning

1. Fray the end of a 2 foot section of old control cable and connect the opposite end to a drill motor (**Figure 38**).

2. Use a glove or shop rag to protect your hand and spin the cable with the drill motor and feed it into the headpipe. Work the cable in and out as you spin it to break the carbon deposits loose. Take your time and do a thorough job.

3. Shake the large pieces out into a trash container. Then, apply compressed air through the tailpipe to blow out the remaining soot.

36

37

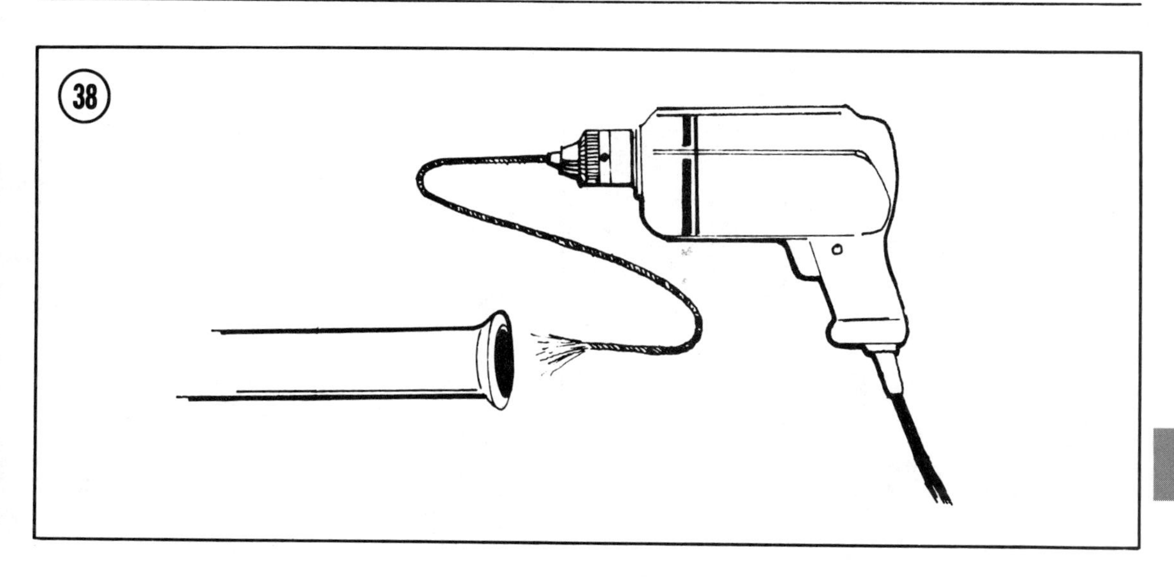

Table 1 CARBURETOR SETTINGS

Model	Idle Speed, rpm	Air Screw, turns out	Throttle Stop Screw, turns out
SA340-A1	2,000	1.5	3
SA440-A1	2,000	1.5	3
SB340-A1	2,000	1.5	3
SB440-A1	2,000	2	6
SA340-A2	2,000	1.5	3
SB440-A2	2,500	1.5	3
SA340-A3 and later	2,000	1.5	6
SB440-A3 and later	2,500	1.5	6
SS340 all	3,000	1.5	3
ST440 all	3,000	1.5	3

Table 2 CARBURETOR SYMPTOMS

Trouble	Check point	Remedy	Adjustment
Poor idling (Relative troubles) • Poor performance at low speeds • Poor acceleration • Slow response to throttle • Engine tends to stall	Improper idling speed adjustment • Pilot screw adjustment	Adjust idling speed	Tighten pilot screw lightly, and check throttle opening. If incorrect, back it out to specification. Start the engine and turn pilot screw in and out ¼ turn each time. When the engine runs faster, back out throttle stop screw so the engine idles at specified speed.
	• Throttle stop screw	Adjust	Tightened too much — Engine speed is higher. Backed out too much — Engine does not idle.
	Damaged pilot screw	Replace pilot screw	
	Clogged bypass hole	Clean	
	Clogged or loose slow jet	Clean and retighten	Remove slow jet, and blow it out with compressed air.
	Air leaking into carburetor joint Defective starter valve seat	Retighten band screw Clean or replace	
	Overflow	Correct	See "Overflow"
Poor performance at mid-range speeds (Relative troubles) • Momentary slow response to throttle • Poor acceleration	Clogged or loose slow jet	Clean and retighten	Remove slow jet, and blow it out with compressed air
	Lean mixtures	Overhaul carburetor	See "Overflow"
Poor performance at normal speeds (Relative troubles) • Excess fuel consumption • Poor acceleration	Clogged air vent	Clean	Remove the air vent pipe, and clean
	Clogged main jet	Clean and retighten	Remove main jet, and blow it out with compressed air
	Overflow	Check float and float valve and clean	See "Overflow"

Table 2 CARBURETOR SYMPTOMS (continued)

Trouble	Check point	Remedy	Adjustment
Poor performance at high speeds	Starter valve is left open	Fully close valve	Return starter lever to its home position
(Relative trouble)	Clogged air vent pipe	Remove and clean	
• Power loss • Poor acceleration	Clogged or loose main jet	Clean and retighten	Remove main jet, and clean with compressed air, then install.
	Clogged power jet	Clean	
	Clogged fuel pipe	Clean or replace	
	Dirty fuel tank	Clean fuel tank	
	Air leaking into fuel line	Check joint and retighten	
	Low fuel pump performance	Repair pump or replace	
	Clogged fuel filter	Replace	
	Clogged silencer outlet	Check for ice, and remove	
Abnormal combustion	Lean mixtures	Clean carburetor and adjust	
(Mainly backfire)	Dirty carburetor	Clean carburetor	
	Dirty or clogged fuel pipe	Clean or replace fuel pipe	
Overflow	Clogged air vent	Clean	
(Relative troubles) • Power idling	Clogged float valve	Disassemble and clean	Clean while taking care not to scratch valve seat.
• Poor performance at low. mid-range, and high speeds • Excessive fuel consumption	Scratched or unevenly worn float valve or valve seat	Clean or replace float valve and valve seat	Valve seat is press-fitted to body. So body must be replaced if seat is damaged
• Hard starting	Broken float	Replace float	
• Power loss • Poor acceleration	Incorrect float level	If not within the specified range, check the following parts and replace any defective part	
	• Worn float tang	Replace float	Replace float assembly
	• Worn arm pin	Replace arm pin	
	• Deformed float arm	Replace float	

5

CHAPTER SIX

ELECTRICAL SYSTEM

The electrical system consists of an ignition system, lighting system, and, on some models, an electric starting system.

Two types of ignition systems are used — a contact breaker flywheel magneto and a CDI (capacitive discharge ignition) electronic system.

The lighting system consists of a headlight, brake/taillight, and console lights.

The electric starting system on some models consists of a battery, starter and solenoid, and an alternator.

Ignition and starter applications are shown in **Table 1**. All tables are at the end of the chapter.

This chapter includes testing and repair of some components of the ignition, lighting, and charging systems. Some testing and repair tasks require special test equipment and tools, and are best left to a dealer or competent auto electric shop.

Wiring diagrams for all models covered are included at the end of the book.

CAPACITIVE DISCHARGE IGNITION (CDI)

The capacitive discharge ignition (CDI) system consists of a flywheel-mounted alternator and a control "black box" that houses the solid-state capacitor and trigger and amplifier devices.

In addition to the ignition chores, the alternator also powers the lights.

In operation, a permanent-magnet rotor attached to the end of the crankshaft revolves around a fixed stator which holds the generating coils and the ignition timing coils.

As the magnets in the flywheel/rotor pass the coils they induce a voltage in them which flows to the electronic control module (as well as to the lights). The voltage is stepped up, or increased, in the module and is stored in the capacitor.

The rotor also induces voltage in the timing coil and this voltage flows to an electronic "switch" that releases the stored energy in the capacitor. This energy discharges to the primary winding in the ignition coil which induces high voltage in the secondary coil winding. This high voltage then jumps the spark plug electrodes and ignites the air/fuel mixture in the combustion chamber.

The primary feature of electronic ignition is that it requires virtually no maintenance other than timing adjustment following installation and a periodic check (annually) to ensure that the timing has not changed.

The system has a drawback and that is that the electronic control module is not repairable; failure of just one component in the module re-

quires replacement of the entire unit. However, the likelihood of an ignition failure is remote and the unit will probably last for the life of the machine.

Removal

It is not necessary to remove the engine from the machine to remove the alternator/ignition assembly.

1. Raise the hood. Disconnect the springs that attach the exhaust system headpipe to the manifold and remove the springs that hold the muffler in place **(Figure 1)** and remove the exhaust system. Remove the starter assembly.
2. Remove the fan cover **(Figure 2)**.
3. Unscrew the bolts that attach the starter pulley to the flywheel/rotor **(Figure 3)** and remove the pulley and fan belt.
4. Unscrew the flywheel/rotor bolt and attach a puller to the flywheel **(Figure 4** — see **Table 2** for puller identification). Hold the puller to prevent it from turning and tighten the puller bolt to break the flywheel/rotor loose from the crankshaft. It may be necessary to rap sharply on the head of the puller to shock the flywheel loose.

CAUTION

Do not strike the flywheel/rotor with a hammer and do not heat it with a torch, or it's likely to be damaged.

5. Mark the stator and crankcase for reference during installation **(Figure 5)**. Disconnect the ignition wiring harness, unscrew the stator screws **(Figure 6)**, and remove the stator.

Testing

Have the stator tested by your dealer. Circuit resistances are critical and require the use of a bench tester to accurately assess the condition of the ignition and timing coils.

Installation

1. Set the stator in place and install the bolts finger-tight. Line up the reference mark on the stator with the reference mark on the crankcase **(Figure 5)** and tighten the stator mounting screws.

2. Line up the keyway in the flywheel/rotor with the key in the crankshaft and press the flywheel onto the crankshaft. Screw the flywheel/rotor nut onto the crankshaft and tighten it to 8.3 mkg (60 ft.-lb.).

3. Install the fan belt and starter pulley and screw in and tighten the 3 bolts (**Figure 7**).

4. Install the fan cover (**Figure 8**).

5. Install the starter assembly and the exhaust system (**Figure 9**). Connect the headpipe-to-manifold springs first, then the long muffler hold-down springs.

6. Refer to *Timing/Adjustment* and *Strobe Timing* in Chapter Two and check and adjust the ignition timing as described.

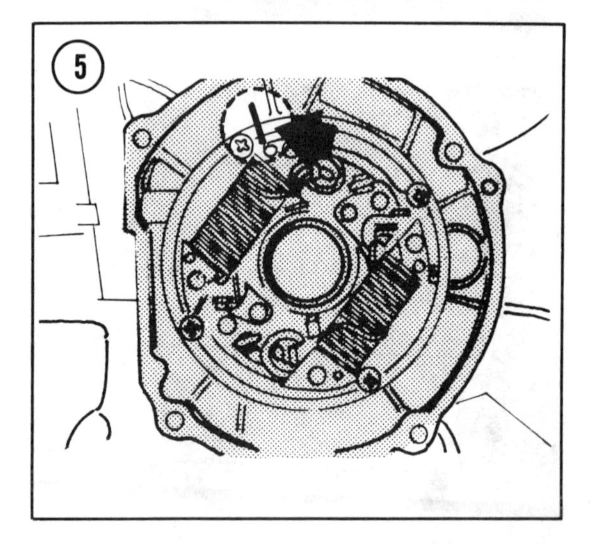

FLYWHEEL MAGNETO

Several models covered in this handbook are equipped with flywheel magneto ignition (see **Table 1**).

The system consists of a magnetic rotor, stationary ignition source coil, lighting coils, contact breaker assembly (points) ignition coil, and ignition switch, safety switch, and kill switch.

The contact breaker gap should be checked and adjusted if necessary every 20 hours or 300 miles (480 km) of operation, and the breaker should be replaced annually. Contact breaker adjustment should be done in conjunction with a timing check and adjustment (see *Timing/ Adjustment* in Chapter Two).

Removal/Installation

The flywheel magneto is removed and installed in the same manner as the CDI unit.

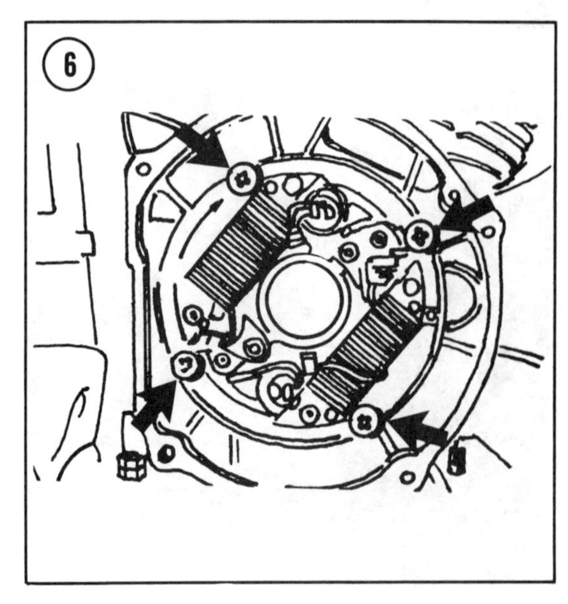

TESTING

Simple resistance tests can reveal much about the condition of electrical components such as the lighting and exciter coils and the ignition coil.

An ohmmeter, or a VOM like the one described in Chapter One, is required.

Lighting/Exciter Coil Resistance

Unplug the connector from the alternator (**Figure 10**). Refer to **Table 3** and check the

resistance between the leads described on the alternator half of the connector.

If the actual resistance is zero there is an open in the coil being tested, and if the resistance is much greater than specified, a short is likely. In either case, the affected coil must be replaced.

Ignition Coil Resistance

Refer to **Table 3** and check the resistance of the ignition coil. If resistance is zero there is an open, and if the resistance is much greater than specified, a short is likely and the coil must be replaced.

LIGHTING SYSTEM

The lighting system consists of a headlight, brake/taillight unit, instrument lights, and an AC (alternating current) generating device. Switches control the circuits.

Testing

If the lights fail to work, do not immediately assume the worst — a major failure in the alternator, regulator, or lighting coils. Very often the problem can be found in the lamps; look for burned out filaments or bulbs that are loose in their sockets. Also check the harness connectors to ensure they are clean, dry, and tight and have not come disconnected.

Four things are required for the lights to function. You must have current and an uninterrupted, non-shorted path for it to follow.

The switch must work correctly. A good ground is required. The bulbs must be in good condition and correctly installed in clean, dry sockets.

A lighting coil resistance test and a regulator test can confirm both good and bad conditions.

Lighting Coil Resistance Test

1. Unplug the main wiring harness connector at the engine (**Figure 11**).

2. Connect an ohmmeter to the leads on the engine side of the connector and measure the resistance (see **Table 3**).

If the resistance is not correct, replace the lighting coil on the stator. If it is correct, remove the regulator and have it tested by your dealer. The test requires only a few minutes of your dealer's time and is very inexpensive. The equipment required for the test (two 12-volt batteries and a regulator checker) is costly and would probably not be used more than once.

11

12

HEADLIGHT BULB REPLACEMENT

1. Raise the hood to gain access to the rear of the headlight.

2. Unplug the harness connector by pulling on the connector — not the wiring harness (**Figure 12**).

3. Turn the retaining ring counterclockwise to release it from the housing (**Figure 13**).

4. Remove the old bulb and install a new one of the same rating. The rating appears on the bulb. Set the retaining ring in place and turn it clockwise to lock it. Reconnect the wiring harness.

13

HEADLIGHT AIMING

1. Set the snowmobile on a level surface, 25 feet (7.6 m) from a vertical surface such as a wall (**Figure 14**).

2. Measure the distance from the floor to the center of the headlight lens. Make a mark on the wall the same distance from the floor. For instance, if the center of the headlight lens is 2 feet above the floor, mark "A" in **Figure 14** should also be 2 feet above the floor.

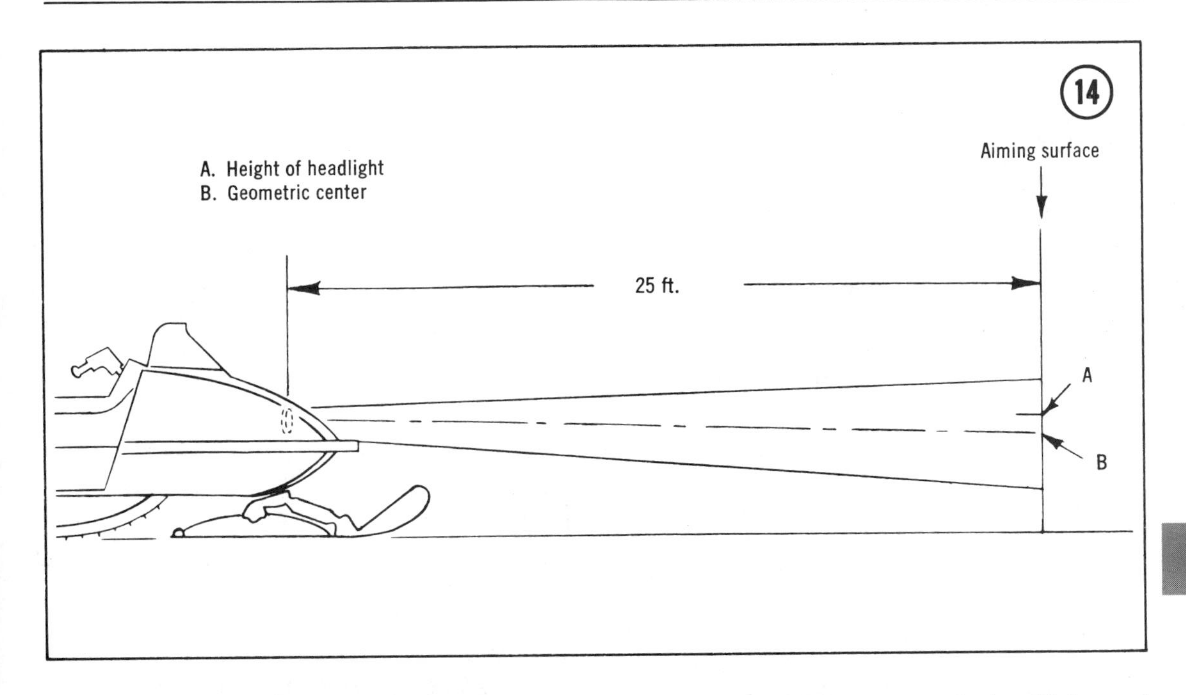

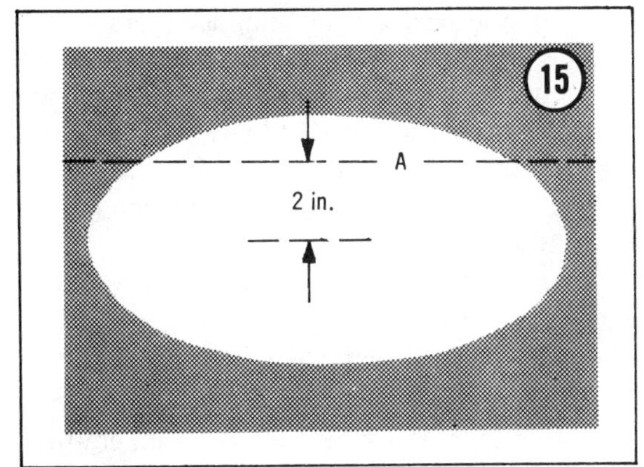

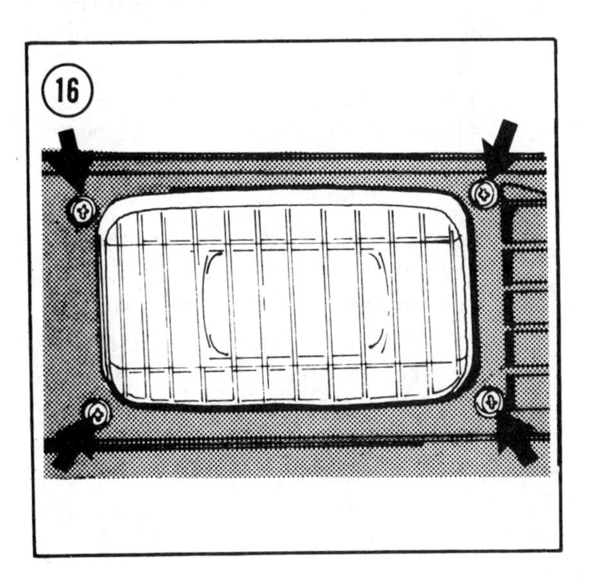

3. Start the engine, turn on the headlight, and set the beam selector at HIGH; do not adjust the headlight beam with the selector set at LOW.

4. The most intense area of the beam on the wall should be 2 in. (50 mm) below the "A" mark (**Figure 15**) and in line with imaginary vertical centerlines from the headlight to the wall.

5. To raise the beam, tighten the top adjuster screws and loosen the bottom screws (**Figure 16**). To lower the beam, tighten the bottom screws and loosen the top screws. To move the beam to the right, tighten the right screws and loosen the left. To move the beam to the left, tighten the left screws and loosen the right.

BRAKE/TAILLIGHT BULB REPLACEMENT

1. Refer to **Figure 17** (a typical brake/taillight assembly) and remove the screws that attach the lens to the seat or toolbox door. Remove the lens and remove the defective bulb by pressing in on it, turning it counterclockwise, and pulling it out of the socket.

2. Clean the socket to remove any corrosion, dirt, and moisture.

3. Line up the guide pins of the new bulb with the slots in the socket, press the bulb in, and turn it clockwise to lock it in place.

NOTE: *On dual filament bulbs (combination brake/taillight) the pins are at different distances from the base of the bulb. Make sure they correspond with the different length slots in the socket.*

4. Clean the inside of the lens with a mild detergent and warm water before installing it. Also make sure that the lens is completely dry. Install all of the screws finger-tight before tightening them, and then tighten them securely but not so tight that they damage the lens.

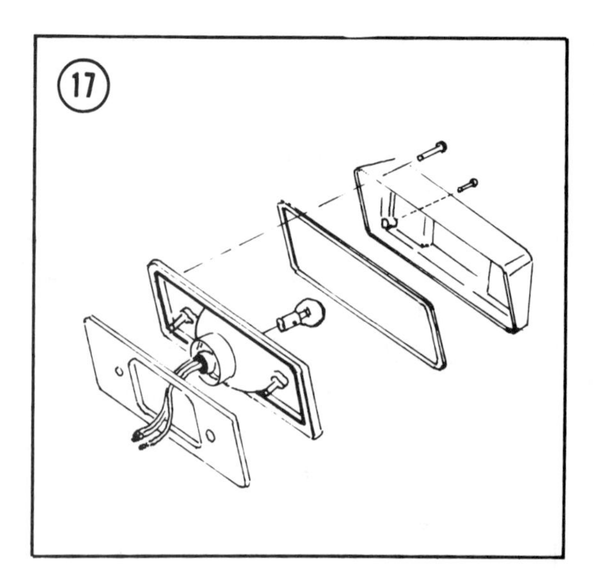

ELECTRIC START SYSTEM

The electric start system, optional on some models consists of a 12-volt battery, starter motor, starter solenoid, AC to DC rectifier, and fuse.

The starter motor pinion gear engages a ring gear on the flywheel to turn the engine over. The battery is charged by the alternator. The rectifier changes the alternating current (AC) produced by the alternator to direct current (DC) which is stored in the battery. The fuse protects the system from overloads and short circuits.

If difficulty is experienced with the electric start system, check the obvious and most likely cause of trouble — the battery. Make certain the battery and starter connections are clean and tight. Check and service the battery as described below. If the battery has sufficient charge and the connections are good, test the starter and solenoid as described. Refer to the wiring diagrams for electric start models at the end of the book for reference when troubleshooting the starter system.

BATTERY

With proper care, the battery should last for 2 to 3 years. Incorrect care or neglect (described below) can cut battery service life short and lead to chronic starting problems.

a. ***Electrolyte level:*** Correct electrolyte level must be maintained at all times for the battery to be able to accept and maintain a full charge (see *Cleaning and Service*).
b. ***Charge level:*** The battery must be maintained at full charge all the time to prevent premature sulfation which will result in internal shorts that destroy the battery. This is a point that is often overlooked during summer storage. The battery should be removed from the machine and cleaned and serviced and fully charged. It should be stored in a dry, warm place, and should be periodically tested and recharged to prevent it from sulfating.
c. ***Overcharging:*** Overcharging or charging the battery at too high a rate creates excessive heat that will destroy the battery.
d. ***Freezing:*** If the machine is left outdoors for long periods in freezing temperatures, freezing of the electrolyte is likely to occur, resulting in deterioration of the battery's service life. Guard against this by removing the battery and storing it in a warm place until the machine will be used.

Removal/Installation

1. Disconnect the cables from the battery — first the negative (−) then the positive (+).
2. Disconnect the battery hold-down strap.
3. Slide the battery part way and note the routing of the vent tube for reference during installation, then remove the battery.
4. Reverse the above to install the battery. Make sure the vent tube is correctly routed and is not kinked. Connect the positive cable first, then the negative.

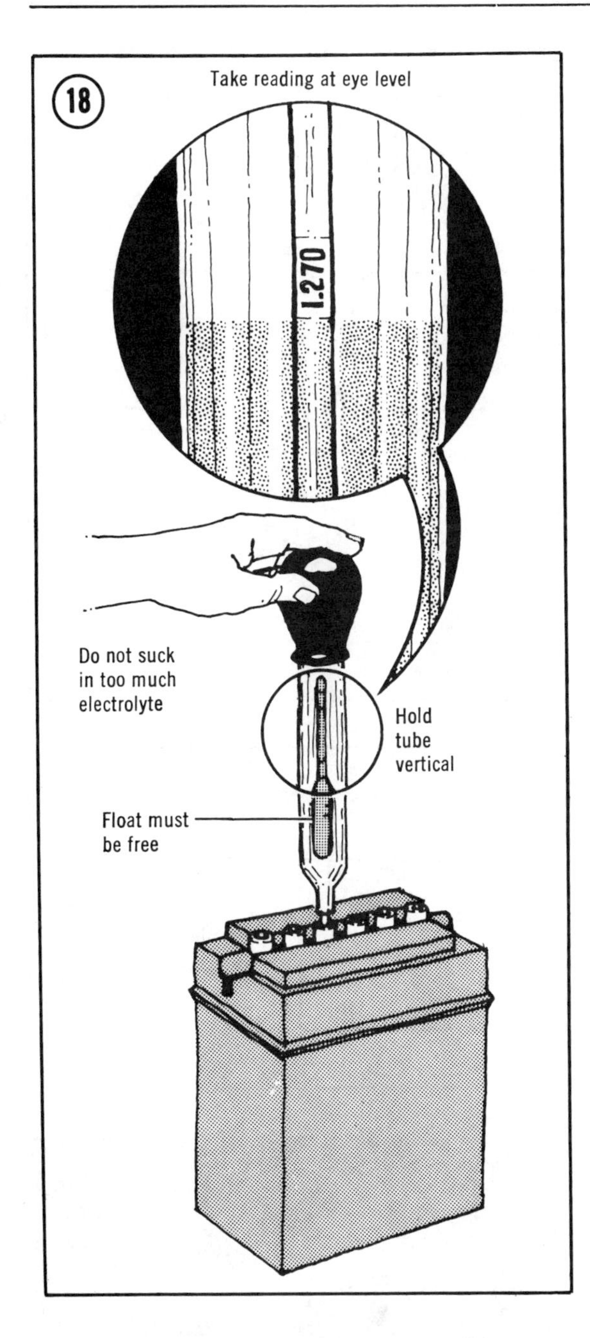

CAUTION

Be sure the battery connections are correct or serious damage to electrical components will occur.

Cleaning and Service

Electrolyte level in the battery should be checked frequently, especially during periods of regular, sustained operation. Use only distilled water and top off the level to the ring in the bottom of the filler neck. The tops of the plates should be covered but take care not to overfill the battery.

Corrosion of the terminals is normal but it must be removed before it creates excessive electrical resistance. Wash the terminals and cable ends with a solution of 4 parts water and 1 part baking soda.

CAUTION

Keep the baking soda solution out of the battery cells or it will dilute the electrolyte and damage the battery.

Clean the battery carrier with baking soda solution and rinse the battery and the case with fresh water. Dry them with an old towel or shop rag.

6

Specific Gravity Test

Use a hydrometer to check the state of charge of the battery. Place the suction tube of the hydrometer in one of the filler openings (**Figure 18**) and draw off just enough electrolyte to lift the float. Hold the hydrometer vertical and read the specific gravity level. Then, return the electrolyte to the cell from which it was removed and check the next cell. Continue until all cells have been tested.

Specific gravity for a battery that is fully charged is 1.260. If the specific gravity is below 1.220, recharge the battery as described below.

NOTE: *Specific gravity varies with temperature. For each 10° F that the battery temperature is less than 80° F, subtract 0.004 from the value shown on the hydrometer. For every 10° F that the battery temperature exceeds 80° F, add 0.004 to the indicated reading.*

Battery Charging

WARNING

Charge the battery in a well-ventilated location and do not smoke or permit any open flame near the battery while it is being charged. Highly explosive hydrogen gas is formed during charg ing. Finally, take care not to arc the terminals during or immediately after charging.

Connect the battery to a charger — positive to positive, negative to negative (**Figure 19**). If the charger is equipped with a variable-rate selector, select a low setting (1-1½ amps). Turn the charger on or plug it in and allow the battery to charge for at least 8 hours.

> NOTE: *Remove the caps from the cells during charging and periodically check and correct the electrolyte level if necessary.*

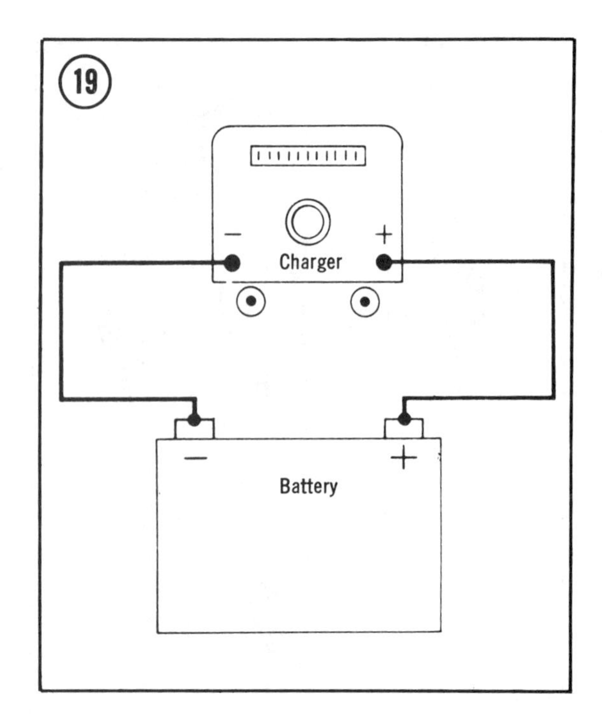

After the battery has had a sufficient time to charge, turn off the charger, disconnect the cables, and check the specific gravity as described above. Then, allow the battery to sit for about an hour and check the specific gravity again. If the level has not dropped, the battery can be considered to be fully charged and in good condition.

STARTER

If the starter fails to crank the engine, or if it cranks it very slowly, test the starter as described below.

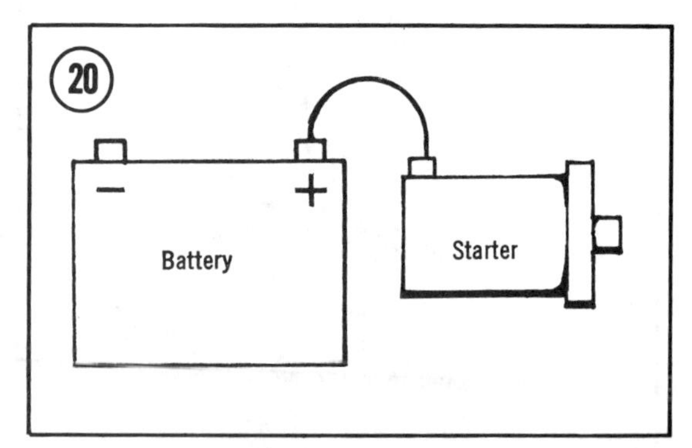

Testing

1. Inspect the starter wiring for loose and corroded terminals and connections and for damaged insulation. Clean and tighten any corroded connections and replace any damaged wires. Inspect the fuse.
2. Perform *Specific Gravity Test* to make sure the battery is fully charged and not defective.
3. Crank the engine with the recoil starter to make sure the engine turns freely and is not seized.
4. Disconnect the high-tension leads from the sparkplugs. Connect a heavy jumper wire from the positive (+) terminal of the battery directly to the starter terminal (**Figure 20**) to bypass the starter switch and solenoid. If the starter turns, it can be assumed to be all right. In such case, check the continuity of the switch with an ohmmeter. Also test the solenoid as described below.

If the starter does not turn, it's defective and should be replaced (see *Starter Removal/Installation*).

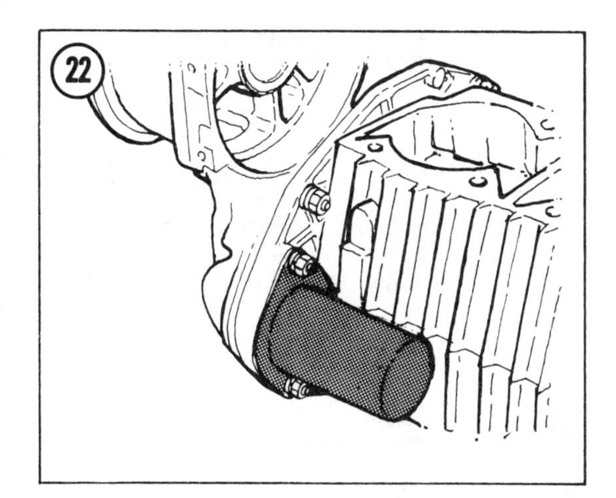

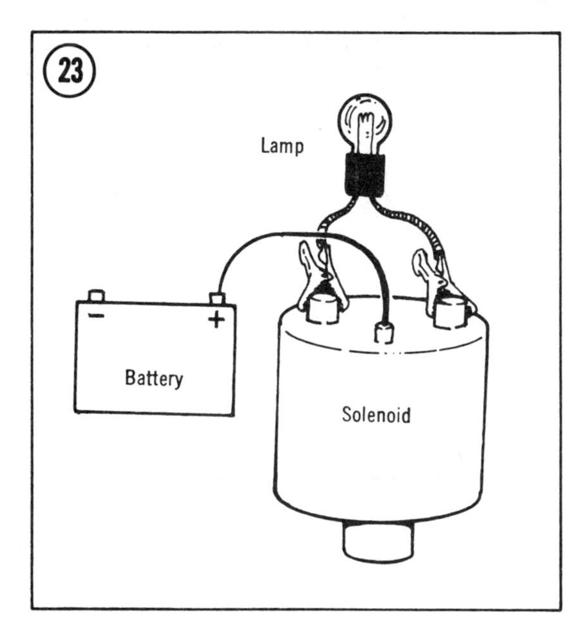

Starter Removal/Installation

1. Disconnect the ground cable from the battery.

2. Disconnect headpipe-to-manifold springs and the long hold-down springs on the exhaust system (**Figure 21**) and remove the system.

3. Disconnect the cable from the starter, unscrew the starter mounting bolts (**Figure 22**), and remove the starter.

4. Reverse the above steps to install the starter. Clean the threads of the starter mounting bolts and coat them with Loctite before installing and tightening them. After all installation work is complete, connect the ground cable.

Starter Solenoid Test

The starter solenoid is a sealed magnetic switch that cannot be repaired. If it's defective, replace it.

1. Disconnect the leads from the solenoid and connect a test light between the 2 large terminals (**Figure 23**).

2. Connect a jumper wire between the small post on the solenoid and the positive terminal of the battery. The solenoid plunger should snap in, light the test light, and hold until the jumper wire is disconnected. If this does not occur, the solenoid is defective and must be replaced.

Table 1 IGNITION/STARTER APPLICATIONS

Model	Ignition Type	Starter Type
SA340-A1	Flywheel magneto	Manual recoil
SA340-A2 and later	Flywheel magneto	Manual recoil and optional electric
SA440-A1	Flywheel magneto	Manual recoil
SB440-A2 and later	CDI	Manual recoil
SS340 all	CDI	Manual recoil
SS440 all	CDI	Manual recoil
SST440 all	CDI	Manual recoil

Table 2 FLYWHEEL/ROTOR PULLER IDENTIFICATION

Model	Puller (Kawasaki Part No.)
SA340-A1, A2	56019-011
SA340-A3 and later	57001-3509
SA440	56019-011
SB440-A2	56019-011
SB440-A3 and later	57001-3509
ST440	56019-01
SS340-A1	56019-01
SS340-A2 and later	57001-3509
SS440-A1	56019-01
SS440-A2 and later	57001-3509

Table 3 RESISTANCE TEST VALUES

Model	Lighting Coil Leads/Resistance	Exciter Coil Leads/Resistance	Ignition Coil Primary Leads/Resistance	Ignition Coil Secondary Leads/Resistance
SA340-A1 SS440-A1	Yellow-Brown/ 0.18 ohm ± 10%	Red, white-ground/ 1.13 ohm ± 10%	Red, white-ground/ 1.61 ohm ± 10%	High-tension-ground/ 5,165 ohm ± 10%
SA340-A2	Yellow-Yellow/ 0.14 ± 20%	Red-ground/ 1.10 ohm ± 20%	Red, white-ground/ 1.61 ohm ± 20%	High-tension-ground/ 5,000 ohm ± 20%
SB440-A2	Yellow-Yellow/ 0.18 ohm ± 20%	Red-ground/ 128 ohm ± 20%	Blue-ground/ 0.37 ohm ± 20%	High-tension-ground/ 10,200 ohm ± 20%
SA340-A3	Yellow-Yellow/ 0.14 ohm ± 20%	Red-ground/ 1.10 ohm ± 20%	Red, white-ground/ 1.61 ohm ± 20%	High-tension-ground/ 5,000 ohm ± 20%
SB440-A3	Yellow-Yellow/ 0.18 ohm ± 20%	Red-ground/ 128 ohm ± 20%	Blue-ground/ 0.37 ohm ± 20%	High-tension-ground/ 10,200 ohm ± 20%
SS and ST—all	Yellow-Yellow/ 0.18 ohm ± 20%	Red-ground/ 128 ohm ± 20%	Blue-ground/ 0.37 ohm ± 20%	High-tension-ground/ 10,200 ohm ± 20%

CHAPTER SEVEN

DRIVE TRAIN

The drive train consists of a drive sheave on the end of the crankshaft, connected to a drive sheave on the chaincase by a belt; a drive chain and sprockets inside the chaincase; a drive shaft with 2 drive sprockets; and a brake. A typical drive train is shown in **Figure 1**.

Some of the procedures in this chapter require the use of special tools and considerable experience. For instance, work on the drive and drive sheaves is not within the abilities of the average hobbyist-mechanic and this work should be entrusted to a dealer.

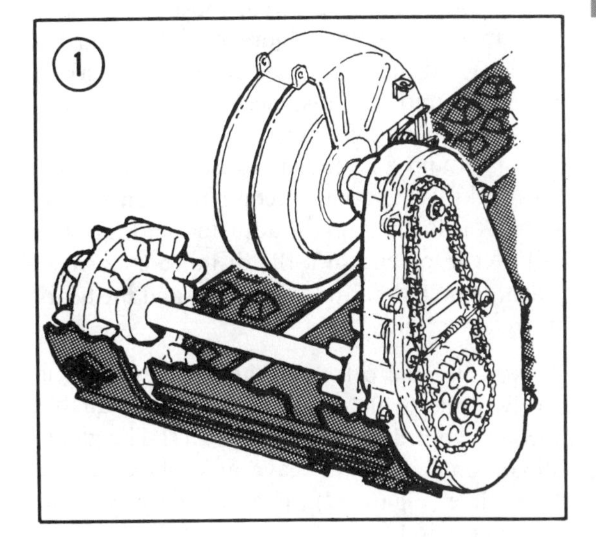

DRIVE BELT

The drive belt should be inspected every 20 hours or 300 miles (480 km) of operation for wear, cracking, damage, or deterioration. If the belt is in poor condition, full power won't be transmitted from the drive sheave to the driven sheave.

Measure the width of the belt (**Figure 2**). A new belt is 31.7 mm (1¼ in.) wide. If the belt is less than 26 mm (1.02 in.) wide at any one spot, replace it.

Removal/Installation

1. Remove the belt guard (**Figure 3**).
2. Lock the brake to prevent the driven sheave from turning.

3. Push against the moveable side of the driven sheave and turn it clockwise. Roll the belt over the sheave **(Figure 4)** to remove it.
4. Remove the belt from the drive sheave.
5. Reverse the above to install the belt.

Drive Belt Adjustment

The center-to-center distance from the drive sheave to the driven sheave and the offset of the sheaves must be correctly maintained for good performance and long belt life.

1. Remove the belt guard **(Figures 5 and 6)**.
2. Measure the distance between the sheave centers **(Figure 7)**. For SA and SB models, the distance should be 262 mm (10.3 in.). For SS and ST models, the distance should be 304 mm (12 in.).
3. If adjustment is required, loosen the chaincase mounting bolts **(Figure 8** — SA and SB models) and turn the clevis adjuster **(Figure 9)** in or out until the distance is correct. Then, tighten the chaincase mounting bolts.

For SS and ST models, loosen the bolts on the jackshaft bearing retainer **(Figure 10)**, loosen the locknut on the adjuster, and turn the adjuster in or out until the distance is correct. Then, tighten the adjuster locknut and the bearing retainer bolts.

4. Measure the offset of the sheaves. This is easily done with a sheave gauge available through your dealer. See **Figure 11**. The gauge should contact each sheave at 2 places at the same time **(Figure 12)**. The offset for SA and SB models is 11.5 mm (0.454 in.). For SS and ST models, the offset is 14.9 mm (0.588 in.).
5. If adjustment is required, loosen the engine mounting bolts **(Figure 13** — SA and SB models) and move the engine right or left until the offset is correct. Then, tighten the engine mounting bolts.

For SS and ST models, remove the trim piece from the body **(Figure 14)** and unscrew the bolt that attaches the driven sheave to the jackshaft. Carefully draw the sheave off the shaft and collect the shims from behind it **(Figure 15)**. Add or subtract shims to correct the offset. Then, install the driven sheave and the trim piece.

3

4

5

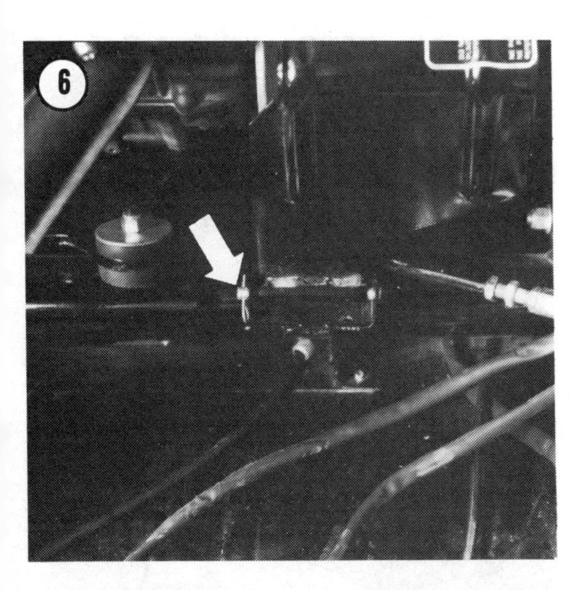

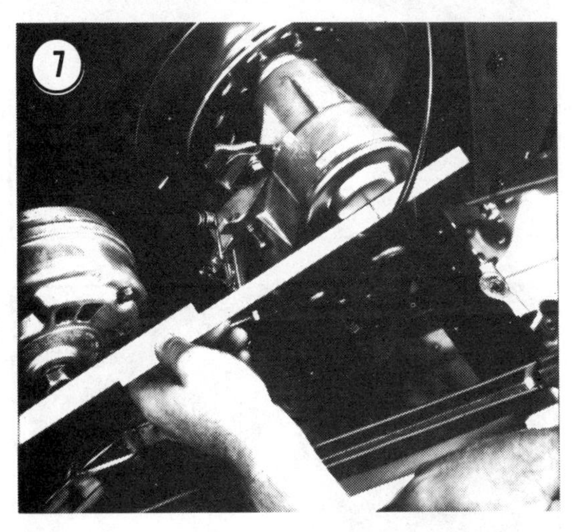

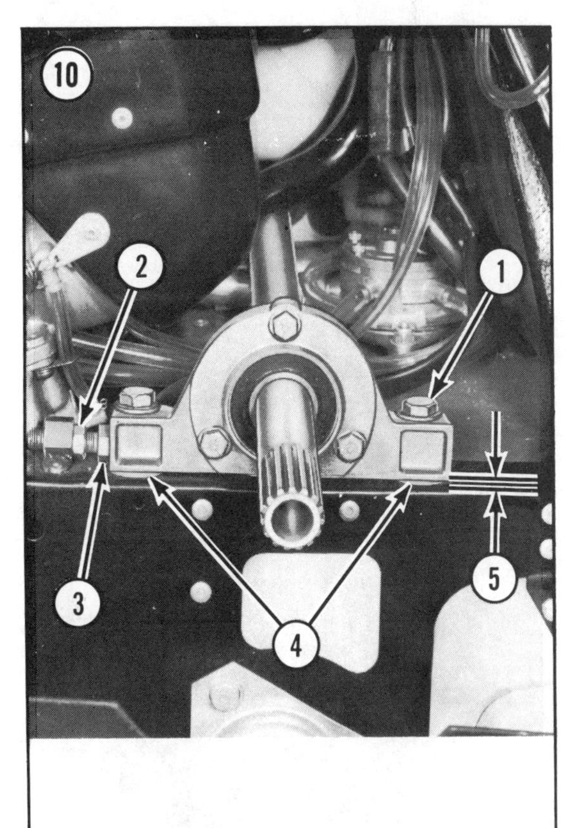

1. Bearing retainer mounting bolts
2. Adjusting bolt locknut
3. Adjusting bolt
4. Bearing retainer shims
5. Parallel

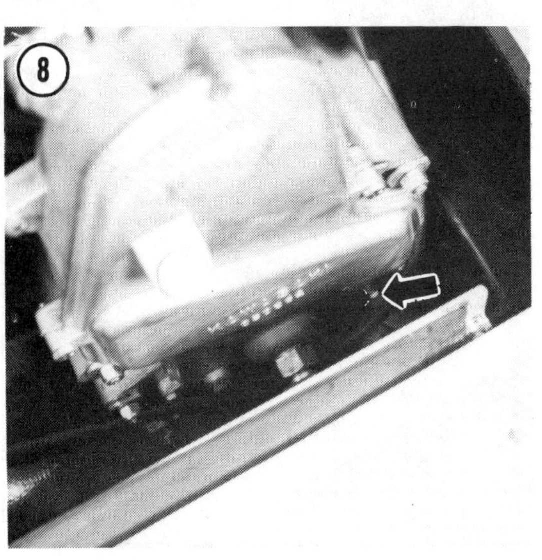

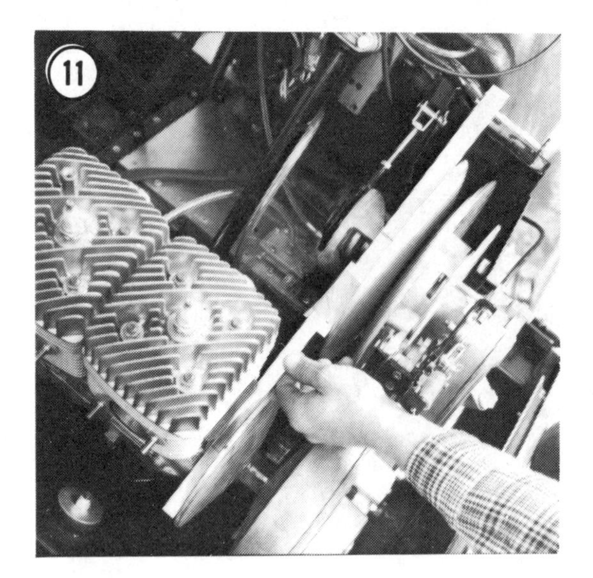

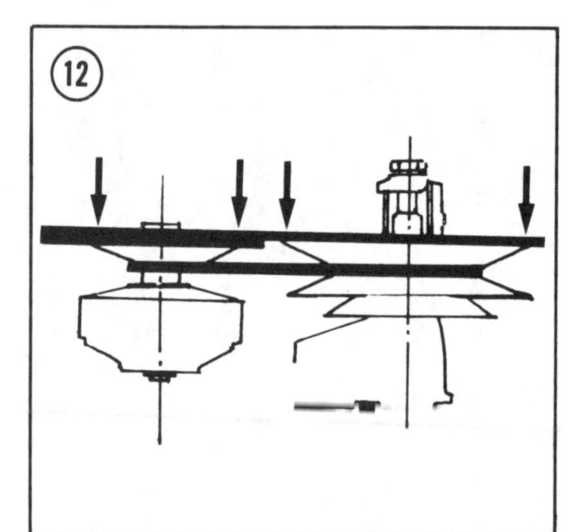

1. Aluminum trim
2. Retaining bolt

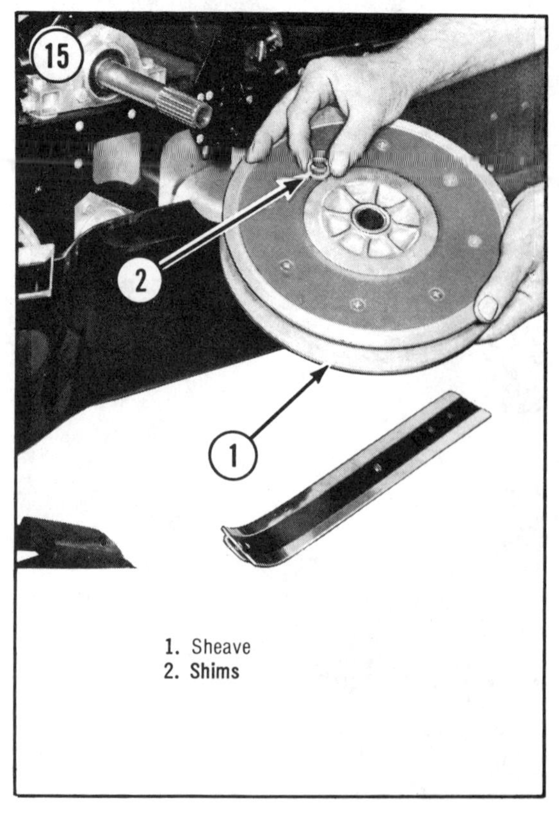

1. Sheave
2. Shims

16

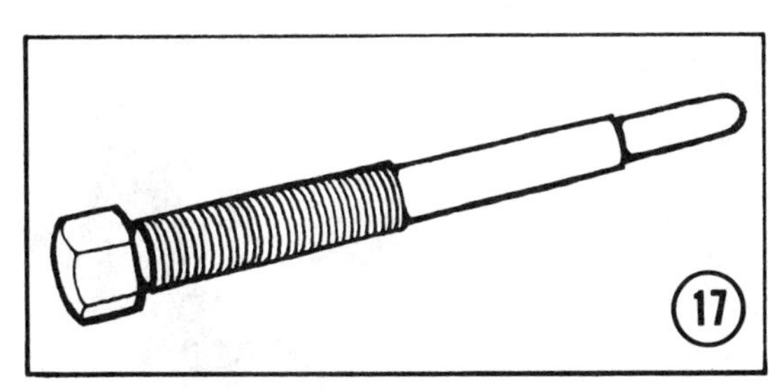
17

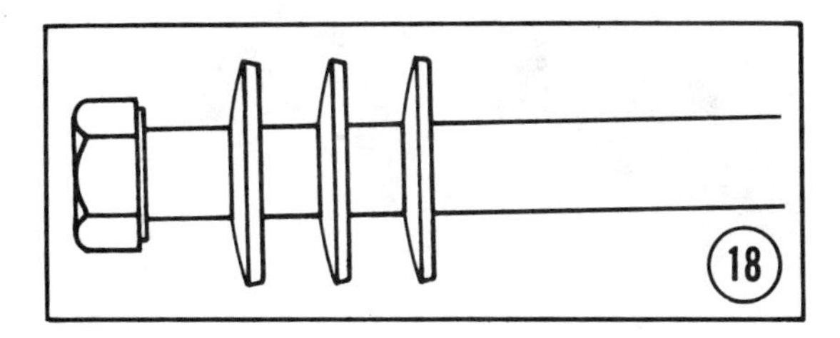
18

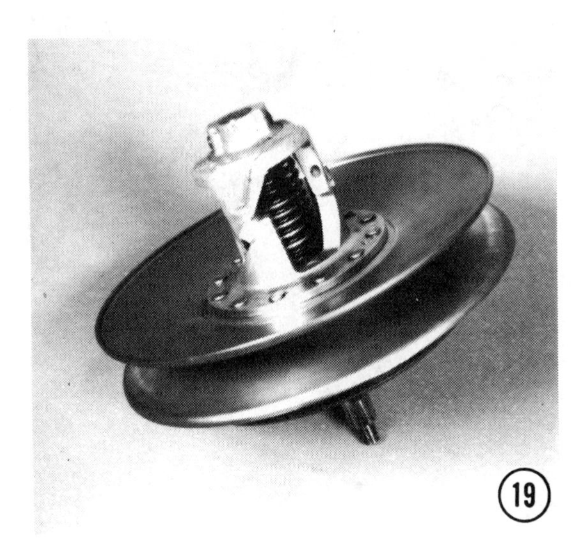
19

NOTE: *The sheave gauge for SA and SB models is identified with Kawasaki part No. 205207. The gauge for SS and ST models is part No. 57001-3503.*

DRIVE SHEAVE REMOVAL/INSTALLATION

Work on the drive sheave should be entrusted to your dealer. The drive sheave can be removed without removing the engine from the machine.

1. Remove the drive belt as described earlier.
2. Unscrew the bolt from the end of the crankshaft (**Figure 16**) and install a puller (**Figure 17**).
3. Hold the sheave to prevent it from turning and tighten the puller bolt.

NOTE: *It may be necessary to rap sharply on the head of the puller bolt to shock the drive sheave loose from the crankshaft.*

4. To install the sheave, first make sure the tapers on the crankshaft and inside the sheave are clean, dry, and oil-free. Install the sheave on the shaft. Install the washers on the bolt as shown in **Figure 18** and screw in and tighten the bolt to 8.3-9.0 mkg (60-65 ft.-lb.)
5. Install the drive belt as described earlier. Refer to *Drive Belt Adjustment* in this chapter and check and adjust the center-to-center distance and the offset of the drive and driven sheaves.

DRIVEN SHEAVE

Service to the driven sheave should be entrusted to your dealer. However, you can save time and money by inspecting, removing, and installing the assembly yourself.

Inspection

1. Press the moveable driven sheave in and rotate it clockwise. Check the feel of the spring. It should be extremely stiff. If it compresses without much resistance, it should be replaced.
2. Check the ramp shoes for wear (**Figure 19**) and have them replaced if they are worn thin or damaged.

3. Inspect the contact surfaces of the sheaves for wear and scoring. If the surfaces are in poor condition they will rapidly wear the drive belt.

4. Check the sheaves and brake disc for warping and replace them if they are not true.

Removal/Installation — SA and SB Models

1. Remove the drive belt as described earlier.

2. Turn the machine onto its side and rest it on cardboard to protect the finish. Remove the speedometer drive (**Figure 20**).

3. Remove the chaincase cover (**Figure 21**).

4. Disconnect the chain tensioner spring (**Figure 22**). Lock the brake and loosen the bolt in the top sprocket. Unscrew the brake mounting bolts and remove the caliper from the chaincase.

5. Unscrew the top sprocket bolt and remove the sprocket from the driven sheave shaft.

6. Tap the driven sheave out of the bearing in the chaincase (**Figure 23**).

7. Reverse the above to install the driven sheave. Adjust the brake as described in Chapter Three (see *Brake Adjustment*), lock the sheave with the brake, install the gear, and screw in and tighten the bolt to 3.9 mkg (28 ft.-lb.). Reconnect the tensioner spring.

Removal/Installation — SS and ST Models

1. Remove the trim piece from the body (**Figure 24**).

2. Remove the drive belt as described earlier.

3. Lock the sheave to prevent it from turning and unscrew the bolt that attaches the sheave to the jackshaft.

4. Carefully draw the sheave off the shaft and collect the shims from behind it (**Figure 25**).

5. Reverse the above to install the sheave. Refer to *Drive Sheave Adjustment* in Chapter Three and check the offset of the sheaves. Lock the sheave to prevent it from turning and tighten the jackshaft bolt to 5.5-6.9 mkg (40-50 ft.-lb.).

CHAINCASE

The chaincase houses sprockets and a drive chain that connect the driven sheave to the

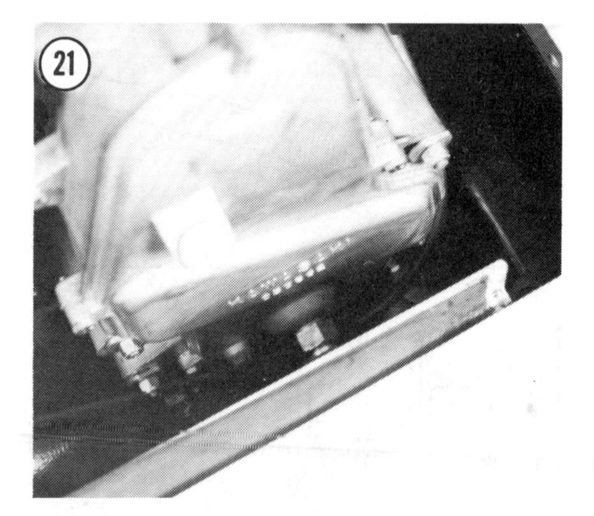

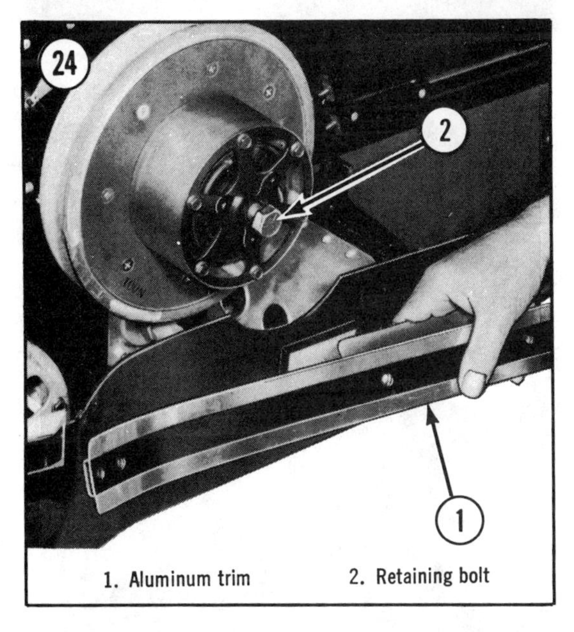

1. Aluminum trim 2. Retaining bolt

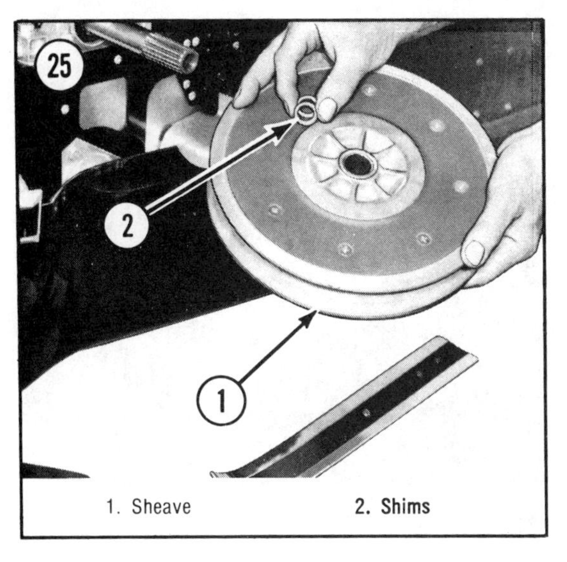

1. Sheave 2. Shims

drive axle. The case is sealed to contain an oil bath in which the chain and sprockets run.

The only service required of the chaincase assembly is a periodic check of the oil level. Chain tension is maintained automatically. Annually, the chaincase should be drained and refilled and the chain tensioner pads should be checked for wear and replaced if necessary. Checking and changing the oil are described in Chapter Two, *Chaincase*.

1. Remove the drive belt as described earlier.
2. Tilt the machine onto its right side and rest it on cardboard to protect the finish.
3. Disconnect the speedometer drive from the bottom of the chaincase cover (**Figure 26**) and remove the cover. Collect the speedometer drive extension (**Figure 27**) from the cover.
4. Drain the oil from the chaincase. A turkey baster, available in markets and variety stores for about a dollar, is handy for siphoning oil from the case.
5. Lock the brake and unscrew the sprocket bolts. Remove the spring from the chain tensioner (**Figure 28**). Disconnect the cable from the brake and remove the brake from the chaincase.
6. Remove the sprockets and chain together.
7. Loosen the locknuts on the track tension bolts (**Figure 29**) and slack off the bolts several turns to relax the track tension on the drive axle.

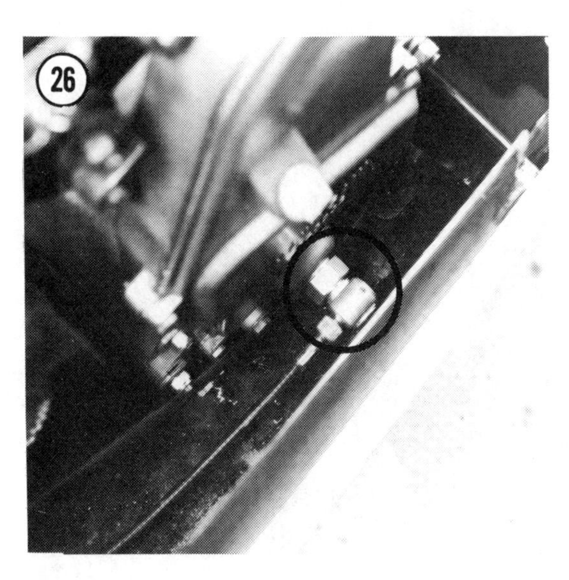

NOTE: *Loosen the track tension bolts an equal number of turns and write down the number of turns for reference during assembly; it will save time when the tension is adjusted later on.*

8. Unscrew the chaincase mounting bolts (**Figure 30**) and remove the chaincase. Pull out on the case to disengage it from the drive axle.

Inspection

1. Clean the case, cover, chain, and sprockets with solvent and dry them with compressed air.
2. Inspect the case for cracks and damage. Minor cracks or porous areas where seeping could occur can be repaired with epoxy or a "liquid aluminum" type filler.
3. Inspect the seals for wear and damage and replace them if they are unsatisfactory.
4. Inspect the sprockets for wear and undercutting of the teeth and replace them if they are not satisfactory.

NOTE: *Unless the chaincase has been run dry, chain and sprocket wear will be negligible.*

5. Check the tensioner pads for wear (**Figure 31**) and replace them if they are worn to within 3 mm of the bolt.

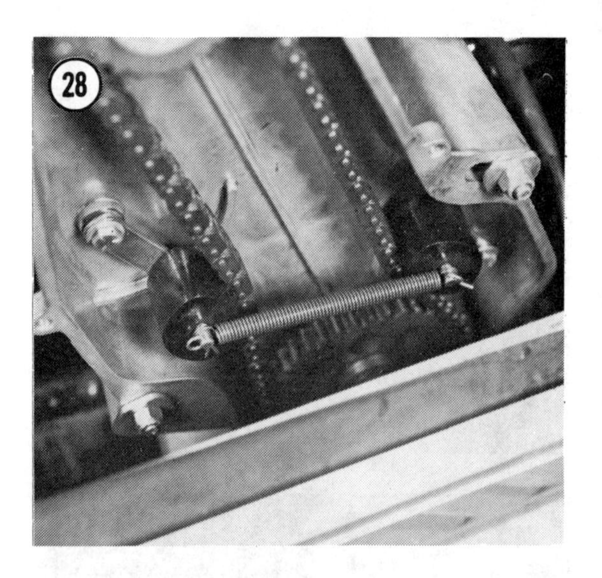

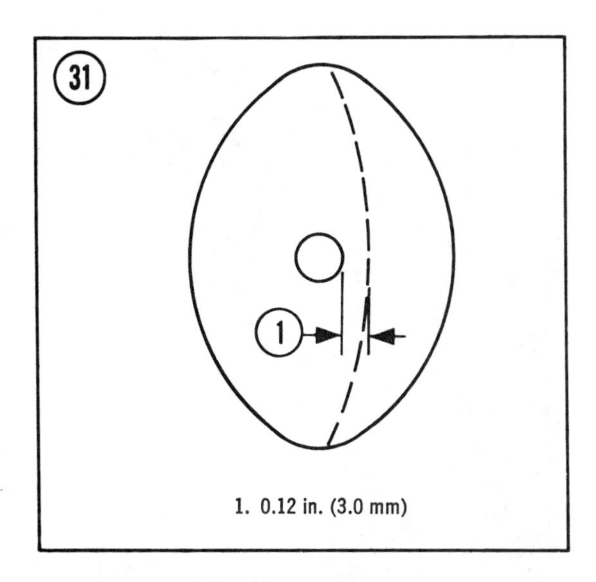

Installation

1. Install the chaincase on the drive axle. Install the mounting bolts finger-tight.
2. Install the sprockets and chain together.
3. Install the brake on the case and connect the brake cable. Lock the brake and screw in and tighten the sprocket bolts. Connect the tensioner spring (**Figure 32**).
4. Position the cover gasket so the break in the gasket is at the top (**Figure 33**). Install the cover.
5. Install the speedometer drive extension in the case and install the speedometer drive (**Figure 34**).
6. Set the machine upright and adjust the center-to-center distance and the offset of the sheaves as described in Chapter Two, *Drive Sheave Adjustment*. When the adjustment is correct, tighten the chaincase mounting bolts securely. Adjust the track tension and alignment as described in Chapter Two (see *Track Tension and Alignment*).
7. Adjust the brake as described in Chapter Three.
8. Remove the level plug and the fill plug (**Figure 35**) and pour fresh Dexron II automatic transmission fluid into the case until it just begins to run out the level hole. Then, install both level and fill plugs. After the machine has been run, check the chaincase cover gasket for leaks and retighten the cover bolts if necessary.

DRIVE AXLE

Service to the drive axle includes drive wheel and bearing replacement. Both of these tasks are within the abilities of the hobbyist mechanic.

Removal

1. Refer to Chapter Nine (*Suspension Removal/Installation*) and remove the rear suspension as described.
2. Remove the drive axle bearing retainer from the right side of the machine (**Figure 36**).
3. Disconnect the speedometer drive from the chaincase cover (**Figure 37**) and remove the chaincase cover.
4. Disconnect the chain tensioner spring (**Figure 38**), lock the driven sheave with the brake, and unscrew the bolt from the bottom sprocket.
5. Pull the drive axle to the right to disengage it from the chaincase, then remove it from the machine.

Inspection

1. Inspect the lugs on the drive wheels for damage or wear. If they are in poor condition, replace them as described later (see *Drive Wheel Replacement*).
2. Turn the right bearing by hand and check it for roughness and radial play (**Figure 39**). If the bearing is in poor condition, press it off the shaft and press on a new bearing.

Installation

1. Guide the right end of the axle through the hole in the body, then move the axle to the left and insert it into the chaincase. Install the sprocket and bolt.
2. Line up the right end of the axle with hole in the body and install the bearing retainer (**Figure 40**). Line up the retainer with the blank grease fitting knob at the top (**Figure 41**) and screw in and tighten the 3 bolts.
3. Engage the track lugs with the lugs on the drive wheels.
4. Refer to Chapter Nine, *Suspension Removal/Installation*, and install the suspension as described.

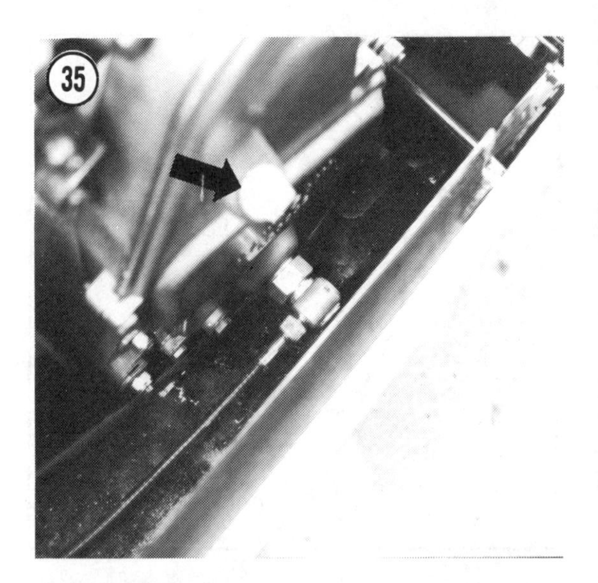
35

36

5. Set the machine upright and connect the chain tensioner spring (**Figure 42**). Tighten the sprocket bolt. Install the chaincase cover and connect the speedometer drive (**Figure 43**).
6. Fill the chaincase with oil as described earlier (see *Chaincase, Installation*.
7. Adjust the track as described in Chapter Two, *Track Tension and Alignment*.

Drive Wheel Replacement

If the drive wheel lugs are severely worn or damaged, the drive wheels should be replaced before they damage the track lugs by "ratcheting" during hard acceleration. Usually, if the

37

38

39

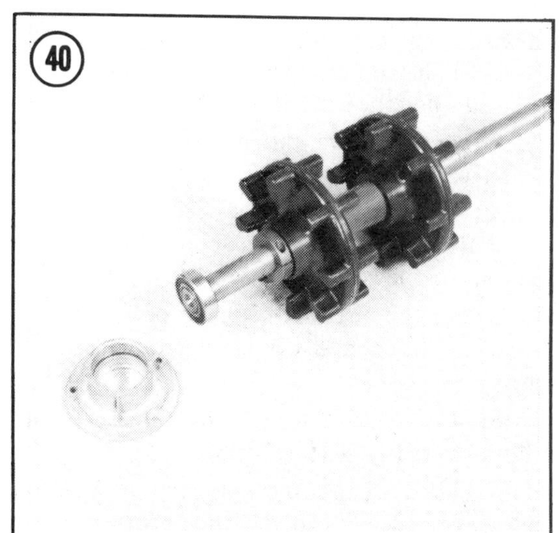
40

41

42

track lugs show ratcheting damage, the drive wheels will be similarly damaged and should be replaced when the track is replaced.

1. Remove the drive axle as described earlier (see *Drive Axle, Removal*).

2. Press the bearing off the right end of the axle shaft and loosen the set screw in the collar and remove the collar (**Figure 44**).

3. Remove the circlips from the outer end of each drive wheel and tap the drive wheels off the axle.

4. Install new drive wheels, making sure the drive lugs are in line (**Figure 45**).

5. Install the circlips on the axle and make sure they are completely seated in their grooves.

6. Install the collars, press them snugly against the drive wheels, and tighten the set screws.

7. Press the bearing onto the right end of the axle.

8. Install the drive axle as described earlier (see *Drive Axle, Installation*).

BRAKE

Kawasaki snowmobiles are equipped with a mechanical disc brake that is mounted on the chaincase and acts on the driven sheave.

Adjustment

Adjustment of the brake is described in Chapter Two, *Brake Adjustment*.

Pad Replacement

To check for brake pad wear and service ability, adjust the brake as described in Chapter Two. Then, measure the distance the threads on the adjuster bolt protrude from the adjuster nut (**Figure 46**). If the threads protrude 7/16 in. (11.1 mm) or more, replace the pads.

1. Loosen the adjuster nuts on the cable adjuster and run the adjuster all the way in until the brake cable can be disconnected from the brake arm (**Figure 47**). Unscrew the brake mounting bolts and remove the brake caliper from the chaincase.

2. Unscrew the nuts from the caliper slide bolts (**Figure 48**) and disassemble the caliper. Remove the old pads and discard them.

3. Refer to **Figure 49** for the arrangement of the caliper parts.

4. Install the slider bolts and the inside caliper half with the spacers as shown (**Figure 50**).

5. Assemble the remaining parts and screw on and tighten the self-locking nuts on the slide bolts.

6. When assembly is complete, install the rubber retractor as shown (**Figure 51**). This band separates the pads from the disc when the brake is released and prevents the pads from dragging on the disc.

7. Install the caliper on the chaincase and connect the cable (**Figure 52**).

8. Adjust the brake as described in Chapter Two, *Brake Adjustment*.

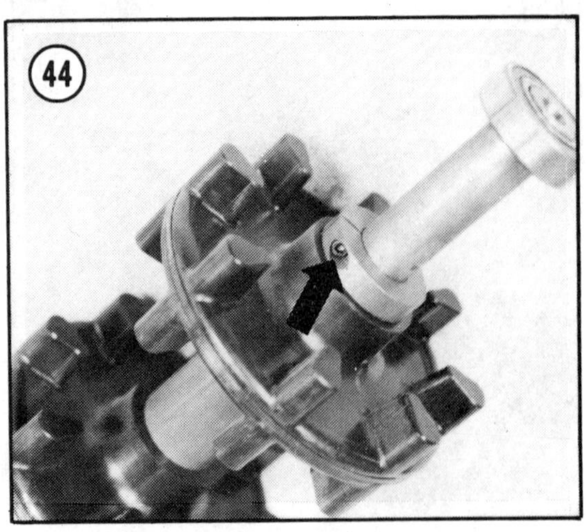

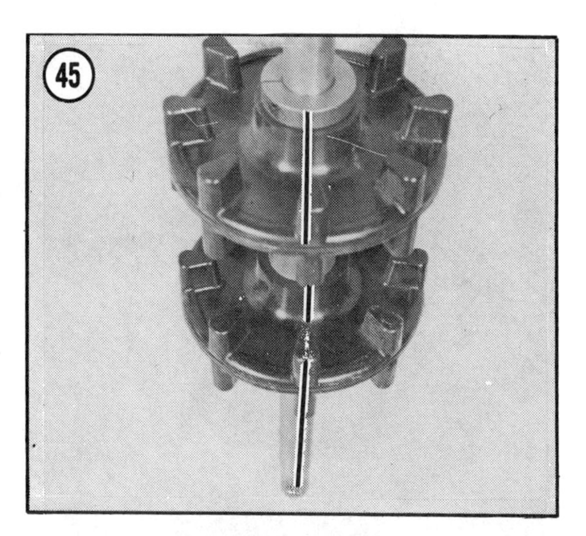
45

48

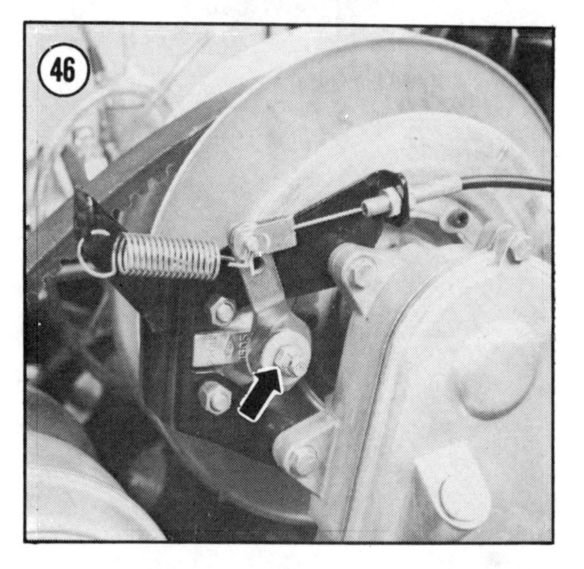
46

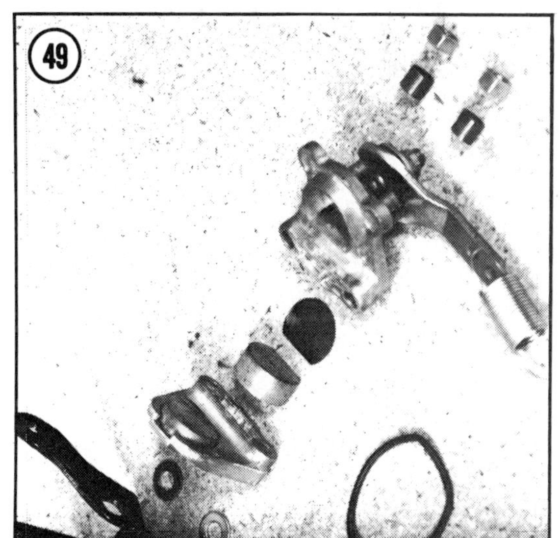
49

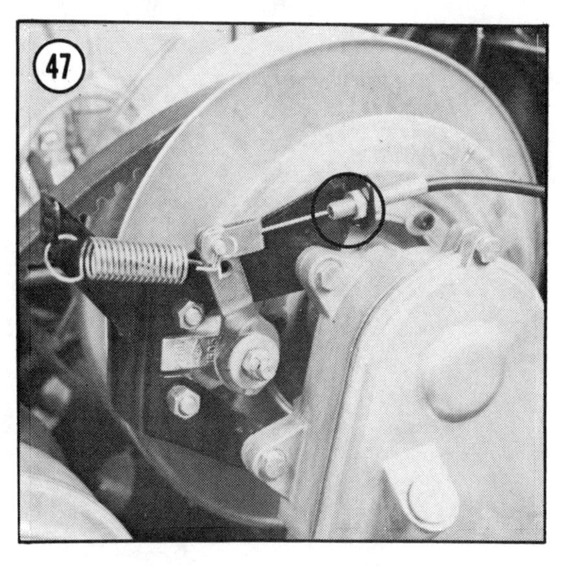
47

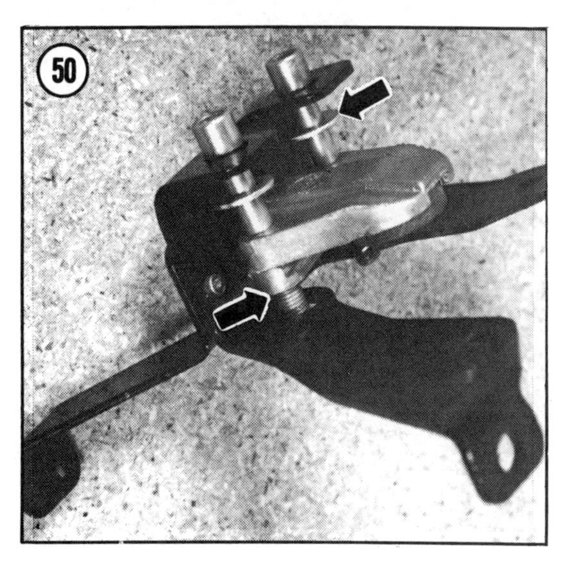
50

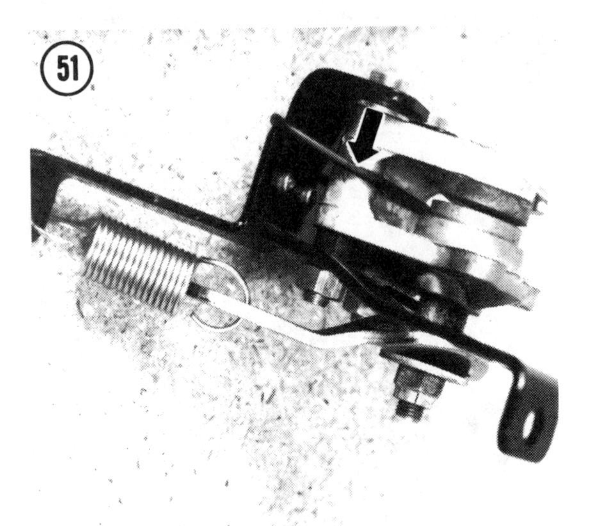
51

52

CHAPTER EIGHT

FRONT SUSPENSION AND STEERING

The front suspension and steering consists of spring-mounted skis that are connected to the steering column by a drag link and to each other with a tie rod.

The front suspension utilizes single-leaf springs and hydraulic shock absorbers.

The steering column is mounted in nylon bushings that also act as steering dampers.

The tie rod and drag link are fitted with adjustable ends to accommodate ski alignment.

SKIS

The skis are equipped with wear bars, or skags, that aid turning and protect the bottoms of the skis from wear and damage caused by road crossings and bare terrain. The bars are expendable and should be checked periodically for wear and damage and replaced when they are worn to the point they no longer protect the skis or aid turning.

Removal/Installation

1. Support the front of the machine so both skis are off the ground.
2. Unscrew the ski-to-spindle bolt (**Figure 1**) and remove the ski assembly.
3. Unscrew the spring-to-ski mounting bolts and remove the spring from the ski.
4. Reverse the above to install the skis. Clean and grease the spring-to-ski bolts and eyes with low-temperature grease.

Wear Bar (Skag) Replacement

1. Remove the skis as described above.
2. Unscrew the locknuts that attach the skag to the ski (**Figure 2**) and remove the skag.
3. Install a new skag and screw on and tighten the locknuts (**Figure 3**).

NOTE: *If the locking inserts in the nuts are worn or extruded, replace the nuts.*

Ski Spindle Lubrication

1. Support the front of the machine so the skis are off the ground.
2. Loosen the pinch bolt on the spindle arm (**Figure 4**). Scribe an alignment mark on the arm and the top of the spindle for reference during installation.
3. Tap the spindle out of the arm with a soft mallet and remove the ski.
4. Clean the spindle and the spindle bore with solvent and dry them with compressed air.
5. Grease the spindle with low-temperature grease.
6. Install the spindle in the bore, align the reference marks on the top of the spindle and the arm and install the arm. Tighten the pinch bolt.
7. Refer to *Ski Alignment* in this chapter and align the skis as described.

STEERING

Steering service includes lubrication and replacement of the steering column bushings, lubrication of the tie rod and drag link ends, and ski alignment.

Steering Column Lubrication

1. Disconnect the drag link from the end of the steering column and unscrew the top and bottom column mounting bolts (**Figure 5**).

> NOTE: *It's not necessary to disturb the hand controls or remove the handlebar.*

2

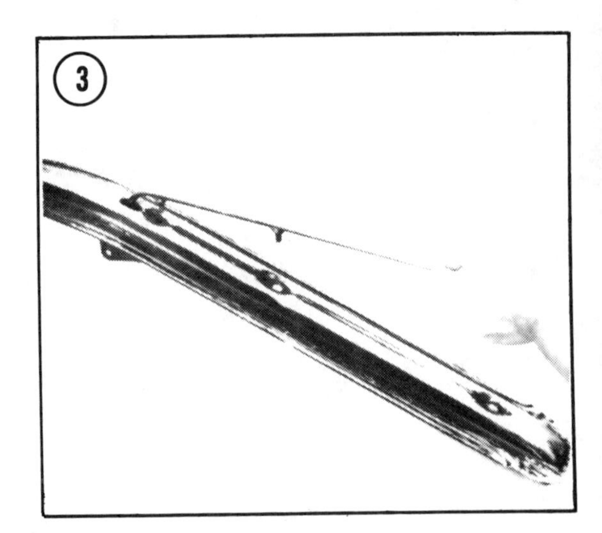
3

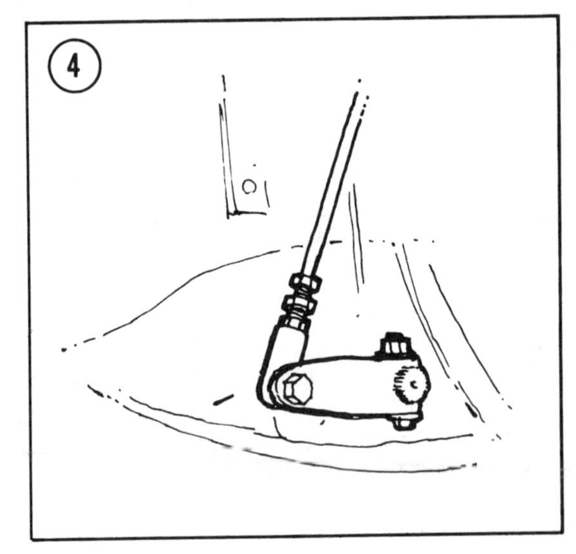
4

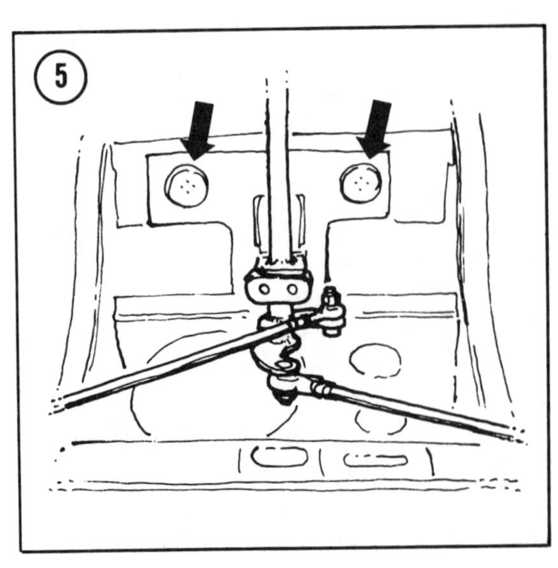
5

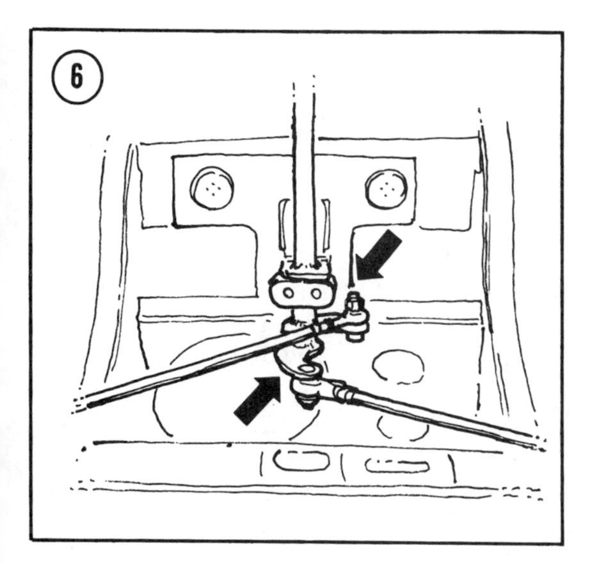

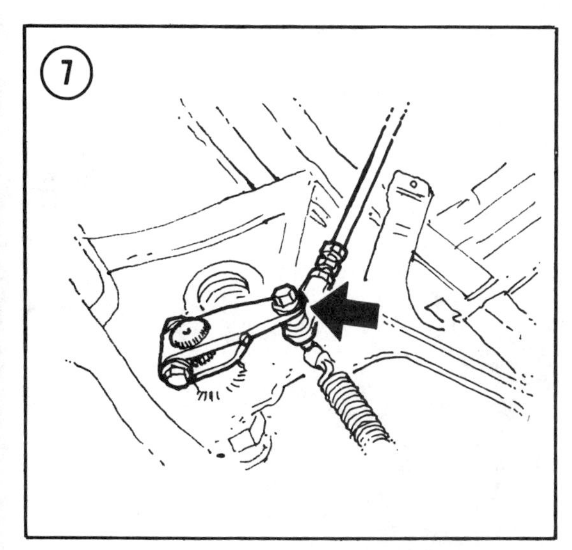

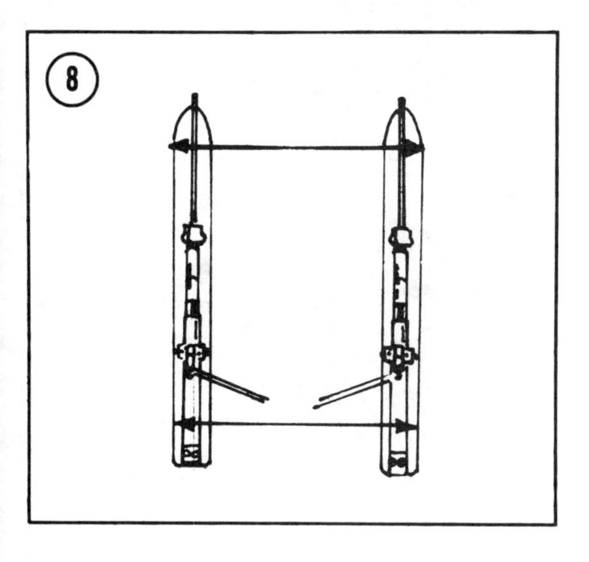

2. Remove the clamps and bushings from the steering column.

3. Clean the column and bushings with solvent and dry them with compressed air.

4. Inspect the bushings for wear and damage and replace them if they are scored or if they are sloppy on the column.

5. Lubricate the inside of the bushings with low-temperature grease.

6. Install the bushings on the column and assemble the clamps with the knob on each bushing engaging with the recess in the outer half of each clamp. Screw in and tighten the column mounting bolts.

7. Connect the drag link to the steering column. Refer to *Ski Alignment* and check and correct the alignment of the skis.

Tie Rod and Drag Link Removal/Installation

1. Disconnect the tie rods/drag links from the bottom of the steering column (**Figure 6**).

2. Unscrew the bolts from the spindle arms (**Figure 7**) and remove the links.

3. Reverse the above to install the links. Refer to *Ski Alignment*; align skis as described.

8

SKI ALIGNMENT

1. Check the alignment of the skis (**Figure 8**). Place a straight board against the edge of the track. Position the left ski parallel to the board and check the handlebar to see if it's centered. If it's not, loosen the locknut (**Figure 9**) and turn the adjuster as required. When the handlebar is centered, tighten the locknuts.

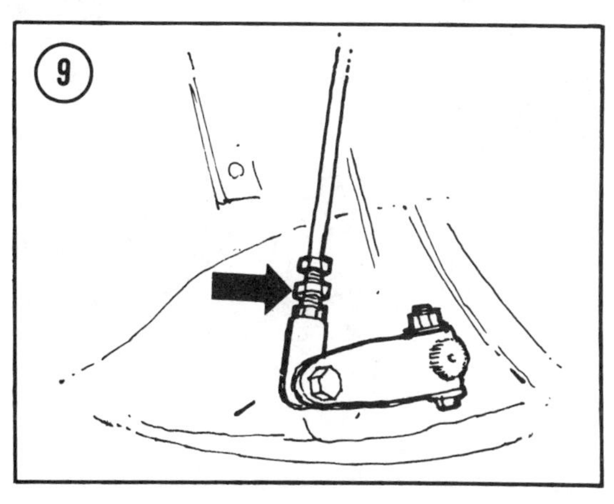

2. Measure the distance between the skis at the front and at the rear. If the 2 measurements are not the same, loosen the locknut on the right drag link and turn the adjuster until the skis are parallel. Then, tighten the locknut.

CONTROL CABLE REPLACEMENT

Throttle Cable

1. Unscrew the top from the carburetor (**Figure 10**) and pull out the slide assembly. Disconnect the cable from the slide and remove the cable and sheath from the carburetor top.
2. Disconnect the cable from the hand control (**Figure 11**).
3. Connect a new cable to the hand control and route it alongside the old cable. Then, remove the old cable from the machine. Connect the new cable to the carburetor slide, making sure the lock plate fits down into the slide and is held in place with the spring to prevent the cable from being disconnected.
4. Adjust the free play of the new cable as described in Chapter Two, *Throttle Cable Free Play*.

Brake Cable

1. Loosen the locknut on the brake cable adjuster at the brake (**Figure 12**) and run the adjuster all the way in and disconnect it from the brake arm.
2. Disconnect the cable from the hand control (**Figure 13**).
3. Connect a new cable to the hand control and route it alongside the old cable. Then, remove the old cable from the machine and connect the new cable to the brake arm.
4. Refer to Chapter Two, *Brake Adjustment*, and adjust the brake as described.

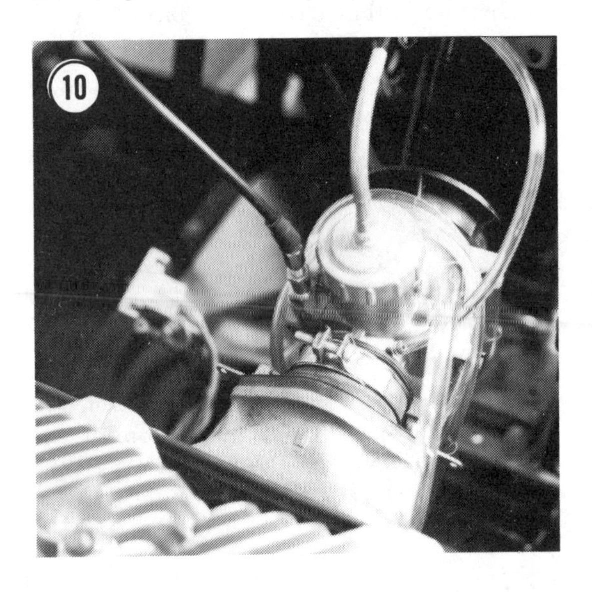

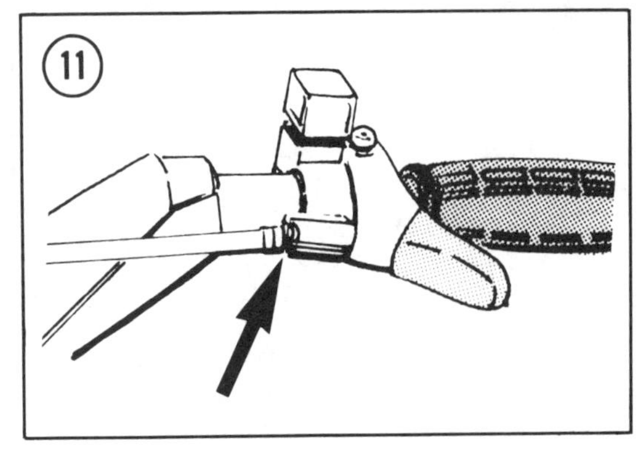

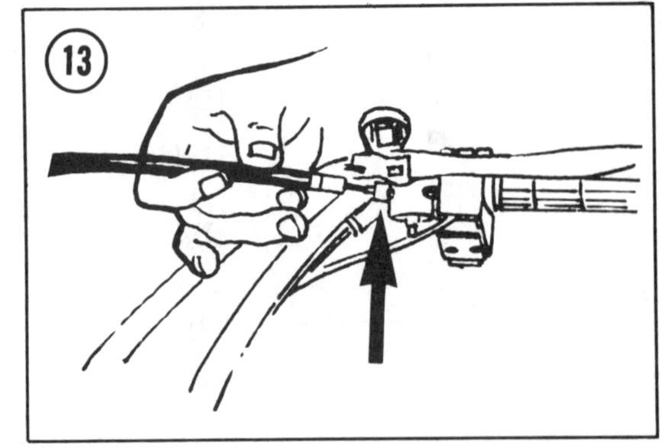

CHAPTER NINE

REAR SUSPENSION AND TRACK

Kawasaki snowmobiles are equipped with slide-rail suspension which utilizes a rear idler wheel assembly to control track tension and alignment.

Two designs are used — one for SA and SB models, and another for SS and ST models. Both systems are equipped with adjustable springs, shock absorbers, and replaceable wear bars.

REAR SUSPENSION

Removal/Installation

1. Loosen the locknuts on the track adjusting bolts **(Figure 1)** and back the bolts out to relieve the tension on the track.
2. Disconnect the rear springs from their hangers **(Figure 2)**.
3. Unscrew the 2 bolts that connect the rear of the suspension to the body **(Figure 3** — one on each side). Then, unscrew the 2 bolts that attach the forward end of the suspension to the body **(Figure 4** — one on each side).
4. Turn the machine onto its side and rest it on a sheet of cardboard to protect the finish.
5. Swing the rear of the suspension and track out of the body and lift the suspension assembly out of the track.
6. Reverse the above steps to install the suspension. Set the suspension into the track, making sure the lugs in the track are inboard of the idler wheels **(Figure 5)**.

Screw in all 4 suspension mounting bolts before tightening them. Then, connect the rear springs **(Figure 6)**. Refer to *Track Tension and Alignment* in Chapter Two and adjust the track as described.

2

3

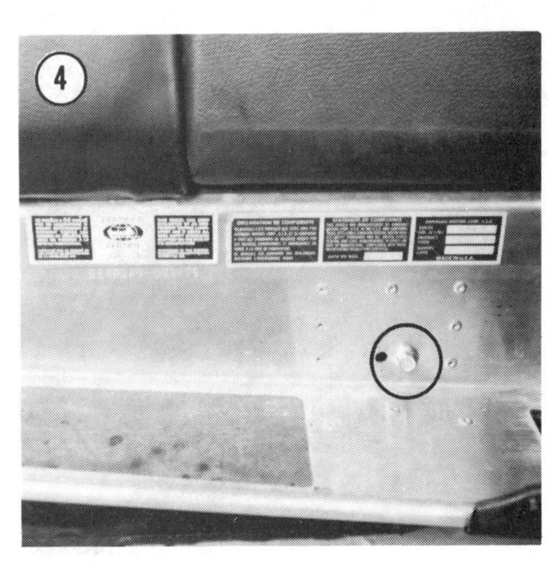
4

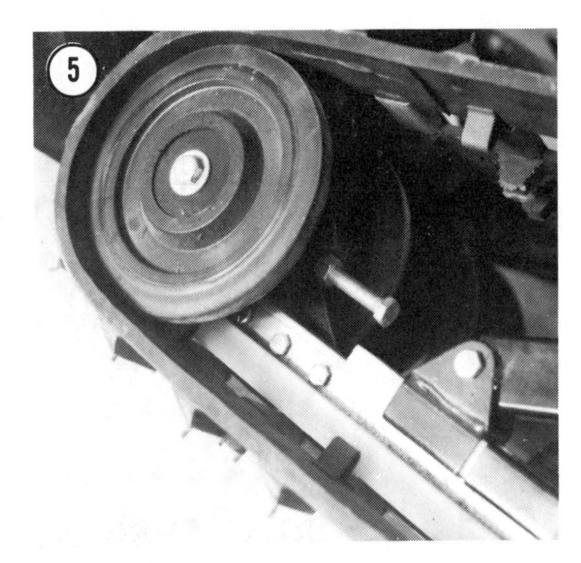
5

6

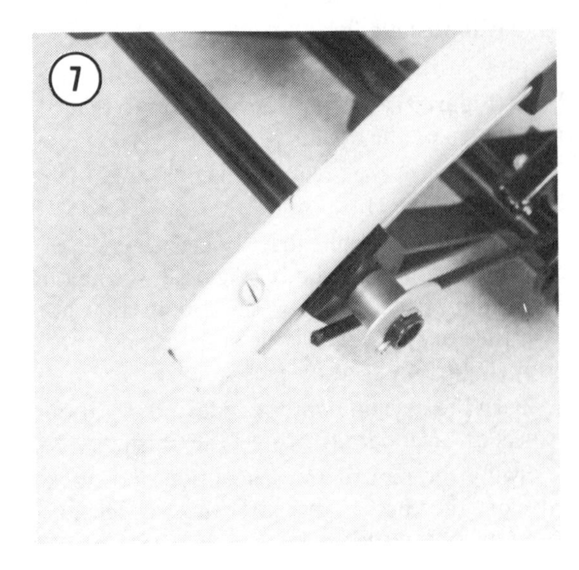
7

Cleaning/Inspection

1. Thoroughly wash the suspension with detergent and warm water and dry it with clean rags and compressed air.

2. Inspect the slide runner wear strips for wear and damage (**Figure 7**). If the wear strips are worn to ⅜ in. (9.5 mm) or less at any point, they must be replaced after the suspension has been cleaned and painted. See *Slide Runner Replacement*.

3. Remove the axles (**Figure 8**) and clean them and the axle sleeves with solvent. Dry them with compressed air.

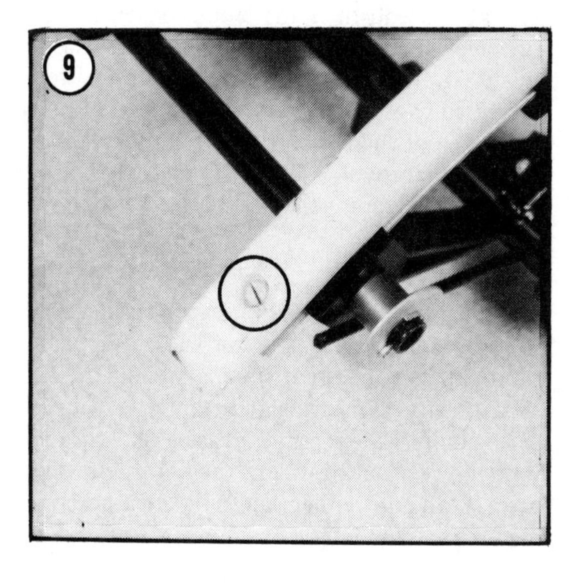

4. Inspect the idler wheels for wear and damage and replace them if the rubber is severely worn or abraded.

5. Lightly sand and feather all chipped painted areas, using a 320 grit wet-or-dry paper. Wash away the sanding residue with water and dry the suspension thoroughly. Paint the sanded areas with a rust-inhibiting paint, such as Rustoleum Gloss Black. Spray on several light coats, allowing the paint to dry between coats to prevent it from running.

6. After the suspension frame has been painted (and the slide runners have been replaced, if required), lightly coat the axles with low-temperature grease and install them in the axle sleeves.

7. Install the suspension as described earlier and adjust the track as described in Chapter Two, *Track Tension and Alignment*.

Slide Runner Replacement

1. Remove the screws that attach the slide runners to the suspension (**Figure 9**). It may be necessary to break the screws loose with an impact driver.

2. Remove the slide runners from the suspension by tapping the runners forward with a block of wood placed against the rear end of the runner.

3. Thoroughly clean the suspension where the runners contact it to remove grit, corrosion, and old paint. Sand and paint the affected area as described above (see *Cleaning/Inspection*).

4. Slide the new runner material onto the suspension. Coat the screw threads with Loctite before installing the screws.

TRACK REMOVAL/INSTALLATION

If track damage or wear is experieced (see Chapter Three, *Track Wear Analysis*) the track should be replaced.

1. Refer to *Suspension Removal/Installation* at the beginning of this chapter and remove the rear suspension assembly from the machine.

2. Refer to *Drive Axle, Removal*, in Chapter Seven and remove the drive axle.

3. Remove the track from the machine.

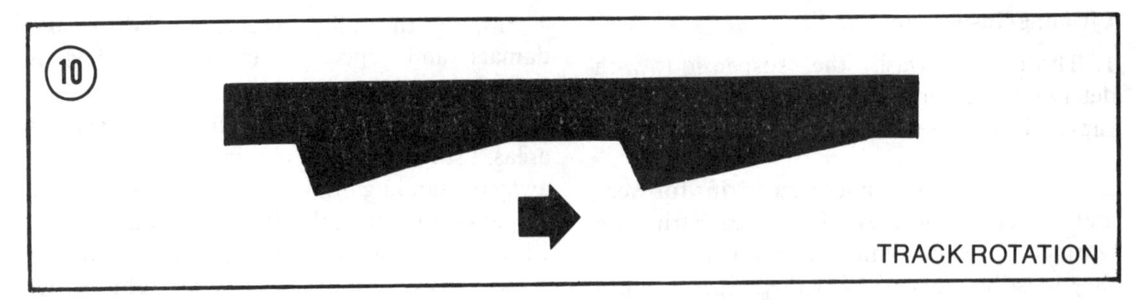

4. Install a new track by reversing the above steps.

> NOTE: *Orient the lugs on the new track as shown in **Figure 10**.*

5. After installation is complete, adjust the track as described in Chapter Two, *Track Tension and Alignment*. Recheck and correct the adjustment after 5 hours and then 10 hours of operation, and then at the intervals recommended in Chapter Two.

SUSPENSION ADJUSTMENT

The suspension can be adjusted to accommodate rider weight (rear springs) and driving conditions (front springs and snubber straps).

Correct suspension adjustment is arrived at mostly through a matter of trial-and-error "tuning." There are several fundamental points that must be understood and applied before the suspension can be successfully adjusted to your needs.

Ski pressure — the load on the skis relative to the load on the track — is the primary factor controlling handling performance. If the ski pressure is too light, the front of the machine tends to float and steering control becomes vague, with the machine tending to drive ahead rather than turn and wander when running straight at steady throttle.

On the other hand, if ski pressure is too heavy, the machine tends to plow during cornering and the skis dig in during straight-line running rather than stay on top of the snow.

Ski pressure for one snow condition is not necessarily good for another condition. For instance, if the surface is very hard and offers little steering traction, added ski pressure — to permit the skis to gain a bite — is desireable; also, the hard surface will support the skis and

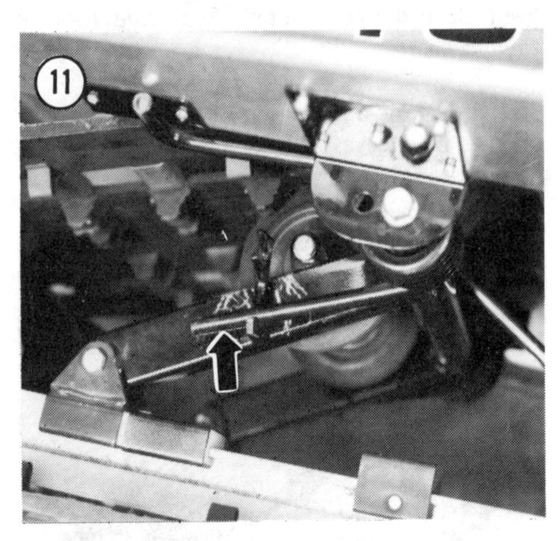

not allow them to penetrate when the machine is running in a straight line under power.

If the surface is soft and tacky, lighter ski pressure is desireable to prevent the skis from digging in, and the increased traction will allow the skis to turn although they are lightly weighted.

It's apparent, then, that good suspension adjustment involves some thought and analysis to relate ski pressure to conditions. Ski pressure is increased by softening the front springs or pulling the front end down with the snubber straps, and it's decreased by stiffening the front springs or extending the snubbers.

Load Compensation — SA and SB Models

If the machine is heavily loaded, either with the addition of a passenger or with equipment, the rear tends to squat, taking weight off the skis and making steering uncertain.

To compensate for an increased load, disconnect the rear springs from their hangers (**Figure 11**), and move them up to the next notch to increase spring tension.

Ski Traction Compensation — SA and SB Models

On ice and hard-packed snow, increased ski pressure is desireable. This could be achieved by stiffening rear suspension as just described, but if the machine is not heavily laden, the ride will be hard and uncomfortable. Instead, the front suspension can be pulled down by shortening the snubber straps (**Figure 12**) and loosening the front spring adjusters (**Figure 13**).

If the snow surface is soft, reduce ski pressure by extending the snubbers and tightening the front spring adjusters.

Load Compensation — SS and ST Models

To compensate for an increased load, increase the spring tension by rotating the adjuster (**Figure 14**) counterclockwise (viewed from the front) to engage a shallower notch. Decrease the tension by rotating the adjuster clockwise to a deeper notch.

Ski Traction Compensation — SS and ST Models

Ski pressure is controlled by shortening the front spring tensioners (**Figure 15**). To increase ski pressure, loosen the adjuster nuts equally.

Decrease ski pressure by tightening the adjuster nuts.

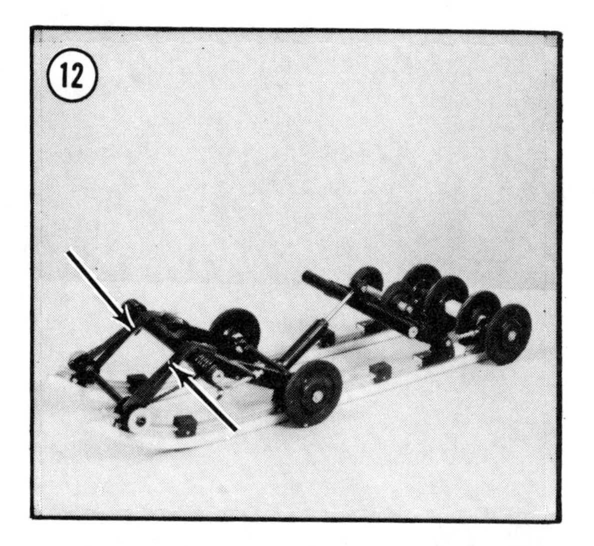

12

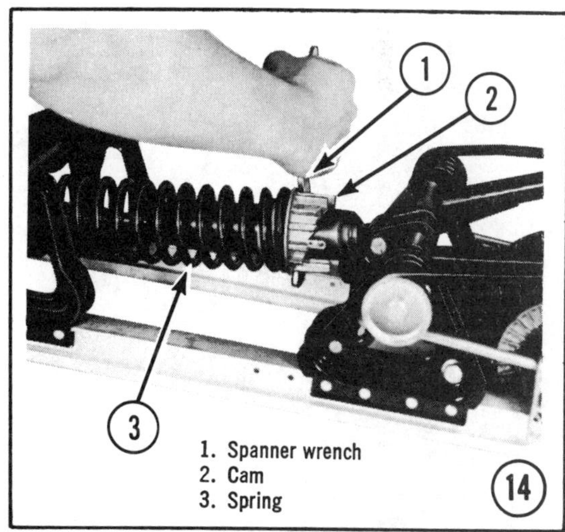

1. Spanner wrench
2. Cam
3. Spring

14

13

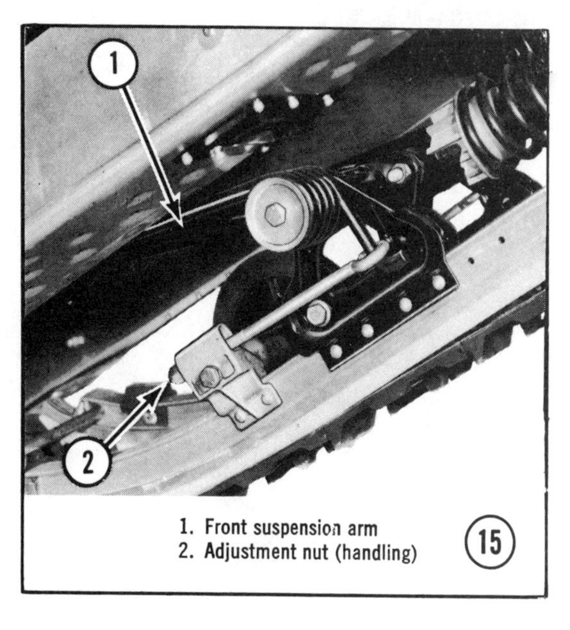

1. Front suspension arm
2. Adjustment nut (handling)

15

INDEX

L

M

O

P

R

S

T

W

SA 340-A1
SA 440-A1
MODELS

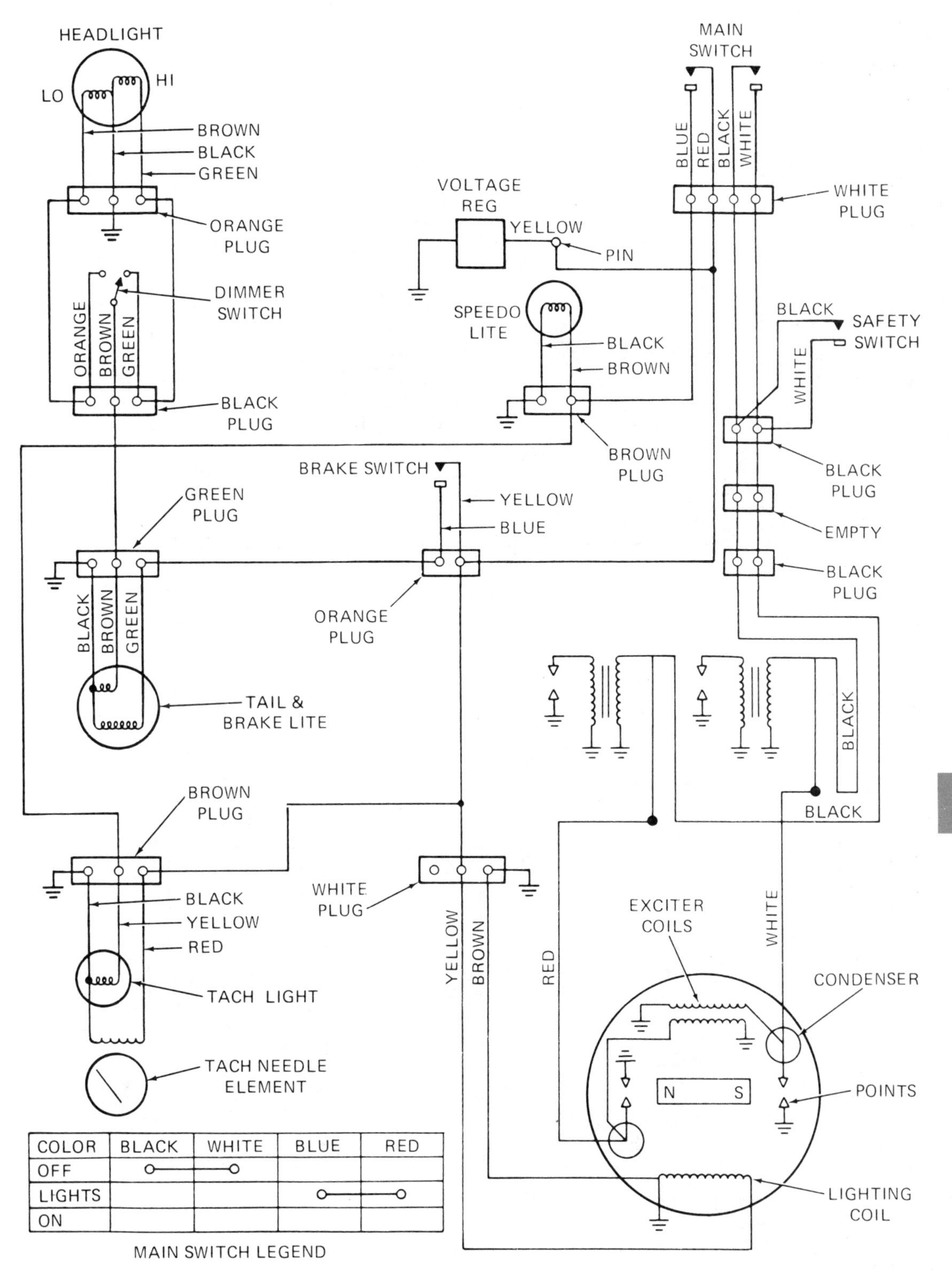

COLOR	BLACK	WHITE	BLUE	RED
OFF	o—	—o		
LIGHTS			o—	—o
ON				

MAIN SWITCH LEGEND

SB 340-A1
SB 440-A1
MODELS

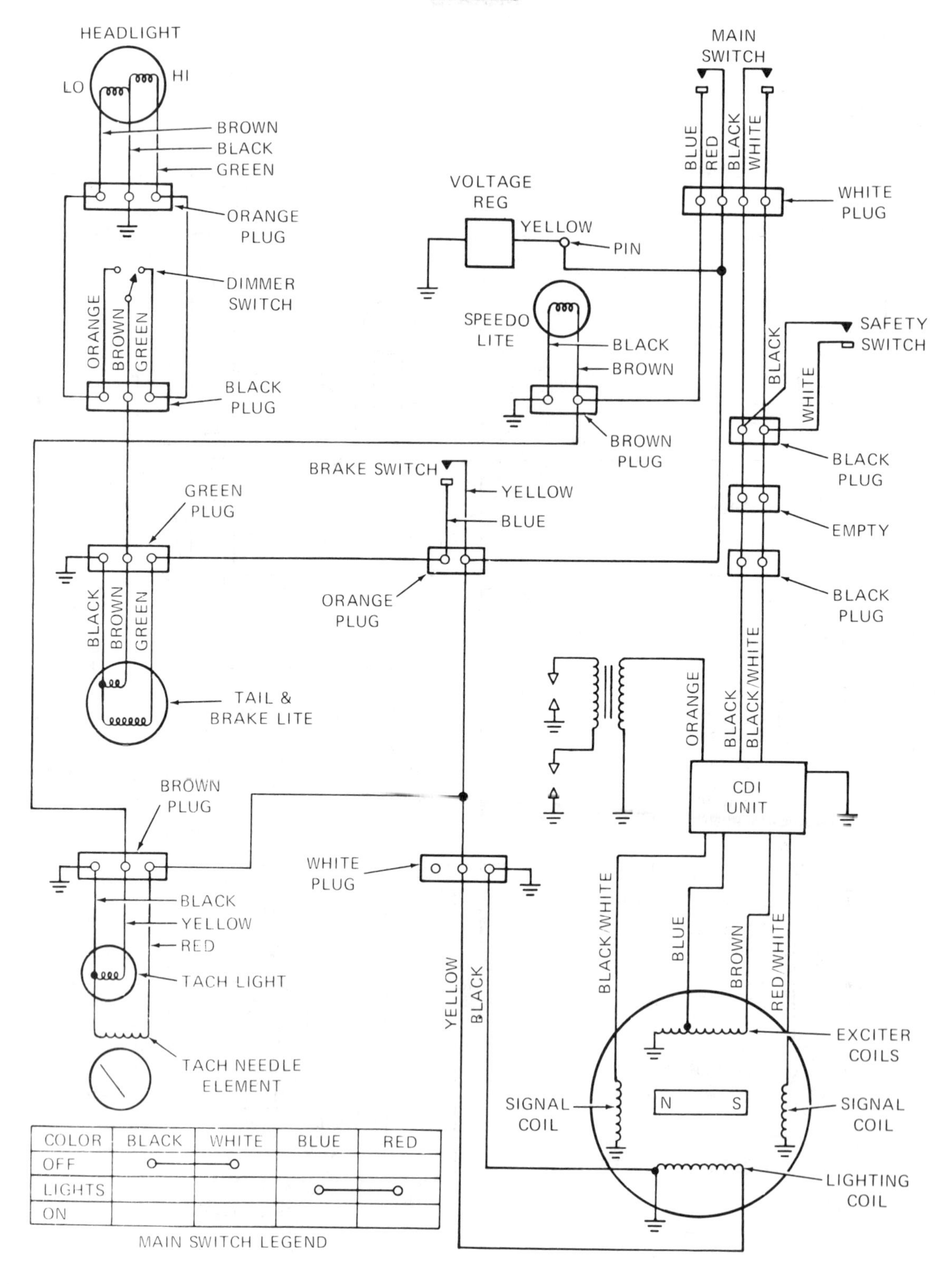

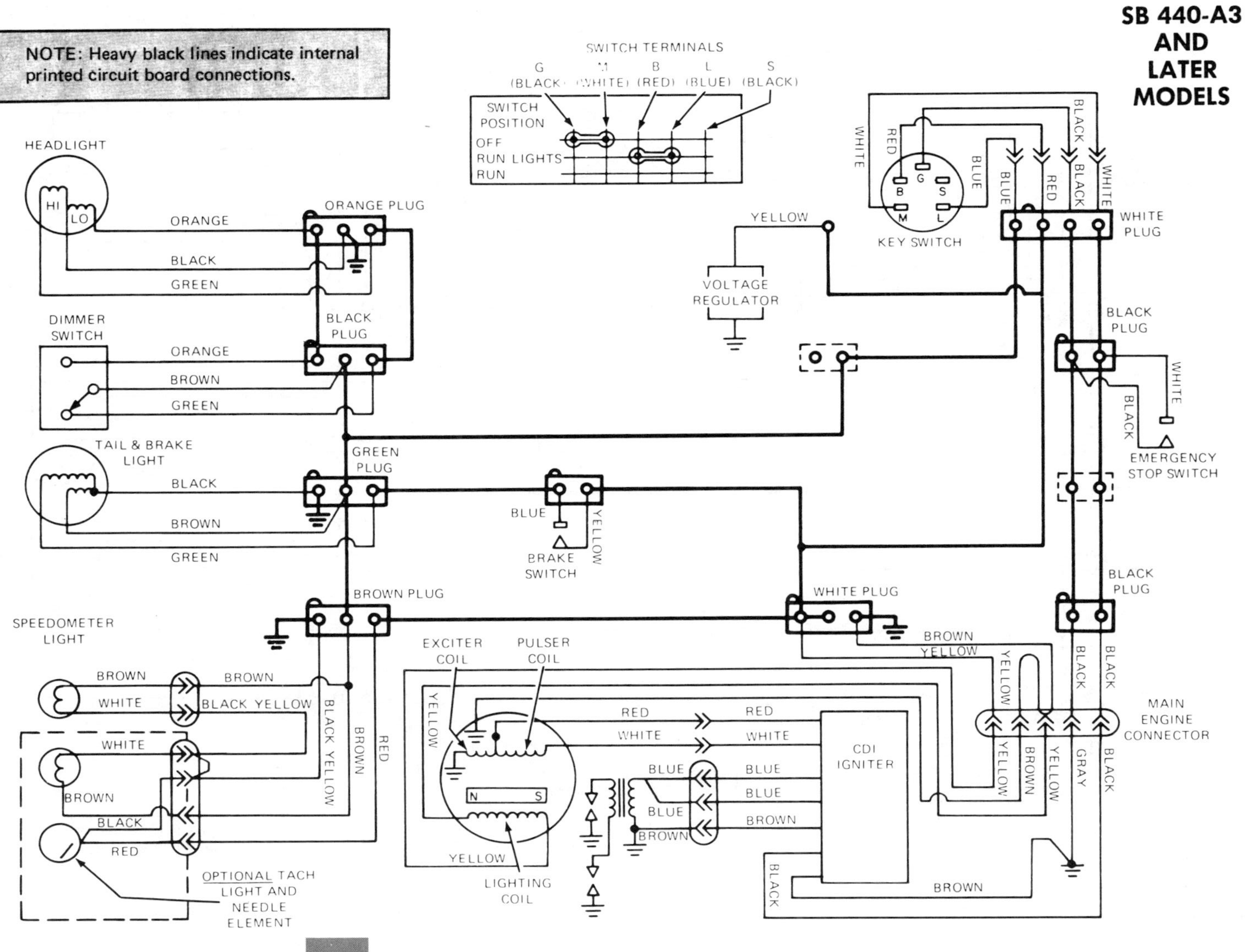
SB 440-A3
AND
LATER
MODELS
NOTE: Heavy black lines indicate internal printed circuit board connections.
SWITCH TERMINALS
G (BLACK)
M (WHITE)
B (RED)
L (BLUE)
S (BLACK)
SWITCH POSITION
OFF
RUN LIGHTS
RUN
HEADLIGHT
HI
LO
ORANGE PLUG
ORANGE
BLACK
GREEN
DIMMER SWITCH
BLACK PLUG
BROWN
TAIL & BRAKE LIGHT
GREEN PLUG
BLUE
YELLOW
BRAKE SWITCH
BROWN PLUG
SPEEDOMETER LIGHT
WHITE
BLACK YELLOW
RED
OPTIONAL TACH LIGHT AND NEEDLE ELEMENT
KEY SWITCH
WHITE PLUG
VOLTAGE REGULATOR
EMERGENCY STOP SWITCH
EXCITER COIL
PULSER COIL
LIGHTING COIL
CDI IGNITER
MAIN ENGINE CONNECTOR
GRAY

SA 340-A3 AND LATER MODELS

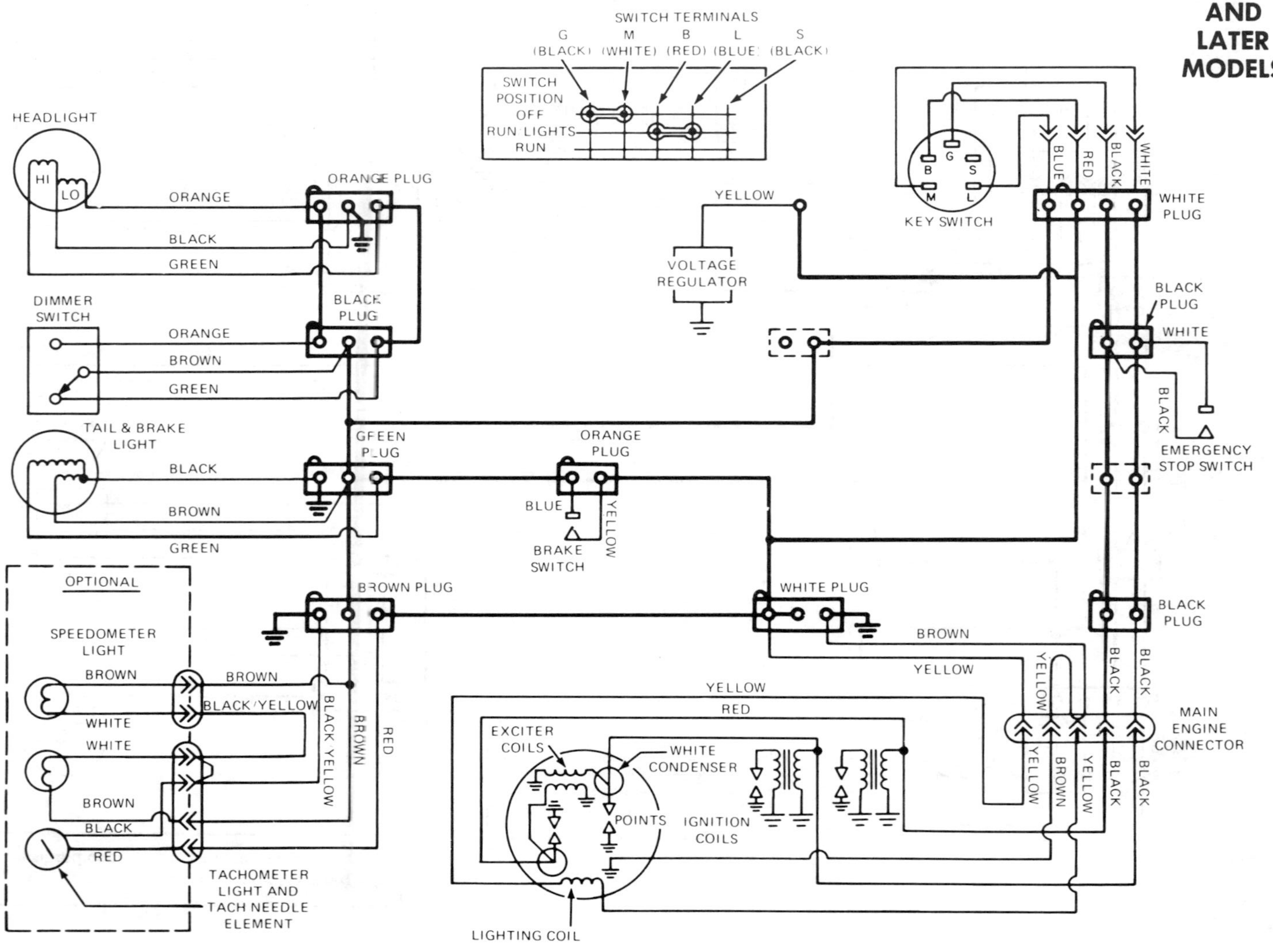

SA 340-A2 MODELS

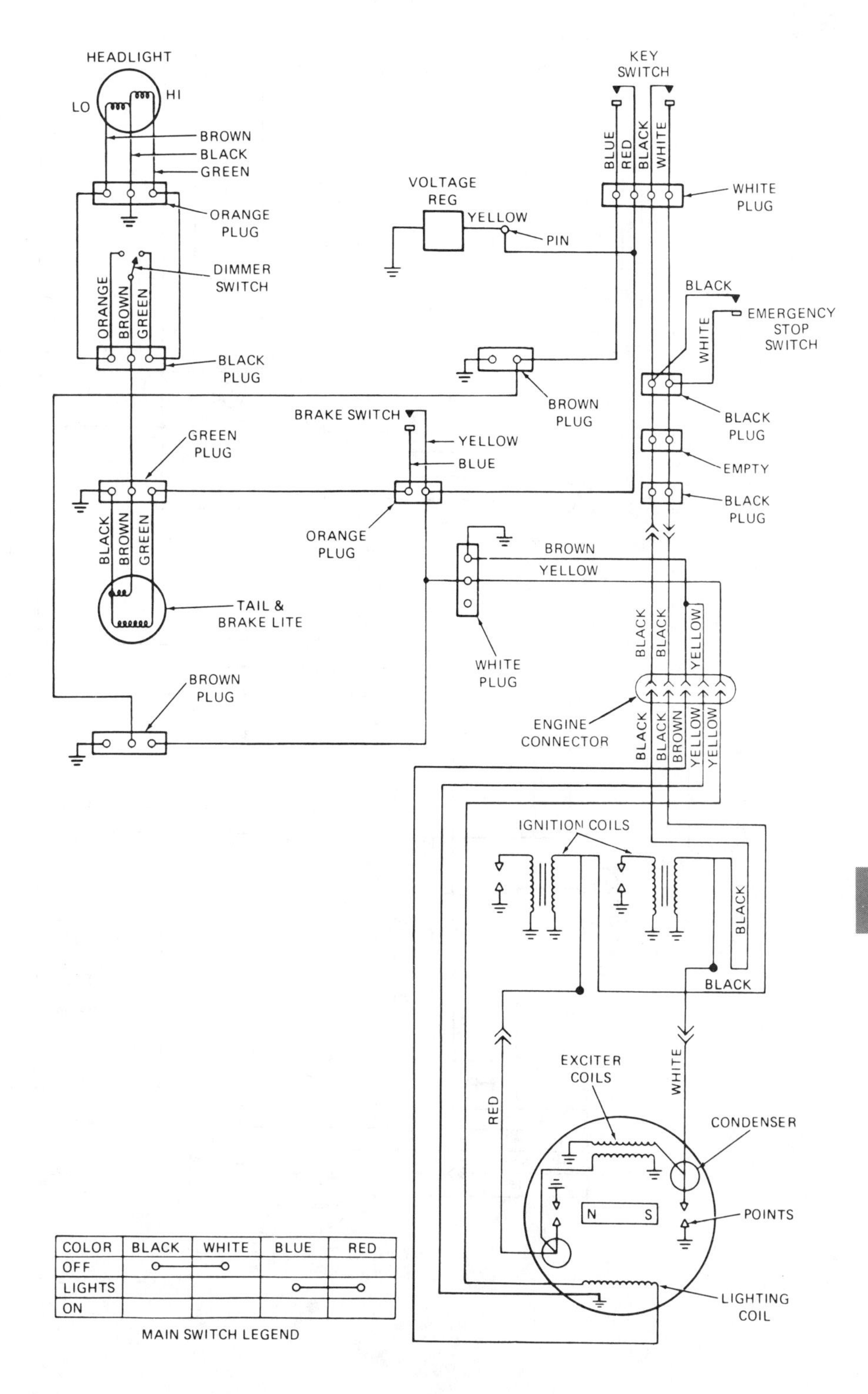

COLOR	BLACK	WHITE	BLUE	RED
OFF	o————	————o		
LIGHTS			o————	————o
ON				

MAIN SWITCH LEGEND

11

ST 440 MODELS

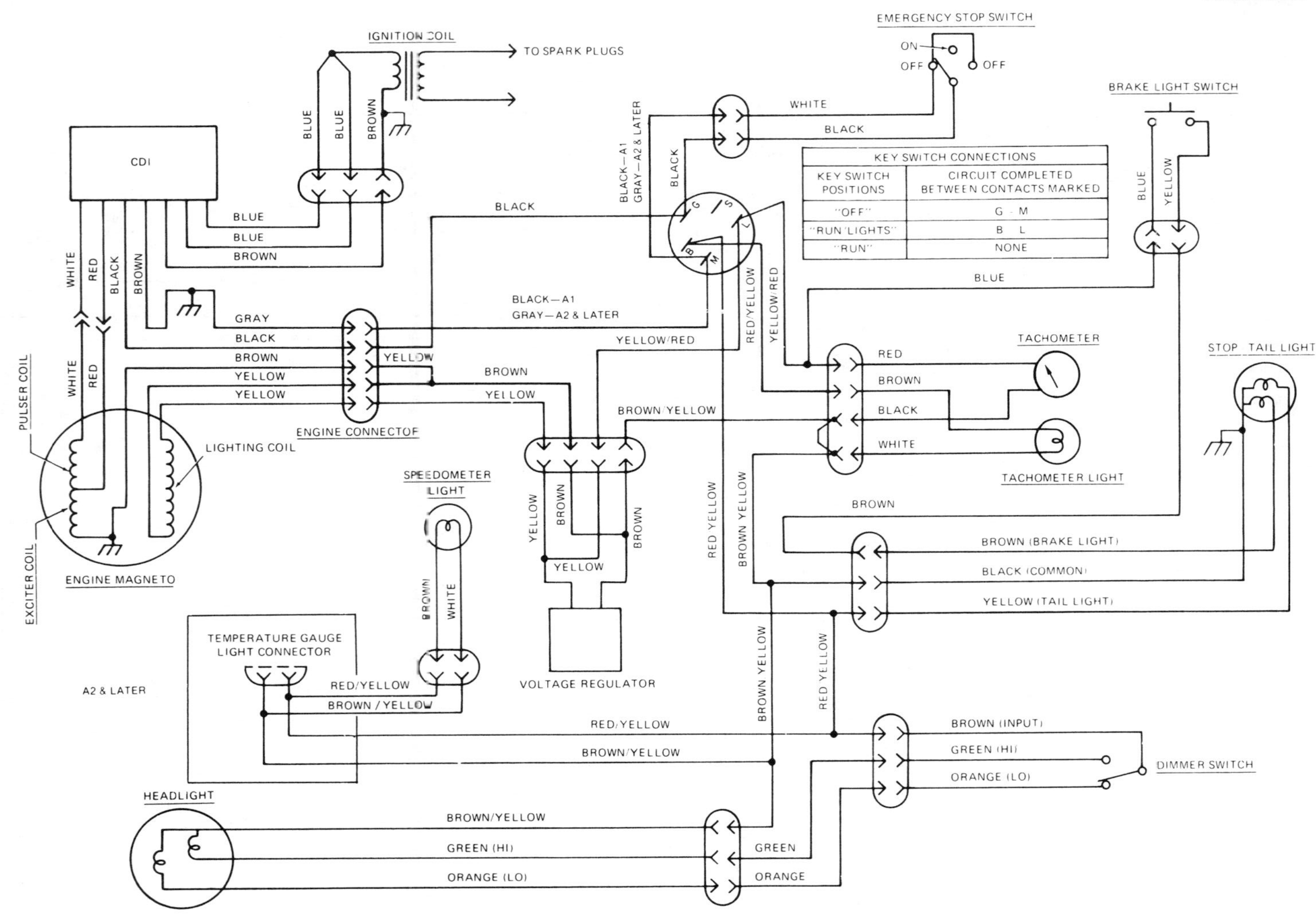

SS 340 AND SS 440 MODELS

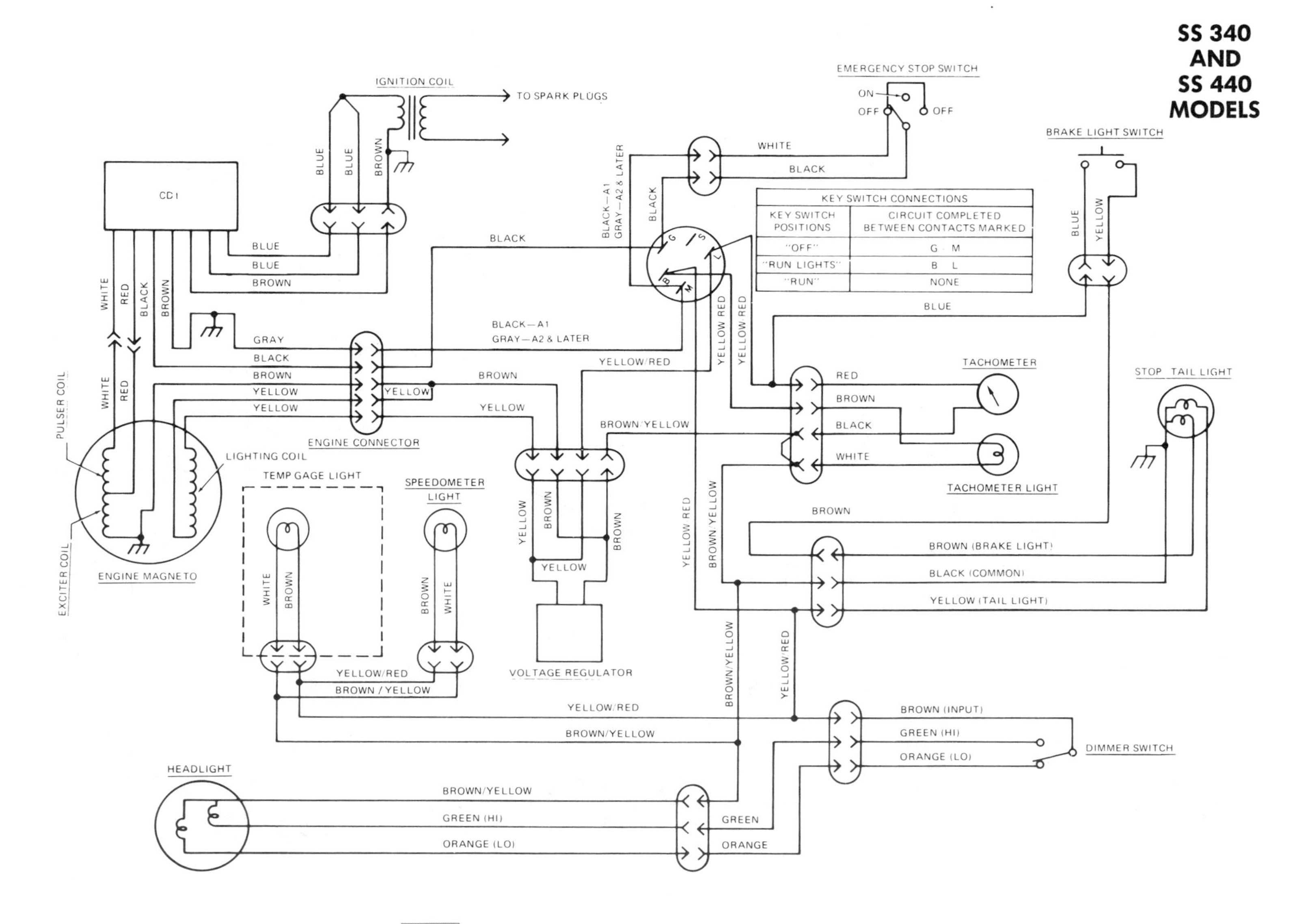

KEY SWITCH CONNECTIONS	
KEY SWITCH POSITIONS	CIRCUIT COMPLETED BETWEEN CONTACTS MARKED
"OFF"	G M
"RUN LIGHTS"	B L
"RUN"	NONE

SB 440-A2 MODELS

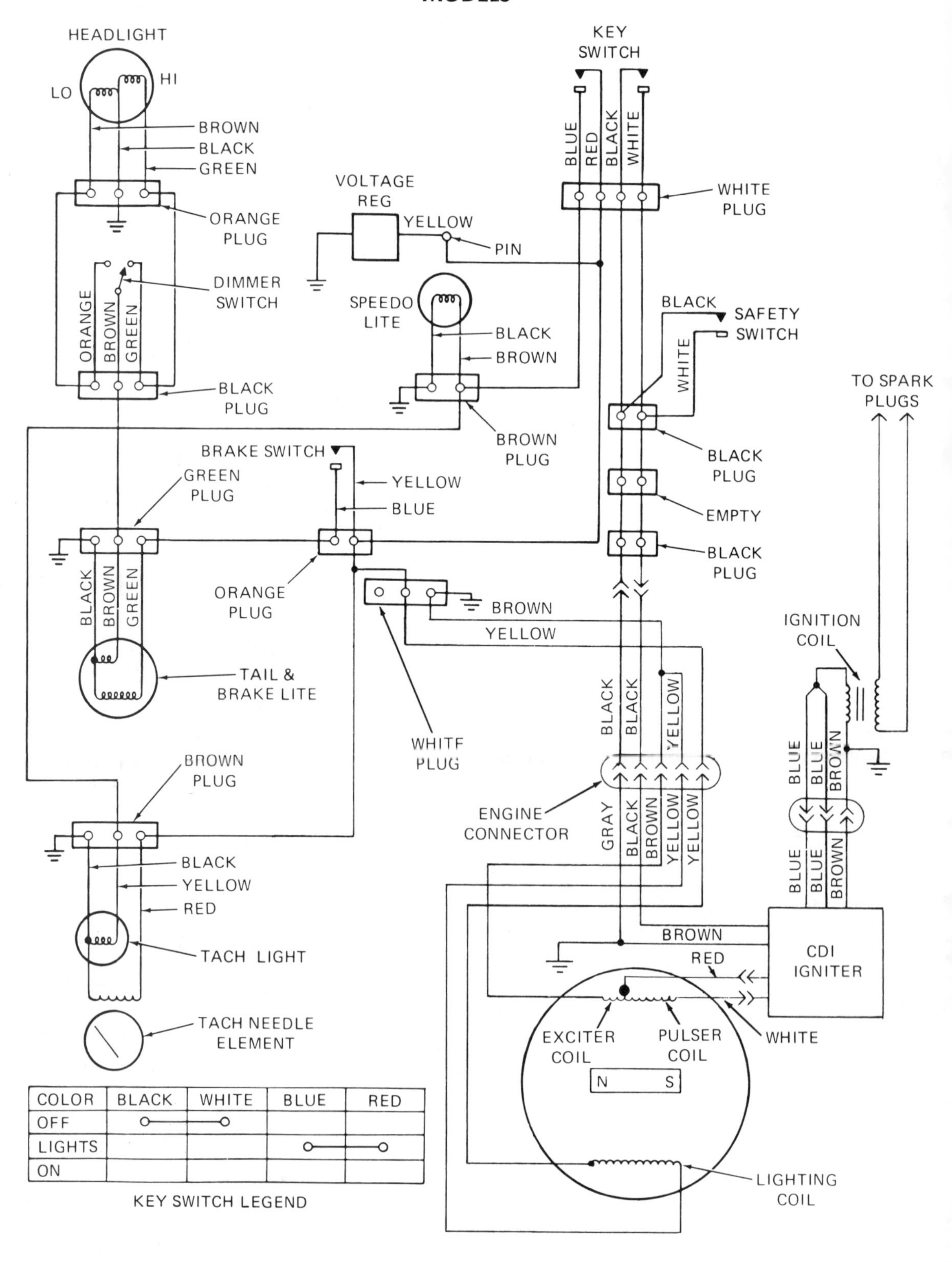

COLOR	BLACK	WHITE	BLUE	RED
OFF	o—	—o		
LIGHTS			o—	—o
ON				

KEY SWITCH LEGEND

Clymer Collection Series

VINTAGE

SNOWMOBILES

VOLUME I

Arctic Cat, 1974-1979

Kawasaki, 1976-1980

➢ John Deere, 1972-1977

Photos and illustrations courtesy of Deere & Co., Moline, Illinois

CONTENTS

Paula Daniels

From: "Matt Spiece" <jdsleds66@nac.net>
To: "Paula Daniels" <jdpd@wamego.net>
Sent: Monday, January 29, 2007 9:32 PM
Subject: Re: cdi box

http://www.cdibox.com - They make an excellent new replacement.

Matt

Paula Daniels wrote:
> Hi Matt
>
> .
>
> I am looking for some help, I am trying to find a cdi box
> for my 77 john deere cyclone
>
> Any help
>
>
> Thanks J.D.
>
>
>

1/30/2007

QUICK REFERENCE DATA

IGNITION SPECIFICATIONS

Spark plug gap	0.020 in. (0.51mm)
Breaker point gap	
CCW & Kioritz engines	0.014 ± 0.002 in. (0.356 ± 0.050mm)
Kohler engines	0.016 ± 0.002 in. (0.406 ± 0.05mm)
Magneto ignition timing	
CCW engines (snowmobile serial No. up to 20,000)	0.023 ± 0.005 in. BTDC (0.584 ± 0.127mm BTDC)
CCW engines (snowmobile serial No. 20,001-30,000)	
Reed valve engines	0.009 ± 0.003 in. BTDC (0.229 ± 0.076mm BTDC)
Piston port engines	0.015 ± 0.002 in. BTDC (0.381 ± 0.050mm BTDC)
Kioritz engines (snowmobile serial No. 30,001 and subsequent)	
Reed valve engines	0.009 ± 0.003 in. BTDC (0.229 ± 0.076mm BTDC)
Piston port engines	0.015 ± 0.002 in. BTDC (0.381 ± 0.050mm BTDC)
Kohler engines	0.090 ± 0.005 in. BTDC (2.286 ± 0.127mm BTDC) —with cam in advanced position
CDI timing	
JD295/S	0.118 in. (2.99mm) BTDC @ 4,000 rpm
JD340/S	0.090 in. (2.28mm) BTDC @ 4,000 rpm
340 Liquidator	0.065 in. (1.65mm) BTDC @ idle
Cyclone and Liquifire	0.096 in. (2.44mm) BTDC @ 4,000 rpm

SPARK PLUG APPLICATION

Serial No. 2551 to 20000

Model	Brand	Plug Number
JDX4	AC	S42XL
	Champion	N-3, N-3G
	NGK	B8ES, B8EV
	Bosch	W260T2, W260S2S
400, 500, 600	AC	S-42F
	Champion	L-81, L-6G
	NGK	B8HZ, B6HV
	Bosch	W225T1
JDX8	AC	S40F
	Champion	L-78, L-4G
	NGK	B9HC, B7HV
	Bosch	W280MZ1, W280S1S

Serial No. 20001 to 30000

Model	Brand	Plug Number
300	AC	S42XLR
	Champion	RN3, RN3G
	NGK	BR8ES, BR8ESA
	Bosch	W260T2, W260S2S
400, 500, 600, JDX4 Special, JDX6, JDX8	AC	S42XLR
	Champion	RN2, N2G*
JD295/S	AC	SV4XL
	Champion	N-19V
	NGK	BR9ES
	Bosch	W280MZ2, W280S2S

Serial No. 30001 and Subsequent

Model	Brand	Plug Number
300	AC	S42XLR
	Champion	RN3
	NGK	BR8ES, BR8ESA
	Bosch	W260T2, W260S2S
400, 600, JDX4, JDX6	AC	S42XLR
	Champion	RN2, N2G
	NGK	BR9ES
	Bosch	W280MZ2, W280S2S
800, JDX8	AC	S41XLR
	Champion	RN2, N2G*
	NGK	BR9ES
	Bosch	W280MZ2, W280S2S
Cyclone, Liquifire, 340 Liquidator	Champion	QN-19V
	NGK	BUE, BUEW
	Bosch	WET2
JD340/S	Champion	N-19V
	NGK	BUE, BUEW
	Bosch	WET2

*For continuous high-speed operation or high performance on 800 and JDX8 models, substitute Champion N59G or NGK B9EV for Champion N2G.

ADJUSTMENTS

Bogie suspension	
To Serial No. 20,000	3/8-1/2 in. (9.5-12.7mm) clearance
After Serial No. 20,001	3/4-1 in. (19.05-25.4mm) clearance
Slide suspension	
Cyclone and Liquifire models with molded grouser bars	0-3/8 in. (0.0-9.6mm) clearance between slide wear bar and top of drive lug
Cyclone and Liquifire models with riveted grouser bars	0-1/4 in. (0.0-6.4mm) clearance between slide wear bar and top of track
All other models	1-1 1/2 in. (25.4-38.1mm) clearance between grouser bars and slide rail
Brake adjustment	1-1 1/2 in. (25-38mm) between brake control lever and handgrip

DRIVE BELT APPLICATION

Model	Width
JDX and 400 (up to serial No. 20,000)	1 3/16 in. (30.16mm)
Model 300 (serial No. 20,001 and subsequent)	1 3/16 in. (30.16mm)
All other models	1 1/4 in. (31.75mm)

NOTE: Replace drive belt when overall width is reduced 1/8 in. (3mm)

FUEL, LUBRICATION, AND CAPACITIES

Fuel	Octane rating 90 or higher. Use premium in all high performance and racing machines.
Engine oil	
Type	John Deere snowmobile oil rated BIA TC-W and BIA-312-69.
Mixture ratio	50:1 (all models except JD295/S) 20:1 (JD295/S)
Chaincase oil	
Cyclone and Liquifire	SAE 90
All other models	SAE 30
Fuel capacities	
Front engine models	8.0 U.S. gal. (30.3 liters)
Mid-engine (plastic tank)	6.5 U.S. gal. (24.6 liters)
Mid-engine (steel tank)	5.5 U.S. gal. (20.9 liters)
JD295/S	2.2 U.S. gal. (8.3 liters)

CARBURETOR TUNING SPECIFICATIONS

Idle speed		
All models with Walbro and Bendix carburetors	2,200-2,600 rpm	
All models with Mikuni carburetors	1,800-2,400 rpm (3,000 rpm on 340 Cyclone and Liquifire models at 8,000-12,000 ft. altitude)	
Air screw turns (Mikuni carburetors)		
340 Liquidator	1½	
JD295/S	1	
JD340/S	1⅛	
440 Liquifire (all models)	1½	
(Serial No. 50,001-70,000)	Sea level to 2,000 ft.	8,000 to 12,000 ft.
340 Cyclone	1	2
440 Cyclone	1	1
340 Liquifire	2	1
(Serial No. 70,001 and subsequent)		
340 Cyclone	2	2
440 Cyclone	2	1½
340 Liquifire	1	½

SCHEDULED MAINTENANCE

Check the following items at indicated intervals:	**Annually**	**Monthly (or 40 hrs. operation)**	**Weekly (or 10 hrs. operation)**	**Daily**
Windshield and coolant level	X	X	X	X
Condition of skis and steering components	X	X	X	X
Track condition and tension	X	X	X	X
Throttle control	X	X	X	X
Brake	X	X	X	X
Emergency stop switch and "tether" switch	X	X	X	X
Lighting system	X	X	X	X
Chain case oil level	X	X	X	
In-line filter for contamination	X	X	X	
Drive belt	X	X	X	
Carburetor adjustments	X	X		
Ski alignment	X	X		
Coolant pump belt or fan belt tension	X	X		
Headlight adjustment	X	X		
Ski wear rods and wear plates	X	X		
Slide suspension wear bars	X	X		
All components for condition and tightness	X			
Drive and driven sheaves	X			

CHAPTER ONE

GENERAL INFORMATION

Snowmobiling has in recent years become one of the most popular outdoor winter recreational past times. It provides an opportunity for an entire family to experience the splendor of winter and enjoy a season previously regarded by many as miserable.

Snowmobiles also provide an invaluable service in the form of rescue and utility vehicles in areas that would otherwise be inaccessible.

As with all sophisticated pieces of machinery, snowmobiles require specific periodic maintenance and repair to ensure their reliability and usefulness.

MANUAL ORGANIZATION

This manual provides periodic maintenance, tune-up, and general repair procedures for John Deere snowmobiles manufactured in 1972 and later.

This chapter provides general information and hints to make all snowmobile work easier and more rewarding. Additional sections cover snowmobile operation, safety, and survival techniques.

Refer to **Table 1**, at the end of the chapter, for general model specifications.

Chapter Two provides all tune-up and periodic maintenance required to keep your snowmobile in top running condition.

Chapter Three provides numerous methods and suggestions for finding and fixing troubles fast. The chapter also describes how a 2-cycle engine works, to help you analyze troubles logically. Troubleshooting procedures discuss typical symptoms and logical methods to pinpoint the trouble.

Subsequent chapters describe specific systems such as engine, fuel system, and electrical system. Each provides disassembly, repair, and reassembly procedures in easy to follow, step-by-step form. If a repair is impractical for the owner/mechanic, it is so indicated. Usually, such repairs are quicker and more economically done by a John Deere dealer or other competent snowmobile repair shop.

Some of the procedures in this manual specify special tools. In all cases, the tool is illustrated in actual use or alone.

The terms NOTE, CAUTION, and WARNING have specific meaning in this book. A NOTE provides additional information to make a step or procedure easier or clearer. Disregarding a NOTE could cause inconvenience, but would not cause damage or personal injury.

A CAUTION emphasizes areas where equipment damage could result. Disregarding a CAUTION could cause permanent mechanical damage; however, personal injury is unlikely.

A WARNING emphasizes areas where personal injury or death could result from negligence. Mechanical damage may also occur. WARNINGS are to be taken seriously. In some cases serious injury or death has been caused by mechanics disregarding similar warnings.

MACHINE IDENTIFICATION AND PARTS REPLACEMENT

Each snowmobile has a serial number applicable to the machine and a model and serial number for the engine.

Figure 1 shows the location of the machine serial number on the right side of the tunnel. **Figure 2** shows the location of engine model and serial numbers on Kohler engines and C.C.W. engines up to serial No. 20,000. **Figure 3** illustrates the location of engine numbers for later C.C.W. and Kioritz engines. Liquid cooled engine numbers are shown in **Figure 4**. Unless specifically noted, all serial numbers called out in procedures in this manual are machine serial numbers and not engine numbers.

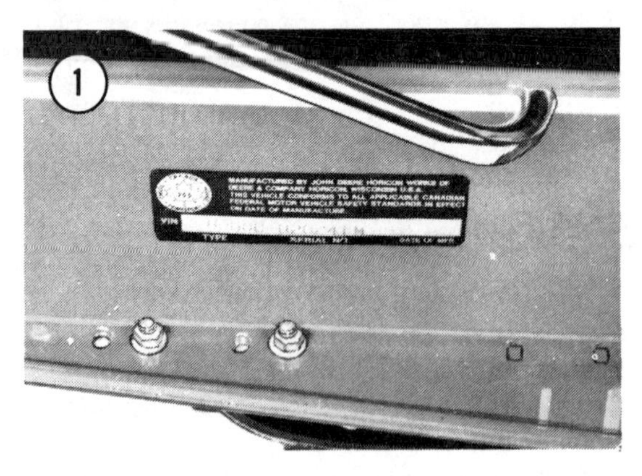

Write down all serial and model numbers applicable to your machine and carry the numbers with you. When you order parts from a dealer, always order by year and engine and machine numbers. If possible, compare old parts to the new ones before purchasing them. If parts are not alike, have the parts manager explain the reason for the difference.

OPERATION

Fuel Mixing

WARNING

Serious fire hazards always exist around gasoline. Do not allow any smoking in areas where fuel is mixed or when refueling your snowmobile.

Always use fresh fuel. Gasoline loses its potency after sitting for a period of time. Old fuel can cause engine failure and leave you stranded in severe weather.

Proper fuel mixing is very important for the life and efficiency of the engine. All engine lubrication is provided by the oil mixed with the gasoline. Always mix fuel in exact proportions. A "too lean" mixture can cause serious and expensive damage. A "too rich" mixture can cause poor performance and fouled spark plugs which can make an engine difficult or impossible to start.

Use a gasoline with an octane rating of 90 or higher. Use premium grade gasoline in all high performance racing machines. Mix gasoline in a separate tank, not the snowmobile fuel tank. Use a tank with a larger volume than necessary to allow room for the fuel to agitate and mix completely.

Use John Deere Snowmobile oil or an equivalent that meets Boating Industry Association (BIA) test qualification TC-W and test procedure BIA-312-69. All machines except JD295/S use a 50:1 ratio fuel/oil mixture. Model JD295/S machines use a 20:1 ratio.

1. Pour required amount of oil into a *clean* container.
2. Add ½ the necessary gasoline and mix thoroughly.
3. Add remainder of gasoline and mix entire contents thoroughly.
4. Always use a funnel equipped with a fine screen when adding fuel to the snowmobile.

Pre-start Inspection

1. Familiarize yourself with your machine, the owner's manual, and all decals on the snowmobile.
2. Clean the windshield with a clean damp cloth. *Do not* use gasoline, solvents, or abrasive cleaners.
3. Check all ski and steering components for wear and loose parts. Correct as necessary.
4. Check track tension.
5. Check operation of throttle and brake controls and ensure that they are free and properly adjusted.
6. Check fuel level. Open fuel shut-off valve on models so equipped.
7. Check coolant level on liquid cooled models.

WARNING

Before starting engine, be sure no bystanders are in front of or behind the snowmobile or a sudden lurch may cause serious injuries.

8. Start engine and test operation of emergency kill switch and "tether" switch. Check that all lights are working.

Emergency Starting

Always carry a small tool kit with you. Carry an extra starting rope for emergency starting or use the recoil starter rope.

1. Remove hood (front engine models) or right access panel (mid-engine models).
2. Remove recoil starter with a 10mm wrench.
3. Wind rope around starter pulley and pull to crank engine (**Figure 5**).

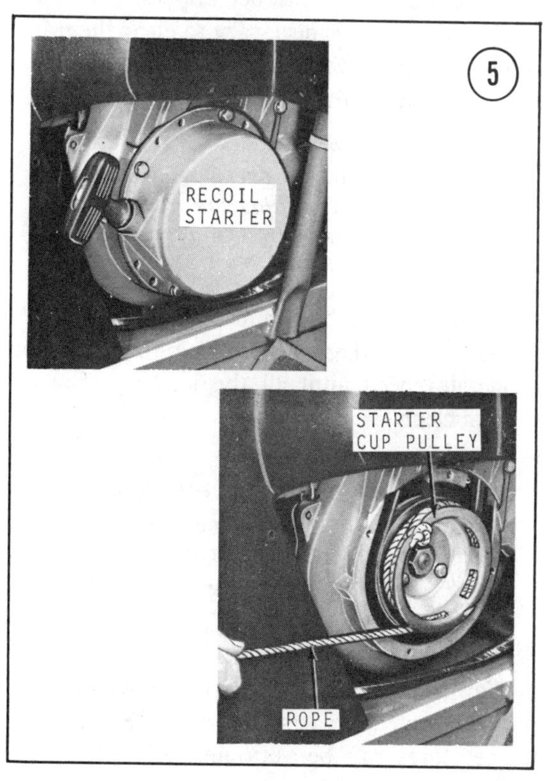

Emergency Stopping

To stop the engine in case of an emergency, pull "tether" string or switch emergency kill switch to STOP or OFF position.

Towing

When preparing for a long trip, pack extra equipment in a sled, do not try to haul it on the snowmobile. A sled is also ideal for transporting small children.

WARNING

Never *tow a sled with ropes or pull straps,* always *use a solid tow bar. Use of ropes or flexible straps could result in a tailgate accident, when the snowmobile is stopped, with subsequent serious injury.*

If it is necessary to tow a disabled snowmobile, securely fasten the disabled machine's skis to the hitch of the tow machine. Remove the drive belt from the disabled machine before towing.

Clearing the Track

If the snowmobile has been operated in deep or slushy snow, it is necessary to clear the track after stopping or the track may freeze, making starting the next time difficult.

WARNING

Always be sure no one is behind the machine when clearing the track. Ice and rocks thrown from the track can cause serious injury.

Tip the snowmobile on its side until the track clears the ground *completely*. Run the track at a moderate speed until all the ice and snow is thrown clear.

CAUTION

If track does freeze it must be broken loose manually. Attempting to force a frozen track with the engine running will burn and damage the drive belt.

Proper Clothing

Warm and comfortable clothing are a must to provide protection from frostbite. Even mild temperatures can be very uncomfortable and dangerous when combined with a strong wind or when traveling at high speed. See **Table 2** for wind chill factors. Always dress according to what the wind chill factor is, not the temperature. Check with an authorized dealer for suggested types of snowmobile clothing.

WARNING

To provide additional warmth as well as protection against head injury, always wear an approved helmet when snowmobiling.

SERVICE HINTS

All procedures described in this book can be performed by anyone reasonably handy with tools. Special tools are required for some procedures; their operation is described and illustrated. These may be purchased at John Deere dealers. If you are on good terms with the dealer's service department, you may be able to borrow from them, however, it should be borne in mind that many of these tools will pay for themselves after the first or second use. If special tools are required, make arrangements to get them before starting. It is frustrating and sometimes expensive to get under way and then find that you are unable to finish up.

Service will be far easier if the machine is clean before beginning work. There are special cleaners for washing the engine and related parts. Just brush or spray on the cleaning solution, let it stand, then rinse it away with a garden hose. Clean all oily or greasy parts with cleaning solvent as they are removed.

WARNING

Never use gasoline as a cleaning agent, as it presents an extreme fire hazard. Be sure to work in a well ventilated area when using cleaning solvent. Keep a fire extinguisher handy, just in case.

Observing the following practices will save time, effort, and frustration as well as prevent possible expensive damage:

1. Tag all similar internal parts for location and mark all mating parts for position. Small parts such as bolts can be identified by placing them in plastic sandwich bags and sealing and labeling the bags with masking tape.

Table 2 WIND CHILL FACTORS

Estimated Wind Speed in MPH	Actual Thermometer Reading (° F)											
	50	40	30	20	10	0	—10	—20	—30	—40	—50	—60
	Equivalent Temperature (° F)											
Calm	50	40	30	20	10	0	—10	—20	—30	—40	—50	—60
5	48	37	27	16	6	—5	—15	—26	—36	—47	—57	—68
10	40	28	16	4	—9	—21	—33	—46	—58	—70	—83	—95
15	36	22	9	—5	—18	—36	—45	—58	—72	—85	—99	—112
20	32	18	4	—10	—25	—39	—53	—67	—82	—96	—110	—124
25	30	16	0	—15	—29	—44	—59	—74	—88	—104	—118	—133
30	28	13	—2	—18	—33	—48	—63	—79	—94	—109	—125	—140
35	27	11	—4	—20	—35	—49	—67	—82	—98	—113	—129	—145
40	26	10	—6	—21	—37	—53	—69	—85	—100	—116	—132	—148
*	**Little Danger** (for properly clothed person)				**Increasing Danger**			**Great Danger**				
					• Danger from freezing of exposed flesh •							
*Wind speeds greater than 40 mph have little additional effect.												

2. Frozen or very tight bolts and screws can often be loosened by soaking with penetrating oil such as WD-40®, then sharply striking the bolt head a few times with a hammer and punch (or screwdriver for screws). A hammer driven impact tool can also be very effective. However, ensure tool is seated squarely on the bolt or nut before striking. Avoid heat unless absolutely necessary, since it may melt, warp, or remove the temper from many parts.

3. Avoid flames or sparks when working near flammable liquids such as gasoline.

4. No parts, except those assembled with a press fit, require unusual force during assembly. If a part is hard to remove or install, find out why before proceeding.

5. Cover all openings after removing parts to keep dirt, small tools, etc., from falling in.

6. Clean all parts as you go along and keep them separated into subassemblies. The use of trays, jars, or cans will make reassembly that much easier.

7. Make diagrams whenever similar-appearing parts are found. You may *think* you can remember where everything came from — but mistakes are costly. There is also the possibility you may be sidetracked and not return to work for days or even weeks — in which interval carefully laid out parts may have become disturbed.

8. Wiring should be tagged with masking tape and marked as each wire is removed. Again, do not rely on memory alone.

9. When reassembling parts, be sure all shims and washers are replaced exactly as they came out. Whenever a rotating part butts against a stationary part, look for a shim or washer. Use new gaskets if there is any doubt about the condition of old ones. Generally, you should apply gasket cement to only one mating surface so the parts may be easily disassembled in the future. A thin coat of oil on gaskets helps them seal effectively.

10. Heavy grease can be used to hold small parts in place if they tend to fall out during assembly. However, keep grease and oil away from electrical and brake components.

11. High spots may be sanded off a piston with sandpaper, but emery cloth and oil do a much more professional job.

12. Carburetors are best cleaned by disassembling them and soaking the parts in a commercial carburetor cleaner. Never soak gaskets and

rubber parts in these cleaners. Never use wire to clean out jets and air passages; they are easily damaged. Use compressed air to blow out the carburetor only if the float has been removed first.

13. Take your time and do the job right. Do not forget that a newly rebuilt snowmobile engine must be broken in the same as a new one. Keep rpm's within the limits given in your owner's manual when you get back on the snow.

14. Work safely in a good work area with adequate lighting and allow sufficient time for a repair task.

15. When assembling 2 parts, start all fasteners, then tighten evenly.

16. Before undertaking a job, read the entire section in this manual which pertains to it. Study the illustrations and text until you have a good idea of what is involved. Many procedures are complicated and errors can be disastrous. When you thoroughly understand what is to be done, follow the prescribed procedure step-by-step.

TOOLS

Every snowmobiler should carry a small tool kit to help make minor adjustments as well as perform emergency repairs.

A normal assortment of ordinary hand tools is required to perform the repair tasks outlined in this manual. The following list represents the minimum requirement:

a. American and metric combination wrenches
b. American and metric socket wrenches
c. Assorted screwdrivers
d. Pliers
e. Feeler gauges
f. Spark plug wrench
g. Small hammer
h. Plastic or rubber mallet
i. Parts cleaning brush

When purchasing tools, always get quality tools. They cost more initially but in most cases will last a life time. Remember, the initial expense of new tools is easily offset by the money saved on a few repair jobs.

Tune-up and troubleshooting require a few special tools. All of the following special tools are used in this manual, however all tools are not necessary for all machines. Read the procedures applicable to your machine to determine what your special tool requirements are.

1. *Ignition gauge* (**Figure 6**). This tool combines round wire spark plug gap gauges with narrow breaker point feeler gauges. The device costs about $3 at auto accessory stores.

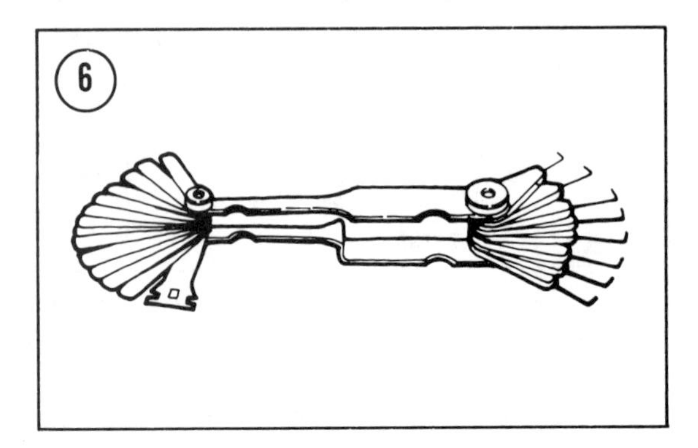

2. *Impact driver* (**Figure 7**). This tool might have been designed with the snowmobiler in mind. It makes removal of screws easy, and eliminates damaged screw slots. Good ones run about $12 at larger hardware stores.

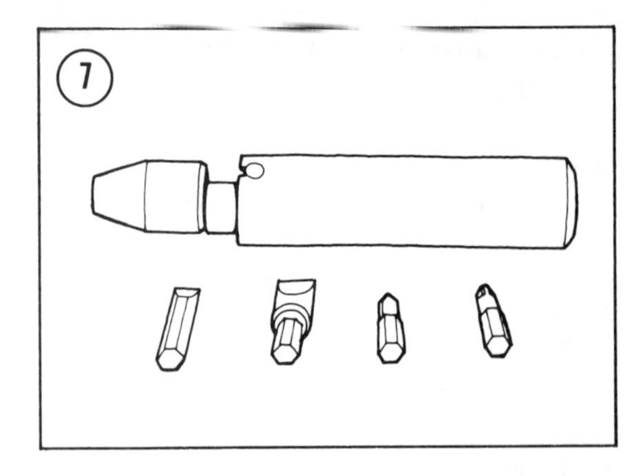

3. *Hydrometer* (**Figure 8**). This instrument measures state of charge of the battery, and tells much about battery condition. Such an instrument is available at any auto parts store and through most larger mail order outlets. Satisfactory ones cost as little as $3.

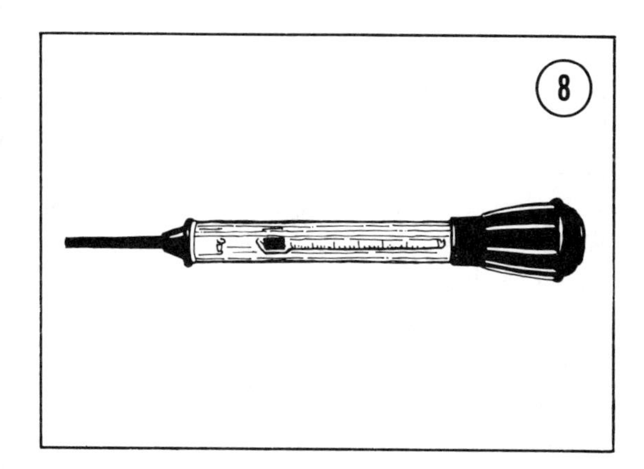

4. *Multimeter or* VOM (**Figure 9**). This instrument is invaluable for electrical system troubleshooting and service. A few of its functions may be duplicated by locally fabricated substitutes, but for the serious hobbyist, it is a must. Its uses are described in the applicable sections of this book. Prices start at around $10 at electronics hobbyists stores and mail order outlets.

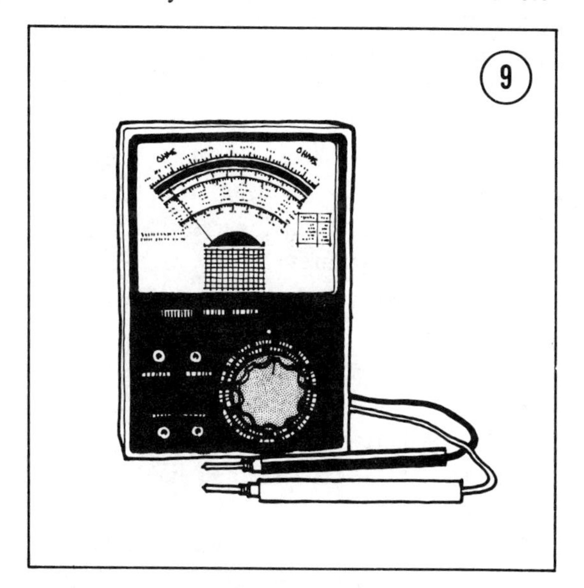

5. *Timing gauge* (**Figure 10**). This device is used to precisely locate the position of the piston before top dead center to achieve the most accurate ignition timing. The instrument is screwed into the spark plug hole and indicates inches and/or millimeters. The tool shown costs about $20 and is available from most dealers and mail order houses. Less expensive tools, which use a vernier scale instead of a dial indicator, are also available.

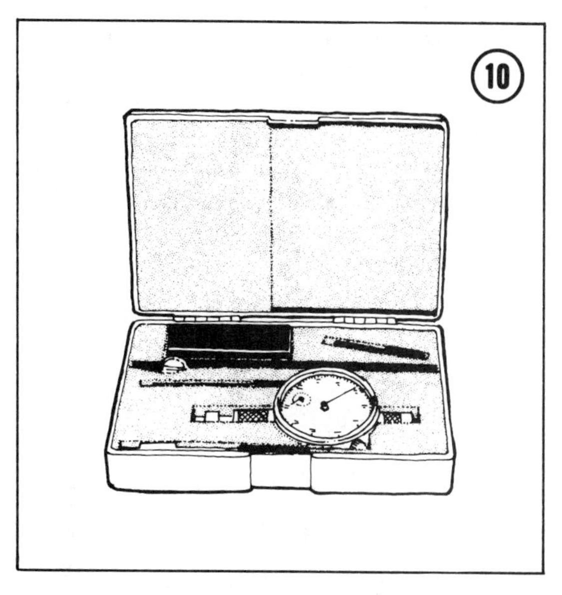

6. *Air flow meter* or *carburetor synchronizer* (**Figure 11**). This device is used on engines with multiple carburetors to fine tune the synchronization and idle speed. The tool shown costs about $10-15 at most dealers, auto parts stores, and mail order houses.

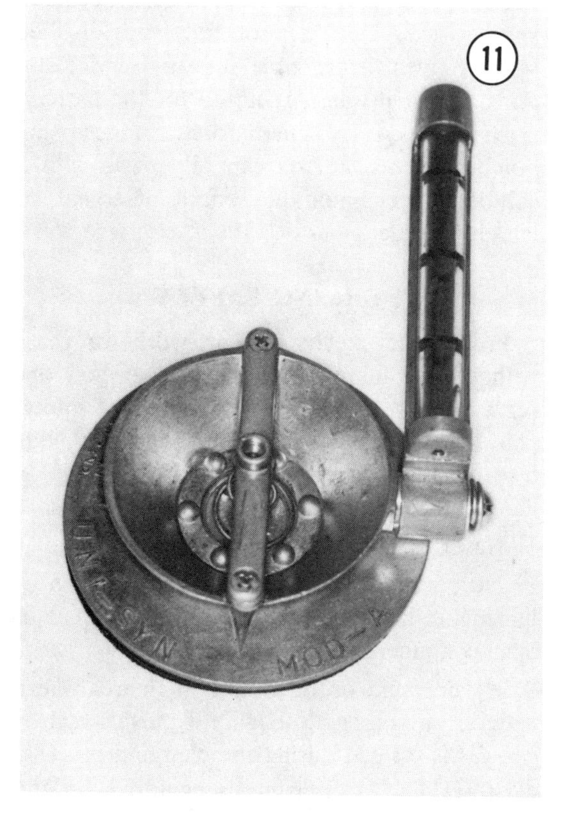

7. *Compression gauge* (**Figure 12**). The compression gauge measures the compression pressure built up in each cylinder. The results, when properly interpreted, indicate general piston, cylinder, ring, and head gasket condition. Gauges are available with or without the flexible hose. Prices start around $5 at most auto parts stores and mail order outlets.

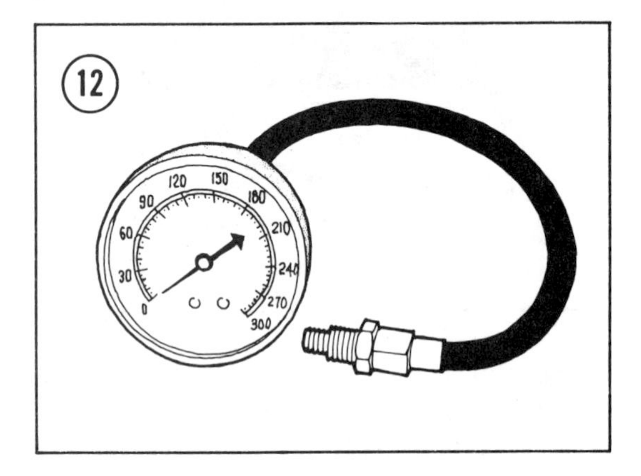

EXPENDABLE SUPPLIES

Certain expendable supplies are also required. These include grease, oil, gasket cement, wiping rags, cleaning solvent, and distilled water. Solvent is available at many service stations. Distilled water, required for the battery, is available at every supermarket. An increasing number of mechanics clean oily parts with a solution of common household detergent or laundry powder.

WORKING SAFELY

Professional mechanics can work for years without sustaining serious injury. If you observe a few rules of common sense and safety, you can enjoy many safe hours servicing your own machine. You can also hurt yourself or damage the machine if you ignore these rules.

1. Never use gasoline as a cleaning solvent.
2. Never smoke or use a torch in the area of flammable liquids, such as cleaning solvent in open containers.
3. Never smoke or use a torch in an area where batteries are charging. Highly explosive hydrogen gas is formed during the charging process.
4. If welding or brazing is required on the machine, remove the fuel tank to a safe distance, at least 50 feet away.
5. Be sure to use properly sized wrenches for nut turning.
6. If a nut is tight, think for a moment what would happen to your hand should the wrench slip. Be guided accordingly.
7. Keep your work area clean and uncluttered.
8. Wear safety goggles in all operations involving drilling, grinding, or use of a chisel.
9. Never use worn tools.
10. Keep a fire extinguisher handy. Be sure it is rated for gasoline and electrical fires.

SNOWMOBILE CODE OF ETHICS

When snowmobiling, always observe the following code of ethics as provided by the International Snowmobile Industry Association.

1. I will be a good sportsman. I recognize that people judge all snowmobile owners by my actions. I will use my influence with other snowmobile owners to promote sportsmanlike conduct.
2. I will not litter trails or camping areas. I will not pollute streams or lakes.
3. I will not damage living trees, shrubs, or other natural features.
4. I will respect other people's property and rights.
5. I will lend a helping hand when I see someone in distress.
6. I will make myself and my vehicle available to assist search and rescue parties.
7. I will not interfere with or harass hikers, skiers, snowshoers, ice fishermen, or other winter sportsmen. I will respect their rights to enjoy our recreation facilities.
8. I will know and obey all federal, state, and local rules regulating the operation of snowmobiles in areas where I use my vehicle. I will inform public officials when using public lands.
9. I will not harass wildlife. I will avoid areas posted for the protection or feeding of wildlife.

10. I will stay on marked trails or marked roads open to snowmobiles. I will avoid country travel unless specifically authorized.

SNOWMOBILE SAFETY

General Tips

1. Read your owner's manual and know your machine.
2. Check throttle and brake controls before starting the engine. Frozen controls can cause serious injury.
3. Know how to make an emergency stop.
4. Know all state, provincial, federal, and local laws concerning snowmobiling. Respect private property.
5. Never add fuel while smoking or when engine is running. Always use fresh, properly mixed fuel. Improper fuel mixtures can cause engine failure, and can leave you stranded in severe weather.
6. Wear adequate clothing to avoid frostbite. Never wear any loose scarves or belts that could catch in moving parts or on tree limbs.
7. Wear eye and head protection. Wear tinted goggles or face shields to guard against snow-blindness. Never wear yellow eye protection.
8. Never allow anyone to operate the snowmobile without proper instruction.
9. Use the "buddy system" for long trips. A snowmobile travels farther in 30 minutes than you can walk in a day.
10. Take along sufficient tools and spare parts for emergency field repairs.
11. Use a sled with a stiff tow bar for carrying extra supplies. Do not overload your snowmobile.
12. Carry emergency survival supplies when going on long trips. Notify friends and relatives of your destination and expected arrival time.
13. Never attempt to repair your machine while the engine is running.
14. Check all machine components and hardware frequently, especially skis and steering.
15. Never lift rear of machine to clear the track. Tip machine on its side and be sure no one is behind machine.
16. Winch snowmobile onto a tilt-bed trailer, never drive it on. Secure machine firmly to trailer and ensure trailer lights operate.

Operating Tips

1. Never operate the vehicle in crowded areas, or steer toward persons.
2. Avoid avalanche areas and other unsafe terrain.
3. Cross highways (where permitted) at a 90 degree angle after looking in both directions. Post traffic guards if crossing in groups.
4. Do not ride snowmobile on or near railroad tracks. The snowmobile engine can drown out the sound of an approaching train. It is difficult to maneuver the snowmobile from between the tracks.
5. Do not ride snowmobile on ski slope areas with skiers.
6. Always check the thickness of the ice before riding on frozen lakes or rivers. Do not panic if you go through the ice; conserve energy.
7. Keep headlight and taillight areas free of snow and never ride at night without lights.
8. Do not ride snowmobile without shields, guards, and protective hoods.
9. Do not attempt to open new trails at night. Follow established trails or unseen barbed wire or guy wires may cause serious injury or death.
10. Always steer with both hands.
11. Be aware of terrain and avoid operating snowmobile at excessive speed.
12. Do not panic if throttle sticks. Pull "tether" string or push emergency stop switch.
13. Drive more slowly when carrying a passenger, especially a child.
14. Always allow adequate stopping distance based on ground cover conditions. Ice requires a greater stopping distance to avoid skidding. Apply brakes gradually on ice.
15. Do not speed through wooded areas. Hidden obstructions, hanging limbs, unseen ditches, and even wild animals can cause accidents.
16. Do not tailgate. Rear end collisions can cause injury and machine damage.

17. Do not mix alcoholic beverages with snowmobiling.

18. Keep feet on footrests at all times. Do not permit feet to hang over sides or attempt to stabilize machine with feet when making turns or in near-spill situations; broken limbs could result.

19. Do not stand on seat, stunt, or show-off.

20. Do not jump snowmobile. Injury or machine damage could result.

21. Always keep hands and feet out of the track area when engine is running. Use extra care when freeing snowmobile from deep snow.

22. Check fuel supply regularly. Do not travel further than your fuel will permit you to return.

23. Whenever you leave your machine unattended, remove the "tether" switch.

Preparing for a Trip

1. Check all bolts and fasteners for tightness. Do not operate your snowmobile unless it is in top operating condition.

2. Check weather forecasts before starting out on a trip. Cancel your plans if a storm is possible.

3. Study maps of the area before the trip and know where help is located. Note locations of phones, resorts, shelters, towns, farms, and ranches. Know where fuel is available. If possible, use the buddy system.

4. Do not overload your snowmobile. Use a sled with a stiff tow bar to haul extra supplies.

5. Do not risk a heart attack if your snowmobile gets stuck in deep snow. Carry a small block and tackle for such situations. Never allow anyone to manually pull on the skis while you attempt to drive machine out.

6. Do not ride beyond one-half the round trip cruising range of your fuel supply. Keep in mind how far it is home.

7. Always carry emergency survival supplies when going on long trips or traveling in unknown territory. Notify friends and relatives of your destination and expected arrival time.

8. Carry adequate eating and cooking utensils (small pans, kettle, plates, cups, etc.) on longer trips. Carry matches in a waterproof container, candles for building a fire, and easy-to-pack food that will not be damaged by freezing. Carry dry food or space energy sticks for emergency rations.

9. Pack extra clothing, a tent, sleeping bag, hand axe, and compass. A first aid kit and snow shoes may also come in handy. Space age blankets (one side silverfoil) furnish warmth and can be used as heat reflectors or signaling devices for aerial search parties.

Emergency Survival Techniques

1. Do not panic in the event of an emergency. Relax, think the situation over, then decide on a course of action. You may be within a short distance of help. If possible, repair your snowmobile so you can drive to safety. Conserve your energy and stay warm.

2. Keep hands and feet active to promote circulation and avoid frostbite while servicing your machine.

3. Mentally retrace your route. Where was the last point where help could be located? Do not attempt to walk long distances in deep snow. Make yourself comfortable until help arrives.

4. If you are properly equipped for your trip you can turn any undesirable area into a suitable campsite.

5. If necessary, build a small shelter with tree branches or evergreen boughs. Look for a cave or sheltered area against a hill or cliff. Even burrowing in the snow offers protection from the cold and wind.

6. Prepare a signal fire using evergreen boughs and snowmobile oil. If you cannot build a fire, make an S-O-S in the snow.

7. Use a policeman's whistle or beat cooking utensils to attract attention or frighten off wild animals.

8. When your camp is established, climb the nearest hill and determine your whereabouts. Observe landmarks on the way, so you can find your way back to your campsite. Do not rely on your footprints. They may be covered by blowing snow.

Table 1 GENERAL SPECIFICATIONS

Engine	2 cylinder, 2-cycle, fan cooled (340 Liquidator and all Liquifire models — liquid cooled)
Electrical system	
Charging and ignition system	
Cyclone, Liquifire 340 Liquidator, JD340/S, JD295/S	Flywheel alternator and CDI (capacitor discharge ignition)
All other models	Flywheel alternator and magneto
Electric starter	Optional on all models
Carburetion	
Front engine models and JD295/S	Mikuni
Mid-engine models	Walbro or Bendix
Power train	
Transmission	Variable drive and driven sheaves
Final drive	Enclosed chain
Brake — front engine models and JD295/S	Mechanical disc (hydraulic disc on JD340/S)
Brake — mid-engine models	External band
Suspension	
Models 300, 400, 500, 600, 800. Models JDX4, JDX8 up to Serial No.20,000. Optional on JD295/S	Trailing arm bogie with 15 wheels
All other models	Slide rail suspension
Track width	
Model 600	18 in. (45.7 cm)
All other models	15.5 in. (39.4 cm)
Body	
Length	
340 Liquidator	107.5 in. (273.0 cm)
Mid-engine models	103.4 in. (262.6 cm)
Front engine models	104.0 in. (264.2 cm)
Width	
Cyclone and Liquifire	39.0 in. (99.0 cm)
JD340/S and 340 Liquidator	40.0 in. (101.6 cm)
JD295/S	35.5 in. (90.2 cm)
All other models	34.5 in. (87.6 cm)

CHAPTER TWO

PERIODIC MAINTENANCE AND TUNE-UP

To gain the utmost in safety, performance, and useful life from your machine, it is necessary to make periodic inspections and adjustments. It frequently happens that minor problems are found during such inspections that are simple and inexpensive to correct at the time, but which could lead to major problems later.

This chapter includes routine maintenance and inspections as well as complete tune-up procedures for all models. **Table 1** summarizes this important information. Keep detailed records of inspections, adjustments, and tune-ups. Such records can help identify recurring trouble areas as well as ensure that required maintenance and tune-up items are accomplished as recommended by the manufacturer.

INLINE FUEL FILTER

Replace the inline fuel filter at the beginning of each season's operation. Examine the filter periodically as specified in **Table 1** and replace it if there is evidence of fuel line contamination.

FAN BELT TENSION

Check fan belt tension at specified intervals (**Table 1**).

1. Remove the fan cover and recoil starter mechanism.
2. Deflect belt with your fingers as shown in **Figure 1**. Examine belt for signs of fraying or deterioration. Adjust belt if deflection is more than ⅜ in. (9.5mm) as outlined under *Fan Belt Adjustment,* Chapter Four. Replace belt if necessary.

COOLANT PUMP DRIVE BELT TENSION

Check coolant pump drive belt tension at specified intervals (**Table 1**).

Table 1 SCHEDULED MAINTENANCE

Check the following items at indicated intervals:	Annually	Monthly (or 40 hrs. operation)	Weekly (or 10 hrs. operation)	Daily
Windshield and coolant level	X	X	X	X
Condition of skis and steering components	X	X	X	X
Track condition and tension	X	X	X	X
Throttle control	X	X	X	X
Brake	X	X	X	X
Emergency stop switch and "tether" switch	X	X	X	X
Lighting system	X	X	X	X
Chain case oil level	X	X	X	
In-line filter for contamination	X	X	X	
Drive belt	X	X	X	
Carburetor adjustments	X	X		
Ski alignment	X	X		
Coolant pump belt or fan belt tension	X	X		
Headlight adjustment	X	X		
Ski wear rods and wear plates	X	X		
Slide suspension wear bars	X	X		
All components for condition and tightness	X			
Drive and driven sheaves	X			

1. Remove belt cover.

2. Deflect belt with your fingers as shown in **Figure 2**. Examine belt for signs of fraying or deterioration. If deflection is more than ⅜ in. (9.5mm), loosen 3 nuts securing pump body and gently pry up on pump body. Tighten 3 nuts securely and recheck belt tension.

DRAINING AND FILLING LIQUID COOLING SYSTEM

CAUTION

Never exceed recommended 50/50 mixture of ethylene glycol anti-freeze. Never use an anti-freeze containing a stop leak or add a radiator stop leak solution to cooling system.

Drain and refill cooling system with new coolant every 2 years. Check coolant level daily during periods of operation. Refer to Chapter Six for draining and filling procedure.

DRIVE AND DRIVEN SHEAVES

All drive and driven sheaves should be removed, disassembled, cleaned, and inspected for worn parts annually. The majority of work on these components requires special tools and expertise. Refer to Chapter Eight for work you can perform. Refer all other work to an authorized dealer.

DRIVE BELT

Examine drive belt periodically as specified in **Table 1** earlier in this chapter. If belt shows unusual signs of wear, refer to Chapter Three for drive belt analysis and troubleshooting. Replace drive belt if its width is reduced by ⅛ in. (3mm). Refer to Chapter Eight for standard width and drive belt replacement for your model.

BRAKES

Check brake operation as scheduled in **Table 1** earlier in this chapter. Brakes are operating properly if the track is locked when the brake control lever is gripped to within one inch of the hand grip. If brake control lever movement is excessive, perform brake adjustment.

Band Brake Adjustment

1. Firmly apply brake and measure distance between brake control lever and handgrip (**Figure 3**). Distance should be 1-1½ in. (25-38mm).

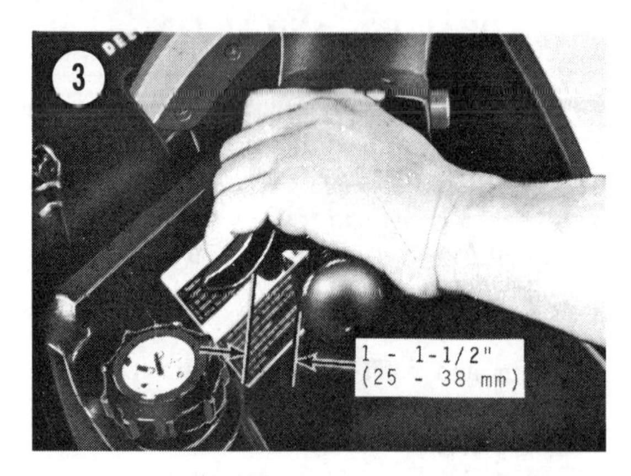

2. To adjust, back off nut "A" (**Figure 4**) several turns and tighten jam nut located behind bracket.

3. Recheck brake operation and readjust if necessary.

> NOTE: *Be sure dowel on end of cable is located properly in recess of brake control lever* (**Figure 4**).

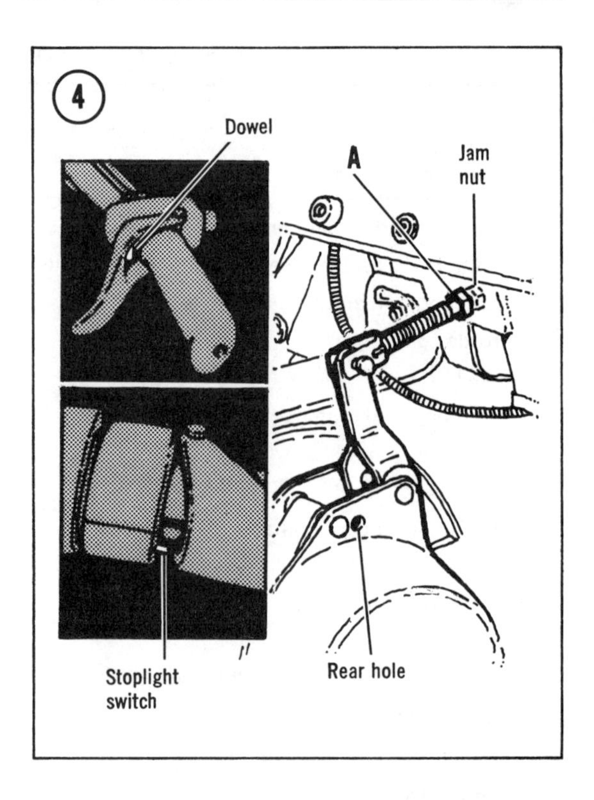

4. Check operation of brake light switch.

5. If brake cable adjustment is used up, perform the following:

   a. Loosen brake cable adjustment.

   b. Move brake band anchor pin to rear hole and readjust brake cables for proper brake operation.

   c. Replace brake band if brakes cannot be adjusted with anchor pin in rear hole.

Mechanical Disc Brake Adjustment

1. Firmly apply brakes and measure distance between brake control lever and handgrip (**Figure 3**). Distance should be 1-1½ in. (25-38mm).

2. Remove cotter pin from brake adjusting nut (**Figure 5**).

3. Tighten adjusting nut until specified brake control lever clearance is achieved. Install cotter pin in adjusting nut.

> NOTE: *Be sure dowel on end of brake cable is located properly in recess of brake control lever.*

4. Check operation of brake light switch.

A. Cotter pin B. Adjusting nut

Hydraulic Disc Brake Bleeding

Check that fluid level is within ⅛ in. (3.2mm) from the top of master cylinder reservoir. Use only brake fluid specified SAE 70R3, DOT 3, or DOT 4 for automotive disc brake application.

If brake work has been performed or if brake operation is "spongy," bleeding may be necessary to expel any air from the system.

> NOTE: *During bleeding operation, be sure master cylinder reservoir is kept topped up to the specified level. If level is allowed to drop too low, air may be ingested, requiring a complete re-bleeding.*

1. Connect a plastic or rubber hose to brake bleeder nipple (**Figure 6**). Place other end of hose in a container with a few inches of clean brake fluid. Keep hose end below the level of the brake fluid.
2. Open brake bleed valve slightly.
3. Operate brake lever and note air bubbles released in jar. Continue operating brakes until all air is expelled. Be sure to keep master cylinder level topped off.
4. After all air has been expelled, close bleeder valve while slowly squeezing brake lever. Check all connections for leaks and remove bleeder hose.

CAUTION

Do not use brake fluid from bleed jar to top off reservoir as the fluid is already aerated.

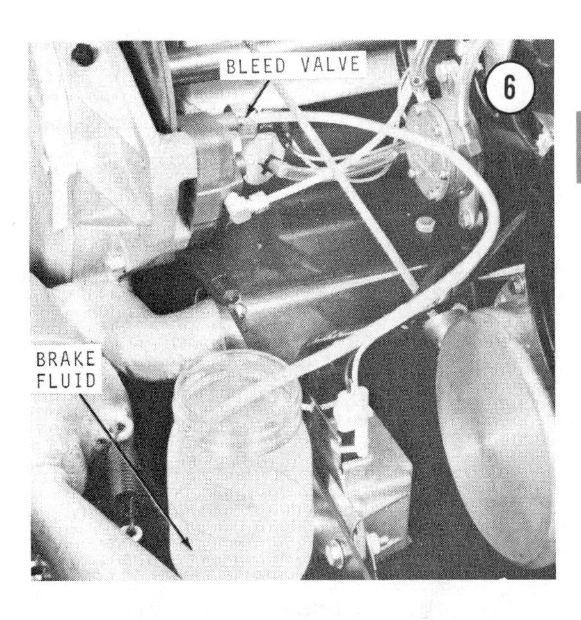

CHAINCASE OIL LEVEL

Check level of chaincase oil at intervals as specified in **Table 1** earlier in this chapter.

On machines with large round access plug, remove plug and check that oil level is ¼-½ in. (6.4-13.0mm) below access hole (**Figure 7**).

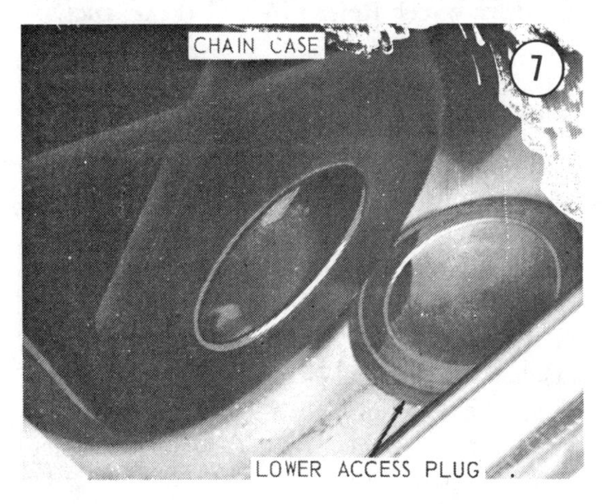

On machines with fill plugs and lower plugs, remove lower plug and check that oil runs from the lower hole (**Figure 8**).

Top up oil level if necessary with SAE 30W oil on all machines except Cyclone and Liquifire. Top off Cyclone and Liquifire models with SAE 90W oil.

Use a syringe or oil suction device to remove old oil when changing oil for machine storage preparation.

A. Lower plug B. Fill plug

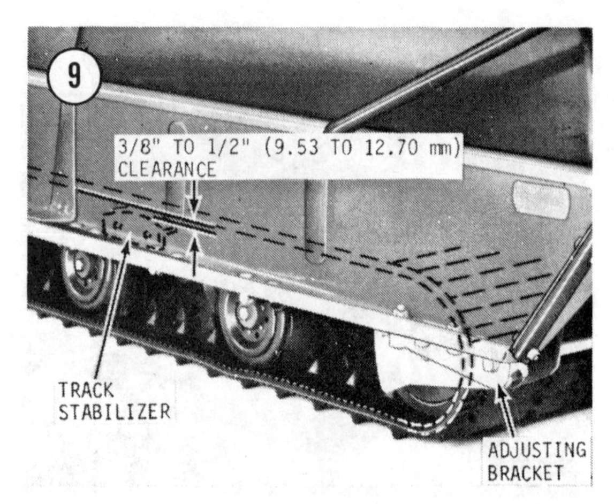

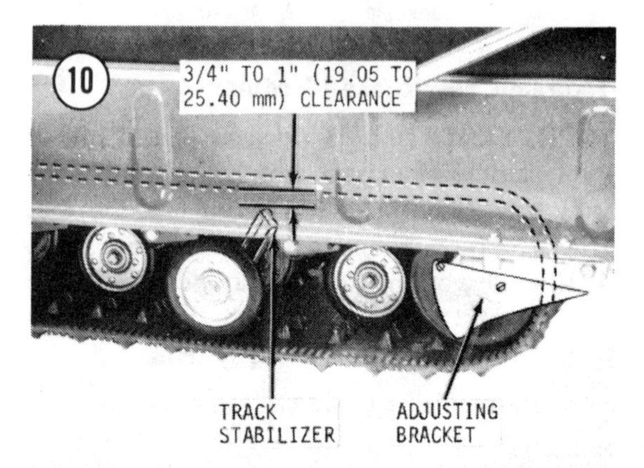

TRACK TENSION ADJUSTMENT

Proper track tension is very important to obtain maximum life and service from the track. Check for track "ratcheting" and proper tension at intervals specified in **Table 1** earlier in this chapter.

Track "ratcheting" occurs if track is too loose and drive lugs on the track slip over the cogs on the drive wheel. Refer to *Track Wear Analysis,* Chapter Three.

Bogie Suspension

1. Position snowmobile on a level surface with an operator on the seat.
2. Measure clearance between track and track stabilizers on each side. On machines up to serial No. 20,000, the clearance should be ⅜-½ in. (9.5-12.70mm) as shown in **Figure 9**. On machines after serial No. 20,001, the clearance should be ¾-1 in. (19.05-25.40mm) as shown in **Figure 10**. If clearance is not as specified, adjust as follows:
   a. Loosen rear bolts and trunnion bolts on both sides of the machine.
   b. Rotate adjusting bolts into trunnion bolts to increase track tension. Both adjusting bolts must be turned an *equal* amount (**Figure 11**). Tighten trunnion and rear bolts.

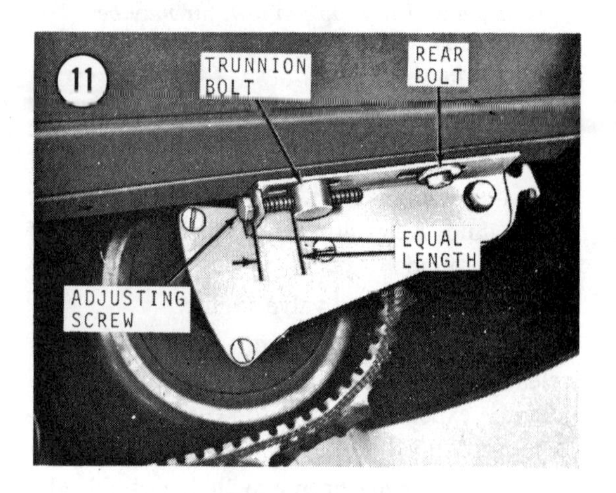

CAUTION
Both sides must be adjusted equally with equal distance between adjusting screw and trunnion bolt (**Figure 11**). *Unequal adjustment will cause track damage.*

NOTE: *If adjustment is used up on adjusting screws, move rear bolts then trunnion bolts to the rear holes* (**Figure 12**) *and adjust the track.*

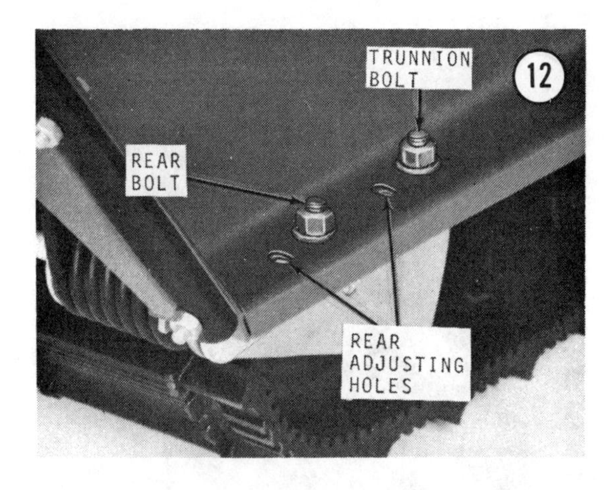

Slide Suspension

1. Tip snowmobile onto its right side or suspend and support rear of machine until it clears the ground.

2. Measure clearance between track and slide bar. Clearance should be as follows:

   a. All models, except Cyclone and Liquifire, 1-1½ in. (25.4-38.1mm) between one of the grouser bars and the slide rail. See **Figure 13**.

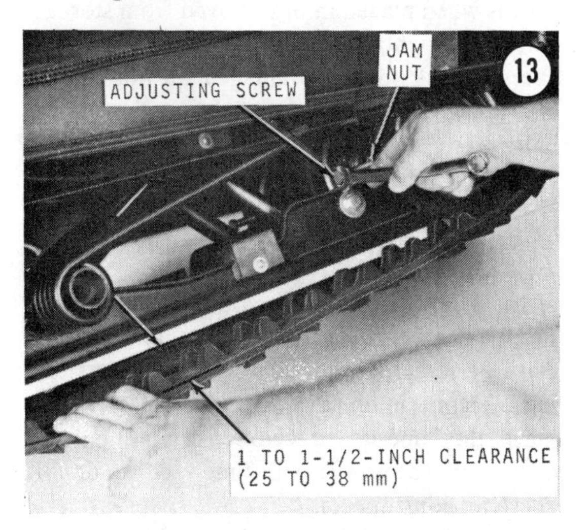

   b. On Cyclone and Liquifire models with molded grouser bars, 0-⅜ in. (0.0-9.6mm) between slide wear bar and top of drive lug (**Figure 14**). On riveted grouser bars, 0-¼ in. (0.0-6.4mm) between slide wear bar and top of track (**Figure 15**).

3. If clearance is not as specified, adjust as follows:

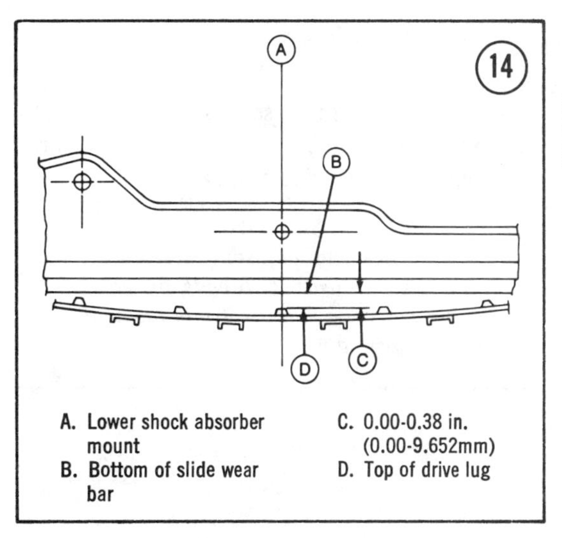

A. Lower shock absorber mount
B. Bottom of slide wear bar
C. 0.00-0.38 in. (0.00-9.652mm)
D. Top of drive lug

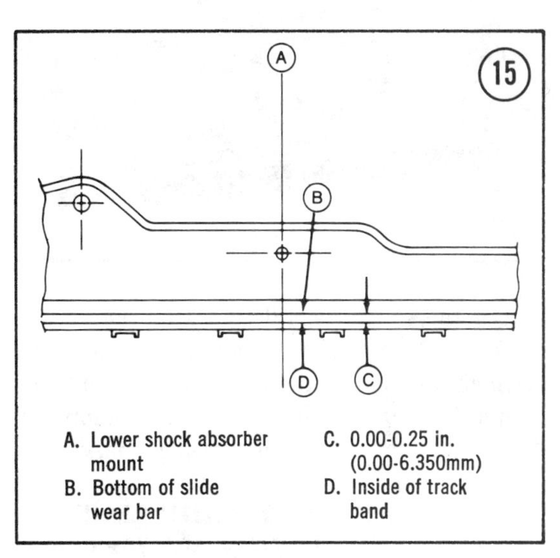

A. Lower shock absorber mount
B. Bottom of slide wear bar
C. 0.00-0.25 in. (0.00-6.350mm)
D. Inside of track band

   a. Loosen jam nut securing adjusting screw.
   b. Turn adjusting screw in to increase tension, out to decrease tension. Check clearance and tighten jam nuts.
   c. Measure between adjusting screw head and jam nut on both sides. Adjustments must be equal.
   d. Start engine and slowly rotate track a few times. Shut off engine and allow track to coast to a stop; do not use the brake.
   e. Recheck track alignment and readjust if necessary. Track will run to the loose side. Tighten loose side slightly to align track. Recheck track tension and readjust if necessary.

SLIDE SUSPENSION SPRING ADJUSTMENT

Adjust front and rear screws equally on both sides for desired ride. Turn screws in for a firm ride; out for a soft ride (**Figure 16**).

CAUTION

Never back out adjusting screws all the way. At least 2 threads on each adjusting screw must protrude through each adjusting nut.

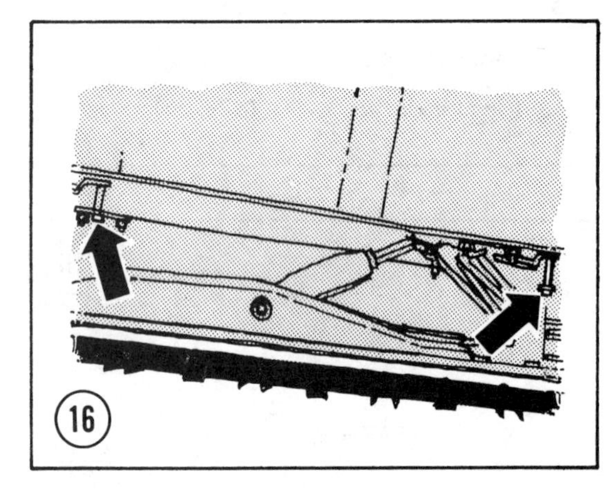

For the smoothest ride, adjust front screws to leave 2 threads protruding through nuts. Tighten rear screws to prevent suspension "bottoming" on all but the most severe bumps. A firmer setting is required if carrying a passenger.

STEERING RESPONSE ADJUSTMENT (SLIDE SUSPENSION SKI LIFT)

Back out adjusting screws to decrease steering response and give lift to skis (**Figure 17**). Turn in adjusting screws to increase steering response and decrease ski lift.

CAUTION

Never back out screws more than "flush" with weld nut. On Liquifire models to serial No. 70,000, never back out adjusting screws more than 0.37 in. (9.39mm) when front pivot arm is in upper hole. Measure from underside of adjusting screw to top of locking nut.

WARNING

When adjusting screws are backed out maximum amount, sudden starts will

lift skis completely off snow. Avoid sudden starts or serious injuries may result.

HARDWARE AND COMPONENT TIGHTNESS CHECK

Hardware and components on all machines should be checked at least once a year. An ideal time is when placed in or removed from storage. Check the tightness of all bolts, nuts, and fasteners. Check for any damaged or worn parts, and areas that require special attention or repair. Refer to **Figure 18** for forward engine models and **Figure 19** for mid-engine models.

ENGINE TUNE-UP

In order to maintain your snowmobile in proper running condition, the engine must receive periodic tune-ups. Since different systems in an engine interact to affect overall performance, the tune-up procedures should be performed in a sequence with time spent to double check all adjustments.

Normal tune-up procedures should begin with ignition adjustment, then followed by carburetor adjustment. Since all adjustments interact, recheck items like idle adjustments after completing the entire tune-up procedure.

Always check the condition of spark plug wires, ignition wires, and fuel lines for splitting, loose connections, hardness, and other signs of deterioration. Check that all manifold nuts and

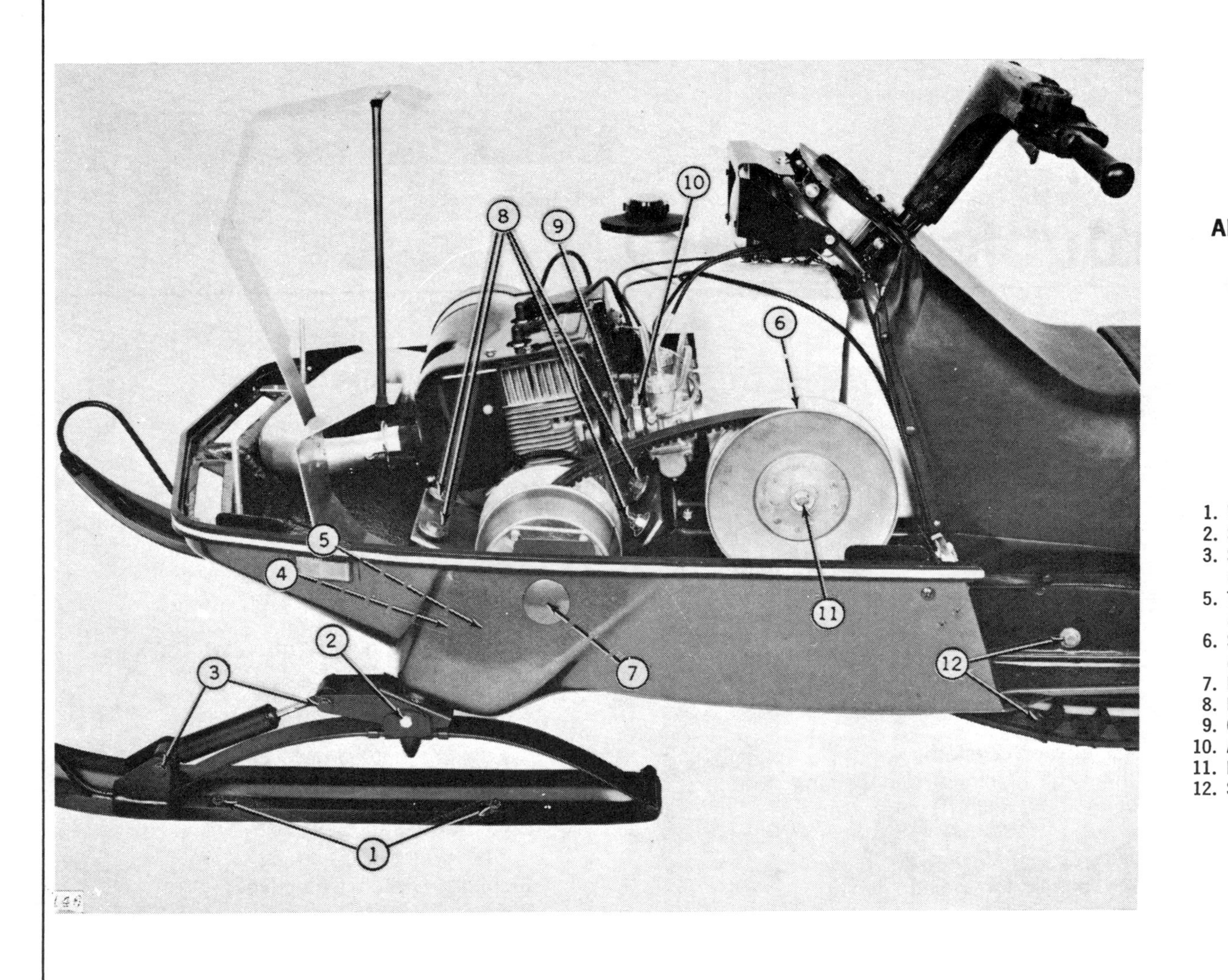

ANNUAL COMPONENT CHECKS FORWARD ENGINE MODELS

1. Wear rod nuts
2. Ski bolts
3. Shock absorber attaching cap screws
5. Tie rod end cap screws and jam nuts
6. Secondary shaft bearing cap screws
7. Drive sheave retaining cap screw
8. Engine mounting bolts
9. Carburetor attaching band
10. Air intake clamps and fuel lines
11. Driven sheave retaining cap screw
12. Suspension components

(18)

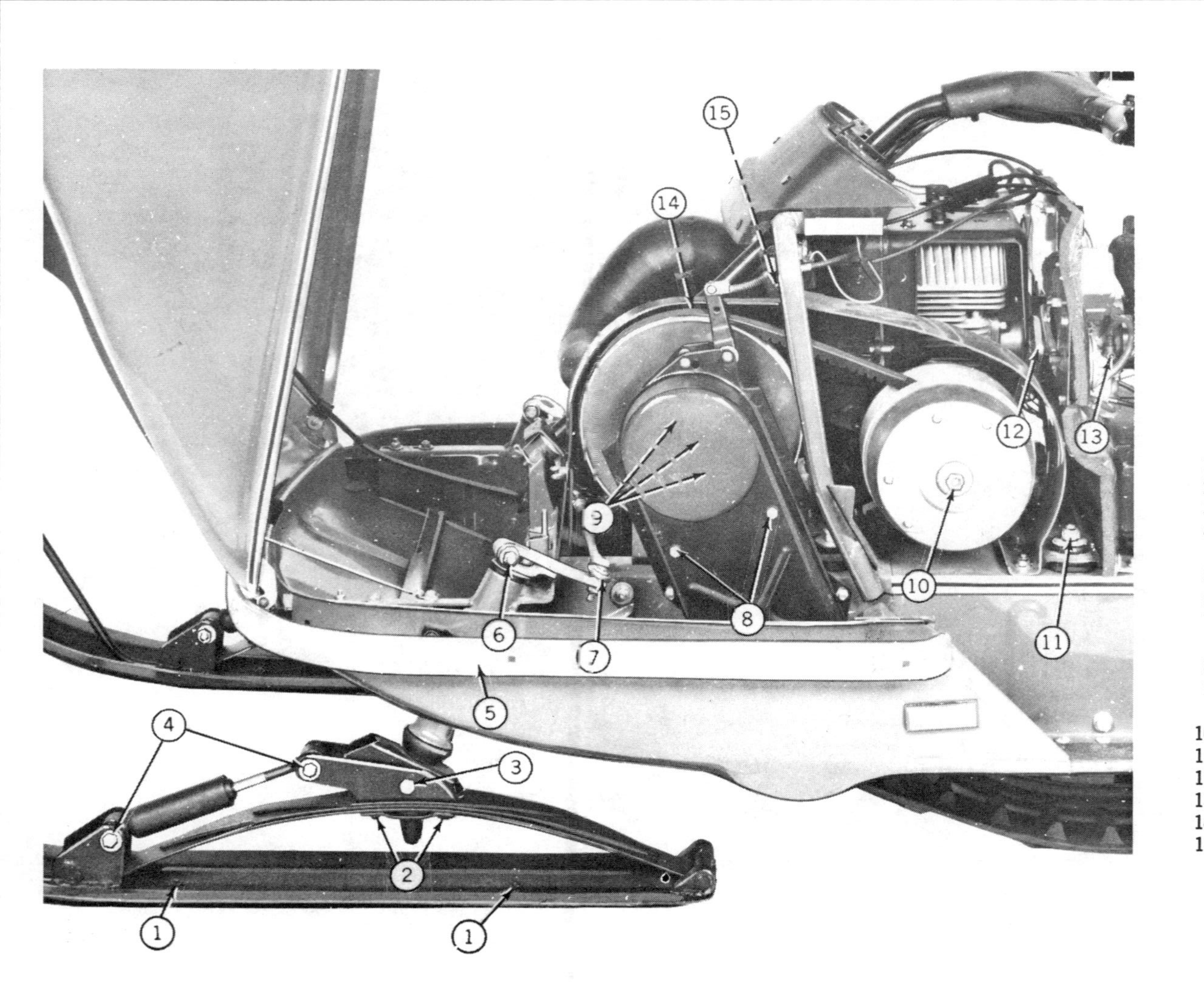

19

ANNUAL COMPONENT CHECKS MID-ENGINE MODELS

1. Wear rod nuts
2. Ski saddle cap screws
3. Ski bolts
4. Shock absorber attaching cap screws
5. Bumper attaching cap screws
6. Steering arm cap screws
7. Tie rod end cap screws and jam nuts
8. Tension block cap screws
9. Secondary shaft bearing cap screws
10. Drive sheave retaining cap screw
11. Engine mounting bolts
12. Carburetor attaching nuts
13. Air intake clamps and fuel lines
14. Driven sheave retaining cap screw
15. Muffler clamp

Also check suspension components.

carburetor nuts are tight and no crankcase leaks are present. A small air leak can make a good tune-up impossible as well as affect performance. A small air leak can also cause serious damage by allowing the engine to run on a "too-lean" fuel mixture.

Tune-up Hints

The following list of general hints will help make a tune-up easier and more successful:

1. Always use good tools and tune-up equipment. The money saved from one or two home tune-ups will more than pay for good tools, from that point you're money ahead. Refer to Chapter One for suitable types of tune-up/test equipment.
2. The purchase of a small set of ignition wrenches and one or two "screwholding" or magnetic screwdrivers will ease the work in replacing breaker points and help eliminate losing small screws.
3. Always purchase quality ignition components.
4. When using a feeler gauge to set breaker points, ensure the blade is wiped clean before inserting between the points.
5. Ensure points are fully open when setting gap with a feeler gauge.
6. Be sure feeler gauge is not tilted or twisted when it is inserted between the contacts. Closely observe the points and withdraw the feeler gauge slowly and carefully. A slight resistance should be felt, however, the movable contact point must *not* "spring back" even slightly when the feeler gauge blade is removed.
7. If breaker points are only slightly pitted, they can be "dressed down" lightly with a small ignition point file. *Do not* use sandpaper as it leaves a residue on the points.
8. After points have been installed, always ensure that they are properly aligned, or premature pitting and burning will result. See **Figure 20**. Bend only the *fixed* half of the points; not the movable arm.
9. When point gap has been set, spring points open and insert a piece of clean paper or cardboard between the contacts. Wipe the contact a few times to remove any trace of oil or grease. A small amount of oil or grease on the contact surfaces will cause the points to prematurely burn or arc.
10. When connecting a timing light or timing tester, always follow the manufacturer's instructions.

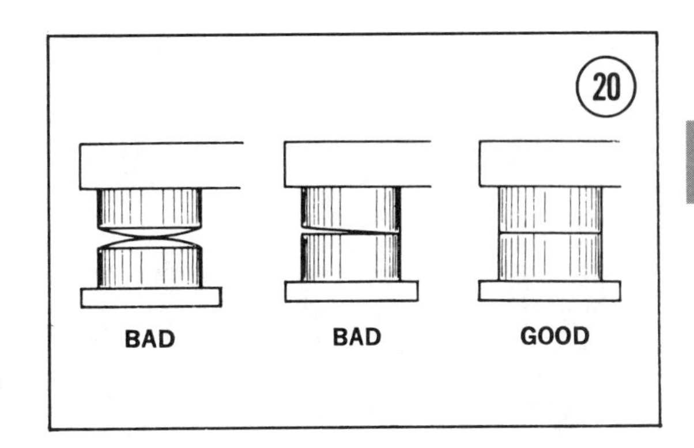

Spark Plugs

Among the first steps to be done during any tune-up is to remove and examine the spark plug. Condition of a used spark plug can tell much about engine condition and carburetion to a trained observer.

To remove the spark plug, first clean the area around its base to prevent dirt or other foreign material from entering the cylinder. Next, unscrew the spark plug, using a 13/16 in. deep socket. If difficulty is encountered removing a spark plug, apply penetrating oil to its base and allow some 20 minutes for the oil to work in. It may also be helpful to rap the cylinder head lightly with a rubber or plastic mallet; this procedure sets up vibrations which helps the penetrating oil to work in.

The proper heat range for spark plugs is determined by the requirement that the plugs operate hot enough to burn off unwanted deposits, but not so hot that they burn themselves or cause preignition. A spark plug of the correct heat range will show a light tan color on the portion of the insulator within the cylinder after the plug has been in service. **Figure 21** illustrates the different construction of the various heat ranges.

Figure 22 illustrates various conditions which might be encountered upon plug removal.

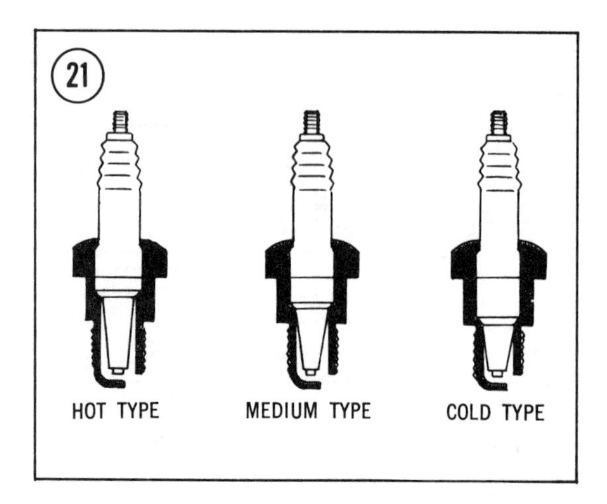

1. *Normal condition*—If plugs have a light tan or gray colored deposit and no abnormal gap wear or erosion, good engine, carburetion, and ignition condition are indicated. The plug in use is of the proper heat range, and may be serviced and returned to use.

2. *Oil fouled*—This plug exhibits a black insulator tip, damp oily film over the firing end, and a carbon layer over the entire nose. Electrodes will not be worn. Common causes for this condition are listed in **Table 2**.

Table 2 CAUSES OF FOULED PLUGS

• Improper fuel/oil mixture	• Weak ignition
• Wrong type of oil	• Excessive idling
• Idle speed too low	• Wrong spark plugs (too cold)
• Clogged air silencer	

Oil fouled spark plugs may be cleaned in a pinch, but it is better to replace them. It is important to correct the cause of fouling before the engine is returned to service.

3. *Overheated*—Overheated spark plugs exhibit burned electrodes. The insulator tip will be light gray or even chalk white. The most common cause for this condition is using a spark plug of the wrong heat range (too hot). If it is known that the correct plug is used, other causes are lean fuel mixture, engine overloading or lugging, loose carburetor mounting, or timing advanced too far. Always correct the fault before putting the snowmobile back into service. Such plugs cannot be salvaged; replace with new ones.

4. *Preignition*—If electrodes are melted, preignition is almost certainly the cause. Check for carburetor mounting or intake manifold leaks, also overadvanced ignition timing. It is also possible that a plug of the wrong heat range (too hot) is being used. Find the cause of preignition before placing the engine back into service.

5. *Carbon fouled*—Soft, dry sooty deposits are evidence of incomplete combustion and can usually be attributed to rich carburetion. This condition is also sometimes caused by weak ignition, retarded timing, or low compression. Such a plug may usually be cleaned and returned to service, but the condition which causes fouling should be corrected.

6. *Gap bridging*—Plugs with this condition exhibit gaps shorted out by combustion chamber deposits used between electrodes. On 2-stroke engines any of the following may be the cause:

a. Improper fuel/oil mixture

b. Clogged exhaust

Be sure to locate and correct the cause of this spark plug condition. Such plugs must be replaced with new ones.

7. *Worn out*—Corrosive gases formed by combustion and high voltage sparks have eroded the electrodes. Spark plugs in this condition require more voltage to fire under hard acceleration; often more than the ignition system can supply. Replace them with new spark plugs of the same heat range.

The spark plugs recommended by the factory are usually the most suitable for your machine. If riding conditions are mild, it may be advisable to go to spark plugs one step hotter than normal. Unusually severe riding conditions may require slightly colder plugs. See **Table 3**.

CAUTION

Ensure that spark plugs used have the correct thread reach. A thread reach too short will cause the exposed threads in the cylinder head to accumulate carbon, resulting in stripped cylinder head threads when the proper plug is installed. A thread reach too long will cause the exposed spark plug

SPARK PLUG CONDITIONS

(22)

NORMAL USE

OIL FOULED

CARBON FOULED

OVERHEATED

GAP BRIDGED

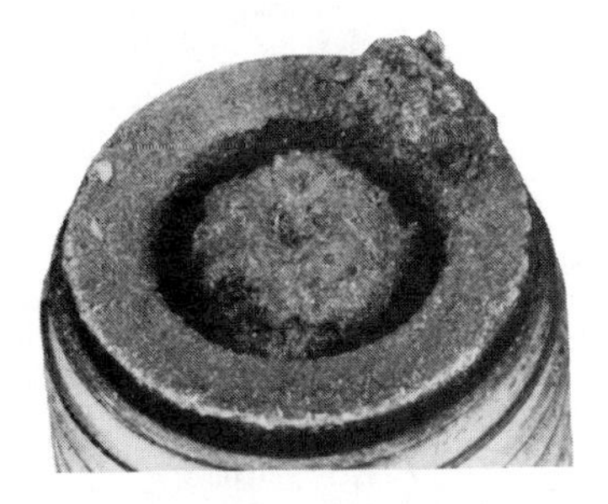

SUSTAINED PREIGNITION

WORN OUT

Photos courtesy of Champion Spark Plug Company.

Table 3 SPARK PLUG APPLICATION

Model	Brand	Plug Number
Serial No. 2551 to 20000		
JDX4	AC	S42XL
	Champion	N-3, N-3G
	NGK	B8ES, B8EV
	Bosch	W260T2, W260S2S
400, 500, 600	AC	S-42F
	Champion	L-81, L-6G
	NGK	B8HZ, B6HV
	Bosch	W225T1
JDX8	AC	S40F
	Champion	L-78, L-4G
	NGK	B9HC, B7HV
	Bosch	W280MZ1, W280S1S
Serial No. 20001 to 30000		
Model	**Brand**	**Plug Number**
300	AC	S42XLR
	Champion	RN3, RN3G
	NGK	BR8ES, BR8ESA
	Bosch	W260T2, W260S2S
400, 500, 600, JDX4 Special, JDX6, JDX8	AC	S42XLR
	Champion	RN2, N2G*
JD295/S	AC	SV4XL
	Champion	N-19V
	NGK	BR9ES
	Bosch	W280MZ2, W280S2S
Serial No. 30001 and Subsequent		
Model	**Brand**	**Plug Number**
300	AC	S42XLR
	Champion	RN3
	NGK	BR8ES, BR8ESA
	Bosch	W260T2, W260S2S
400, 600, JDX4, JDX6	AC	S42XLR
	Champion	RN2, N2G
	NGK	BR9ES
	Bosch	W280MZ2, W280S2S
800, JDX8	AC	S41XLR
	Champion	RN2, N2G*
	NGK	BR9ES
	Bosch	W280MZ2, W280S2S
Cyclone, Liquifire, 340 Liquidator	Champion	QN-19V
	NGK	BUE, BUEW
	Bosch	WET2
JD340/S	Champion	N-19V
	NGK	BUE, BUEW
	Bosch	WET2

*For continuous high-speed operation or high performance on 800 and JDX8 models, substitute Champion N59G or NGK B9EV for Champion N2G.

threads to accumulate carbon resulting in stripped cylinder head threads when the plug is removed.

It may take some experimentation to arrive at the proper plug heat range for your type of riding. As a general rule, use as cold a spark plug as possible without fouling. This will give the best performance.

Remove and clean spark plugs at least once a season. After cleaning, inspect them for worn or eroded electrodes. Replace them if in doubt about their condition. If the plugs are serviceable, file the center electrodes square, then adjust the gaps by bending the outer electrodes only. Measure the gap with a round wire spark plug gauge only; a flat gauge will yield an incorrect reading. **Figure 23** illustrates proper spark plug gap measurement. Gap should be 0.020 in. (0.51mm).

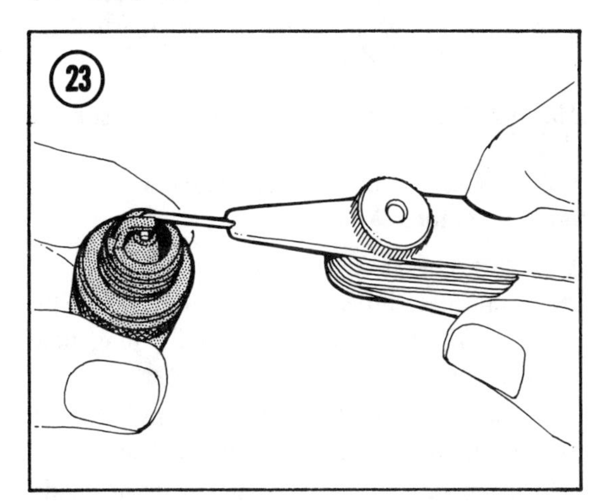

Be sure to clean the seating area on the cylinder head and use a new gasket whenever you replace a spark plug. Install the plug finger-tight, then tighten it an additional ½ turn. If using a torque wrench, torque spark plugs to 14 ft.-lb. (1.9 mkg).

C.C.W. and Kioritz Engine Breaker Point Adjustment

Refer to list of general tune-up hints as outlined under *Engine Tune-up.*

1. Remove recoil starter.
2. Remove starter cup, lower fan belt pulley, and window plate.

NOTE: *On all models except those with a Denso ignition, the flywheel must be installed to adjust points because the breaker point cam is part of the flywheel. On Denso ignition models, the cam is separate and may be installed for point adjustment without the flywheel.*

3. Remove spark plugs. Install dial indicator type timing gauge with a one inch adapter in No. 1 cylinder spark plug hole (flywheel side). Refer to Chapter One for a description of dial indicator type timing gauge.
4. Rotate crankshaft clockwise until No. 1 piston is just past TDC (top dead center) and No. 1 set of breaker points is completely open.

NOTE: *Breaker points for No. 1 cylinder have white wires and points for No. 2 have red wires.*

5. Loosen screw securing No. 1 set of breaker points. Carefully examine points and replace if necessary.
6. Using a feeler gauge, set breaker point gap to 0.014 in. (0.36mm) and tighten breaker point retaining screw (**Figure 24**). Recheck gap and readjust if necessary.

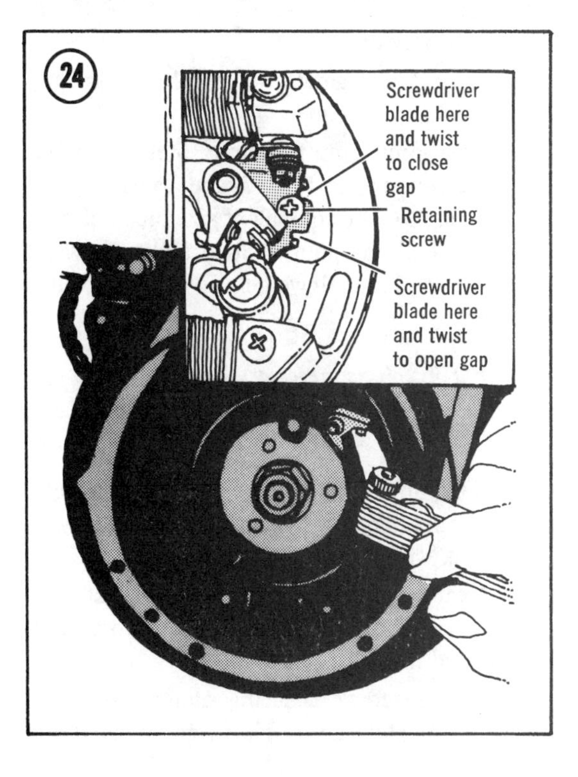

7. Install dial timing gauge in No. 2 spark plug hole and repeat procedure for No. 2 breaker points.

Kohler Engine Breaker Point Adjustment and Timing Adjustment

Refer to list of general tune-up hints as outlined under *Engine Tune-up*. Refer to **Figure 25** for this procedure.

1. Remove recoil starter.
2. Remove fan belt pulley and belt.
3. Remove spark plugs.

> NOTE: *Breaker point and timing adjustment can be performed with flywheel removed if cam is installed over crankshaft.*

4. Install dial indicator type timing gauge with a one inch adapter in No. 1 cylinder on the power take-off (PTO) side. Refer to Chapter One for a description of dial indicator type timing gauge.
5. Rotate engine until No. 1 piston is at TDC (top dead center). Set dial indicator to "zero."
6. Loosen screw securing No. 1 set of breaker points (lower set). Carefully examine points and replace if necessary.
7. Using a feeler gauge, set breaker point gap to 0.016 in. (0.41mm).
8. Move point cam flyweight (behind larger flywheel window) to full advance position. Install special tool or equivalent size piece of wire through small hole immediately above flywheel window to hold flyweight in full advance position.

> NOTE: *It is necessary to hold flyweight in full advance position or incorrect timing will result.*

9. Using paper clips in wire coupler, connect a battery operated test light between brown wire and black wire for No. 1 cylinder.
10. Rotate flywheel clockwise until test light brightens (points just start closing). Note reading on dial indicator.
11. Loosen screws securing stator plate and move stator plate until light brightens at 0.090 ± 0.005 in. (2.29 ± 0.13mm) indication on dial indicator. Tighten stator plate and recheck setting. Readjust if necessary.
12. Install dial indicator in No. 2 spark plug hole.
13. Move test light lead to black wire for No. 2 cylinder. Leave other lead connected to the brown wire.
14. Rotate flywheel counterclockwise. Test light should brighten at 0.090 ± 0.005 in. (2.29 ± 0.13mm) indication on dial indicator. If indication is not as specified, adjust No. 2 breaker point gap (upper set) to 0.014-0.018 in. (0.36-0.46mm) until specified indication is achieved.

> NOTE: Do not *move stator plate when adjusting No. 2 breaker points or No. 1 point setting will be incorrect. Both pistons should be within 0.005 in. (0.13mm) of each other when adjustment is completed.*

15. Install fan belt, pulley, and recoil starter.

C.C.W. and Kioritz Engines Magneto Ignition Timing Adjustment

1. Perform breaker point adjustment.
2. Install dial indicator type timing gauge with a one inch adapter in No. 1 spark plug hole (flywheel side). See **Figure 26**.
3. Rotate crankshaft clockwise until No. 1 piston is at TDC. Set dial indicator to "zero."
4. Disconnect red and white ignition leads between coupler and ignition coils. Connect one lead of a battery powered test light to white lead and other lead to ground (**Figure 27**).

> NOTE: *If engine is not removed from snowmobile, disconnect large white connector from engine to wiring harness.*

5. Rotate crankshaft clockwise until dial indicator indicates proper timing position as specified in **Table 4**. Breaker points should just start to open. If points do not just start to open, loosen screws securing stator plate and move stator plate until test light dims. Tighten stator plate screws (**Figure 28**).

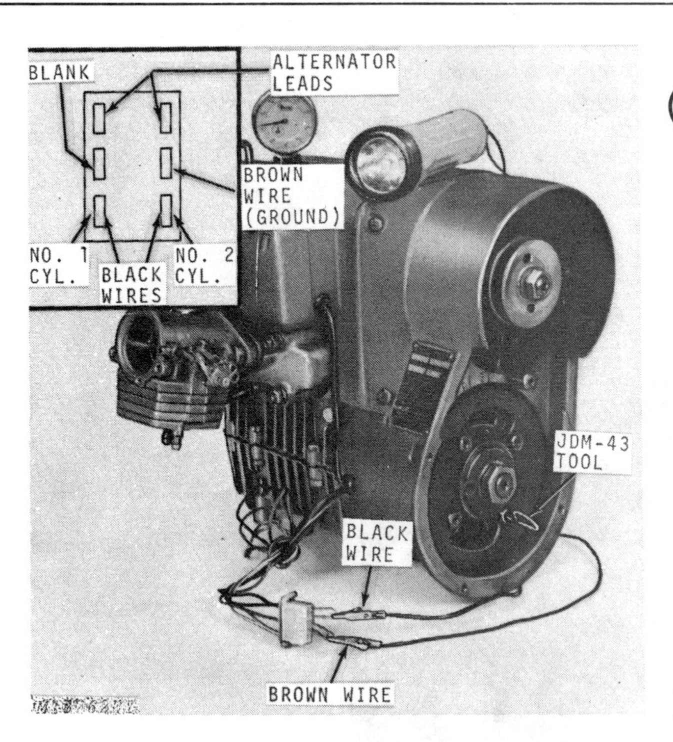
BLANK
ALTERNATOR LEADS
BROWN WIRE (GROUND)
NO. 1 CYL.
BLACK WIRES
NO. 2 CYL.
JDM-43 TOOL
BLACK WIRE
BROWN WIRE

25

DIAL INDICATOR
"ZERO" NEEDLE AT TDC

26

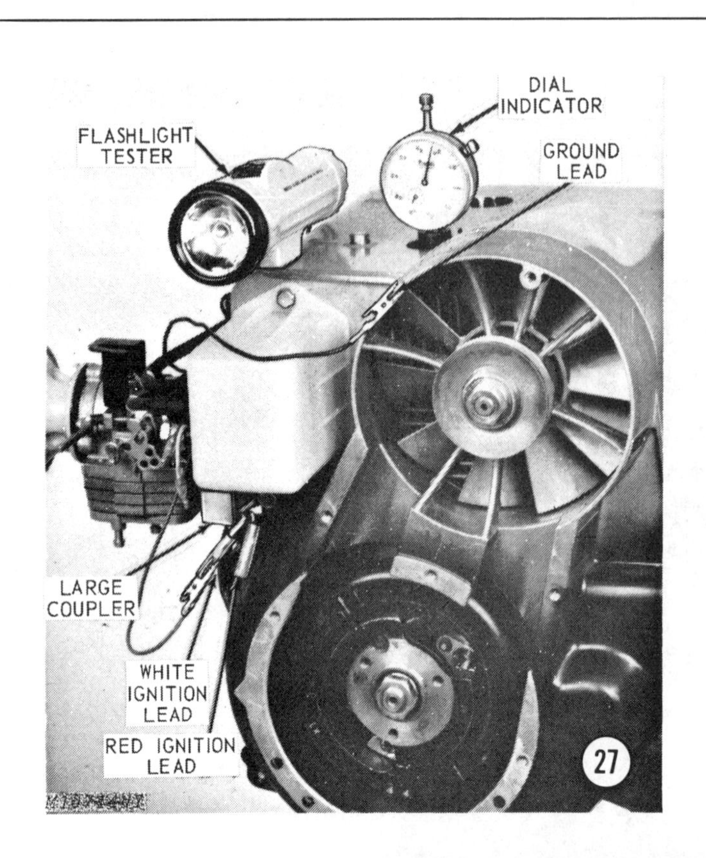

NOTE: *Loosening and tightening stator plate screws can affect breaker point adjustment. Recheck point gap. Timing dimension must be within tolerance specified in* **Table 4**. *The higher side of tolerance (more advanced) will give slightly better performance in higher altitudes.*

Table 4 IGNITION SPECIFICATIONS

Spark plug gap	0.020 in. (0.51mm)
Breaker point gap	
CCW & Kioritz engines	0.014 ± 0.002 in. (0.356 ± 0.050mm)
Kohler engines	0.016 ± 0.002 in. (0.406 ± 0.05mm)
Magneto ignition timing	
CCW engines (snowmobile serial No. up to 20,000)	0.023 ± 0.005 in. BTDC (0.584 ± 0.127mm BTDC)
CCW engines (snowmobile serial No. 20,001-30,000)	
Reed valve engines	0.009 ± 0.003 in. BTDC (0.229 ± 0.076mm BTDC)
Piston port engines	0.015 ± 0.002 in. BTDC (0.381 ± 0.050mm BTDC)
Kioritz engines (snowmobile serial No. 30,001 and subsequent)	
Reed valve engines	0.009 ± 0.003 in. BTDC (0.229 ± 0.076mm BTDC)
Piston port engines	0.015 ± 0.002 in. BTDC (0.381 ± 0.050mm BTDC)
Kohler engines	0.090 ± 0.005 in. BTDC (2.286 ± 0.127mm BTDC) —with cam in advanced position
CDI timing	
JD295/S	0.118 in. (2.99mm) BTDC @ 4,000 rpm
JD340/S	0.090 in. (2.28mm) BTDC @ 4,000 rpm
340 Liquidator	0.065 in. (1.65mm) BTDC @ idle
Cyclone and Liquifire	0.096 in. (2.44mm) BTDC @ 4,000 rpm

6. Connect test light lead to red lead and repeat procedure for No. 2 cylinder.

> NOTE: Do not *move stator plate when timing No. 2 cylinder or No. 1 setting will be incorrect.*

7. Readjust No. 2 cylinder breaker points, if necessary, to get test light to dim just as points open with No. 2 piston at specified position.

> NOTE: *Closing breaker point gap retards timing and opening gap advances ignition timing. For smoothest engine operation, time both cylinders to exactly the same dimension as specified in* **Table 4**.

8. Install window plate fan belt pulley, recoil starter cup, and recoil starter. Install spark plugs and reconnect ignition wires (red to red, and white to white). Run engine and check for proper operation. Perform carburetor adjustments.

CDI Ignition Timing

1. Remove recoil starter.
2. Using a small piece of wire, bend a pointer and secure to flywheel housing. On some models it may be easier to use a felt marker and place a reference mark on the flywheel housing.
3. Remove No. 1 spark plug and install a dial indicator type timing gauge in No. 1 spark plug hole (**Figure 29**). Refer to Chapter One for a description of dial indicator type timing gauge.
4. Rotate crankshaft until piston is at TDC (top dead center). Set dial indicator to "zero."
5. Rotate crankshaft counterclockwise (opposite of normal rotation) until specified timing BTDC (before top dead center) is reached. See **Table 4**.
6. Place a mark on flywheel opposite pointer or earlier placed mark on flywheel housing.
7. Remove dial indicator and install spark plug.
8. Connect a timing light to No. 1 cylinder spark plug wire as shown in **Figure 30**.

A. Dial indicator
B. Brass pointer
C. Mark on flywheel sheave

A. Timing light
B. Clamp on pick-up
C. Pointer
D. Mark on sheave

NOTE: *If a* DC *powered timing light is used, it must be connected to a 12 volt battery.*

9. Raise rear of snowmobile enough to clear track of the ground. Start engine and run at approximately 4,000 rpm.

10. Point timing light at marks on flywheel and flywheel housing. Marks should line up if timing is correct. If timing marks are not aligned, perform the following:

   a. Remove flywheel housing, flywheel, and alternator/stator.

   b. Loosen 4 screws securing timing ring. Rotate timing ring clockwise to advance timing and counterclockwise to retard timing. Tighten timing ring retaining screws.

11. Install alternator/stator, flywheel housing, and flywheel. Recheck timing and readjust if necessary.

Throttle Cable Adjustment (Walbro and Bendix Carburetors)

Remove windshield and console. Adjust throttle cable so throttle lever on carburetor is fully open when the throttle control on the handlebar is in the wide-open throttle position.

Choke Cable Adjustment (Walbro and Bendix Carburetors)

Remove windshield and console. Adjust choke cable so that choke lever on carburetor is fully open when the choke button on the instrument panel is pushed in all the way.

Walbro Carburetor Mixture and Idle Speed Adjustment

Refer to **Figure 31** for this procedure.

1. Remove console and windshield if not previously removed. Set high-speed mixture needle ⅞ turn open.

2. Set idle mixture needle one turn open.

NOTE: *On 800 and JDX8 models Serial No. 30,001 and subsequent, open high-speed and idle mixture needles 1⅛ turns.*

CAUTION

Do not attempt to adjust carburetor without intake silencer installed. A too lean fuel mixture with subsequent serious engine damage may result.

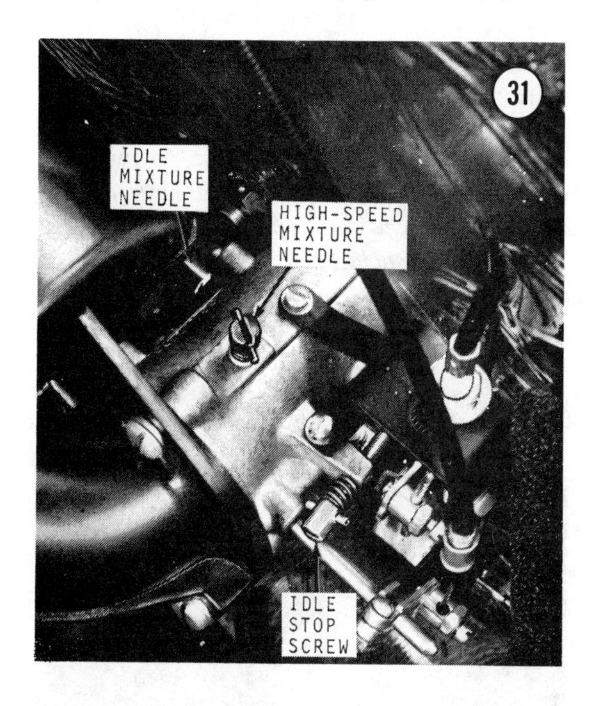

3. Block up rear of snowmobile so track is clear of the ground.

4. Start and warm up engine. Adjust idle mixture needle for highest engine rpm without opening the throttle.

5. Adjust idle stop screw for an idle of 2,200-2,600 rpm.

6. Make several short trial runs with snowmobile for final fine tuning. Adjust high-speed needle for maximum engine performance, then back out high-speed needle additional ⅛ turn.

CAUTION

A too lean fuel mixture can cause serious engine damage. On reed valve engines, mixture adjustment must never be less than ⅞ turn open.

7. Install console and windshield.

Bendix Carburetor Mixture and Idle Speed Adjustment

Refer to **Figure 32** for this procedure.

1. Remove console and windshield if not previously removed.

2. Screw in low-and high-speed needles until they are *lightly* seated. *Do not* force or needles may be damaged. Back high-speed needle 1½ turns and low-speed needle 1¾ turns.

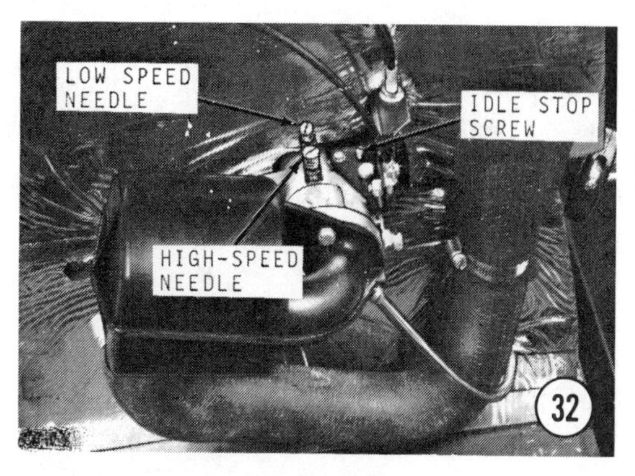

3. Start engine and warm up by driving snowmobile. Adjust low-speed needle slowly clockwise until engine stumbles (loses rpm) then back out needle ¼ turn.

4. Check that idle speed is 2,200-2,600 rpm. Adjust idle stop screw, if necessary, for proper idle speed. If idle stop screw is adjusted, it is necessary to readjust low-speed needle as performed in Step 3.

5. Start engine and make several high speed trial runs. Turn high-speed needle ⅛ turn in or out on each run until optimum performance is achieved. After maximum performance is obtained, back out needle ¼ turn.

CAUTION

Never attempt to adjust needle with track off the ground. The engine must be adjusted in a "loaded" condition or a too lean fuel mixture with subsequent serious engine damage will result.

NOTE: *Above 5,000 feet altitude, turn high-speed needle one turn open.*

6. Recheck machine performance. If engine stumbles or hesitates on acceleration, it may be necessary to lean or enrichen low-speed needle slightly.

7. Install console and windshield.

Mikuni Carburetor Starter (Choke) Adjustment

The starting (choke) system on Mikuni carburetors is controlled by a starter plunger with separate metering of fuel/air mixture through independent jets. When the engine is started, the throttle valve must be closed or the starting mixture will be made too lean for engine starting.

1. Place starter lever on instrument panel in the down position. Lever should have slight free play.

2. Look through starter plunger air hole (3 o'clock position in the carburetor bore). Check that starter plunger is all the way down in its bore (**Figure 33**).

3. To adjust starter plunger, perform the following:

   a. Loosen jam nut securing adjusting nut (**Figure 33**).

   b. Rotate adjusting nut clockwise to lower starter plunger down in its bore. Tighten jam nut.

NOTE: *If starter plunger is not fully down in its bore, the carburetor will run rich and affect entire engine performance level.*

Mikuni Carburetor Adjustment and Synchronization

This procedure includes throttle cable adjustments and idle speed adjustments for all models equipped with Mikuni carburetors.

On models equipped with 2 carburetors, more precise synchronization can be achieved with an air flow meter as described in Chapter One. If such a device is available, perform the following procedure as a preliminary adjustment and proceed to *Mikuni Carburetor Air Flow Meter Synchronziation* for the final fine tuning.

Refer to **Figure 34** for this procedure.

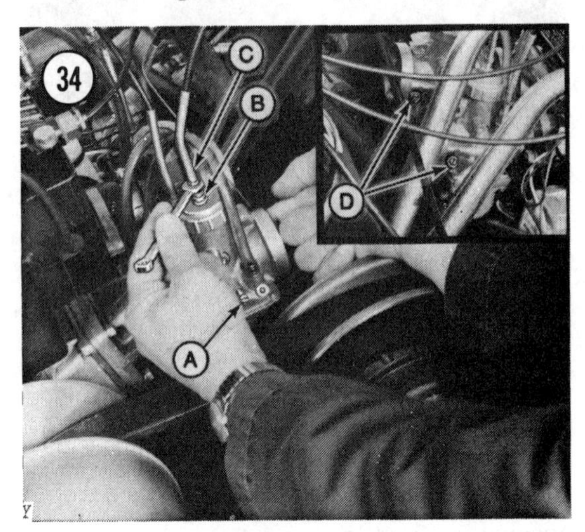

A. Throttle stop screw
B. Jam nut
C. Adjusting sleeve
D. Pilot air screw

1. Remove air intake silencer.

2. Use a strong rubber band and clamp throttle lever to handlebar grip in the wide-open-throttle position.

3. Loosen jam nut securing adjusting sleeve. Feel inside carburetor bore and turn adjusting sleeve until cut-out portion of throttle valve is flush with inside of carburetor bore.

4. Turn adjusting sleeve counterclockwise the required number of additional turns to position the backside of the throttle valve flush with the carburetor bore.

   a. 340 Cyclone and 440 Liquifire—4 turns

   b. 340 Liquifire, JD340/S, JD295/S—3½ turns

   c. 440 Cyclone and 340 Liquidator—5 turns

NOTE: *The specified additional turns on the adjusting sleeve should position the throttle valve flush with or slightly above the carburetor bore. If any part of the throttle slide protrudes into the carburetor bore, turn adjusting sleeve until throttle slide is flush.*

5. Rotate throttle stop screw counterclockwise until the tip is flush with inside of carburetor bore.

6. Remove rubber band clamp from handlebar and allow throttle to return to idle position.

7. Turn in throttle stop screw until tip just contacts throttle slide valve. Turn in stop screw 2 additional turns for a preliminary idle setting.

8. Slowly operate throttle lever on handlebar and observe that throttle valve begins to rise. On models with 2 carburetors, ensure that throttle valves move an equal amount together. Readjust throttle cables if necessary.

9. Slowly turn in pilot air screw until light seating is felt. *Do not* force or air screw may be damaged. Back out pilot air screw number of turns specified in **Table 5**.

10. Install air intake silencer and start engine. Warm up engine to operating temperature and check that idle speed is 1,800-2,400 rpm. Adjust throttle stop screw as necessary for specified idle speed. On 2 carburetor models, ensure that both throttle stop screws are adjusted an equal amount.

NOTE: Do not *use pilot air screws to attempt to set engine idle speed. Pilot air screws must be set as specified in* **Table 5**.

Mikuni Carburetor Air Flow Meter Synchronization

To obtain a precise synchronization of twin carburetor models, use an air flow meter device as described in Chapter One. Perform *Mikuni Carburetor Adjustment and Synchronization* to obtain proper preliminary adjustments.

Refer to **Figure 35** for this procedure.

WARNING

The following procedure is performed with the engine running. Ensure that arms and clothing are clear of drive belt or serious injury may result.

1. Raise and support rear of snowmobile so track is clear of the ground.

2. Start engine. Wedge in throttle lever to maintain engine speed at 4,000 rpm.

Table 5 CARBURETOR TUNING SPECIFICATIONS

Idle speed		
All models with Walbro and Bendix carburetors	2,200-2,600 rpm	
All models with Mikuni carburetors	1,800-2,400 rpm (3,000 rpm on 340 Cyclone and Liquifire models at 8,000-12,000 ft. altitude)	
Air screw turns (Mikuni carburetors)		
340 Liquidator	1½	
JD295/S	1	
JD340/S	1⅛	
440 Liquifire (all models)	1½	
(Serial No. 50,001-70,000)	Sea level to 2,000 ft.	8,000 to 12,000 ft.
340 Cyclone	1	2
440 Cyclone	1	1
340 Liquifire	2	1
(Serial No. 70,001 and subsequent)		
340 Cyclone	2	2
440 Cyclone	2	1½
340 Liquifire	1	½

A. Idle adjusting screw
B. JDM-64-2 air flow meter
C. Tube in vertical position
D. Float
E. Air flow control

3. Open air flow control of air flow meter and place meter over right carburetor throat. Tube on meter must be vertical.

4. Slowly close air flow control until float in tube aligns with a graduated mark on tube.

5. Without changing adjustment of air flow control, place air flow meter on left carburetor. If carburetors are equal, no adjustment is necessary.

6. If adjustment is necessary, loosen jam nut on carburetor with lowest float level and turn adjusting sleeve until float level matches other carburetor.

7. Return engine to idle and repeat Steps 3, 4, and 5. Adjust throttle stop screws as necessary for a balanced idle.

Mikuni Carburetor Main Jet Selection

The main jet controls the fuel metering when the carburetor is operating in the ½ to full throttle range. Since temperature and altitude affect the air density, each snowmobile owner will have to perform the following trial and error method of jet selection to obtain peak engine efficiency and performance for his own particular area of operation.

CAUTION
Air intake silencer must be installed during the following procedure or a "too lean" mixture may result. A "too lean" fuel mixture can cause engine overheating and subsequent serious damage.

NOTE: *Snowmobile must be operated on a flat well-packed area for best results.*

1. Operate machine at wide open throttle for several minutes. If peak rpm cannot be achieved or engine appears to be laboring, main jet needs to be changed.

2. Make another trial run and shut off ignition while throttle is still wide open. Examine the exhaust and spark plugs to determine if mixture is too rich or too lean. Mixture is too rich if exhaust manifold or spark plug insulator is dark brown or black. Refer to *Spark Plugs* in this chapter. Decrease jet size if mixture is too rich.

NOTE: *Change jet sizes* one *increment at a time and test after each change to obtain best results.*

If manifold or spark plug insulator is a very light color, mixture is too lean. Correct by increasing jet size.

3. If state of fuel/air mixture cannot be determined by color of exhaust manifold or spark plug insulator, assume mixture is too lean and increase jet size. If operation improves, continue increasing jet size until maximum performance is achieved. If operation gets worse, decrease jet size until best results are obtained.

OFF-SEASON STORAGE

Proper storage techniques are essential to help maintain your snowmobile's life and usefulness. The off-season is also an excellent time to perform any maintenance and repair tasks that are necessary.

Placing in Storage

1. Use soap and water to thoroughly clean the exterior of your snowmobile. Use a hose to remove rocks, dirt, and debris from the track area. Clean all dirt and debris from the hood and console areas.

CAUTION

Do not spray water around the carburetor or engine. Be sure you allow sufficient time for all components to dry.

2. Use a good automotive type cleaner wax and polish the hood, pan, and tunnel. Use a suitable type of upholstery cleaner on the seat. Touch up any scratched or bare metal parts with paint. Paint or oil the skis to prevent rust.

3. Drain the fuel tank. Start the engine and run it at idle to burn off all fuel left in the carburetor. Check the fuel filter and replace if contaminated.

4. Wrap up carburetor(s) and intake manifold in plastic and tie securely.

5. Remove spark plugs and add a teaspoon of snowmobile oil to each cylinder. Pull the engine over several times with the starter rope to spread the oil over the cylinder walls. Replace the spark plugs.

6. Remove the drive belt. Apply a film of light grease to drive and driven sheaves to prevent rust and corrosion.

7. Change chaincase oil.

8. Raise rear of snowmobile off the ground. Loosen the track adjusting screws to remove any tension on the track.

9. Carefully examine all components and assemblies. Make a note of immediate and future maintenance and repair items and order the necessary parts. Perform *Hardware and Component Tightness Check.*

10. Cover snowmobile and store inside if possible.

Removing From Storage

1. Perform *Hardware and Component Tightness Check.*

2. Remove grease from the drive and driven sheaves and install the drive belt.

3. Fill the fuel tank with new gasoline/oil mixture. Refer to Chapter One.

4. Check throttle and brake controls for proper operation and adjust if necessary.

5. Adjust the track to proper tension.

6. Familiarize yourself with all safety and operating instructions.

7. Start the engine and check the operation of the emergency stop switch and "tether" switch. Check that all lights and switches operate properly. Replace any burned out bulbs.

8. Start out slowly on short rides until you are sure your machine is operating properly and is dependable.

CHAPTER THREE

TROUBLESHOOTING

Diagnosing snowmobile ills is relatively simple if you use orderly procedures and keep a few basic principles in mind.

Never assume anything. Do not overlook the obvious. If you are riding along and the snowmobile suddenly quits, check the easiest most accessible problem spots first. Is there gasoline in the tank? Has a spark plug wire fallen off? Check the ignition switch. Maybe that last mogul caused you to accidentally switch the emergency switch to OFF or pull the emergency stop "tether" string.

If nothing obvious turns up in a cursory check, look a little further. Learning to recognize and describe symptoms will make repairs easier for you or a mechanic at the shop. Describe problems accurately and fully. Saying that "it won't run" isn't the same as saying "it quit at high speed and wouldn't start," or that "it sat in my garage for 3 months and then wouldn't start."

Gather as many symptoms together as possible to aid in diagnosis. Note whether the engine lost power gradually or all at once, what color smoke (if any) came from the exhaust, and so on. Remember that the more complicated a machine is, the easier it is to troubleshoot because symptoms point to specific problems.

You do not need fancy equipment or complicated test gear to determine whether repairs can be attempted at home. A few simple checks could save a large repair bill and time lost while the snowmobile sits in a dealer's service department. On the other hand, be realistic and do not attempt repairs beyond your abilities. Service departments tend to charge heavily for putting together disassembled components that may have been abused. Some will not even take on such a job—so use common sense; do not get in over your head.

OPERATING REQUIREMENTS

An engine needs three basics to run properly: correct gas/air mixture, compression, and a spark at the right time. If one or more are missing, the engine will not run. The electrical system is the weakest link of the three. More problems result from electrical breakdowns than from any other source; keep this in mind before you begin tampering with carburetor adjustments.

If the snowmobile has been sitting for any length of time and refuses to start, check the battery (if the machine is so equipped) for a charged condition first, and then look to the gasoline delivery system. This includes the tank, fuel petcocks, lines, and the carburetor. Rust may have formed in the tank, obstructing fuel flow. Gasoline deposits may have gummed up carburetor jets and air passages. Gasoline tends to lose its potency after standing for long periods. Condensation may contaminate it with water. Drain old gas and try starting with a fresh tankful.

Compression, or the lack of it, usually enters the picture only in the case of older machines. Worn or broken pistons, rings, and cylinder bores could prevent starting. Commonly, a gradual power loss and harder and harder starting will be readily apparent in this case.

PRINCIPLES OF 2-CYCLE ENGINES

The following is a general discussion of a typical 2-cycle piston-ported engine. The same principles apply to reed valve engines except that during the intake cycle, the fuel/air mixture passes through a reed valve assembly into the crankcase. During this discussion, assume that the crankshaft is rotating counterclockwise.

In **Figure 1**, as the piston travels downward, a scavenging port (A) between the crankcase and the cylinder is uncovered. Exhaust gases leave the cylinder through the exhaust port (B), which is also opened by downward movement of the piston. A fresh fuel/air charge, which has previously been compressed slightly, travels from the crankcase (C) to the cylinder through scavenging port (A) as the port opens. Since the incoming charge is under pressure, it rushes into the cylinder quickly and helps to expel exhaust gases from the previous cycle.

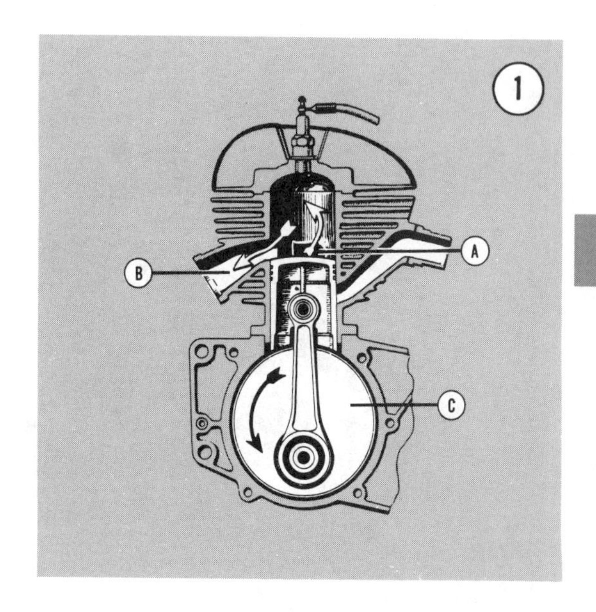

Figure 2 illustrates the next phase of the cycle. As the crankshaft continues to rotate, the piston moves upward, closing the exhaust and scavenging ports. As the piston continues upward, the air/fuel mixture in the cylinder is compressed. Notice also that a low pressure area is created in the crankcase at the same time. Further upward movement of the piston uncovers intake port (D). A fresh fuel/air charge is then drawn into the crankcase through the intake port because of the low pressure created by the upward piston movement.

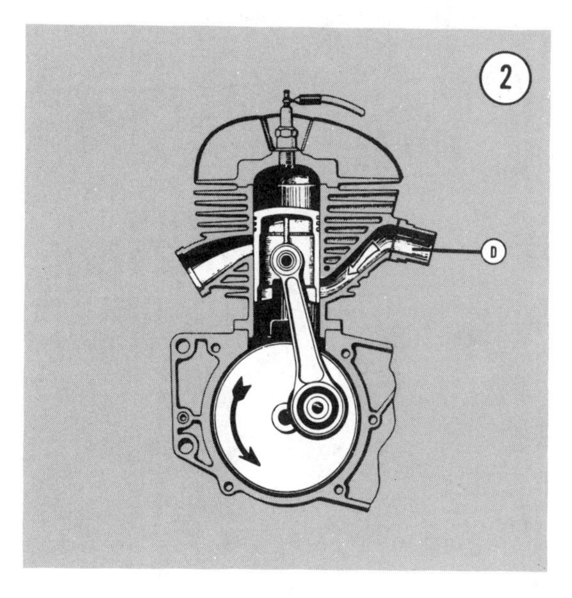

The third phase is shown in **Figure 3**. As the piston approaches top dead center, the spark plug fires, igniting the compressed mixture. The piston is then driven downward by the expanding gases.

When the top of the piston uncovers the exhaust port, the fourth phase begins, as shown in **Figure 4**. The exhaust gases leave the cylinder through the exhaust port. As the piston continues downward, the intake port is closed and the mixture in the crankcase is compressed in preparation for the next cycle. Every downward stroke of the piston is a power stroke.

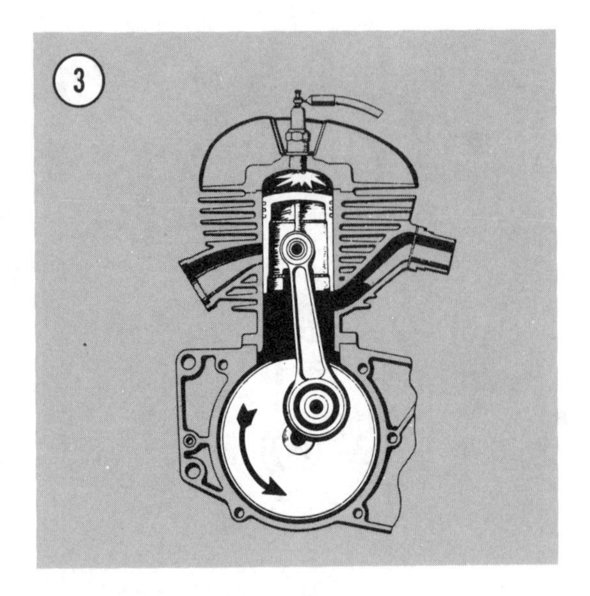

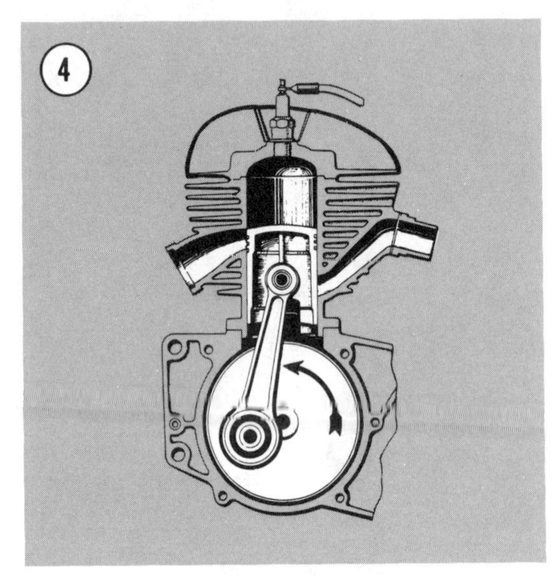

ENGINE STARTING

An engine that refuses to start or is difficult to start can try anyone's patience. More often than not, the problem is very minor and can be found with a simple and logical troubleshooting approach.

The following items provide a beginning point from which to isolate an engine starting problem.

Engine Fails to Start

Perform the following spark test to determine if the ignition system is operating properly.

1. Remove a spark plug.
2. Connect spark plug connector to spark plug and clamp base of spark plug to a good grounding point on the engine. A large alligator clip makes an ideal clamp. Position spark plug so you can observe the electrode.
3. Turn on ignition and crank engine over. A fat blue spark should be evident across spark plug electrode.

WARNING

On machines equipped with CDI *(capacitor discharge ignition), do not hold spark plug, wire, or connector or a serious electrical shock may result.*

4. If spark is good, check for one or more of the following possible malfunctions:
   a. Fouled or defective spark plugs
   b. Obstructed fuel filter or fuel line
   c. Defective fuel pump
   d. Leaking head gasket (*see Compression Test*)
5. If spark is not good, check for one or more of the following:
   a. Burned, pitted, or improperly gapped breaker points
   b. Weak ignition coil or condenser
   c. Loose electrical connections
   d. Defective CDI components—have CDI system checked by an authorized dealer

Engine Difficult to Start

Check for one or more of the following possible malfunctions:

a. Fouled spark plugs
b. Improperly adjusted choke
c. Defective or improperly adjusted breaker points
d. Contaminated fuel system
e. Improperly adjusted carburetor
f. Weak ignition coil
g. Incorrect fuel mixture
h. Defective reed valve
i. Crankcase drain plugs loose or missing
j. Poor compression (see *Compression Test*)

Engine Will Not Crank

Check for one or more of the following possible malfunctions:

a. Defective recoil starter
b. Seized piston
c. Seized crankshaft bearings
d. Broken connecting rod

Compression Test

Perform compression test to determine condition of piston ring sealing qualities, piston wear, and condition of head gasket seal.

1. Remove spark plugs. Insert a compression gauge in one spark plug hole (**Figure 5**). Refer to Chapter One for a suitable type of compression tester.

2. Crank engine vigorously and record compression reading. Repeat for other cylinder. Compression readings should be from 120-175 psi (8.4-12.30 kg/cm²). Maximum allowable variation between cylinders is 10 psi (0.70 kg/cm²).

3. If compression is low or variance between cylinders is excessive, check for defective head gaskets, damaged cylinders and pistons, or stuck piston rings.

ENGINE PERFORMANCE

In the following discussion, it is assumed that the engine runs, but is not operating at peak efficiency. This will serve as a starting point from which to isolate a performance malfunction.

The possible causes for each malfunction are listed in a logical sequence and in order of probability.

Engine Will Not Idle

a. Carburetor incorrectly adjusted
b. Fouled or improperly gapped spark plugs
c. Head gasket leaking—perform compression test
d. Fuel mixture incorrect
e. Spark advance mechanism not retarding
f. Obstructed fuel pump impulse tube
g. Crankcase drain plugs loose or missing

Engine Misses at High Speed

a. Fouled or improperly gapped spark plugs
b. Defective or improperly gapped breaker points
c. Improper ignition timing
d. Defective fuel pump
e. Improper carburetor high-speed adjustment (Walbro and Bendix carburetors) or improper main jet selection (Mikuni carburetor)
f. Weak ignition coil
g. Obstructed fuel pump impulse tube

Engine Overheating

a. Too lean fuel mixture—incorrect carburetor adjustment or jet selection
b. Improper ignition timing
c. Incorrect spark plug heat range
d. Intake system or crankcase air leak
e. Cooling fan belt or coolant pump drive belt broken or slipping
f. Cooling fan or coolant pump defective
g. Leak in liquid cooling system
h. Damaged or blocked cooling fins

Smoky Exhaust and Engine Runs Rough

a. Carburetor adjusted incorrectly—mixture too rich
b. Incorrect fuel/oil mixture

c. Choke not operating properly
d. Obstructed muffler
e. Water or other contaminants in fuel

Engine Loses Power

a. Carburetor incorrectly adjusted
b. Engine overheating
c. Defective or improperly gapped breaker points
d. Improper ignition timing
e. Incorrectly gapped spark plugs
f. Weak ignition coil
g. Obstructed muffler
h. Defective reed valve

Engine Lacks Acceleration

a. Carburetor mixture too lean
b. Defective fuel pump
c. Incorrect fuel/oil mixture
d. Defective or improperly gapped breaker points
e. Improper ignition timing
f. Defective reed valve

ENGINE FAILURE ANALYSIS

Overheating is the major cause of serious and expensive engine failures. It is important that each snowmobile owner understands all the causes of engine overheating and takes the necessary precautions to avoid expensive overheating damage. Proper preventive maintenance and careful attention to all potential problem areas can often eliminate a serious malfunction before it happens.

Fuel

All John Deere snowmobile engines rely on a proper fuel/oil mixture for engine lubrication. Always use an approved oil and mix the fuel carefully as described in Chapter One.

Gasoline must be of sufficiently high octane (90 or higher) to avoid "knocking" and "detonation."

Fuel/Air Mixture

Fuel/air mixture is determined by carburetor adjustment (Walbro and Bendix) or main jet selection (Mikuni). Always adjust carburetors carefully and pay particular attention to avoiding a "too lean" mixture.

Heat

Excessive external heat on the engine can be caused by the following:

a. Hood louvers plugged with snow
b. Damaged or plugged cylinder and head cooling fins
c. Slipping or broken fan or coolant pump belt
d. Damaged cooling fan or coolant pump
e. Operating snowmobile in hot weather
f. Plugged or restricted exhaust system
g. Liquid cooling system low on coolant
h. Defective thermostat in liquid cooling system

See **Figures 6 and 7** for examples of cylinder and piston scuffing caused by excessive heat.

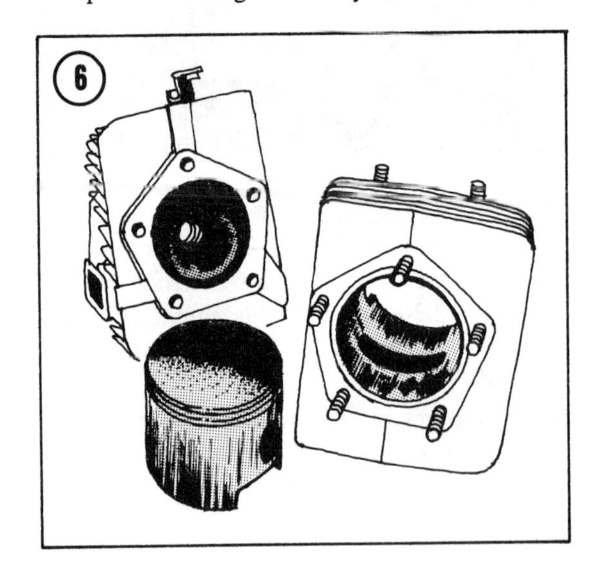

Dirt

Dirt is a potential problem for all snowmobiles. The air intake silencers on all models are not designed to filter incoming air. Avoid running snowmobiles in areas that are not completely snow covered.

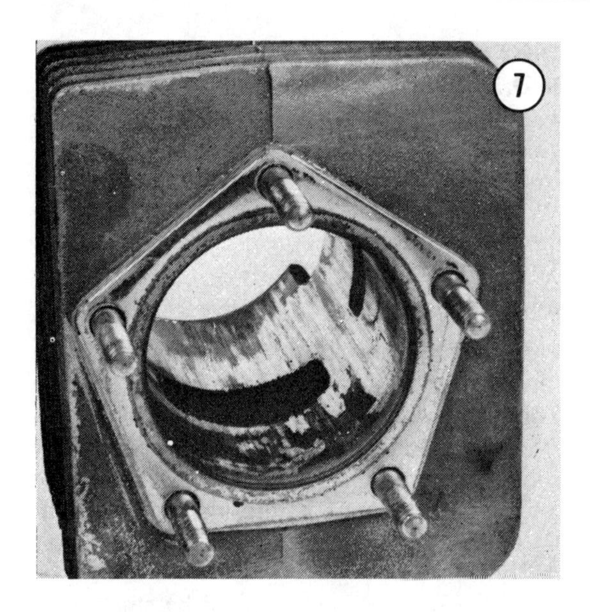

Ignition Timing

Ignition timing that is too far advanced can cause "knocking" or "detonation." Timing that is too retarded causes excessive heat buildup in the cylinder exhaust port areas.

Spark Plugs

Spark plugs must be of a correct heat range. Too hot a heat range can cause preignition and detonation which can ultimately result in piston burn-through as shown in **Figure 8**.

Refer to Chapter Two for recommended spark plugs.

Preignition

Preignition is caused by excessive heat in the combustion chamber due to a spark plug of improper heat range and/or too lean a fuel mixture. See **Figure 9** for an example of a melted and scuffed piston caused by preignition.

Detonation (Knocking)

Knocking is caused by a too-lean fuel mixture and/or too-low octane fuel.

ELECTRICAL SYSTEM

The following items provide a starting point from which to troubleshoot electrical system malfunctions. The possible causes for each malfunction are listed in a logical sequence and in order of probability.

Ignition system malfunctions are outlined under *Engine Starting* and *Engine Performance,* covered earlier.

Lights Will Not Light

a. Bulbs are burned out
b. Loose electrical connections
c. Defective switch
d. Defective lighting coil or alternator
e. Defective voltage regulator

Bulbs Burn Out Rapidly

a. Incorrect bulb type
b. Defective voltage regulator

Lights Too Bright or Too Dim

a. Defective voltage regulator
b. Defective alternator

Discharged Battery

a. Defective battery
b. Low electrolyte level
c. Dirty or loose electrical connections
d. Defective voltage regulator
e. Defective lighting coil
f. Defective rectifier
g. Defective circuit breaker

Cracked Battery Case

a. Discharged battery allowed to freeze
b. Improperly installed hold-down clamp
c. Improperly attached battery cables

Starter Motor Does Not Operate

a. Loose electrical connections
b. Discharged battery
c. Defective starter solenoid
d. Defective starter motor
e. Defective circuit breaker
f. Defective ignition switch

Poor Starter Performance

a. Commutator or brushes worn, dirty, or oil soaked
b. Binding armature
c. Weak brush springs
d. Armature open, shorted, or grounded

POWER TRAIN

The following items provide a starting point from which to troubleshoot power train malfunctions. The possible causes for each malfunction are listed in a logical sequence and in order of probability. Also refer to *Drive Belt Wear Analysis,* later in this chapter.

Drive Belt Too Tight at Idle

a. Engine idle speed too fast
b. Distance between sheaves incorrect
c. Belt length incorrect

Drive Belt Not Operating Smoothly in Drive Sheave

a. Face of drive sheave is rough, grooved, pitted, or scored
b. Defective drive belt

Uneven Drive Belt Wear

a. Misaligned drive and driven sheaves
b. Loose engine mounts

Glazed Drive Belt

a. Excessive slippage
b. Oil on sheave surfaces

Drive Belt Worn Narrow in One Place

a. Excessive slippage caused by stuck track
b. Too high engine idle speed

Drive Belt Edge Cord Failure

a. Misaligned sheaves
b. Loose engine mounting bolts

Brake Not Holding Properly

a. Incorrect brake cable adjustment or air in hydraulic brake system
b. Brake lining or pucks worn
c. Oil saturated brake lining or pucks
d. Sheared key on brake pulley or disc

Brake Not Releasing Properly

a. Weak or broken return spring
b. Bent or damaged brake lever

Leaking Chaincase

a. Gaskets on drive shaft bearing flangettes or secondary shaft bearing flangettes damaged
b. Damaged O-ring on drive shaft or secondary shaft
c. Cracked or broken chaincase

Rapid Chain and Sprocket Wear

a. Insufficient chaincase oil
b. Misaligned sprockets
c. Broken chain tension blocks

DRIVE BELT WEAR ANALYSIS

Frayed Edge

A rapidly wearing drive belt with a frayed edge cord indicates the drive belt is misaligned. See **Figure 10**. Also, check for loose engine mounting bolts.

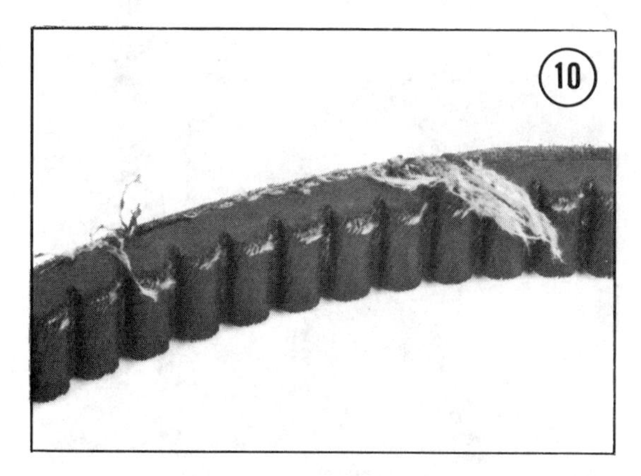

Worn Narrow in One Section

Excessive slippage due to a stuck track or too high an engine idle speed will cause the drive belt to be worn narrow in one section. See **Figure 11**.

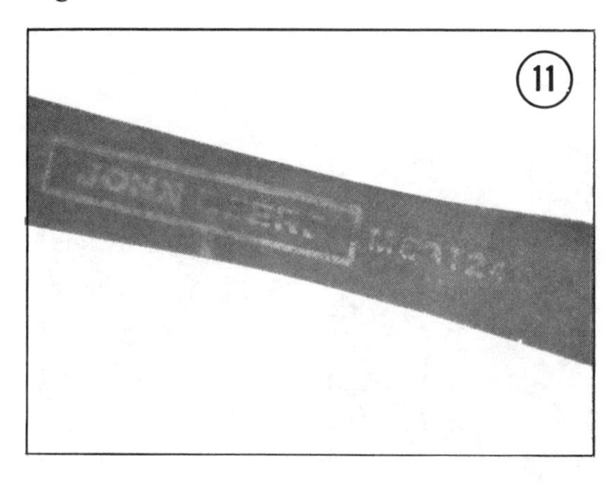

Belt Disintegration

Drive belt disintegration is usually caused by misalignment. Disintegration can also be caused by using an incorrect belt or oil on sheave surfaces. See **Figure 12**.

Sheared Cogs

Sheared cogs as shown in **Figure 13** are usually caused by violent drive sheave engagement.

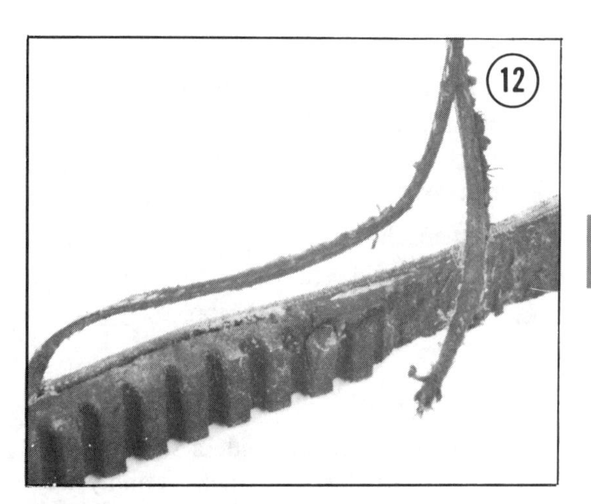

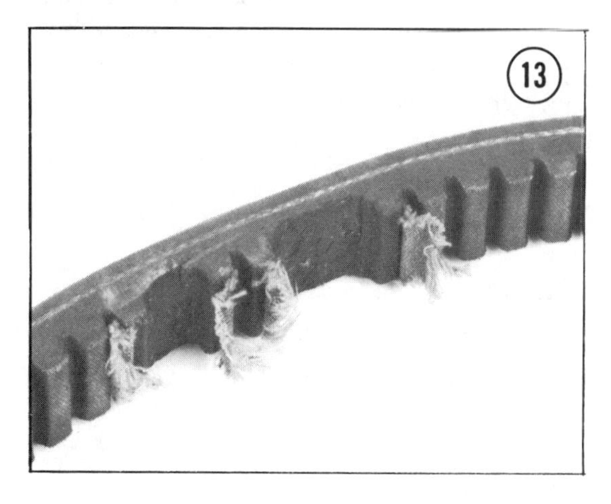

This is an indication of a defective or improperly installed drive sheave.

SKIS AND STEERING

The following items provide a starting point from which to troubleshoot ski and steering malfunctions. The possible causes for each malfunction are listed in a logical sequence and in order of probability.

Loose Steering

a. Loose steering post bushing (models to Serial No. 20,000)

b. Loose steering post or steering column cap screw (models to Serial No. 20,000)

c. Loose tie rod ends

d. Worn spindle bushings

e. Stripped spindle splines

Unequal Steering

a. Improperly adjusted tie rods
b. Improperly installed steering arms

Rapid Ski Wear

a. Skis misaligned
b. Worn out ski wear rods
c. Worn out spring wear plate

TRACK ASSEMBLY

The following items provide a starting point from which to troubleshoot track assembly malfunctions. The possible causes for each malfunction are listed in a logical sequence and in order of probability. Also refer to next section, *Track Wear Analysis.*

Frayed Track Edge

Track is misaligned.

Track Grooved on Inner Surface

a. Track too tight
b. Frozen bogie wheel(s)
c. Frozen rear idle shaft bearing

Track Drive Ratcheting

Track is too loose.

Rear Idlers Turning on Shaft

Rear idler shaft bearings are frozen.

Bogie Wheels Not Turning Freely

Bogie wheel bearing is defective.

Bogie Assemblies Not Pivoting Freely

Bogie tube and axle are bent.

TRACK WEAR ANALYSIS

The majority of track failures and abnormal wear patterns are caused by negligence, abuse, and poor maintenance. The following items illustrate typical examples. In all cases the damage could have been avoided by proper maintenance and good operator technique.

Obstruction Damage

Cuts, slashes, and gouges in the track surface are caused by hitting obstructions such as broken glass, sharp rocks, or buried steel. See **Figure 14**.

Worn Grouser Bars

Excessively worn grouser bars are caused by snowmobile operation over rough and non-snow covered terrain such as gravel roads and highway roadsides (**Figure 15**).

Lug Damage

Lug damage as shown in **Figure 16** is caused by lack of snow lubrication.

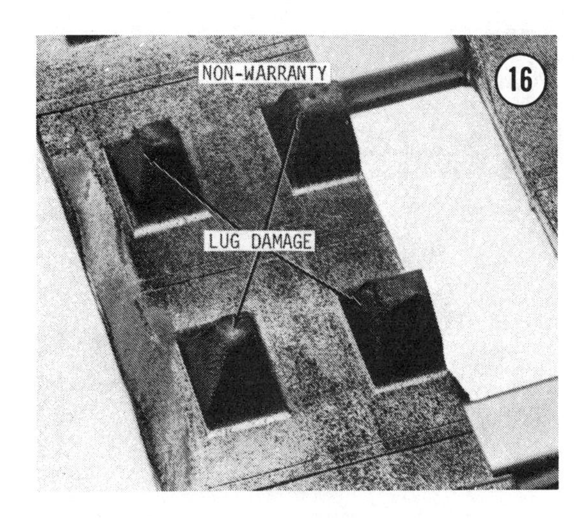

Ratcheting Damage

Insufficient track tension is a major cause of ratcheting damage to the top of the lugs. See **Figure 17**. Ratcheting damage can also be caused by too great a load and constant "jack-rabbit" starts.

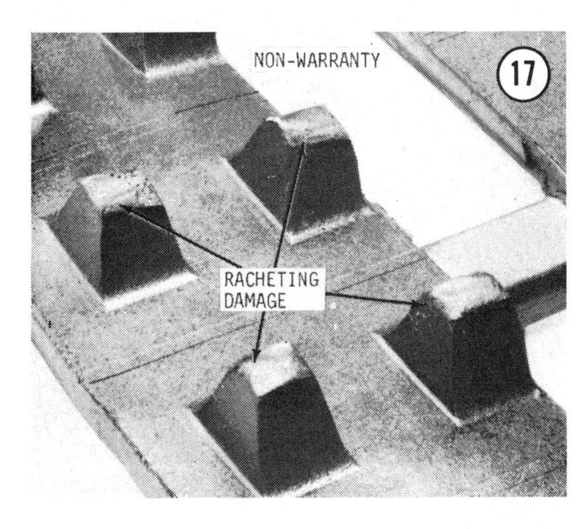

Overtension Damage

Excessive track tension can cause too much friction on the wear bars. This friction causes the wear bars to melt and adhere to the track grouser bars. See **Figure 18**. An indication of this condition is a "sticky" track that has a tendency to "lock up."

Loose Track Damage

A track adjusted too loosely can cause the outer edge to flex excessively. This results in

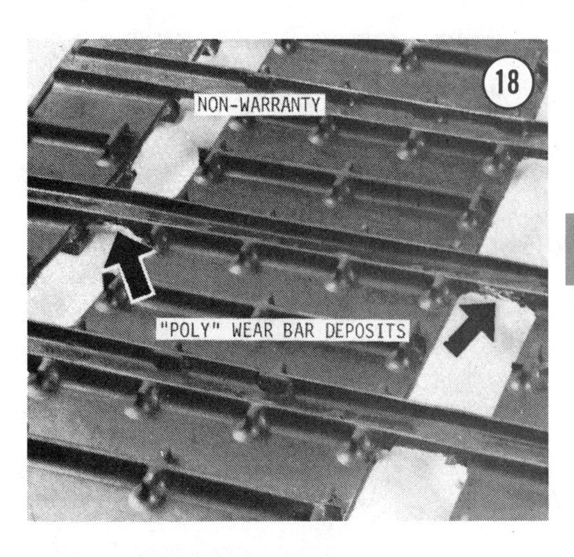

the type of damage shown in **Figure 19**. Excessive weight can also contribute to the damage.

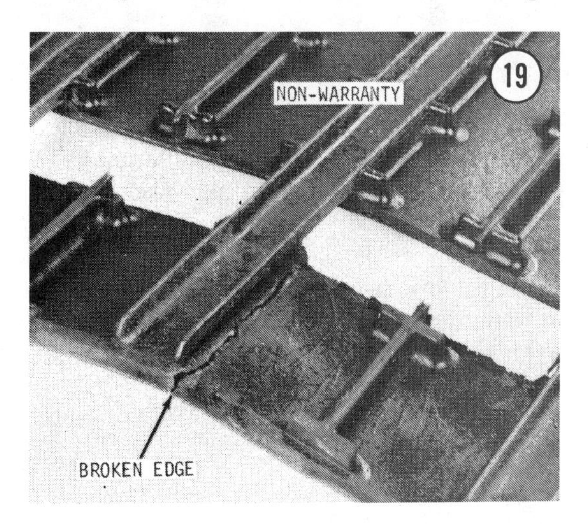

Impact Damage

Impact damage as shown in **Figure 20** causes the track rubber to open and expose the cord. This frequently happens in more than one place. Impact damage is usually caused by riding on rough or frozen ground or ice. Insufficient track tension can allow the track to pound against the track stabilizers inside the tunnel.

Edge Damage

Edge damage as shown in **Figure 21** is usually caused by tipping the snowmobile on its side to clear the track and allowing the track edge to contact an abrasive surface.

20
NON-WARRANTY
IMPACT DAMAGE

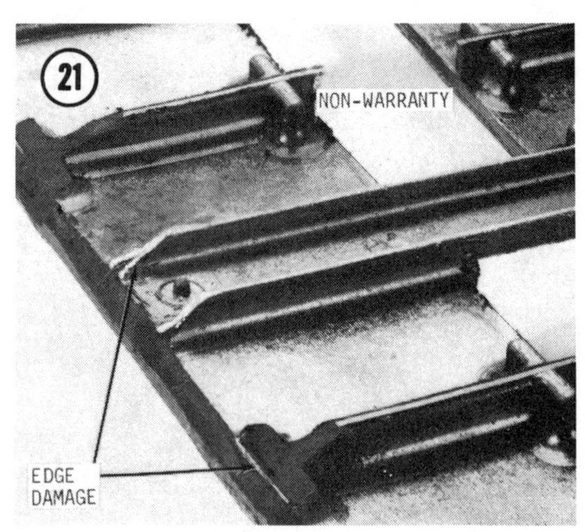
21
NON-WARRANTY
EDGE
DAMAGE

CHAPTER FOUR

4

ENGINE

All John Deere snowmobiles are equipped with one of three types of engines: Kohler, C.C.W. (Canadian-Curtis-Wright), or Kioritz. All are 2-cylinder, 2-cycle engines. All Kohler, certain models of C.C.W. and air-cooled Kioritz are piston-ported. Other models of C.C.W. and Kioritz, including liquid-cooled models, are equipped with reed valves. Refer to **Table 1** to determine what engine your machine is equipped with. Refer to Chapter Three for the *Principles of Operation* of 2-cycle piston-port and reed valve engines.

All engines have ball bearing main crankshaft bearings and needle bearings on the lower and upper bearings of the connecting rods. C.C.W. and Kioritz engines are equipped with 5 main bearings: 2 between crank throws, one on the flywheel end, and 2 on the PTO (power take off) end of the crankshaft. Kohler engines use 4 main bearings: 2 between crank throws, one on the flywheel end, and one on the PTO end. The crankshaft, ball bearings, seals, connecting rod and wrist pin bearings are available only as a complete assembly. Outer crankshaft bearings and seals are available separately.

This chapter includes removal and repair procedures for most engine components. However, due to the special tools and expertise required, all crankshaft assembly inspection and alignment should be performed by an authorized dealer or competent machine shop. Some procedures in this chapter require the use of special tools. In all cases the special tools are illustrated and in many cases can be easily fabricated or substituted by a well-equipped home mechanic. However, each snowmobile owner must be honest with himself about his own supply of tools and expertise and avoid repair procedures that are not within his capabilities. It is often less expensive and easier in the long run to remove the engine and take it to an authorized dealer for required service and repair than to risk expensive damage if you do not have the proper tools and facilities for the necessary work.

TOP END AND COMPLETE OVERHAUL

The following is an orderly sequence for removing and disassembling the engine to perform a top end overhaul or complete overhaul. Proceed to the applicable engine section and perform the procedures necessary in the order indicated, to achieve desired level of disassembly for the necessary repairs. Refer to **Figures 1, 2, 3, or 4** during engine disassembly and repair. Refer to **Table 2** at the end of the chapter for torque specifications.

Table 1 ENGINE APPLICATION

Snowmobile Model	**300**	**300**	**400**	**400**	**400**
Serial No.	(20,001-30,000)	J300D 030001M	(2,551-11,000)	(11,001-20,000)	(20,001-30,000)
Engine manufacturer	Kohler	Kohler	CCW	CCW	CCW
Engine model No.	K295-2AX	K295-2AX	KEC-340	KEC-340/4	KEC-340/5
	Piston port	Piston port	Piston port	Piston port	Piston port
Snowmobile Model	**400**	**500**	**500**	**500**	**600**
Serial No.	J400D 030001M	(2,551-11,000)	(11,001-20,000)	(20,001-30,000)	(2,551-20,000)
Engine manufacturer	Kioritz	CCW	CCW	CCW	CCW
Engine model No.	KEC-340/5	KEC-440	KEC-440/4	KEC-440/5	KEC-440/4
	Piston port	Piston port	Piston port	Piston port	Piston port
Snowmobile Model	**600**	**600**	**800**	**JDX4**	**JDX4**
Serial No.	(20,001-30,000)	J600D 030001M	J800D 030001M	(2,551-20,000)	JDX4D 030001M
Engine manufacturer	CCW	Kioritz	Kioritz	Kohler	Kioritz
Engine model No.	KEC-440/5	KEC-440/5	KEC-440/22	K295-2AX	KEC-340/22A
	Piston port	Piston port	Reed valve	Piston port	Reed valve
Snowmobile Model	**JD295/S**	**JDX4 Special**	**JDX6**	**JDX6**	**JDX8**
Serial No.	(20,001-30,000)	(20,001-30,000)	(20,001-30,000)	JDX6D 030001M	(2,551-20,000)
Engine manufacturer	Kioritz	CCW	CCW	Kioritz	CCW
Engine model No.	KEC-295RS/2	KEC-340/5	KEC-400/22	KEC-400/22	KEC-440/21
	Piston port	Piston port	Reed valve	Reed valve	Reed valve
Snowmobile Model	**JDX8**	**JDX8**	**340 Cyclone**	**340 Cyclone**	**340 Liquifire**
Serial No.	(20,001-30,000)	JDX8D 030001M	(55,001-70,000)	(70,001-)	(70,001-)
Engine manufacturer	CCW	Kioritz	Kioritz	Kioritz	Kioritz
Engine model No.	KEC-440/22	KEC-440/22A	340/22A	340/22B	340/23ALC
	Reed valve	Reed valve	Reed valve	Reed valve	Reed valve
Snowmobile Model	**440 Cyclone**	**440 Cyclone**	**340 Liquifier**	**440 Liquifire**	**440 Liquifire**
Serial No.	(55,001-70,000)	(70,001-)	(55,001-70,000)	(55,001-70,000)	(70,001-)
Engine manufacturer	Kioritz	Kioritz	Kioritz	Kioritz	Kioritz
Engine model No.	440/22A	440/22B	340/23LC	440/23LC	440/23ALC
	Reed valve	Reed valve	Reed valve	Reed valve	Reed valve

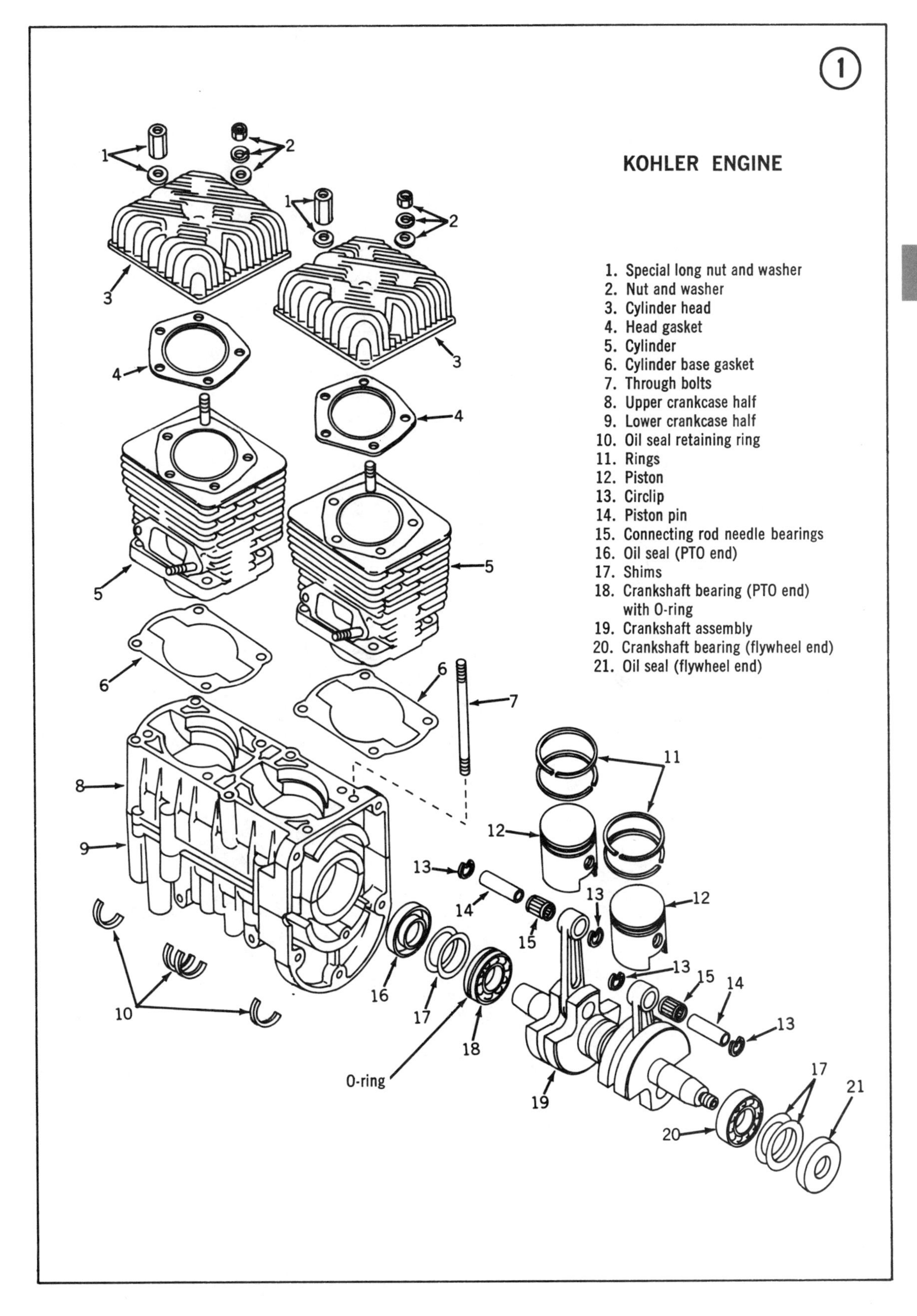
1
KOHLER ENGINE
1. Special long nut and washer
2. Nut and washer
3. Cylinder head
4. Head gasket
5. Cylinder
6. Cylinder base gasket
7. Through bolts
8. Upper crankcase half
9. Lower crankcase half
10. Oil seal retaining ring
11. Rings
12. Piston
13. Circlip
14. Piston pin
15. Connecting rod needle bearings
16. Oil seal (PTO end)
17. Shims
18. Crankshaft bearing (PTO end) with O-ring
19. Crankshaft assembly
20. Crankshaft bearing (flywheel end)
21. Oil seal (flywheel end)
O-ring

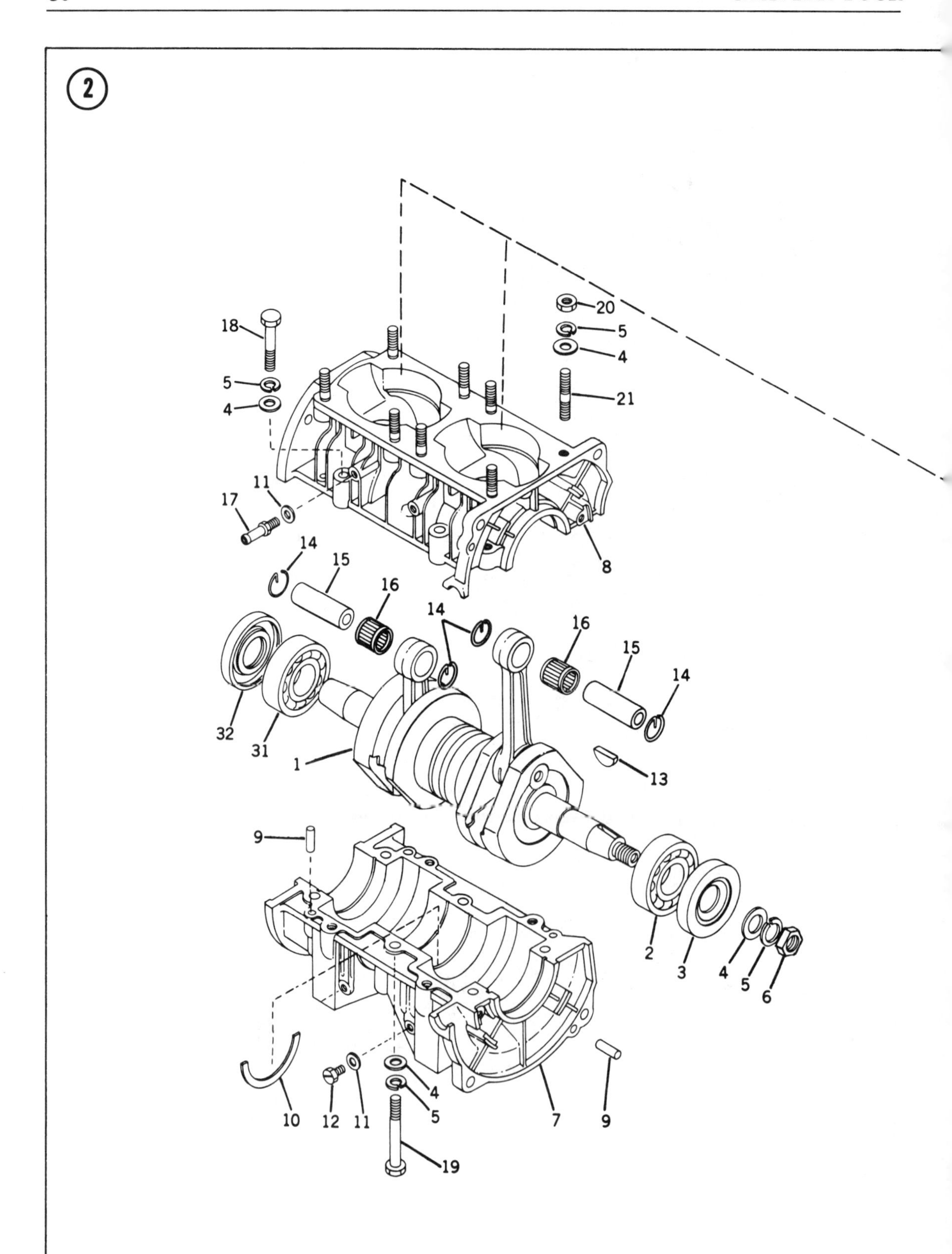
2
18
5
4
11
17
20
5
4
21
8
14
15
16
14
16
15
14
32
31
1
13
9
2
3
4
5
6
10
12
11
4
5
7
9
19

C.C.W. AND KIORITZ PISTON-PORTED ENGINES

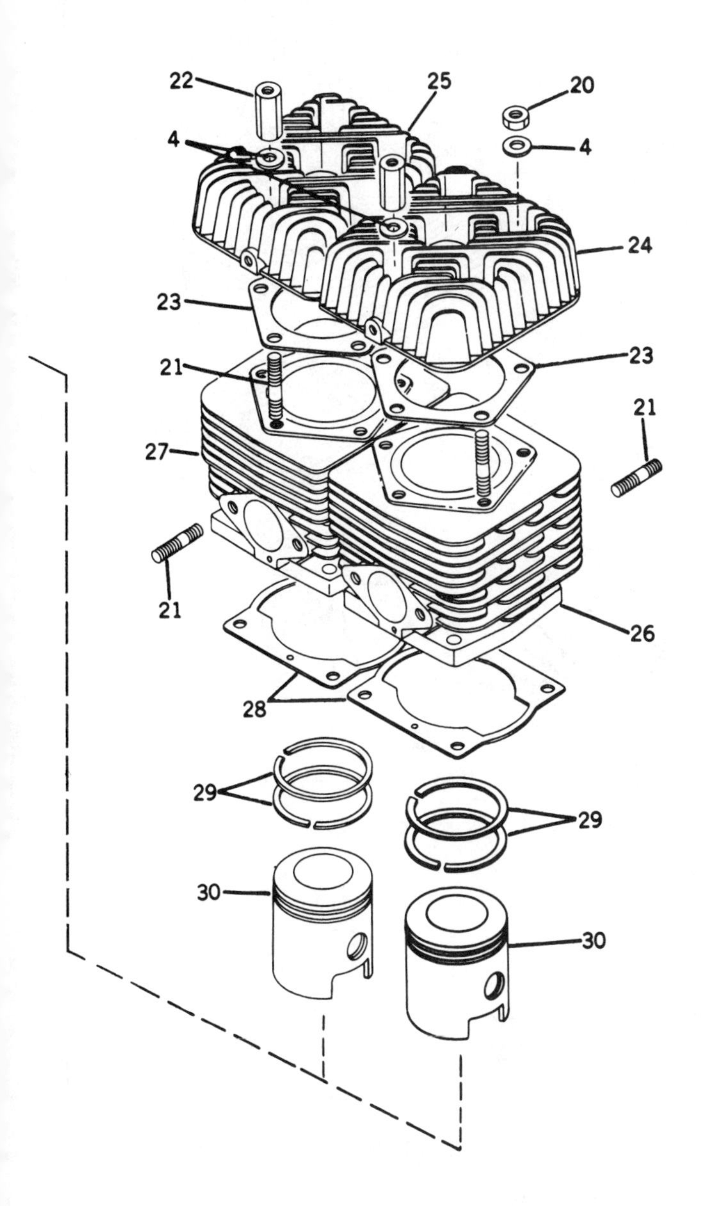

1. Crankshaft assembly
2. Ball bearing (flywheel end)
3. Oil seal (flywheel end)
4. Washer
5. Lockwasher
6. Flywheel nut
7. Lower crankcase half
8. Upper crankcase half
9. Dowel pin (4)
10. Bearing retaining clip (4)
11. Gasket
12. Drain screw
13. Woodruff key
14. Circlip (4)
15. Piston pin
16. Needle bearing
17. Impulse tube fitting
18. Cap screw, 8 x 65mm (4)
19. Cap screw, 8 x 45mm (6)
20. Nut, 8mm
21. Stud (8)
22. Special nut (2)
23. Head gasket (2)
24. Right cylinder head
25. Left cylinder head
26. Right cylinder
27. Left cylinder
28. Cylinder base gasket (2)
29. Ring set
30. Piston
31. Ball bearing, PTO end (2)
32. Oil seal, PTO end

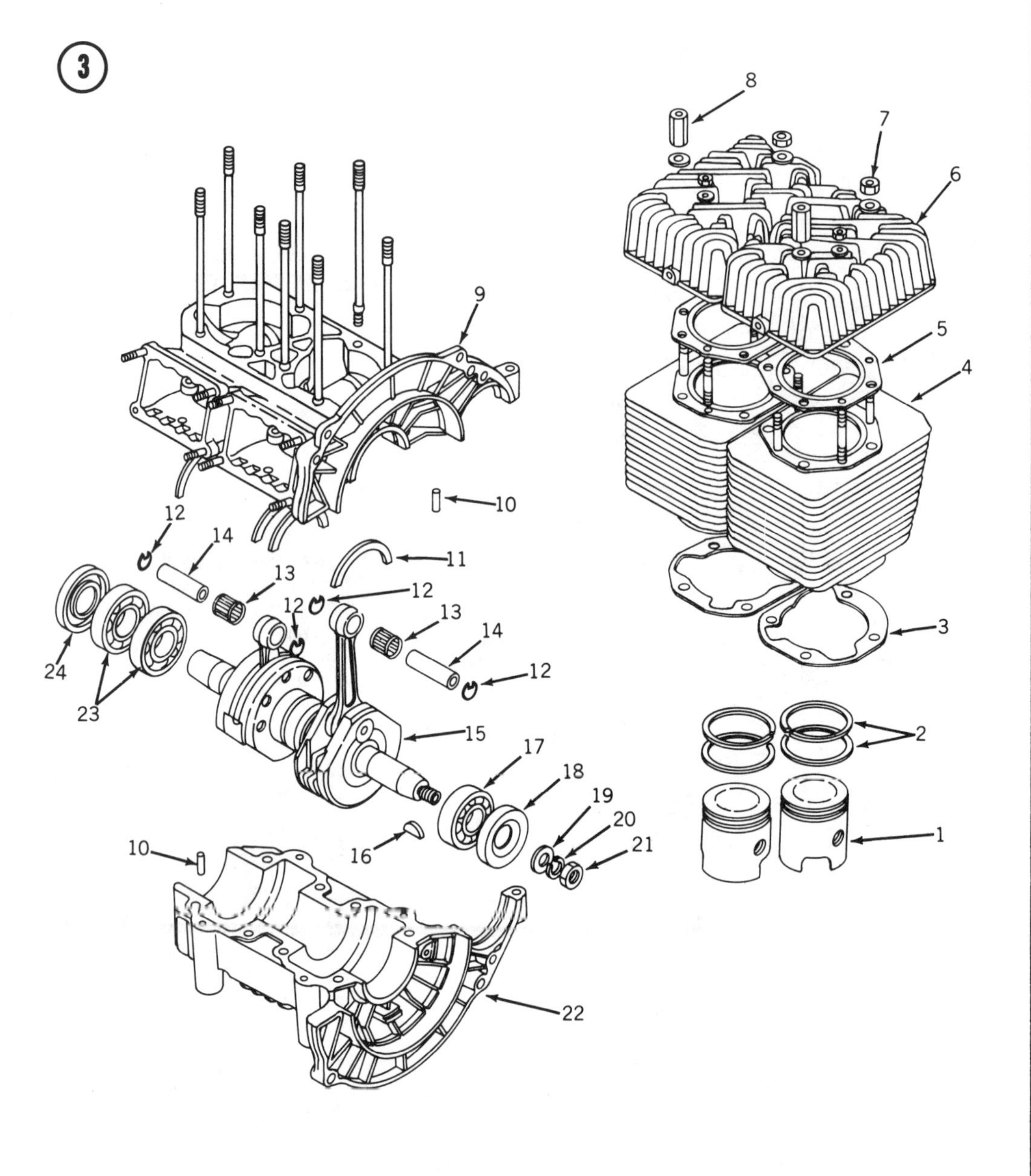

C.C.W. REED VALVE ENGINE

1. Piston
2. Rings
3. Cylinder base gasket
4. Cylinder
5. Head gasket
6. Cylinder head
7. Nut
8. Special long nut
9. Upper crankcase half
10. Dowel
11. Bearing retaining clip
12. Circlip
13. Connecting rod needle bearings
14. Piston pin
15. Crankshaft assembly
16. Woodruff key
17. Crankshaft bearing (fan end)
18. Oil seal (fan end)
19. Washer
20. Lockwasher
21. Nut
22. Lower crankcase half
23. Crankshaft bearing (PTO end)
24. Oil seal (PTO end)

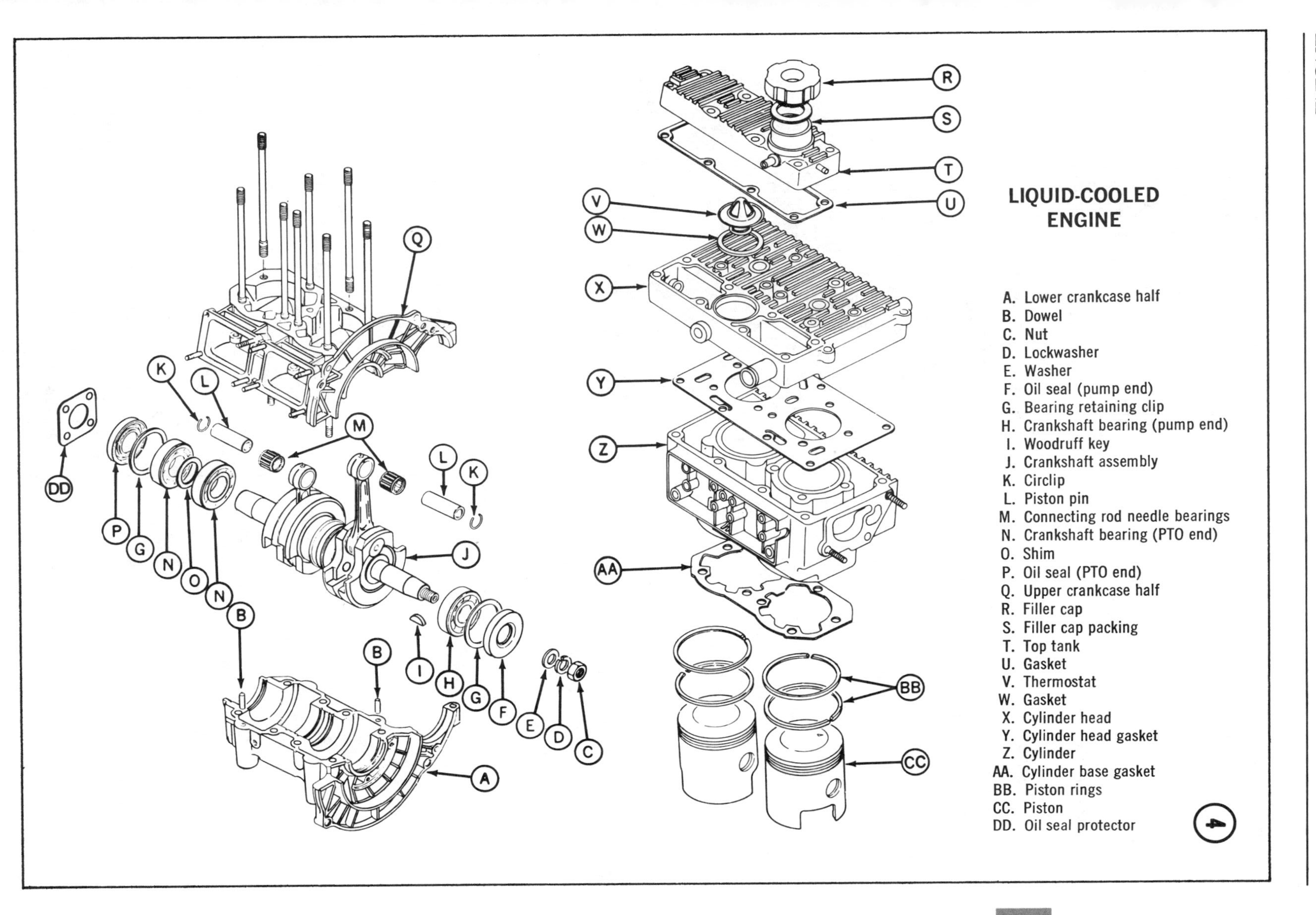
LIQUID-COOLED ENGINE
A. Lower crankcase half
B. Dowel
C. Nut
D. Lockwasher
E. Washer
F. Oil seal (pump end)
G. Bearing retaining clip
H. Crankshaft bearing (pump end)
I. Woodruff key
J. Crankshaft assembly
K. Circlip
L. Piston pin
M. Connecting rod needle bearings
N. Crankshaft bearing (PTO end)
O. Shim
P. Oil seal (PTO end)
Q. Upper crankcase half
R. Filler cap
S. Filler cap packing
T. Top tank
U. Gasket
V. Thermostat
W. Gasket
X. Cylinder head
Y. Cylinder head gasket
Z. Cylinder
AA. Cylinder base gasket
BB. Piston rings
CC. Piston
DD. Oil seal protector
4

Top End Overhaul

a. Remove engine
b. Remove exterior components
c. Remove fan cover
d. Remove coolant pump and flywheel housing (liquid-cooled models)
e. Remove flywheel
f. Remove stator
g. Remove cylinder head
h. Remove cylinder, piston, and rings
i. Perform component inspection

Complete Overhaul

a. Perform top end overhaul
b. Remove reed valve assembly (if so equipped)
c. Remove crankshaft assembly
d. Perform component inspection

KOHLER ENGINE

Refer to *Top End and Complete Overhaul* for proper engine disassembly sequence.

Engine Removal/Installation

1. Remove access panel from top of console and remove windshield by removing 6 attaching screws (**Figure 5**).

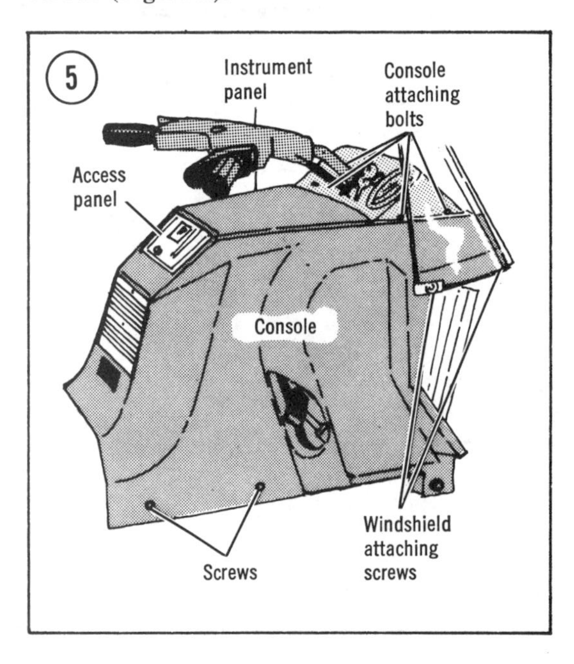

2. Remove left-hand access panel.
3. Remove screws securing lower console to tunnel.
4. Loosen 4 console attaching bolts, located under the console, on each side of instrument panel, and remove console by lifting it up and sliding it rearward. Instrument panel remains in place and need not be removed.
5. Disconnect choke and throttle cables. See **Figure 6**. Remove silencer and air hose.

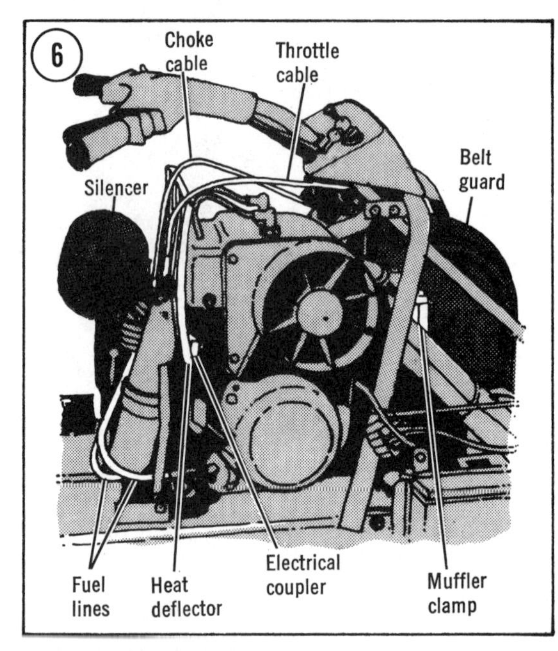

6. Disconnect fuel lines from the carburetor.
7. Remove heat deflector.
8. Loosen clamp securing muffler to engine.
9. Remove drive belt guard and drive belt.
10. On electric starter-equipped models, remove lead from starter terminal and disconnect ground strap from engine.
11. Mark crossbars (**Figure 7**) to aid drive and driven sheaves alignment when engine is installed. Mark both front and rear crossbars.
12. Disconnect muffler bracket from front of engine base. Remove 4 bolts securing engine base to crossbars. Gently lift engine from snowmobile and remove engine base from engine.
13. Installation is the reverse of these steps. Ensure drive and driven sheaves are aligned correctly as outlined in Chapter Eight.

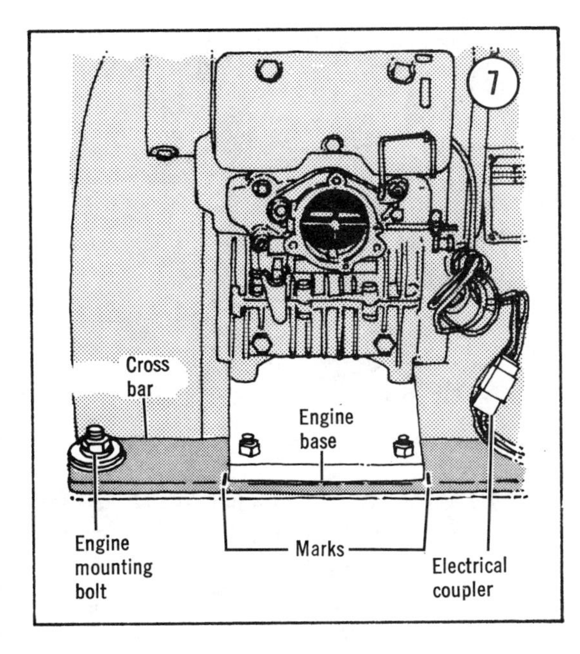

CAUTION
A rebuilt or overhauled engine must be broken-in like a new machine or serious overheating and damage may result.

Exterior Components Removal/Installation

1. Remove recoil starter assembly.
2. Remove drive sheave
3. Remove electric starter, if engine is so equipped.
4. Remove lower fan sheave and fan belt.
5. Installation is the reverse of these steps.

Ring Gear and Hub Removal/Installation

1. Install special flywheel holding tool or locally fabricated equivalent as shown in **Figure 8** to hold crankshaft from turning.
2. Remove 4 ring gear retaining screws and remove ring gear.
3. Install puller to ring gear hub and tighten puller to remove hub (**Figure 9**).

NOTE: *On some models, ring gear and hub are one piece and are removed as one unit.*

4. Installation is the reverse of these steps. Keep the following points in mind:

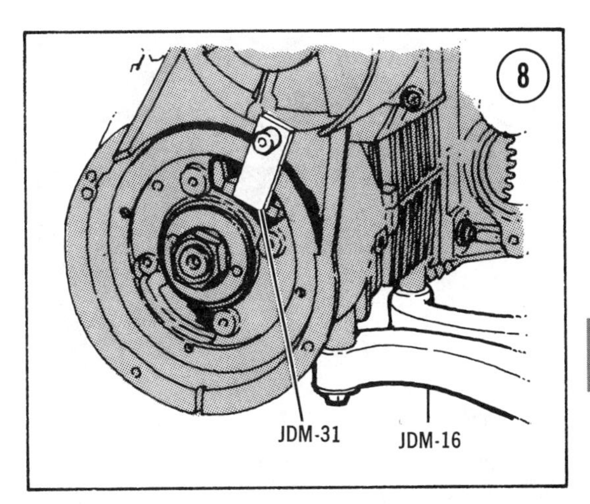

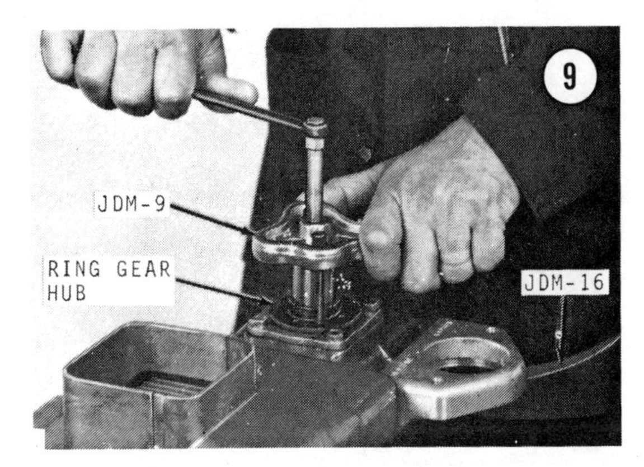

a. Use a hammer and block of wood to tap ring gear hub in place. Ensure hub is properly aligned and is not cocked.

NOTE: *On one-piece ring gear and hub units, install unit with recess in gear toward the inside and flat side of gear toward outside.*

b. Be sure ring gear hub is fully in contact with crankcase and secure with 4 screws.

c. Install lower sheave, fan belt, and starter cup.

Flywheel Removal/Installation

1. Install special flywheel holding tool or locally fabricated equivalent as shown in **Figure 10**.
2. Remove flywheel retaining nut and washers. Install flywheel puller and tighten puller center bolt to break flywheel loose. Remove flywheel holding tool and remove flywheel.

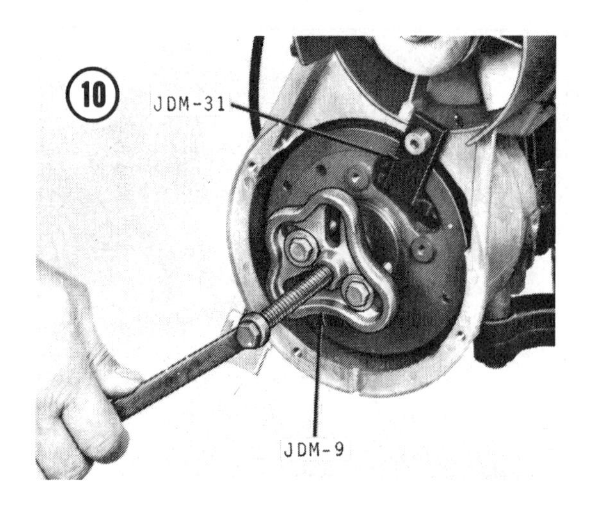

CAUTION

Do not strike puller center bolt with hammer or crankshaft and/or bearing damage may occur.

3. Installation is the reverse of these steps. Keep the following points in mind.

   a. Ensure flywheel keyway is aligned with key on crankshaft.

   b. Use flywheel holding tool to keep flywheel from turning and torque flywheel retaining nut to 90 ft.-lb. (12.4 mkg).

CAUTION

Do not tighten flywheel retaining nut with an impact wrench or flywheel and/or engine damage may occur..

Shrouds and Fan Housing Removal/Installation

1. Remove flywheel.
2. Remove coil cover and disconnect stator leads from coils. Remove coils.
3. Remove intake and exhaust manifolds.
4. Remove front and rear engine shrouds.
5. Remove fan housing and stator as an assembly. See **Figure 11**.
6. Inspect fan, fan bearings, and belt as outlined in *Component Inspection*.
7. Installation is the reverse of these steps. Keep the following points in mind:

   a. Use new O-ring between fan housing and crankcase.

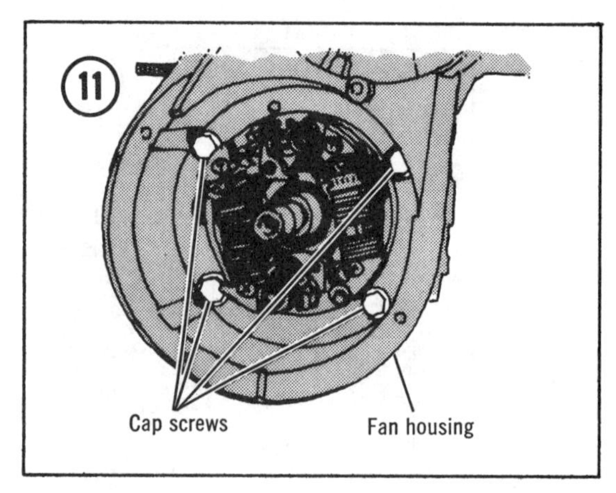

   b. Be sure correct number of shims are placed and seated in crankcase when fan housing is installed.

   c. Torque 4 cap screws securing fan housing to 15-18 ft.-lb. (2.0-2.5 mkg).

   d. Use new gaskets on intake and exhaust manifolds.

   e. Blue/red wires goes to No. 1 coil (closest to ring gear) and blue wire to No. 2 coil.

Cylinder Head, Cylinder, Piston, and Ring Removal

1. Remove nuts securing cylinder heads. Note location of long nuts to aid installation. To aid head removal, gently tap head with a rubber mallet. Remove heads and discard old gaskets.
2. Remove nuts and washers securing cylinders to crankcase (**Figure 12**).

3. Gently slide cylinders up over pistons.

4. Note mark on each piston indicating exhaust side of engine (**Figure 13**). If no marks are visible, inscribe them accordingly. Also make sure that pistons are marked 1 and 2 since they are not interchangeable.

> NOTE: *If rings are going to be replaced, but not pistons, rings may now be removed. However, if pistons are going to be removed, leave old rings on pistons to protect ring grooves until new rings are to be installed.*

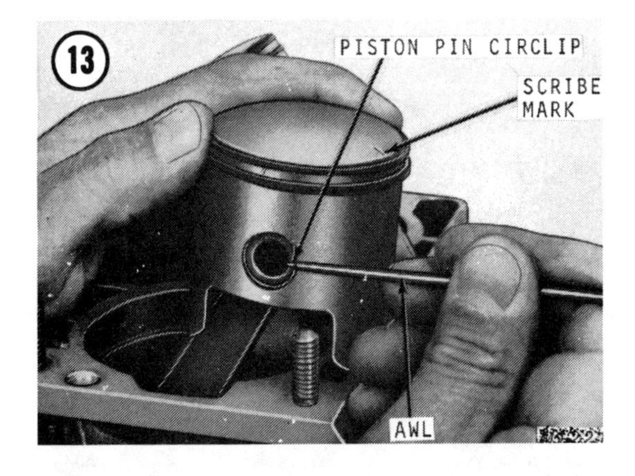

5. Using a ring expander tool or your thumbs on each end of piston ring, gently expand ring and slide up and off piston (**Figure 14**).

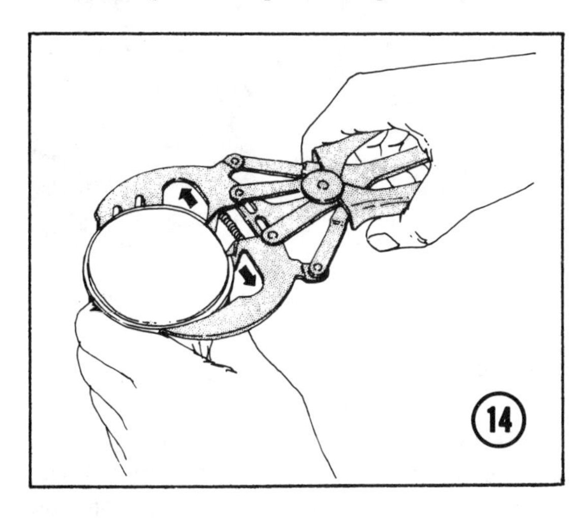

6. Be sure pistons are appropriately marked. Remember, pistons are *not* interchangeable. Remove circlips from each end of piston pins (**Figure 15**).

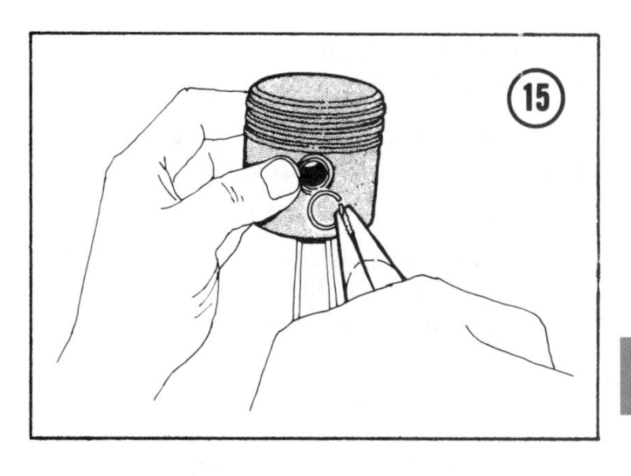

> NOTE: *Stuff clean rags around connecting rods in crankcase to help prevent circlips from dropping into the crankcase.*

Using a piston pin removal tool or an appropriately sized wooden dowel, gently remove pins from piston and connecting rod (**Figure 16**).

CAUTION

Exercise care when removing pins to avoid damaging connecting rod needle bearings. If a wooden dowel is used to drive out piston pins, ensure piston is properly supported so lateral shock is not transmitted to lower connecting rod bearing, otherwise rod and/or bearing damage may occur.

Remove needle bearings from connecting rods.

7. *Refer to Component Inspection* and inspect cylinder heads, cylinders, pistons, pins, and rings.

Cylinder Head, Cylinder, Piston, and Ring Installation

1. Lubricate piston pin needle bearings with oil and insert bearings into connecting rods.

2. Slide piston over connecting rod.

3. Using piston pin installation tool or appropriately sized wooden dowel, install piston pins through piston and rod ends (**Figure 16**).

CAUTION

Exercise care when installing pins to avoid damage to connecting rod needle bearings. If a wooden dowel is used to drive in piston pins, ensure piston is properly supported so lateral shock is not transmitted to lower connecting rod bearing, otherwise rod and/or bearing damage may occur.

4. Secure piston pins to pistons with circlips (**Figure 15**).

NOTE: *Stuff clean rags around connecting rods in crankcase to help prevent circlip from dropping into the crankcase.*

CAUTION

If possible, use new circlips to secure piston pins. If old circlips are used, they must snap securely into grooves in pistons. A weak circlip could become disengaged during engine operation and cause severe engine damage.

5. Using a ring expander tool or your thumbs on each end of piston ring, gently expand ring and slide over piston in ring groove. Install ring in bottom groove first. Be sure ring groove clearance is within tolerance as outlined in *Component Inspection.*

NOTE: *Be sure ring end gap is properly positioned around locating tab* (**Figure 17**). *Install L-ring in top groove.*

6. Install new cylinder base gaskets over cylinder studs.

7. Thoroughly lubricate pistons, rings, and cylinder cores with engine oil.

8. Position a suitable wooden block under pistons for piston support (**Figure 18**).

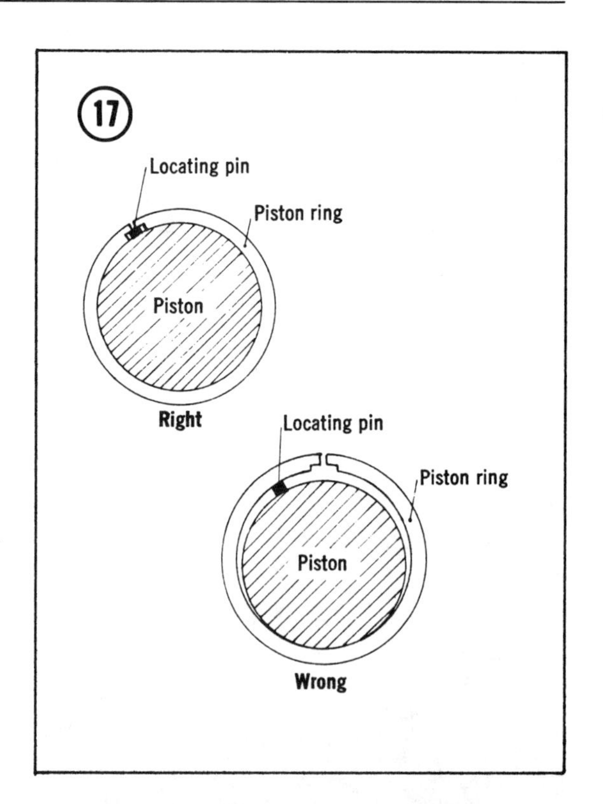

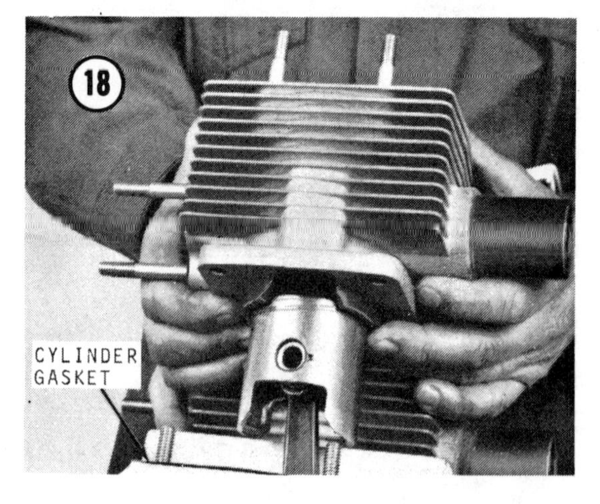

9. Compress rings with a suitable ring compressor or your fingers and carefully slide cylinder down over piston. Ensure rings are properly centered around locating tabs (**Figure 17**).

10. Install cylinder bolts, washers, and nuts. *Do not* tighten at this time. Place special serrated lockwashers on 2 stud nuts on fan housing end.

11. Using new gaskets, install intake manifold to cylinders and tighten nuts to 15-18 ft.-lb. (2.0-2.5 mkg).

CAUTION
Intake manifold must be installed and secured to cylinders before cylinder nuts are tightened in order to align cylinders. Misalignment of cylinders will cause air leaks resulting in serious engine damage.

12. Tighten center crankcase bolts first, then tighten end bolts. Torque bolts to 15-18 ft.-lb. (2.0-2.5 mkg). Torque 4 crankcase side bolts and PTO end bearing plate cap screws to 8-10 ft.-lb. (1.1-1.4 mkg).

13. Install cylinder heads with raised bosses toward exhaust side of engine. Tighten head nuts evenly in sequence shown in **Figure 19**. Torque nuts to 15-18 ft.-lb. (2.0-2.5 mkg). Be sure long head nuts are installed in proper location.

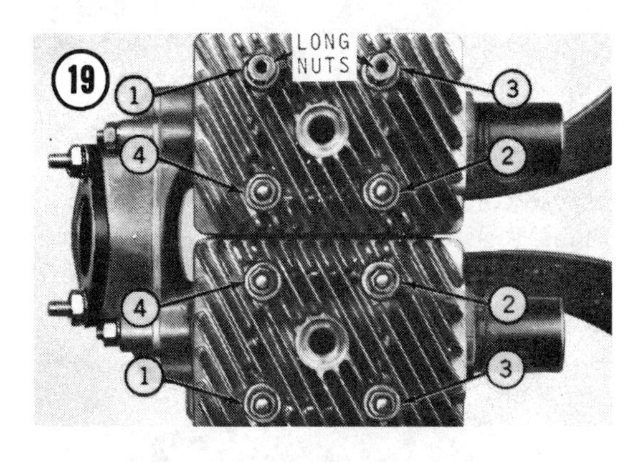

Crankshaft Assembly Removal

1. Remove cylinder heads, cylinders, and pistons.

2. Remove PTO end bearing plate. Remove shims and save for installation (**Figure 20**).

3. Remove crankcase retaining cap screws. Tap upper crankcase half with soft mallet (**Figure 21**) to separate crankcase.

CAUTION
Never attempt to pry crankcase halves apart with a screwdriver or similar object or crankcase sealing surfaces will be damaged.

4. Carefully lift out crankshaft assembly.

5. Perform crankshaft assembly and crankcase inspection as outlined later in *Component*

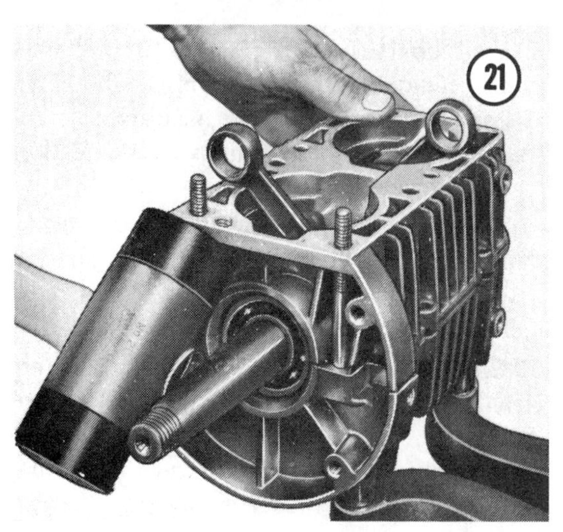

Inspection. Refer all crankshaft assembly repair and service to an authorized dealer or competent machine shop. They are equipped with the necessary special tools and expertise for the work.

Crankshaft Assembly Installation

1. Place crankshaft in lower crankcase half and install fan housing and PTO end bearing plate with new gasket (**Figure 22**). Be sure O-ring is around PTO end bearing.

2. Push crankshaft toward fan housing and measure clearance between PTO crankshaft bearing and PTO end bearing plate with a feeler gauge. Correct crankshaft end play is 0.006-0.012 in. (0.15-0.30mm).

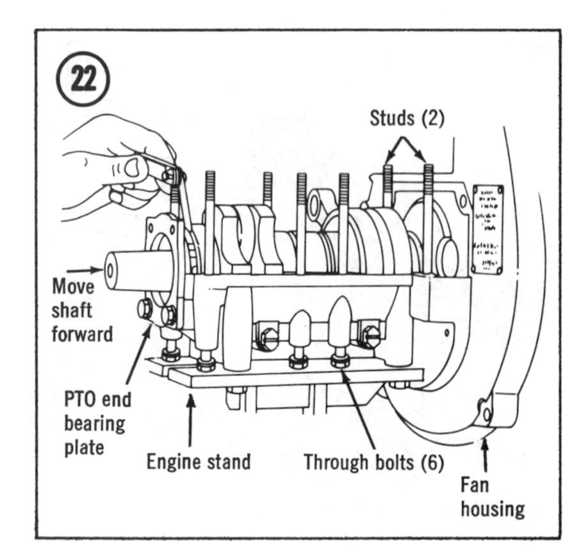

3. To determine number of 0.006 in. (0.15mm) shims required, subtract a nominal 0.009 in. (0.23mm) from feeler gauge measurement. Install an equal number of shims at each end of crankshaft whenever possible.

4. After shim requirements have been determined, remove PTO end bearing plate and fan housing. Leave crankshaft in lower crankcase half.

5. Check that crankcase sealing surfaces are clean and not damaged. Apply an even coat of silicone rubber adhesive to sealing surfaces on both crankcase halves. Make sure no rubber adhesive runs into crankcase.

6. Install upper crankcase half. Loosely install crankcase side bolts. Do not tighten at this time.

7. Install PTO end bearing plate and shims. Do not tighten end bearing plate cap screws at this time.

8. Install pistons, cylinders, and cylinder heads.

CAUTION

Do not tighten PTO end bearing plate cap screws until cylinders have been installed and torqued. Crankcase misalignment will occur resulting in a serious air leak in crankcase.

9. Tighten center crankcase bolts first, then tighten end bolts. Torque bolts to 15-18 ft.-lb. (2.0-2.5 mkg). Torque 4 crankcase side bolts and PTO end bearing plate cap screws to 8-10 ft.-lb. (1.1-1.4 mkg).

Fan Belt Tension Adjustment

1. Check fan belt for proper tension (**Figure 23**). Properly adjusted fan belt should deflect approximately ⅜ in. (9mm) when flexed at a point near center of belt span.

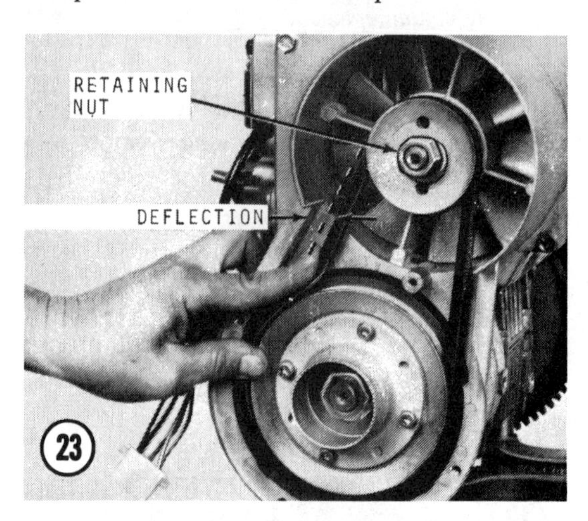

2. To adjust belt, remove fan retaining nut. Remove outer fan sheave half (**Figure 24**) to expose adjusting shims. Remove one or more shims. Reassemble and check for ⅜ in. (9mm) belt deflection. Repeat until proper tension is achieved.

3. Install retaining nut and torque to 35 ft.-lb. (4.8 mkg).

COMPONENT INSPECTION

This procedure applies to all 3 engines, and should be performed during each overhaul.

Some of the following inspection procedures require the use of micrometers and dial indicators for precise wear analysis. If such precision tools are not available, refer inspection procedures to an authorized dealer or competent machine shop. Refer to **Table 3** at the end of the chapter for engine component dimensions and wear tolerances.

Cooling Fan and Belt Inspection

1. Inspect fan (**Figure 25**) for cracked, broken, or damaged fins. Dress nicks or dents with a file. If fins are cracked or broken, fan must be replaced.

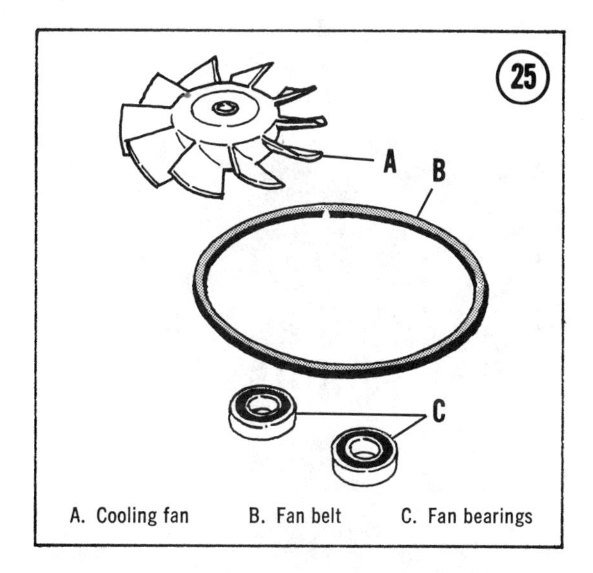

A. Cooling fan B. Fan belt C. Fan bearings

2. Inspect fan bearings for wear or looseness. Replace if necessary.

3. Inspect fan belt and replace if frayed, stretched, or deteriorated.

Coolant Pump and Belt Inspection

1. Inspect O-ring (**Figure 26**) for damage, cracks, or breaks. Replace if necessary.

2. Inspect coolant pump for internal wear caused by cooling system contamination or dirt.

3. Inspect coolant pump belt and replace if frayed, stretched, or deteriorated.

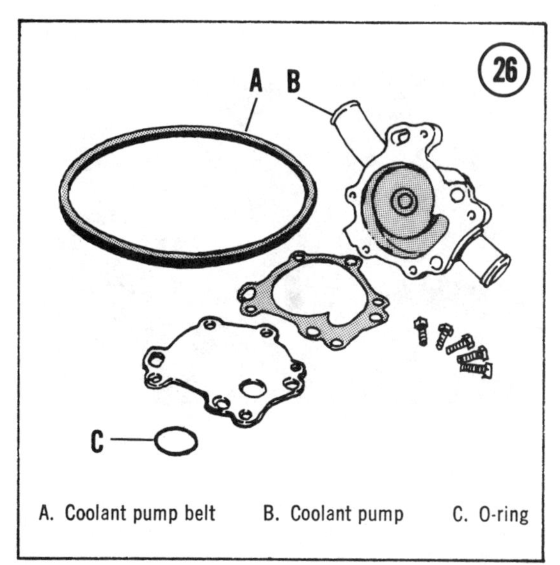

A. Coolant pump belt B. Coolant pump C. O-ring

Liquid Cooled Cylinder Inspection

1. Inspect cylinder assembly for scoring or scuffing (**Figure 27**).

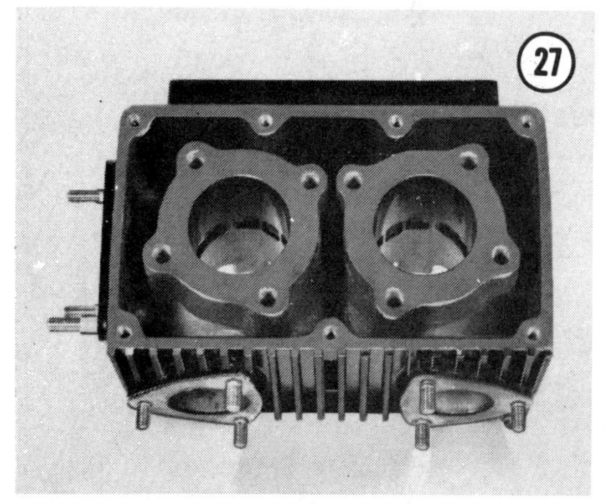

2. Inspect all mating surfaces for nicks, scratches, or other damage that could prevent proper sealing of cylinder head.

Liquid Cooled Cylinder Head Inspection

Inspect head (**Figure 28**) for excessive carbon buildup in combustion chambers. Ensure all coolant passages are clean and open.

Air Cooled Cylinder Inspection

1. Install a serviceable cylinder head on cylinders. Install without gasket or hold-down nuts.

2. Measure for distortion between studs with a feeler gauge (**Figure 29**). Maximum allowable distortion is 0.002 in. (0.05mm).

29

A
B
C

A. Cylinder B. Cylinder head C. Feeler gauge

3. Measure cylinders with an inside micrometer (**Figure 30**). Measure parallel with crankshaft and at right angles to crankshaft at both top and bottom of ring travel. See **Table 3** at end of chapter for wear tolerances.

Air Cooled Cylinder Head Inspection

1. Carefully scrape carbon from cylinder head and exhaust ports of cylinders. Use a soft metal (nonferrous) scraper to avoid damage. A wooden spatula works well for cleaning exhaust ports.
2. Use a spark plug tap (14mm) to clean carbon from spark plug threads in cylinder head, if required.

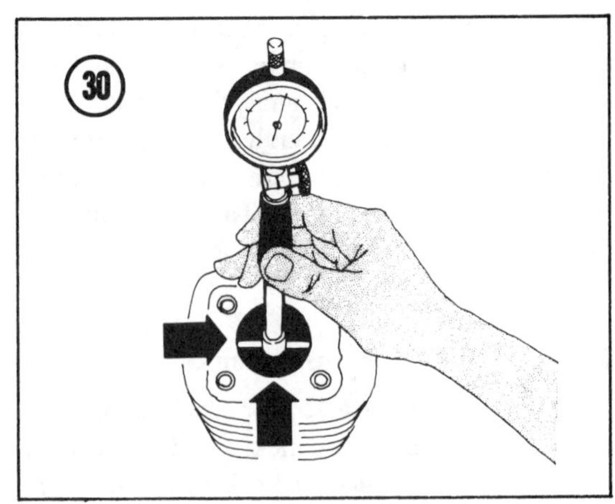

3. Place cylinder head on a surface plate (**Figure 31**) and measure at various points between head and surface plate with a feeler gauge.

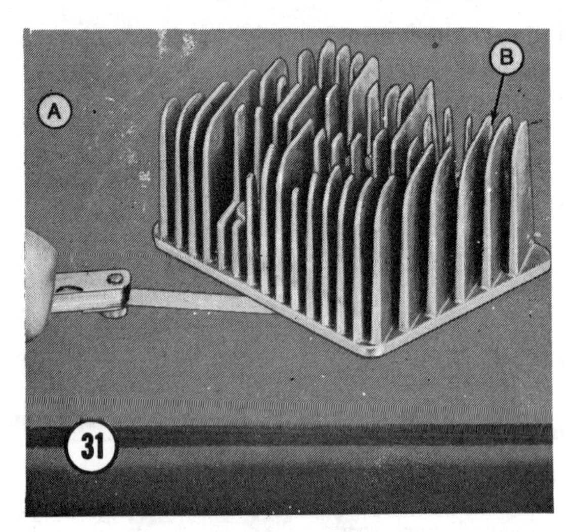

A. Surface plate B. Cylinder head

Honing Cylinder Bore

If cylinder is within wear tolerance (**Table 3**, end of chapter), but lightly scored, hone by running a fine stone cylinder hone lightly in cylinder (**Figure 32**). On C.C.W. and Kioritz engines, do not hone over 0.005 in. (0.127mm) oversize or cylinder will have to be replaced.

Clean thoroughly with detergent and water to remove all particles.

Crankcase Inspection

1. Inspect crankcase sealing surfaces (**Figure 33)** for deep scratches, scoring, or pitting.

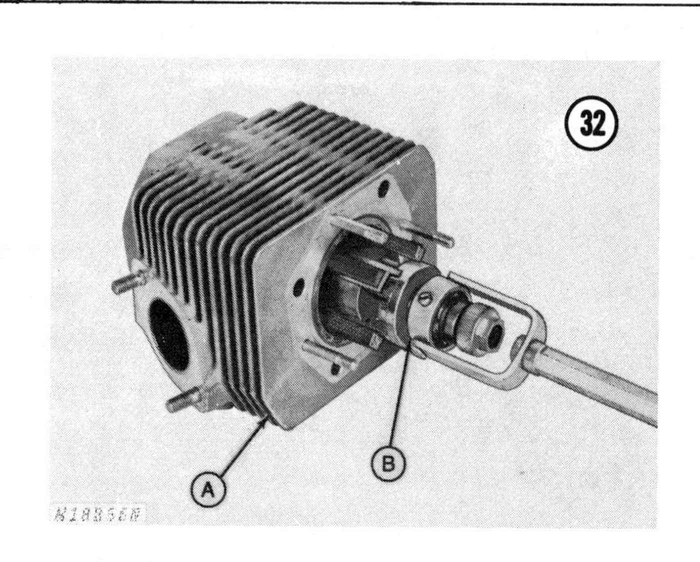

A. Cylinder
B. Fine stone cylinder hone

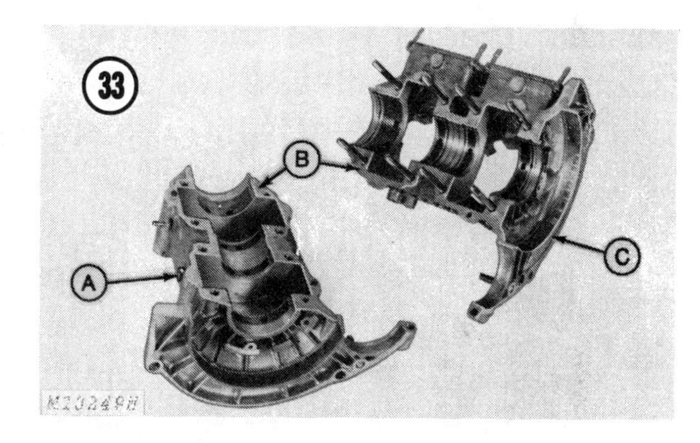

A. Lower crankcases
B. Sealing surfaces
C. Upper crankcase

2. Inspect bearing and oil seal retaining inserts for wear, scoring, or conditions that could cause leaks.

> NOTE: *Minor indication of crankshaft bearing outer race "rotation" in crankcase halves is considered normal.*

3. Replace crankcase halves if damaged. Crankcase halves are available only in a matched set.

Piston and Piston Pin Inspection

1. Clean piston ring grooves with a ring groove cleaner or the broken section of an old ring.
2. Inspect piston for evidence of scoring, pitting, or corrosion.
3. Measure piston at right angles to piston pin for wear at top land and at skirt (**Figure 34**). Measure piston pin bore with an inside micrometer.

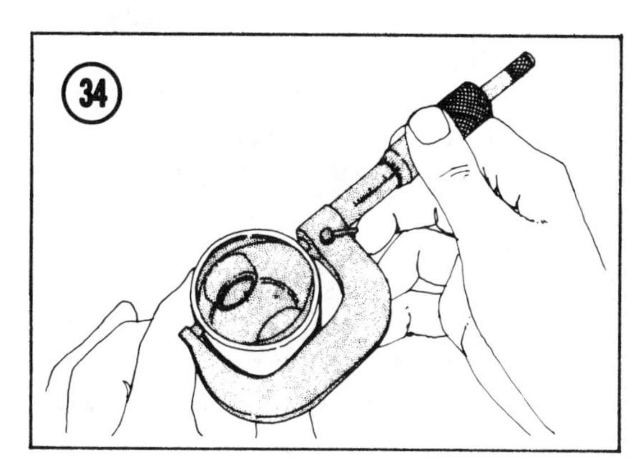

4. Install a new ring into ring groove and measure side clearance with a feeler gauge as shown in **Figure 35**.

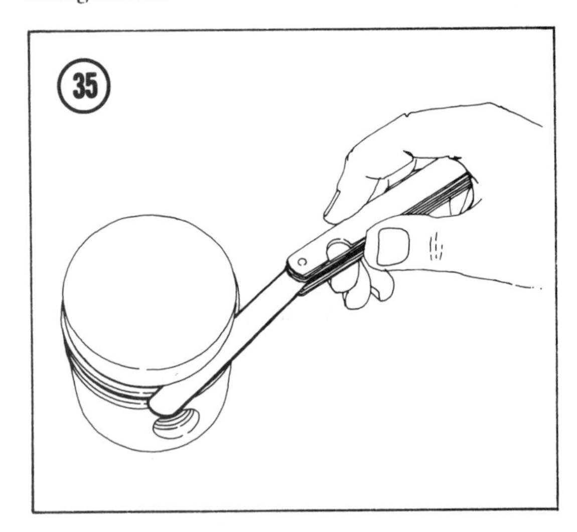

5. Measure piston pin with a micrometer (**Figure 36**).

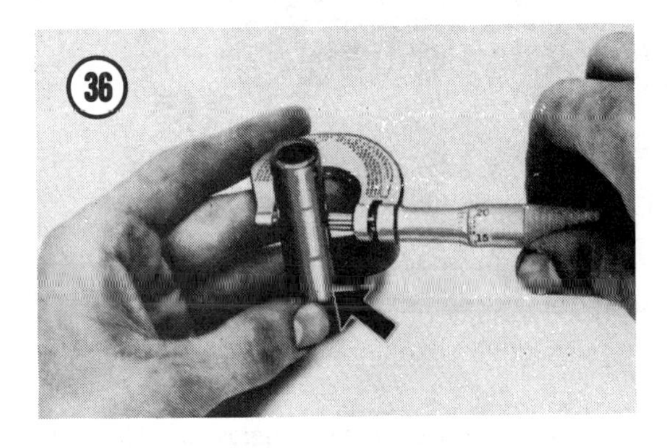

Crankshaft and Connecting Rod Inspection

Refer all allowable service and repair work on crankshaft assembly to an authorized dealer or competent machine shop.

1. Inspect threads on each end of crankshaft. Inspect keyway on flywheel end and taper on each end of crankshaft for scoring or wear (**Figure 37**).

2. Inspect ball bearing for wear, free movement, and security. Outer bearings are replaceable; however, if inner bearings are worn, crankshaft assembly must be replaced.

3. Inspect seals for wear or damage. Outer seals are replaceable; however, inner seals are

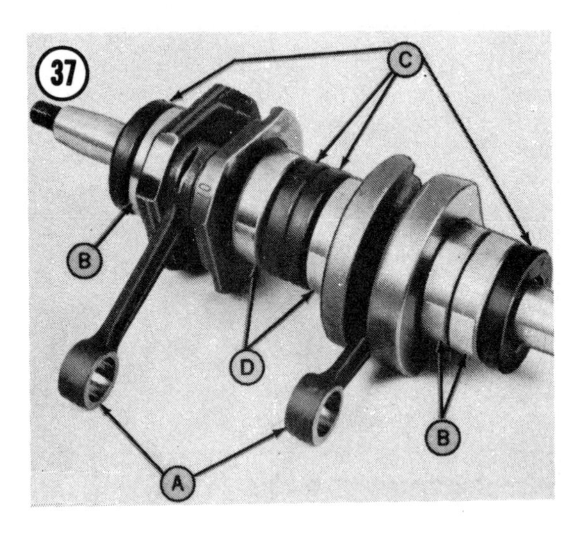

A. Connecting rods
B. Outer bearings
C. Seals
D. Inner bearings

available only with the complete crankshaft assembly.

4. Measure connecting rod side clearance with a feeler gauge (**Figure 38**).

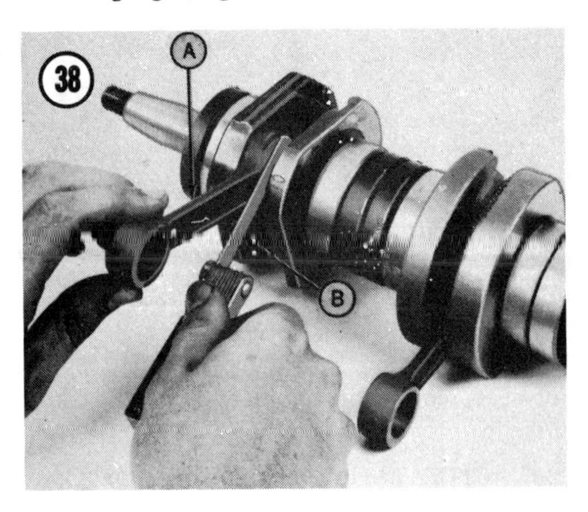

A. Connecting rod
B. Feeler gauge

5. Inspect crankpin (**Figure 39**) and piston pin needle bearings for wear or looseness. New piston pin bearings are available while crankpin bearings are available only with complete crankshaft assembly.

6. Measure connecting rod small end diameter with an inside micrometer (**Figure 39**). Replace crankshaft assembly if badly scored or not within tolerance (**Table 3**, end of chapter).

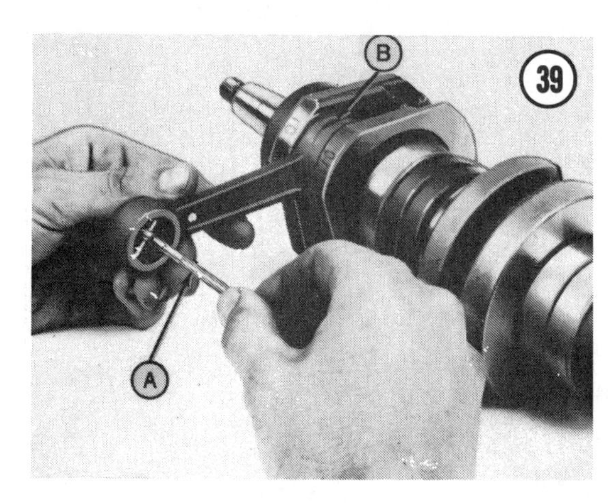

A. Inside micrometer B. Crankpin bearing

Crankshaft Runout

1. Set up a dial indicator (**Figure 40**) against crankshaft.

2. Rotate crankshaft. Refer to **Table 3** (end of chapter) for maximum permissible runout. Replace crankshaft assembly if not in tolerance. Check both ends of crankshaft.

Crankshaft Twist

1. Install dial indicator in No. 1 cylinder (**Figure 41**) to determine TDC (top dead center).
2. With No. 1 cylinder at TDC, proceed as follows:

   a. On liquid-cooled models, fasten a pointer on intake manifold stud and point it to a mark on tooth of flywheel (**Figure 42**).

   b. On air-cooled models, mark flywheel cover to align with "T" mark on flywheel (**Figure 41**).

3. Install dial indicator in No. 2 cylinder. Rotate flywheel 180° to other mark on flywheel.
4. With marks aligned, dial indicator must indicate No. 2 piston is at TDC or within 0.003 in. (0.762mm) either way of TDC. Replace crankshaft assembly if it is not within tolerance.

C.C.W. AND KIORITZ ENGINES

The basic procedures for removal, disassembly, and repair of C.C.W. and Kioritz engines are the same for piston port and reed valve engines, both air and liquid cooled. Specific differences will be noted in the procedures where necessary.

Refer to *Top End and Complete Overhaul,* at beginning of chapter, for proper engine disassembly sequence.

Engine Removal/Installation (Cyclone and Liquifire Models)

Refer to **Figure 43** for Cyclone models and **Figure 44** for Liquifire models.

A. Air silencer
B. Expansion chamber
C. Recoil starter rope
D. Fuel line
E. Engine coupler
F. Choke plunger and cable

1. On Liquifire models, remove drain screw and cylinder head vent screw and drain coolant

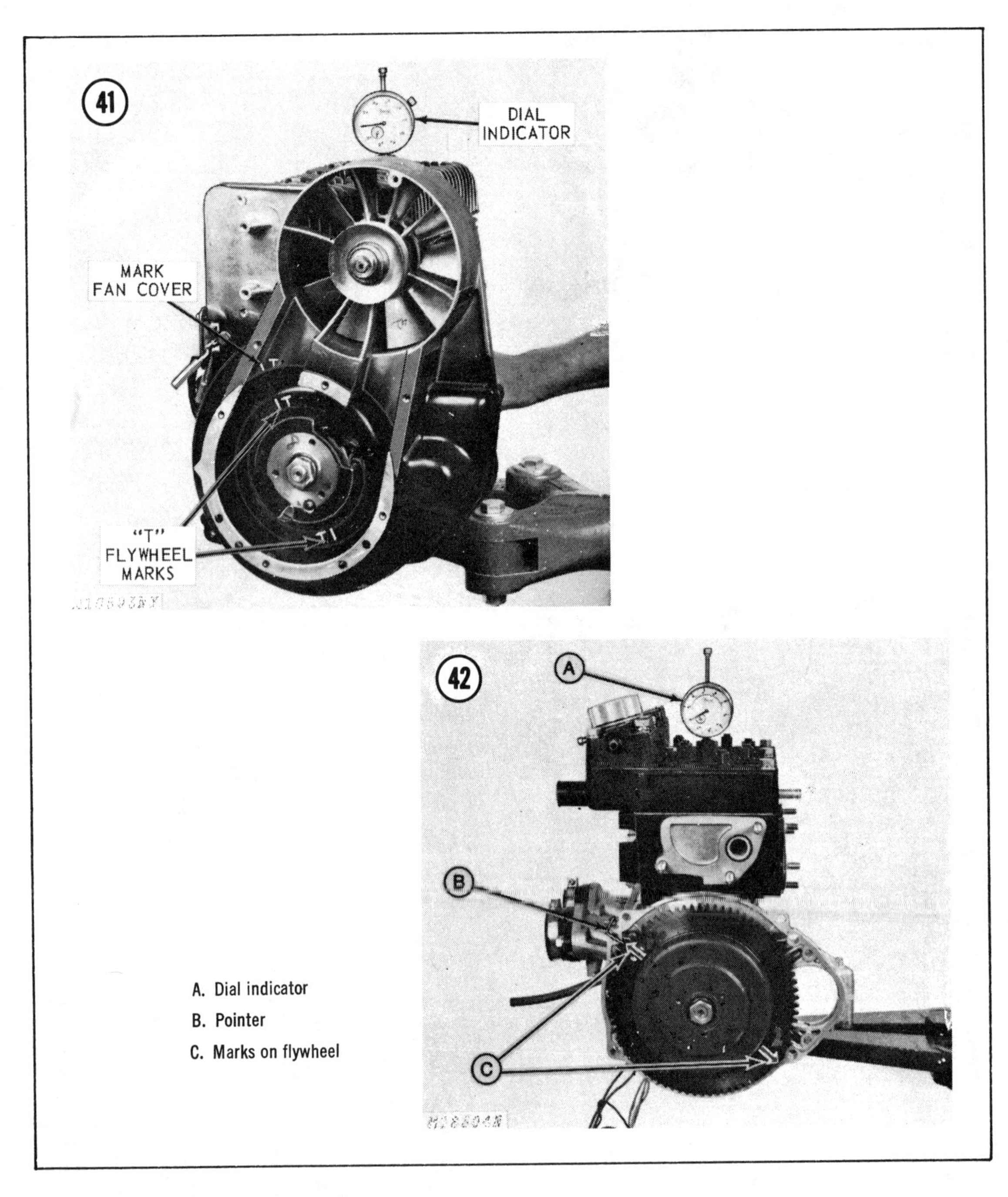

A. Dial indicator

B. Pointer

C. Marks on flywheel

from engine block (**Figure 45**). If coolant is to be reused, ensure drain pan is free of dirt and contamination.

2. Remove hood support and 2 attaching cap screws on hinge plate and lift off hood.

3. Remove spring on air silencer strap. Loosen clamps on carburetor boots and remove air silencer.

4. Remove springs securing expansion chamber to exhaust manifold and supporting bracket. Remove expansion chamber. Loosen hose clamps on Liquifire models.

5. Untie knot in starter handle and allow starter rope to partially rewind into starter housing. Tie a slip knot in starter rope to prevent rope from winding all the way into starter housing.

A. Air silencer
B. Hose clamps
C. Expansion chamber
D. Rewind starter rope
E. Fuel line
F. Electrical coupler

A. Drive belt
B. Water temperature sender
C. Carburetors

A. Cylinder head vent screw B. Drain screw

A. Drive belt B. Throttle valve and cable

6. Close fuel shutoff valve. Disconnect fuel line between fuel shutoff valve and fuel pump.

7. Remove plastic strap lock and disconnect electrical coupler.

8. On Cyclone models, remove choke plunger and cable.

9. On Liquifire models, remove water temperature sender from cylinder head (**Figure 46**).

10. Remove drive belt (**Figure 47**).

11. On Cyclone models, remove throttle valve and cable (**Figure 47**).

12. On Liquifire models, loosen clamps securing carburetors and remove carburetors. See **Figure 46**.

13. Remove 4 locknuts (**Figure 48**) securing engine base assembly to rubber mounts. Gently lift engine and base from snowmobile.

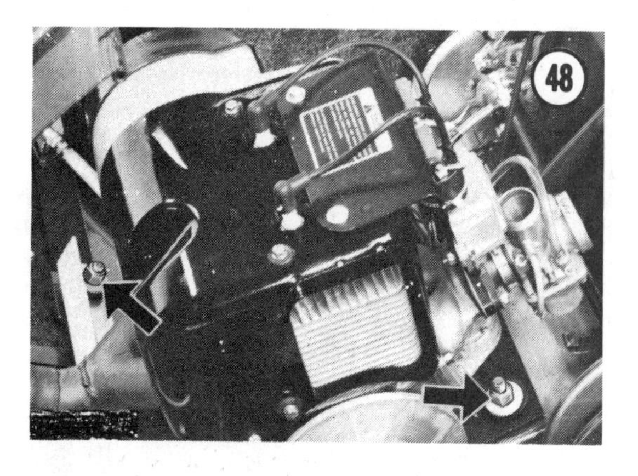

14. Installation is the reverse of these steps. On Liquifire models, fill cooling system as outlined in Chapter Six; check all connections for leaks.

CAUTION

A rebuilt or overhauled engine must be broke-in like a new machine.

Engine Removal/Installation (All Models Except Cyclone and Liquifire)

1. Remove 6 attaching screws and remove windshield (**Figure 49**). Remove access panel from top of console.
2. Remove left-hand access panel.

> NOTE: *Disconnect fuel lines from primer and wiring from heat gauge on models so equipped.*

3. Remove screws securing lower console to tunnel.
4. Loosen 4 console attaching bolts, located under console, on each side of instrument panel, and remove console by lifting it up and sliding it rearward. Instrument panel remains in place and need not be removed.
5. Disconnect choke and throttle cables. See **Figure 50**. Remove silencer and air hose.

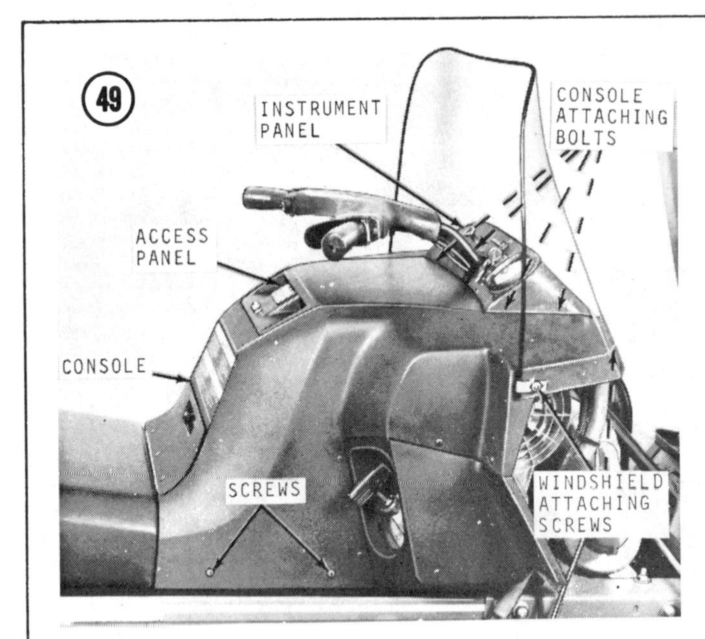

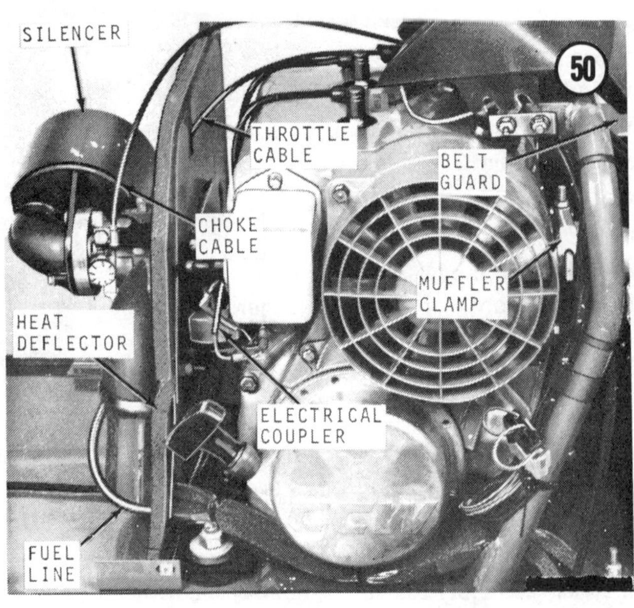

6. Disconnect fuel lines from carburetor.

7. Remove heat deflector and fuel pump (on models so equipped). Disconnect electrical coupler from engine.

8. Loosen clamp securing muffler to engine.

9. Remove drive belt guard and drive belt.

10. On electric starter equipped models, remove lead from starter terminal and disconnect ground strap from engine.

11. Mark crossbars (**Figure 51**) to aid drive and driven sheaves alignment when engine is installed. Mark both front and rear crossbars.

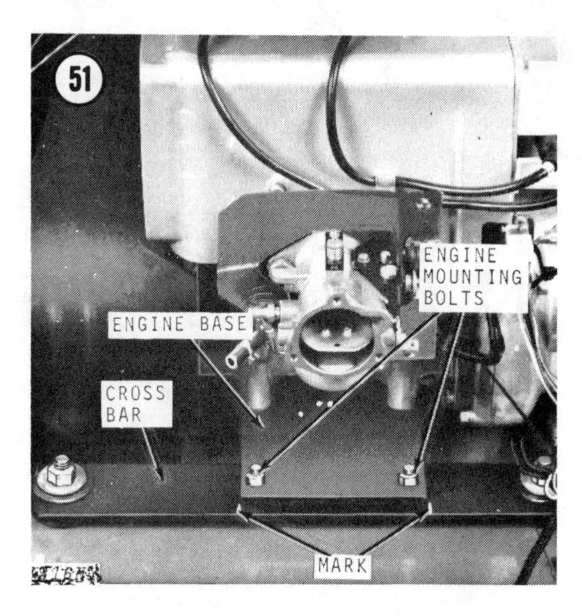

12. Disconnect muffler bracket from front of engine base. Remove 4 bolts securing engine base to crossbars. Gently lift engine from snowmobile and remove engine base from engine.

13. Installation is the reverse of these steps. Ensure drive and driven sheaves are aligned correctly as outlined in Chapter Eight.

CAUTION

A rebuilt or overhauled engine must be broke-in like a new machine or serious overheating and damage may result.

Exterior Components Removal/Installation (Cyclone Models)

Refer to **Figure 52** for this procedure.

1. Remove carburetor.

2. Disconnect ignition leads from coils and disconnect electronic pack coupler. Remove electronic pack.

3. Remove spark plug cables and coil bracket.

4. Remove bolts securing cylinder cover to engine. Remove cylinder cover.

5. Remove bolts securing fan guard to engine and remove fan guard.

6. Remove recoil starter assembly.

7. Installation is the reverse of these steps. Keep the following points in mind:

  a. Use silicone rubber adhesive on exhaust manifold gaskets.
  b. Torque all fasteners as specified in **Table 2** (end of chapter).

Exterior Components Removal/Installation (Liquifire Models)

Refer to **Figure 53** for this procedure.

1. Disconnect coupler and remove electronic pack.

2. Remove coil cover and disconnect ignition leads. Remove coils along with spark plug cables.

3. Remove coolant pump cover. Remove 6 nuts securing exhaust manifold to cylinder assembly and remove exhaust manifold.

4. Remove recoil starter.

5. Remove intake manifold complete with fuel pump.

6. Installation is the reverse of these steps. Keep the following points in mind:

  a. Use silicone rubber adhesive on exhaust manifold gaskets.
  b. Torque all fasteners as specified in **Table 2**.

Exterior Components Removal/Installation (All Models Except Cyclone and Liquifire)

Refer to **Figure 54** for this procedure.

1. Disconnect spark plug cables and remove plugs. Disconnect ignition leads between ignition coil and coupler. Remove coil cover and high tension coils with spark plug cables.

2. Remove bolts securing cylinder cover to engine and remove cover. Remove exhaust mani-

4

EXTERIOR COMPONENTS

CYCLONE MODELS

A. Carburetor
B. Electronic pack
C. Ignition leads
D. Spark plug cables
E. Coil bracket
F. Cylinder cover
G. Fan guard
H. Recoil starter assembly

LIQUIFIRE MODELS

A. Electronic pack
B. Coupler
C. Fuel pump
D. Intake manifold
E. Coil cover and coils
F. Spark plug cables
G. Pump cover
H. Exhaust manifold
I. Recoil starter

folds. Remove starter motor, if so equipped. Remove intake manifold on piston-port engines.

3. Remove recoil starter assembly, starter cup, belt sheave, and window plate from flywheel.

4. Remove drive sheave as outlined in Chapter Eight.

5. Installation is the reverse of these steps. Keep the following points in mind:

   a. Use silicone rubber adhesive on exhaust gaskets.

   b. Torque all fasteners as specified in **Table 2** (end of chapter).

   c. Ensure plug wires are connected correctly. No. 1 cylinder is nearest fan cover.

Fan Cover Removal/Installation

Air cooled engines are equipped with one of 2 types of fan covers; one with the cooling fan mounted inside the cover (**Figure 55**) and one with the fan mounted outside the cover (**Fig-**

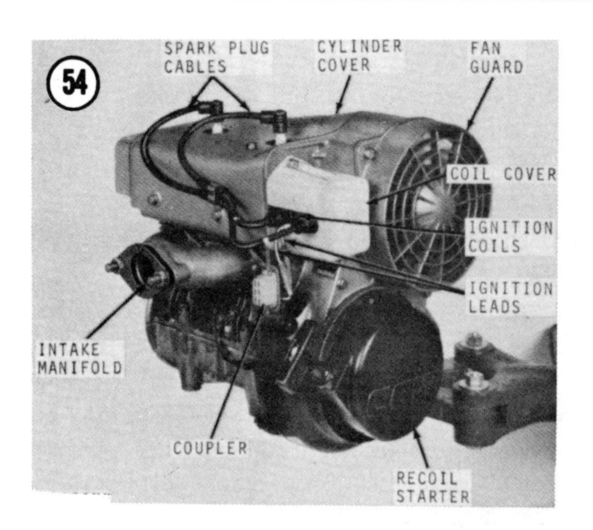

ure 56). Be sure to refer to correct illustration during removal and installation.

1. Remove retaining nut and washers. Remove pulley halves and spacer from fan shaft. Remove fan from units with outside fan (**Figure 56**).

2. Remove ignition coupler (**Figure 57**). Remove 4 bolts securing fan cover to backing plate and remove cover.

3. To disassemble fan assembly, gently tap assembly outer bearings with a soft mallet or hammer and block of wood. Inner bearing should remain on fan shaft and can be pulled off shaft. Gently drive outer bearing out of fan cover.

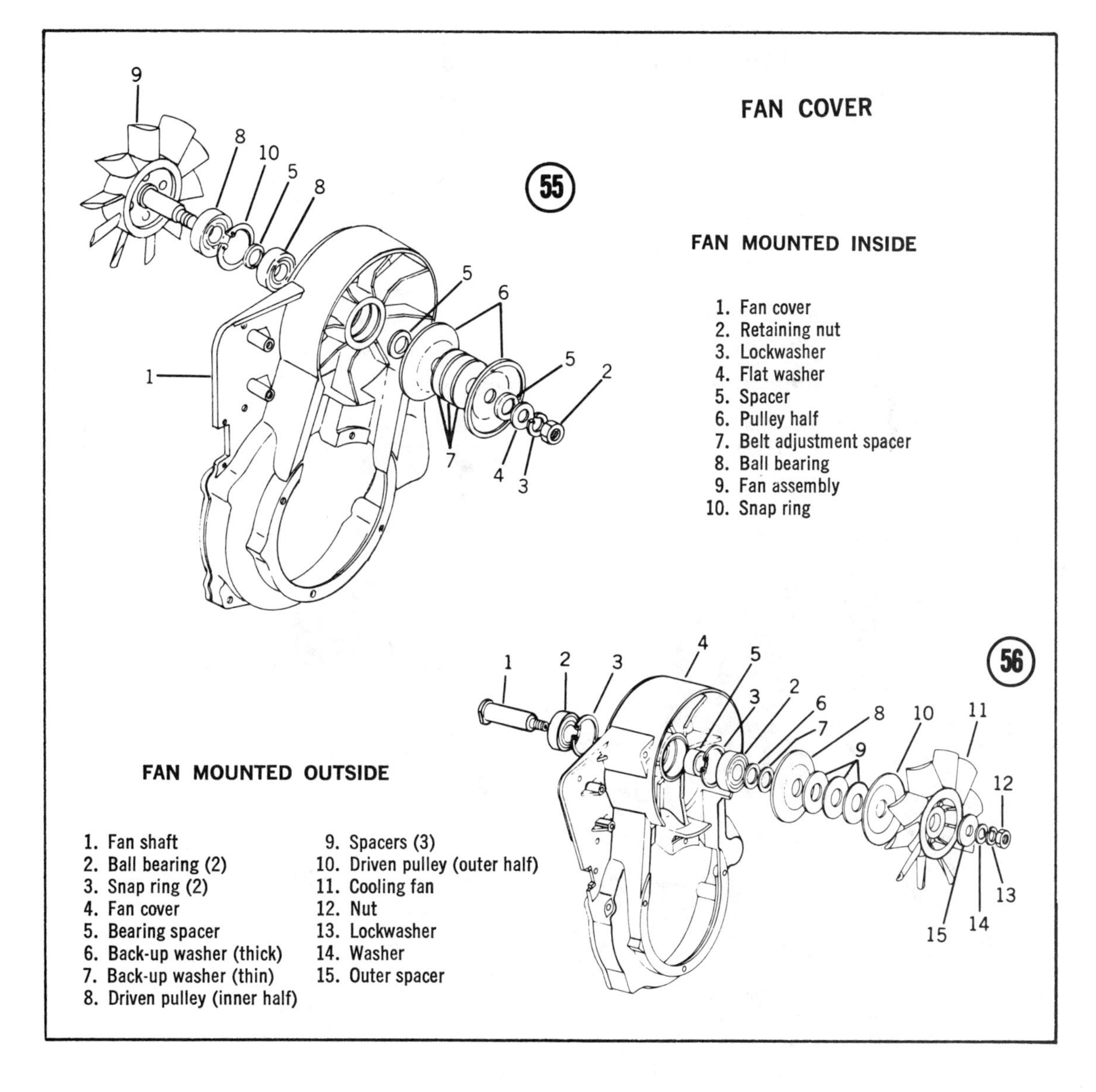

4. Inspect fan components and bearings as outlined under *Component Inspection* and replace defective components.

5. Remove bolts securing fan cover to engine and remove cover.

6. Installation is the reverse of these steps. Perform *Fan Belt Tension Adjustment*. Torque fan retaining nut to 28-31 ft.-lb. (3.9-4.3 mkg).

Coolant Pump and Flywheel Housing Removal/Installation (Liquifire Models)

Refer to **Figure 58** for this procedure.

1. Remove 3 screws securing starter cup and lower pulley to flywheel. Remove starter cup, pulley, and drive belt.

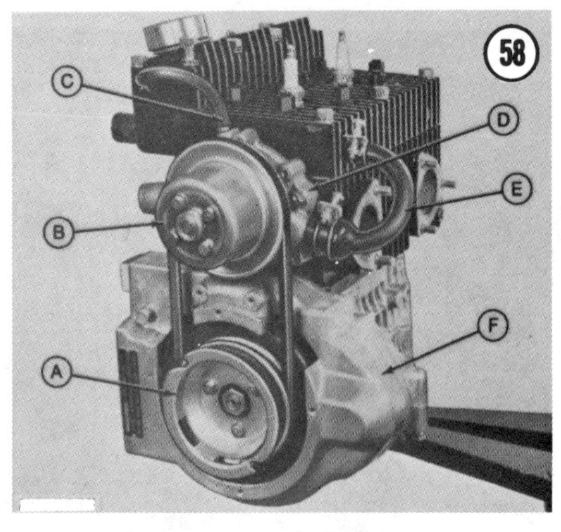

A. Starter cup
B. Pump pulley
C. Bleed hose
D. Coolant pump
E. Head-to-pump hose
F. Flywheel housing

2. Remove 4 screws securing pump drive pulley and remove pulley.

3. Remove head-to-pump and bleed hoses.

4. Remove 3 nuts securing coolant pump and remove pump.

5. Remove 4 bolts and remove flywheel housing.

6. Inspect pump and belt as outlined under *Component Inspection*.

7. Installation is the reverse of these steps. Ensure O-ring on pump is not defective and is properly located. Adjust pump for approximately ⅜ in. (9mm) deflection on drive belt when flexed near center of belt.

> NOTE: *It may be necessary to grind wrench thinner in order to tighten pump securing nuts behind pump pulley.*

Flywheel Removal/Installation

1. Remove fan cover or coolant pump and flywheel housing.

2. Install special flywheel holding tool or locally fabricated equivalent. See **Figure 59** for Cyclone, Liquifire, and JD295/S models. See **Figure 60** for other models.

3. While holding flywheel, remove retaining nut, lockwasher, and flat washer securing flywheel to crankshaft.

4. Install puller to flywheel using 3 tapped holes in flywheel. Hold flywheel and tighten puller center bolt (**Figure 61**) to 35-40 ft.-lb. (4.8-5.5 mkg).

CAUTION

Do not overtorque or hammer on center bolt of puller, otherwise, crankshaft and/or main bearing may be damaged.

If flywheel does not break loose, leave tension on puller and strike flywheel with a wood or plastic mallet in line with keyway.

CAUTION

Do not strike flywheel with a steel hammer or flywheel may be damaged.

5. Installation is the reverse of these steps. Ensure stator plate is securely in place before installing flywheel. Torque flywheel retaining nut (**Figure 62**) to 60 ft.-lb. (8.3 mkg) for JD295/S models and 45-50 ft.-lb. (6.2-6.9 mkg) for all other models.

Stator Assembly Removal/Installation

Refer to **Figure 63** for magneto equipped models and **Figure 64** for CDI equipped models.

1. Remove flywheel.

A. Stator assembly B. Trigger ring

2. On magneto models, remove 4 bolts securing backing plate to crankcase and remove backing plate.

3. On CDI models, remove 4 screws securing stator and trigger ring assembly to crankcase and remove stator and trigger ring.

> NOTE: *Store stator assembly inside flywheel to retain magnetic properties.*

5. Installation is the reverse of these steps. On magneto models, be sure wires and rubber grommets are installed in recess in crankcase flanges.

Cylinder Head Removal/Installation

Refer to **Figure 65** for air cooled models and **Figure 66** for liquid cooled models.

> NOTE: *Mark location of long head nuts to ensure they are installed in the correct position.*

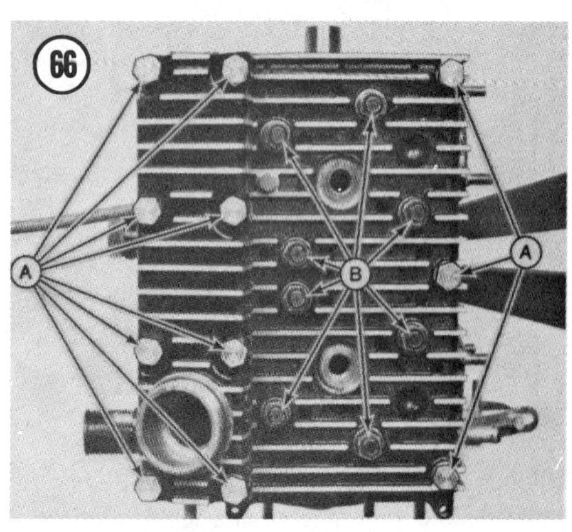

A. Head bolts B. Cylinder stud nuts

1. On air cooled models, remove 8 head retaining nuts. On liquid cooled models, remove 11 head bolts and 8 retaining nuts. To aid head removal, gently tap head with a rubber mallet. Remove head and discard old gaskets.
2. Refer to *Component Inspection* and inspect cylinder heads.
3. Use new head gaskets when installing cylinder heads. *Do not* use gasket sealant on head gaskets. Torque cylinder head nuts in sequence as shown in **Figure 67** for air cooled piston-port models, **Figure 68** for air cooled reed valve models, and **Figure 69** for liquid cooled models. Torque head nuts as follows:

   a. Air cooled piston-port models: 15-18 ft.-lb. (2.0-2.5 mkg)
   b. Air cooled reed valve models:
      Nuts A, B, C, and D:
      11.5-14.5 ft.-lb. (1.6-2.0 mkg)
      Nuts E, F, G, and H:
      5.0-6.5 ft.-lb. (0.69-0.89 mkg)
   c. Liquid cooled models:
      Nuts A through H:
      21-23 ft.-lb. (2.9-3.2 mkg)
      Nuts I through O:
      14-16 ft.-lb. (1.9-2.2 mkg)
      Nuts P through S:
      13-14 ft.-lb. (1.8-1.9 mkg)

Cylinder, Piston, and Ring Removal

1. Remove nuts and washers securing cylinders to crankcase (**Figure 70**).
2. Gently slide cylinders up over pistons.
3. Note mark on each piston indicating exhaust side of engine (**Figure 71**). If no marks are visi-

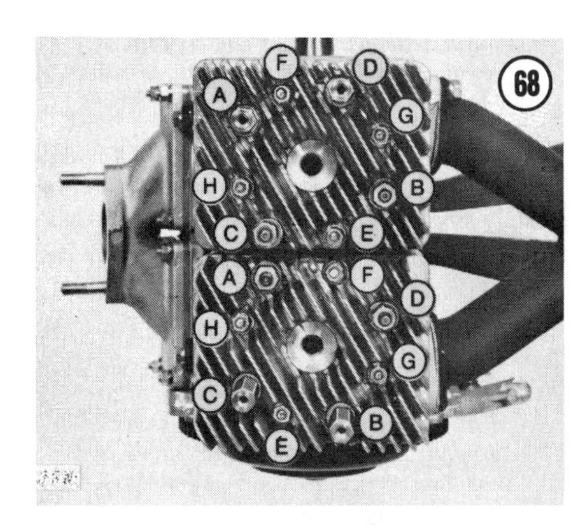

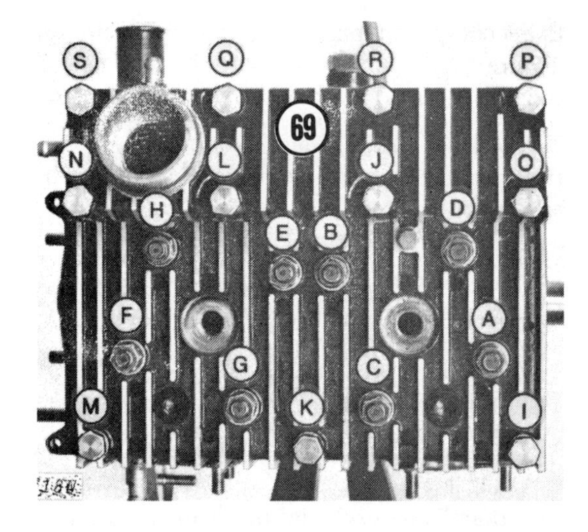

A. Circlip C. No. 1 piston
B. Mark D. No. 2 piston

ble, inscribe them accordingly. Also ensure pistons are marked 1 and 2 since they are not interchangeable.

NOTE: *If rings are going to be changed, but not pistons, rings may now be removed. However, if pistons are going to be removed, leave old rings on pistons to protect ring grooves until new rings are to be installed.*

4. Using a ring expander tool or your thumbs on each end of piston ring, gently expand ring and slide up and off piston (**Figure 72**).

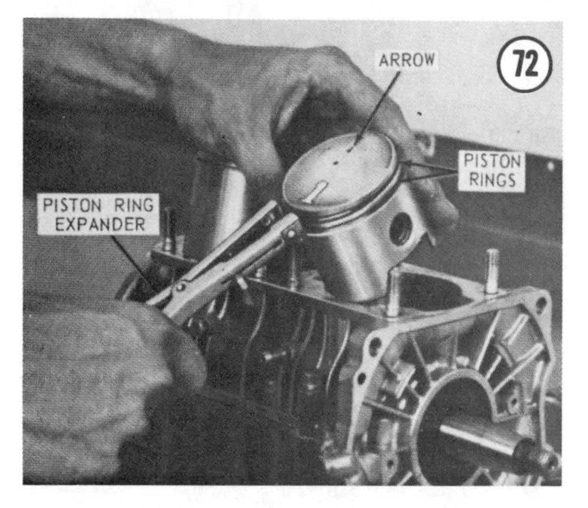

5. Be sure pistons are appropriately marked. Remember, pistons are *not* interchangeable. Remove circlips from each end of piston pins (**Figure 73**).

NOTE: *Stuff clean rags around connecting rods in crankcase to help prevent circlips from dropping into the crankcase.*

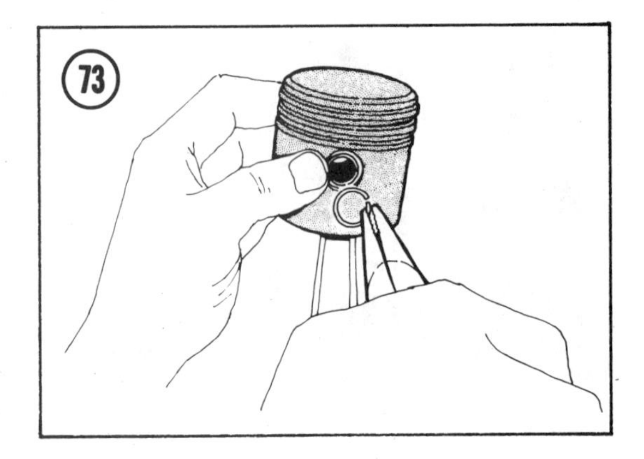

Using a piston pin removal tool or an appropriately sized wooden dowel, gently remove pins from piston and connecting rod (**Figure 74**).

CAUTION
Exercise care when removing pins to avoid damaging connecting rod needle bearings. If a wooden dowel is used to drive out piston pins, ensure piston is properly supported so lateral shock is not transmitted to lower connecting rod bearing, otherwise rod and/or bearing damage may occur.

Remove needle bearings from connecting rods.

6. Refer to *Component Inspection* and inspect pistons, pins, and rings.

Cylinder, Piston, and Ring Installation

1. Lubricate piston pin needle bearing with oil and insert bearing into connecting rods.
2. Slide piston over connecting rod.

CAUTION
Be sure pistons are installed in proper order with piston crowns forward or toward exhaust side of engine. Pistons are not interchangeable and improper installation will result in engine damage.

3. Using piston pin installation tool or appropriately sized wooden dowel, install piston pins through pistons and rod ends (**Figure 74**).

CAUTION
Exercise care when installing pins to avoid damage to connecting rod needle bearings. If a wooden dowel is used to drive in piston pins, ensure piston is properly supported so lateral shock is not transmitted to lower connecting rod bearing, otherwise rod and/or bearing damage may occur.

4. Secure piston pins to pistons with circlips (**Figure 73**).

NOTE: *Stuff clean rags around connecting rods in crankcase to help prevent circlips from dropping into the crankcase.*

CAUTION
If possible, use new circlips to secure piston pins. If old circlips are used, they must snap securely into grooves in pistons. A weak circlip could become disengaged during engine operation and cause severe engine damage.

5. Using a ring expander tool (**Figure 75**) or your thumbs on each end of piston ring, gently expand ring and slide over piston in ring groove. Install ring in bottom groove first. Be sure ring groove clearance is within tolerance as outlined in *Component Inspection.*

NOTE: *Be sure ring end gap is properly positioned around locating tab* (**Figure 75**). *On earlier models, chrome rings or Keystone style rings must be installed in top ring groove. Later model engines use Keystone rings in both piston grooves. See* **Figure 76** *for Keystone ring installation.*

6. Install new cylinder base gaskets over cylinder studs.
7. Thoroughly lubricate pistons, rings, and cylinder bores with engine oil.

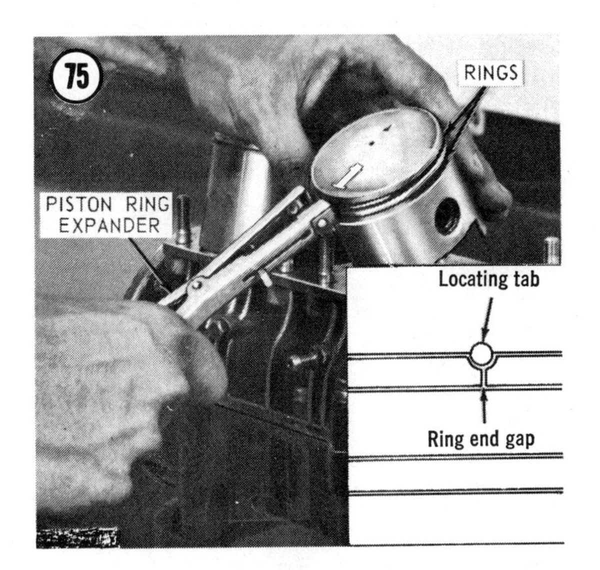

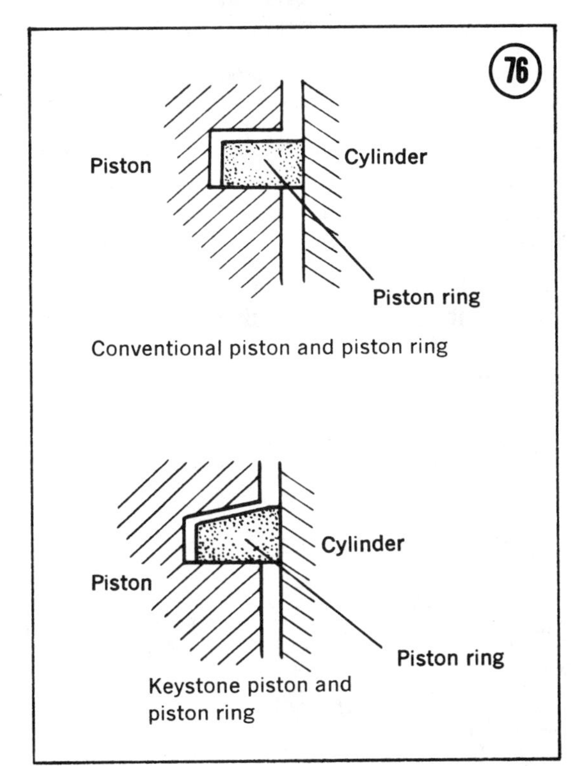

8. Position a suitable wooden block under pistons for piston support (**Figure 77**).

9. Compress rings with a suitable ring compressor or your fingers and carefully slide cylinder down over pistons. Ensure rings are properly centered around locating tabs (**Figure 75**).

> NOTE: *On some engines, cylinders are marked "R" and "L" for right and left. Install "R" cylinder on flywheel end of engine and "L" cylinder on* PTO *end* (**Figure 77**).

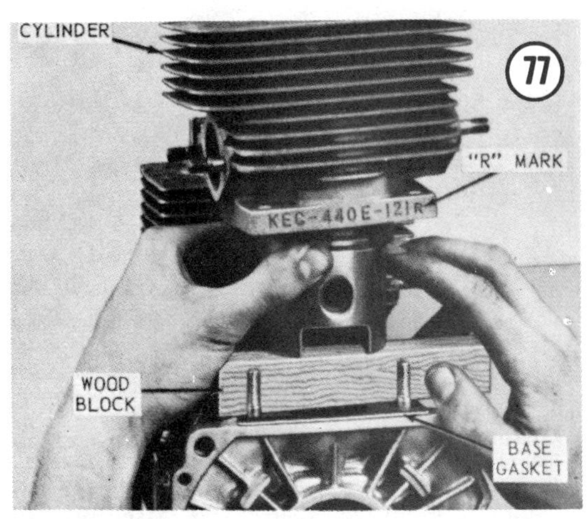

10. Install washers, lockwashers, and nuts on cylinder retaining studs. *Do not* tighten nuts at this time.

11. On piston-ported engines, install intake manifold and tighten nuts securing manifold.

12. On reed valve engines, use silicone rubber adhesive on exhaust manifold gaskets and install exhaust manifold. Tighten nuts securing exhaust manifold.

CAUTION

The appropriate manifold must be installed and tightened to align cylinders. Misalignment of cylinders will cause air leaks, resulting in serious engine damage.

13. On piston-ported engines, tighten 6 nuts securing cylinders to crankcase. Torque nuts to 15-18 ft.-lb. (2.0-2.5 mkg) as shown in **Figure 78**. Remove intake manifold to gain access to 2 remaining nuts and tighten to specified torque.

14. On reed valve engines, install cylinder heads.

Reed Valve Removal/Installation

1. Remove 4 nuts securing each reed valve assembly to crankcase and carefully remove reed valve assembly (**Figure 79**).

2. Carefully inspect reed valve assembly for warped or cracked reeds (**Figure 80**). Measure "hang open" between reed valve and valve body. Maximum "hang open" is 0.060 in. (1.524mm). Reed valve and body must have a "light-tight" seal when pinched together with finger pressure. Replace reeds that are damaged or out of tolerance.

CAUTION
Operating an engine with a faulty reed valve assembly could cause serious damage due to excessive lean mixture and improper fuel transfer.

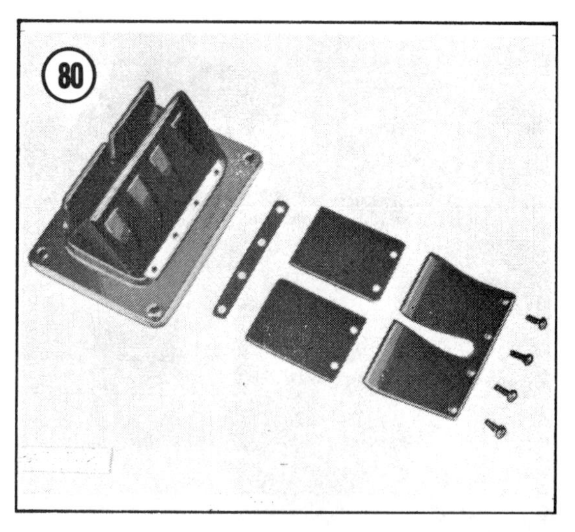

Crankshaft Assembly Removal (Reed Valve Engines)

1. Remove cylinder heads, cylinders, and pistons.

2. Remove 8 nuts and washers securing crankcase halves (**Figure 81**). Remove starter motor cover plate and cap screw. Remove 2 screws securing stator plate if not previously removed.

3. Lightly tap crankcase halves at joint with a soft hammer to separate halves.

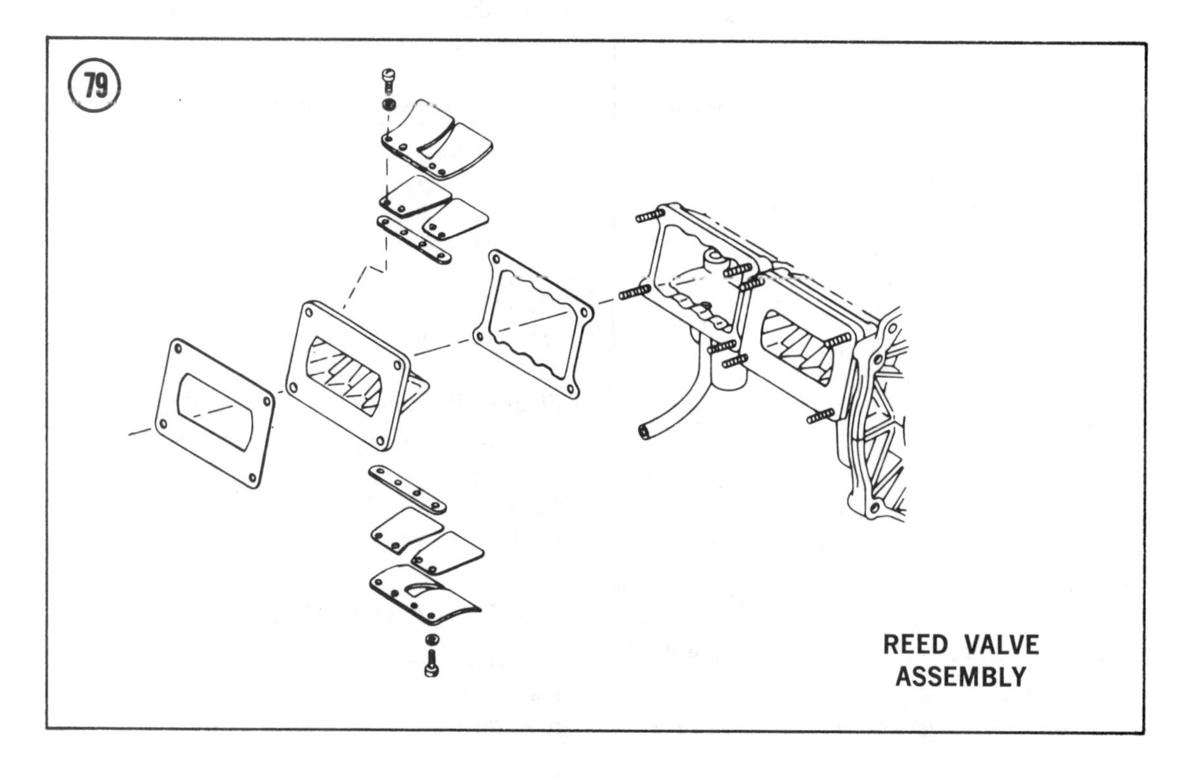

A. Nuts on bottom of crankcase
B. Starter motor cover plate
C. Cap screw

CAUTION

Never attempt to pry crankcase halves apart with a screwdriver or similar object or crankcase sealing surfaces will be damaged.

4. Gently lift up and remove crankshaft assembly. Ensure 4 bearing retaining clips in upper crankcase are not lost.

5. Perform crankshaft assembly and crankcase inspection outlined in *Component Inspection*. Refer all crankshaft assembly repair and service to an authorized dealer or competent machine shop. They are equipped with the necessary special tools and expertise for the work.

Crankshaft Assembly Installation (Reed Valve Engines)

1. Stand upper crankcase half on stud bolts. Install bearing retaining clips in crankcase (**Figure 82**).

2. Thoroughly lubricate crankshaft and bearings with engine oil and install crankshaft in upper crankcase half. Ensure bearing clips are correctly positioned.

3. Check that crankcase sealing surfaces are clean and not damaged. Apply an even coat of silicone rubber adhesive to sealing surfaces on both crankcase halves. Make sure no rubber adhesive runs into crankcase.

4. Install stator assembly over crankshaft. Position rubber grommet in recess between upper and lower crankcase halves.

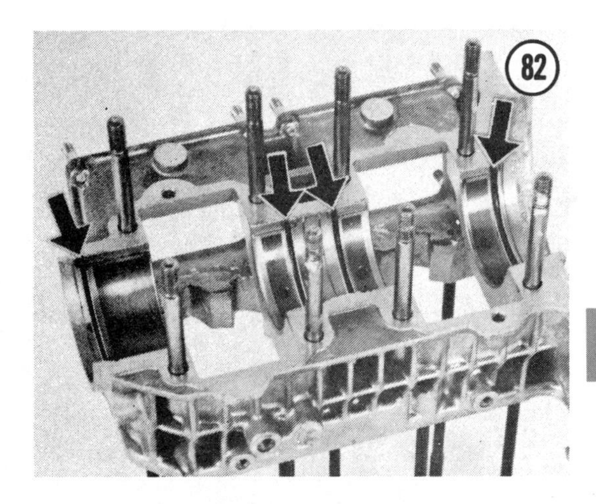

5. Install lower crankcase half over upper crankcase half. Ensure 2 dowel pins engage holes on opposite crankcase half.

6. Install flat washers, lockwashers and nuts to crankcase studs. Tighten nuts evenly in sequence shown in **Figure 83**. Torque nuts to 15-18 ft.-lb. (2.0-2.5 mkg).

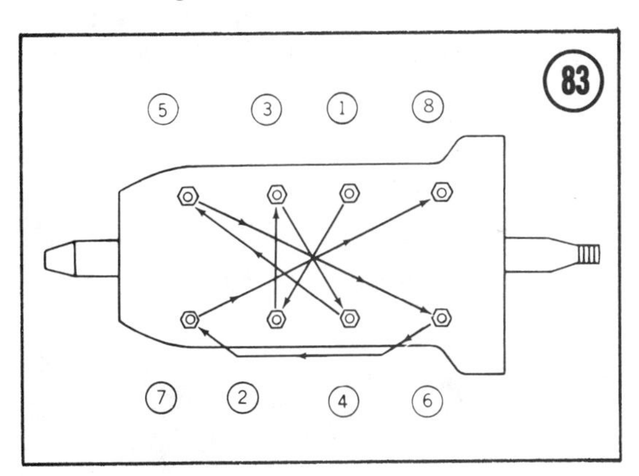

Crankshaft Assembly Removal (Piston-Ported Engines)

1. Remove the cylinder heads, cylinders, and pistons.

2. Remove cap screws securing crankcase together.

3. Tap upper crankcase half lightly with a soft mallet to separate crankcase (**Figure 84**).

CAUTION

Never attempt to pry crankcase halves apart with screwdriver or similar object or crankcase sealing surface will be damaged.

4. Gently lift up and remove crankshaft assembly. Ensure 4 bearing retaining clips in lower crankcase half are not lost.

5. Perform crankshaft assembly and crankcase inspection as outlined in *Component Inspection*. Refer all crankcase assembly repair and service work to an authorized dealer or competent machine shop. They are equipped with the necessary special tools and expertise for the task.

Crankshaft Assembly Installation (Piston-Ported Engines)

1. Install 4 bearing retaining clips in lower crankcase half as shown in **Figure 85**.

2. Install crankshaft assembly into lower crankcase half (**Figure 86**). Check that bearing retaining clips are correctly positioned. Position seals

tightly against clips. Ensure flywheel end of crankshaft is positioned on the same end as flanged end of crankcase. Thoroughly lubricate crankshaft and bearings with engine oil.

3. Check that crankcase sealing surfaces are clean and not damaged. Apply an even coat of silicone rubber adhesive to sealing surfaces on both crankcase halves. Make sure no rubber adhesive runs into crankcase.

4. Install upper crankcase half. Check that seals are correctly positioned and are not cocked.

5. Install cap screws, flat washers and lockwashers and tightenly evenly in sequence shown in **Figure 87**. Torque cap screws to 15-18 ft.-lb. (2.0-2.5 mkg). Letter "A" in **Figure 87** identifies upper 4 cap screws.

PTO End Bearing Seal Removal/Installation

On early models engines, if drive belt breaks and winds around back side of clutch fixed face, check PTO end bearing seal for damage. Later model engines are equipped with a PTO end seal guard (**Figure 88**) to prevent damage from a broken belt.

The PTO end bearing seal can be replaced without removing the engine.

1. Remove clutch.

2. Remove PTO end seal guard if engine is so equipped.

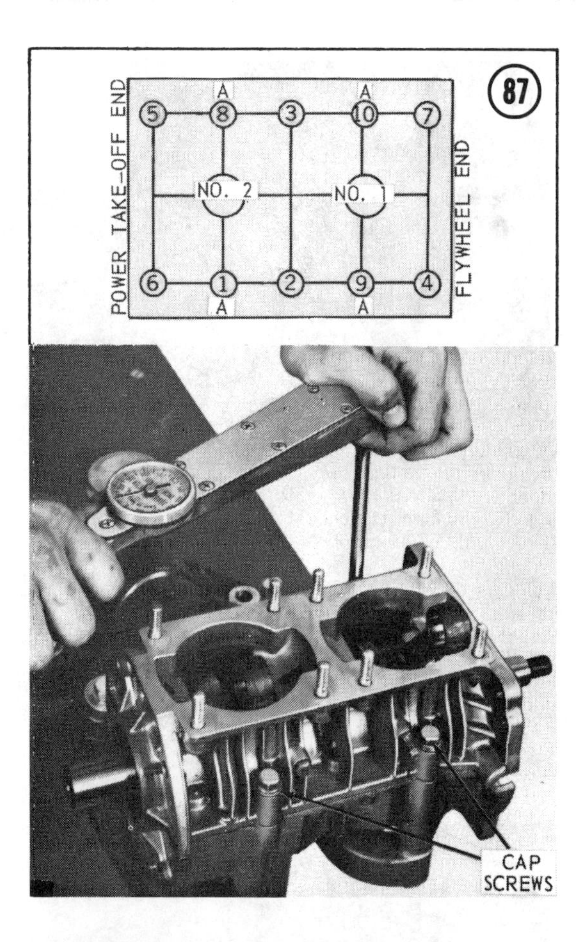

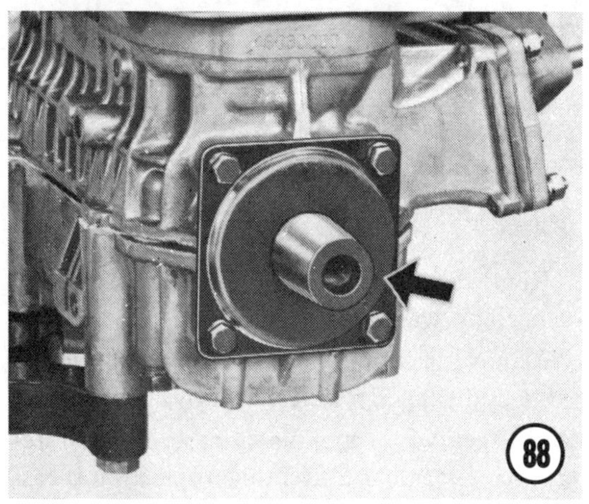

3. Using a cold chisel make a slot in old seal and pry old seal out with a long screwdriver blade (**Figure 89**).

CAUTION

Exercise care not to chisel too deep or damage to crankshaft and/or bearing may result.

A. Wooden block B. PTO end bearing seal C. Screwdriver

4. Clean crankshaft surface and coat inner and outer seal surfaces with STP or equivalent to ease seal installation. Do not use snowmobile oil as a lubricant.

5. Install new seal with a 4x4 in. (10.1x10.1cm) block of hard wood approximately 12 in. (30.5 cm) long with a 1¼ in. (3.14 cm) diameter hole drilled 3 in. (7.6 cm) deep in end of block. Place block over crankshaft, align seal, and carefully drive seal into place flush with crankshaft. Ensure seal is properly aligned.

6. To install seal guard (if engine is so equipped), center hole in guard around crankshaft before tightening screws to prevent guard from contacting crankshaft.

Fan Belt Tension Adjustment

1. Install fan belt window plate belt sheave and starter cup if components were removed. Secure to flywheel with three 6x15mm cap screws and lockwashers.

CAUTION

Use 6 x 22mm cap screws from recoil starter to secure starter cup to flywheel. Longer cap screws will make centrifugal advance mechanism inoperative.

2. Check fan belt for proper tension as shown in **Figure 90**. Properly adjusted fan belt should deflect approximately ⅜ in. (9mm) when flexed at a point near center of belt span.

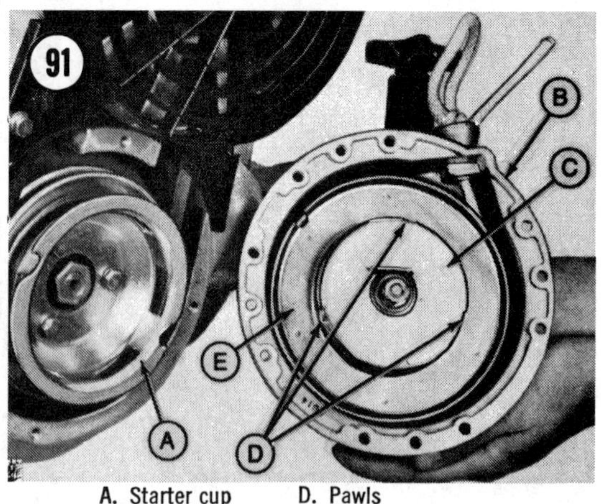

A. Starter cup
B. Recoil starter
C. Friction plate
D. Pawls
E. Reel

3. To adjust belt, remove fan retaining nut and proceed as follows:

a. On engines with fan outside belt (**Figure 56**) remove fan, belt and outer sheave half. Remove one of the larger spacers from between pulley halves and place it outside outer pulley half. Reassemble and torque retaining nut to 28-31 ft.-lb. (2.5-4.3 mkg).

b. On engines with belt outside fan (**Figure 55**), remove outer pulley half. Remove one of the large spacers from between pulley halves and place on outside of outer pulley. Install retaining nut and torque to 28-31 ft.-lb. (2.5-4.3 mkg).

c. Recheck the belt tension and readjust if necessary.

RECOIL STARTERS (C.C.W. and Kioritz Engines)

Removal/Installation

Remove bolts securing recoil starter assembly to engine and lift off starter (**Figure 91**). Installation is the reverse of removal.

Disassembly

Refer to **Figure 92** for this procedure.

1. Untie knot in recoil starter handle and let rope recoil into case. Unscrew rope guide (**Figure 93**) out of case.

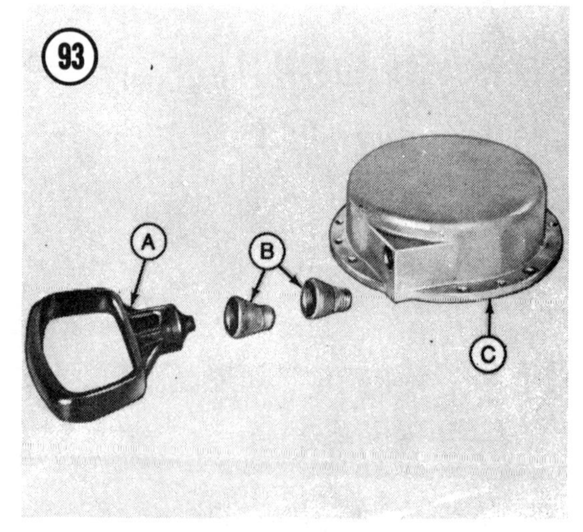

A. Handle B. Rope guides C. Case

2. Remove retaining nut (**Figure 94**), lockwasher and washer from reel shaft.
3. Remove friction plate, return spring, friction spring and cupped washer. Remove 3 pawls.
4. Lift reel, with rope, out of case. Remove recoil spring from case. Remove rope from reel if rope is to be replaced.

Inspection

1. Thoroughly clean all parts, except rope, in cleaning solvent.
2. Inspect recoil starter cup on engine for excessive wear in rectangular slots as shown in **Figure 95**. Replace if necessary.

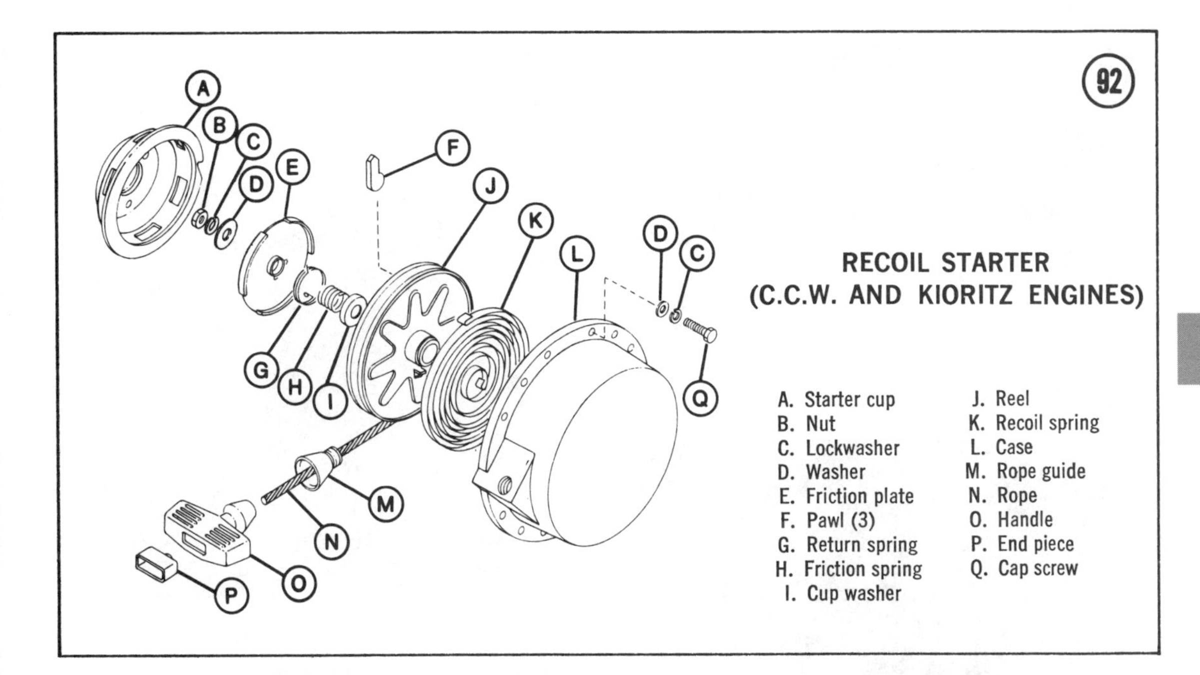

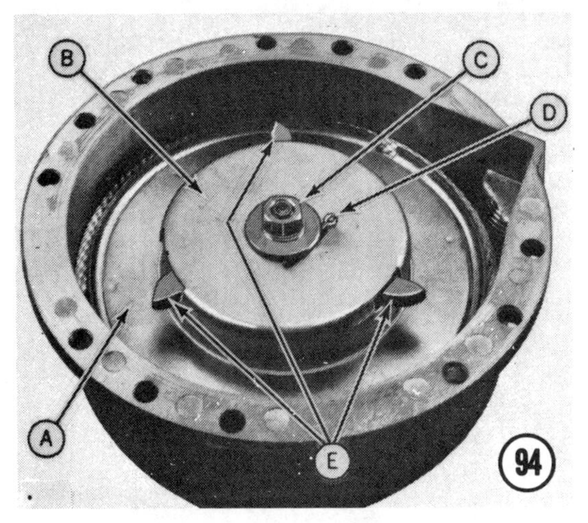

A. Reel
B. Friction plate
C. Retaining nut
D. Return spring
E. Pawls

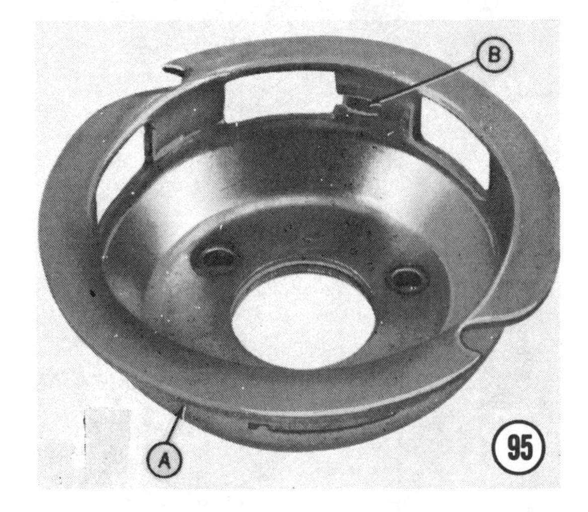

A. Starter cup B. Worn slots

3. Inspect 3 pawls and friction plate for wear (**Figure 96**).

4. Inspect return spring and friction spring and replace if cracked, broken, or distorted.

5. Inspect reel hub and reel shaft for excessive clearance (**Figure 97**). Replace parts as necessary.

6. Replace the rope if frayed or broken. See **Figure 98**. Replace coil spring if broken or distorted.

7. Inspect condition of spring ends, and bend, if necessary, so spring properly engages tabs on reel and case when installed.

Assembly

Refer to **Figure 92** for this procedure.

1. Install rope through hole in reel. Loop and tie knot as shown in **Figure 99**.

2. Place loop around reel hub and pull tight. Wind rope counterclockwise around reel when viewed as shown in **Figure 100**.

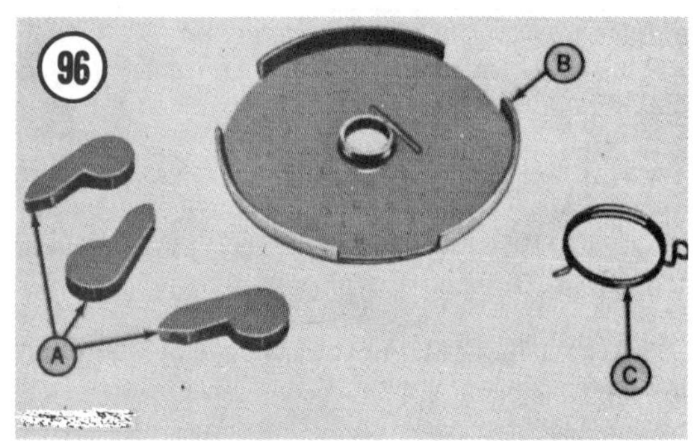

A. Pawls B. Friction plate C. Return spring

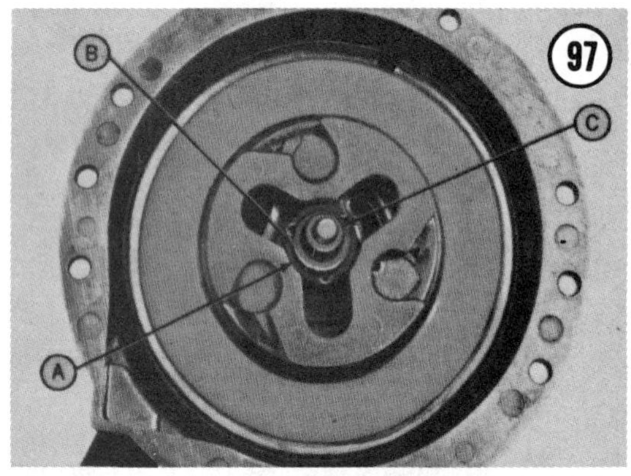

A. Reel hub B. Excessive clearance C. Reel shaft

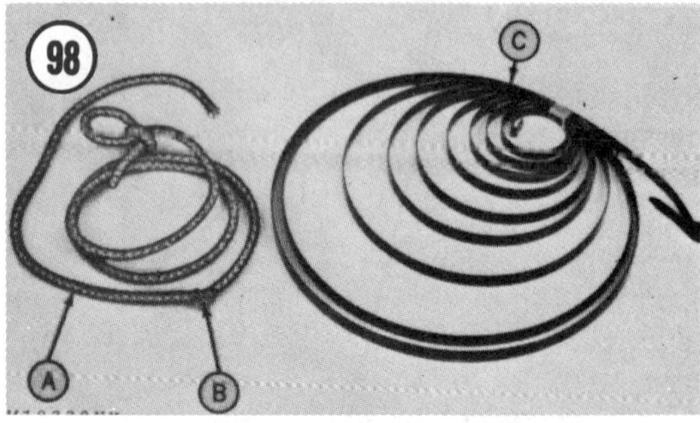

A. Rope B. Frayed C. Recoil spring

3. Place recoil spring into spring winding tool as shown in **Figure 101**, so spring can be wound onto tool by winding clockwise.

4. Assemble winding tool and wind plate, with handle, clockwise to wind spring into tool as shown in **Figure 102**.

5. Let plate with handle unwind and remove bolt and plate.

6. Set spring winding tool with spring into case with loop on spring end positioned as shown in **Figure 103**, around dowel pin.

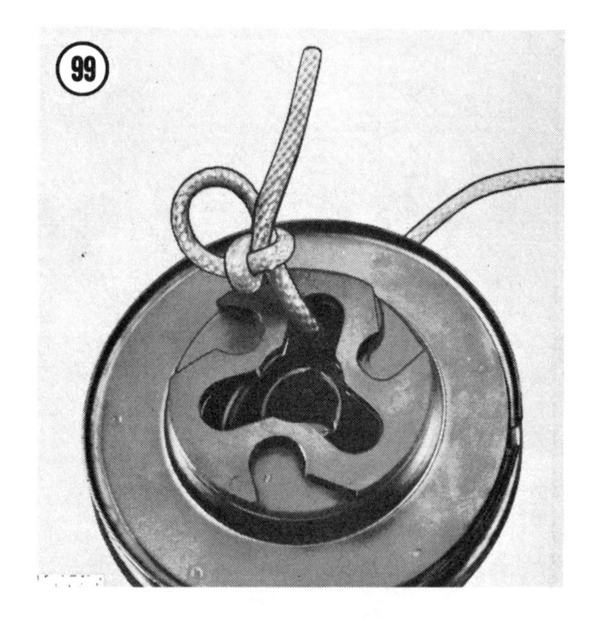

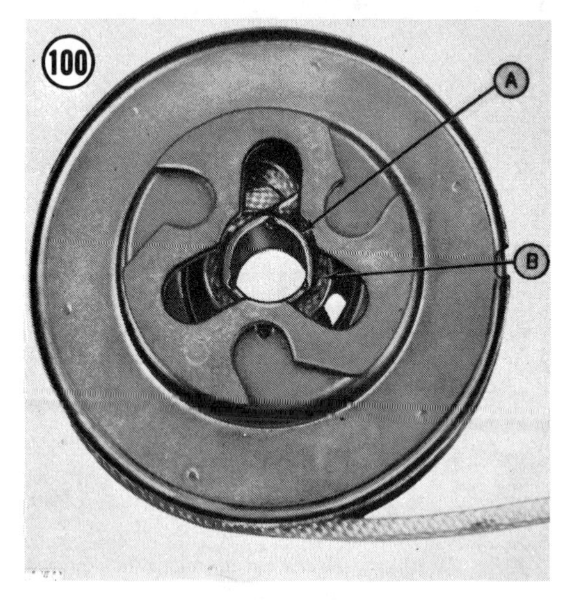

A. Reel hub B. Loop

7. Carefully pull winding tool off recoil spring, leaving spring in position in case. Lightly coat spring with low-temperature grease.

8. Apply a light film of low-temperature grease to reel shaft. Place end of rope through notch in reel and place reel in position on reel shaft (**Figure 104**).

> NOTE: *Position inner end of recoil spring away from reel shaft or reel will not drop into proper position.*

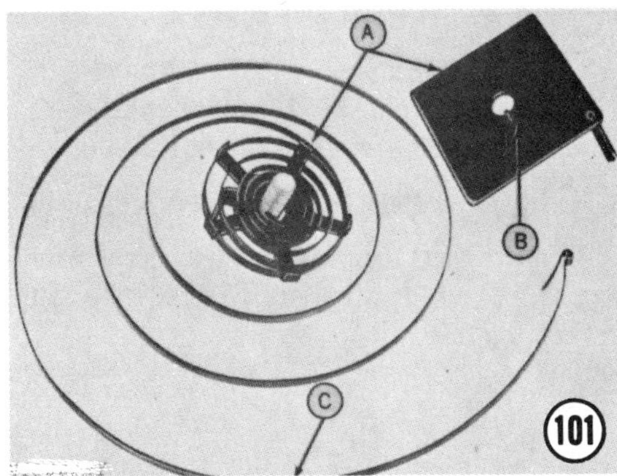

A. Spring winding tool B. Spring pin C. Recoil spring

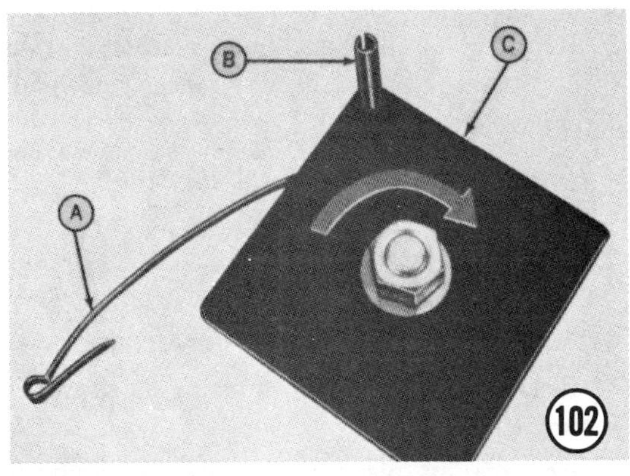

A. Recoil spring B. Handle C. Plate

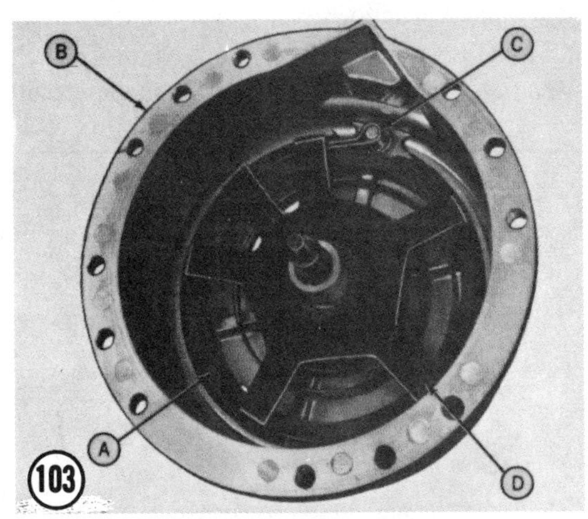

A. Recoil spring
B. Case
C. Dowel pin
D. Spring winding tool

9. Install cupped washer, cup up, onto reel shaft. Place friction spring and return spring into place.

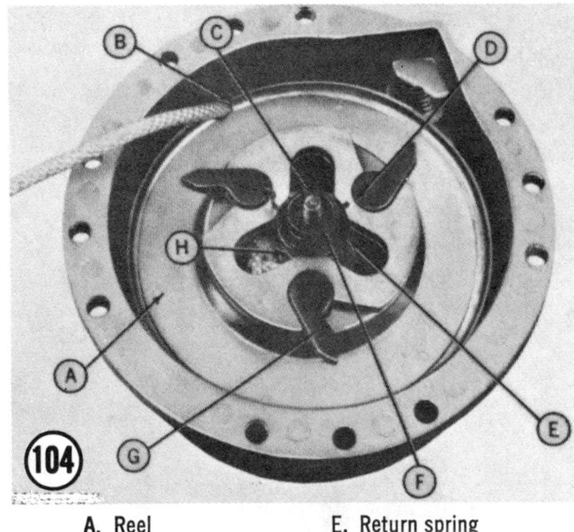

A. Reel
B. Notch
C. Reel shaft
D. End of return spring in retaining hole
E. Return spring
F. Friction spring
G. Pawl
H. Cupped washer

NOTE: *End of return spring must be placed into retaining hole in reel* (**Figure 104**), *or pawls will not retract properly.*

10. Apply a light film of low-temperature grease to 3 pawls and position them as shown in **Figure 104**.

11. Slip eye of return spring (**Figure 105**) through slot in friction plate and position friction plate so notches line up with pawls. Install washer, lockwasher and nut and tighten securely. Lubricate washer with anti-seize compound.

A. Notch
B. Washer
C. Nut
D. Return spring eye
E. Slot
F. Pawl
G. Friction plate

4

12. Tension reel by winding it 3 turns counterclockwise. Insert rope through eye of case and rope guide and tie a slip knot (**Figure 106**). Screw rope guide into case.

> NOTE: *Heat end of rope with a match to fuse strands* (**Figure 106**) *to enable rope to be threaded easily through rope guide and handle.*

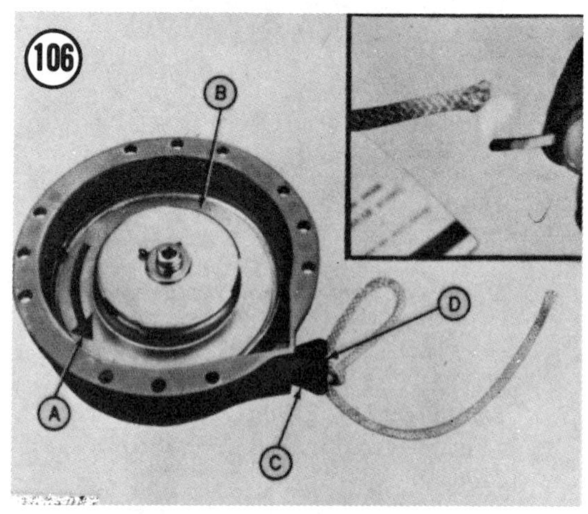

A. 3 turns B. Reel C. Rope Guide D. Slip knot

13. Install end of rope through handle. Tie a knot and pull knot into handle. Untie slip knot and check operation of recoil starter. Pawls should extend when rope is pulled and retract when rope is released. Rope should recoil sharply back into case.

RECOIL STARTER

(Kohler Engines on JDX4 Models Serial No. 2551-20,000 and 300 Models Serial No. 20,001-21,504)

Removal/Installation

Remove bolts securing recoil starter assembly to engine and lift off starter. Installation is the reverse of removal.

Disassembly

Refer to **Figure 107** for this procedure.

1. Remove rope handle and allow pulley to unwind.
2. Remove circlip, retainer washer, brake spring, steel washer, fiber washer, and pawl assembly. Remove recoil spring.

> NOTE: *Recoil spring in steel case is not repairable.*

Assembly

Refer to **Figure 107** for this procedure.

1. Install recoil spring in housing as shown in **Figure 108**.
2. Remove any piece of old rope remaining in pulley. Save rope lock that pulley end of rope is wrapped around.
3. Fuse both ends of new rope with a match. Bend one end of rope into a "U" shape around

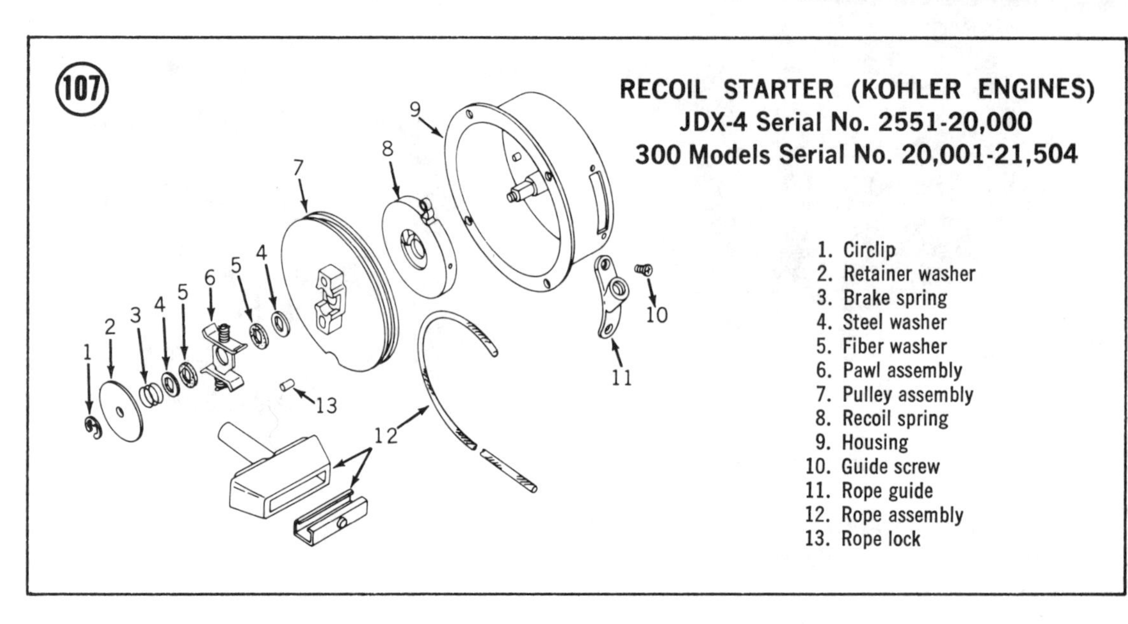

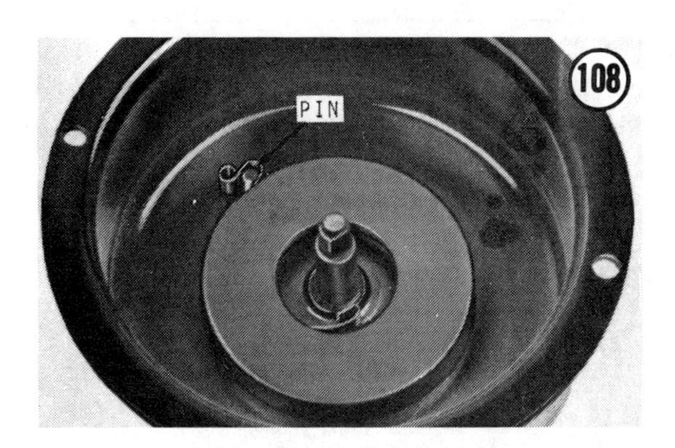

rope lock (**Figure 109**) and insert into hole in pulley hub. Be sure loop is flush with top of pulley hub. Run rope in groove on inside of pulley. Wrap rope around pulley in a counter-clockwise direction (when viewed from outside face of pulley).

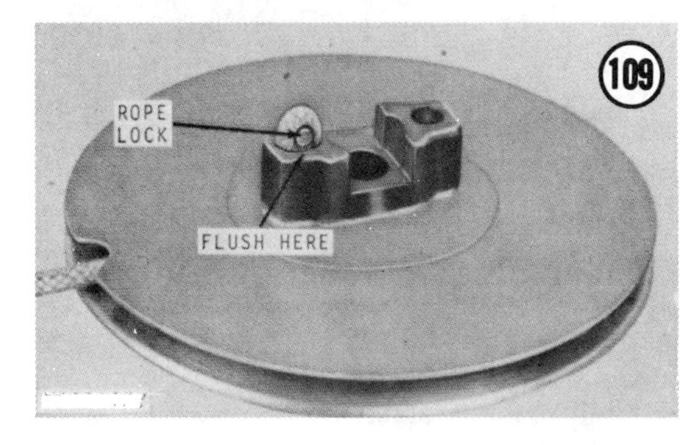

4. Pull open end of rope through notch in pulley. Install pulley with rope in housing.
5. Install steel washer, fiber washer and pawl assembly as shown in **Figure 110**. Install fiber washer, steel washer, brake spring, retainer washer and circlip.

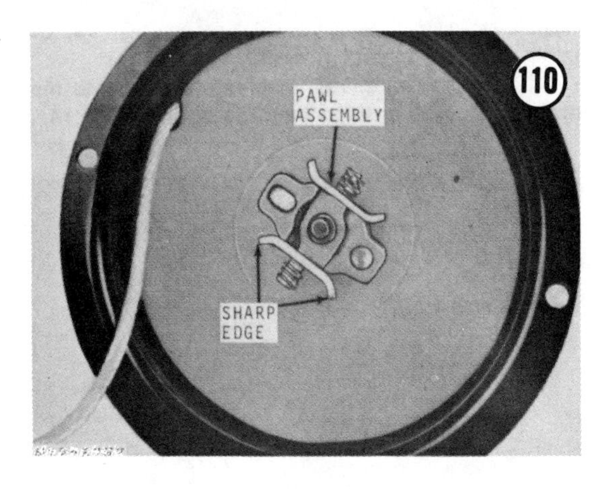

6. Pre-tension pulley by rotating counterclock-wise 3½ to 4 turns. While holding this position, thread rope end through rope guide.
7. Pull rope out about 6 inches and tie a slip knot in rope to prevent it from retracting. Insert rope through handle and hole in retainer. Tie a permanent knot in end of rope and then install retainer in handle.
8. Release slip knot and guide rope slowly back into retracted position.

RECOIL STARTER

(Kohler Engine on 300 Models Serial No. 21,505 and Later)

Removal/Installation

Remove bolts securing recoil starter assembly to engine and lift off starter. Installation is the reverse of removal.

Disassembly

Refer to **Figure 111** for this procedure.

1. Remove rope handle, and allow pulley to unwind.
2. Remove center screw, pawl cam, brake spring and brake washer.
3. Remove retaining rings, pawls and pawl springs.
4. Lift pulley, spring and keeper assembly out of housing.

NOTE: *Recoil spring in steel case is not repairable.*

Inspection

1. Inspect rope and replace if broken or frayed.
2. Check pawls and starter cup engagement surfaces for wear.
3. Check pawl return springs for adequate tension.

Assembly

Refer to **Figure 111** for this procedure.

1. Install spring and keeper assembly into housing, placing loop on end of spring into notch as shown in **Figure 112**. Apply a liberal coating

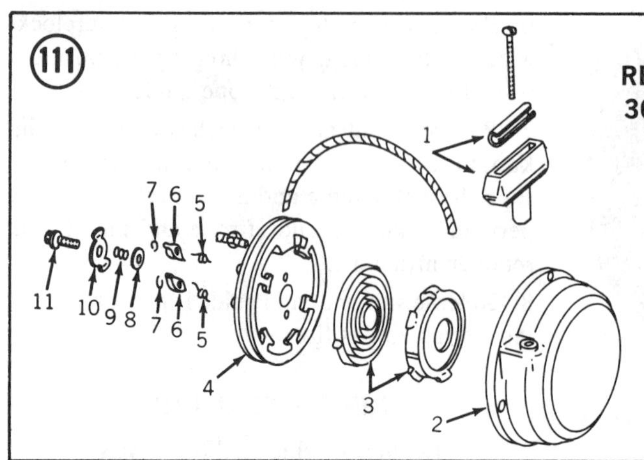

RECOIL STARTER (KOHLER ENGINES)
300 Models Serial No. 21,505 and Later

1. Handle
2. Housing
3. Spring and keeper
4. Pulley
5. Pawl spring
6. Pawl
7. Retaining rings
8. Brake washer
9. Brake spring
10. Pawl cam
11. Center screw

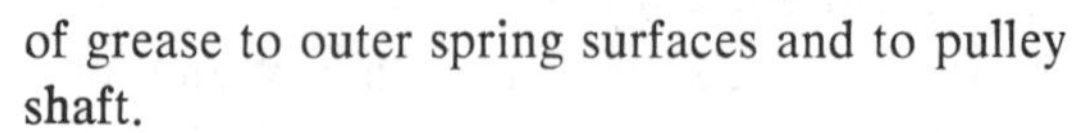
of grease to outer spring surfaces and to pulley shaft.

2. Lubricate pawl pivot shafts with oil. Install pawl springs, pawls, and retaining rings on pulley.

3. Fuse both ends of new rope with a match. Insert rope through hole in pulley and tie a permanent knot in end of rope. Wrap rope around pulley in a counterclockwise direction. File a small notch in pulley rim as shown in **Figure 113** to permit easier spring pretensioning.

4. Hold rope in notch, and install pulley into housing. Rotate pulley until it drops in place (**Figure 113**).

5. Install brake washer, brake spring and pawl cam. Lubricate center screw bearing surfaces with a light coat of grease. Install center screw and torque to 5 ft.-lb. (0.7 mkg).

6. Hold rope in notch and wind pulley counterclockwise 2 turns. Insert rope through hole in housing.

7. Pull rope out about 12 inches and tie a slip knot in rope to prevent it from retracting.

8. Insert rope through handle and hole in retainer. Tie a permanent knot in end of rope and install retainer in handle.

9. Release slip knot and guide rope slowly back into retracted position (**Figure 114**).

Table 2 TORQUE SPECIFICATIONS

	Kioritz and CCW	Kohler
Crankcase	15-18 ft.-lb. (2.0-2.5 mkg)	—
Cylinder-to-crankcase	15-18 ft.-lb. (2.0-2.5 mkg)	15-18 ft.-lb. (2.0-2.5 mkg)
Cylinder head (air cooled)	15-18 ft.-lb. (2.0-2.5 mkg)	15-18 ft.-lb. (2.0-2.5 mkg)
Cylinder head (liquid cooled)	21-23 ft.-lb. (2.9-3.2 mkg)	—
	14-16 ft.-lb. (1.9-2.2 mkg)	—
	13-14 ft.-lb. (1.8-1.9 mkg)	—
	Refer to cylinder head installation	
Intake and exhaust manifold	10-12 ft.-lb. (1.4-1.7 mkg)	15-18 ft.-lb. (2.0-2.5 mkg)
Flywheel to crankshaft	60 ft.-lb. (8.3 mkg)	85-90 ft.-lb. (11.7-12.4 mkg)
Fan pulley retaining nut	28-31 ft.-lb. (3.9-4.3 mkg)	30-40 ft.-lb. (4.1-5.5 mkg)
PTO end bearing plate	—	8-10 ft.-lb. (1.1-1.4 mkg)
Spark plug	14 ft.-lb. (1.9 mkg)	14 ft.-lb. (1.9 mkg)

Table 3 (Dimensions and Wear Tolerances) follows on page 90.

Table 3 DIMENSIONS AND WEAR TOLERANCES

Item	300 JDX4	400 JDX4 Special	500, 600	Wear Tolerance
Cylinder bore	2.2631-2.2651 in. (57.483-57.534mm)	2.3622-2.3629 in. (59.999-60.018mm)	2.6772-2.6779 in. (68.001-68.019mm)	0.005 in. (0.127mm)
Connecting rod small end	0.7872-0.7876 in. (19.995-20.005mm)	0.8664-0.8669 in. 22.0007-22.019mm)	0.8664-0.8669 in. (22.007-22.019mm)	0.0005 in. (0.0127mm)
Connecting rod side clearance	0.008-0.016 in. (0.203-0.406mm)	0.014-0.016 in. (0.356-0.406mm)	0.014-0.016 in. (0.356-0.406mm)	+0.004 in. (+0.1016mm)
Crankshaft runout	0.0012 in. (0.030mm)	0.0035 in. (0.089mm)	0.0035 in. (0.089mm)	
Crankshaft twist				0.003 in. off TDC (0.076mm)
Piston top land	2.2504-2.2516 in. (57.160-57.191mm)	2.3500-2.3510 in. (59.690-59.715mm)	2.6650-2.6660 in. (67.691-67.716mm)	—
Piston at skirt	2.2600-2.2608 in. 57.404-57.424mm)	2.3570-2.3580 in. (59.868-59.893mm)	2.6720-2.6730 in. (67.869-67.894mm)	0.005 in. (0.127mm)
Piston pin bore	0.6300-0.6302 in. (16.002-16.007mm)	0.7083-0.7087 in. (17.991-18.001mm)	0.7083-0.7087 in. (17.991-18.001mm)	No clearance permissible
Piston pin	0.6297-0.6299 in. (15.994-15.999mm)	0.7084-0.7087 in. (17.993-18.001mm)	0.7084-0.7087 in. (17.993-18.001mm)	—
Ring groove clearance (top)	0.0022-0.0037 in. (0.056-0.094mm)	0.004 in. (0.1016mm)	0.004 in. (0.1016mm)	+0.002 in. (+0.0508mm)
Ring groove clearance (bottom)	0.0012-0.0024 in. (0.030-0.061mm)	0.003 in. (0.0762mm)	0.003 in. (0.0762mm)	+0.002 in. (+0.0508mm)
Crankshaft O.D. (PTO end)	30mm	30mm (see note) ①	30mm	—
Crankshaft end play	0.006-0.012 in. (0.1524-0.3048mm)	0.003 in. (0.0762mm)	0.003 in. (0.0762mm)	—
		(continued)		

Table 3 **DIMENSIONS AND WEAR TOLERANCES** (continued)

Item	JDX6, 800, JDX8	JD295/S	Wear Tolerance
Cylinder bore	2.6000-2.6007 in. (66.04-66.058mm)	2.2047-2.2050 in. (55.999-56.007mm)	0.005 in. (0.127mm)
Connecting rod small end	0.7870-0.7878 in. (19.990-20.010mm)	0.8660-0.8667 in. (21.996-22.014mm)	0.0005 in. (0.0127mm)
Connecting rod side clearance	0.014-0.016 in. (0.3556-0.4064mm)	0.014-0.016 in. (0.3556-0.4064mm)	+0.004 in. (0.1016mm)
Crankshaft runout	0.0035 in. (0.0889mm)	0.0008 in. (0.0203mm)	
Crankshaft twist			0.003 in. off TDC (0.0762mm)
Piston at top land	2.5880-2.6000 in. (65.735-66.04mm)	2.1870-2.1880 in. (55.550-55.575mm)	—
Piston at skirt	2.5953-2.6000 in. (65.921-66.04mm)	2.1950-2.1960 in. (55.753-55.778mm)	0.005 in. (0.127mm)
Piston pin bore (see note) ②	0.7083-0.7087 in. (17.991-18.001mm)	0.7080-0.7087 in. (17.983-18.001mm)	No clearance permissible
Piston pin (see note) ②	0.7084-0.7087 in. (17.993-18.001mm)	0.7083-0.7087 in. (17.991-18.001mm)	—
Ring groove clearance (top)	0.004 in. (0.1016mm)	0.002-0.0035 in. (0.0508-0.0889mm)	+0.002 in. (0.0508mm)
Ring groove clearance (bottom)	0.003 in. (0.0762mm)	0.002-0.0035 in. (0.0508-0.0889mm)	+0.002 in. (0.0508mm)
Crankshaft O.D. (PTO end)	30mm	30mm	—
Crankshaft end play	0.003 in. (0.0762mm)	0.008 in. (0.2032mm)	—

(continued)

4

Table 3 **DIMENSIONS AND WEAR TOLERANCES** (continued)

Item	340 Cyclone, Liquifire	440 Cyclone, Liquifire	Wear Tolerance
Cylinder bore	2.2843-2.2850 in. (58.00-58.01mm)	2.5998-2.6006 in. (66.03-66.05mm) See Note ③	0.005 in. (0.127mm)
Connecting rod small end	0.8664-0.8669 in. (22.00-22.01mm)	0.8664-0.8669 in. (22.00-22.01mm)	0.0008 in. (0.0203mm)
Connecting rod side clearance	0.0078-0.0118 in. (0.198-0.298mm)	0.0078-0.0118 in. (0.198-0.298mm)	—
Crankshaft runout	0.0008-0.0020 in. (0.020-0.050mm)	0.0008-0.0020 in. (0.020-0.050mm)	—
Crankshaft twist	—	—	0.003 in. (0.076mm) off TDC
Crankshaft end play	0.0098-0.0197 in. (0.249-0.500mm)	0.0098-0.0197 in. (0.249-0.500mm)	—
Piston at top land	2.2699-2.2711 in. (57.65-57.69mm)	2.5850-2.5858 in. (65.66-65.68mm)	—
Piston at skirt 0.255 in. (6.5mm) up from bottom	2.2780-2.2787 in. (57.86-57.87mm)	2.5929-2.5937 in. (65.86-65.88mm)	0.008 in. (0.203mm)
Piston pin bore	0.7084-0.7087 in. (17.00-18.00mm)	0.7084-0.7087 in. (17.99-18.00mm)	—
Piston pin	0.7083-0.7087 in. (17.99-18.00mm)	0.7083-0.7087 in. (17.99-18.00mm)	—
Ring groove clearance (top)	0.0012-0.0028 in. (0.030-0.071mm)	0.0012-0.0028 in. (0.030-0.071mm)	+0.004 in. (0.102mm)
Ring groove clearance (bottom)	0.0012-0.0028 in. (0.030-0.071mm)	0.0012-0.0028 in. (0.030-0.071mm)	+0.004 in. (0.102mm)
Crankshaft O.D. (PTO) end	30mm	30mm	—

NOTES: ① Crankshaft dimension is 25mm on model 400 machines up to Serial No. 11,000.

② Dimension is 0.6296-0.6300 in. (15.992-16.002mm) on JDX8 and 800 snowmobiles before Serial No. 20,001.

③ Dimension for 440 Liquifire is 2.5990-2.6000 in. (66.02-66.04mm).

CHAPTER FIVE

FUEL SYSTEM

The fuel system consists of a fuel tank, fuel line or lines, inline fuel filter, and carburetor. See **Figure 1** for fuel system on front engine models and **Figure 2** for fuel system on mid-engine models.

The Walbro carburetor has an integral fuel pump. An auxiliary impulse pump is provided on models equipped with Bendix or Mikuni carburetors. The fuel pumps operate off differential pressure in the engine crankcase. An air intake silencer is used on all models to quiet incoming air and to catch fuel that may spit back out of the carburetor.

This chapter covers removal, installation, and replacement and/or repair of carburetors, fuel pumps, inline filters, and fuel tanks. Carburetor tuning is covered in Chapter Two.

See **Table 1** for carburetor application.

WALBRO CARBURETORS

Removal

1. Remove windshield and console.
2. Disconnect choke and throttle cables. Remove fuel inlet, vapor return lines, and fuel pump impulse tube from carburetor.
3. Remove silencer and throttle and choke brackets. Remove carburetor from the intake manifold.

Disassembly

1. Clean exterior of carburetor with a nonflammable solvent.

CAUTION

Never use compressed air to clean an assembled carburetor or diaphragm may be damaged.

2. Carefully disassemble carburetor according to **Figure 3**. Pay particular attention to the location of different sized screws and springs.

Cleaning and Inspection

WARNING

Most carburetor cleaners are highly caustic. They must be handled with extreme care or skin burns and possible eye injury may result.

1. Clean all metallic parts in carburetor cleaning solvent. Do not place gaskets or diaphragms in solvent or they will be destroyed.

CAUTION

Never clean holes or passages with small drill bits or wire or a slight enlargement or burring of holes will result, drastically affecting carburetor performance.

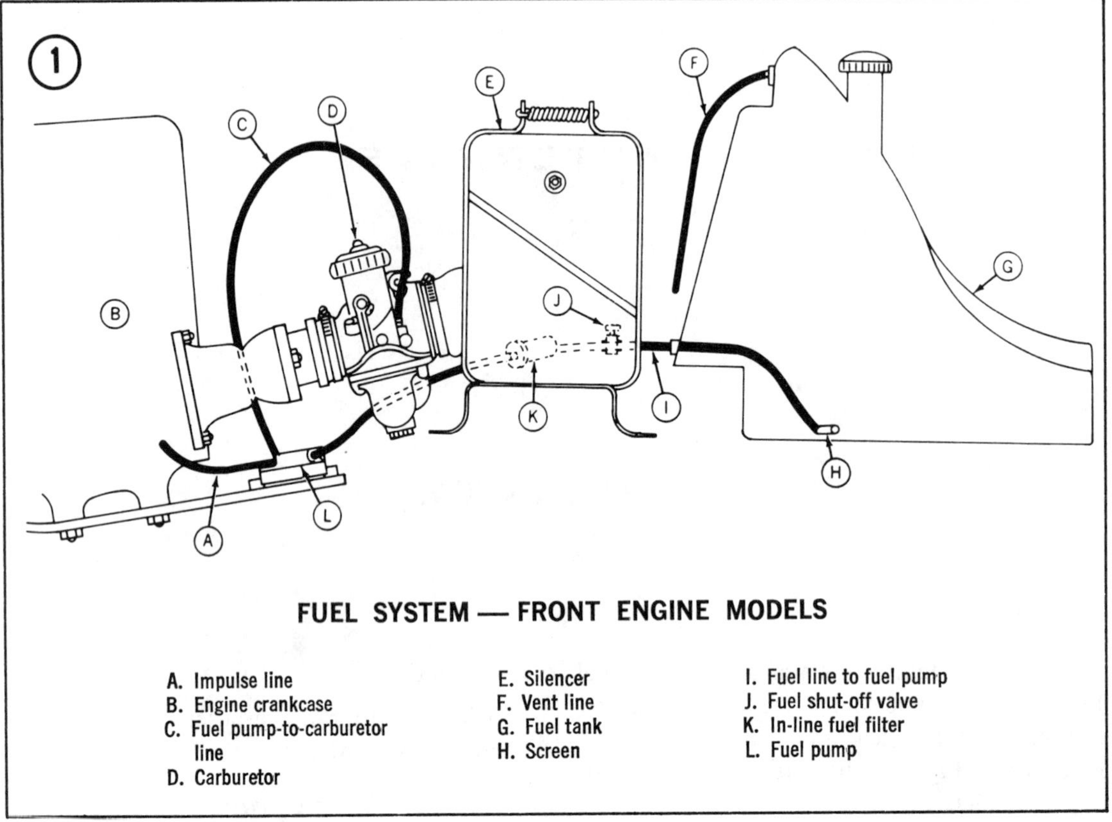

FUEL SYSTEM — FRONT ENGINE MODELS

A. Impulse line
B. Engine crankcase
C. Fuel pump-to-carburetor line
D. Carburetor
E. Silencer
F. Vent line
G. Fuel tank
H. Screen
I. Fuel line to fuel pump
J. Fuel shut-off valve
K. In-line fuel filter
L. Fuel pump

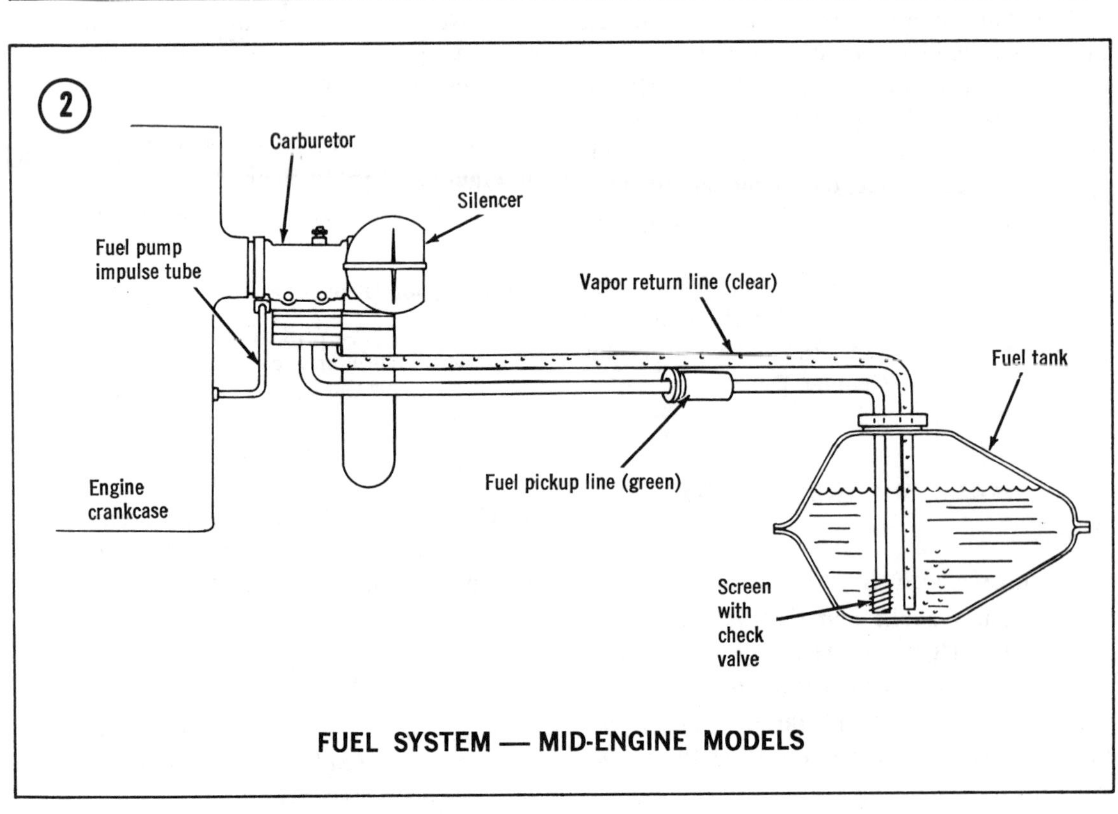

FUEL SYSTEM — MID-ENGINE MODELS

Table 1 CARBURETOR APPLICATION

Snowmobile		Carburetor	
Model	Serial Number	Manufacturer	Series
400	(-11000)	Walbro	WR-7
500	(-11000)	Walbro	WD-8
400, 500	(11001-20000)	Walbro	WDA-32
600	(-20000)	Walbro	WDA-32
JDX4	(-20000)	Walbro	WDA-37
JDX8	(-20000)	Walbro	WDA-34
300	(20001-)	Walbro	WDA-31
400, 500, 600, JDX4 Special	(20001-30000)	Bendix	1612
JDX6, JDX8	(20001-30000)	Walbro	WRA-31
JD295/S	(20001-30000)	Mikuni (2 used)	VM34-55
400, 600	(30001-)	Bendix	1612
800, JDX4, JDX6, JDX8	(30001-)	Walbro	WRA-31

Serial No. (55,001 to 70,000)

Snowmobile Model	Mikuni Carburetor
340 Cyclone	VM34-83
440 Cyclone	VM34-84
340 Liquifire	VM34-79 (2 used)
440 Liquifire	VM34-80 (2 used)

Serial No. (70,001 and Subsequent)

Snowmobile Model	Mikuni Carburetor
340 Cyclone	VM34-123
440 Cyclone	VM34-124
340 Liquifire	VM34-122 (2 used)
440 Liquifire	VM34-125 (2 used)
JD340/S	VM34-55 (2 used)
340 Liquidator	VM36-45 (2 used)

5

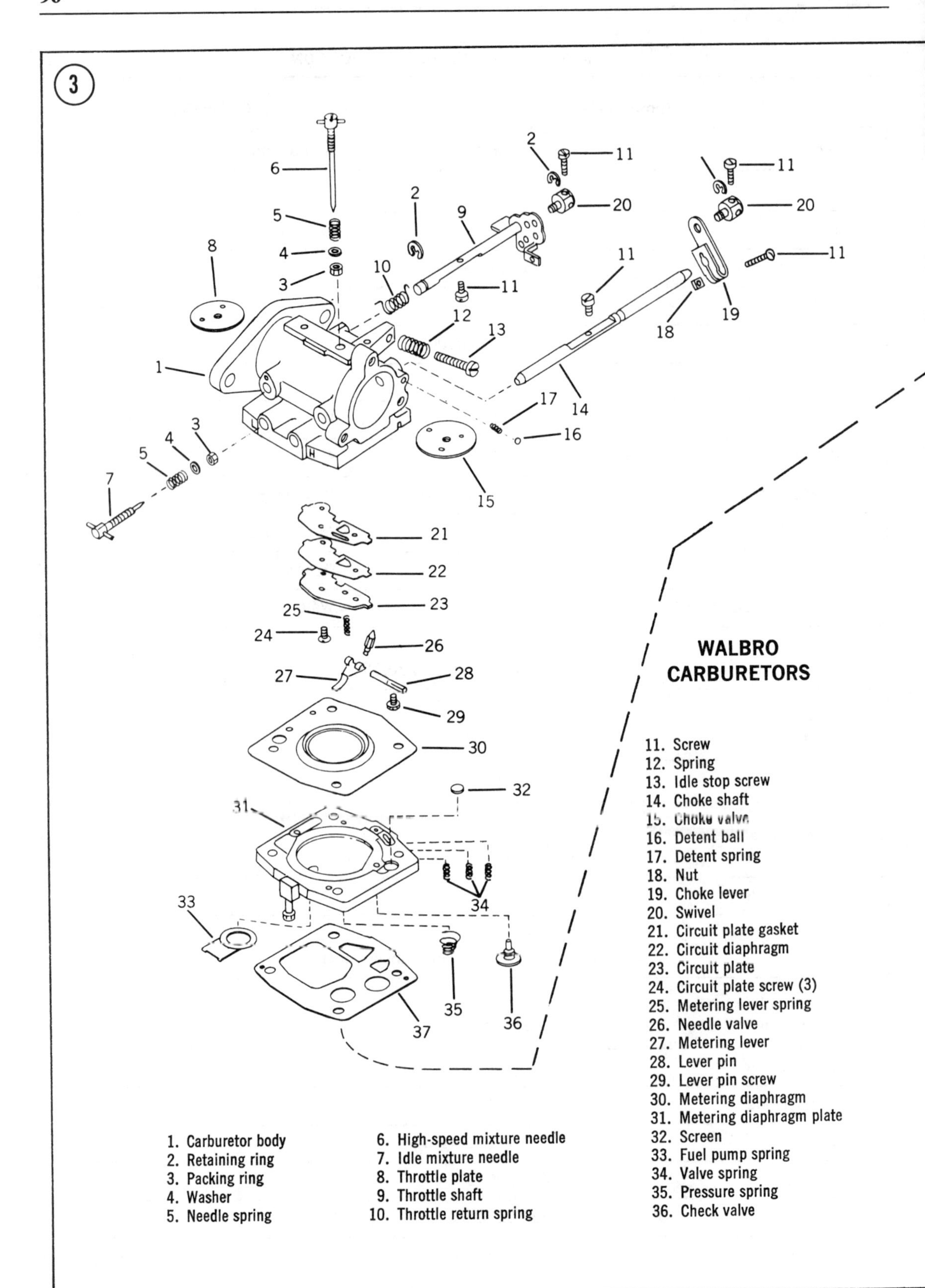
3
WALBRO
CARBURETORS
1. Carburetor body
2. Retaining ring
3. Packing ring
4. Washer
5. Needle spring
6. High-speed mixture needle
7. Idle mixture needle
8. Throttle plate
9. Throttle shaft
10. Throttle return spring
11. Screw
12. Spring
13. Idle stop screw
14. Choke shaft
15. Choke valve
16. Detent ball
17. Detent spring
18. Nut
19. Choke lever
20. Swivel
21. Circuit plate gasket
22. Circuit diaphragm
23. Circuit plate
24. Circuit plate screw (3)
25. Metering lever spring
26. Needle valve
27. Metering lever
28. Lever pin
29. Lever pin screw
30. Metering diaphragm
31. Metering diaphragm plate
32. Screen
33. Fuel pump spring
34. Valve spring
35. Pressure spring
36. Check valve
37

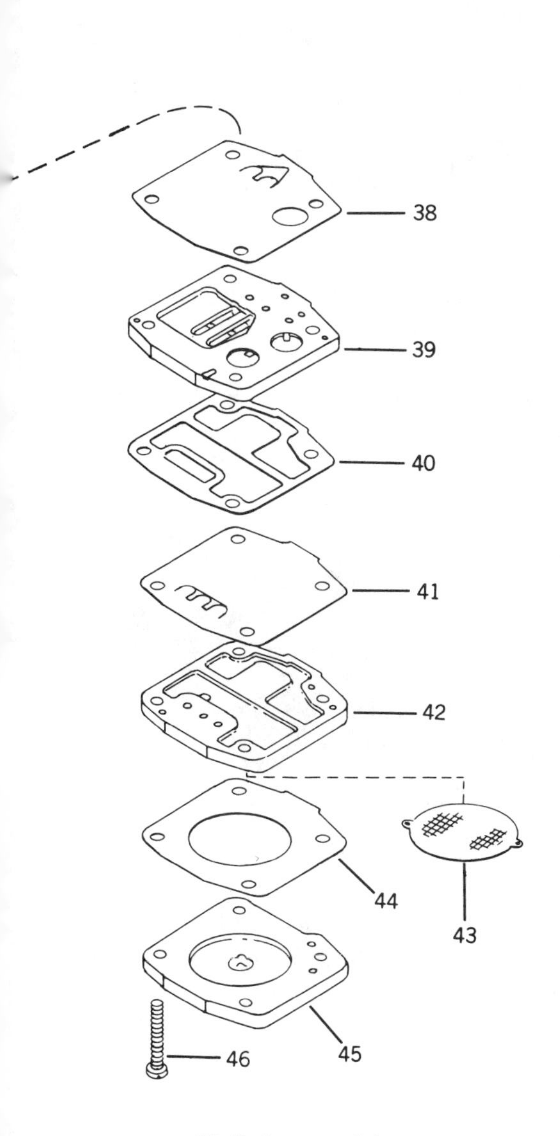

37. Fuel pump gasket
38. Fuel pump diaphragm
39. Fuel pump plate
40. Check valve gasket
41. Check valve diaphragm
42. Filter plate
43. Filter screen
44. Fuel inlet gasket
45. Cover plate
46. Cover screw (4)

2. After cleaning carburetor parts, dry with compressed air. Make sure all holes are open and free of carbon and dirt.

> NOTE: *Do not use rags or wastepaper to dry parts. Lint may plug jets or channels and affect carburetor operation.*

3. Inspect shaft bearing surfaces in carburetor body (**Figure 4**) for excessive wear.

CAUTION

If excessive clearance is found between shafts and carburetor body, worn parts must be replaced. Excessive clearance will allow air to enter, causing a damaging lean mixture.

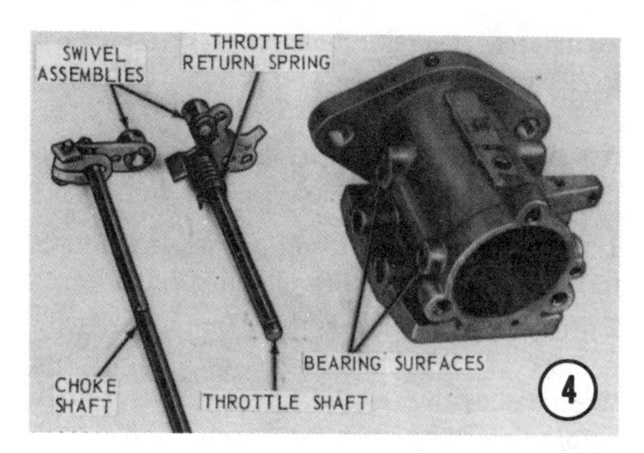

4. Inspect choke and throttle plates for damage. Inspect swivel assemblies on choke and throttle levers for wear. Inspect condition of throttle return spring. Replace all worn parts.

5. Inspect mixture needles and needle valve seating surfaces for pitting or wear (**Figure 5**), and replace if worn or damaged.

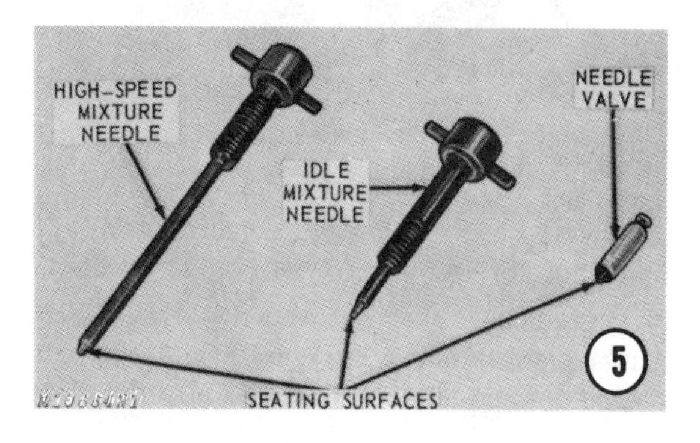

6. Inspect diaphragms for distortion, cracks or punctures (**Figure 6**).

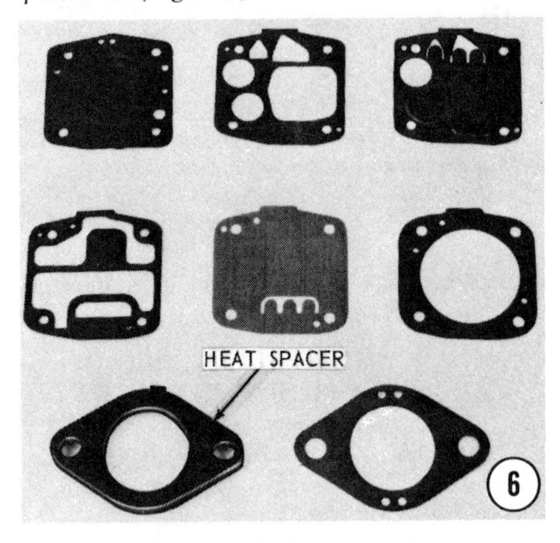

7. Inspect carburetor mounting gasket and heat spacer gasket.

Assembly

Refer to **Figure 3** for this procedure.

1. Install throttle shaft, with spring, in carburetor body and secure with retaining ring.
2. Insert throttle plate as shown in **Figure 7** with numbers facing out and below shaft. Close throttle plate to center plate in carburetor body before tightening screw, and secure screw with Loctite Lock N' Seal.

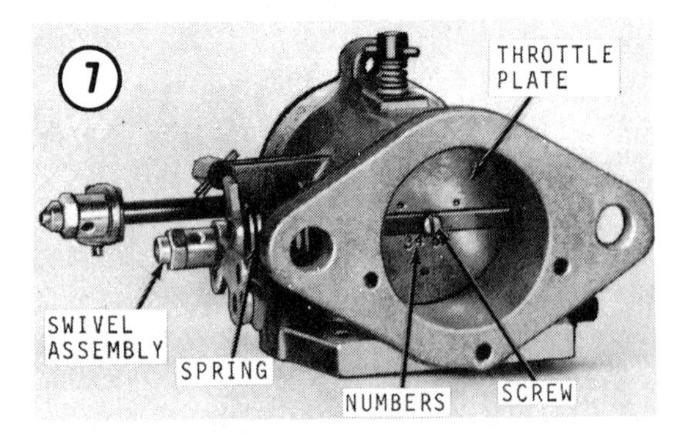

3. Install detent spring and detent ball in hole in carburetor body (**Figure 8**).
4. Install choke shaft in carburetor body. Insert choke plate as shown in **Figure 8** with numbers facing out and notch on top. Close choke plate

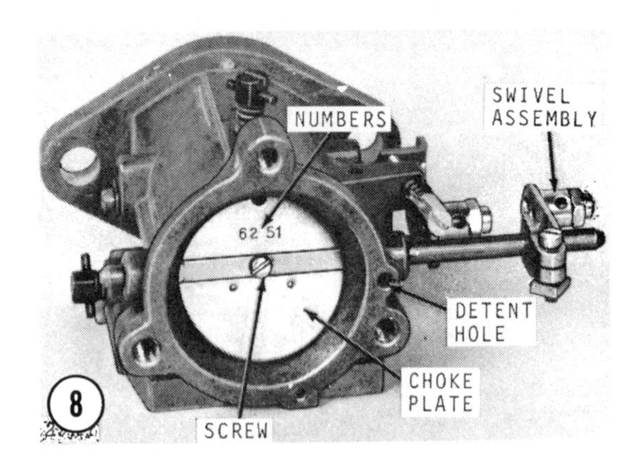

to center plate in carburetor body before tightening screw and secure screw with Loctite Lock N' Seal.

5. Install spring, washer and packing rings on mixture needles (**Figure 9**). Install needles in carburetor body. Install idle stop screw and spring.

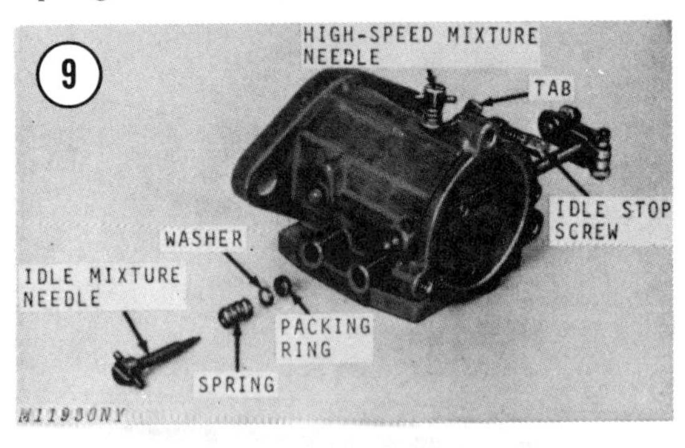

6. Carefully seat both needles. Use finger pressure only. Do not force needles or damage to needle and/or seat will result.
7. Open high speed mixture needle ⅞ turn. Open idle mixture one turn
8. Turn idle stop screw in until it contacts tab on throttle lever, then turn in one additional turn.

> NOTE: *The preceding are preliminary carburetor adjustments. Complete carburetor adjustment must be performed as outlined in Chapter Two after carburetors are installed.*

9. Install thin and thick gaskets, circuit diaphragm and circuit plate as shown in **Figure 10**. Secure with 3 screws.

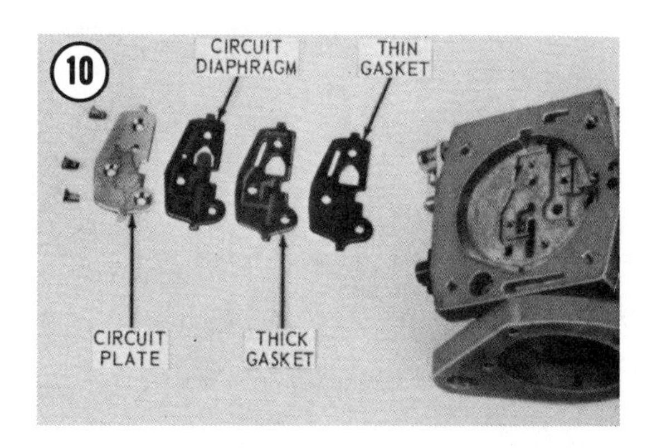

10. Place spring in cavity in carburetor body, and needle valve in tab on metering lever. Install lever pin through metering lever (**Figure 11**).

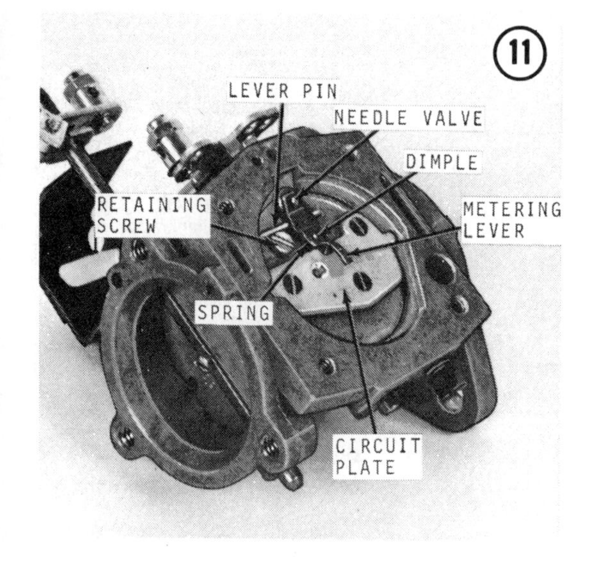

11. Install assembly in carburetor body and secure with screw. Ensure spring is positioned around dimple in metering lever.

12. Invert carburetor and lay straight edge across carburetor body. Adjust needle valve tab so lever is 0.005 to 0.020 in. (0.127-0.508mm) above carburetor body as shown in **Figure 12**.

13. Install valve springs, pressure spring, check valve and fuel pump spring in metering diaphragm plate as shown in **Figure 13**.

14. Install small screen (32, **Figure 3**) in other side of metering diaphragm plate.

15. Install diaphragms, gaskets, filter screen, and plates in carburetor body as shown in **Figure 14**. Install and tighten 4 cover screws.

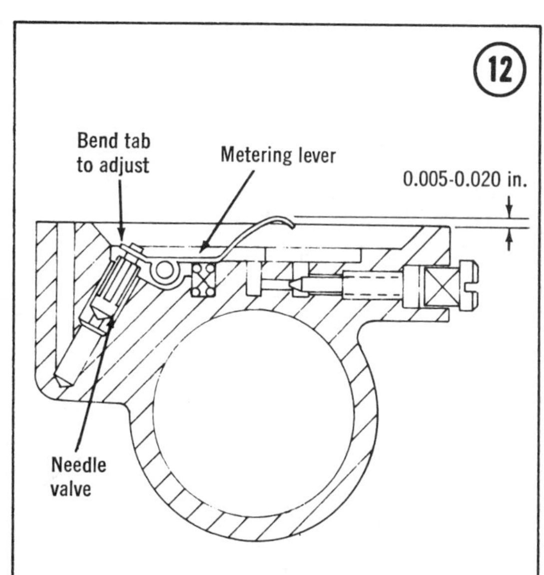

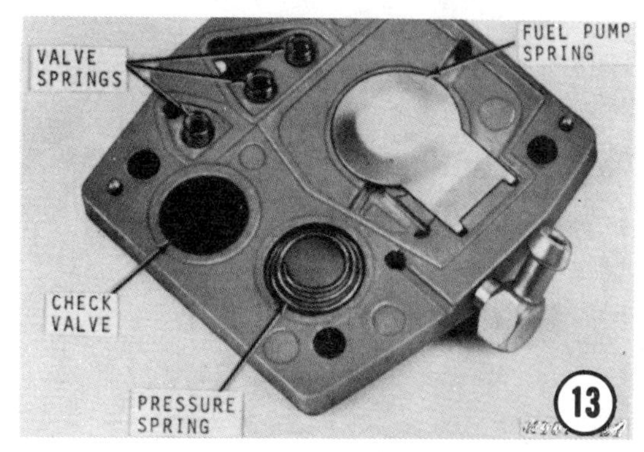

Installation

Refer to **Figure 15** for this procedure.

1. Install new gasket, heat spacer, new gasket, small square heat shield, new gasket and carburetor, and tighten nuts securely.
2. Install silencer and throttle bracket.
3. Connect fuel pickup line (green) to center fitting under carburetor. Install vapor return line (clear) on other lower fitting.
4. Connect impulse tube to angle fitting on left-hand side of carburetor.
5. Connect choke cable to choke lever. Ensure choke is in full open position when choke button on instrument panel is pushed in.
6. Connect throttle cable to throttle lever and adjust so throttle lever fully opens when throttle control on handlebar is fully actuated.

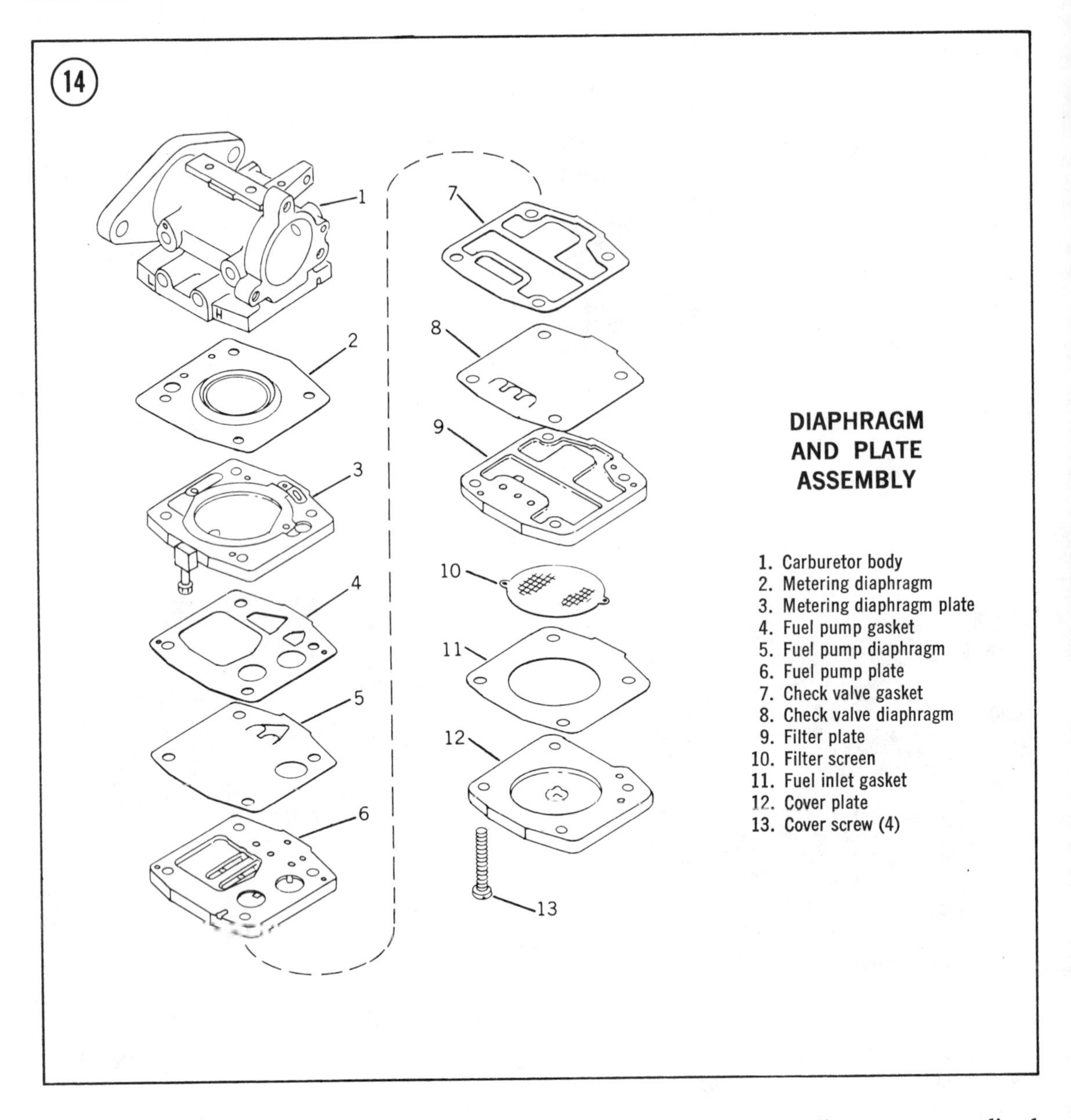

NOTE: *Ensure dowel on end of throttle cable is properly positioned in recess of handlebar throttle control.*

7. Make preliminary carburetor adjustments as follows:

a. High speed adjustment needle ⅞ turn open.
b. Idle mixture needle one turn open.

NOTE: *On 800 and JDX8 snowmobiles, serial No. 30,001 and subsequent, back out high speed and idle mixture needles 1⅛ turns.*

8. Perform carburetor adjustments as outlined in Chapter Two.

BENDIX CARBURETOR

Removal

1. Remove windshield and console.
2. Remove air intake silencer complete with air intake tube and spacer.
3. Disconnect choke and throttle cables.
4. Disconnect fuel inlet line from carburetor.
5. Disconnect impulse line from fuel pump.

1. Choke cable
2. Silencer
3. Heat spacer and gaskets
4. Throttle bracket
5. Fuel line (green)
6. Fuel line (clear)

6. Remove heat deflector and fuel pump.

7. Remove choke and throttle cable bracket from carburetor.

8. Remove carburetor complete with heat deflector.

Disassembly

1. Clean exterior of carburetor with a non-flammable solvent.

2. Carefully disassemble carburetor according to **Figure 16**. Pay particular attention to the location of different size screws and-springs.

Cleaning and Inspection

WARNING

Most carburetor cleaners are highly caustic. They must be handled with extreme care or skin burns and possible eye injury may result.

1. Clean all metallic parts in carburetor cleaning solvent. Do not place gaskets in solvent or they may be destroyed.

CAUTION

Never clean holes or passages with small drill bits or wire or a slight enlargement or burring of hole will result, drastically affecting carburetor performance.

2. After cleaning carburetor parts, dry with compressed air. Make sure all holes are open and free of carbon and dirt.

NOTE: *Do not use rags or wastepaper to dry parts. Lint may plug jets or channels and affect carburetor operation.*

3. Inspect shaft bearing in carburetor body for excessive wear. Inspect bearing surfaces on choke and throttle shaft for wear.

4. Inspect choke and throttle plates for damage. Inspect condition of throttle return spring.

5. Inspect mixture needles and needle valve seating surfaces for pitting or wear and replace if worn or damaged.

6. Inspect all gaskets for damage or tears. Inspect carburetor mounting gasket and heat spacer gasket.

Assembly

Refer to **Figure 16** for this procedure.

1. Install a new throttle shaft seal into a new seal retainer (lip on seal must face into retainer) **Figure 17**. Position seal in retainer over boss on carburetor body (**Figure 18**). Use a plastic hammer and gently tap seal and retainer into place.

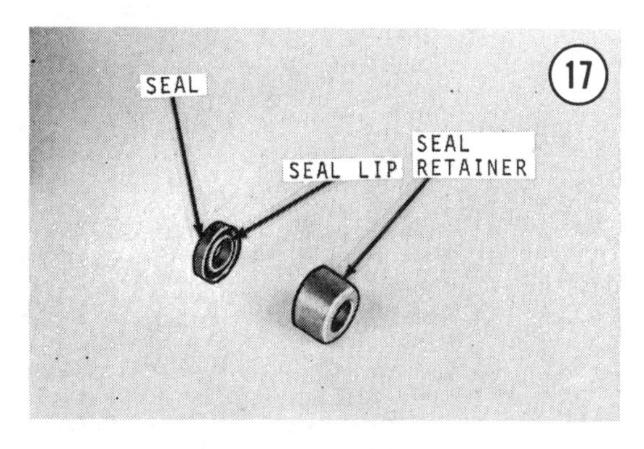

2. Install throttle return spring on throttle shaft. Lightly oil throttle shaft seal and install throttle shaft and lever (**Figure 19**). Locate long leg of

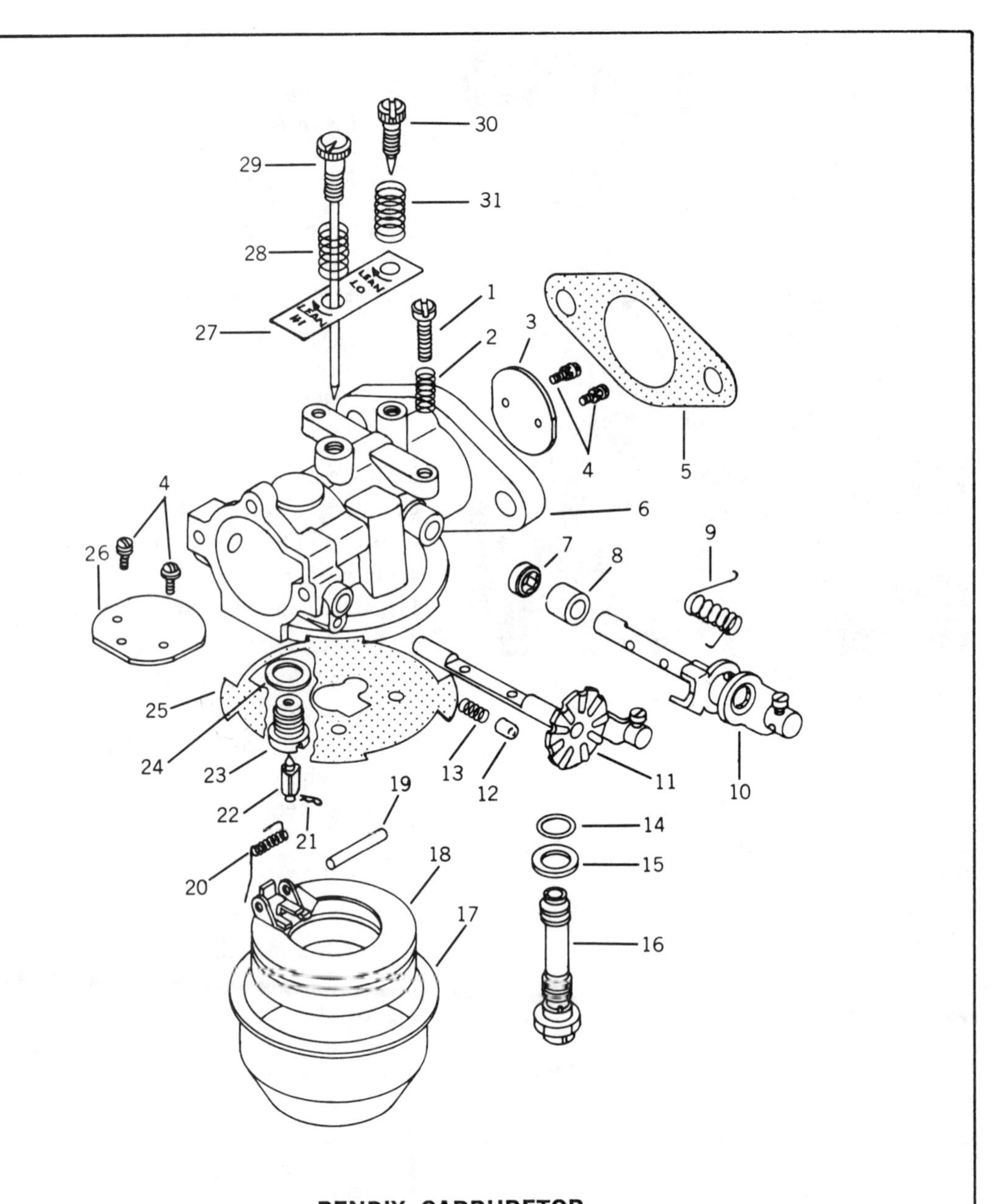

BENDIX CARBURETOR

1. Throttle stop screw
2. Spring
3. Throttle plate
4. Screws and lockwashers
5. Flange gasket
6. Carburetor body
7. Throttle shaft seal
8. Seal retainer
9. Throttle return spring
10. Throttle lever
11. Choke shaft and lever
12. Friction pin
13. Friction pin spring
14. O-ring
15. Washer
16. Main jet and discharge tube
17. Fuel bowl
18. Float
19. Float axle
20. Float spring
21. Fuel valve spring clip
22. Fuel valve
23. Fuel valve seat
24. Gasket
25. Bowl-to-body gasket
26. Choke plate
27. Identification plate
28. Spring
29. High-speed needle (idle tube)
30. Low-speed needle
31. Spring

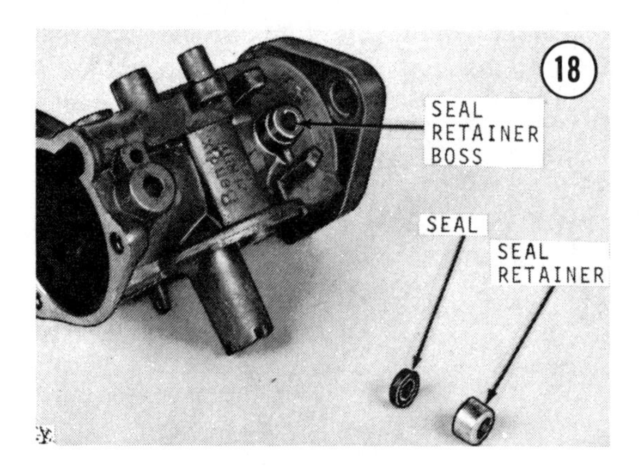

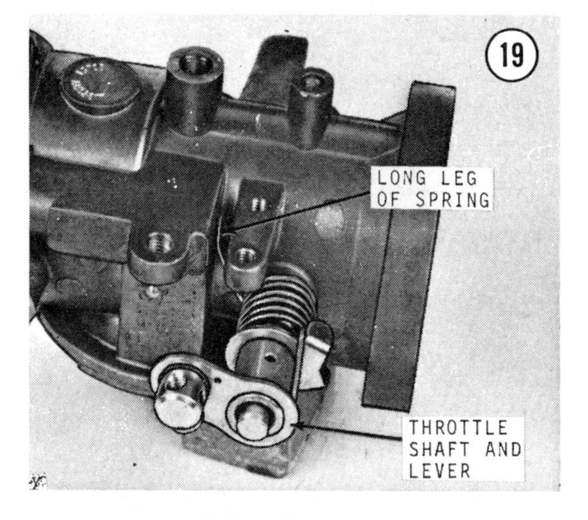

throttle return spring up between bosses on carburetor body.

CAUTION
Install throttle shaft gently or sharp edge on throttle shaft may damage shaft seal.

3. Rotate throttle shaft counterclockwise 1/3 turn until flat center section faces manifold opening. Install throttle plate to shaft with stamped marking facing *out* and towards top as shown in **Figure 20**. Snap throttle plate open and shut several times to center the plate. Tighten screws and secure with Loctite.

4. Insert spring and friction pin in hole in body (**Figure 21**). Install choke shaft and lever.

5. Rotate choke shaft until flat center section faces towards manifold opening. Loosely secure choke plate to shaft. Ensure plate is installed with stamped markings *in* and towards the top

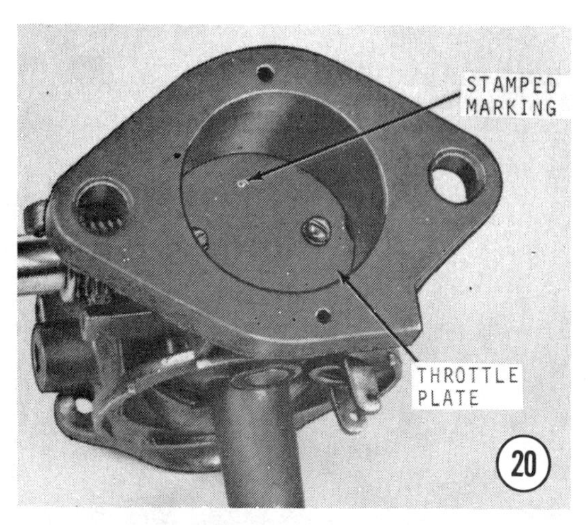

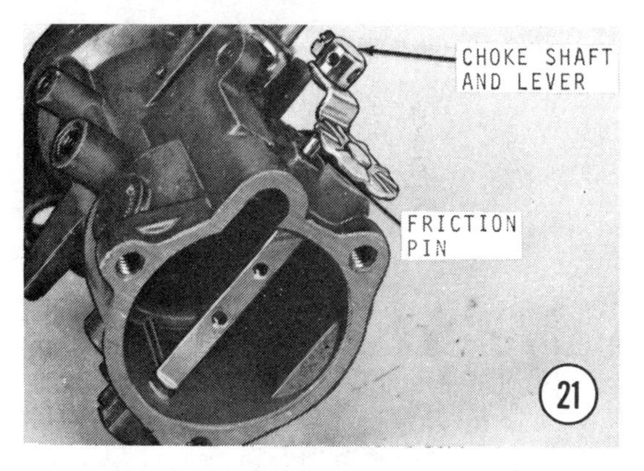

as shown in **Figure 22**. Snap choke plate open and shut several times to center the plate. Push in on shaft, and tighten screws. Secure screws with Loctite Lock N' Seal.

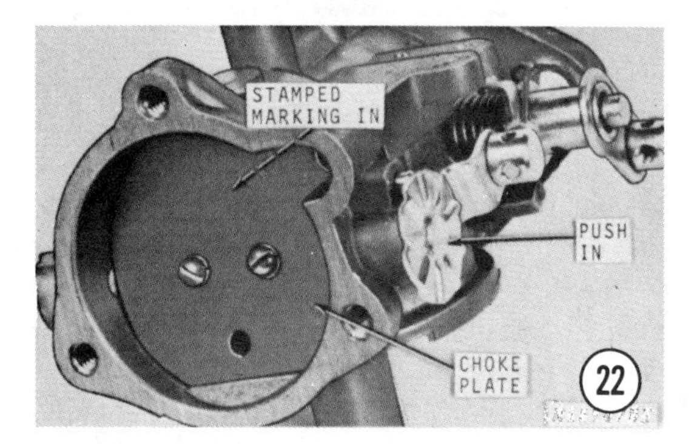

6. Invert carburetor body and install bowl-to-body gasket (**Figure 23**). Install gasket on fuel valve seat and install seat in body. Tighten valve seat. Place spring clip on fuel valve and install valve in seat.

5

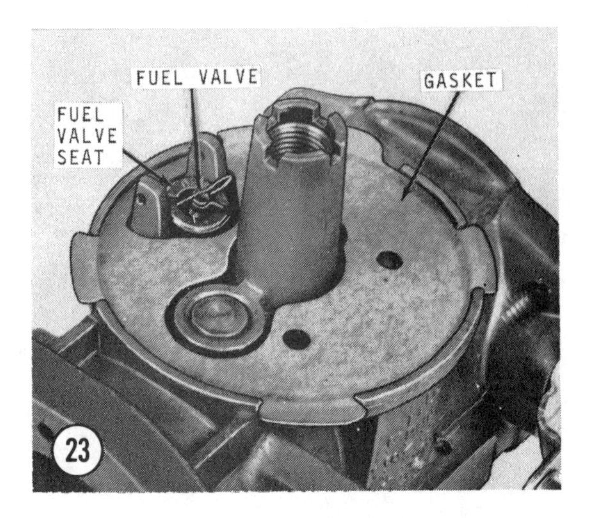

7. Install float spring, float and float axle.

NOTE: *Tang on float must be positioned between spring clip and fuel valve. Ensure tang fits loosely between spring clip and fuel valve as shown in* **Figure 24**.

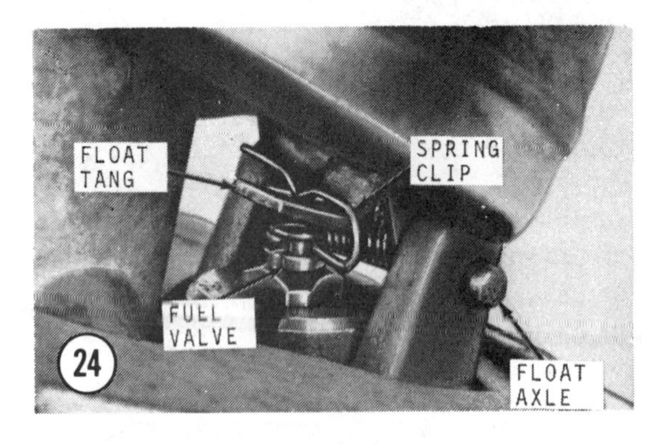

8. Lay a 1/8 in. drill on flat surface of gasket as shown in **Figure 25**. Bottom surface of float should be 1/8 in. (3.18mm) from gasket surface at point on float furthest from hinge. Bend tang that contacts fuel valve to adjust float.

9. Install fiber washer on main jet and discharge tube. Lightly lubricate O-ring with oil and install O-ring on main jet.

10. Position fuel bowl on carburetor body. Make sure long leg of float spring is inside the fuel bowl. Install main jet in discharge tube through hole in fuel bowl and tighten securely (**Figure 26**).

11. Install throttle stop screw and spring (**Figure 27**). Adjust screw to open throttle plate

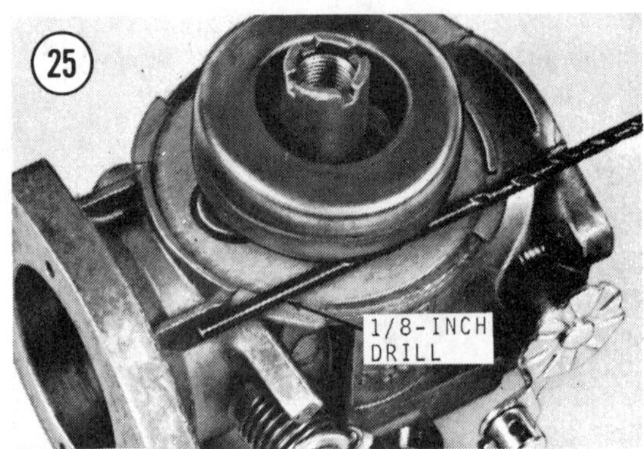

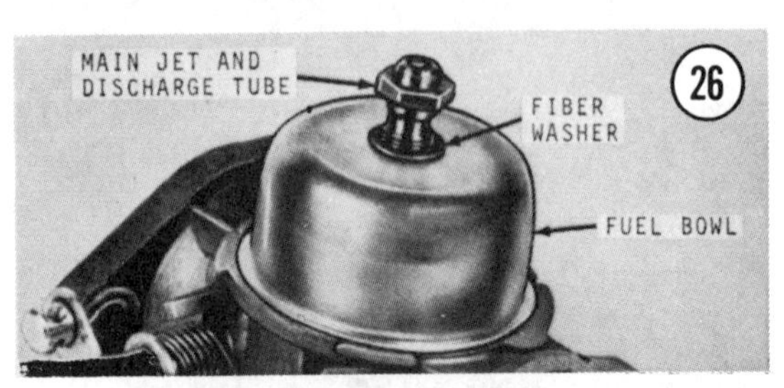

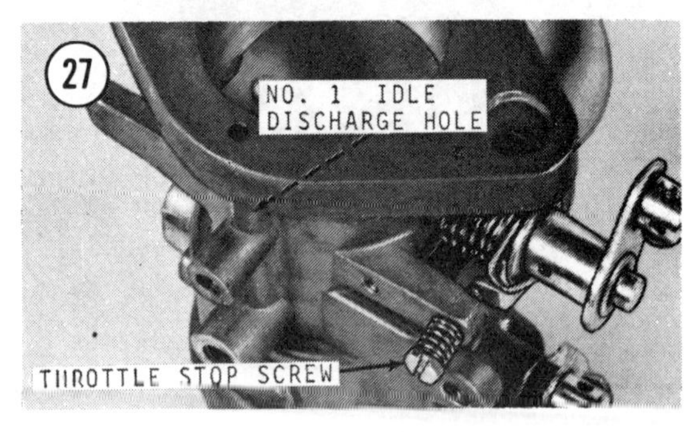

slightly but not enough to uncover No. 2 idle discharge hole (**Figure 28**).

12. Refer to **Figure 29** and install identification plate. Install low speed needle and spring. Screw needle in until it seats *lightly* against No. 1 discharge hole. Back needle out 1¾ turns as a preliminary adjustment.

13. Place spring on high-speed needle (**Figure 30**) and install needle through identification plate into discharge tube on opposite side of venturi. Screw needle in until it seats *lightly* then back out 1½ turns as a preliminary adjustment.

Installation

Refer to **Figure 31** for this procedure.

1. Install carburetor on engine.

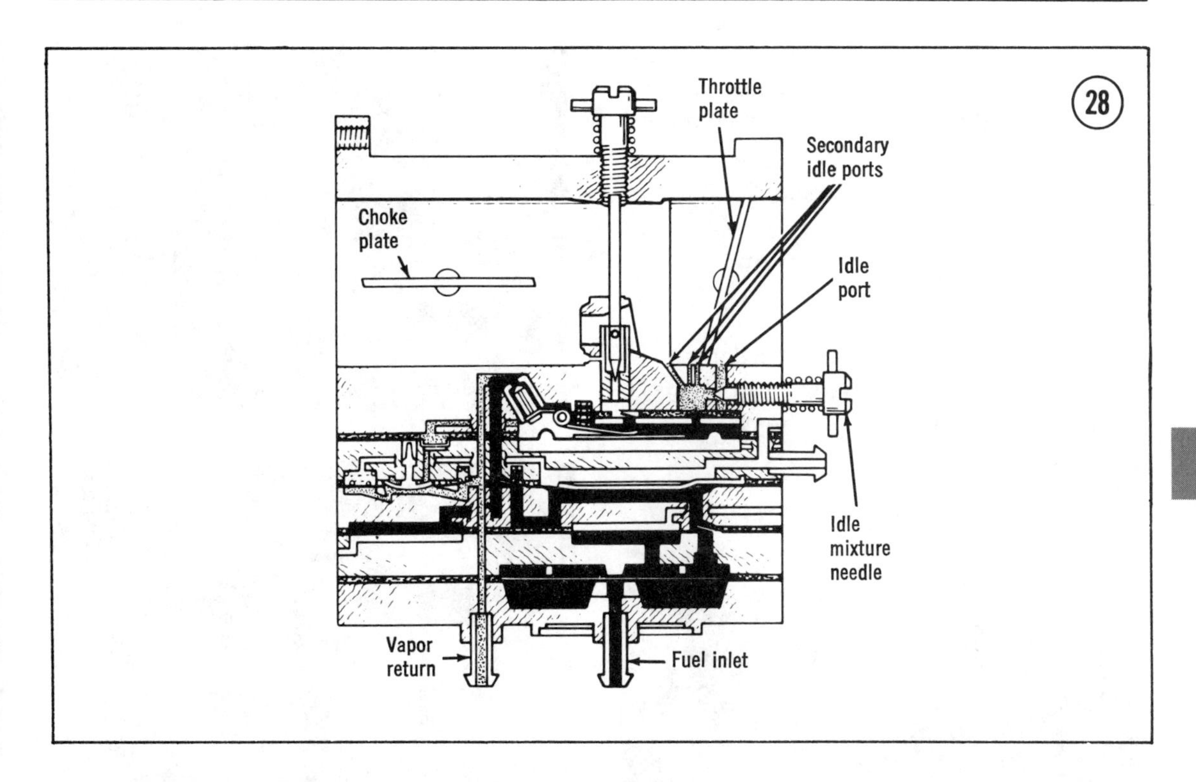

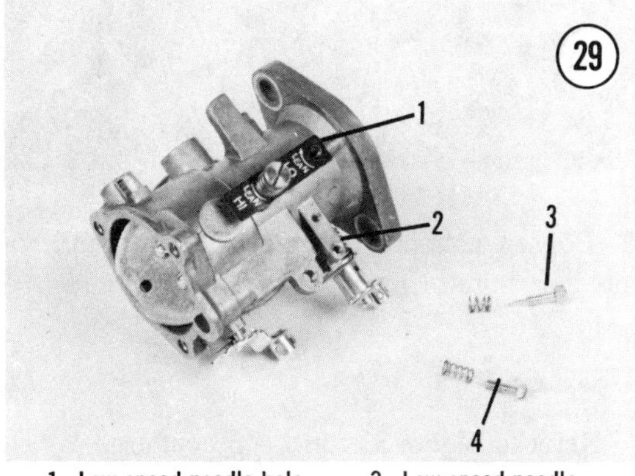

1. Low-speed needle hole
2. Throttle stop screw hole
3. Low-speed needle
4. Throttle stop screw

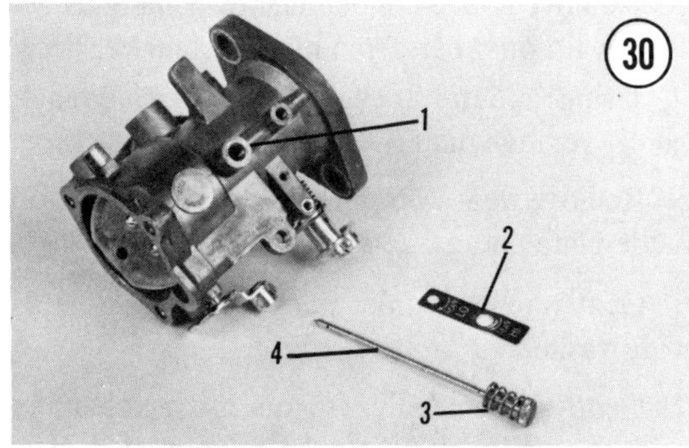

1. High-speed needle hole
2. Identification plate
3. Spring
4. High-speed needle

1. Choke and throttle bracket
2. Impulse pump
3. Spacer

2. Install choke and throttle cable bracket to carburetor.

3. Install heat deflector and fuel pump. Connect impulse line to fuel pump. Connect fuel line from fuel pump to carburetor.

4. Connect choke cable to choke lever. Ensure choke plate is fully open (choke button on instrument panel is pushed in).

5. Connect throttle cable to throttle lever and adjust so throttle fully opens when throttle control on handlebar is fully actuated.

6. Install spacer and silencer with air intake tube.

7. Perform carburetor adjustments as outlined in Chapter Two. Check all fuel line connections and carburetor mounting connections for leaks.
8. Install windshield and console.

MIKUNI CARBURETOR

Removal

Refer to **Figure 32** for this procedure.

1. Remove console and windshield on JD295/S models and top half of air intake silencer assembly.

A. Spring
B. Top half
C. Clamp
D. Fuel line
E. Clamp
F. Carburetor
G. Lower half

2. Disconnect fuel lines from carburetor.
3. Loosen locknut on starter cable. Lift starter plunger assembly from mixing chamber body as shown in **Figure 33**.
4. Unscrew mixing chamber cap and gently lift out throttle valve assembly from carburetor body. See **Figure 34**.
5. Remove drain plug from bottom of float chamber. Drain fuel into suitable container. Install drain plug.

WARNING
Handle and dispose of drained fuel carefully or a serious or fatal fire may occur.

6. Loosen clamp securing carburetor and remove carburetor from rubber mount.

Disassembly

Refer to **Figure 35** for this procedure.

1. Remove throttle stop screw and spring.
2. Remove air screw and spring.
3. Remove float chamber as shown in **Figure 36**. Gently lift out floats from mixing chamber body.
4. Using a 6mm socket or box end wrench, gently remove main jet and ring.
5. Remove float arm pin and float arm. Lift off baffle plate and gaskets (**Figure 36**).
6. Gently remove inlet needle valve assembly with washer.
7. Gently push needle jet from mixing chamber using an awl or similar sharp pointed device. See **Figure 37**.

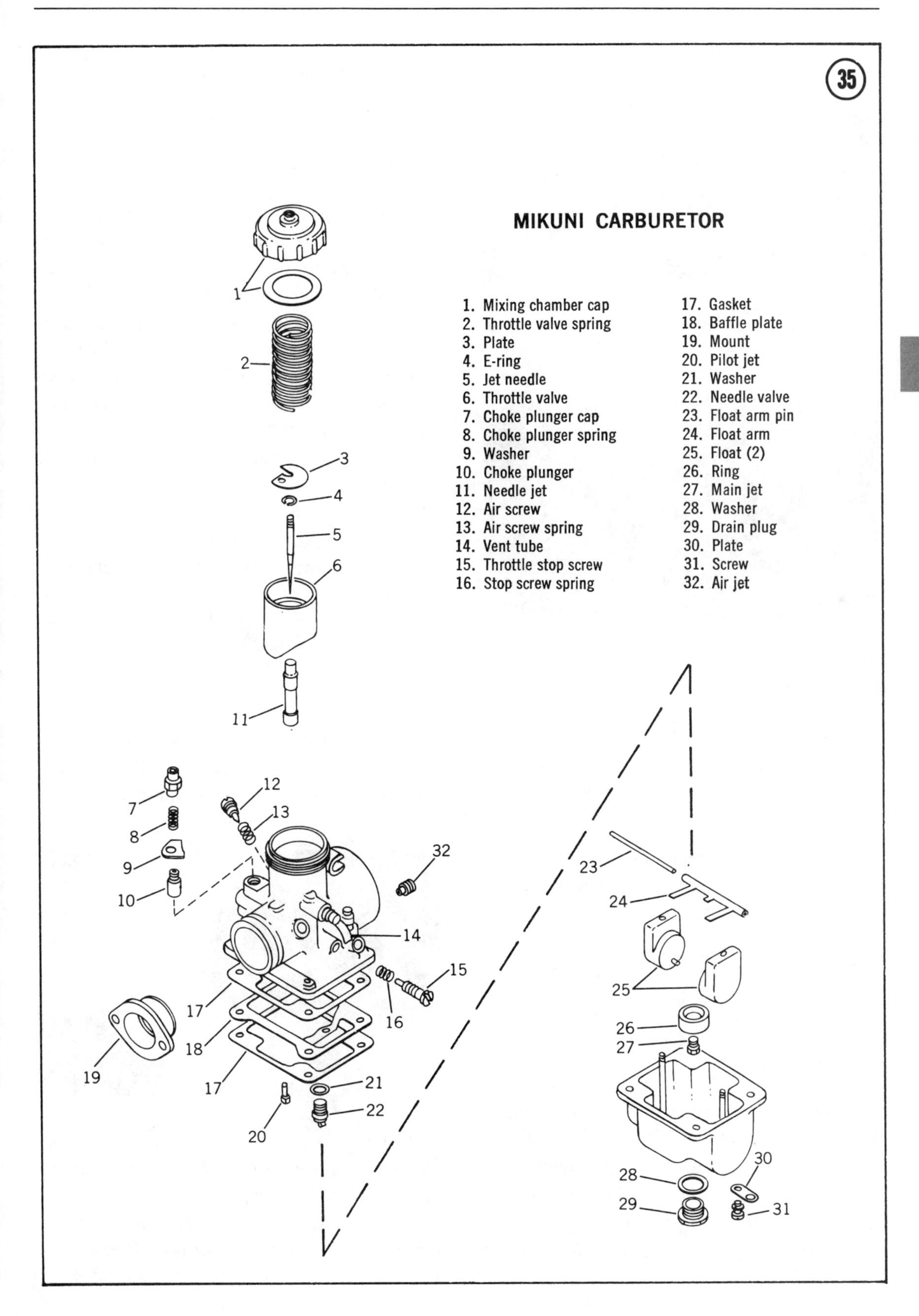
35
MIKUNI CARBURETOR
1. Mixing chamber cap
2. Throttle valve spring
3. Plate
4. E-ring
5. Jet needle
6. Throttle valve
7. Choke plunger cap
8. Choke plunger spring
9. Washer
10. Choke plunger
11. Needle jet
12. Air screw
13. Air screw spring
14. Vent tube
15. Throttle stop screw
16. Stop screw spring
17. Gasket
18. Baffle plate
19. Mount
20. Pilot jet
21. Washer
22. Needle valve
23. Float arm pin
24. Float arm
25. Float (2)
26. Ring
27. Main jet
28. Washer
29. Drain plug
30. Plate
31. Screw
32. Air jet
1
2
3
4
5
6
11
7
8
9
10
12
13
32
14
15
16
17
18
17
19
20
21
22
23
24
25
26
27
28
29
30
31

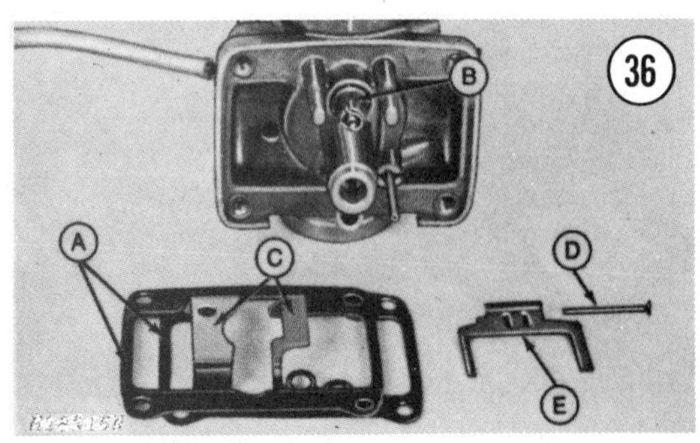

A. Gaskets
B. Inlet needle valve assembly
C. Baffle plate
D. Float air pin
E. Float arm

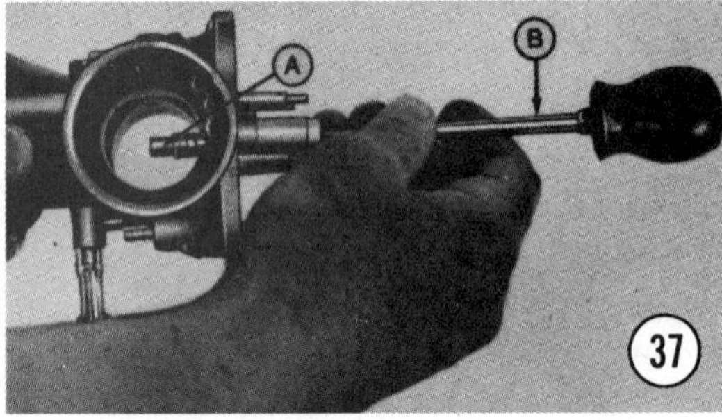

A. Needle jet B. Awl

Cleaning and Inspection

WARNING

Most carburetor cleaners are highly caustic. They must be handled with extreme care or skin burns and possible eye injury may result.

1. Clean all metallic parts in carburetor cleaning solvent. Do not place gaskets in solvent or they will be destroyed.

CAUTION

Never clean holes or passages with small drill bits or wire or a slight enlargement or burring of hole will result, drastically affecting carburetor performance.

2. Inspect float chamber and mixing chamber body for fine cracks or evidence of fuel leaks.
3. Check spring for distortion or damage.
4. Inspect air screw and throttle stop screw for surface damage or stripped threads.
5. Inspect pilot jet and main jet for damage or stripped threads.

CAUTION

Pilot jet and main jet must be scrupulously clean and shiny. Any burring, roughness, or abrasion will cause a lean fuel and air mixture and possible engine damage.

6. Remove retainer and inlet valve from valve seat. Carefully examine seating surface on inlet valve and seat for damage. Ensure retainer does not bind and hinder movement of inlet valve.
7. Inspect jet needle and needle jet for damage. Jet needle must slide freely within needle jet.
8. Install float guides in float chamber. Move floats up and down several times to ensure they are not binding on float guides.
9. Inspect float arm and float pin to ensure float arm does not bind on pin.
10. Inspect choke plunger. Plunger must move freely in passage of mixing chamber.
11. Install throttle valve in mixing chamber body and move several times up and down to check for sticking motion or wear. Ensure guide pin in mixing chamber body is not broken.

Assembly

Refer to **Figure 35** for this procedure.

1. Using a small screwdriver, install pilot jet in carburetor body as shown in **Figure 38**.

2. Install gaskets and baffle plate on mixing chamber surface (**Figure 39**). Install second gasket on top of baffle plate.
3. Place washer on inlet needle valve seat and install seat in mixing chamber body (**Figure 39**). Install inlet valve (point down) and retainer.

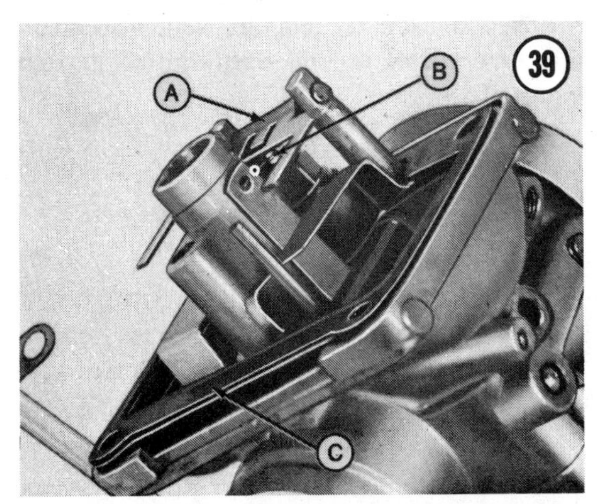

A. Float arm B. Inlet valve C. Baffle plate and gaskets

4. Install float arm and secure float arm with float arm pin.

5. Invert carburetor body. Edge of mixing chamber (**Figure 40**) must be parallel with float arm. Adjust if necessary by bending float arm actuating tab.

A. Mixing chamber B. Float arm

6. Install needle jet. Make sure notch on needle jet is correctly aligned with pin on bore of mixing chamber (**Figure 41**). Install ring over needle jet bore (recess in ring next to bore) and screw main jet into needle jet.

7. Slide floats over float pin. Pins on float must be down and point to inside of float chamber as shown in **Figure 42**.

8. Install float chamber to mixing chamber body and secure with 4 screws.

A. Pin B. Notch C. Needle jet

9. Slide air screw spring over air screw and install air screw gently.

CAUTION

Do not force air screw or seat damage may occur.

10. Install throttle stop screw with spring. Install screw until it is just flush with inside of bore.

Installation

1. Position carburetor in rubber mount and secure carburetor with clamp.

2. Connect fuel line from pump to carburetor.

3. Install E-ring in middle groove on Cyclone models and 4th groove from top for Liquifire models. See **Figure 43**.

4. Route throttle cable end button through cap, spring, and slot in throttle valve as shown in **Figure 43**.

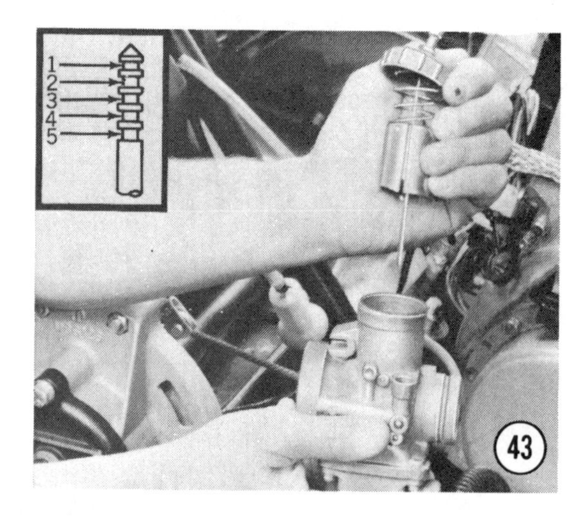

5. Slide cable into narrow part of slot in throttle valve. Install plate between spring and throttle valve with tab on plate in slot of throttle valve.

6. Compress throttle valve spring and tighten cap on mixing chamber.

7. Place choke lever on instrument panel in down position. Route choke as shown in **Figure 44** and place washer on mixing chamber body. Install assembly and tighten cap.

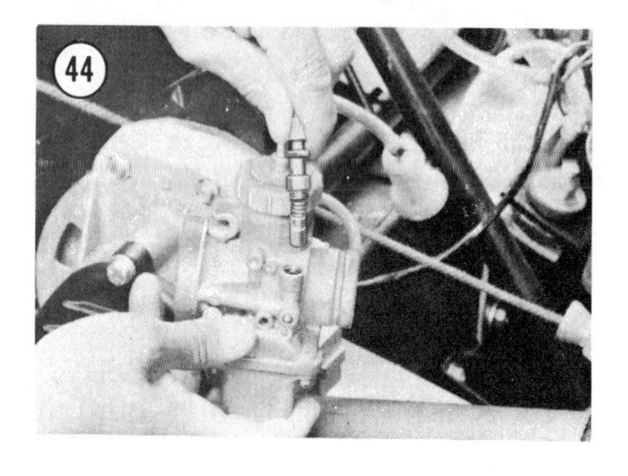

8. Perform carburetor adjustment as outlined in Chapter Two.

9. Install air intake silencer.

10. Install windshield and console if removed.

INTAKE SILENCERS

Intake silencers are installed on snowmobiles to quiet the sound of rushing air and to catch fuel that spits back out of the carburetor throat.

The silencer is not intended to filter incoming air. Operate snowmobiles only in clean, snow covered areas.

CAUTION

Never operate snowmobile with silencer removed. Loss of power and engine damage will result due to lean mixture.

Service of air intake silencers is limited to removal and cleaning of components. Refer to **Figures 45 through 51** and **Table 2** for air intake silencers used on various models.

Table 2 AIR SILENCER APPLICATION

Model/Series	Illustration
400 and 500 (up to Serial No. 11,000)	Figure 45
400 and 500 (Serial No. 11,001 and subsequent); 300, 600, 800, JDX4, JDX4 Special, JDX6, and JDX8	Figure 46
JD295/S (Serial No. 20,001 to 30,000)	Figure 47
340 and 440 Cyclone (Serial No. 55,001 to 70,000)	Figure 48
340 and 440 Cyclone (Serial No. 70,001 and subsequent)	Figure 49
340 and 440 Liquifire (Serial No. 55,001 to 70,000)	Figure 50
340 and 440 Liquifire (Serial No. 70,001 and subsequent)	Figure 51

FUEL TANK (FRONT ENGINE MODELS)

The fuel tank (**Figure 52**) incorporates a fuel gauge in the filler cap and a spill ledge to prevent fuel from spilling on to the seat. The fuel tank cap is sealed and the tank is vented by a line to the top of the tank.

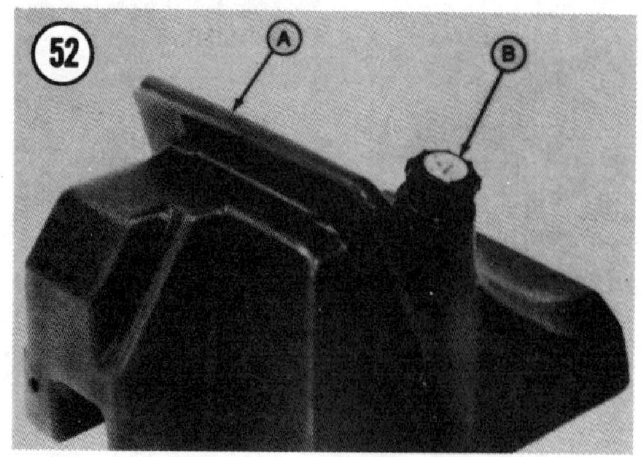

A. Fuel tank B. Fuel cap and gauge

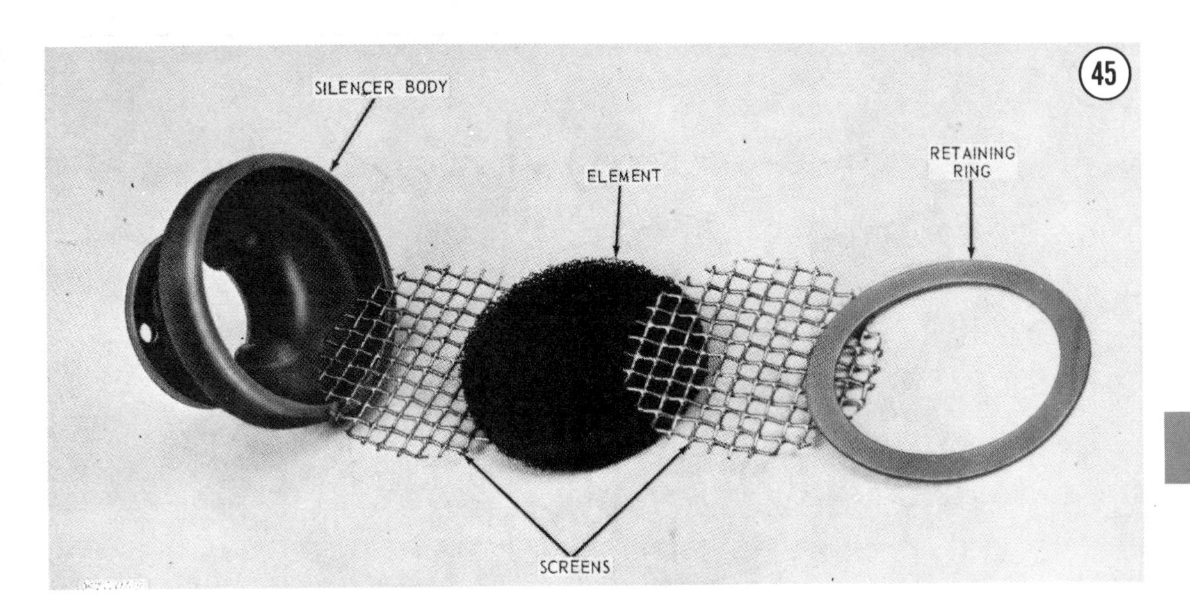

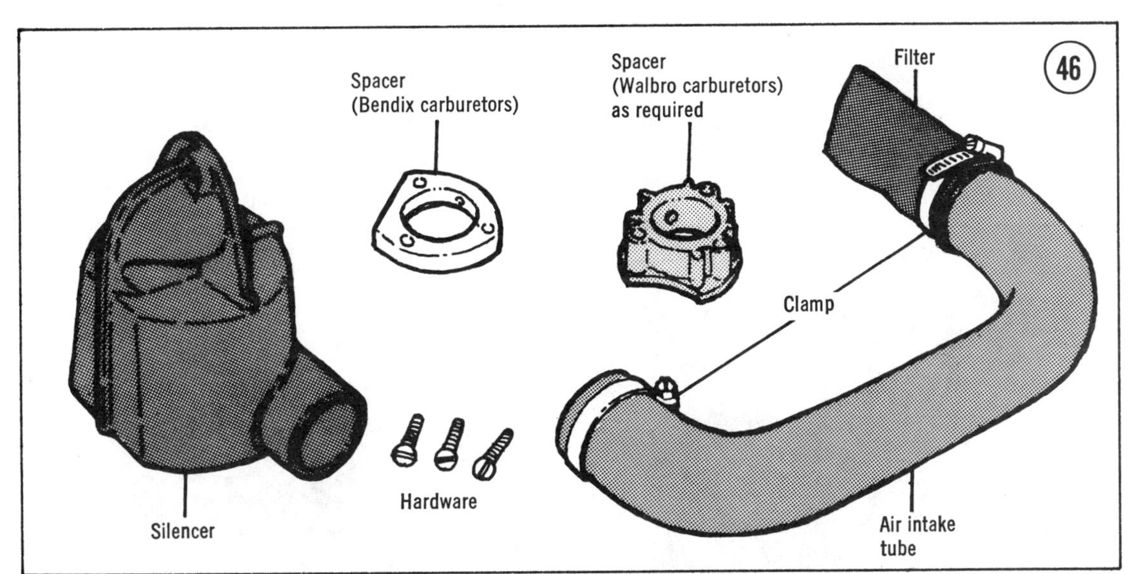

A fuel shut-off valve is located between the fuel pickup line and the inline fuel filter.

> NOTE: *If snowmobile has been transported on a trailer without the fuel valve shut off, the engine may be flooded.*

Removal/Installation

1. Disconnect fuel lines and vent lines.
2. Remove seat and tank hold-on clip.
3. Remove seat by sliding tank rearward.
4. Installation is the reverse of these steps.

Pick-up Screen Cleaning

1. Disconnect fuel lines from fitting and remove fitting from fuel tank (**Figure 53**).
2. Remove pickup screen from end of line.
3. Rinse screen carefully in solvent and blow dry with compressed air. Replace screen if damaged. Replace gasket on fuel line fitting if necessary.

FUEL TANK (MID-ENGINE MODELS)

The fuel tank (**Figure 54**) incorporates a fuel gauge and a separate filler opening. A spill tray

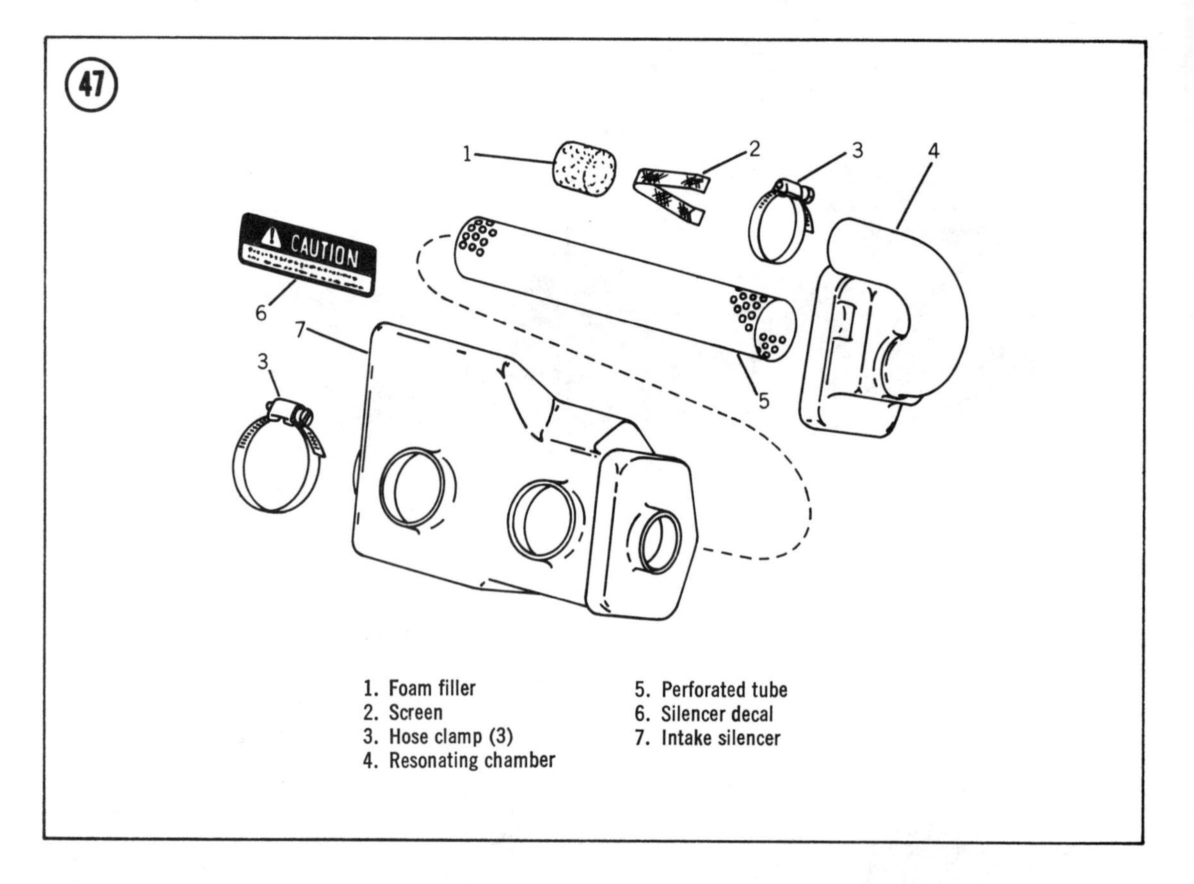

1. Foam filler
2. Screen
3. Hose clamp (3)
4. Resonating chamber
5. Perforated tube
6. Silencer decal
7. Intake silencer

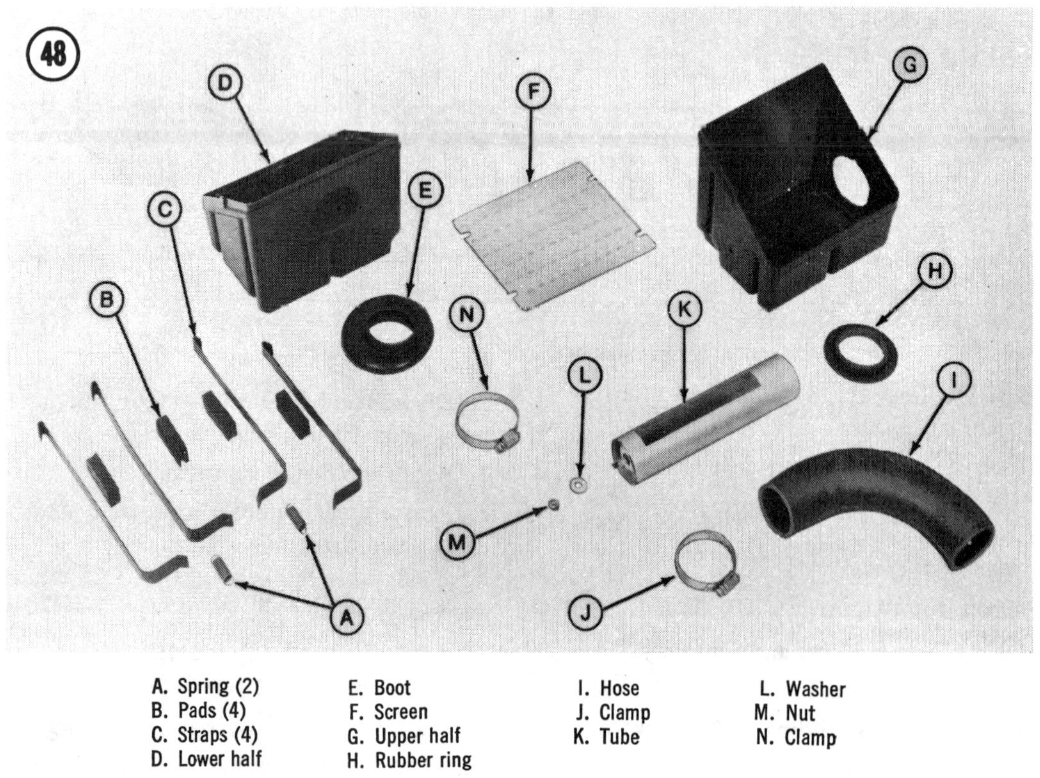

A. Spring (2)
B. Pads (4)
C. Straps (4)
D. Lower half
E. Boot
F. Screen
G. Upper half
H. Rubber ring
I. Hose
J. Clamp
K. Tube
L. Washer
M. Nut
N. Clamp

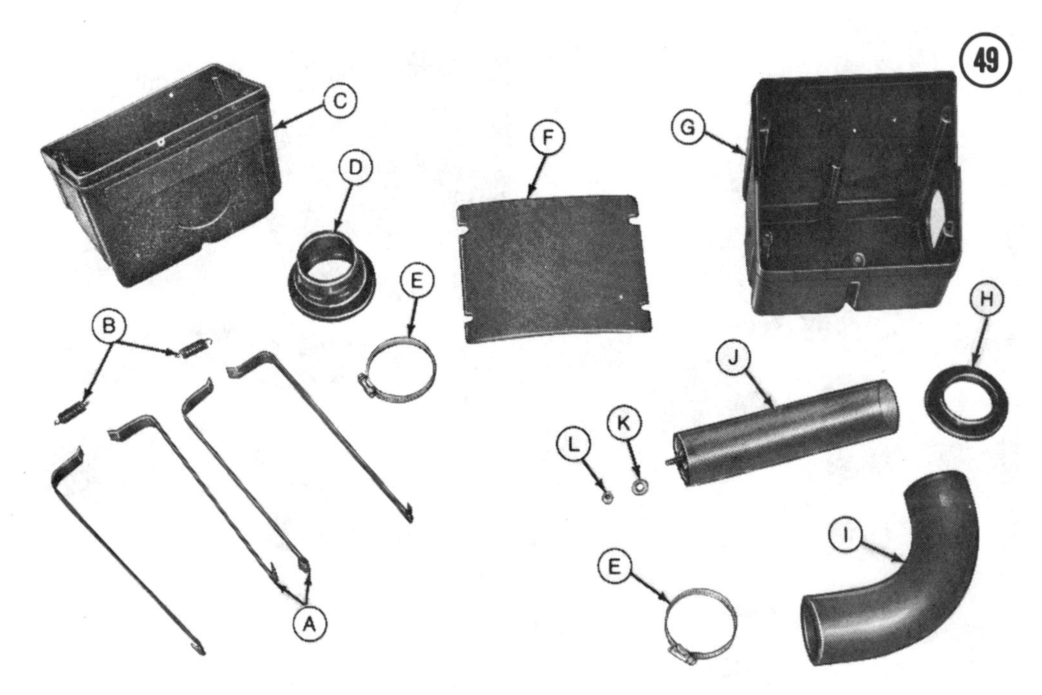

A. Straps (4)
B. Spring (2)
C. Lower half
D. Boot
E. Clamp (2)
F. Screen
G. Upper half
H. Rubber ring
I. Hose
J. Screen
K. Washer
L. Nut

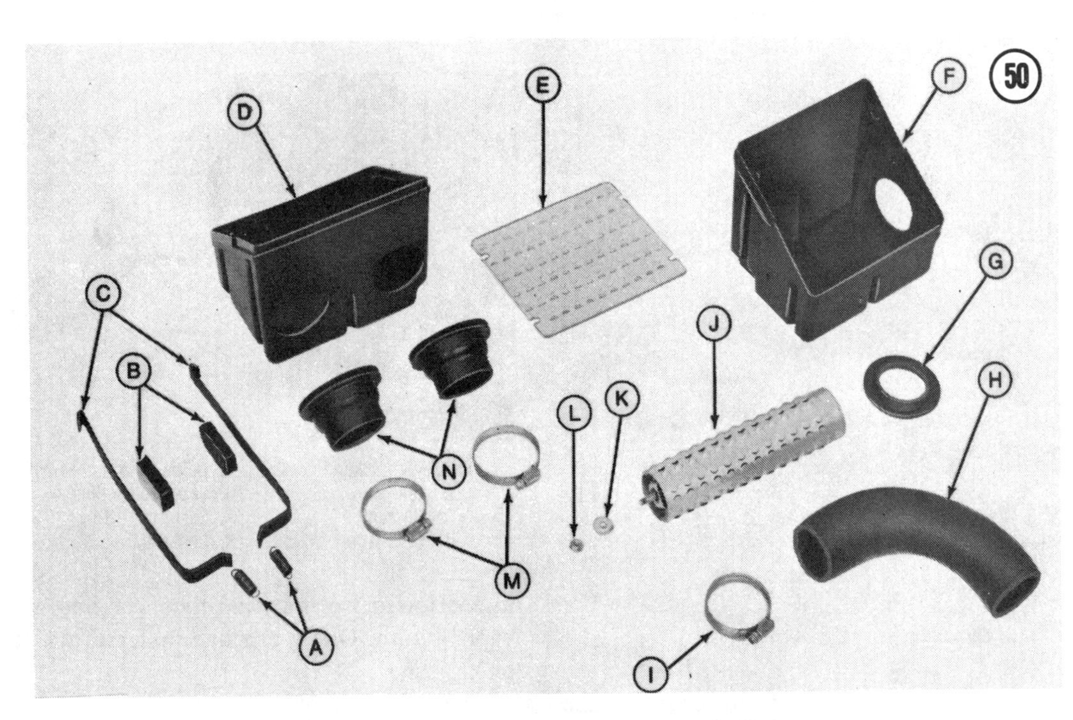

A. Springs (2)
B. Pads (2)
C. Straps (2)
D. Lower half
E. Screen
F. Upper half
G. Rubber ring
H. Hose
I. Clamp
J. Tube
K. Washer
L. Nut
M. Clamps (2)
N. Boots (2)

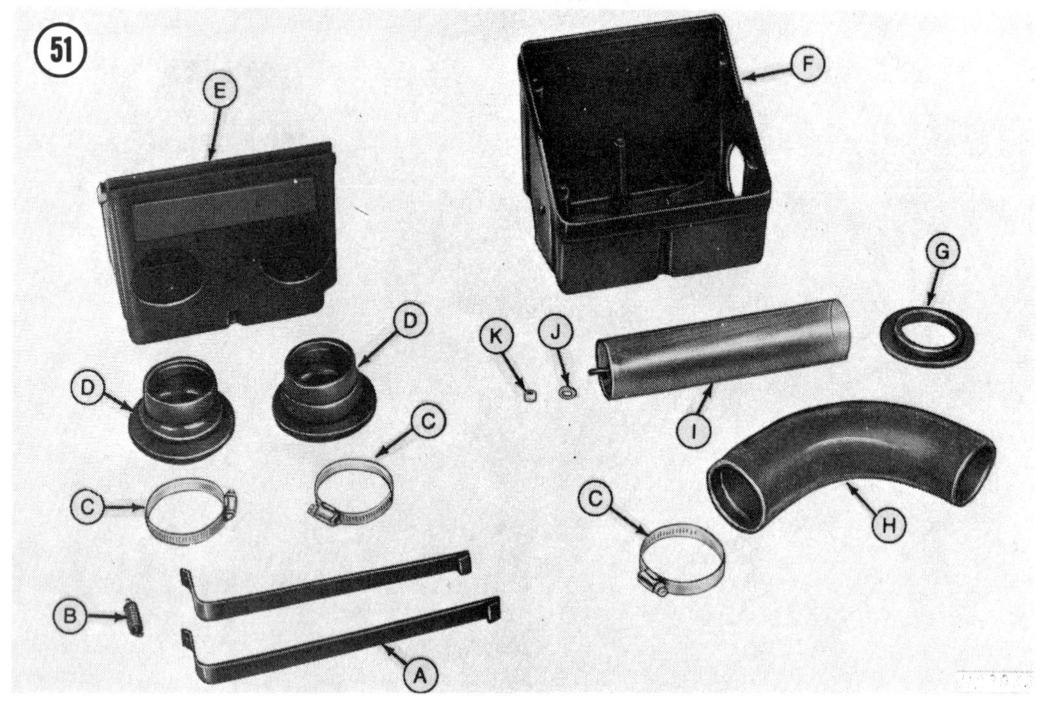

A. Straps (2)
B. Spring
C. Clamp (3)
D. Boots (2)
E. Lower half
F. Upper half
G. Rubber ring
H. Hose
I. Screen
J. Washer
K. Nut

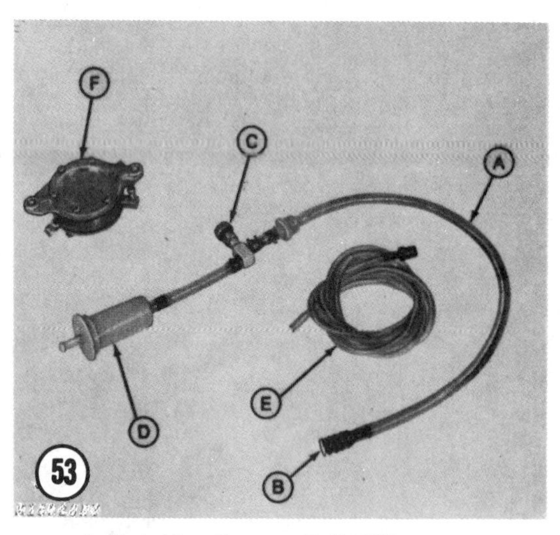

A. Fuel pick-up line
B. Screen
C. Shut-off valve
D. Fuel filter
E. Vent line
F. Fuel pump

prevents spilled gasoline from entering the snowmobile pan area.

The fuel tank will have 2 fuel lines with Walbro carburetors or one fuel line with Bendix carburetors. When 2 lines are used, the green line is for fuel pickup and the clear line is for vapor return. One green line indicates the fuel pickup line.

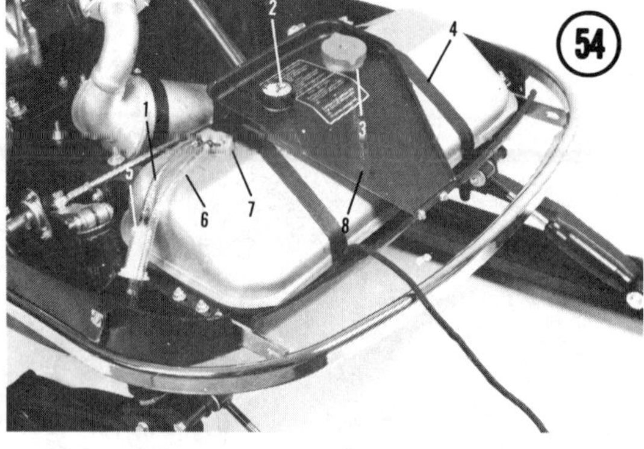

1. Fuel pickup line
2. Fuel gauge
3. Fill cap
4. Tank strap
5. Inline filter
6. Vapor return line
7. Fuel tank fitting
8. Spill tray

The fuel pickup line is connected to a self-cleaning screen with an internal ball check valve as shown in **Figure 55**. The internal check valve prevents fuel from running back out of carburetors and lines when the engine is stopped.

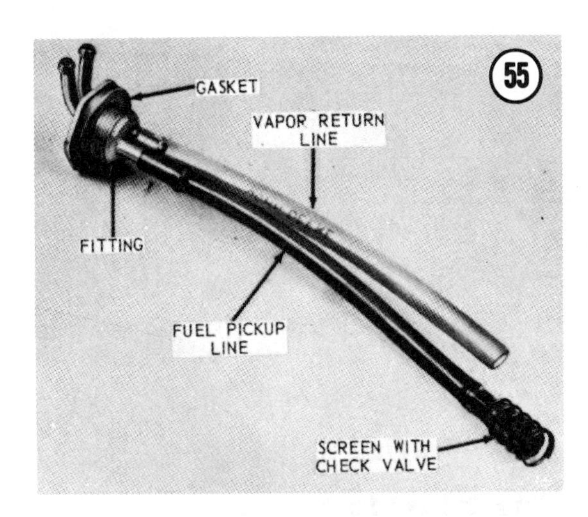

The fuel tank filler cap is sealed. The vent for the fuel tank is in the gas gauge cap.

Removal/Installation

1. Disconnect fuel lines.
2. Unbolt tank straps (**Figure 54**) and fold straps forward.
3. Remove 2 bolts securing spill tray to front of pan. Lift out fuel tank.
4. Installation is the reverse of these steps.

Pickup Screen Cleaning

1. Disconnect fuel lines from fitting.
2. Remove fitting from fuel tank (**Figure 54**).
3. Remove pickup screen with check valve from end of line.
4. Rinse screen in solvent and blow dry with compressed air. Replace assembly if screen is damaged or if check valve does not restrict flow of fuel backward. Replace gasket on fuel line fitting if necessary.

INLINE FUEL FILTER

Service of inline fuel filter is limited to annual replacement or replacement when contamination builds up at the base of the cone in the filter unit.

FUEL PUMP

To check fuel pump operation, disconnect fuel line from pump to carburetor at the carburetor. Make sure ignition switch is off and pull recoil starter handle and check for fuel flow at fuel line. If fuel flow from pump is unsatisfactory, replace pump.

CHAPTER SIX

LIQUID COOLING SYSTEM

A liquid cooling system is used on all Liquifire and Liquidator models. The cooling system consists of a drive belt, coolant pump, thermostat, heat exchangers on Liquifire models, and a radiator on Liquidator models. The thermostat maintains uniform engine temperatures throughout the engine's operating range. Refer to **Figure 1** for a typical liquid cooling system.

The pressure cap maintains the cooling system under pressure to achieve a higher potential coolant boiling point. Coolant is a 50/50 mixture of water and ethylene glycol anti-freeze. The coolant recovery tank holds any possible system overflow. Coolant captured in the recovery tank is siphoned back into the cooling system when engine cools. See **Table 1** (end of chapter) for cooling system specifications.

COOLING SYSTEM PRESSURE TESTING

Special pressure testing tools and adapters are required for system pressure tests. For this reason, have an authorized dealer perform any necessary cooling system tests.

DRAINING AND FILLING COOLING SYSTEM

Drain and refill cooling system with new coolant at 2-year intervals.

Refer to **Figure 2** for Liquifire models and **Figure 3** for Liquidator models.

A. Hose to heat exchangers
B. Filler cap
C. Thermostat
D. Cylinder head air vent screw
E. Drain screw

1. Remove drain screw and filler cap.

NOTE: *On Liquidator models, remove lower hose from connection between exhaust silencers.*

2. Remove hoses from engine and elevate rear of snowmobile to drain system.
3. Rinse engine and engine compartment with clean water.

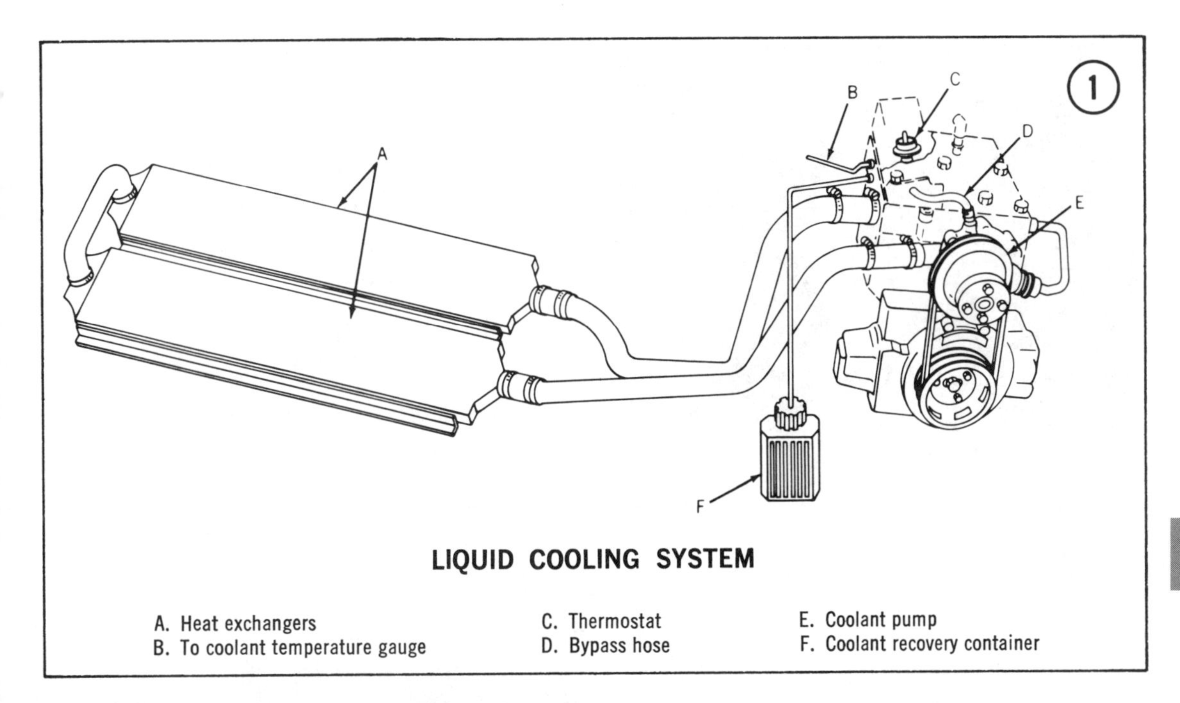

LIQUID COOLING SYSTEM

A. Heat exchangers
B. To coolant temperature gauge
C. Thermostat
D. Bypass hose
E. Coolant pump
F. Coolant recovery container

A. Pressure cap
B. Pump air vent screw
C. Cylinder block air vent screw
D. Thermostat

4. Position machine on a level surface.
5. Remove air vent.
6. Start filling system with a 50/50 mixture of water and ethylene glycol anti-freeze. Fill system until solution flows from air vent. Install air vent washer and screw.

CAUTION

Never exceed 50/50 mixture of anti-freeze and water. Never use anti-freeze containing a radiator stop leak or add a radiator stop leak to cooling system.

7. Continue filling cooling system until top tank or radiator is full. Liquifire models hold approximately 4½ quarts (4.25 liters). Liquidator models hold approximately 4.0 quarts (3.78 liters). Add coolant to recovery tank until approximately one inch covers end of hose in tank.
8. Check all hose connections for leaks. Install filler cap.
9. Block up rear of machine to clear track off the ground.
10. Start engine and warm up to operating temperature. Check entire cooling system for leaks.
11. Shut off engine and let cool. Recheck coolant level.

THERMOSTAT REMOVAL/INSTALLATION

Refer to **Figure 4** for Liquifire models and **Figure 5** for Liquidator models.

1. Drain cooling system.
2. Remove bolts securing thermostat housing and remove housing. Lift out thermostat. If engine has been running too cold or overheating, replace thermostat.
3. Installation is the reverse of these steps. Use a new gasket on thermostat housing. Fill cooling

A. Top tank B. Thermostat

system as outlined under *Draining and Filling Cooling System.*

COOLANT PUMP REMOVAL/INSTALLATION

Refer to **Figure 6** for this procedure.

1. Drain cooling system.

NOTE: *On Liquifire models it is not necessary to drain heat exchangers.*

2. Remove drive belt cover and loosen screws securing pump. Remove drive belt.

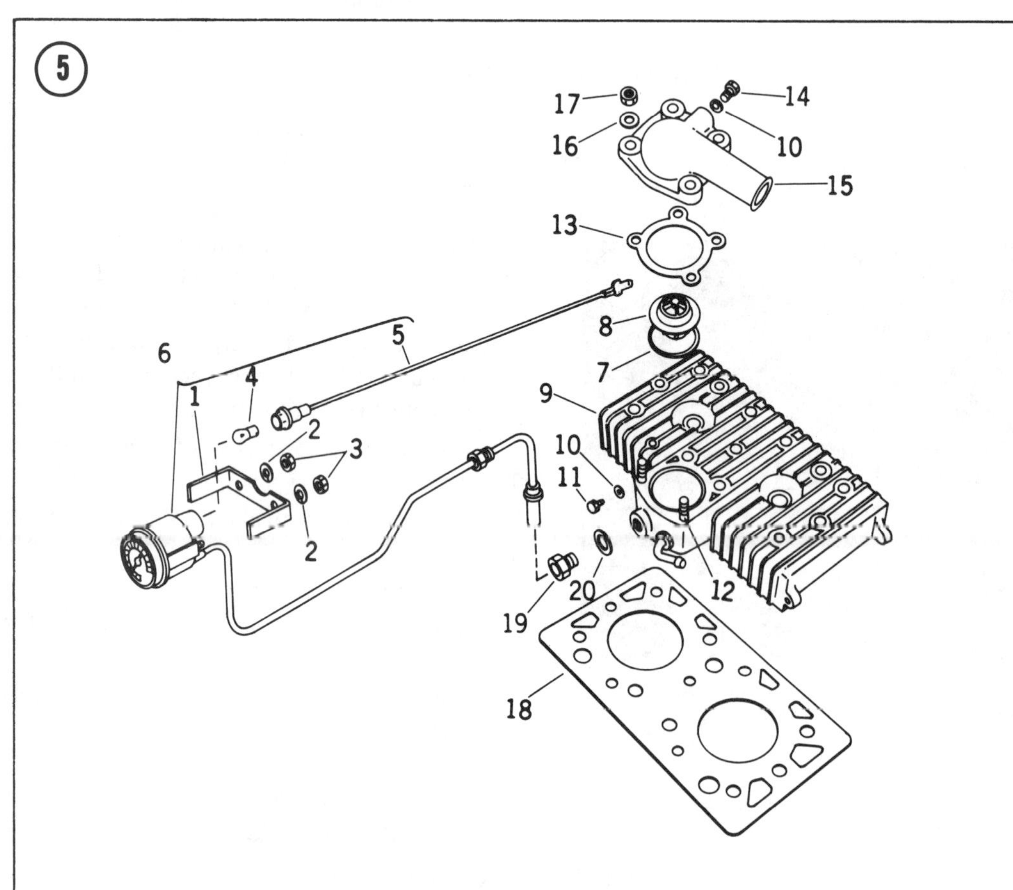

THERMOSTAT — LIQUIDATOR MODELS

1. Temperature gauge bracket
2. Lockwasher
3. Nut
4. Bulb
5. Lamp holder
6. Temperature gauge
7. Thermostat gasket
8. Thermostat
9. Cylinder head
10. Bleeder screw gasket
11. Bleeder screw
12. Outlet adapter stud
13. Outlet adapter gasket
14. Outlet adapter screw
15. Outlet adapter
16. Washer
17. Nut
18. Cylinder gasket
19. Adapter
20. Adapter to engine packing

A. Starter cup
B. Pump pulley
C. Bleed hose
D. Head-to-pump hose
E. Coolant pump
F. Flywheel housing

3. Remove all hoses connected to pump.

4. Remove pulley from pump bracket and remove pump from engine.

5. Installation is the reverse of these steps. If pump repair is required, refer work to an authorized dealer. Keep the following points in mind:

   a. Leave pump mounting nuts loose.

   b. Gently pry up on pump until drive belt deflects about ⅜ in. (9.5mm) as shown in **Figure 7**. Tighten nuts securely.

   c. Fill cooling system as outlined in *Draining and Filling Cooling System.*

LIQUIFIRE HEAT EXCHANGER REMOVAL/INSTALLATION

1. Drain cooling system.

2. Remove slide suspension.

3. Disconnect heat exchanger hoses (**Figure 8**).

4. Using a cold chisel, gently remove rivets securing heat exchangers.

5. Installation is the reverse of these steps. Pop rivet heat exchangers to tunnel from the bottom. Refer heat exchanger repair to an authorized dealer. Fill cooling system as outlined under *Draining and Filling Cooling System.*

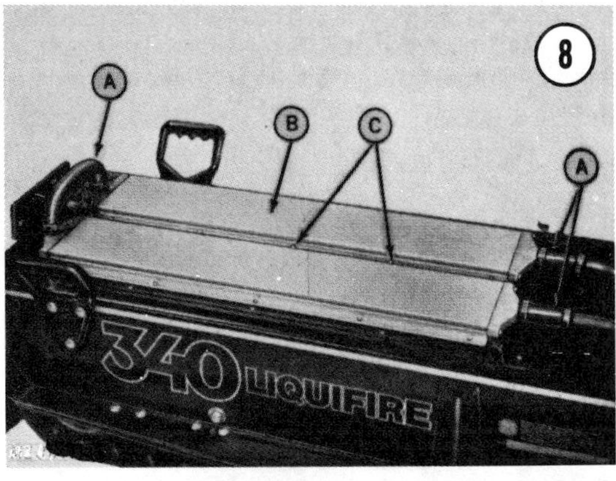

A. Hoses
B. Heat exchanger
C. Rivets

LIQUIDATOR RADIATOR REMOVAL/INSTALLATION

Refer to **Figure 9** for this procedure.

1. Drain cooling system.

2. Remove inlet and outlet hoses from radiator.

3. Remove nuts securing radiator to mounting bracket and remove radiator.

4. Installation is the reverse of these steps. Refer all radiator repair work to an authorized dealer or competent radiator repair shop. Fill cooling system as outlined under *Draining and Filling Cooling System.*

COOLANT TEMPERATURE GAUGE

Coolant temperature gauge is replaceable only as a complete unit. If faulty, remove gauge from instrument panel and sender from cylinder and replace with a new unit.

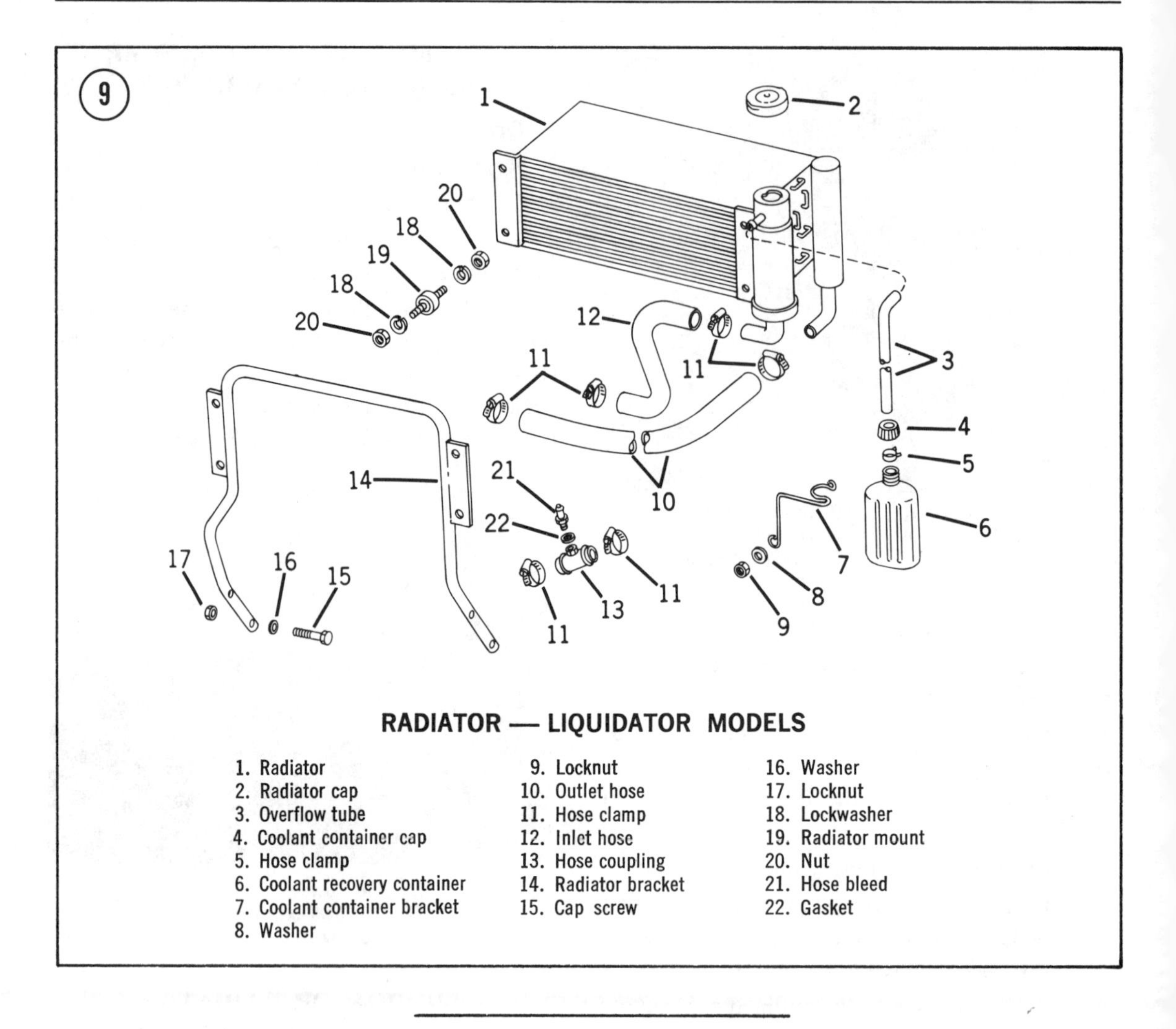

RADIATOR — LIQUIDATOR MODELS

1. Radiator
2. Radiator cap
3. Overflow tube
4. Coolant container cap
5. Hose clamp
6. Coolant recovery container
7. Coolant container bracket
8. Washer
9. Locknut
10. Outlet hose
11. Hose clamp
12. Inlet hose
13. Hose coupling
14. Radiator bracket
15. Cap screw
16. Washer
17. Locknut
18. Lockwasher
19. Radiator mount
20. Nut
21. Hose bleed
22. Gasket

Table 1 COOLING SYSTEM SPECIFICATIONS

Coolant capacity	
Liquifire	4½ quarts (4.25 liters)
Liquidator	4.0 quarts (3.78 liters)
Pressure cap pressure	15 psi (1.1 kg/cm^2)
Coolant pump belt deflection	⅜ in. (9.5mm)
Thermostat opening temperature	144°F (61°C)

CHAPTER SEVEN

ELECTRICAL SYSTEM

The electrical system on John Deere snowmobiles consists of an ignition system, lighting system, and an optional electric starting system.

Two types of ignition systems are used; a magneto and capacitor discharge ignition (CDI). Refer to **Table 1** to determine ignition system application.

Table 1 IGNITION SYSTEM APPLICATION

Model	Ignition
300	Robert Bosch
400	Kokusan
500 (up to serial No. 20,000)	Denso
500 (serial No. 20,001 and subsequent)	Kokusan
600 (up to serial No. 20,000)	Denso
600 (serial No. 20,001 and subsequent)	Kokusan
JDX4 (up to serial No. 20,000)	Robert Bosch
JDX4 (serial No. 20,001 and subsequent, and JDX4 Special)	Kokusan
800, JDX6, and JDX8	Denso
Cyclone, Liquifire, JD295/S, 340 Liquidator, and JD340/S	CDI (capacitor discharge ignition)

The lighting system consists of a headlight, brake/taillight, and console lights.

The electric starting system is an optional package consisting of a battery, a starter, and solenoid, and charging equipment.

This chapter includes testing and repair of some components of the ignition, lighting, and charging systems. Some testing and repair tasks referenced in this chapter require special testing equipment and tools. These tasks are best accomplished by an authorized dealer or competent auto electric shop.

Wiring diagrams for all models are included at the end of the chapter.

Refer to Chapter Two for ignition timing and breaker point adjustment.

CDI IGNITION

The capacitor discharge ignition system, **Figure 1**, consists of a permanent magnet flywheel, alternator, and solid-state capacitor. The system supplies high voltage for ignition and generates current required for the lighting system.

The flywheel incorporates a special flexible magnet and is mounted on the engine crankshaft. The flywheel and magnet revolve around the stator assembly, which is fixed to the engine.

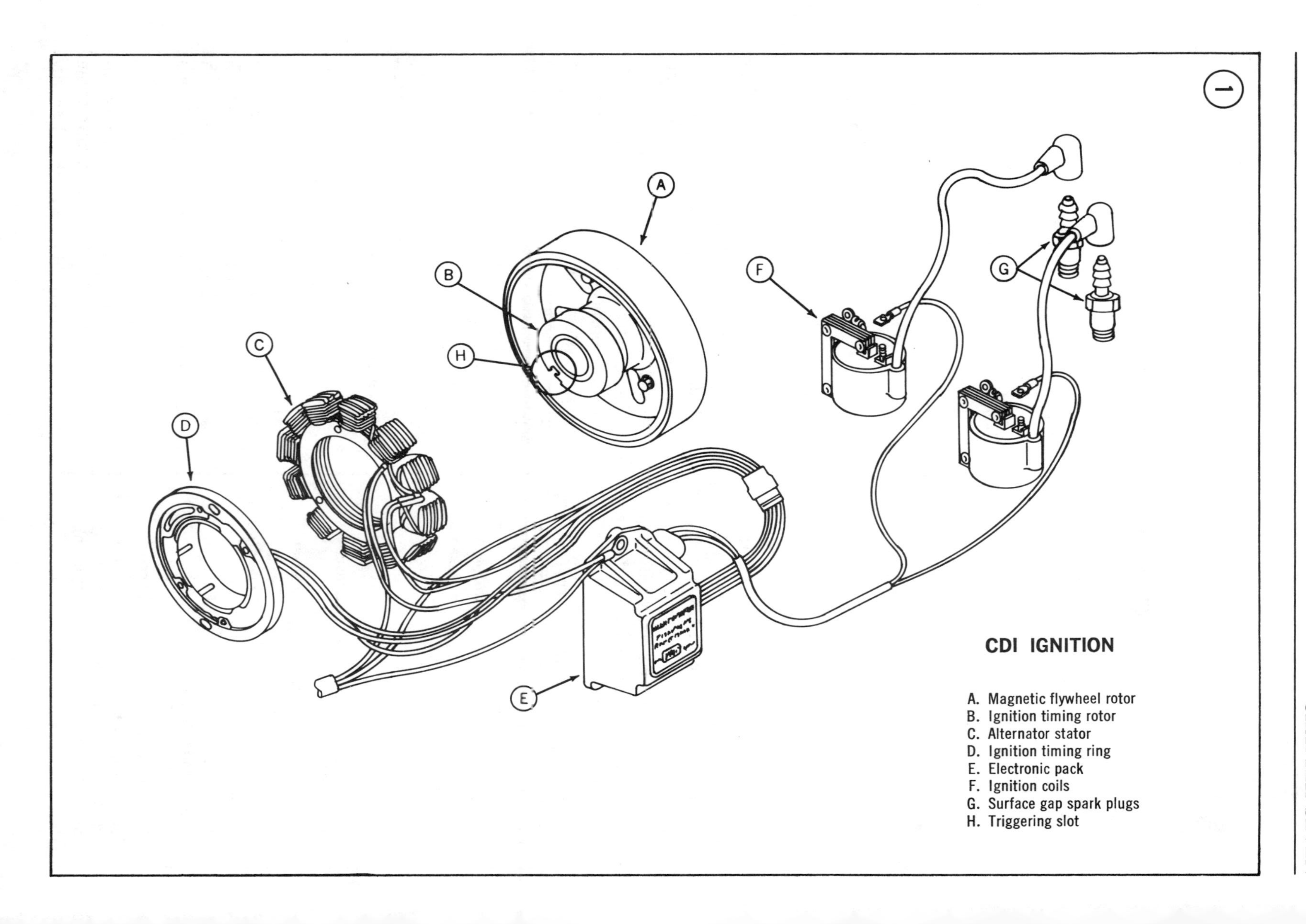
1
A
B
C
D
E
F
G
H
CDI IGNITION
A. Magnetic flywheel rotor
B. Ignition timing rotor
C. Alternator stator
D. Ignition timing ring
E. Electronic pack
F. Ignition coils
G. Surface gap spark plugs
H. Triggering slot

Current is generated in the 12 pole windings of the stator.

Nine poles supply power for the lighting system and 3 poles supply power for the ignition.

The ignition timing ring, alternator stator, and electronic pack require special test equipment to troubleshoot malfunctions and to monitor performance. If trouble exists in any of these units, refer the testing and repair to an authorized dealer.

Testing Engine Kill Switch (Serial No. 50,001-70,000)

Refer to **Figure 2** for this procedure.

1. Disconnect white and black wires leading to terminal block.
2. Connect test leads of a continuity test light to white and black wires.
3. Turn ignition switch to ON position.
4. With kill switch in the OFF position, test light should light. Test light should not light with kill switch in ON position. Replace kill switch if defective.

Ignition Switch Test (Serial No. 50,001-70,000)

Refer to **Figure 3** for this procedure.

1. Disconnect white and black wires leading to terminal block.
2. Connect test leads of a continuity test light to white and black wires.
3. Place engine kill switch in OFF position.
4. With ignition switch in OFF position, test light should light.

Ignition and Kill Switches Test (Serial No. 70,001 and Subsequent)

Refer to **Figure 4** for this procedure.

1. Disconnect white and black wires between electronic pack and ignition and kill switches (2 wires are taped together).
2. Connect test leads of a continuity test light to white wire and ground.
3. Turn ignition switch to ON position.
4. With kill switch in OFF position, test light should light. Test light should not light with kill switch in ON position.
5. Place kill switch in OFF position.
6. With ignition switch in OFF position, test light should light.

Flywheel, Alternator Stator, and Trigger Removal/Installation

It is not necessary to remove engine from snowmobile to perform the following procedure.

1. Remove recoil starter, flywheel housing, and lower fan sheave.

CAUTION

Do not strike flywheel with a steel hammer or serious damage to flywheel may occur.

2. Install special flywheel holding tool, or locally fabricated equivalent, to flywheel (**Figure 5**).

3. While holding flywheel, remove retaining nut.
4. Install puller to flywheel as shown in **Figure 6**. Tighten puller and remove flywheel.

7

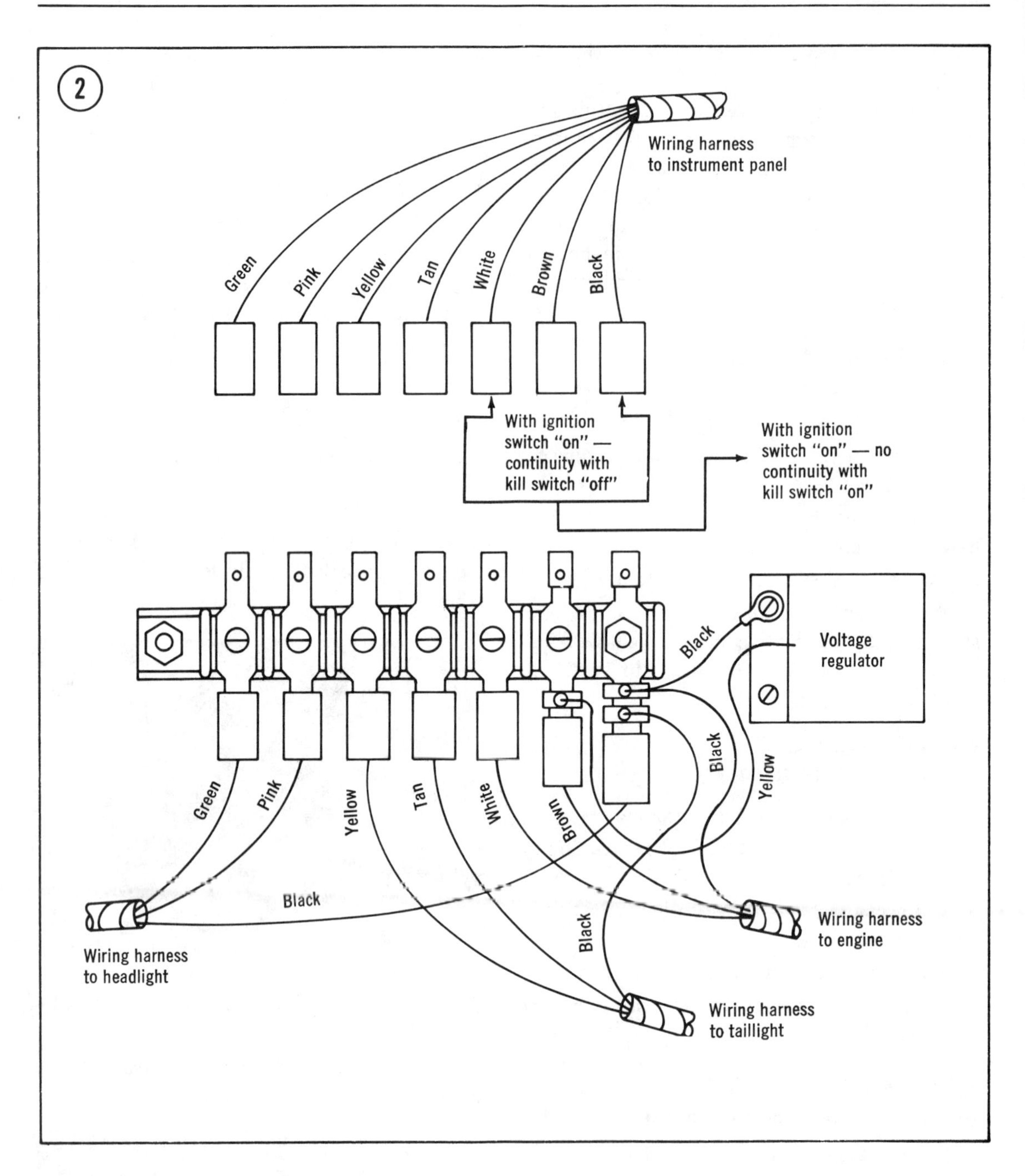

5. Remove stator and trigger assembly (**Figure 7**).

> NOTE: *Screws securing stator and trigger assembly are secured with Loctite. Tap screwdriver with hammer to break Loctite loose.*

6. Installation is the reverse of these steps. Keep the following points in mind:

a. Use Loctite to secure stator and trigger screws.
b. Position 4 screws on trigger assembly in the center of their slots for preliminary ignition timing.
c. Torque flywheel retaining nut to 60 ft.-lb. (8.3 mkg).
d. Refer to Chapter Two and perform ignition timing.

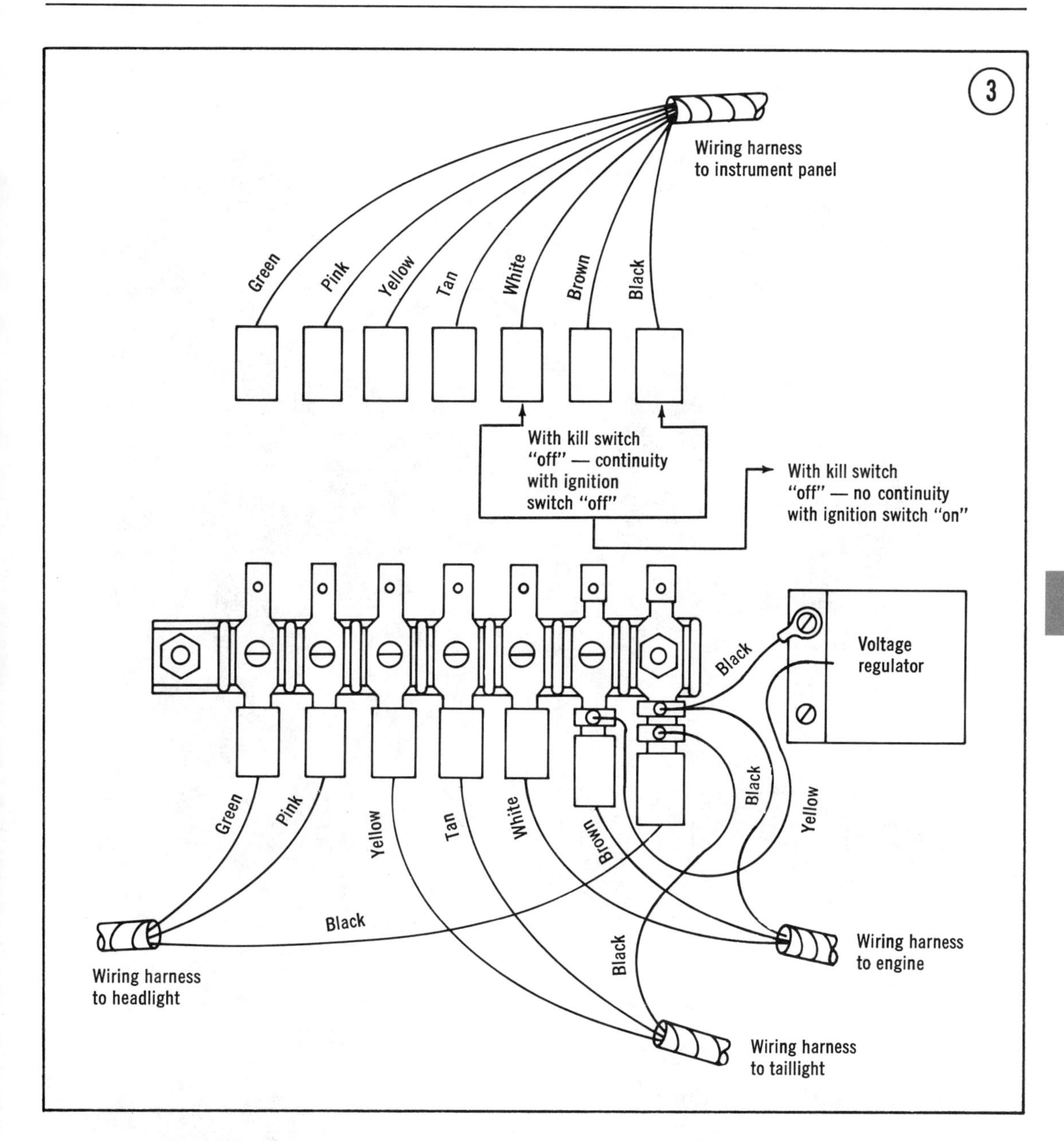

MAGNETO IGNITION

The magneto equipped snowmobiles feature an energy transfer ignition system (**Figure 8**). The Kohler engine has a Robert Bosch ignition system. The C.C.W. and Kioritz engines utilize either a Kokusan or Denso ignition system. See **Table 1** for systems used on each machine. All systems function in the same manner and similar procedures can be used for testing and repair.

The ignition system consists of 2 ignition generating coils, 2 sets of breaker points, 2 condensers, and 2 ignition coils. Each cylinder has separate components.

Refer to Chapter Two for breaker point and timing adjustments.

Emergency Stop Switch Test

1. Disconnect coupler to emergency stop switch.

2. Connect a light type continuity tester between terminals in coupler as shown in **Figure 9**.

4

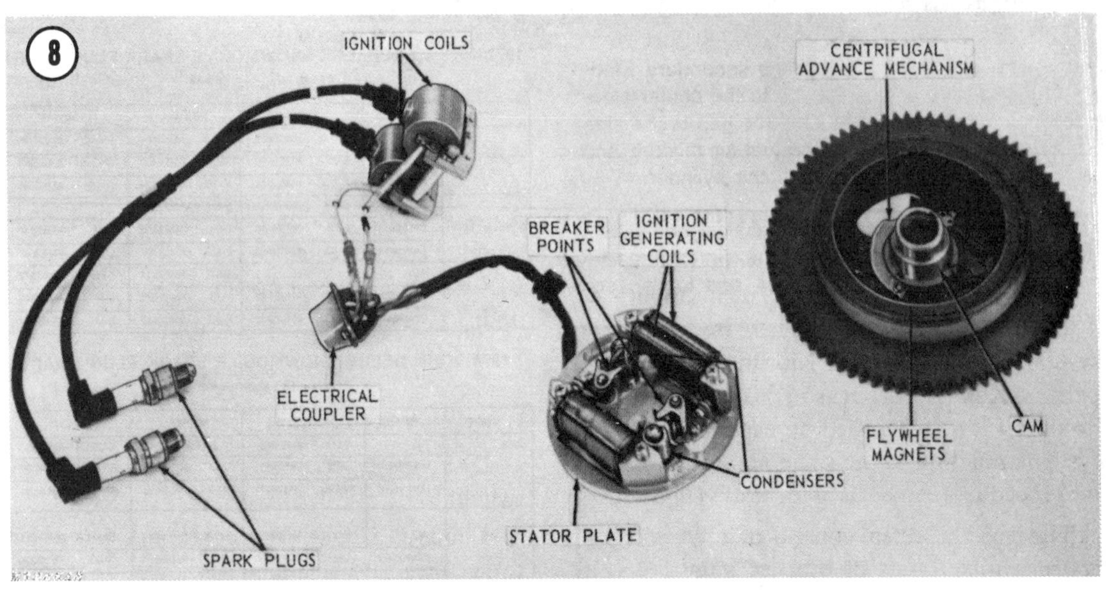
8
IGNITION COILS
CENTRIFUGAL ADVANCE MECHANISM
BREAKER POINTS
IGNITION GENERATING COILS
ELECTRICAL COUPLER
FLYWHEEL MAGNETS
CAM
CONDENSERS
STATOR PLATE
SPARK PLUGS

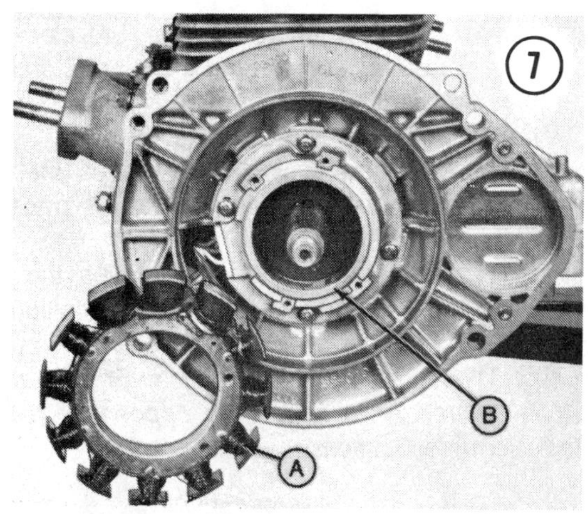

A. Alternator stator B. Trigger assembly

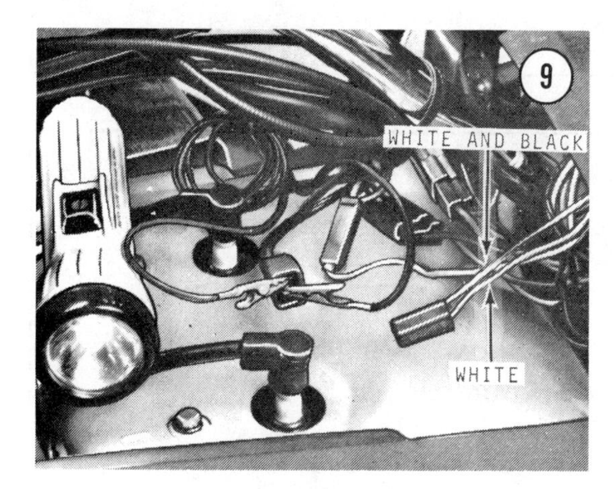

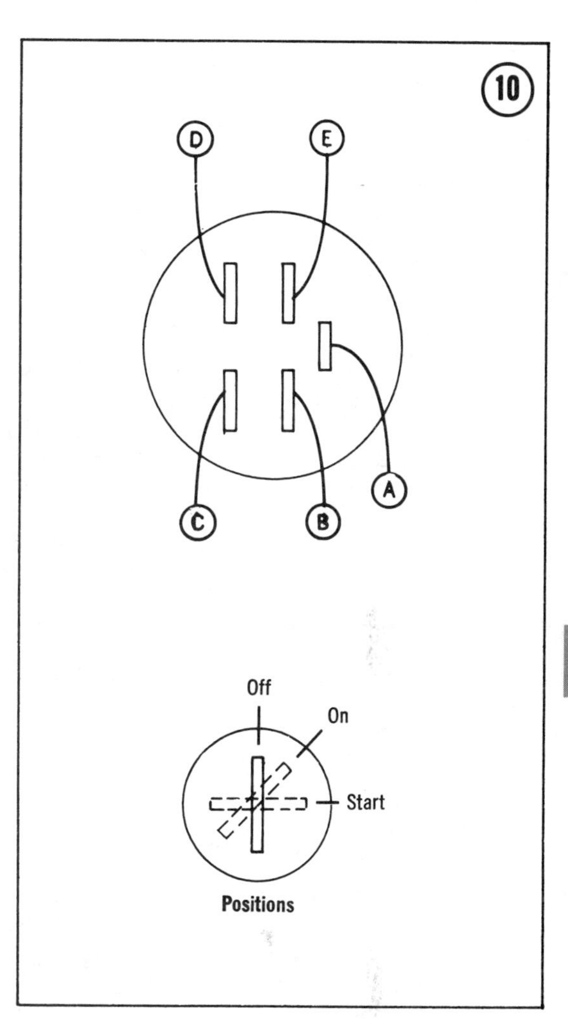

3. Test light must light when switch is pressed and go out when stop switch is in operating position.

Ignition Switch Test

Check ignition switch with a light type continuity tester. Refer to **Figures 10** and **Table 2** and check for continuity between switch terminals.

Table 2 IGNITION SWITCH TEST

Position	Closed	Open
Off	A & D	B, C, E
On	C & E	A, B, D
Start	B & E	A, C, D

C.C.W. ENGINE MAGNETO STATOR ASSEMBLY

Remove recoil starter, fan cover, and flywheel to gain access to stator assembly. Remove backing plate if stator assembly is to be removed.

NOTE: *Breaker points, ignition generating coil and condenser with white wires are No. 1 cylinder components; red leads are for No. 2 cylinder.*

Refer to **Figures 11 and 12**, Kokusan and Denso stator assemblies, for the following procedures.

Condenser

1. Loosen soldered leads on condenser terminal.
2. Remove screw securing condenser to stator plate and remove condenser.

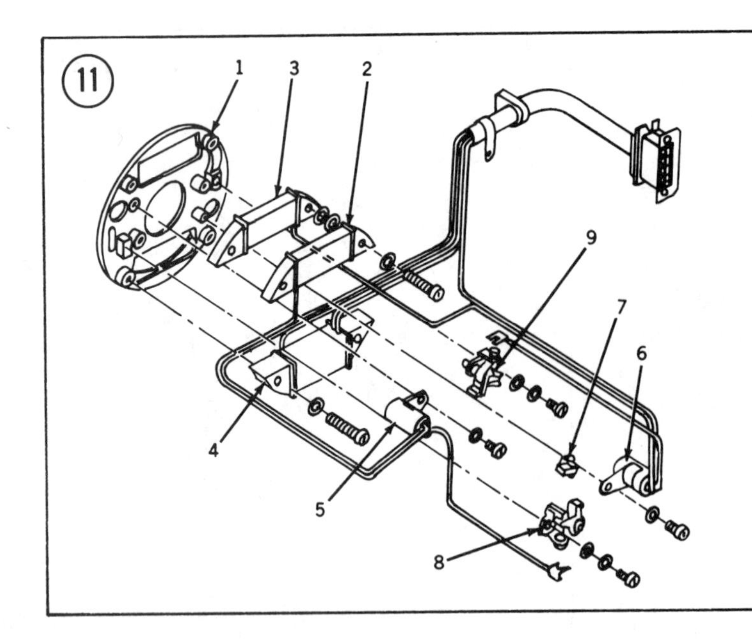

KOKUSAN STARTER

1. Stator plate
2. No. 1 ignition generating coil (white leads)
3. No. 2 ignition generating coil (red leads)
4. Lighting coil
5. Condenser (white leads)
6. Condenser (red leads)
7. Felt oil pad and holder
8. Breaker points (white leads)
9. Breaker points (red leads)

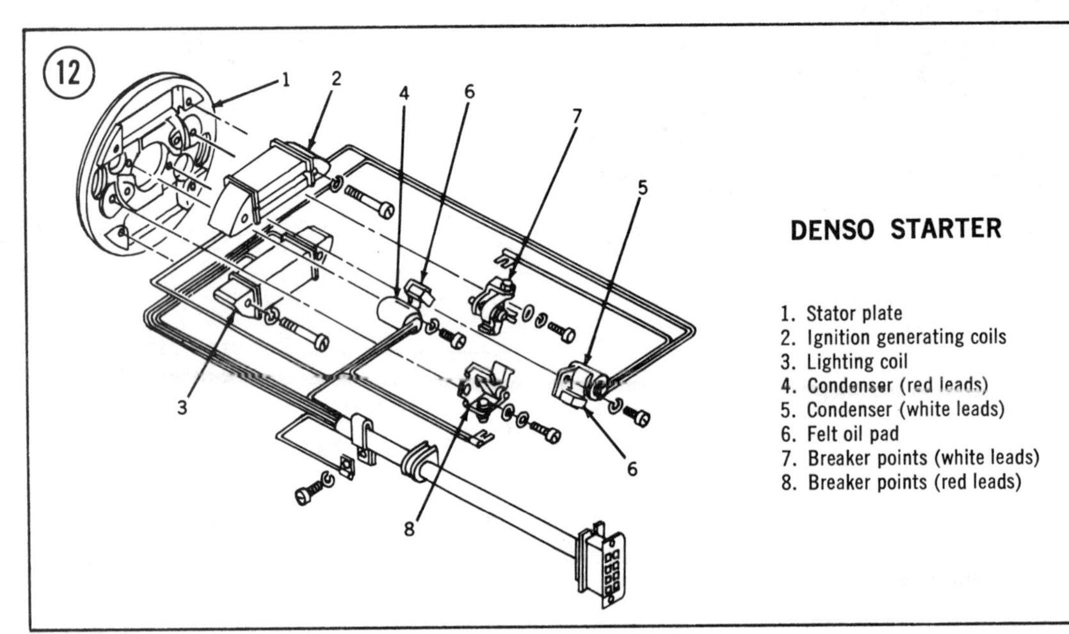

3. Install new condenser and solder leads to terminal. Repeat for other condenser.

CAUTION

Exercise care when resoldering wire to condenser as too much heat can destroy condenser.

Lighting Coil

1. Remove 2 screws securing lighting coil to stator plate (**Figure 13**). Remove 2 yellow lead connectors from coupler and pull wires through protector.

2. Thread new yellow coil wires through protector and secure new lighting coil to stator plate.

3. Install new connectors to yellow wires and insert into coupler.

NOTE: *On snowmobile models 400, 500, 600, and JDX4 Serial No. 20,001 and subsequent, the production light*

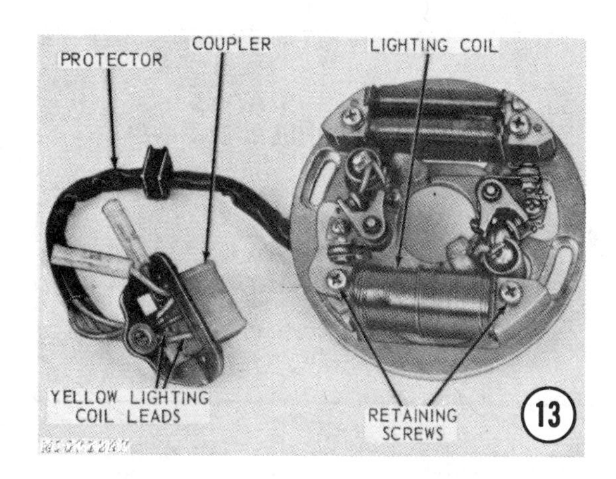

ing coil has 2 yellow leads and one brown lead. Replacement parts have 2 green and 2 black leads. Connect lighting coil as follows:

a. *Black leads together to one yellow lead.*

b. *Green lead to remaining yellow lead.*

c. *Brown lead grounded to stator plate.*

Kohler engines (Serial No. 722605694) are equipped with 2 lighting coils, a 75-watt coil and a 23-watt coil in parallel. Engines (Serial No. 722605695-732701000) are equipped with a single 100-watt lighting coil. For replacement, should the 75-watt or 23-watt coil fail, remove both and replace with single 100-watt lighting coil. Beginning with Serial No. 732701001, 23-watt and 100-watt lighting coils are again used.

Ignition Generating Coils

NOTE: *Both ignition generating coils must be replaced as a unit on Denso system. Kokusan system coils can be replaced separately.*

1. Loosen each coil lead from condenser terminal with a soldering gun.
2. Remove 2 screws securing coils to stator plate and remove coils.
3. Install new coil(s) and secure stator plate.
4. Solder leads to condenser terminals noting color code.

NOTE: *It may be necessary to remove breaker points to gain access to ignition generating coil leads.*

Breaker Points

1. Loosen breaker point terminal and remove leads.
2. Remove screw securing breaker points to stator plate and remove breaker points.
3. Install new breaker points and attach leads. Flywheel must be installed to adjust points. Perform breaker point adjustment as outlined in Chapter Two.

Felt Oil Pads

Replace felt oil pads, if their lubricating capacity is questionable. Oil pads with one or two drops of light oil whenever breaker points are replaced.

KOHLER ENGINE STATOR ASSEMBLY

See **Figure 14** for the following procedures.

Stator Assembly Removal/Installation

1. Remove recoil starter, fan drive sheave, fan belt, and flywheel.
2. Remove coil cover and disconnect blue/red and blue wire from coils. Remove coils.
3. Remove ventilator housing complete with stator assembly and wiring harness.
4. Remove stator from ventilator housing.

Condensers

1. Unsolder leads on condenser. Locally fabricate a condenser driver from a ½ in. hardwood dowel by drilling a ¼ in. hole ¼ in. deep in one end.
2. Drill a ¾ in. hole in a 2 x 4 x 6 in. block of wood. Place condenser over hole and drive out with wooden driver.
3. Insert new condenser and use wood driver to install condenser flush with bottom of stator plate. Solder leads to condenser.

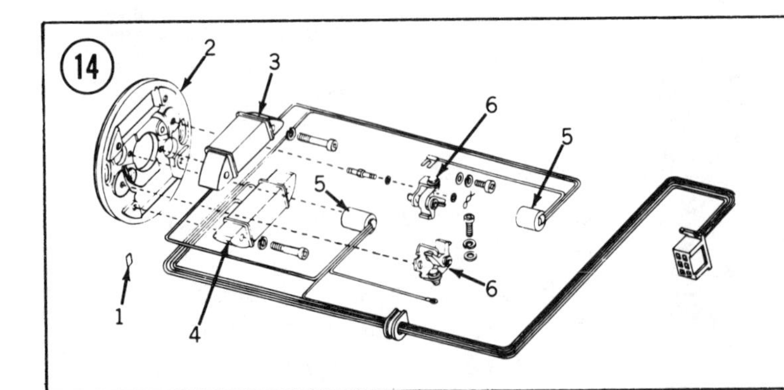

STATOR (KOHLER ENGINE)

1. Felt oil pad
2. Stator plate
3. Lighting coil
4. Ignition generating coil
5. Condenser
6. Breaker points

CAUTION

Use minimum heat to solder condenser lead or condenser will be destroyed.

Lighting Coil

Refer to *C.C.W. Engine Magneto Stator Assembly* for lighting coil replacement.

Ignition Generating Coils

1. Loosen each coil from condenser terminals with a soldering gun. Remove coil from stator plate.

NOTE: *On engines equipped with 23- and 100-watt lighting coils, remove 23-watt coil to gain access to lighting coil.*

2. Install new coil and secure to stator plate. Solder leads to condenser terminals. Install 23-watt lighting coil if removed.

NOTE: *It may be necessary to remove breaker points to gain access to ignition generating coil leads.*

Breaker Points

1. Remove leads from breaker point terminal and remove breaker points from stator plate.
2. Install new breaker points and attach leads. Perform breaker point adjustment as outlined in Chapter Two.

Felt Oil Pads

Replace felt oil pad if lubricating capacity is questionable. Oil pad with one or two drops of light oil whenever breaker points are replaced.

KOKUSAN FLYWHEEL ASSEMBLY

1. Remove flywheel.
2. Remove snap ring (**Figure 15**) and slide cam off collar. Examine cam and collar for scoring or wear and replace if necessary.
3. Inspect governor springs. If stretched or broken, remove retaining rings and replace springs.

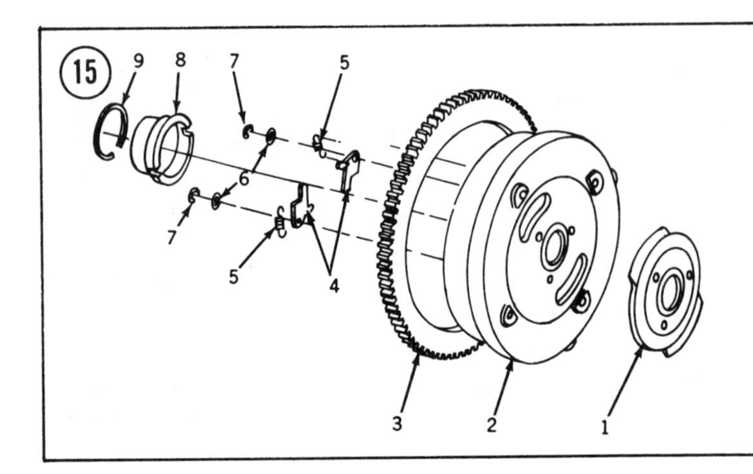

KOKUSAN FLYWHEEL ASSEMBLY

1. Window plate
2. Flywheel
3. Ring gear
4. Governor weight
5. Governor spring
6. Washer
7. Retaining ring
8. Cam
9. Snap ring

4. Lubricate flywheel collar and governor weight mechanism with a light film of grease and install cam.

NOTE: *Do not over-lubricate or excess grease will contaminate breaker points and cause ignition failure.*

5. Install governor weights and springs and secure with retaining rings.
6. Install cam and secure with snap ring.

NOTE: *Install cam with cam lobe positioned slightly to left of keyway.*

DENSO FLYWHEEL ASSEMBLY

1. Remove flywheel.
2. Remove retaining ring (**Figure 16**) and governor weight. Lift cam off flywheel collar.
3. Inspect cam and collar for scoring and wear. Inspect governor spring. Replace defective parts.
4. Lubricate flywheel collar and governor weight mechanism with a light film of grease prior to installing cam.

NOTE: *Do not over-lubricate or excess grease will contaminate breaker points and cause ignition failure.*

5. Install cam onto flywheel collar. Install governor weight and spring and secure with retaining ring.

NOTE: *Align keyways in flywheel collar and cam.*

BOSCH FLYWHEEL ASSEMBLY

1. Remove flywheel.
2. Remove screws and washers, spring retainer, and cam retainers securing cam, and lift out cam (**Figure 17**). Inspect cam, spring retainer, and cam retainers for wear or damage and replace as necessary.

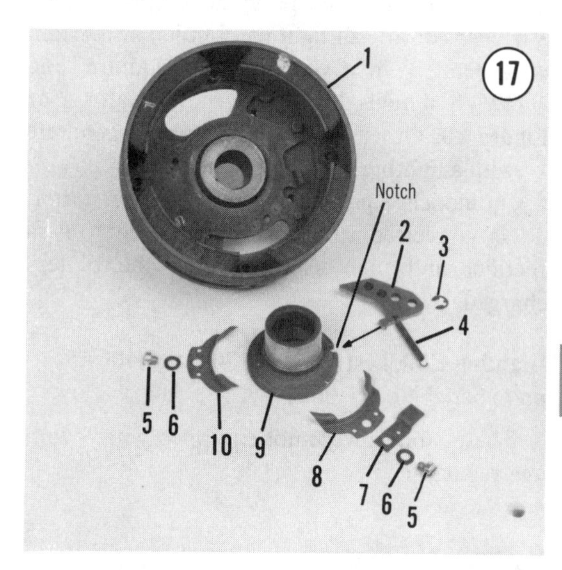

1. Flywheel
2. Flyweight
3. Retaining clip
4. Spring
5. Screw
6. Washer
7. Spring retainer
8. Cam retainer
9. Cam
10. Cam retainer

3. Remove retaining clip and disconnect spring from pin flywheel. Remove flyweight and spring. Inspect flyweight and spring for wear or damage and replace as necessary.

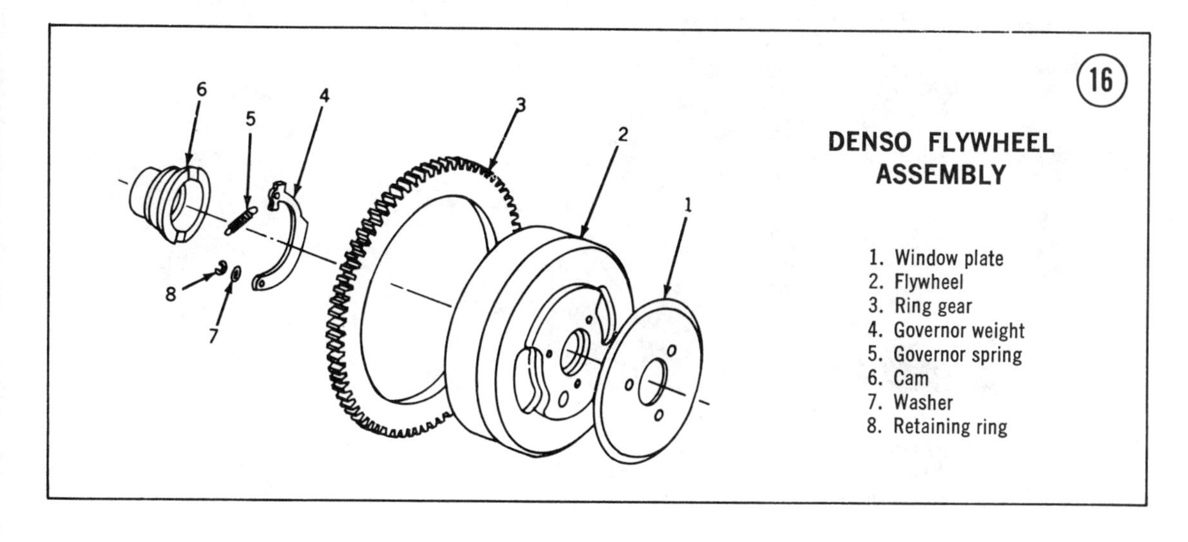

DENSO FLYWHEEL ASSEMBLY

1. Window plate
2. Flywheel
3. Ring gear
4. Governor weight
5. Governor spring
6. Cam
7. Washer
8. Retaining ring

NOTE: *When installing cam or flyweight be sure protrusion of flyweight is properly positioned in notch of cam.*

LIGHTING SYSTEM

The lighting system consists of a headlight and brake/taillight unit, instrument lights, and an AC (alternating current) generating device. Switches control all lighting circuits. AC current is generated in Cyclone and Liquifire and JD295/S models by an alternator-stator. See **Figure 18**. Other model snowmobiles generate AC with a lighting coil. See **Figure 19**.

On models equipped with an electric starter, AC is converted to DC (direct current) by a rectifier and then used to keep the battery charged.

Lighting Coil Test (Models 300 and 400 up to Serial No. 11,000)

These models were not equipped with a voltage regulator.

1. Block up track securely so engine can be safely run at operating speeds.

CAUTION

Do not operate engine with drive belt removed or engine over-revving and subsequent damage may occur.

2. If machine is equipped with electric starter, disconnect starting system coupler (**Figure 20**).

3. Connect AC voltmeter to orange and black terminals in coupler (**Figure 20**).

4. Turn on lights and start engine; AC voltage output should be from 12 to 15 volts at 6,000 rpm. If voltage is lower, lighting coil may be defective. Make continuity test of coil. If test proves coil satisfactory, check wiring and connections.

5. If voltage output is higher than specified, check light bulbs for operation. Output voltage will build as load is decreased.

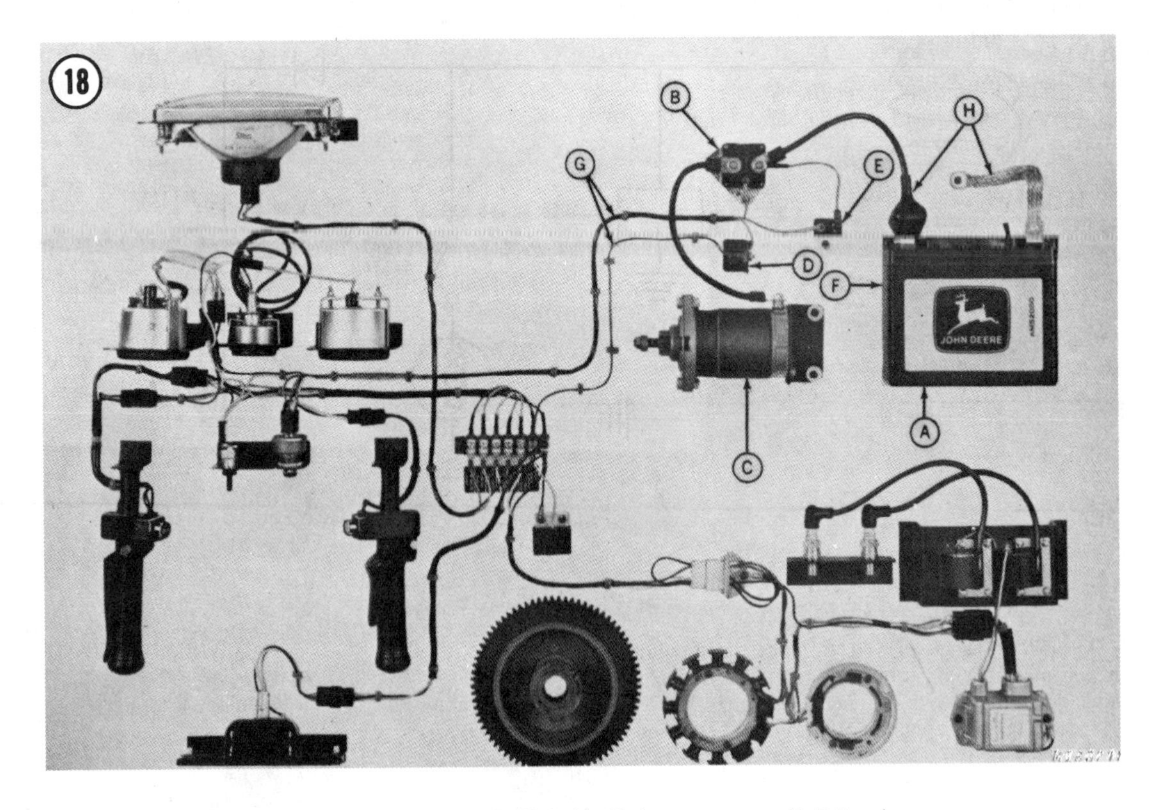

A. Battery
B. Solenoid
C. Starter motor
D. Diode (rectifier)
E. Circuit breaker
F. Battery box
G. Wiring harness
H. Battery cables

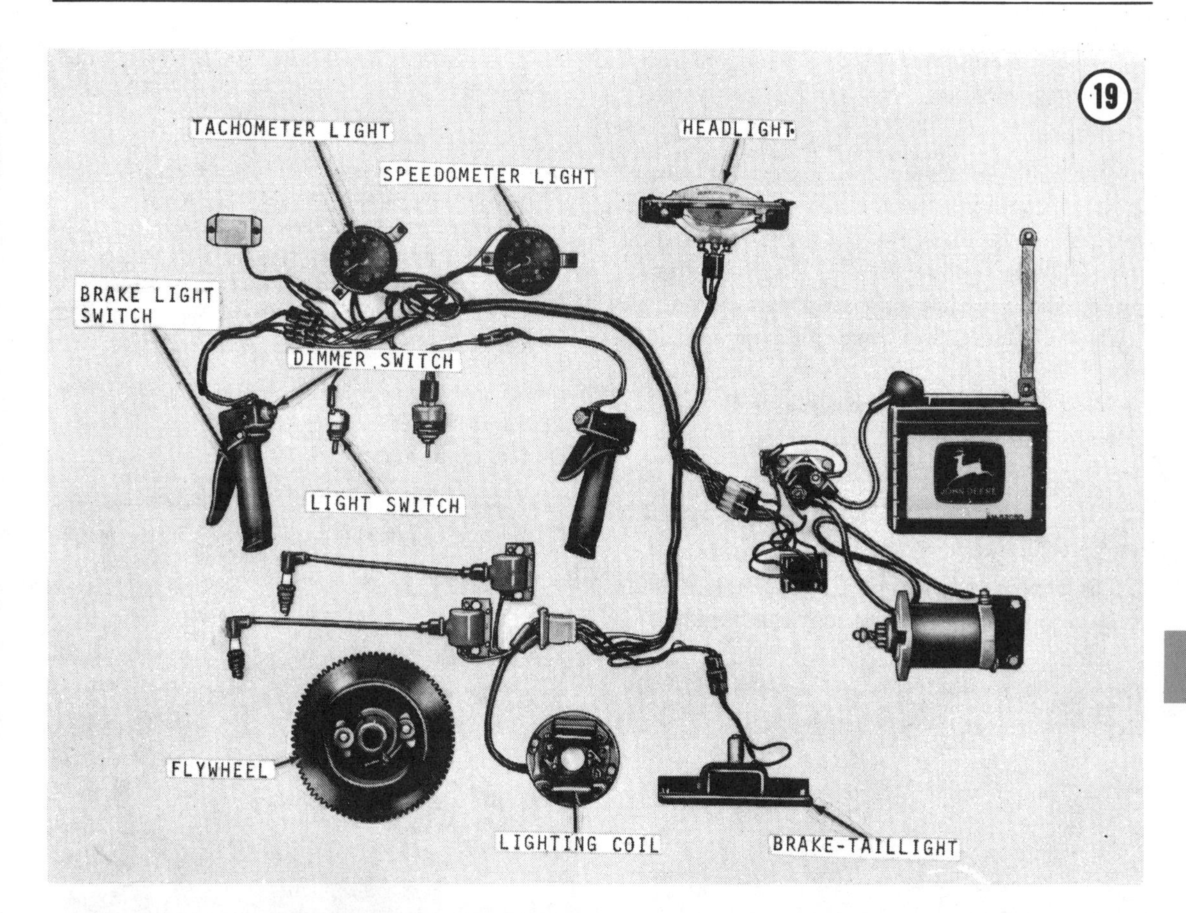

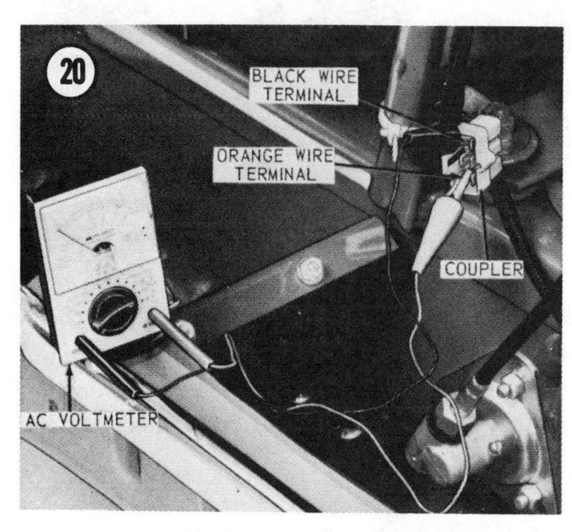

Lighting Coil and Voltage Regulator Test (Manual Start Models)

The voltage regulator will malfunction in only 2 ways: an "open" condition will cause all lights in system to burn out (at engine speeds over idling); and a "shorted" condition will prevent any lights from lighting. A defective lighting coil will also prevent the lights from lighting.

1. Block up track securely so engine can be safely run at operating speeds.

CAUTION
Do not operate engine with drive belt removed or engine over-revving and subsequent damage may occur.

2. Install a *new* taillight bulb.

3. With the light OFF, start and run engine at approximately 6,000 rpm; then, apply brake. Taillight bulb will burn out immediately if voltage regulator is defective. Replace voltage regulator. If taillight does not burn out, voltage regulator is working properly.

4. Disconnect voltage regulator from electrical system.

CAUTION
When checking headlight intensity with voltage regulator disconnected, run engine at idle speed only. Lighting sys-

tem is unregulated and a faster engine speed will cause lighting coil output to burn out the lights.

5. Start and run engine at *idle speed*. Turn on lights. If headlight lights and has good brilliance, lighting coil is functioning. If headlight is dim or will not light at all, lighting coil is malfunctioning and must be replaced. Headlight intensity should remain uniform at all engine speeds.

Lighting Coil and Voltage Regulator Test (Electric Start Models)

On electric start models, the lights are powered by the battery. The lighting coil is used to charge the battery.

1. Disconnect wires to circuit breaker and connect to a DC ammeter as shown in **Figure 21**.

2. Block up track securely so engine can be safely run at operating speeds. Run engine at 6,000-7,000 rpm. On 1973 snowmobiles, output should be 5-10 amps with lights off and 1-2 amps with lights on. On 1974 and 1975 snowmobiles, output should be 2-4 amps with lights off and 3-5 amps with lights on.
3. If output is not as specified, disconnect voltage regulator and recheck. If output increases, voltage regulator is defective. If output does not increase, check key switch, rectifier, and wiring. If still no output is indicated, the lighting coil could be defective.
4. Disconnect electric start and reinstall manual start coupler. Test as outlined under *Lighting Coil and Voltage Regulator Test (Manual Start Models).*

NOTE: *Snowmobiles (1973-75) with electric start can be operated with voltage regulator temporarily disconnected if voltage regulator is proven defective. However, higher charging rates will result with the lights off when operating the snowmobile. Check the battery water level frequently.*

Light Switch Test (Models up to Serial No. 20,000)

Refer to **Figure 22** for this procedure.

1. Remove windshield and loosen bolts securing instrument panel.

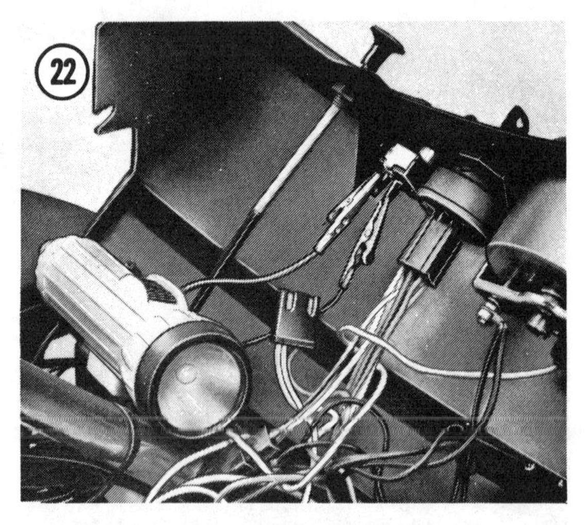

2. Lift panel up to provide access to back side of light switch.
3. Remove coupler from light switch.
4. Connect an ohmmeter or light type tester between light switch terminals in on position. If switch does not show continuity, replace switch.

Light Switch Test (Models Serial No. 20,001 and Subsequent Except Cyclone and Liquifire)

Refer to **Figure 23** for this procedure.

1. Remove windshield and loosen bolts securing instrument panel.
2. Lift panel up to provide access to back side of light switch.

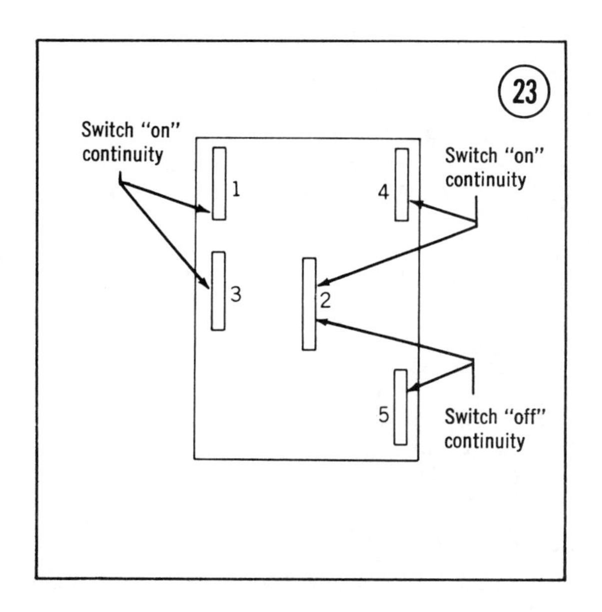

3. Remove coupler from light switch. Connect an ohmmeter or light type tester across terminals 1 and 3 and 2 and 4 with switch in ON position. Continuity should be present.

4. Turn switch off and check for continuity across terminals 2 and 5. If continuity is not present, replace switch.

> NOTE: *Continuity should not be present except as specified.*

Light Switch Test (Cyclone and Liquifire Models)

Refer to **Figure 24** for this procedure.

1. Disconnect yellow and brown wire from terminal block.

2. Connect leads of continuity test light across 2 wires.

3. With light switch in ON position, test light should light. With switch off, test light should not light. Replace switch if defective.

Alternator Output Test

Refer to **Figure 25** for this procedure.

1. Connect AC voltmeter to brown and black leads at terminal block. Disconnect yellow voltage regulator lead from terminal block.

2. Start and run the engine at approximately 4,000 rpm. AC voltmeter should indicate 40-70 volts. Replace stator if voltage is not as specified.

> NOTE: *Wiring harness coupler must be connected to engine coupler during test. Engine will run with harness plug disconnected but* cannot be stopped.

Headlight Adjustment (Front Engine Models)

1. Position snowmobile on a flat surface with headlight 25 ft. (7.6m) from a vertical surface. With an operator on the seat and headlight on high beam, light beam centerline should be straight ahead of machine and 27¼ in. (69.2 cm) above ground level (**Figure 26**).

2. If headlight needs to be adjusted, proceed as follows:

   a. Refer to **Figure 27** and adjust 2 top screws clockwise to lower beam and counterclockwise to raise beam.

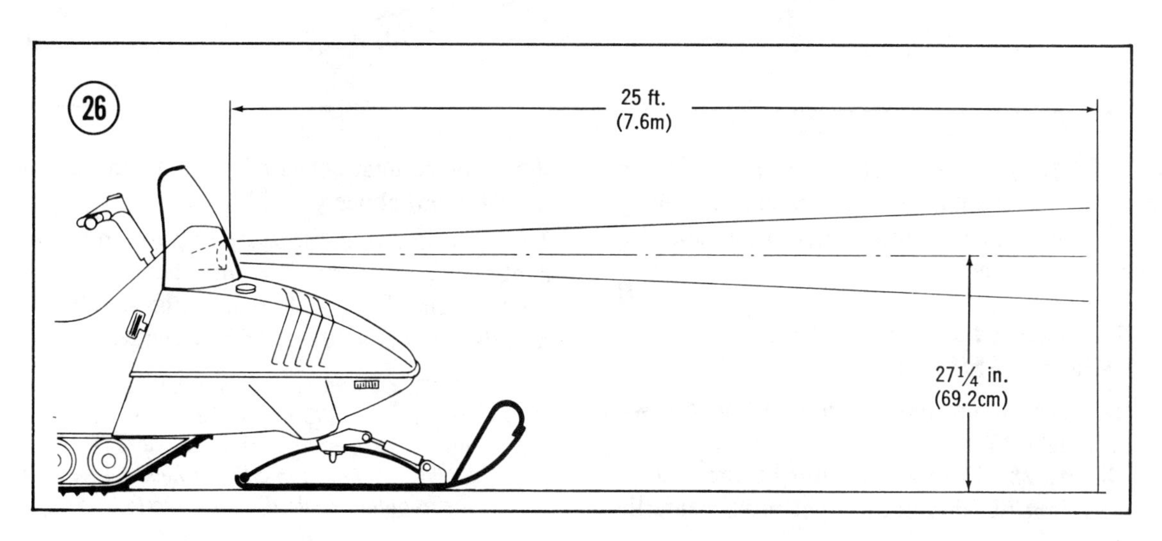

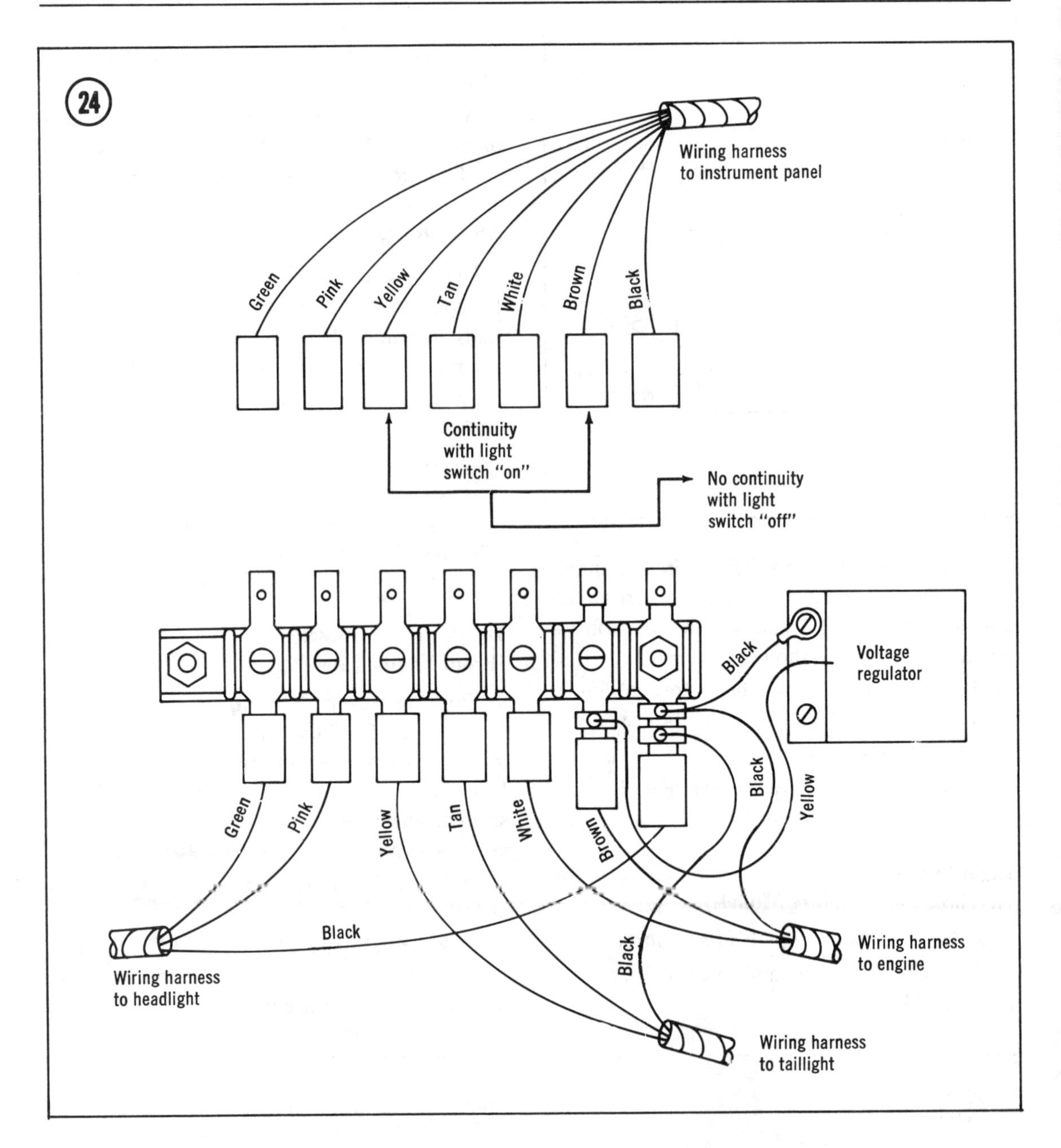

b. To move light beam right or left, loosen 2 side adjustment screws and tighten appropriate one to center beam ahead of machine.

Headlight Adjustment (Mid-engine Models)

1. Position snowmobile on a flat surface with headlight 25 ft. (7.6 m) from a vertical surface (**Figure 28**). With an operator on the seat and headlight on high beam, light beam centerline should be straight ahead of machine and 21½ in. (54.6 cm) above ground level.

2. Adjust screws as necessary to position headlight correctly. Refer to **Figure 29** for models up to Serial No. 30,000 and **Figure 30** for models Serial No. 30,001 and subsequent.

> NOTE: *On 400 and 500 models up to Serial No. 11,000 it may be necessary to loosen 2 headlight attaching screws and shim behind top edge of headlight unit to obtain desired light elevation.*

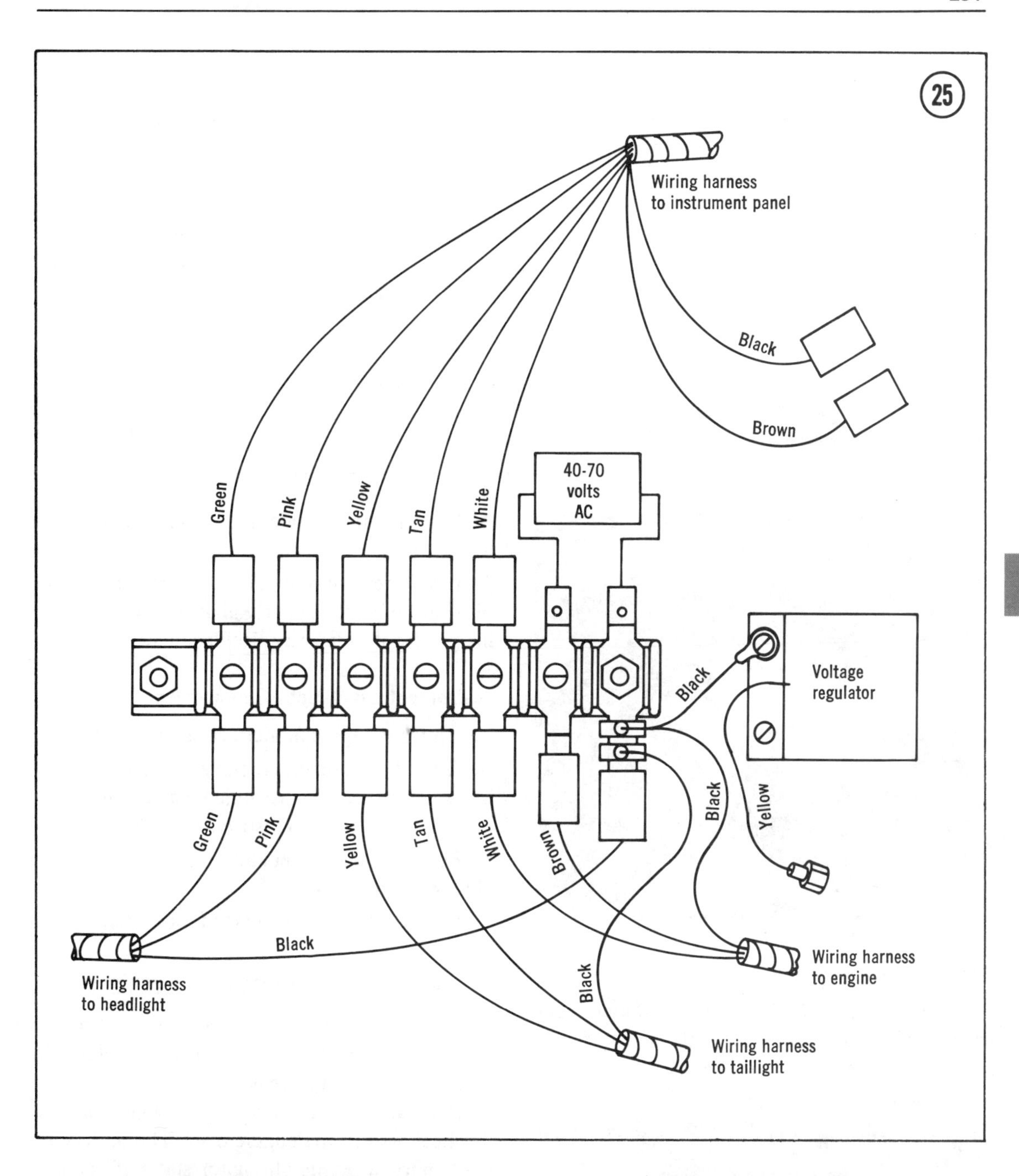
25
Wiring harness
to instrument panel
Black
Brown
40-70
volts
AC
Green
Pink
Yellow
Tan
White
Black
Voltage
regulator
Black
Yellow
Green
Pink
Yellow
Tan
White
Brown
Black
Wiring harness
to headlight
Black
Wiring harness
to engine
Wiring harness
to taillight

27
1
2
3

TIGHTEN TO MOVE
BEAM RIGHT
LOOSEN TO MOVE
BEAM LEFT
TIGHTEN TO
LOWER BEAM
TIGHTEN TO
RAISE BEAM
29

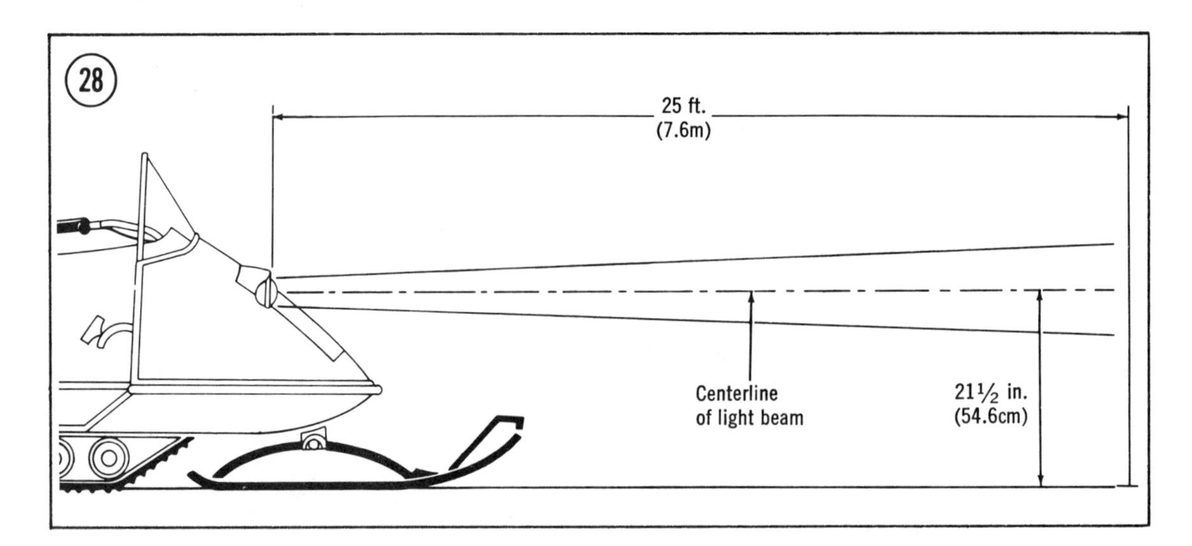

Headlight Replacement

1. Open hood and disconnect headlight connector. Note position of old light unit to facilitate installation of new unit.
2. Unhook headlight retaining wire end from slot and remove wire. Remove old headlight unit.
3. Install new unit making sure it is right side up.
4. Secure headlight with retaining wire. Wire end placed in slot must pass over other part of wire.

Brake/Taillight Bulb Replacement

1. Remove 2 screws securing light lens and remove lens.
2. Push in and rotate bulb counterclockwise to remove.
3. Install new bulb making sure that alignment pins on bulb are properly aligned. Bulb can be installed only one way. Replace lens.

Instrument Light Bulb Replacement

1. On front engine models, raise the hood.
2. On other models, remove windshield and loosen nuts securing instrument panel. Tip panel back to gain access to instrument bulbs.
3. Pull out bulb socket and rotate bulb counterclockwise to remove.
4. Installation is the reverse of these steps.

ELECTRIC STARTING SYSTEM

The electric starting system consists of a 12-volt battery, starter motor, starter solenoid, rectifier and circuit breaker.

The starter motor engages a ring gear on the flywheel to turn the engine over. The battery is kept charged by the alternator stator (Cyclone, Liquifire, and JD295/S models) or the lighting coil (other models). An automatic reset type of circuit breaker is used to protect the system from overloads or short circuits.

Battery Removal/Installation

1. Disconnect negative (—) cable (**Figure 31**). Remove rubber boot and disconnect positive (+) cable.

A. Battery box
B. Negative terminal
C. Positive terminal
D. Hold-down
E. Hold-down bolt

2. Loosen hold-down bolts and unhook bolts from battery box. Remove hold-down clamp.
3. Disconnect vent tube from battery. Carefully lift battery out of battery box.
4. Installation is the reverse of these steps. Keep the following points in mind:

CAUTION
Be sure battery connections are correct or serious damage to electrical components will occur.

a. Be sure exterior of battery and terminals are clean and free from corrosion.
b. Connect positive (+) cable to battery first.

Battery Cleaning and Service

Electrolyte level in the battery should be checked periodically, especially during periods of regular operation. Use only distilled water and top off battery to bottom of ring (filler neck) so the tops of the plates are covered. *Do not* overfill.

Battery corrosion is a normal reaction; however, it should be cleaned off periodically to keep battery deterioration to a minimum.

Remove battery and wire brush terminals and cable ends. Wash terminals and exterior of battery with about a 4:1 solution of warm water and baking soda.

CAUTION
Do not allow any baking soda solution to enter battery cells or serious battery damage may result.

Wash battery box and hold-down bolts with baking soda solution. Rinse all parts in clear water and wipe dry.

In freezing weather, never add water to a battery unless the machine will be operated for a period of time to mix electrolyte and water.

CAUTION
Keep battery fully charged. A discharged battery will freeze causing the battery case to break.

Remove the battery from the machine during extended non-use periods and keep battery fully charged. Perform periodic specific gravity tests with a hydrometer to determine the level of charge and how long charge stays up before it starts to deteriorate.

Battery Specific Gravity Test

Determine the state of charge of the battery with a hydrometer. To use this instrument, place the suction tube (**Figure 32**) into the filler opening and draw in just enough electrolyte to lift the float. Hold the instrument in a vertical position and take the reading at eye level.

Specific gravity of electrolyte varies with temperature, so it is necessary to apply a temperature correction to the reading you obtain. For each 10° that the battery temperature exceeds 80°F, add 0.004 to the indicated specific gravity. Subtract 0.004 from the indicated value for each 10° that the battery temperature is below 80°F.

WARNING
Do not smoke or permit any open flame in any area where batteries are being charged. Highly explosive hydrogen gas is formed during the charging process.

The specific gravity of a fully charged battery is 1.260. If the specific gravity is below 1.220, recharge the battery (**Figure 33**).

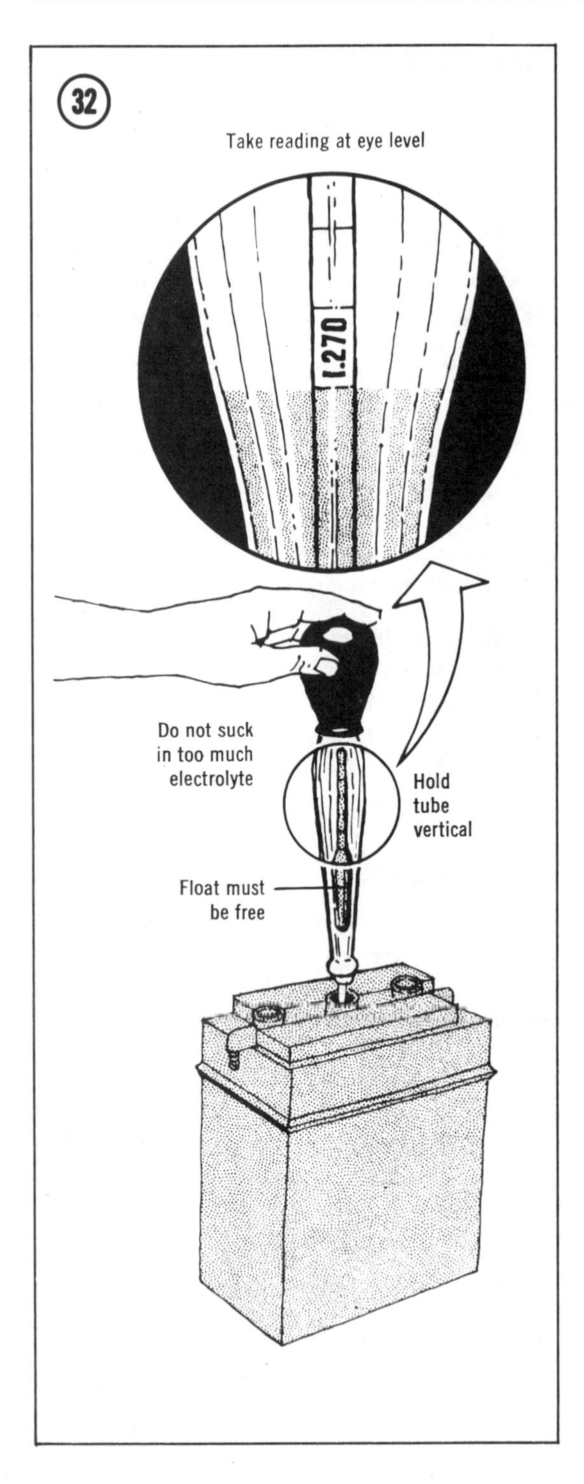

Starter Test

If starter fails to crank engine or cranks engine very slowly, perform the following:

1. Inspect cranking circuit wiring for loose or badly corroded connections or damaged wiring.
2. Perform *Battery Specific Gravity Test* to be certain battery is charged and not defective.
3. Crank engine with recoil starter to make sure engine turns freely and is not seized.

> NOTE: *Remove spark plug wires. The following bypasses the ignition switch.*

4. If starter still will not crank engine, place a heavy jumper lead from positive (+) battery terminal directly to starter terminal (**Figure 34**). This bypasses ignition switch, circuit breaker, and starter solenoid. If starter now cranks the engine, then one of these items is defective. If starter will still not crank engine, starter is defective.

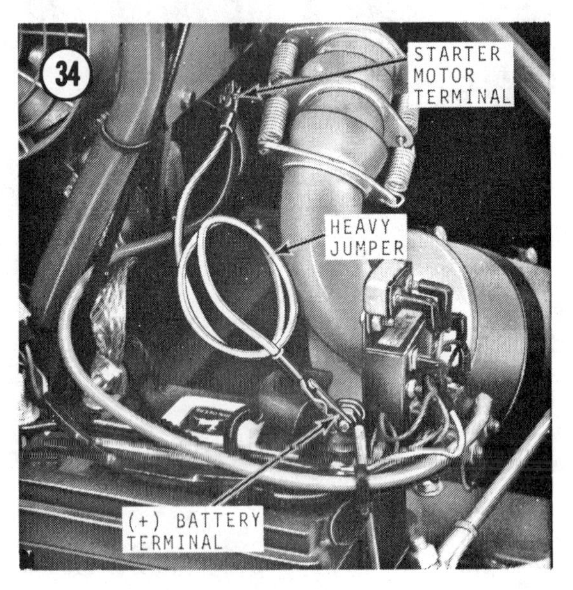

Starter Removal/Installation

Refer to **Figure 35** for this procedure. Starter repair consists of armature and/or brush replacement. It is recommended that all starter service and repair be referred to an authorized dealer or competent auto electric shop. However, if disassembly is desired, refer to **Figures 36, 37, and 38**.

1. Disconnect ground cable from battery.
2. Disconnect solenoid-to-starter cable from starter terminal.
3. Remove mounting bolts securing starter to engine and mounting bracket to engine. Remove starter and mounting bracket.

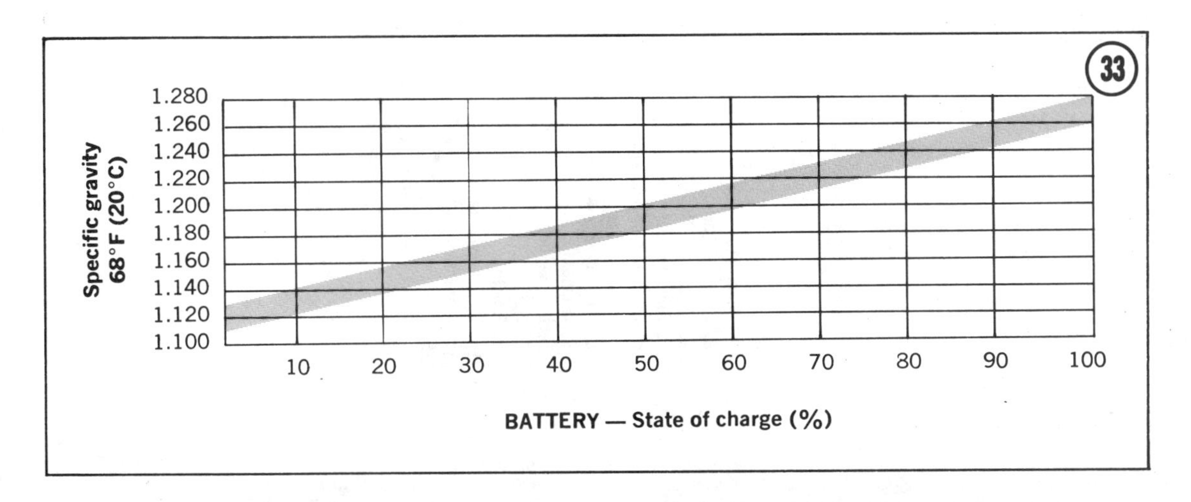

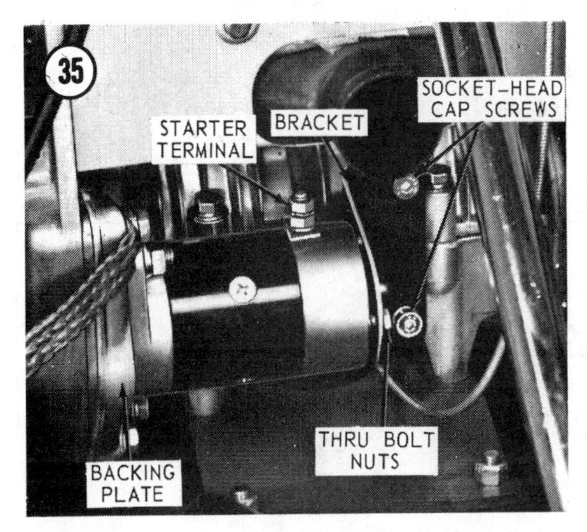

A. Positive battery terminal
B. Heavy jumper lead
C. Small terminal
D. Large terminals

4. Remove mounting bracket from starter.
5. Installation is the reverse of these steps. Keep the following points in mind:
   a. Tighten 3 bolts securing bracket last to prevent misalignment of starter.
   b. Torque 3 bolts to 6 ft.-lb. (0.8 mkg).

Starter Solenoid Test

1. Starter solenoid is a sealed magnetic switch and cannot be repaired. If defective, it must be replaced.
2. Remove and insulate cable from starter terminal. Connect test light across 2 large terminals (**Figure 39**) of starter solenoid.
3. With a jumper lead, connect positive (+) battery post to small terminal on solenoid. Solenoid plunger should snap in, light the test lamp, and hold until the jumper is removed. If not, solenoid is defective.

Circuit Breaker Test

Test light must light when connected across circuit breaker terminals (**Figure 40**). Replace if defective. The circuit breaker terminals are designated BAT and AUX. Short red lead from starter solenoid must be connected to BAT terminal of circuit breaker.

Ignition Switch Test

Perform *Ignition Switch Test* as outlined under *Magneto Ignition.*

Rectifier Test (1972 Models)

1. Disconnect 3 connectors from rectifier (**Figure 41**).

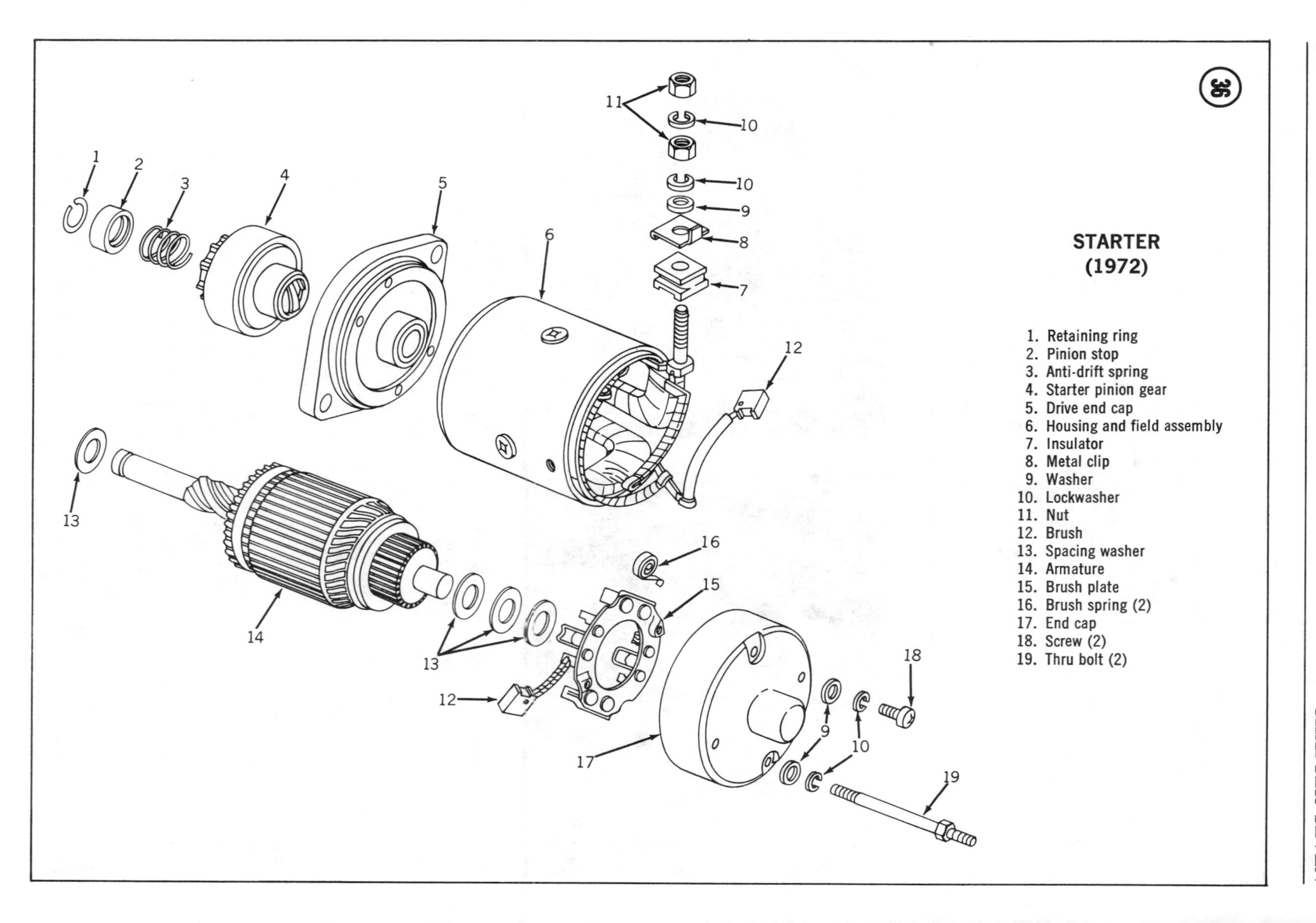
36
STARTER
(1972)
1. Retaining ring
2. Pinion stop
3. Anti-drift spring
4. Starter pinion gear
5. Drive end cap
6. Housing and field assembly
7. Insulator
8. Metal clip
9. Washer
10. Lockwasher
11. Nut
12. Brush
13. Spacing washer
14. Armature
15. Brush plate
16. Brush spring (2)
17. End cap
18. Screw (2)
19. Thru bolt (2)

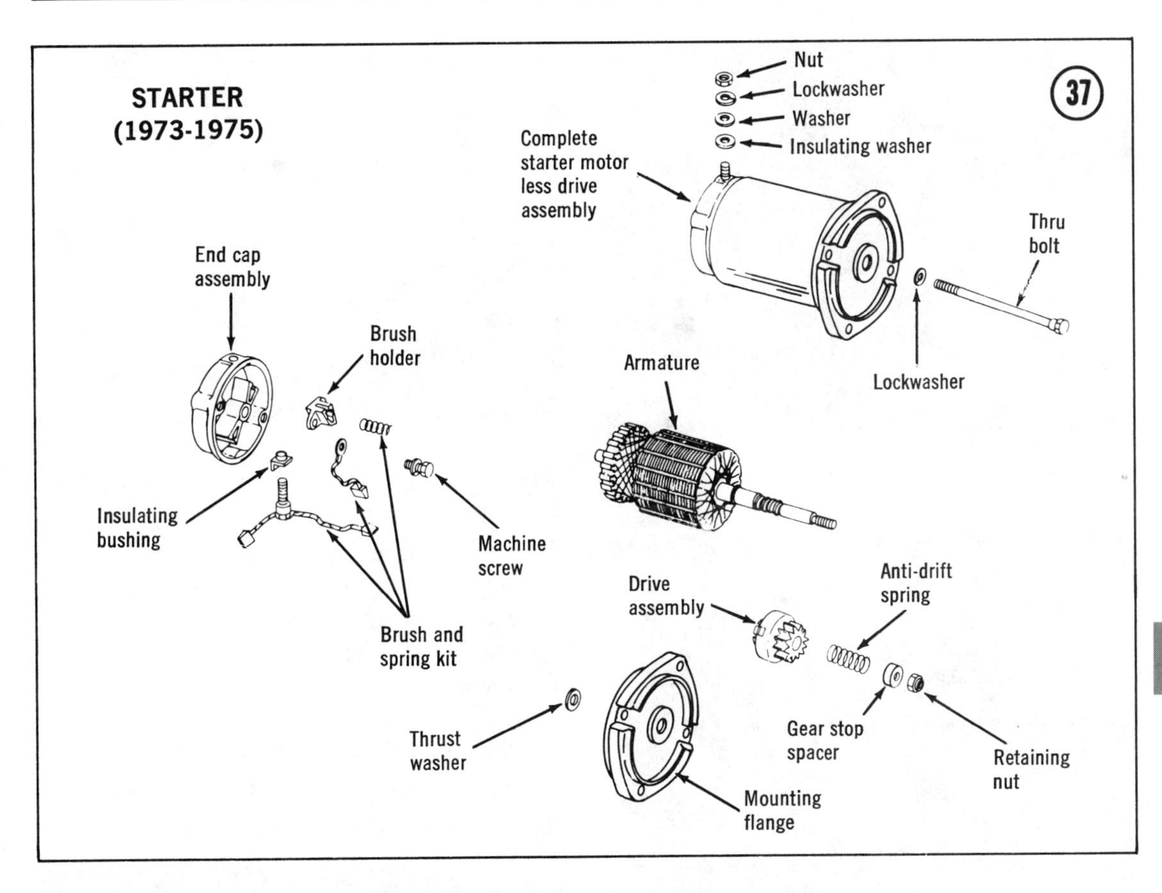

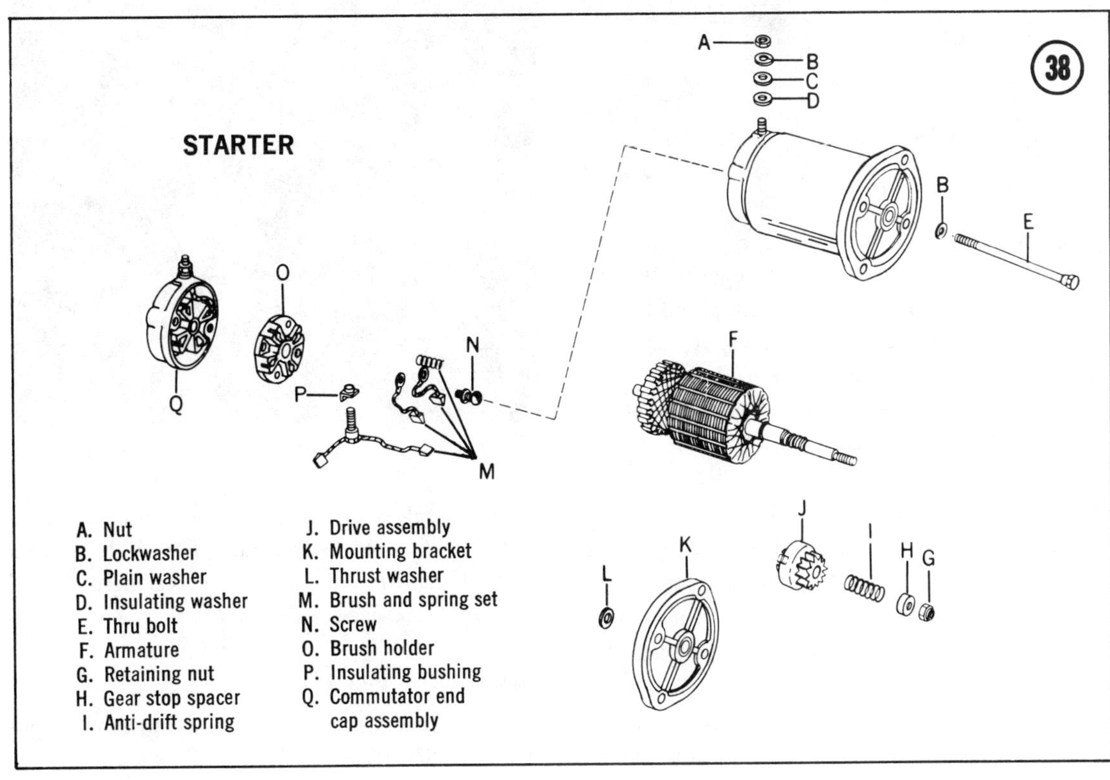

A. Nut
B. Lockwasher
C. Plain washer
D. Insulating washer
E. Thru bolt
F. Armature
G. Retaining nut
H. Gear stop spacer
I. Anti-drift spring
J. Drive assembly
K. Mounting bracket
L. Thrust washer
M. Brush and spring set
N. Screw
O. Brush holder
P. Insulating bushing
Q. Commutator end cap assembly

A. Circuit breaker B. Flashlight tester

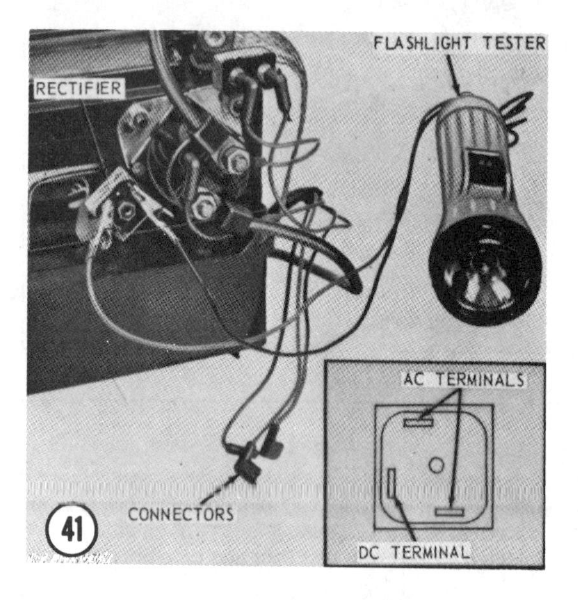

2. Connect positive (red) lead of test light to output terminal of rectifier and negative lead to one of 2 input terminals. Test light should light. Connect negative lead to other input terminal. Test light should light once again. Reverse positive and negative tester leads. Test light should not light. Replace rectifier if defective.

Rectifier Test (1973-1975 Models)

1. Disconnect 4 connectors from rectifier. A diode exists between each of the 4 terminals in the rectifier. Test the 4 diodes one at a time by connecting a test light to 2 adjacent terminals (**Figure 42**).

2. Test with leads on 2 top terminals; 2 bottom terminals, 2 left terminals, and 2 right terminals.

Reverse terminal contacts in each test set-up. Do not test terminals in a diagonal pattern.

3. With leads connected one way, test light should light. With leads reversed, a high resistance or open condition should be indicated. Repeat test for the other 3 diodes. Replace if defective.

Rectifier Test (Cyclone and Liquifire)

1. Connect black test lead (**Figure 43**) from flashlight tester to brown wire side of diode.

A. Diode B. Red lead C. Black lead

2. Connect red lead from flashlight tester to orange wire side of diode. Test light should light. With the leads reversed, test light should not light. Replace if defective.

Charging System Tests

Perform lighting coil test or alternator output test as outlined under *Lighting System*.

WIRING DIAGRAMS

Wiring diagrams are found on the following pages:

WIRING DIAGRAM — ELECTRIC-START EQUIPPED 400 AND 500 (SERIAL NO. 2,551-11,000)

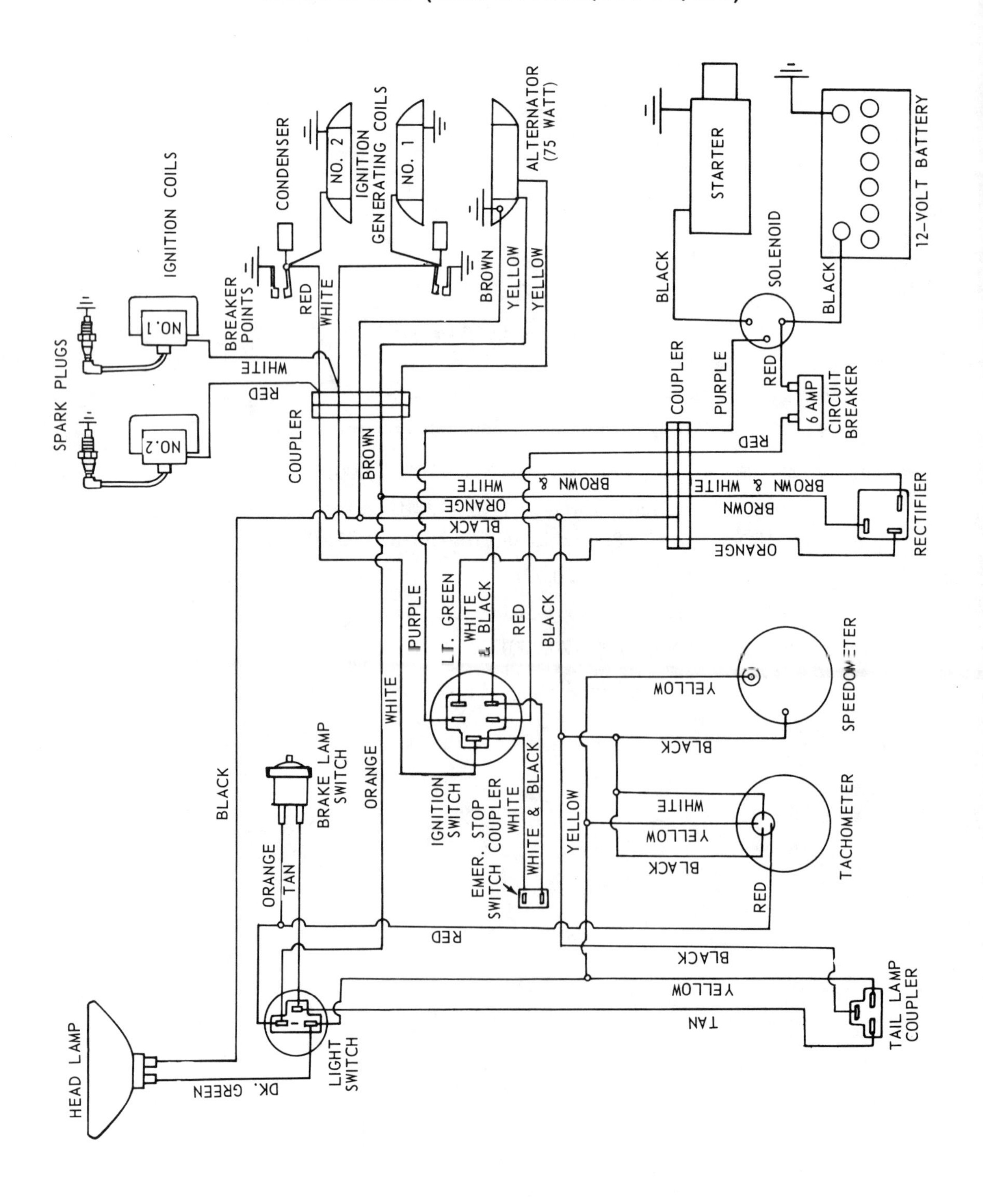

WIRING DIAGRAM — ELECTRIC-START EQUIPPED 400 AND 500 (SERIAL NO. 11,001-22,000) 600, JDX4, AND JDX8 (SERIAL NO. 2551-20,000)

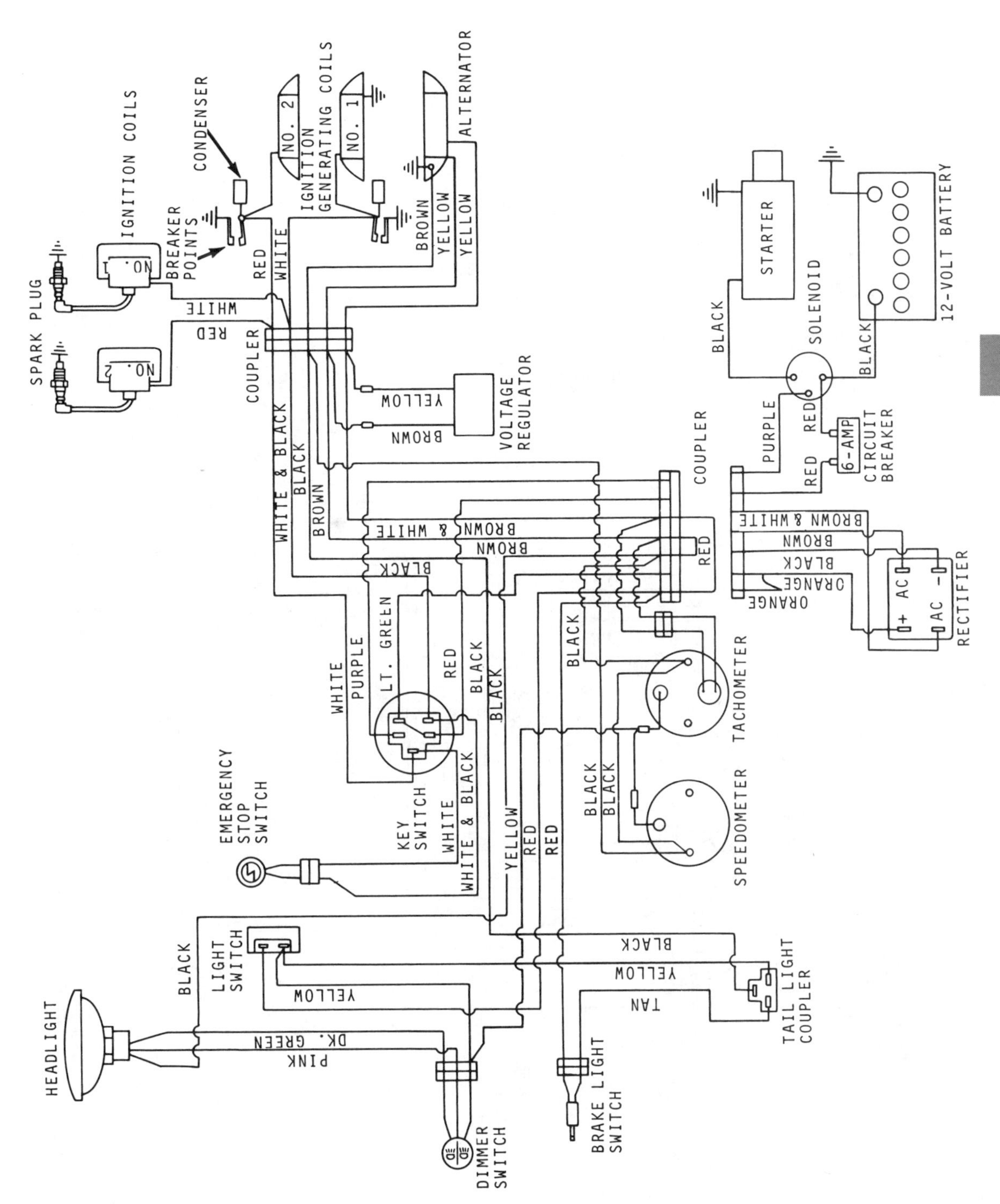

WIRING DIAGRAM — ELECTRIC-START EQUIPPED SERIAL NO. 20,001 ON

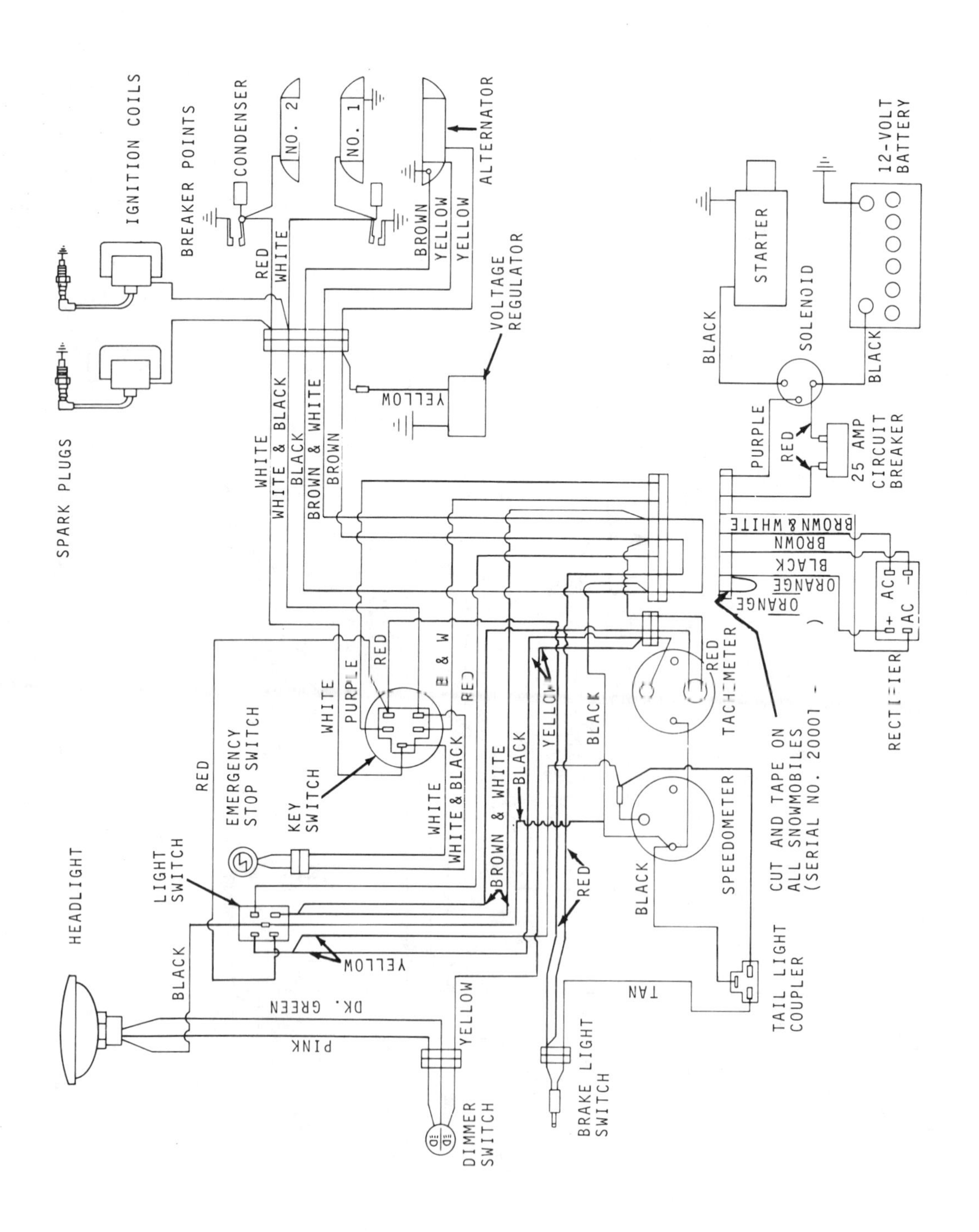

WIRING DIAGRAM — JD295/S (SERIAL NO. 20,001-30,000)

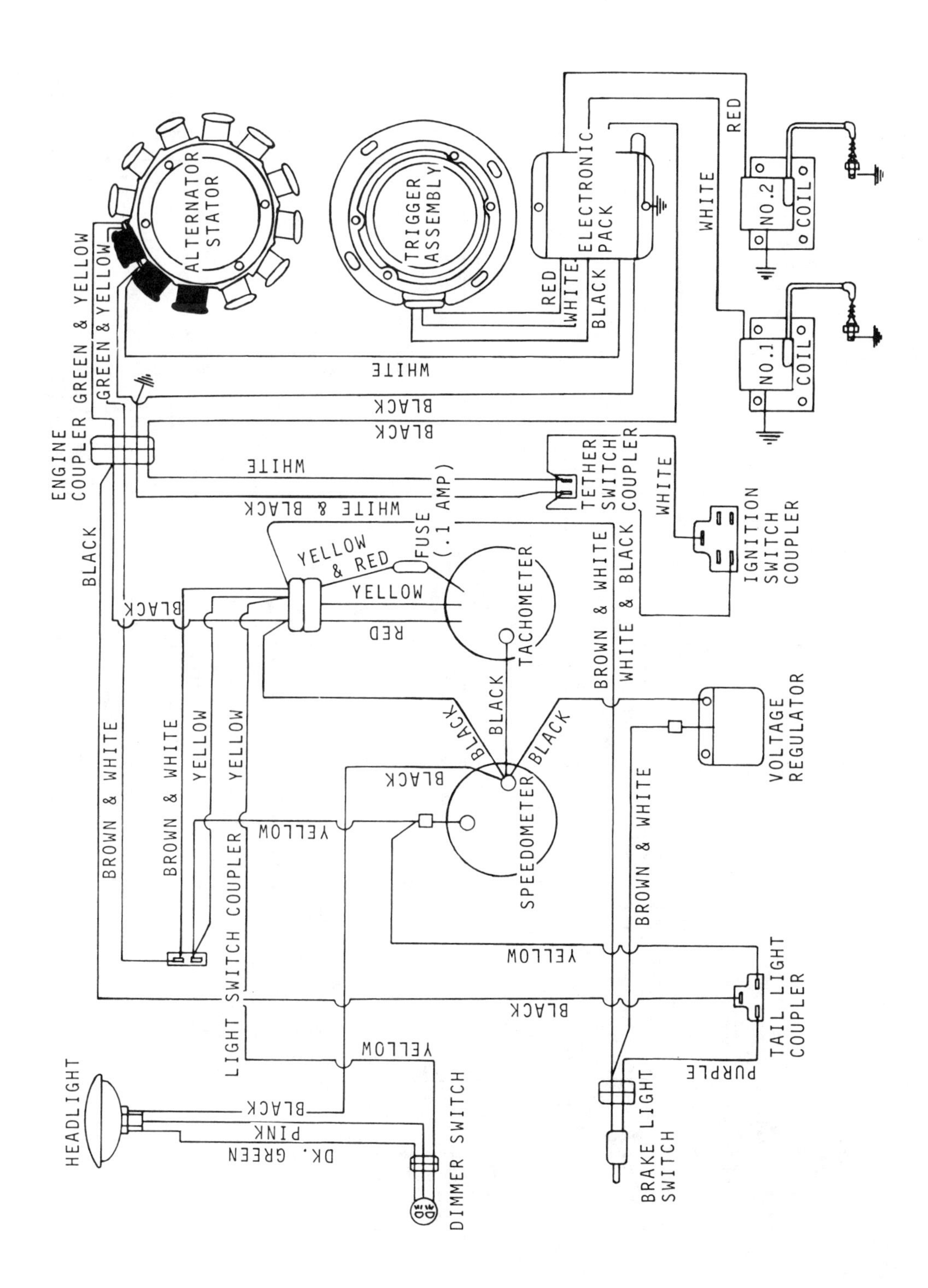

WIRING DIAGRAM — 1976 (SERIAL NO. 50,001-70,000)

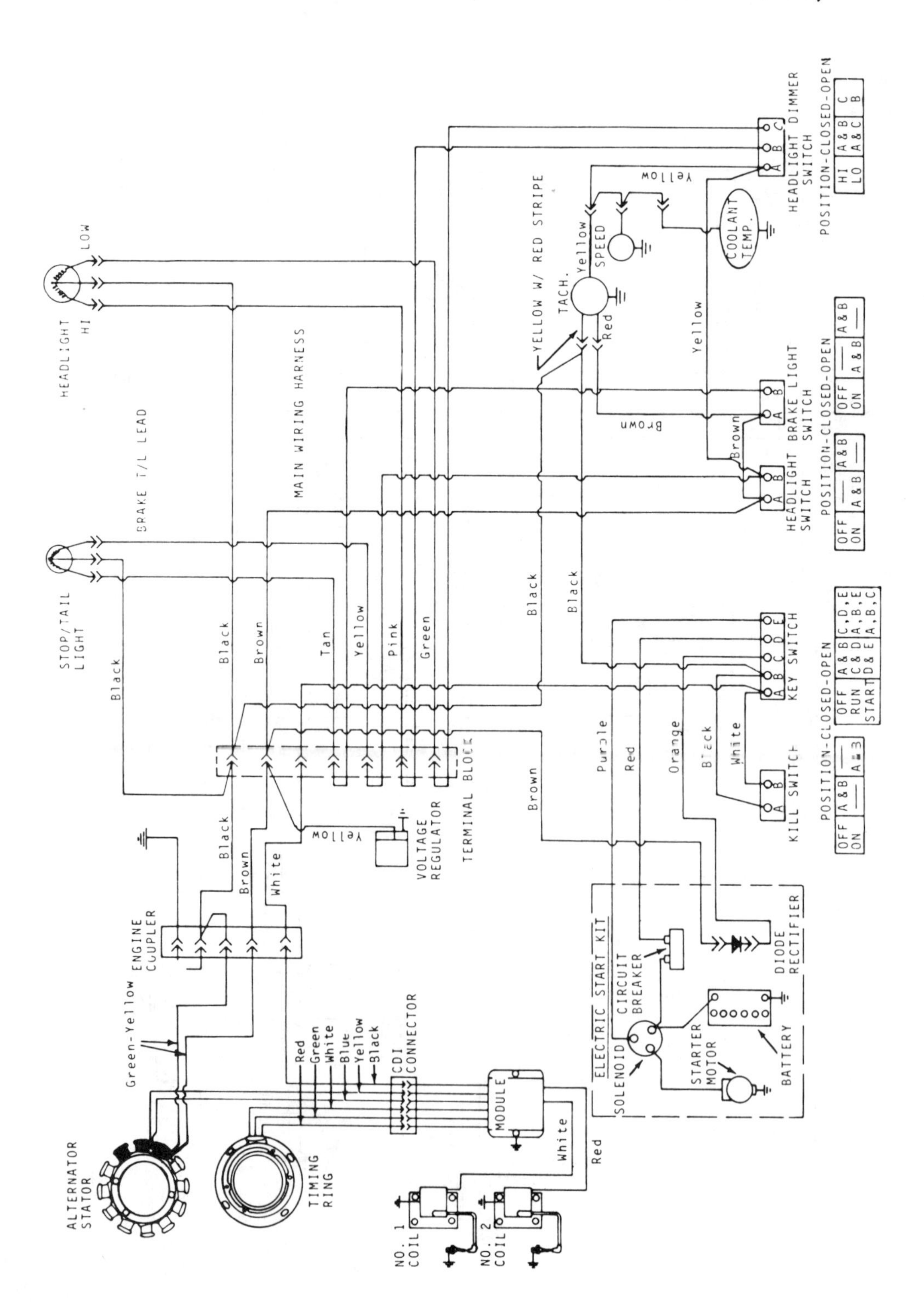

WIRING DIAGRAM — 1977 (SERIAL NO. 70,001 ON)

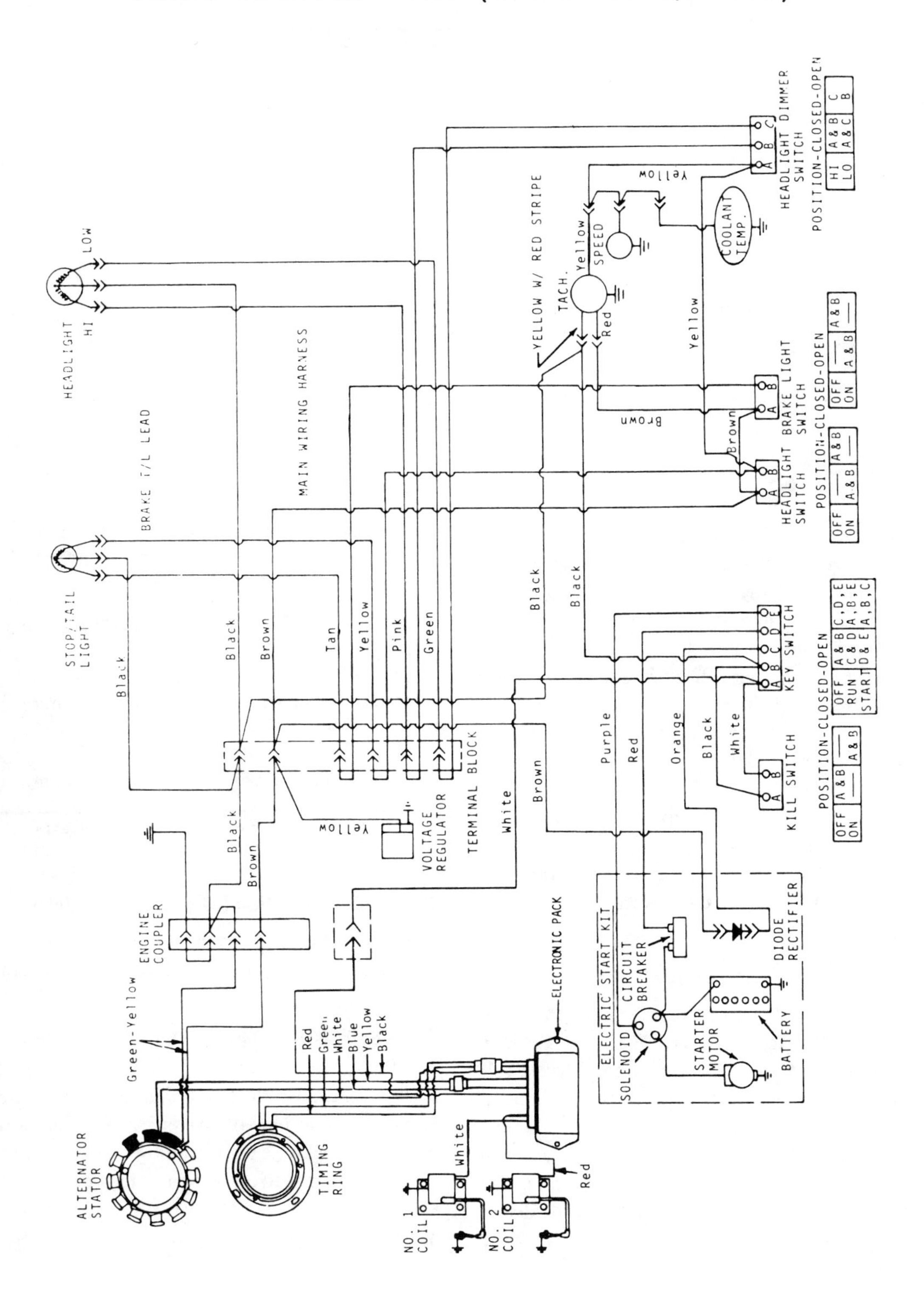

CHAPTER EIGHT

POWER TRAIN

The power train consists of a drive belt, drive and driven sheaves, drive chain and sprockets, secondary shaft, drive shaft, and a brake assembly. Refer to **Figure 1** for a typical example of drive train components.

Earlier model machines are equipped with a band brake while later models use a mechanical disc brake. Model JD340/S was fitted with a hydraulic disc brake.

Some procedures in this chapter require the use of special tools for removal and repair work. If such tools are not available, and substitutes cannot be locally fabricated, refer the removal and repair work to an authorized dealer.

DRIVE BELT

The drive belt transmits power from the drive sheave to the driven sheave. Refer to **Table 1** for drive belt model application. Drive belt should be replaced when its width is reduced by 1/8 in. (3mm). Drive belts are not interchangeable between different models even though belt width is the same. Refer to *Drive Belt Wear Analysis* in Chapter Three to determine causes of drive belt unusual wear or failure.

Table 1 DRIVE BELT APPLICATION

Model	Width
JDX and 400 (up to serial No. 20,000)	1 3/16 in. (30.16mm)
Model 300 (serial No. 20,001 and subsequent)	1 3/16 in. (30.16mm)
All other models	1 1/4 in. (31.75mm)

NOTE: Replace drive belt when overall width is reduced by 1/8 in. (3mm).

Removal/Installation (Mid-engine Models)

Refer to **Figure 2** for this procedure.

1. Loosen wing nuts and remove driven sheave belt shield.

2. Remove access panel from left-hand side of console.

3. Standing on right-hand side of snowmobile, grasp movable half of driven sheave and turn it counterclockwise while pulling to open sheave. Pull belt upward to allow belt to be worked over and off top of driven sheave.

4. Work belt off drive sheave and remove belt from snowmobile.

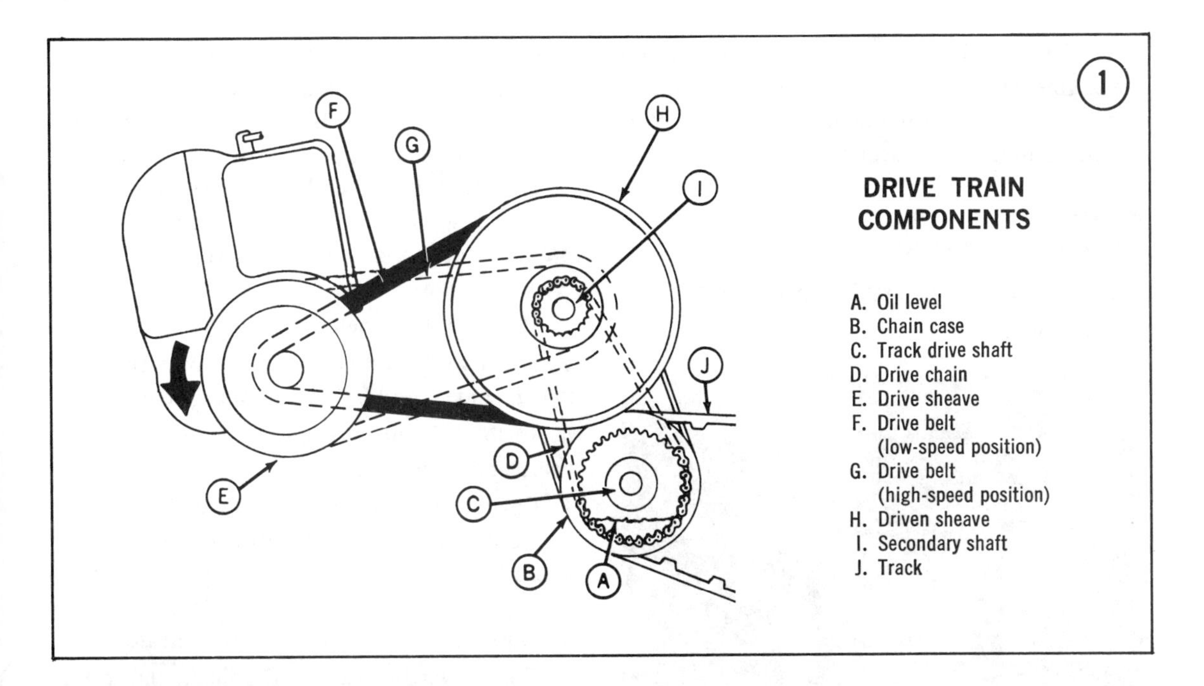

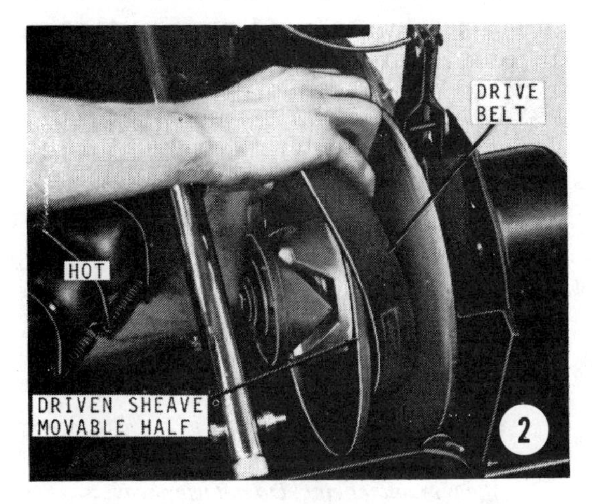

CAUTION

Do not pry belt off over sheaves or belt and/or sheave damage may result.

5. Installation is the reverse of these steps.

Belt Alignment (Mid-engine Models)

CAUTION

Proper belt alignment is critical. A misaligned belt can be destroyed in a few hours of operation.

Check belt alignment whenever engine is installed or rapid belt wear is experienced. Distance between sheave shaft centers is factory set and not adjustable.

The following procedure requires the use of a special alignment tool. If tool is not available, have an authorized dealer perform the task. Refer to **Figure 3**.

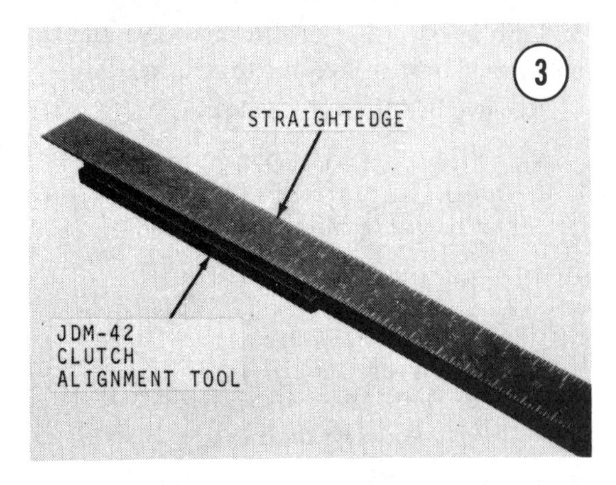

1. Remove drive belt. Hold clutch aligning tool against inside edge of drive and driven sheaves. Tool should contact sheaves at 4 points.

CAUTION

Position tool so reinforcement webs in drive sheave do not hold tool away from either edge of sheave. Be careful not to deflect tool to match sheaves.

2. If alignment is incorrect, loosen bolts securing engine base plate to cross mounts and move engine in slots until alignment is correct. Tighten engine mounting bolts securely.

Removal/Installation (Front Engine Models)

Refer to **Figure 4** for this procedure.

1. Raise hood.

2. Pull spring lock and tip belt guard forward to provide access to drive belt.
3. Push in on center of driven sheave and lift belt up and over sheave half to remove.
4. Remove belt from drive sheave.

CAUTION
Do not pry belt over sheaves or belt and/or sheave damage may result.

WARNING
Keep fingers out of area between center of driven sheave halves when sheave is opened. If driven sheave sticks closed, use care in opening or fingers may be pinched.

5. Installation is the reverse of these steps.

Belt Alignment (Front Engine Models)

CAUTION
Proper belt alignment is critical. A misaligned belt can be destroyed in a few hours of operation.

Check belt alignment whenever engine is installed or rapid belt wear is experienced. The distance between sheave centers is adjustable by moving engine forward or rearward. Lateral adjustment is achieved by adding or subtracting shims from between driven sheave and bearing.

The following procedure requires the use of a special alignment tool. If tool is not available, have an authorized dealer perform the task. Refer to **Figure 5**.

1. Remove drive belt.
2. Remove driven sheave and shims from between sheave and secondary shaft bearing.
3. Install clutch aligning tool on secondary shaft and drive shaft.
4. For lateral adjustment, remove clutch aligning tool and add shims between the tool and secondary shaft bearing to give proper dimension (**Figure 5**). Shims are available in 0.018 or 0.060 in. (0.46 or 1.5mm) thicknesses.
5. For forward or rearward adjustment, loosen engine mounting bolts and move engine as required to achieve proper alignment.

DRIVE SHEAVES

The following procedures require the use of special tools for removal, installation, and repair. If special tools or locally fabricated equivalents are not available, refer work to an authorized dealer. Refer to **Table 2** for drive sheave model application.

CAUTION
Drive sheaves are matched to driven sheaves and engine. Do not use sheaves not designated for your particular machine or improper operation may result.

Salsbury 780, 910, and 850 Sheave Removal/Installation

1. Remove access panel from left-hand side of console and remove drive belt.
2. Hold sheave with drive sheave holding tool (**Figure 6**) and remove retaining cap screw.
3. On 400 and 500 models up to serial No. 11,000 perform the following:
   a. Bend back locking tab on large ramp plate nut, and remove nut.

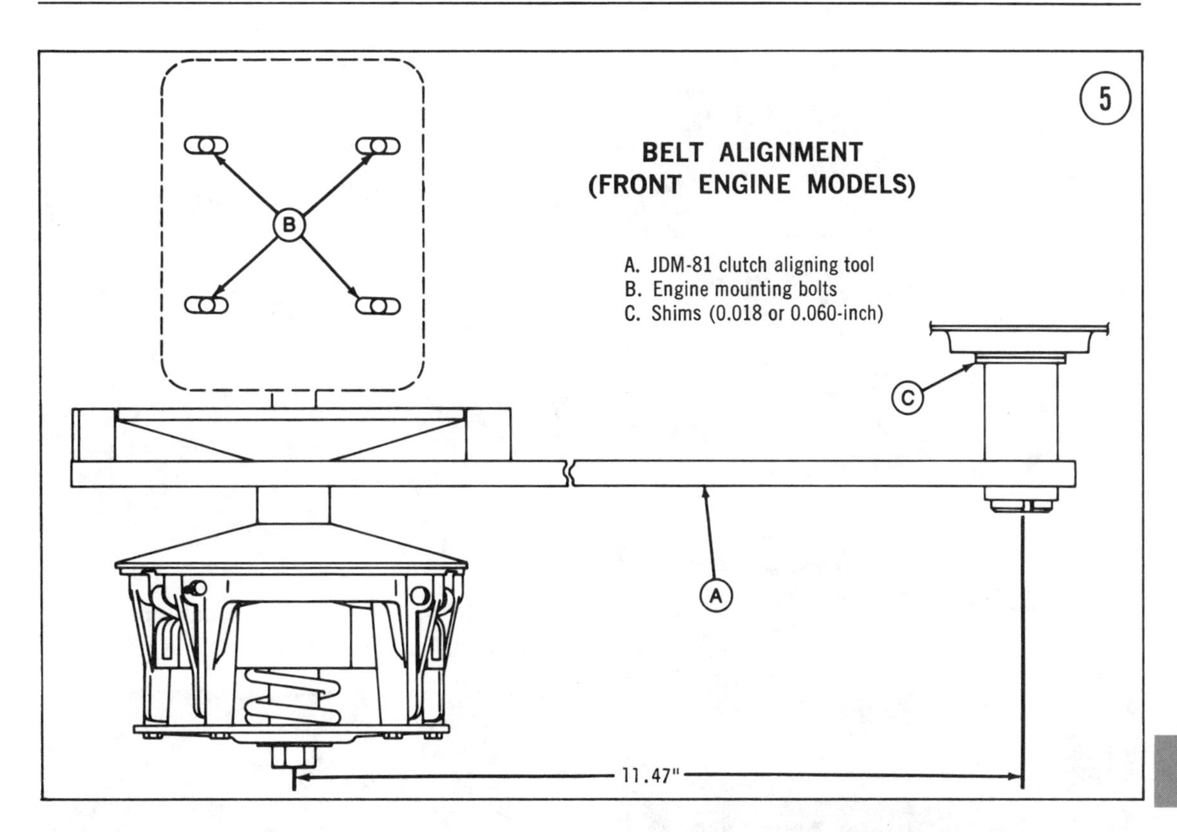

Table 2 DRIVE SHEAVE APPLICATION

Model	Sheave
JDX4 and 400 (up to serial No. 20,000) 300 (serial No. 20,001 and subsequent)	Salsbury 780
500 (up to serial No. 11,000)	Salsbury 910
500 (serial No. 11,001-20,000) 600 and JDX8 (up to serial No. 20,000)	Salsbury 850
400, 500, 600, JDX4 Special, JDX6, and JDX8 (serial No. 20,001-30,000) All JD295/S	John Deere (Comet) 100C
400, 600, 800, JDX4, JDX6, and JDX8 (serial No. 30,001 and subsequent)	John Deere (Comet) 101C
Cyclone, Liquifire, JD340/S, and 340 Liquidator	John Deere (Comet) 102C

b. Thread socket-cap screw into end of crankshaft and reinstall large ramp plate nut (**Figure 7**).

c. Hold socket-head cap screw with wrench and tighten large ramp plate nut with wrench until drive sheave is removed.

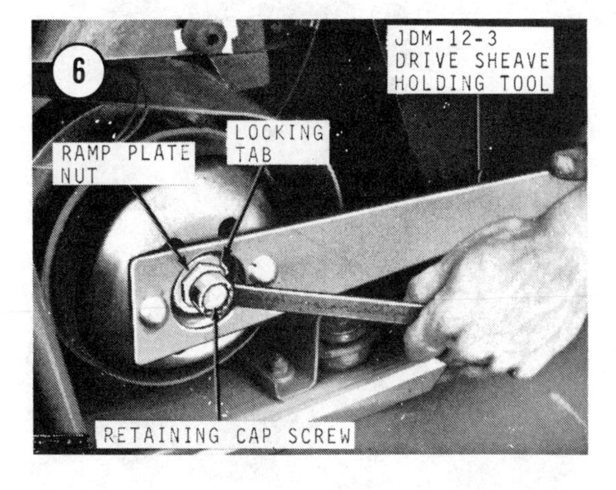

4. On all other models use drive sheave holding tool (**Figure 8**) to prevent drive sheave from turning and screw clutch puller into large ramp plate nut (large special screw on 850 sheaves) until drive sheave comes loose from shaft. See **Figure 9** for 850 sheaves.

5. Refer all necessary inspection and repair to an authorized dealer.

6. Installation is the reverse of these steps. Torque retaining cap screw to 50 ft.-lb. (6.9 mkg).

8

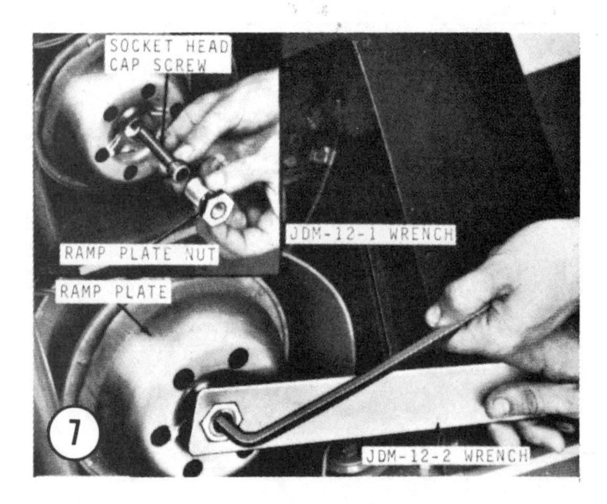

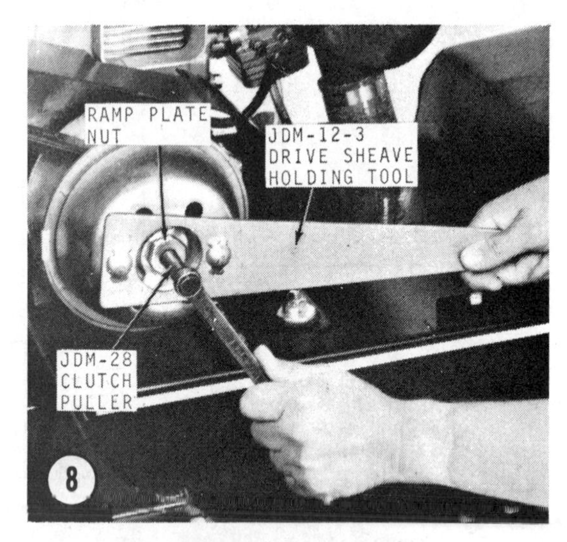

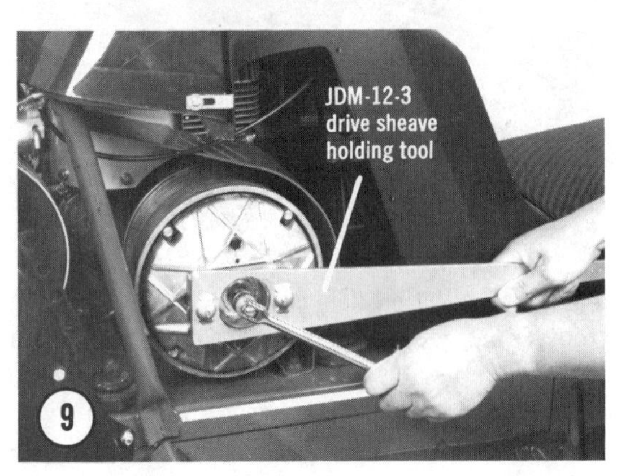

John Deere (Comet 100C, 101C, 102C) Sheave Removal/Installation

1. On mid-engine models remove left-hand access panel. On front engine models raise hood and remove knock-out plug from side of pan.

2. Remove drive belt.

3. Install compressor ring and compress movable face to expose cross hole in hub (**Figure 10**).

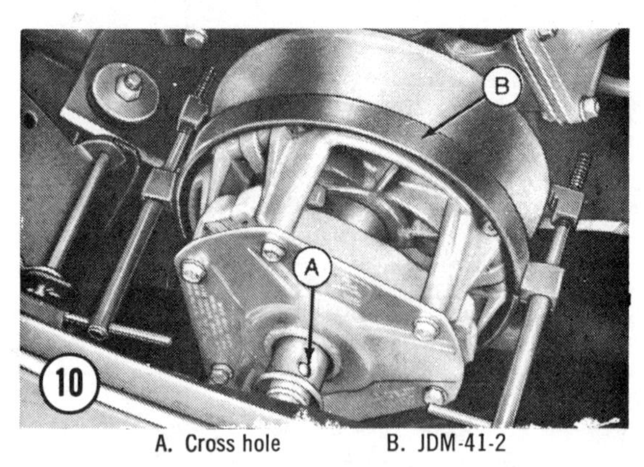

A. Cross hole B. JDM-41-2

4. Install 2-piece nut around hub and engage pins on nut in hub cross hole (**Figure 11**).

5. Hold nut and remove retaining cap screw with lockwasher and pilot washer. See **Figure 12** for mid-engine models and **Figure 13** for front engine models. Install puller bolt into hub and turn until drive sheave is removed from crankshaft.

6. Refer all necessary inspection and repair to an authorized dealer.

7. Installation is the reverse of these steps. Torque retaining cap screw to 50 ft.-lb. (6.9 mkg).

CAUTION

Do not torque cap screw more than 50 ft.-lb. (6.0 mkg) or hub end will "swell" and drive sheave will not operate freely.

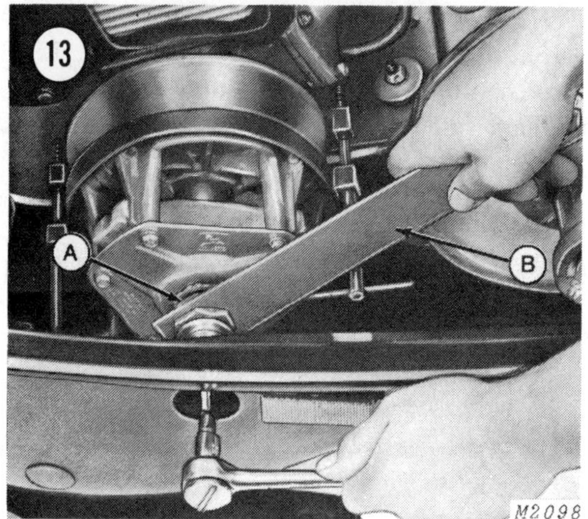

A. JDM-41-4 nut B. JDM-12-2 nut wrench

DRIVEN SHEAVES

CAUTION

Driven sheaves are matched to drive sheaves and engine. Do not use sheaves not designated for your particular machine or improper operation will result. Refer to **Table 3** *for driven sheave model application.*

Removal/Installation

NOTE: *On models 400 and 500 up to serial No. 11,000, perform Step 1; all other models proceed to Step 2.*

1. Remove drive belt. Remove cap screw and tapered sleeves securing steering column to steering post and slide steering column up approximately 6 inches (**Figure 14**). Loosen exhaust pipe clamp and swing exhaust system away from driven sheave.

Table 3 DRIVEN SHEAVE APPLICATION

Model	Sheave
400 and JDX4 (up to serial No. 20,000) 300 (serial No. 20,001 and subsequent)	Salsbury 780
500 (serial No. 11,001-20,000) JDX8 and 600 (serial No. up to 20,000)	Salsbury 850
500 (up to serial No. 11,000)	Salsbury 910
All other mid-engine models	John Deere (Comet)
All front engine models	John Deere

1. Exhaust pipe clamp
2. Steering column
3. Retaining cap screw
4. Steering post

2. Remove drive belt guard and remove drive belt.
3. Remove retaining cap screw (**Figure 15**) and washer. Slide sheave with key off shaft.
4. On mid-engine models, if sheave does not slide off easily, perform the following:
   a. Back out retaining screw approximately ½ in. (13mm) as shown in **Figure 16**.
   b. Use a soft mallet and carefully tap secondary shaft toward chain case while pulling out on sheave.

CAUTION

Secondary shaft can only move approximately ½ in. (13mm). Do not attempt to move shaft more or shaft bearing damage will result.

5. Refer all necessary inspection and repair to an authorized dealer.
6. Installation is the reverse of these steps. Torque retaining cap screw to 20 ft.-lb. (2.8 mkg).

DRIVE TRAIN (MID-ENGINE MODELS)

Refer to **Figure 17** for the following procedures.

Drive Chain Removal/Installation

1. Remove 2 rubber access plugs in chain case.
2. Remove 2 cap screws (**Figure 18**) securing chain tension blocks. Lift tension block assemblies out of chain case.

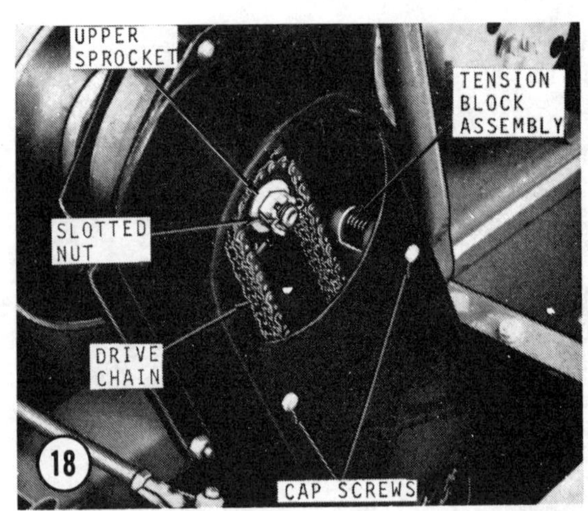

3. Remove cotter pin, slotted nut washer, and upper sprocket from secondary shaft. Remove spacer from behind upper sprocket.
4. Slide drive chain off lower sprocket. Lift chain and upper sprocket out upper access hole in chain case.
5. Installation is the reverse of these steps.

CAUTION

Secondary shaft must receive final torque from driven sheave end of shaft. If chain is installed with driven sheave in place, loosen retaining cap screw securing driven sheave before tightening slotted nut on chain end of shaft, then torque retaining cap screw to 20 ft.-lb. (2.8 mkg).

Drive Chain and Sprocket Inspection

Refer to **Figure 19** for this procedure.

1. Inspect drive chain, and replace if badly worn or broken.

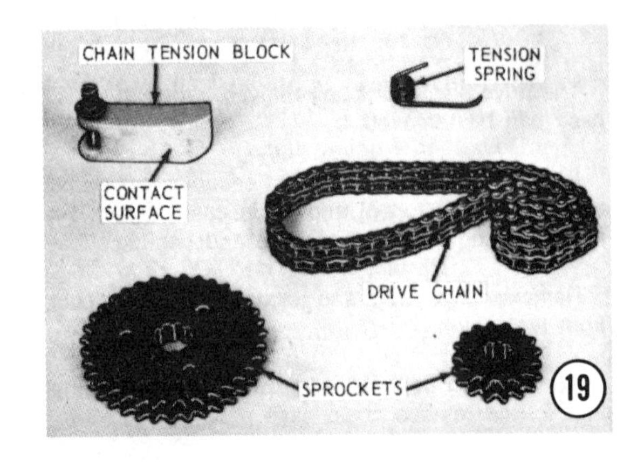

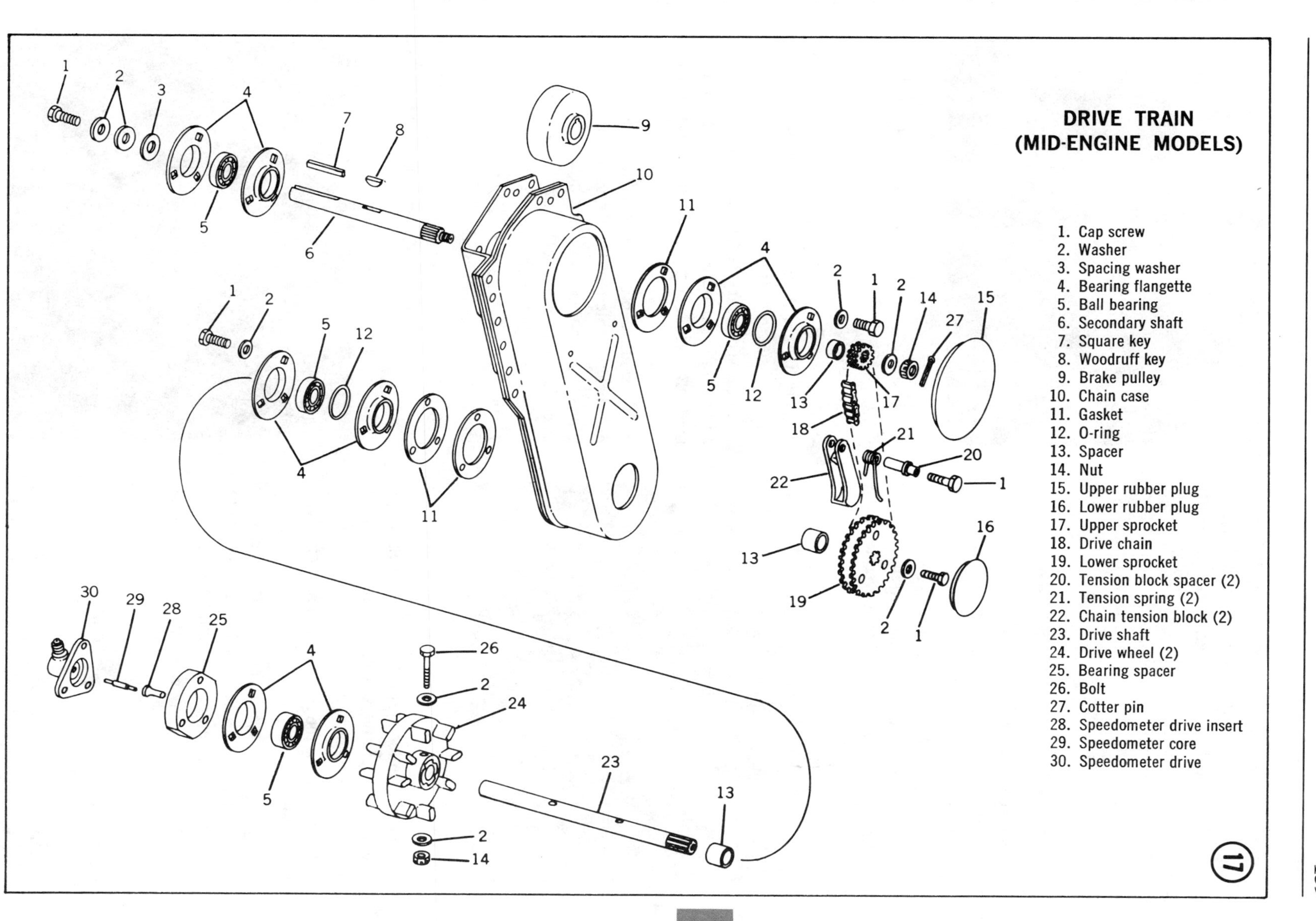
DRIVE TRAIN
(MID-ENGINE MODELS)
1. Cap screw
2. Washer
3. Spacing washer
4. Bearing flangette
5. Ball bearing
6. Secondary shaft
7. Square key
8. Woodruff key
9. Brake pulley
10. Chain case
11. Gasket
12. O-ring
13. Spacer
14. Nut
15. Upper rubber plug
16. Lower rubber plug
17. Upper sprocket
18. Drive chain
19. Lower sprocket
20. Tension block spacer (2)
21. Tension spring (2)
22. Chain tension block (2)
23. Drive shaft
24. Drive wheel (2)
25. Bearing spacer
26. Bolt
27. Cotter pin
28. Speedometer drive insert
29. Speedometer core
30. Speedometer drive
17

2. Inspect sprocket teeth for wear. If a new drive chain is installed, replace sprockets. A new chain will not properly match worn sprockets.

3. Replace chain tension blocks, if contact surfaces are worn deeply. Replace tension springs if cracked, broken, or pitted.

Lower Sprocket Removal/Installation

1. Remove drive chain.

2. Remove cap screw and washer securing lower sprocket to drive shaft.

3. Install puller (**Figure 20**) on drive shaft and sprocket. Tighten cap screws, threaded into sprocket, evenly, until sprocket is free of drive shaft.

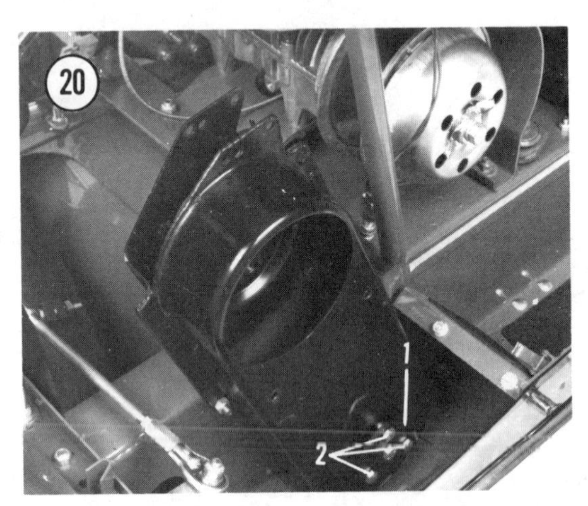

1. JDM-13 puller 2. Cap screws

4. Remove puller and lift sprocket from upper access hole. Remove spacer next to lower bearing from drive shaft.

5. Installation is the reverse of these steps. Keep the following points in mind:

   a. Use a *new* cap screw to retain the lower sprocket. Cap screw has a nylon locking insert and can be used only once.

   b. Torque cap screw to 30-38 ft.-lb. (4.1-4.5 mkg).

Secondary Shaft and Brake Pulley Removal/Installation

Refer to **Figure 21** for this procedure.

1. Remove driven sheave.

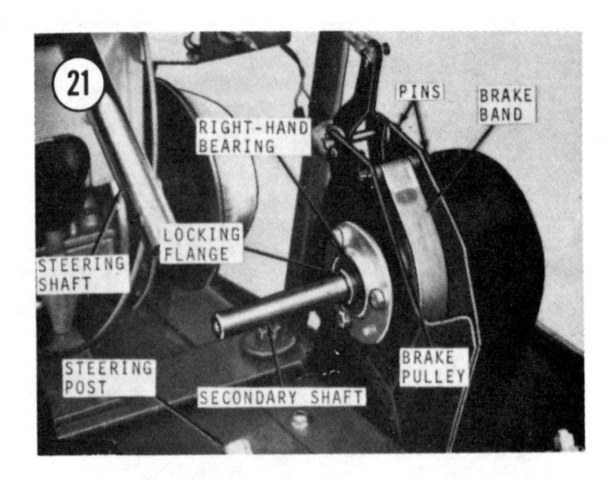

2. Disconnect steering shaft and slide shaft up. Remove square key from secondary shaft.

3. Remove drive chain and upper sprocket.

4. Remove 3 cap screws securing right-hand bearing flangettes to chain case. Remove flangettes and bearing from secondary shaft.

5. Slide secondary shaft out of other bearing and out of brake pulley. Remove left-hand bearing flangettes, bearing and O-ring if replacement is necessary.

6. To remove brake pulley if desired, disconnect brake cable from brake arm and remove 2 pins securing brake band and brake arm to chain case. Do not lose spacers. Lift brake band and brake pulley out of chain case.

7. Installation is the reverse of these steps. Keep the following points in mind:

   a. Use a new gasket and gasket sealer on chain case upper bearing mount (**Figure 22**).

b. Lubricate and install O-ring on bearing.

c. Install inner flangette, bearing with O-ring, and outer flange to chain case but do not tighten cap screws at this time.

CAUTION

Locking flange on bearing must face right-hand side of machine or chain and driven sheave will not align properly.

d. Align secondary shaft perpendicular to chain case and tighten 6 cap screws securely.

e. Perform *Band Brake Adjustment* as outlined in Chapter Two.

Secondary Shaft Inspection

Refer to **Figure 23** for this procedure.

1. Inspect secondary shaft bearing surfaces for evidence of bearings turning on shaft. Inspect condition of splined end and threads. Replace shaft if defective.

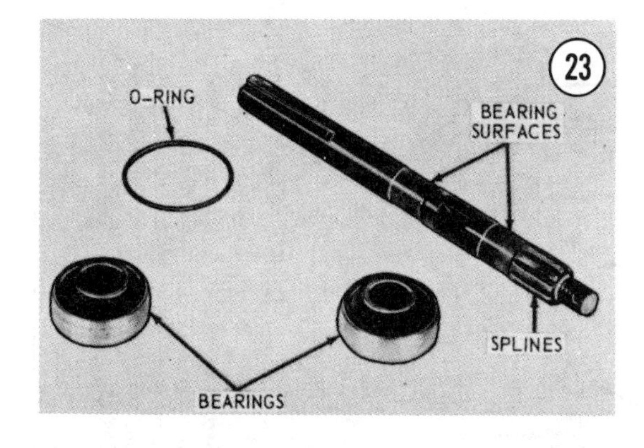

2. Inspect secondary shaft bearings, and replace if binding, worn, or noisy. Install new O-ring on left-hand bearing if it is crimped or damaged.

Chain Case Removal/Installation

1. Remove secondary shaft and brake pulley.
2. Remove lower sprocket.
3. Remove drive shaft and track as outlined in Chapter Ten.
4. Remove cap screws securing lower chain case bearing to tunnel and chain case.
5. Remove the bearing flangettes, O-ring, and bearing.
6. Remove bolts securing chain case to tunnel (**Figure 24**) and lift out chain case.

1. Chain case 2. Bolts

7. Installation is the reverse of these steps. Keep the following points in mind:

a. Install new gaskets with gasket sealer on both sides of tunnel where left-hand drive shaft bearing mounts.

b. Do not tighten chain case retaining bolts until lower bearing has been installed.

Drive Shaft Removal/Installation

Remove drive shaft with track as outlined in Chapter Ten.

DRIVE TRAIN (FRONT ENGINE MODELS)

Refer to **Figure 25** for the following procedures.

Drive Chain and Sprocket Removal

1. Loosen chain case cover screws and allow oil to drain out.
2. Remove chain case cover. Remove cap screws securing upper and lower sprockets.
3. On models, serial No. 55,001-70,000 remove sprockets, drive chain, and chain tensioners as an assembly. On models serial No. 70,001 and

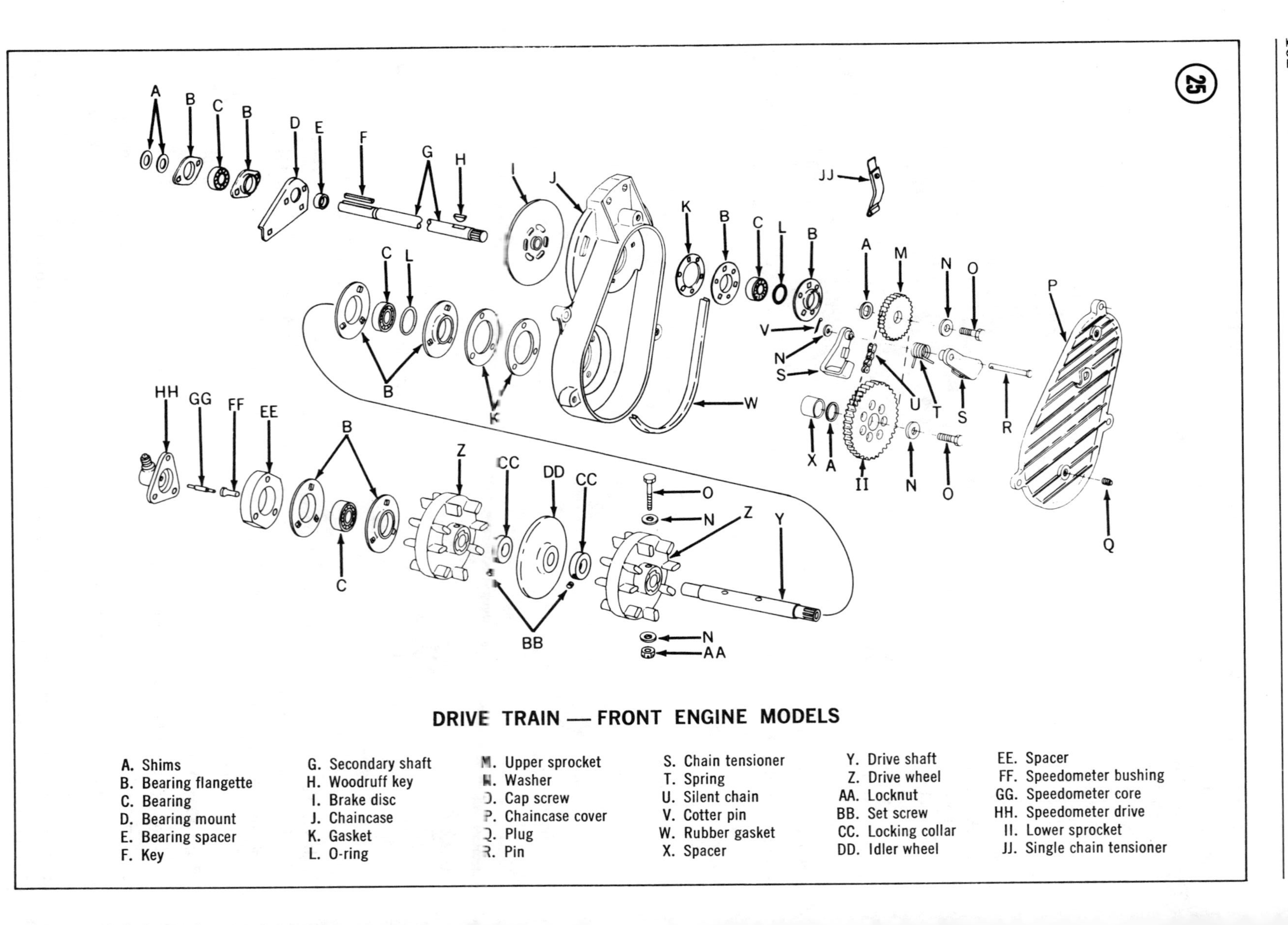

DRIVE TRAIN — FRONT ENGINE MODELS

A. Shims
B. Bearing flangette
C. Bearing
D. Bearing mount
E. Bearing spacer
F. Key
G. Secondary shaft
H. Woodruff key
I. Brake disc
J. Chaincase
K. Gasket
L. O-ring
M. Upper sprocket
N. Washer
O. Cap screw
P. Chaincase cover
Q. Plug
R. Pin
S. Chain tensioner
T. Spring
U. Silent chain
V. Cotter pin
W. Rubber gasket
X. Spacer
Y. Drive shaft
Z. Drive wheel
AA. Locknut
BB. Set screw
CC. Locking collar
DD. Idler wheel
EE. Spacer
FF. Speedometer bushing
GG. Speedometer core
HH. Speedometer drive
II. Lower sprocket
JJ. Single chain tensioner

subsequent, remove chain tensioner first then remove sprockets and drive chain as an assembly.

> NOTE: *Record number of shims between sprockets and shafts to aid reassembly and sprocket alignment.*

Drive Chain, Sprockets, and Chain Tensioner Inspection

Refer to **Figure 26** for this procedure.

1. Inspect drive chain, and replace if badly worn or broken.

2. Inspect sprocket teeth for wear. If a new drive chain is installed, replace sprockets. A new chain will not properly match worn sprockets.
3. Replace chain tensioners, if contact surfaces are worn deeply. Replace tension spring if it is cracked, broken, or pitted.

Drive Chain and Sprocket Installation

1. Check sprocket alignment as follows:
   a. Install four 0.010 in. (0.25mm) shims and spacers on lower shaft and install lower sprockets. Install two 0.018 in. (0.45mm) shims and upper sprocket. Install washers and cap screws and tighten securely.
   b. Place a straight-edge on flat surface of lower sprocket (**Figure 27**) and slide up toward upper sprocket to check alignment. Add or deduct shims as necessary to align sprockets.

CAUTION

Use a maximum of three 0.018 in. (0.45mm) shims behind upper sprocket and ten 0.010 in. (0.25mm) shims behind lower sprocket. Shims on lower sprocket should be between spacer and bearing. When sprockets are shimmed correctly, shafts should be recessed into sprockets. Do not *allow shafts to protrude beyond sprockets.*

   c. Remove sprockets. Leave spacer and shims in place.

2. On models serial No. 50,001 to 70,000 place chain over sprockets and position chain tensioner over chain with head of chain tensioner bolt to the outside. Install spring in grooves of chain tensioner. Hold tensioner in place with chain tensioner tool (**Figure 28**) and install chain and sprocket assembly.

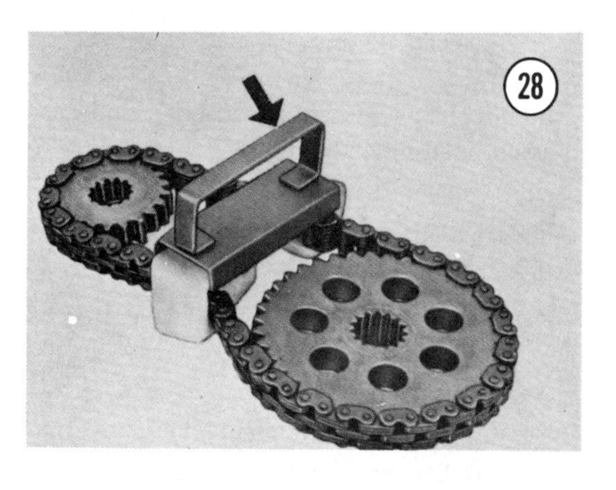

3. On models serial No. 70,001 and subsequent, install chain and sprockets in chain case. Position chain tensioner on pin in chain case as

shown in **Figure 29**. Ensure tensioner is properly aligned.

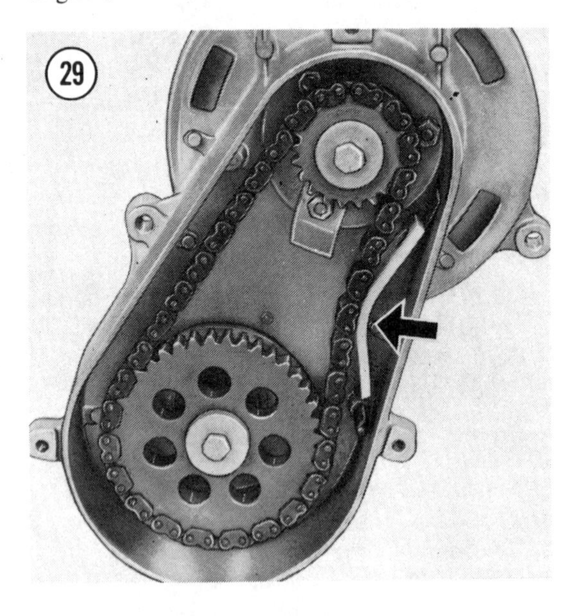

4. Apply Loctite (grade AV, red) to sprocket threads and install cap screws with washers. Tighten securely.
5. Install gasket on chain case with gap in 11 o'clock position. Gap should be 1/16-3/16 in. (1.6-4.8mm) and serves as crankcase vent.

> NOTE: *On models serial No. 70,001 and subsequent, a one-piece gasket is used. Position seam in gasket in 11 o'clock position. If one-piece gasket is used on older machines, drill a 1/16 in. (1.6mm) hole lengthwise through upper plug for a vent hole.*

6. Install chain case cover. Install chain case oil as outlined in Chapter Two.

Secondary Shaft and Brake Disc Removal/Installation

1. Remove drive belt and driven sheave. Record number of shims behind driven sheave to aid reassembly.
2. Remove drive chain and sprockets.
3. Remove flangettes and bearing from left side of tunnel (**Figure 30**).
4. Slide secondary shaft out of chain case bearing and out of brake disc.

5. Remove chain case bearing flangettes, bearing, and O-ring if replacement is necessary.
6. Remove brake puck body, brake puck, and backing plate. Remove brake disc.
7. Remove brake puck from chain case.
8. Remove exhaust silencer. Slide shaft past chain case and out right-hand side.
9. Installation is the reverse of these steps. Keep the following points in mind:
   a. Use an anti-seize compound on secondary shaft in area of brake disc to keep disc from binding or sticking. Ensure no anti-seize compound gets on face of brake disc.
   b. Perform *Mechanical Disc Brake Adjustment* as outlined in Chapter Two.

Secondary Shaft Inspection

Refer to **Figure 31** for this procedure.

1. Inspect secondary shaft bearing surfaces for evidence of bearings turning on shaft. Inspect condition of splined-end threads and replace shaft if defective.

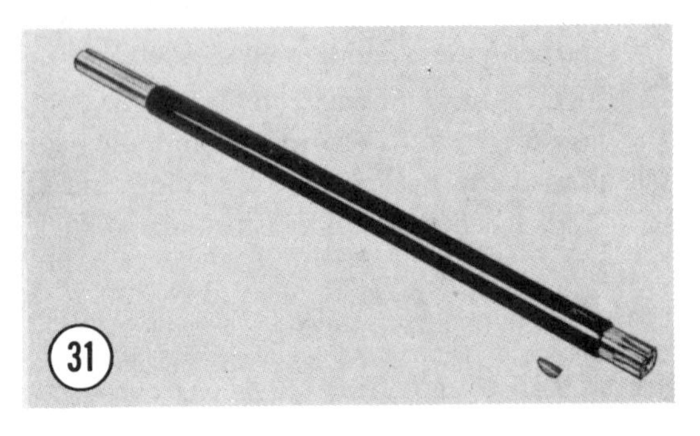

2. Inspect secondary shaft bearings and replace if binding, worn, or noisy. Install a new O-ring on bearing in chain case.

Chain Case Removal/Installation

1. Loosen track tension.
2. Remove drive chain and sprockets.
3. Remove secondary shaft and brake disc.
4. Remove upper and lower bearing flangette nuts.
5. Remove nuts securing chain case to tunnel and lift out chain case (**Figure 32**).
6. Installation is the reverse of these steps.

Drive Shaft Removal/Installation

Remove drive shaft with track as outlined in Chapter Ten.

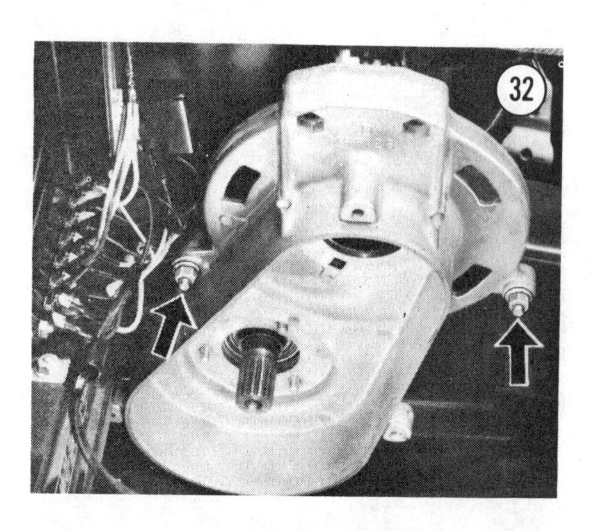

BAND BRAKE

Refer to **Figure 33** for the following procedure.

33

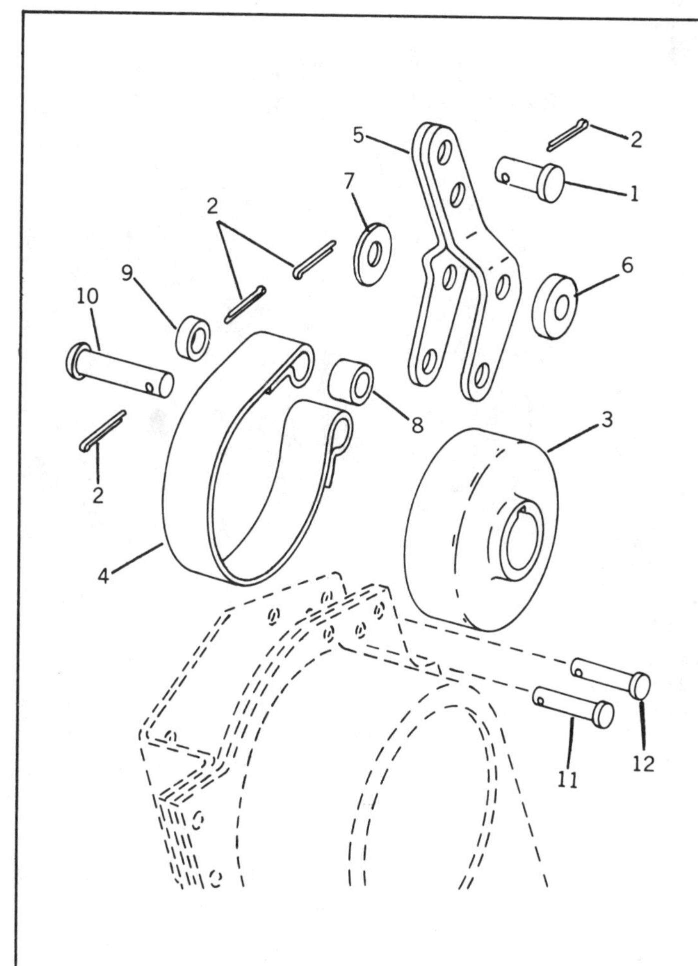

BAND BRAKE

1. Cable attaching pin
2. Cotter pin (4)
3. Brake pulley
4. Brake band with lining
5. Brake arm
6. Thick washer
7. Thin washer
8. Thick spacer
9. Thin spacer
10. Brake arm attaching pin
11. Anchor pin
12. Pivot pin

Removal

1. Remove driven sheave belt guard.
2. Disconnect brake cable from brake arm (**Figure 34**).

3. Remove anchor pin and pivot pin. Ensure spacers and washers are not lost.
4. Rotate brake band forward slightly and remove pin connecting brake arm to brake band.
5. Rotate band backward and out from under brake pulley (**Figure 35**).

Inspection

Refer to **Figure 36** for this procedure.

1. Inspect brake band, and replace if lining is oil-contaminated or if no more adjustment is left.

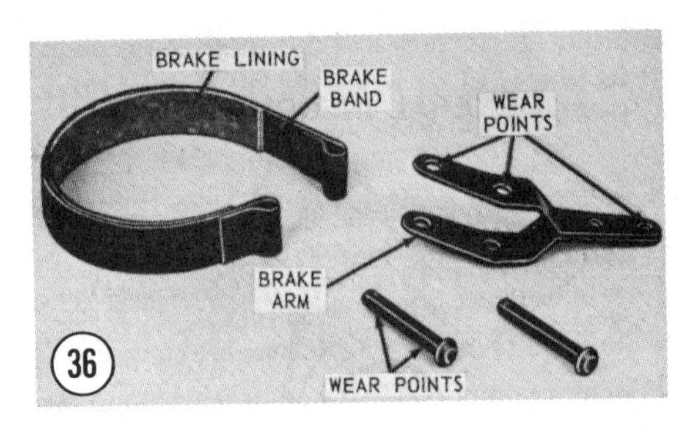

2. Inspect brake arm, anchor pin, and pivot pin, for wear at bearing surfaces. Replace if worn excessively.

Installation

1. Rotate band forward and into position around brake pulley (**Figure 37**). Insert trailing end of brake band first.

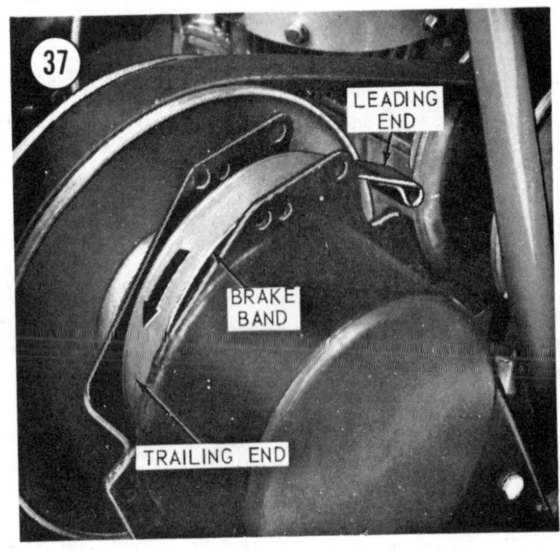

2. Rotate brake band forward sufficiently to allow room to attach brake arm to brake band (**Figure 38**). Secure pin with cotter pin.
3. Install anchor pin through forward hole in chain case, 2 spacers and brake band (**Figure 39**). Thick spacer must be on chain case side of brake band.
4. Install pivot pin through chain case, 2 washers, and brake arm. Thick washer must be on chain case side of brake arm.
5. Connect brake cable to brake arm. Install driven sheave belt guard.

6. Perform *Band Brake Adjustment* as outlined in Chapter Two.

MECHANICAL DISC BRAKE

Removal/Installation

1. Remove air silencer.
2. Remove drive belt and driven sheave. Record number of shims between driven sheave and bearing to aid reassembly.
3. Remove drive chain. Record number of shims between upper sprocket and bearing to aid reassembly. Do not remove spacer and shims from lower drive shaft.
4. Remove bearing flangettes and bearing from secondary shaft on left side of tunnel.
5. Slide secondary shaft out to free shaft from brake disc.
6. Remove cotter pin (**Figure 40**) and adjusting nut from stud of brake puck body.

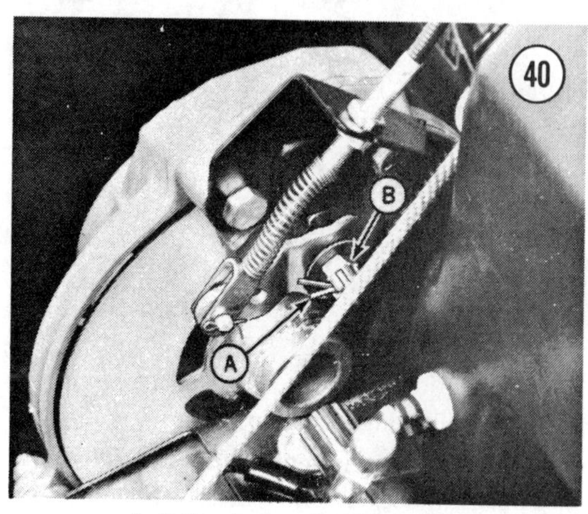

A. Cotter pin B. Adjusting nut

7. Loosen jam nuts securing brake cable to bracket.
8. Remove cam and spring with brake cable and bracket from stud of brake puck body.
9. Remove brake puck body (**Figure 41**), brake puck, backing plate and pins.

NOTE: *On models serial No. 70,001 and subsequent, it may be necessary to heat puck to remove it. Puck was installed with Loctite and 400°F heat is required to loosen it.*

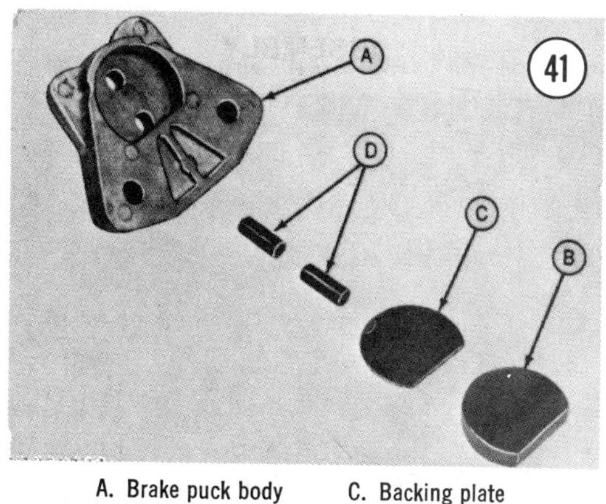

A. Brake puck body C. Backing plate
B. Brake puck D. Pins

10. Remove brake disc (**Figure 42**).
11. Remove brake puck from chain case.

8

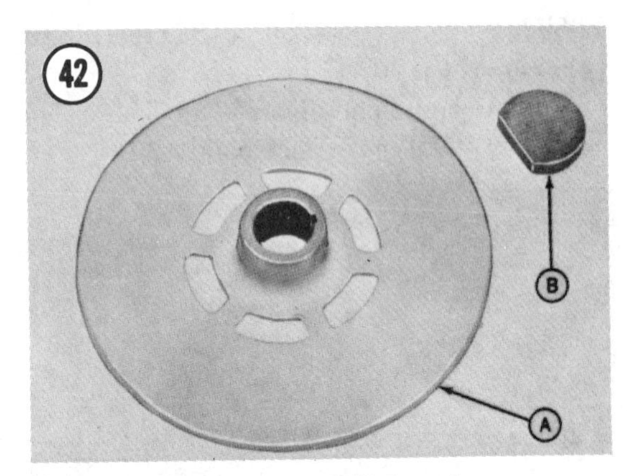

A. Brake disc B. Brake puck

12. Installation is the reverse of these steps. Keep the following points in mind:

a. Use an anti-seize compound on secondary shaft in area of brake disc to prevent disc from seizing or binding on shaft. Make sure that no compound gets on face of brake disc.
b. Place puck and brake disc in chain case as shown in **Figure 43**.
c. Perform *Mechanical Disc Brake Adjustment* as outlined in Chapter Two.

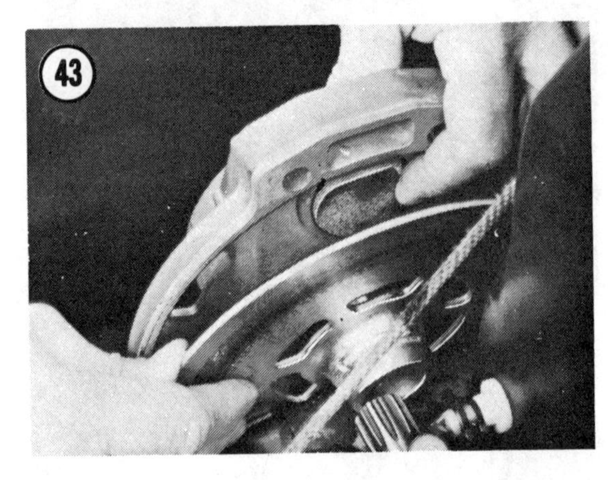

CHAPTER NINE

FRONT SUSPENSION AND STEERING

The front suspension and steering consists of spring mounted skis on spindles connected to the steering column by tie rods. See **Figure 1** for mid-engine models and **Figure 2** for front engine models.

The skis have replaceable wear rods and wear plates. All standard model machines are equipped with multi-leaf springs without shock absorbers. High-performance and racing machines are equipped with either mono-leaf or multi-leaf springs with shock absorbers. Multi-leaf springs are replaceable only as complete units.

Ski spindles are mounted in replaceable bushings pressed into the spindle housing. The handlebar steering column is mounted in rubber at the top with a replaceable nylon bushing at the bottom. All snowmobiles serial No. 20,000 and earlier, are equipped with a steering post and steering column connected by 2 tapered sleeves. All subsequent models use a one piece steering column.

The tie rod and drag link ends are color coded for easy identification. Silver colored ends indicate right-hand threads and gold colored ends indicate left-hand threads. The tie rods, drag link, and spindles are designed to bend rather than break if any extreme shock loads are encountered.

SKIS AND STEERING (CYCLONE AND LIQUIFIRE MODELS)

Refer to **Figure 3** for following procedures.

Ski Wear Rod Removal/Installation

1. Remove locknuts (**Figure 4**) securing wear rod to the ski.

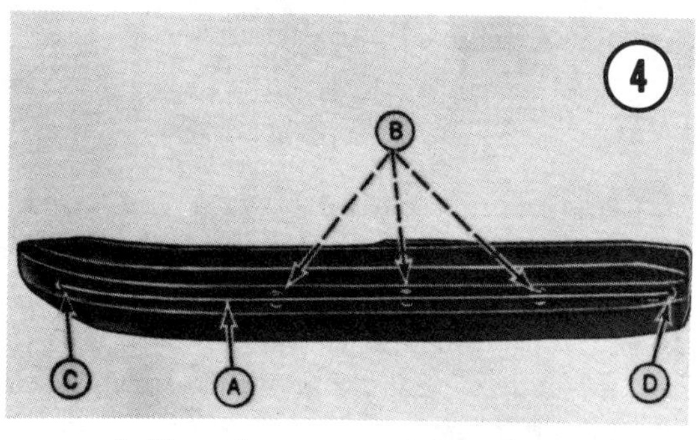

A. Wear rod
B. Locknuts
C. Front hole
D. Rear hole

2. Pry wear rod down to get studs out of holes. Then slide rod forward to remove back of rod from rear hole in ski and remove.

3. Place forward end of new wear rod in position and slide it rearward, positioning back end of rod and studs into hole in ski. Install locknuts and tighten securely.

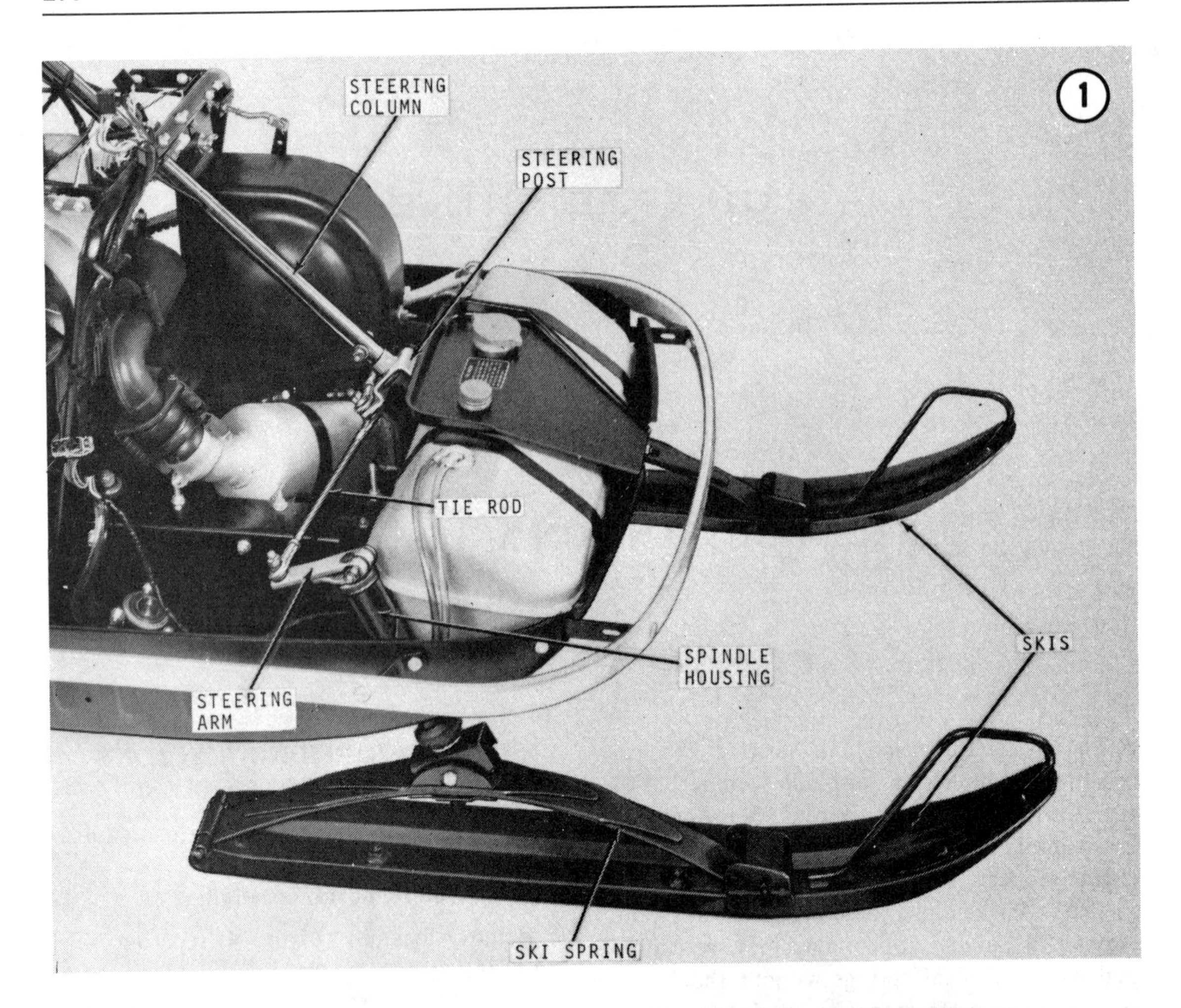

Wear Plate Removal/Installation

NOTE: *It is not necessary to remove ski spring to replace wear plate.*

1. Remove pin securing forward end of ski spring. Lift spring up slightly and slide out old wear plate.
2. Slide new wear plate into position. Lower forward end of spring and secure spring with pin and cotter pin.

Ski Spring Removal/Installation

1. Remove cap screw (**Figure 5**) securing ski to ski spindle. Remove ski stop spring from spindle.
2. Remove cotter pins and drilled pins securing spring assembly to ski.
3. Examine old wear plate and replace if necessary. Install spring assembly to ski and secure with drilled pins and cotter pins. Install spacer in ski spindle and place ski spring stop over

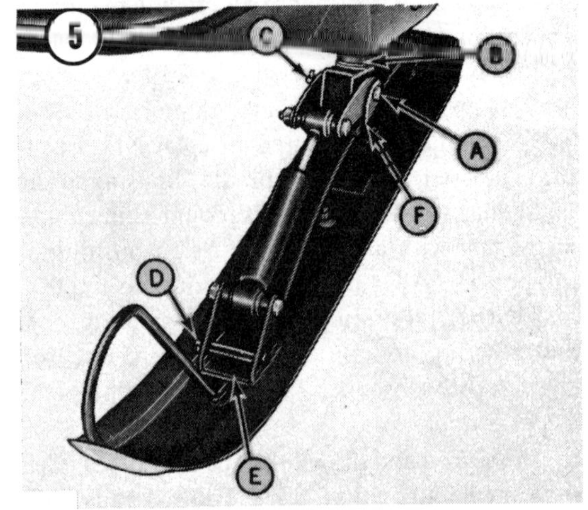

A. Cap screw
B. Ski spindle
C. Cotter pin
D. Drilled pin
E. Wear plate
F. Ski spring stop

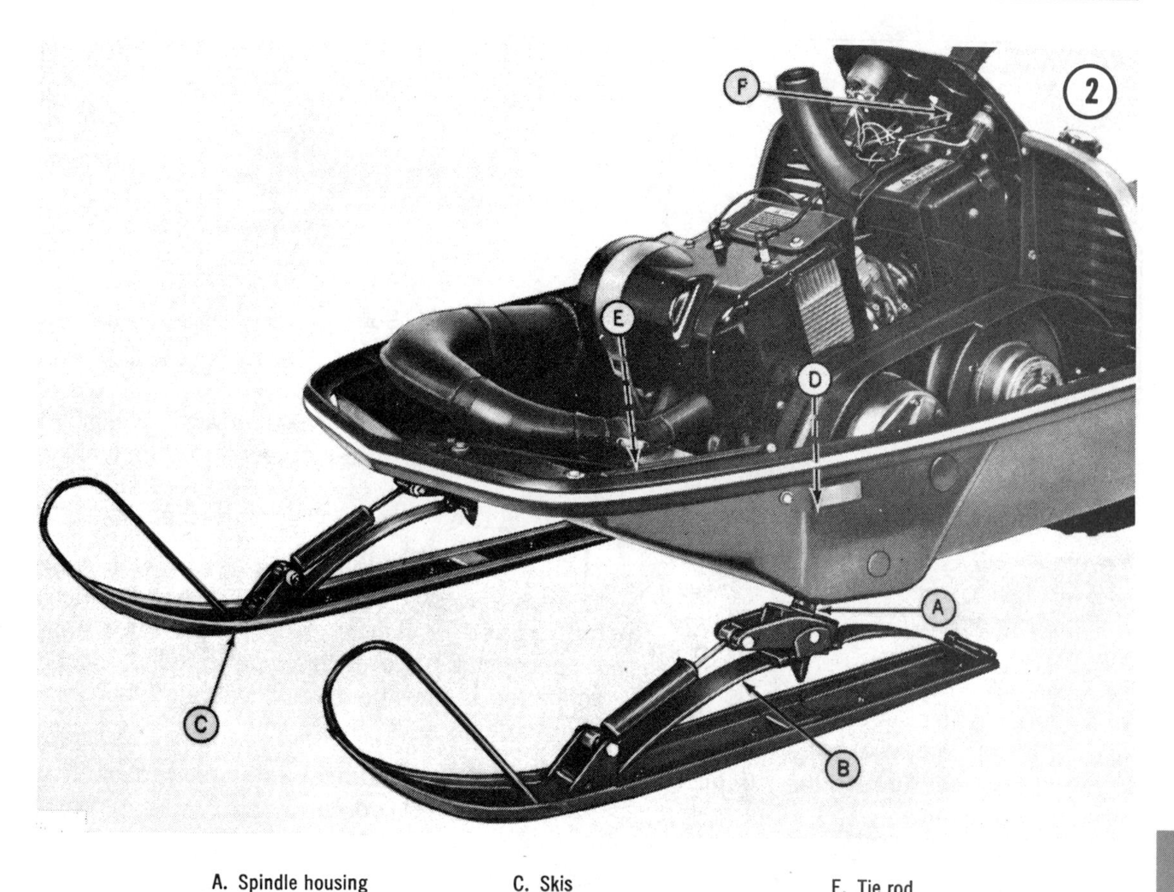

A. Spindle housing
B. Ski spring
C. Skis
D. Steering arm
E. Tie rod
F. Steering post

spindle. Attach ski assembly to spindle with cap screw, washer, and locknut. Torque nut to 39 ft.-lb. (5.4 mkg).

Ski Bumper and Saddle Removal/Installation

NOTE: *Mono-leaf spring and saddles must be replaced as an assembly. Mono-leaf bumpers can be removed by removing screw securing bumper to bottom of saddle.*

Perform the following procedure for 3- and 4-leaf spring units.

1. Remove 2 locknuts securing saddle and bumper retainer to ski spring. See **Figure 6**.
2. Remove bumper retainer. Install bumper and secure with bumper retainer and locknut. Tighten nuts to 35-43 ft.lb. (4.8-5.9 mkg).

NOTE: *Always check condition of 2 cap screws securing saddle and bumper retainer to ski springs and replace if worn. Replace ski saddle if necessary.*

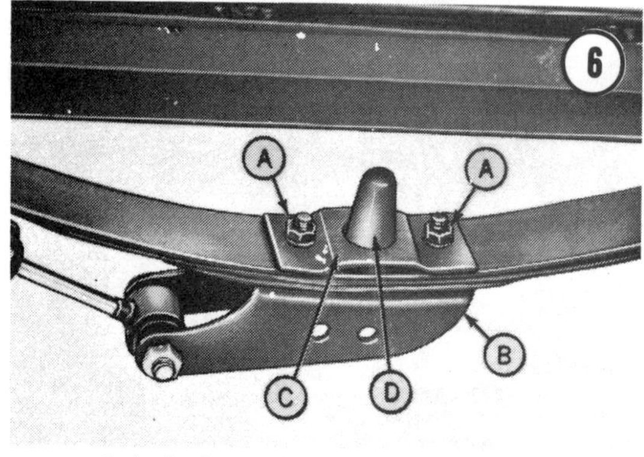

A. Locknuts
B. Saddle
C. Bumper retainer
D. Bumper

Ski Spindle and Bushing Removal/Installation

1. Remove ski and spring assembly and ski spring stop from spindle.

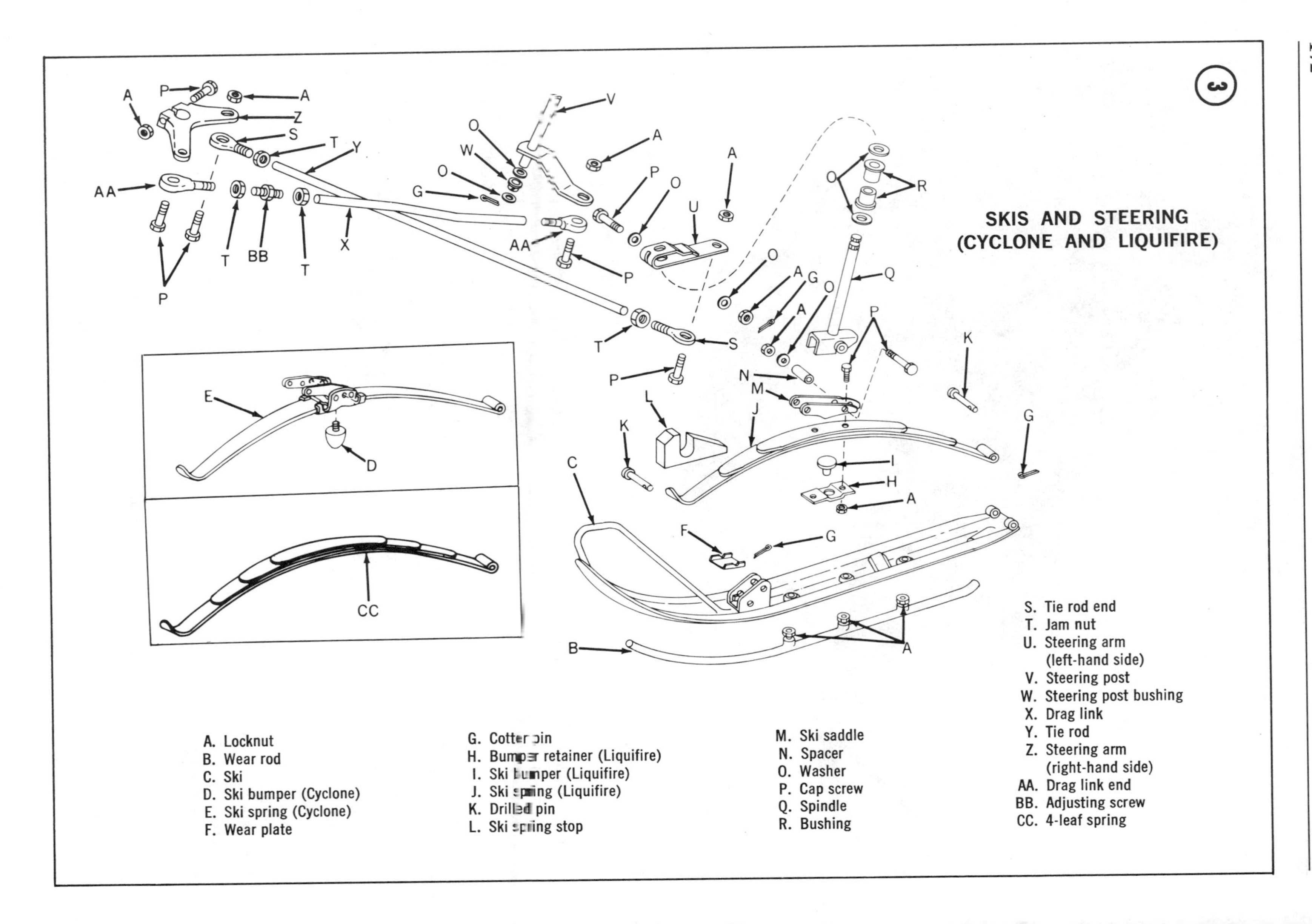
3
SKIS AND STEERING
(CYCLONE AND LIQUIFIRE)
A. Locknut
B. Wear rod
C. Ski
D. Ski bumper (Cyclone)
E. Ski spring (Cyclone)
F. Wear plate
G. Cotter pin
H. Bumper retainer (Liquifire)
I. Ski bumper (Liquifire)
J. Ski spring (Liquifire)
K. Drilled pin
L. Ski spring stop
M. Ski saddle
N. Spacer
O. Washer
P. Cap screw
Q. Spindle
R. Bushing
S. Tie rod end
T. Jam nut
U. Steering arm (left-hand side)
V. Steering post
W. Steering post bushing
X. Drag link
Y. Tie rod
Z. Steering arm (right-hand side)
AA. Drag link end
BB. Adjusting screw
CC. 4-leaf spring

2. Remove spacer from spindle.

3. Remove cap screw, washer, and locknut securing steering arm to spindle. Using a brass drift or a block of wood, gently drive spindle down and out of front frame. Carefully examine bushings. If bushings are satisfactory, proceed to next step. If bushings are worn, damaged, or cracked, replace as follows:

a. Use a punch to remove bushings as shown in **Figure 7**. Drive lower bushing out from the top and upper bushing out from the bottom.

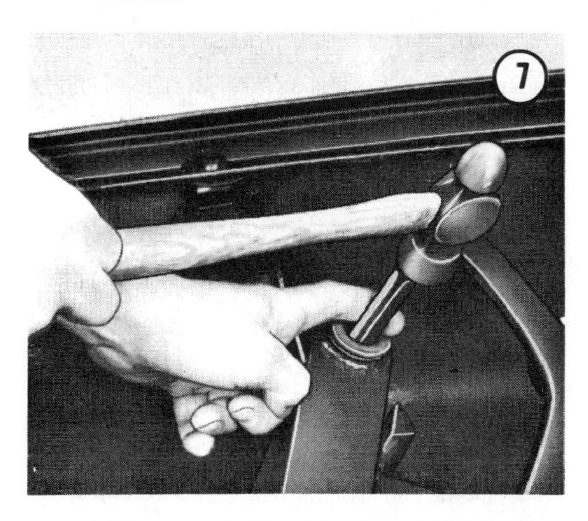

b. Install new bushing until bushing bottoms on front frame. Use care not to crack or distort during installation.

4. Slide ski spindle up into place from bottom.

5. Install spacer in spindle. Place ski spring stop over spindle and install ski in spring assembly.

6. Secure with cap screw, washer, and locknut. Torque nut to 39 ft.-lb. (5.4 mkg).

Steering Arm Removal/Installation

1. Position handlebars and skis to point straight ahead.

2. Remove exhaust silencer.

3. Disconnect tie rod from left- and right-hand steering arm.

4. Disconnect drag link from right-hand steering arm.

5. Remove steering arm retaining cap screws and gently lift arms off steering spindles.

6. Installation is the reverse of these steps. Keep the following points in mind:

a. Ensure steering arm is correctly installed over spindle by measuring from center of tie rod hole in arm to snowmobile frame. The approximate distance should be 5.9 in. (14.9cm) as shown in **Figure 8**. This dimension may vary by ⅛ in. (3mm) due to spline alignment of arm and spindle.

WARNING

The approximate 5.9 in. (14.9cm) dimension from steering arm to snowmobile is critical. Improper alignment of steering arm could cause interference in steering effort.

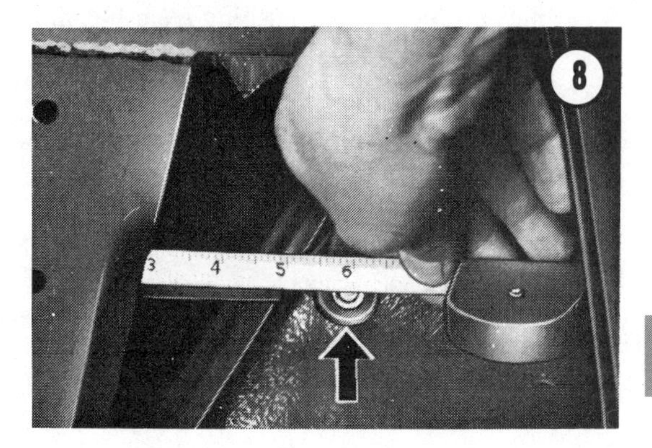

b. When steering arm is positioned correctly on spindle, secure steering arm with cap screw, washer, and locknut.

c. Connect tie rod and drag link to steering arm. Perform *Ski Adjustment and Alignment.*

d. Install exhaust silencer.

Tie Rod and Drag Link Removal/Installation

1. Remove cap screw securing tie rod and drag link.

2. Remove tie rod through knockout hole on left-hand side of pan.

3. Installation is the reverse of these steps. Connect gold colored tie rod end to left-hand steering arm and silver colored tie rod end to right-hand steering arm. Silver color indicates right-hand threads, gold color indicates left-hand threads.

4. Install drag link to steering post and right-hand steering arm. Adjusting screw end of drag link must be attached to steering arm.

5. Perform *Ski Adjustment and Alignment.*

Steering Post Assembly Removal/Installation

1. Remove seat and air intake silencer.

2. Close fuel shut off valve and disconnect fuel line from inline fuel filter.

3. Remove fuel tank.

4. Disconnect drag link from steering post. Remove cotter pin and washer from bottom of steering post.

5. Remove upper rubber bushing from bushing bracket. Remove bracket from dash.

6. Remove brake and throttle grip from handlebar. Pull up and twist and turn steering post to remove.

7. Installation is the reverse of these steps.

Steering Post Upper Bushing Removal/Installation

1. Remove 2 locknuts (**Figure 9**) and clamp securing bushing.

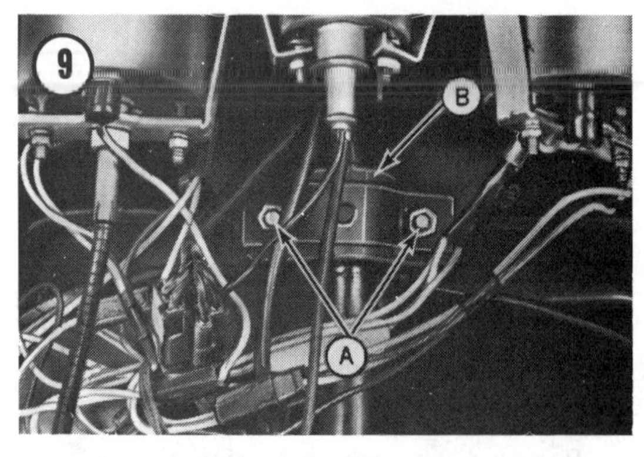

A. Locknuts B. Bushing

2. Open bushing and remove from steering post.

3. Installation is the reverse of these steps.

Steering Post Lower Bushing Removal/Installation

1. Disconnect drag link (**Figure 10**) from steering post.

A. Washer
B. Cotter pin
C. Bushing
D. Washers
E. Drag link

2. Remove cotter pin and washer from end of steering post.

3. Slide steering post up sufficiently to remove washers and bushing.

4. Install bushing. Place washers over steering post and slide back in place. Secure with washer and cotter pin.

Component Inspection

WARNING

Snowmobiles travel at near highway speed. Worn, bent, or damaged ski and steering components must be replaced or fatal accidents may occur.

1. Replace wear rods on skis if they are worn as shown in **Figure 11**. Rapid wear of this type indicates operation without proper snow lubrication. Worn wear rods cause a loss of snowmobile maneuverability.

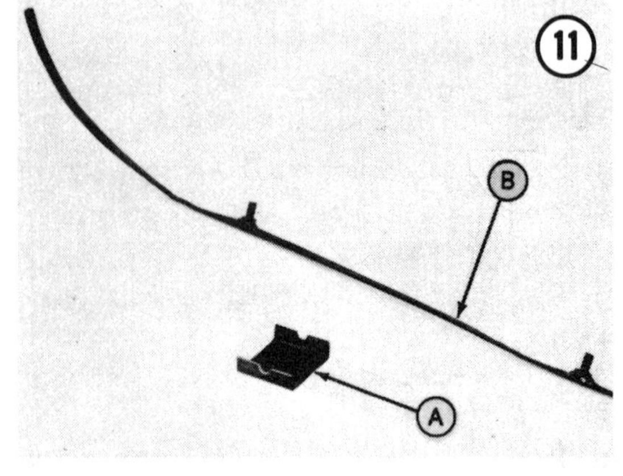

A. Wear plate B. Worn wear rod

2. Examine ski wear plates (**Figure 11**). If wear plates are worn or damaged, they must be replaced or a ski spring will wear through ski making ski replacement necessary.

3. Examine ski springs for fractures or breaks and replace if necessary.

4. Inspect all steering components. A bent or damaged steering component indicates that a snowmobile was subjected to severe forces such as hitting an obstruction at high speed or landing on one ski after jumping. Whenever damage of this type is found, closely examine the rest of the ski and steering mechanism for bends, damage, or cracks. Replace items as necessary.

5. Replace ski attaching pins if worn.

6. Replace tie rod and drag link ends if excessively loose.

Ski Adjustment and Alignment

Refer to **Figure 12** for proper positioning of skis in relationship to steering arms, tie rods, and steering column.

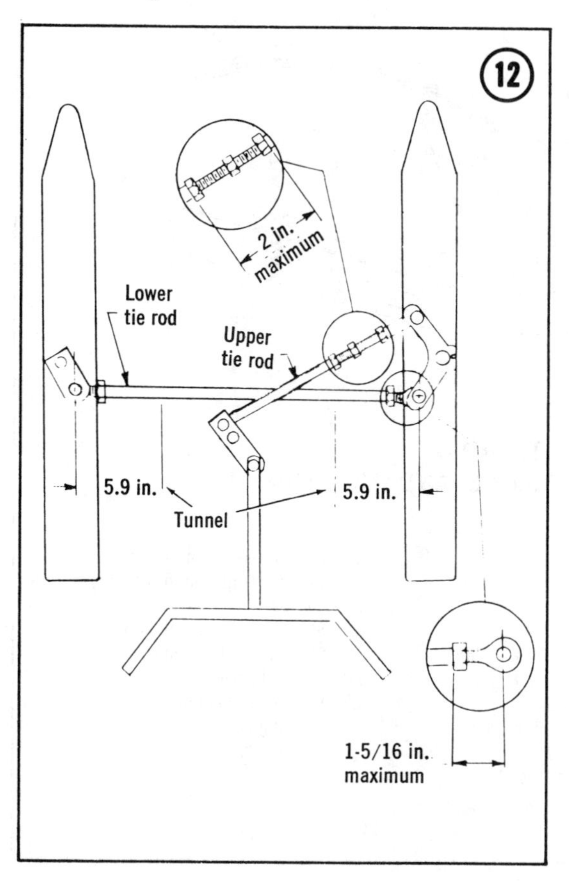

1. Raise front of snowmobile slightly to remove weight from skis.

2. Position handlebars straight ahead.

3. Measure distance over front and rear wear rod nuts. The dimensions should be the same.

4. If adjustments are necessary, remove exhaust silencer for access to tie rods.

5. Loosen jam nuts on each end of lower tie rod. Rotate tie rod until skis are parallel and tighten jam nut.

> NOTE: *Turn tie rod toward front of snowmoile to spread skis apart. Turn tie rod toward rear of snowmoile to bring skis together.*

6. To realign handlebars, loosen jam nuts on both sides of adjusting screw on upper tie rod. Rotate adjusting screw until handlebars are aligned. Tighten jam nuts.

> WARNING
> *Do not exceed 1-5/16 in. (33.34mm) between tie rod and center of tie rod when adjusting tie rod. Do not exceed 2 in. (50.58mm) between the drag link and drag link end when adjusting drag link.*

7. After completing ski alignment, ensure all jam nuts are tight. Install exhaust silencer.

SKIS AND STEERING (ALL MODELS EXCEPT CYCLONE AND LIQUIFIRE)

Refer to **Figure 13** for following procedures.

Ski Wear Rod Removal/Installation

1. Remove 2 locknuts (**Figure 14**) securing wear rod to ski. Pry wear rod down to get studs out of holes. Slide rod forward to remove back of rod from rear hole in ski and remove.

2. Place forward end of wear rod in position and slide it rearward, positioning back end of rod and studs into holes in ski. Install locknuts and tighten securely.

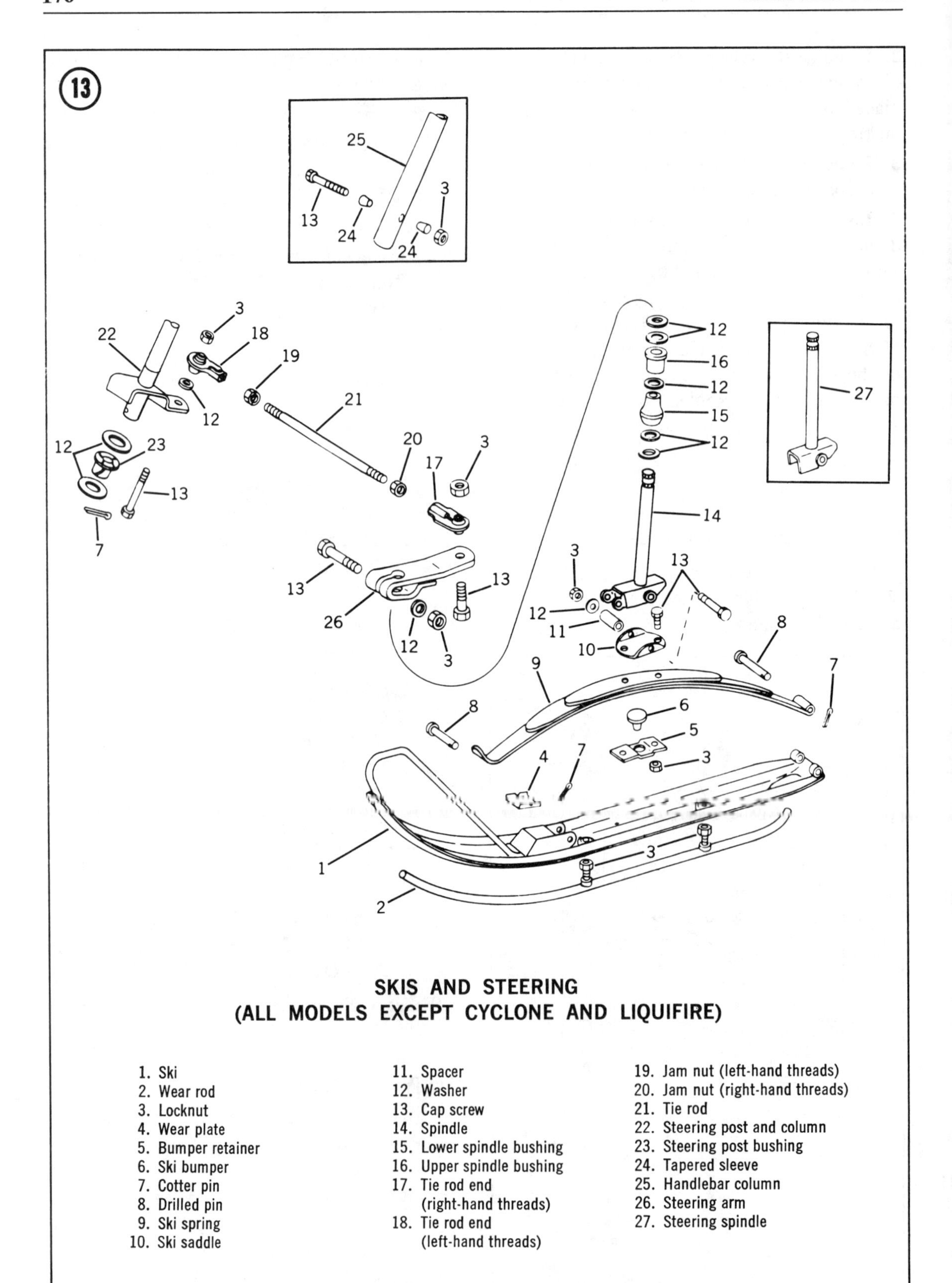

SKIS AND STEERING (ALL MODELS EXCEPT CYCLONE AND LIQUIFIRE)

1. Ski
2. Wear rod
3. Locknut
4. Wear plate
5. Bumper retainer
6. Ski bumper
7. Cotter pin
8. Drilled pin
9. Ski spring
10. Ski saddle
11. Spacer
12. Washer
13. Cap screw
14. Spindle
15. Lower spindle bushing
16. Upper spindle bushing
17. Tie rod end (right-hand threads)
18. Tie rod end (left-hand threads)
19. Jam nut (left-hand threads)
20. Jam nut (right-hand threads)
21. Tie rod
22. Steering post and column
23. Steering post bushing
24. Tapered sleeve
25. Handlebar column
26. Steering arm
27. Steering spindle

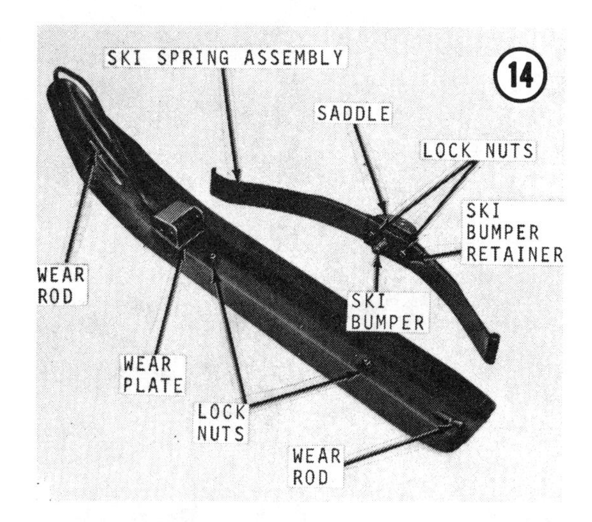

Wear Plate Removal/Installation

NOTE: *Ski springs do not have to be removed to replace wear plates.*

1. Remove pin securing forward end of ski spring. Lift spring up slightly and slide out old wear plate (**Figure 14**).
2. Slide wear plate into position. Lower forward end of spring and secure with pin and cotter pin.

Ski Spring Removal/Installation

1. Remove locknut, washer, and cap screw (**Figure 15**) securing ski and spring assembly to ski spindle.

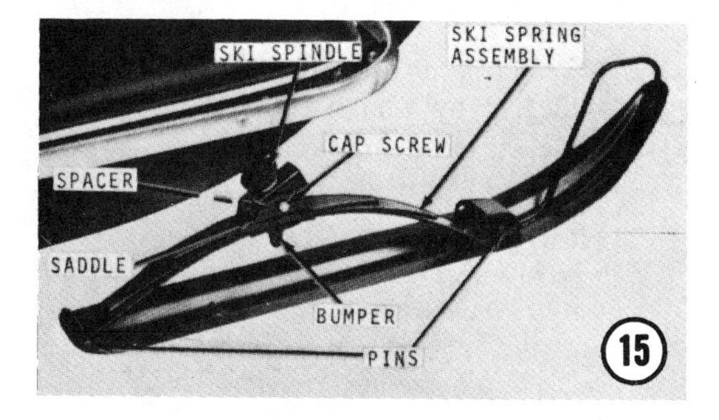

2. Remove cotter pins and pins securing spring assembly to ski and remove spring assembly. Inspect wear plate and replace if necessary.
3. Install spring assembly and secure to ski with pins and cotter pins. With spacer in place in ski spindle, attach ski and spring assembly to spindle with cap screw, washer, and locknut. Torque to 39 ft.-lb. (5.4 mkg).

Ski Bumper and Saddle Removal/Installation

1. Remove locknuts and cap screws securing saddle and bumper retainer to ski springs (**Figure 14**).
2. Remove bumper retainer.
3. To remove saddle from ski spindle, remove cap screw, washer, and locknut.
4. Installation is the reverse of these steps.

NOTE: *Inspect 2 cap screws securing saddle and bumper retainer to ski spring and replace if worn. Torque nuts to 35-40 ft.-lb. (4.8-5.5 mkg).*

Ski Spindle Removal/Installation

1. Remove locknut, washer, and cap screw (**Figure 16**) securing ski and spring assembly to spindle.

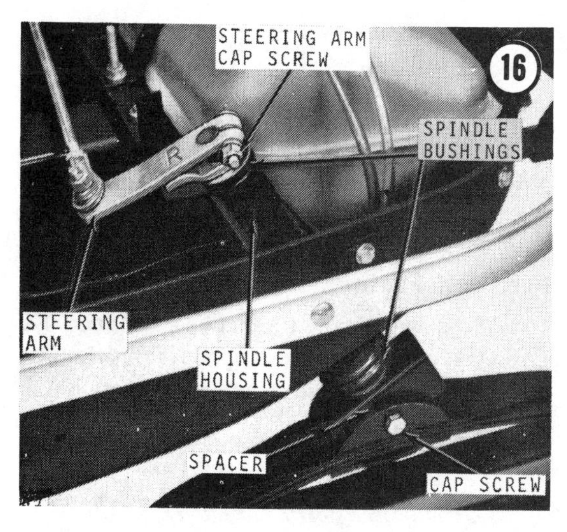

2. Remove ski and spring assembly and remove spacer from ski spindle.
3. Remove cap screw securing steering arm to spindle. Using a brass drift or a block of wood, gently drive spindle down and out of spindle housing. Carefully inspect bushings and replace if worn, damaged, or cracked.
4. Installation is the reverse of these steps. Keep the following points in mind:

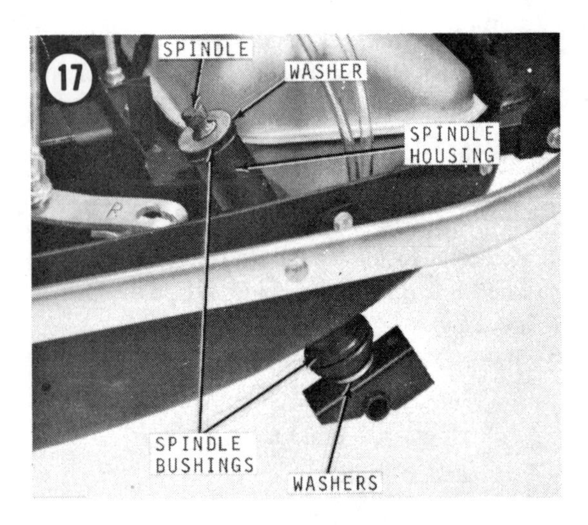

a. Install 2 or more washers over spindle on top of upper bushing (**Figure 17**). Use as many washers in this position as necessary to take play out of spindle and spindle bushings with steering arm installed.

b. After installing cap screw, washer, and locknut, torque to 39 ft.-lb. (5.4 mkg).

Steering Arm Removal/Installation

1. Position handlebars and skis to point straight ahead.

2. Remove tie rod end cap screw and steering arm retaining cap screw and lift steering arm off ski spindle.

3. Installation is the reverse of these steps. Keep the following points in mind:

a. Steering arms are stamped "L" and "R" (**Figure 18**) to indicate proper positioning. Ensure proper steering arm is installed in the correct position.

CAUTION

Do not force steering arm over spindle or spindle splines will be damaged.

b. Perform *Ski Adjustment and Alignment*.

Tie Rod Assembly Removal/Installation

1. Remove cap screw securing tie rod end to steering post and steering arm (**Figure 19**).

2. Installation is the reverse of these steps. Keep the following points in mind:

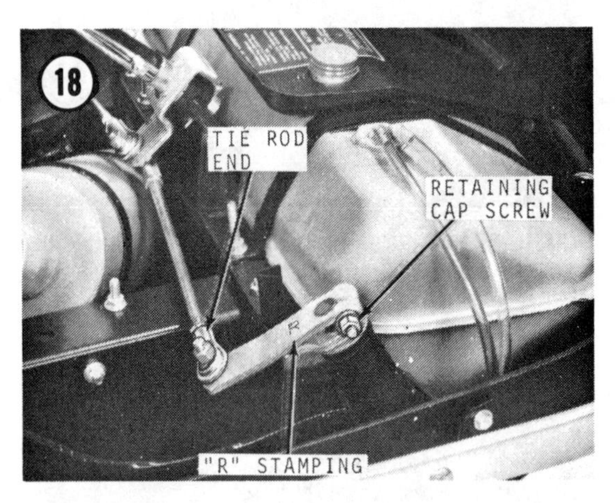

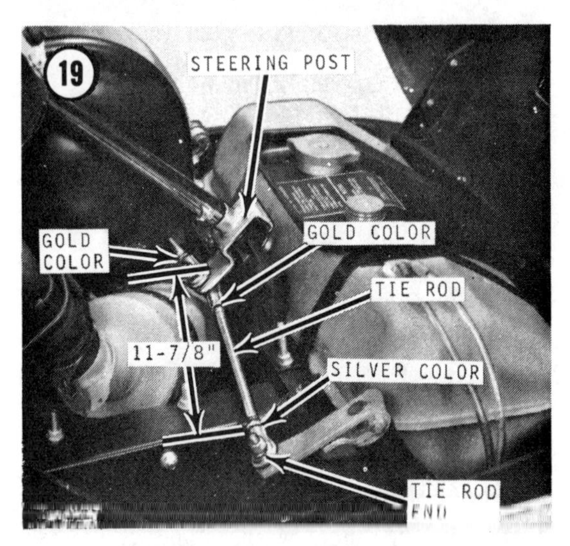

a. Adjust tie rods to 11⅞ in. dimension (hole center to hole center) as shown in **Figure 19**. Each tie rod has one silver colored and one gold colored end. Silver color indicates right-hand threads. Gold color indicates left-hand threads.

NOTE: *Do not at any time adjust tie rod to exceed 12¼ in. from center hole to center hole. If necessary to maintain correct dimension of 11⅞ in., relocate steering arm on spindle.*

b. Attach gold colored tie rod ends to steering post and silver colored ends to steering arm. Secure with cap screws and locknuts, positioning locknuts on top as shown in **Figure 19**. Tighten nuts securely.

NOTE: *A washer must be placed on each side of steering post. Right-hand tie rod must be installed below steering post and left-hand tie rod above post.*

c. Perform *Ski Adjustment and Alignment.*

Steering Post Assembly Removal/Assembly

1. Remove steering column cap screw and nut (**Figure 20**).

NOTE: *On all machines serial No. 20,000 and earlier, it is necessary to remove tapered sleeve from steering column.*

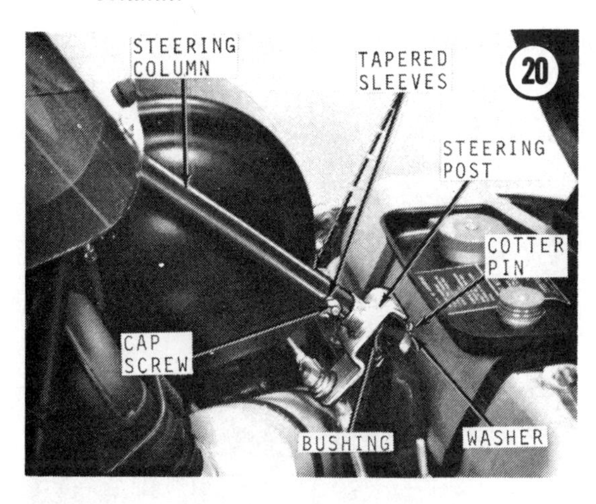

2. Remove tie rods from steering post.
3. Remove cotter pin and washer securing post to frame.
4. Installation is the reverse of these steps.

Steering Column Bushing Removal/Installation

1. Remove locknuts and clamp securing bushing (**Figure 21**).
2. Open bushing and remove from steering column.
3. Installation is the reverse of these steps.

Ski Adjustment and Alignment

Refer to **Figure 22** for proper positioning of skis in relation to steering arms, tie rods, and steering column.

NOTE: *Measure from parallel sides of skis only, not from tapered ends.*

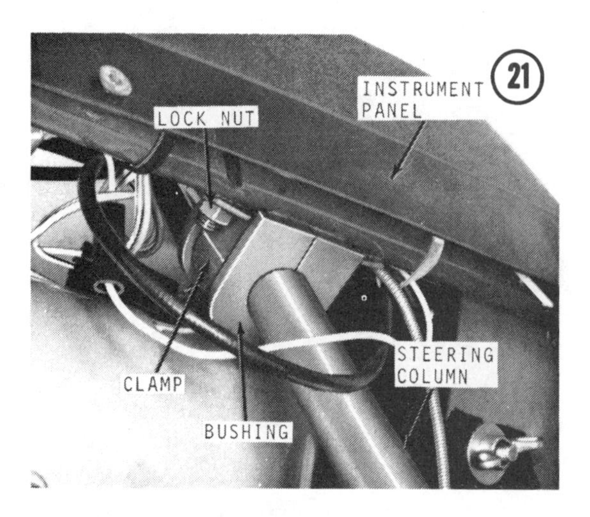

Loosen jam nuts on tie rod ends. Turn tie rods to lengthen or shorten as required to make skis parallel and keep handlebars in alignment with skis. Tighten jam nuts securely.

NOTE: *When tightening jam nuts on tie rods be certain tie rod ends are still free to swivel after jam nuts are tight.*

CAUTION
Hold tie rod with Vise Grips after adjustment is made and while jam nut is being tightened. Damage or stripping of threads may occur within ball-joint if tie rod is not held.

9

CHAPTER TEN

REAR SUSPENSION AND TRACK

John Deere snowmobiles are equipped with either a bogie or slide rail rear suspension. Both suspensions utilize a rear idler assembly.

The bogie suspension is of the trailing arm type and is designed so it cannot come loose from the tunnel even if bent or severely damaged. It consists of 6 independent bogie wheel assemblies: three 2-wheel units, and three 3-wheel units. The bogie wheels are equipped with sealed ball bearings and do not require lubrication.

The slide rail suspension (**Figure 1**) is fitted with adjustable springs, shock absorbers, and replaceable wear bars. The suspension also includes a weight transfer adjustment to vary the amount of ski lift.

This chapter includes removal and installation procedures for suspension components and tracks. Refer to Chapter Two for suspension and track adjustments and Chapter Three for *Track Wear Analysis.*

BOGIE WHEEL SUSPENSION

Bogie Wheel Removal/Installation

1. Inspect bogie wheels periodically for freeness of operation. See **Figures 2 and 3**.

CAUTION
A stuck bogie wheel must be replaced or track damage will result.

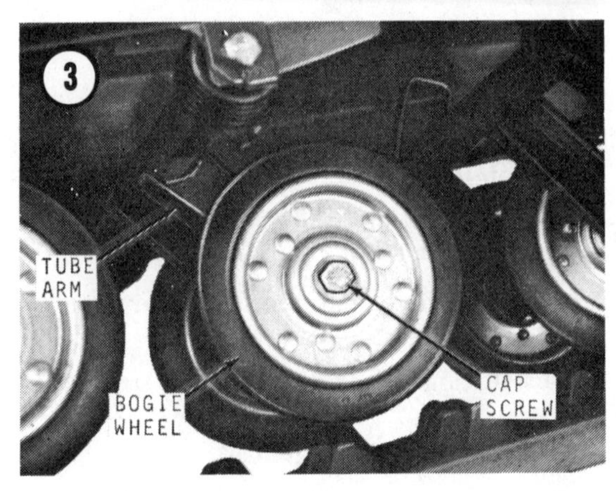

2. Remove cap screw securing bogie wheel and remove wheel.

3. Install *black* bogie wheels with spring washer between wheel and bogie tube arm. Install *silver*

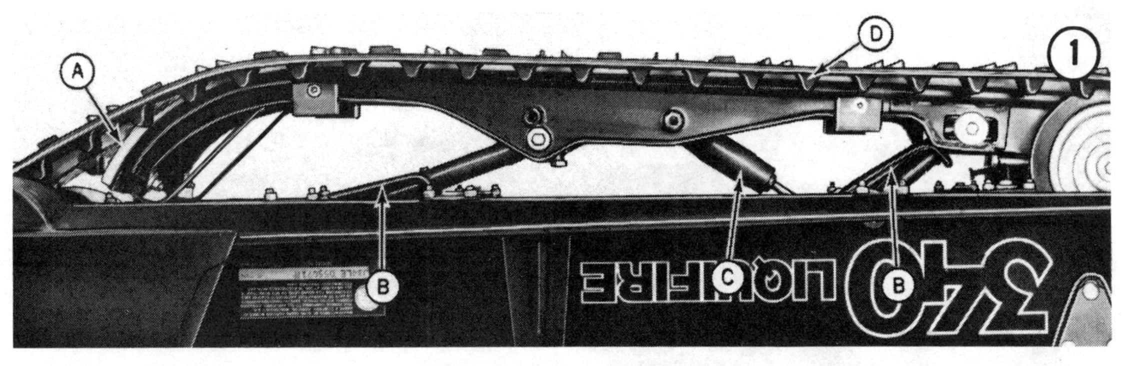

A. Replaceable wear bars B. Suspension springs C. Shock absorber D. Grouser bar track

bogie wheels with shoulder of wheel next to tube arm.

CAUTION

If shoulder on silver *wheel is not against tube arm, wheel will bind and not turn freely.*

Bogie Wheel Assembly Removal/Installation (Models Serial No. to 20,000)

Refer to **Figure 4** for this procedure.

1. Block up rear of machine. Adjust rear idler assembly forward to relieve track tension.

1. Carriage bolts
2. Bogie axle clamp
3. Bogie axle
4. Spacer
5. Notch
6. Spring
7. Bogie axle clip
8. Groove

2. Remove 2 carriage bolts from each side of bogie assembly to be removed.

3. Remove bogie axle clip and clamp from each end of axle. Remove bogie assembly.

4. Remove springs, spacers, axle, and axle bushings if replacement is necessary.

5. Installation is the reverse of these steps. Keep the following points in mind:

a. Install front and rear assemblies first as shown in **Figure 5**. Install 2 middle assemblies last.

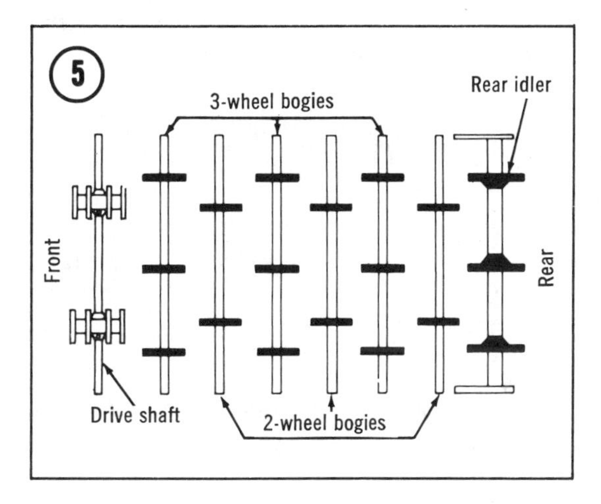

b. Be sure axle clip is positioned in groove on end of bogie axle.

c. Short spring leg must be retained in notch in bogie tube. Long spring leg must be under clamp pointed toward rear of machine.

d. Perform *Track Tension Adjustment* as outlined in Chapter Two.

Bogie Wheel Assembly Removal/Installation (Models Serial No. 20,001 and Subsequent)

Refer to **Figure 6** for this procedure.

1. Block up rear of machine. Adjust rear idler assembly to relieve track tension.

2. Straighten locking tabs.

3. Remove one cap screw while holding opposite cap screw to prevent axle rotation.

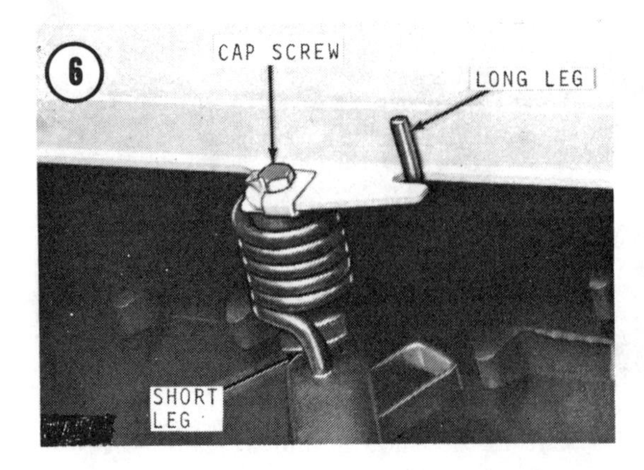

4. Pull bogie assembly down and clear of the tunnel. Use locking pliers to hold axle and remove retaining cap screw.
5. Installation is the reverse of these steps. Keep the following points in mind:

   a. Install front and rear assemblies first as shown in **Figure 5**. Install 2 middle assemblies last.
   b. Position long leg of spring in notch on bogie rail. Position short leg of spring in tube arm.
   c. After tightening cap screw, secure with locking tab.
   d. Perform *Track Tension Alignment* as outlined in Chapter Two.

Bogie Wheel Inspection

Refer to **Figure 7** for this procedure.

1. Inspect wheel ball bearings. Replace wheel if excessively loose or not free.

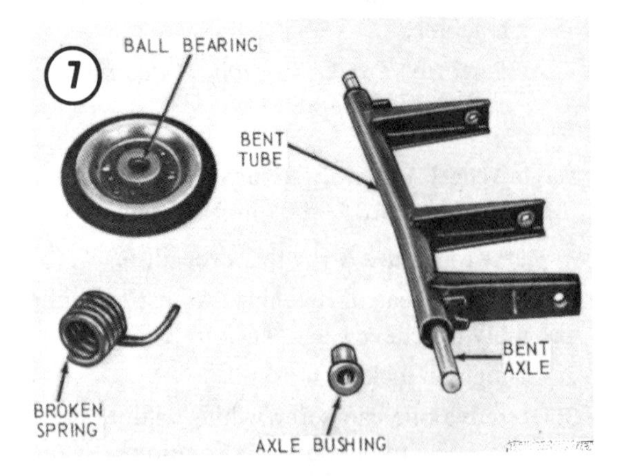

2. Replace bogie springs if cracked or broken.
3. Replace tubes or arms if bent.
4. Replace axle if bent.
5. Replace worn or damaged axle bushings.

> NOTE: *It is not necessary to replace a slightly bent bogie tube if it does not interfere with bogie axle or freedom of operation.*

Rear Idler Assembly Removal/Installation

1. Adjust idler assembly forward to loosen track tension (**Figure 8**).

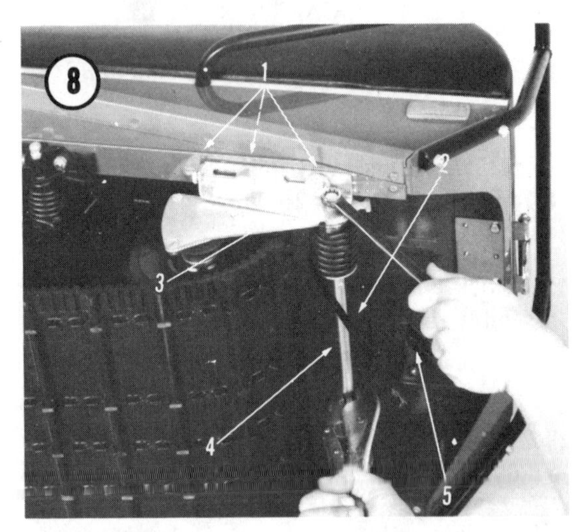

1. Tension adjustment
2. Spring
3. Cap screw
4. Pivot rod
5. Spring bracket

2. Unhook long ends of 2 large springs from spring brackets.
3. Clamp a Vise Grip to pivot rod to keep it from turning and remove cap screws from each end of pivot rod.
4. Remove springs, spacers, and washers and slide idler assembly out from inside the track. Refer rear idler assembly repair to an authorized dealer.
5. Installation is the reverse of these steps. Keep the following points in mind:

   a. Install small spacers over cap screws and large spacers over pivot arm tubes. See **Figure 9**.

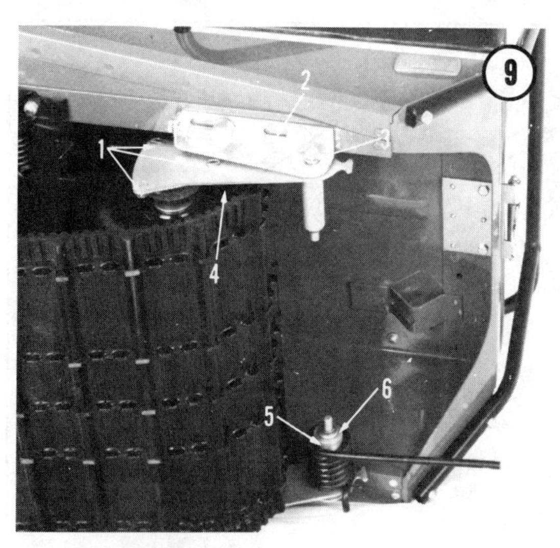

1. Countersunk screws
2. Adjusting bracket
3. Cap screw
4. Pivot arm
5. Large spring spacer
6. Small spacer

b. Perform *Track Tension Alignment* as outlined in Chapter Two.

SLIDE SUSPENSION

Refer to **Figures 10 and 11** for the following procedures.

Suspension Removal

1. Siphon fuel from fuel tank and remove oil from chaincase.

2. Lift or tip snowmobile to gain access to suspension.

3. Back out track adjusting screws to relieve track tension (**Figure 12**).

4. Use a 1¼ in. wrench to relieve spring tension and remove the 2 adjustment screws from each side of suspension (**Figure 13**).

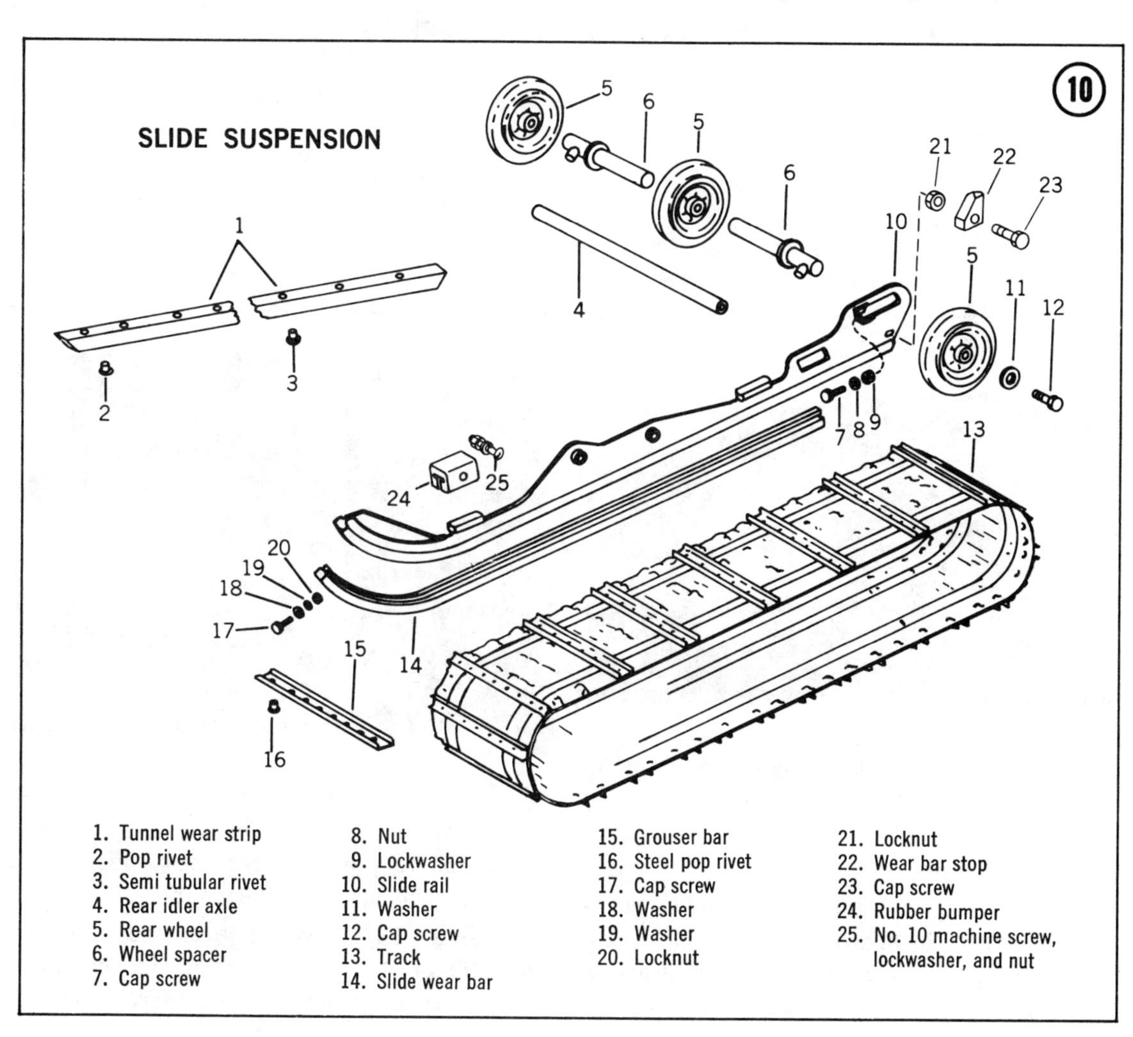

1. Tunnel wear strip
2. Pop rivet
3. Semi tubular rivet
4. Rear idler axle
5. Rear wheel
6. Wheel spacer
7. Cap screw
8. Nut
9. Lockwasher
10. Slide rail
11. Washer
12. Cap screw
13. Track
14. Slide wear bar
15. Grouser bar
16. Steel pop rivet
17. Cap screw
18. Washer
19. Washer
20. Locknut
21. Locknut
22. Wear bar stop
23. Cap screw
24. Rubber bumper
25. No. 10 machine screw, lockwasher, and nut

10

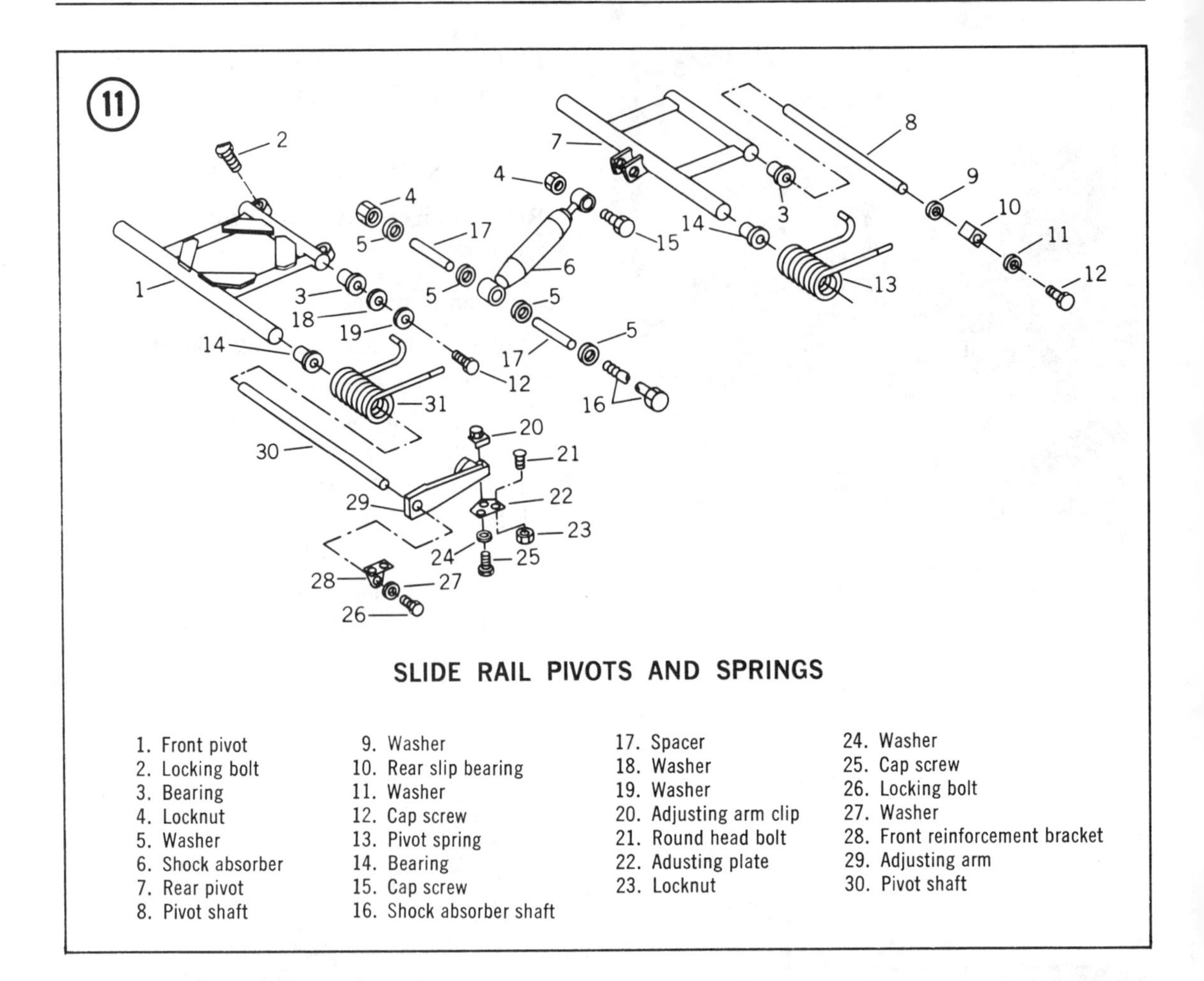

5. Remove 4 cap screws (**Figure 14**) securing suspension and remove suspension.

Suspension Installation

1. Assemble springs and arms to slide suspension as shown in **Figure 15**. Use tape to help hold springs in place and to hold arms to springs.

2. Install suspension inside track and into place in tunnel.

3. Install front then rear cap screws (**Figure 14**). Tighten all 4 cap screws securely.

4. Use a 1¼ in. wrench over end of spring arm (**Figure 13**) to pivot spring arm and allow installation of spring adjustment screw and spe-

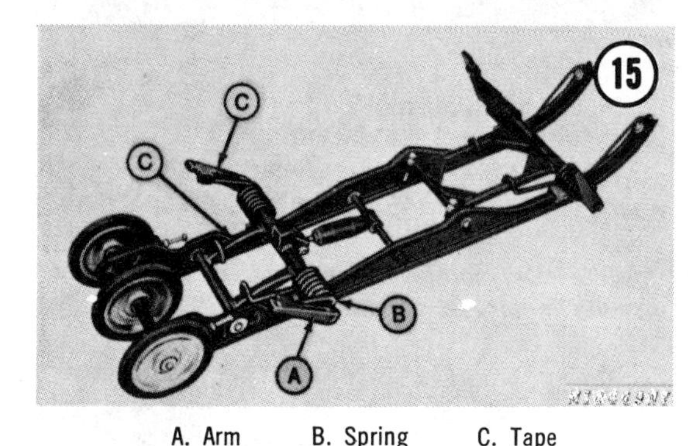

A. Arm B. Spring C. Tape

cial nut. Turn adjustment screw through nut until screw is through by 2 threads. Repeat procedures on remaining adjustment screws.

5. Turn both track adjusting screws (**Figure 12**) in equally until track is fairly tight. *Do not* tighten jam nuts on adjusting screws at this time.

6. Perform *Track Tension Adjustment* and then *Slide Suspension Spring Adjustment* as outlined in Chapter Two.

7. Fill fuel tank. Fill chaincase with oil as outlined in Chapter Two.

Tunnel Wear Bar Replacement

1. On mid-engine models, remove windshield and console to gain access to front rivets.

2. Remove suspension.

3. Lay track over front of machine to gain access to tunnel wear bars (**Figure 16**).

4. Remove cap screws securing seat and remove seat.

5. Using a cold chisel, remove rivets securing wear bars. Remove rivets from the top (seat side).

6. Install new wear bars. Install rivets from

A. Tunnel wear bars B. Seat screws

wear bar side. Use pop rivets in front 3 holes and semi-tubular rivets in remaining holes.

7. Install seat. Install windshield and console if removed.

Suspension Wear Bar Replacement

1. Back out adjusting screws (**Figure 12**) to loosen track tension.

2. Remove wear bar retaining bolt and nut (**Figure 17**).

> NOTE: *On models serial No. 70,000 and subsequent, remove wear bar stop from rear of suspension.*

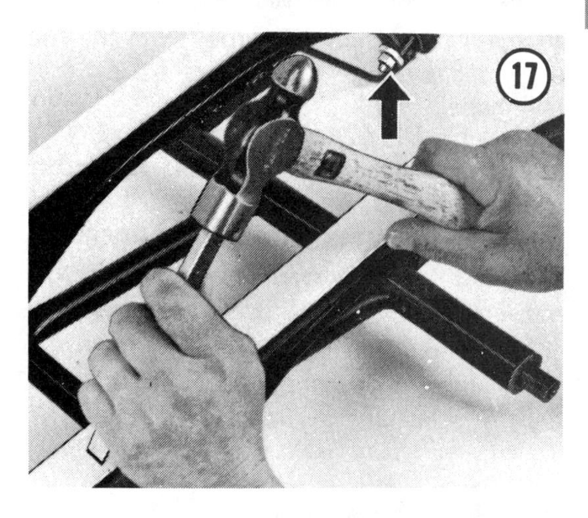

3. Using a hammer and cold chisel, tap wear bar to the rear and out of slide rail grooves (**Figure 17**).

4. If wear bars are difficult to remove, perform the following:

a. Carefully drill a ⅜ in. diameter hole in the center of wear bar approximately 18 inches (45 cm) from rear of wear bar.

CAUTION

Drill hole carefully so drill bit does not damage the metal rail.

b. Install a ⅜ x ¾ in. cap screw in drilled hole.
c. Using a hammer and cold chisel, drive against cap screw and remove wear bar (**Figure 18**).

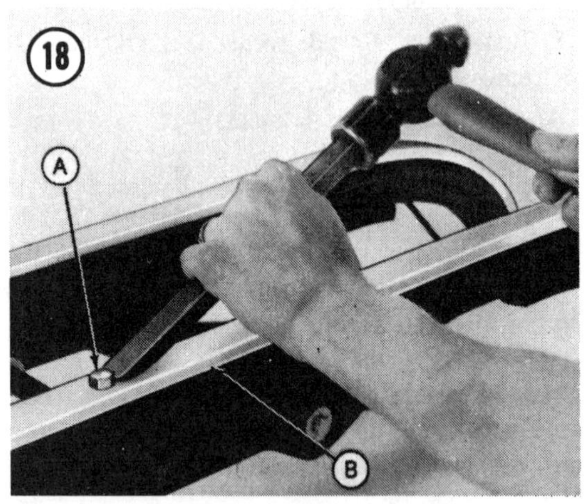

A. ⅜ x ¾ in. cap screw B. Wear bar

5. Measure width of opening in rail. Opening should be 0.44-0.50 in. (11.1-12.7mm) wide. If opening is too narrow, gently pry rail open to specified dimension.

6. Lubricate rail with grease and install new wear bar from the rear. Drive wear bar into place with a soft mallet. *Do not* use a chisel.

7. Install retaining bolt and nut to secure wear bar. Install wear bar stop on models serial No. 70,001 and subsequent.

Wear Block Replacement

1. Remove cap screw, washer, and old wear block from each end of lower pivot shaft. See **Figure 19**.

2. Install new wear block and secure with cap screw and washer.

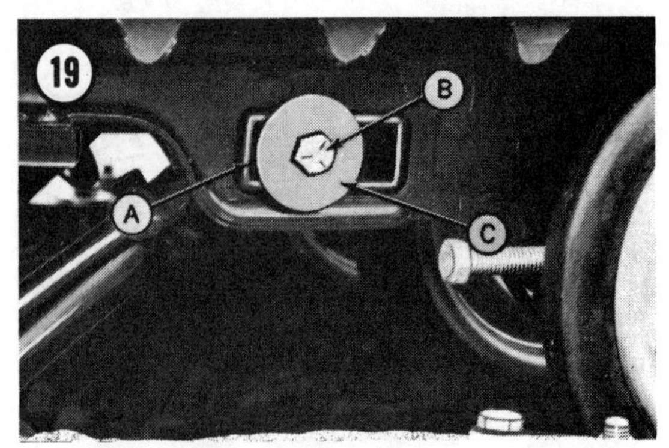

A. Wear block B. Cap screw C. Washer

Rear Idler Wheel Replacement

Refer to **Figure 20** for models up to serial No. 70,000 and **Figure 21** for models serial No. 70,001 and subsequent.

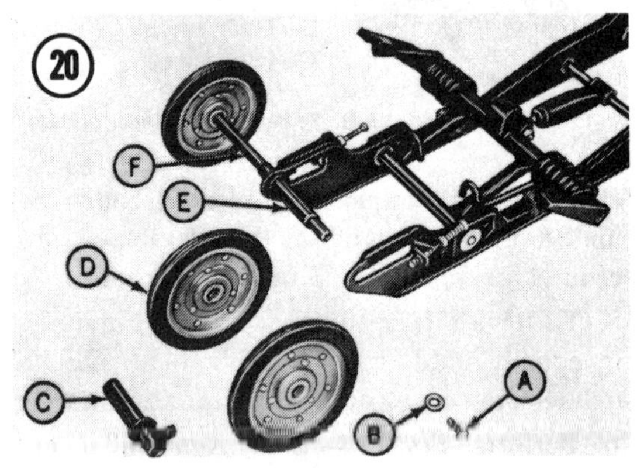

A. Cap screw E. Wheel spacer C. Wheel spacer
B. Washer F. Idler axle D. Idler wheel

1. Remove suspension.

2. Remove cap screw and washer from either end of idler axle.

3. Drive out axle from other side and remove idler wheels and spacers.

4. Assemble idler wheels as shown in **Figure 20 or 21**. Use Loctite on cap screws.

TRACK AND DRIVE SHAFT (BOGIE SUSPENSION)

Removal

1. Remove chaincase oil. Remove tension blocks, drive chain, and lower sprocket.

2. Remove battery and battery box if so equipped. Tip snowmobile on its side.

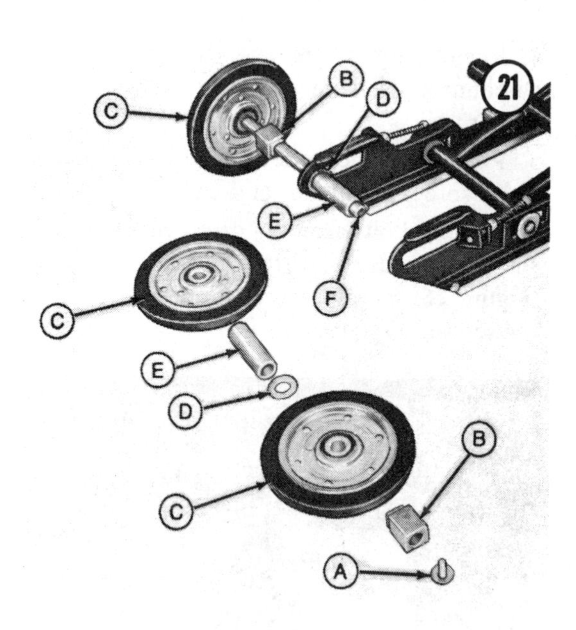

A. Cap screw and washer
B. Outer spacer
C. Idler wheel
D. Washer
E. Inner spacer
F. Idler axle

3. Remove rear idler assembly.

4. Remove all 6 bogie wheel assemblies.

5. Remove 3 cap screws securing right-hand bearing flangettes, spacer, and speedometer drive as shown in **Figure 22**.

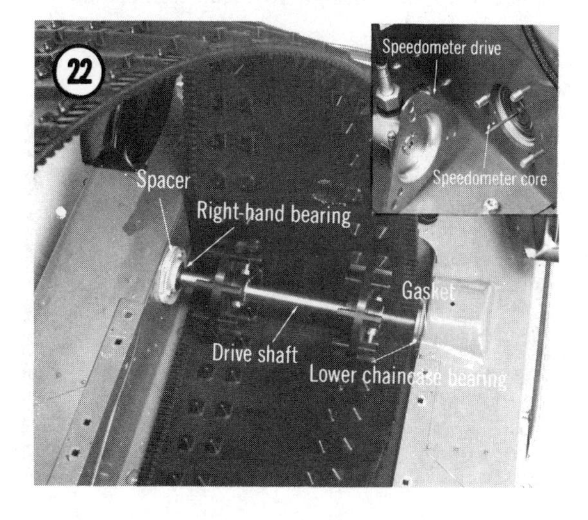

6. Remove speedometer drive with speedometer core.

7. Loosen 3 cap screws securing lower chaincase bearing flangette.

8. Lift up right-hand end of drive shaft enough for drive wheels to clear tunnel (**Figure 23**).

Remove drive shaft with chaincase bearing, flangettes, and O-ring.

9. Remove track. If track is defective, refer to *Track Wear Analysis* in Chapter Three to help determine cause of track failure.

Installation

1. Install track in snowmobile tunnel. Be sure arrow points in direction of track rotation on directional tracks.

2. Assemble spacers, bearings, and bearing flangettes on drive shaft as shown in **Figure 24**. Lubricate and install O-ring on lower chaincase bearing.

CAUTION

Be sure locking flanges on bearings face splined end of drive shaft.

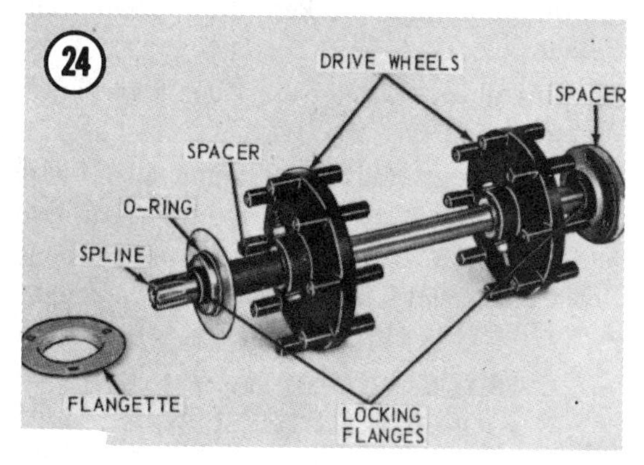

3. Install a new gasket on lower chaincase bearing mounts as shown in **Figure 25**.

4. Install splined end of drive shaft into place and drop other end of drive shaft down into position.

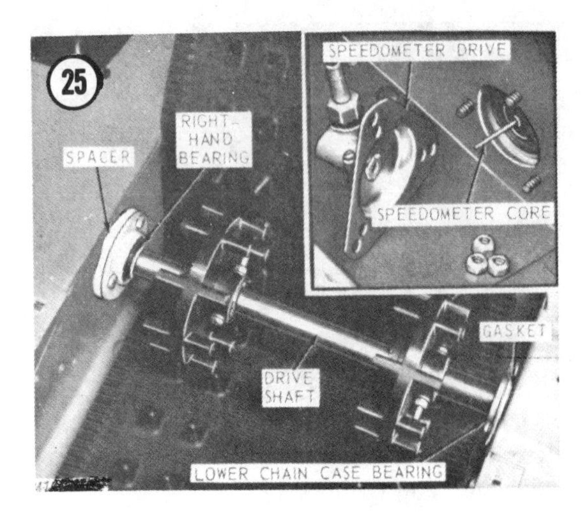

5. Install lower chaincase bearing flangettes to chaincase with 3 cap screws. Do not tighten until right-hand bearing is secured to tunnel.

CAUTION

Be certain O-ring is in correct position on bearing and between flangettes, or oil leak will result. Be sure gasket is in position between flangette and tunnel.

6. Insert speedometer core in right-hand end of drive shaft (**Figure 25**). Install bearing flangettes, spacer, and speedometer drive to tunnel. Tighten all cap screws securely.
7. Install rear idler assembly.
8. Install bogie wheel assemblies.
9. Install lower sprocket, drive chain, and tension blocks.
10. Install battery and battery box if previously removed.
11. Fill chaincase with recommended oil as outlined in Chapter Two.
12. Perform *Track Tension Adjustment* as outlined in Chapter Two.

TRACK AND DRIVE SHAFT (SLIDE SUSPENSION)

Removal

1. Siphon fuel from fuel tank and remove oil from chaincase. Close fuel shut-off valve.
2. Remove battery, if so equipped.
3. Turn the snowmobile on its side and remove suspension.
4. Remove drive belt and secondary sheave. Disconnect speedometer cable from speedometer drive.
5. Remove chaincase cover. Remove upper and lower sprockets, chain, and chain tensioner.
6. Remove bolts securing drive wheels to drive shaft. Slide drive wheels toward the center (**Figure 26**) to gain access to bearing flangette cap screws.

7. Remove cap screws securing drive shaft flangettes to tunnel (**Figure 27**). Slide drive shaft toward chaincase side. Lift opposite end (with spacer) up far enough to clear tunnel and remove drive shaft.
8. Remove track.

Installation

1. Install track in snowmobile tunnel.
2. Assemble bearing flangettes, bearings, and spacer on drive shaft as shown in **Figure 28**. Lubricate and install O-ring on chaincase bearing.

CAUTION

Be sure locking flanges on bearings face splined end of drive shaft.

3. Install new gasket on side of flangette facing chaincase. Gasket sticks to flangette and is between flangette and chaincase.

4. Position drive shaft assembly in tunnel, splined end first (**Figure 29**).

CAUTION

Be certain O-ring on bearing between flangettes on chaincase side is in correct position, or oil leak will result.

5. Install chaincase bearing flangettes to chaincase. Do not tighten cap screws until opposite bearing flangette is secured to tunnel.

6. Install speedometer core into left-hand end of drive shaft. Attach bearing flangettes, spacer, and speedometer drive to tunnel. Tighten all cap screws securely.

7. Slide drive wheels into position and secure with bolt and locknut.

8. Install suspension.

9. Turn snowmobile right side up and install sprockets, chain, and chain tensioner in chaincase.

10. Install secondary sheave and drive belt.

12. Fill chaincase with recommended oil as outlined in Chapter Two.

13. Install battery if so equipped and fill fuel tank.

14. Perform *Track Tension Adjustment* as outlined in Chapter Two.

SUSPENSION AND TRACK

To adjust track tension, slide suspension springs, and slide suspension ski lift, refer to the applicable procedure as outlined in Chapter Two.

INDEX

NOTES

NOTES

MAINTENANCE LOG

Service Performed	Mileage Reading				
Oil change (example)	2,836	5,782	8,601		